E&M Endocrinology and Metabolism

Progress in Research and Clinical Practice

Piero P. Foà
Series Editor

Endocrinology and Metabolism
Progress in Research and Clinical Practice

Piero P. Foà
  Series Editor

Cohen and Foà (eds.): Hormone Resistance and Other Endocrine Paradoxes
  (Vol. 1)
Jovanovic (ed.): Controversies in Diabetes and Pregnancy (Vol. 2)
Cohen and Foà (eds.): The Brain as an Endocrine Organ (Vol. 3)
Ginsberg-Fellner and McEvoy (eds.): Autoimmunity and the Pathogenesis of
  Diabetes (Vol. 4)
Foà (ed.): Humoral Factors in the Regulation of Tissue Growth (Vol. 5)
Foà and Walsh (eds.): Ion Channels and Ion Pumps: Metabolic and Endocrine
  Relationships in Biology and Clinical Medicine (Vol. 6)

Forthcoming volume:

Grossman (ed.): Bilateral Communication Between the Endocrine and Immune
  Systems (Vol. 7)

Piero P. Foà   Mary F. Walsh
Editors

Ion Channels and Ion Pumps

Metabolic and Endocrine Relationships in Biology and Clinical Medicine

With 67 Figures

Springer-Verlag
New York Berlin Heidelberg London Paris
Tokyo Hong Kong Barcelona Budapest

Piero P. Foà, M.D., Sc.D.
Professor Emeritus
Department of Physiology
Wayne State University
Detroit, MI 48202, USA
Mailing address:
  2104 Rhine Road
  West Bloomfield, MI 48323,
  USA

Mary F. Walsh, Ph.D.
Assistant Professor
Section of Endocrinology
Department of Internal Medicine
School of Medicine
Wayne State University
Detroit, MI 48202, USA

Library of Congress Cataloging-in-Publication Data
Ion channels and ion pumps : metabolic and endocrine relationships in
  biology and clinical medicine / Piero P. Foà, Mary F. Walsh, editors.
    p.   cm. — (Endocrinology and metabolism ; v. 6)
  Includes bibliographical references and index.
  ISBN-13:978-1-4612-7599-2
  1. Calcium channels. 2. Ion channels. 3. Ion pumps. I. Foà,
Piero P. (Piero Pio), 1911–   . II. Walsh, Mary F. III. Series:
Endocrinology and metabolism (New York, N.Y.) ; 6.
  [DNLM: 1. Ion Channels—physiology. 2. Ion Pump—physiology.
3. Biological Transport.   W1 EN396SN v. 6 1993 / QH 601 I64665   1993]
QP535.C2I58   1993
612'.01524—dc20
DNLM/DLC
for Library of Congress   93-31926
Printed on acid-free paper.
© 1994 by Springer-Verlag New York, Inc.
Softcover reprint of the hardcover 1st edition 1994

Production coordinated by Chernow Editorial Services, Inc., and managed by.
  Terry Kornak; manufacturing supervised by Vincent Scelta.
Typeset by Best-set Typesetter Ltd, Hong Kong.

9 8 7 6 5 4 3 2 1

ISBN-13:978-1-4612-7599-2   e-ISBN-13:978-1-4612-2596-6
DOI:10.1007/978-1-4612-2596-6

Preface

Omnis cellula e cellula, "every cell from a cell," was dogma to the 19th-century cellular physiologist and the cornerstone of Virchow's *Cellular-pathologie*. "Spread out a cell into a layer and you will find that, in ceasing to be a cell, it has ceased to act as such," wrote the British physiologist G.H. Lewes more than a century age.[1] "The cell remains vital as long as its wall remains intact . . ." keeping its content "pure and clear" and thus preserving the "vital principle" within, echoed Claude Bernard[2] a few years later. The notion of the cell membrane as a protecting envelope held sway until it became clear that it could not account for the "coalescence" of poorly differentiated embryonic "vesicles" and for their transformation into "cell-like structures" capable of auto-regulation and yet subject to what the grandfather of one of us defined as the "federal obligations imposed by the whole organism."[3] A new concept was needed, and soon the membrane was described as a structure capable of uniting as well as separating adjacent cells. Morphologic evidence for this dual function was obtained several years later when the electron microscope revealed the existence of tight and gap junctions which, acting as intercellular bonds and channels, allowed the cells to communicate with one another and thus coordinate their biologic activities.

Clearly intercellular communication and fulfillment of federal obligations requires that signals be generated, transported to, and identified by their targets, and provided with mechanisms needed to generate appropriate responses. A variety of specific receptors and transduction systems make up this complex yet precise system of communication whose function requires, among other conditions, the appropriate concentration of inorganic ions. In turn, this depends upon the timely and controlled activation of specific channels through which these ions are transported from the extra- to the intracellular compartment, and from one intracellular pool to another. This book describes much of what has been learned about the molecular structure of these channels and about their functions, using such techniques as voltage clamping and single-channel recording, measurements of gating currents, molecular cloning, and other genetic manipulations. The book also provides the information needed to understand the role that the movement of ions has in general biologic phenomena as diverse as neuromuscular excitation, stimulus-secretion coupling, and gene expression, and in such diverse events as cell multiplication and differentiation; central and peripheral neurotransmission; contraction of skeletal, myocardial, and smooth muscle; exocrine and endocrine secretion; platelet aggregation; and immune responses—in other words, to understand the ion movements involved in the function and malfunction of essentially all organ systems.

It is obvious that a single book, no matter how large, cannot do justice to this complex subject; we had to be selective, an editorial task further complicated by the fact that most, if not all, ion channels are functionally related. Thus, although we have attempted to arrange chapters dealing with the molecular biology and action of calcium, sodium, potassium, chloride, and other anion channels in a systematic fashion, this has not always been possible, and perhaps would not have been desirable. Indeed, we hope that this book will impress upon the reader the multiple and interdependent responses of ion channels to physiologic stimuli, pharmacologic agonists, or agents of disease; their multiple and interdependent roles in such diverse phenomena as the secretion of vasopression, aldosterone, and atrial natriuretic peptide, and therefore in the regulation of electrolyte and water balance and arterial pressure; the secretion of gonadotropins, thyroid hormones, and the hormones of the islets of Langerhans, and the maintenance of endocrine and metabolic homeostasis; the release of synaptic neurotransmitters and the control of neuromuscular responses and possibly of the cognitive functions of the brain, the action of anesthetic drugs, or the pathogenesis of malignant hyperthermia or ethanol intoxication and dependence; the pathogenesis of the recently described Syndrome X (obesity, hypertension, and diabetes) and of cystic fibrosis; and the transport of metals in bacteria and their antibacterial effectiveness.

We believe that this book and the additional reading suggested by the extensive bibliographies provided at the end of each chapter will prepare the reader for the constant flood of new information relating ion channels to matters as diverse as the mechanism of fertilization, the formation of synaptic circuits from immature neurons, the function of hair cells of the inner ear, the interaction of immune complexes with their receptors and the activation of macrophages, the pathogenesis of learning disabilities (at least in a mouse model), and the role of ion channels in the movement of water.

Detroit, MI   Piero P. Foà
                                                        Mary F. Walsh

References

1. Lewes GH. The Physiology of Common Life, Vol. 1. Leipzig: B Tauchnitz; 1860:xii–315.
2. Bernard C. Lecons sur les Propriétés des Tissus Vivants. Paris: G Baillière; 1866:492.
3. Foà P. Introduzione. Trattato di Anatomia Patologica Generale, Vol. 1. Torino: UTET; 1922:1–35.

Contents

Contributors

AALKJÆR, CHRISTIAN, PH.D.
Institute of Pharmacology, University of Århus, Århus, Denmark

ANDERSSON, KARL-ERIK, M.D., PH.D.
Department of Clinical Pharmacology, Lund University Hospital, 22185 Lund, Sweden

ANHOLT, ROBERT R.H., PH.D.
Department of Neurobiology, Duke University Medical Center, Durham, North Carolina 27710, USA

BATTAINI, FIORENZO, M.D.
Department of Experimental Medicine, Biological Science, University of Rome, Rome, Italy

BELLONE, MATTEO, M.D.
Department of Biochemistry, College of Biological Sciences, University of Minnesota, St. Paul, Minnesota 55108, USA, and H.S. Raffaele Scientific Institute, Milan, Italy

BHAT, MANJUNATHA, B., M.Sc.
Department of Pharmacology, Faculty of Medicine, University of Alberta, Edmonton, Alberta, Canada T6G 2H7

CHEW, CATHERINE S., PH.D.
Departments of Medicine and Cell Biology, Medical College of Georgia, Augusta, Georgia 30912, USA

CONTI-TRONCONI, BIANCA M., M.D.
Department of Biochemistry, College of Biological Sciences, University of Minnesota, St. Paul, Minnesota 55108, USA, and H.S. Raffaele Scientific Institute, Milan, Italy

DORIS, PETER A., PH.D.
Department of Cell Biology and Anatomy, Texas Tech University Health Sciences Center, Lubbock, Texas 79430, USA

DUNBAR, JOSEPH C., PH.D.
Department of Physiology, School of Medicine, Wayne State University, Detroit, Michigan 48201, USA

DUNN, SUSAN M.J., PH.D.
Department of Pharmacology, Faculty of Medicine, University of Alberta, Edmonton, Alberta, Canada T6G 2H7

GOVONI, STEFANO, M.D.
Department of Pharmacology, University of Bari, Bari, Italy

GUNDLACH, ANDREW L., PH.D.
Department of Medicine, Clinical Pharmacology and Therapeutics Unit, University of Melbourne, Austin and Repatriation Hospital, Heidelberg, Victoria, Australia

HIROUCHI, MASAAKI, PH.D.
Department of Pharmacology, Kyoto Prefectural, University of Medicine, Kyoto, 602 Japan

HÖGESTÄTT, EDWARD D., M.D., PH.D.
Department of Clinical Pharmacology, Lund University Hospital, 22185 Lund, Sweden

KJELDSEN, KELD, M.D., PH.D.
Department of Medicine, Division of Cardiology, University of Copenhagen School of Medicine, DK-2100 Copenhagen, Denmark

KOSTOWSKI, WOJCIECH, M.D., PH.D.
Department of Pharmacology and Physiology of the Nervous System, Institute of Psychiatry and Neurology, 02-957 Warsaw, Poland

KURIYAMA, KINYA, M.D., PH.D.
Department of Pharmacology, Kyoto Prefectural, University of Medicine, Kyoto, 602 Japan

LAWSON, DAVID M. PH.D.
Department of Physiology, School of Medicine, Wayne State University, Detroit, Michigan 48201, USA

LEVY, JOSEPH, M.D.
Departments of Internal Medicine and Physiology, Diabetes Section, Division of Endocrinology, Metabolism, and Hypertension, School of Medicine, Wayne State University, Detroit, Michigan 48201, USA

LIEDTKE, CAROLE M., PH.D.
Department of Pediatrics at Rainbow Babies and Children Hospital, Cystic Fibrosis Center, and Department of Physiology and Biophysics, Case Western Reserve University, Cleveland, Ohio 44106, USA

LYNN, ANITA R., PH.D.
Department of Biochemistry, School of Medicine, Wayne State University, Detroit, Michigan 48201, USA

MAESTRONE, EGIDIO, M.D.
Department of Anaesthesia, USSL 22 Ospedale Civile, 23100 Sondrio, Italy

MAGNONI, MARIA SANDRA, M.D.
Institute of Pharmacological Science, University of Milan, Milan, Italy

MANFREDI, ANGELO A., M.D.
Department of Biochemistry, College of Biological Sciences, University of Minnesota, St. Paul, Minnesota 55108, USA, and H.S. Raffaele Scientific Institute, Milan, Italy

MARKWARDT, FRITZ, PH.D.
Julius Bernstein Institute for Physiology, Martin Luther University, Halle, Germany

MENINI, ANNA, PH.D.
Institute of Cybernetics and Biophysics, National Research Council, 16146 Genova, Italy

MOIOLA, LUCIA, M.D.
Department of Biochemistry, College of Biological Sciences, University of Minnesota, St. Paul, Minnesota 55108, USA

NOBILE, MARIO, M.D.
Institute of Cybernetics and Biophysics, National Research Council, 16146 Genova, Italy

ÖZ, MURAT A., PH.D.
Department of Pharmacology, Faculty of Medicine, University of Alberta, Edmonton, Alberta, Canada T6G 2H7

PRESTIPINO, GIANFRANCO, PH.D.
Institute of Cybernetics and Biophysics, National Research Council, 16146 Genova, Italy

PROTTI, MARIA PIA, M.D.
Department of Biochemistry, College of Biological Sciences, University of Minnesota, St. Paul, Minnesota 55108, USA, and H.S. Raffaele Scientific Institute, Milan, Italy

REMBOLD, CHRISTOPHER M., PH.D.
Department of Internal Medicine and Physiology, Division of Cardiology, University of Virginia School of Medicine, Charlottesville, Virginia 22908, USA

ROSEN, BARRY P., PH.D.
Department of Biochemistry, School of Medicine, Wayne State University, Detroit, Michigan 48201, USA

SOWERS, JAMES R., M.D.
Department of Internal Medicine and Physiology, Division of Endocrinology, Metabolism, and Hypertension, School of Medicine, Wayne State University, Detroit, Michigan 48201, USA

STRONG, P.N., M.D.
Department of Paediatrics and Neonatal Medicine, Neuromuscular Unit, Royal Postgraduate Medical School, Hammersmith Hospital, London, WI2 0NN, United Kingdom

TRABUCCHI, MARCO, M.D.
Department of Experimental Medicine, Biological Science, University of Rome, Rome, Italy

USKI, TORE K., M.D., PH.D.
Department of Clinical Pharmacology, Lund University Hospital, 22185 Lund, Sweden

WALSH, MARY F., PH.D.
Department of Pathology, School of Medicine, Wayne State University, Detroit, Michigan 48201, USA

WEIK, RALF, M.D.
Max Planck Institute for Experimental Medicine, Göttingen, Germany

WHITFIELD, JAMES F., PH.D.
Cell Systems Section, Institute for Biological Sciences, National Research Council of Canada, Ottawa, Ontario, Canada K1A OR6

YINGST R. DOUGLAS, PH.D.
Department of Physiology, Wayne State University, Detroit, Michigan 48201, USA

1
The Molecular Structure and Gating of Calcium Channels

Susan M.J. Dunn, Manjunatha B. Bhat, and A. Murat Öz

The concentration of cytosolic free calcium is of critical importance for the control of many essential cellular functions. A diverse array of Ca^{2+} transporting systems acts to maintain the steep concentration gradient between the millimolar concentrations of extracellular Ca^{2+} and resting intracellular concentrations of about $0.1\,\mu M$. These low resting concentrations are maintained by several pumps and exchange mechanisms that transport Ca^{2+} either out of the cell or into intracellular storage sites. A rise in intracellular calcium to micromolar levels initiates many physiologic responses, including excitation-contraction and excitation-secretion coupling. The influx of calcium ions through calcium-selective channels in the plasma membrane plays an important role in these transient increases in concentration. Although the plasma membrane is normally virtually impermeable to Ca^{2+}, the opening of calcium channels allows Ca^{2+} to move into the cell down its steep electrochemical gradient.

Different types of calcium channels can be characterized by differences in their gating properties. Calcium channels that open in response to activation of an associated receptor are referred to as receptor-operated calcium channels. These include smooth muscle channels that are coupled to ATP receptors and neuronal channels that are opened by activation of receptors for the neurotransmitter, N-methyl-D-aspartate (NMDA).[1,2] Other calcium channels are gated primarily by changes in membrane potential.[3-11] These voltage-dependent calcium channels (VDCCs), which are the focus of this chapter, are ubiquitous in the membranes of excitable cells and have also been shown to exist in a number of non-excitable cell types including glial cells, myeloma cells, and fibroblasts.[12-14] As described below, the combined application of advanced biophysical, biochemical, and molecular techniques has contributed greatly to our understanding of the molecular properties of VDCCs.

Multiplicity of Voltage-Dependent Calcium Channels

A significant advance in our understanding of VDCCs was the realization that most excitable cells express several types of voltage-activated calcium channels that can be distinguished by differences in their voltage dependencies, kinetic properties, and pharmacologic sensitivities. At least three different VDCCs, termed L, T, and N, have been described in chick dorsal root ganglion (DRG) neurons[15,16] and, more recently, a fourth type (P) was identified in cerebellar Purkinje cells.[17] Two types of calcium channels, similar to the neuronal L- and T-type, have also been identified in skeletal, cardiac, and smooth muscle.[5] The best characterized of these calcium channels are those that mediate long-lasting depolarization (L-type). As discussed in more detail below, these channels are activated by relatively large depolarization, carry the largest current of the known VDCCs and inactivate slowly. L-type channels are modulated by a variety of organic calcium channel ligands, some of which are used clinically in the treatment of cardiovascular and cerebrovascular diseases.[18] Channel-blocking drugs, which exert their therapeutic effect by reducing the contractility of cardiac and smooth muscle, have diverse chemical structures and fall into three main groups, ie, 1,4-dihydropyridine (DHP) derivatives such as nifedipine, phenylalkylamines such as verapamil, and benzo-thiazepines, exemplified by diltiazem. The binding sites for the three classes of drugs are distinct but interact in an allosteric manner such that the dissociation of radio-labeled DHPs (eg, [^{3}H]-nitrendipine) from the receptor complex is enhanced by verapamil and slowed by diltiazem.[19] Some DHPs, eg, Bay K8644, activate rather than block calcium channels, and both the DHP antagonists and agonists, which bind with high affinity (Kds of nM) to VDCCs, are proving to be invaluable tools in the study of the molecular properties of the channel proteins.[20]

Structure of a Skeletal Muscle L-Type Calcium Channel

Although DHP-sensitive calcium channels have been demonstrated in essentially all excitable cells, the most abundant source of DHP-binding proteins is rabbit skeletal muscle, where they are present in transverse tubule membranes at a density 50 to 100 times greater than in other tissues.[21] Since this is consistent with the transverse tubular localization of skeletal muscle VDCCs, these membranes have been the preferred starting material for purification of the calcium channel protein.

The DHP-binding protein has been purified in a number of laboratories[22] using conventional purification procedures involving detergent solubilization, lectin-Sepharose chromatography, ion exchange chromatography, and sucrose-density gradient sedimentation.[23,24] In most reports, a multi-subunit complex of five distinct subunits (α_1 [170 kDa]; α_2 [140 kDa],

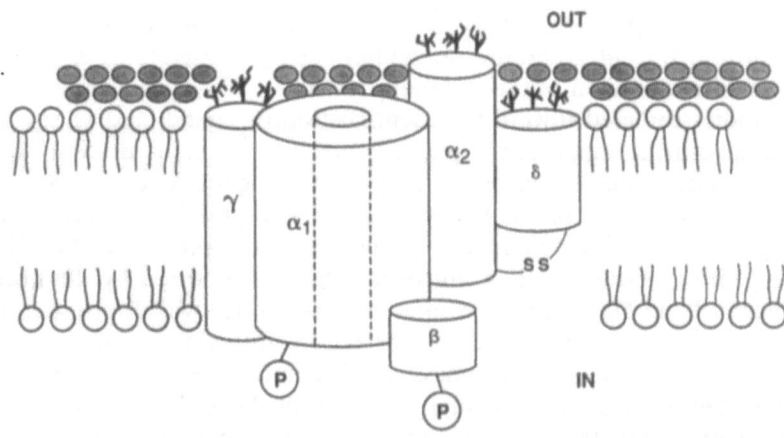

FIGURE 1.1. A model of the subunit structure of the skeletal muscle DHP-sensitive calcium channel, based on previously proposed models[3,7,22] and depicting subunit glycosylation (γ), phosphorylation (P), and disulfide bonds (SS).

linked by a disulfide group to δ [30 kDa]; β [55 kDa]; and γ [32 kDa]) has been isolated. Photoaffinity labeling studies have demonstrated that the α_1 subunit carries binding sites for DHPs and for other classic organic calcium-channel blockers[25-27] and, in expression studies (see below), this subunit has been shown to form a functional voltage-dependent calcium channel. The disulfide-linked α_2 and δ subunits are encoded by the same gene[28,29] and separate as a result of a proteolytic event during processing. These subunits do not have appreciable hydrophobicity and may not, therefore, be integral transmembrane proteins. They are, however, highly glycosylated and this underlies the usefulness of lectin affinity chromatography in the purification of the channel complex. The β subunit is neither hydrophobic nor glycosylated but, like the α_1 subunit, contains a site for phosphorylation by cAMP-dependent protein kinase. The γ subunit is glycosylated and is believed to be a transmembrane protein.[3] A model, based on ones previously described[3,7,22] is shown in Figure 1.1. The exact subunit structure and arrangement in the membrane, however, remain to be established. Recently, for example, it was proposed that the α_2 subunit is a peripheral protein that is exposed on the extracellular surface and is anchored in the membrane by the δ subunit.[29]

All five subunits of the DHP-binding protein from skeletal muscle have now been cloned and their cDNAs have been sequenced.[28,20,31-33] The α_1 subunit displays 30–40% homology to the α_1 subunit of the voltage-dependent sodium channel.[30] Like the sodium channel,[34] this subunit has four internal repeats that exhibit sequence homology and each repeat contains six conserved hydrophobic stretches that are predicted to be transmembrane. The fourth transmembrane helix (S4) of each of these motifs contains positively charged amino acids (arginine or lysine) every third or fourth residue and this region has been proposed to constitute the voltage-sensing region of the channel protein.[3]

In studies directed toward the identification of ligand-binding sites on the α_1 subunit, a region in the putative cytosolic domain adjacent to the sixth transmembrane helix of the fourth homologous repeat (IVS6) was recently shown to be photoaffinity labeled by the DHP analogues, [^{3}H]-azidopine and [^{3}H]-nitrendipine.[35] These results suggest that the DHP-binding site is intracellular, a suggestion that is supported by the hydro-phobicity of the DHPs and is consistent with models which propose that DHPs partition into the lipid bilayer before diffusing to their protein binding sites.[36] However, this possibility is contradicted by electrophysio-logic evidence for an extracellular exposure of the DHP-binding site. In experiments using the ionized and presumably membrane-impermeant DHP derivatives, amlodipine and SDZ-207-180, the drugs were reported to be ineffective when applied intracellularly.[37] In support of this ob-servation, two recent studies employing photolabeling techniques[38,39] and subsequent sequence-directed antibody mapping of the proteolytic frag-ments have implicated putative extracellular domains in forming a DHP-binding site. These regions were identified as a portion of the loop linking the fifth (S5) and sixth helix (S6) of the third domain, and also the sixth helix in repeats III and IV.

While there is still some controversy about the location of high affinity DHP-binding sites, most available evidence suggests that the phenylalkyl-amine binding site is accessible only from the intracellular surface. In reconstitution studies using planar bilayers, it was shown that D890, a charged derivative of verapamil, was effective only when applied to the intracellular-equivalent side.[40] Recent studies have also shown that the peptide labeled by a photoactive derivative of verapamil is located between Glu-1349 and Trp-1391, a segment which lies on the presumed intra-cellular end of the sixth helix in the fourth homologous repeat.[41] Thus, the proposal that the binding sites for DHPs and phenylalkylamines are located at the opposite ends of transmembrane helix IVS6 leads to the suggestion that the allosteric interaction between these two sites may involve conformational changes induced by the movement of this IVS6 segment.[38]

Recently, we have provided evidence for distinct low-affinity binding sites on, or related to, the high-affinity DHP-binding proteins in skeletal muscle transverse tubule membranes. These sites were first revealed by the ability of micromolar concentrations of unlabeled DHPs to accelerate the rate of dissociation of the DHP antagonist, [^{3}H]PN200-110, from its high-affinity sites.[42] More recently the low-affinity sites have been charac-terized by monitoring the fluorescence changes of the fluorescent DHP, felodipine, that accompany its binding to transverse tubule membranes.[43] The binding of felodipine results in a large saturable enhancement of its fluorescence, which can be excited either directly or indirectly via energy transfer from membrane protein. This fluorescence enhancement was inhibited in a competitive fashion by micromolar concentrations of DHP agonists and antagonists. It has been suggested that occupancy of DHP-binding sites with similar low affinity has a functional significance in the modulation of VDCCs in both skeletal[44] and cardiac[45] muscle. Gaining further insight into the detailed molecular properties of both the high and low affinity sites may help understand the process of calcium channel gating and its modulation by various ligands.

The Roles of DHP-Binding Proteins in Skeletal Muscle

Although DHP-binding proteins are abundant in skeletal muscle transverse tubule membranes, their functional significance remains to be established. Skeletal muscle contraction, unlike contraction of cardiac or smooth muscle,[46-50] is not dependent on the influx of extracellular calcium. Also, skeletal muscle L-type calcium channels activate only very slowly, reaching peak amplitude only after 100–200 ms (see below), and thus do not have time to open during an action potential that lasts only about 2 ms.[51] It has been suggested that only a few percent of DHP-binding proteins in skeletal muscle function as VDCCs,[52] although, as recently pointed out,[53] this may be true at a given instant under specific conditions, and all of them may be intrinsically capable of mediating calcium transport. The current view is that DHP-binding proteins in skeletal muscle are likely to play a dual role, acting as calcium channels under some conditions and also as voltage-sensors which, by mechanisms presently unknown, communicate the depolarization of the transverse tubule membrane to the sarcoplasmic reticulum, triggering the release Ca^{2+}.[54] Alternatively, different proteins may carry out the two functions, and it has been suggested that two forms of the α_1 subunit may be responsible for performing the two roles.[55] As described above, the apparent molecular weight of the α_1 subunit in the purified protein is about 170 kDa. However, the molecular weight predicted from cloning studies is 212 kDa. Recently both forms have been identified in skeletal muscle, with approximately 90% having the smaller molecular weight.[55] This has led to the suggestion that the smaller form acts as a voltage-sensor and the larger species as a VDCC. However, this remains to be established, and several lines of evidence suggest that the one protein species may be capable of carrying out both functions. In studies in which the purified DHP-binding protein, which did not contain detectable amounts of the 212 kDa species, were reconstituted into phospholipid vesicles[56] or planar bilayers,[57] voltage-dependent calcium currents could be measured. Furthermore, although measurements of calcium fluxes in reconstituted vesicle preparations originally suggested that only a few percent of the receptors mediated calcium flux,[56] this percentage could be increased about tenfold after protein phosphorylation.[58] Recent $^{45}Ca^{2+}$ flux studies in this laboratory have also shown that isolated transverse tubule vesicles contain functional VDCCs[59] that mediate rapid voltage-dependent flux responses. This implies that many, if not all, DHP-binding proteins can function as calcium channels. It is thus likely that information gathered about the relatively abundant skeletal muscle DHP-binding protein may give us further insight into the molecular properties of DHP-sensitive VDCCs in other tissues.

Homologous Calcium Channels

Following the cloning of the α_1 subunit of the L-type VDCC in skeletal muscle, a homologous protein was identified in cardiac muscle.[60] The cardiac α_1 subunit displays 66% homology with its skeletal muscle counterpart, and was shown to form a functional VDCC when expressed from its encoding mRNA in *Xenopus* oocytes. Snutch et al.[61] isolated four partial cDNA clones, which they termed A, B, C, and D, from rat brain. Full-length cDNAs corresponding to class C clones have since been isolated from rabbit lung,[62] rat brain,[63] and rat aorta,[64] and these are virtually identical to the original rabbit cardiac clone. Complete class D clones have also been isolated from rat brain,[65] human brain,[66] and human pancreas,[67] and these also appear to encode DHP-sensitive L-type calcium channels. Classes A and B are highly homologous to each other but are less closely related to the skeletal and cardiac muscle α_1.[9] A full-length class A clone, designated BI, has been reported from rabbit brain.[68] When expressed in *Xenopus* oocytes, this channel was insensitive to nifedipine and ω-conotoxin, but was inhibited by crude venom from the funnel web spider, suggesting that this clone may be a P-type channel (see below). In addition to the expression of multiple genes, alternative splicing also appears to generate a diversity of calcium channels and variants of brain, skeletal, cardiac, and smooth muscle channels have been identified.[63,64,68,69]

Gating of Voltage-Dependent Calcium Channels

The analysis of whole cell and single-channel current kinetics has provided deep insights into the gating properties of VDCCs. As described above, at least three different types of VDCCs have been identified in chick dorsal root ganglion (DRG) neurons and a fourth, P-type, has been described in cerebellar Purkinje cells. The main features of these different types of VDCCs are summarized in Table 1.1.

TABLE 1.1. Electrical and pharmacologic properties of the four types of voltage-sensitive calcium channels.[16,17]

Channel type	Low voltage activated Ca²⁺ channel	High voltage activated Ca²⁺ channels		
	T	P	N	L
Single channel conductance (110 Ba)	~8 pS	~10–12 pS	~13 pS	~25 pS
Relative conductance	$Ba^{2+} = Ca^{2+}$	$Ba^{2+} > Ca^{2+}$	$Ba^{2+} > Ca^{2+}$	$Ba^{2+} > Ca^{2+}$
Inorganic ion block	$Ni^{2+} > Cd^{2+}$	Cd^{2+} sensitive	$Cd^{2+} > Ni^{2+}$	$Cd^{2+} > Ni^{2+}$
ω-CgTx block	Weak, reversible	Resistant	Sensitive	Resistant
Dihydropyridine sensitivity	Resistant?	Resistant	Resistant	Sensitive
FTX sensitivity	Resistant	Sensitive	Resistant	Resistant
Activation range (for 10 Ca)[a]	Positive to −70 mV	Positive to −50 mV	Positive to −20 mV	Positive to −10 mV4
Inactivation range (for 10 Ca)	−100 to −60 mV	—	−120 to −30 mV	−60 to −10 mV
Inactivation rate (0 mV, 10 Ca, or 10 Ba)	Rapid (tau ~ 20–50 ms)	Very slow (tau ~ 1 s)	Moderate (tau ~ 50–80 ms)	Very slow (tau > 500 ms)

[a] In 80 mM Ba^{2+} for P-type channels.[17]

The L-type VDCC, in addition to being the best characterized in terms of its molecular structure, is also the best characterized in terms of its gating properties. This channel has greater permeability to Ba^{2+} than to Ca^{2+}, and it has a large unitary Ba^{2+} conductance (25 pS in 110 mM Ba^{2+}). These channels have a high-voltage threshold for activation (positive to -30 mV) and conduct maximum current (peak current) in the voltage range of 0 to $+10$ mV.[70] Although L-type channels in different tissues share similar current-voltage-related features such as activation threshold and the voltage range of the peak amplitude,[51] their activation kinetics show significant tissue variability.[71] In a DRG neuron, for example, during depolarizations to $+10$ mV from a holding potential of -80 mV, whole cell currents reach their peak amplitude within 10 ms.[70] Skeletal muscle channels, however, activate much more slowly, and under similar conditions the peak amplitude in frog skeletal muscle fibers was reached only after 100–200 ms.[51]

In addition to their high threshold of activation, L-type VDCCs are also distinguished by their slow and unique type of inactivation, which is both voltage and Ca^{2+} dependent.[72] Depending on membrane potential, temperature, the charge-carrying ion, and the tissue type, the inactivation time constant for L-type VDCCs varies between 30 to 1000 ms. The voltage-dependent nature of this inactivation is evident in the decay from its peak during a sustained depolarizing step. This voltage-dependence of inactivation of L-type channels is less pronounced than that of T currents (see below). However, unlike T channels, L channels can also be inactivated by a rise in intracellular free Ca^{2+}. Thus in the presence of Ba^{2+} as the charge carrier and/or in the cells in which intracellular Ca^{2+} is chelated by EGTA or BAPTA, the inactivation of L currents can be dramatically reduced, leaving "long-lasting" L currents.[72]

T-type calcium channels have also been identified and characterized in neurons and muscle cells. However, study of these channels is more difficult since their pharmacologic properties are not well known. In general they are resistant to organic Ca^{2+} channel antagonists,[73] although they are blocked by various inorganic blockers, and are particularly sensitive to Ni^{2+}, which blocks at low concentrations (40 μM).[74] The T current is characterized by a small single channel conductance of only about 5–10 pS (in 100 mM Ba^{2+}); indeed, this channel was originally named because of its "tiny" conductance.[15] These channels are equally permeable to Ca^{2+} and Ba^{2+}; they can be activated at a low threshold range of -50 to -70 mV, and usually reach their peak amplitude in the range of -20 to -10 mV. Their activation kinetics are faster than those of L-type channels, leading to their description as "fast" Ca^{2+} channels by some authors.[5,6] Unlike the L-type current, the T-current shows purely voltage- and time-dependent inactivation kinetics. Their inactivation time constant is in the range of 20–50 ms in DRG neurons[16] and because of this fast inactivation these channels have also been referred to as "transient" Ca^{2+} channels.[75] In skeletal muscle fibers a component of Ca^{2+} current has been described which, like T-type channels in cardiac muscle and neurons, activates rapidly with a low threshold.[5] However, this current in muscle fiber inactivates only very slowly, over several seconds.[76]

N-type Ca^{2+} channels, first described in chick DRG cells by Nowycky and colleagues,[15] are largely restricted to neurons, but their relative prominence varies significantly from one neuron to another. Although N-type channels are the dominant Ca^{2+} entry pathway in sensory, sympathetic, and myenteric plexus neurons,[5] they are mostly absent in cerebellar Purkinje cells.[9] These channels have a conductance that is intermediate between T- and L-type channels (13 to 17 pS in 110 mM Ba^{2+}). N-type channels, like L channels, become activated at high thresholds. From a holding potential of -100 mV, strong depolarizations to -20 mV or more positive elicit rapidly activating N currents in chick DRG.[15] The N-type current differs from the L-type mainly by its greater sensitivity to holding potentials, and faster and strictly voltage-dependent inactivation kinetics.[15] It is inactivated in a range between T- and L-type channels and thus can be largely eliminated by using holding potentials greater than -40 mV. Although in chick DRG cells the inactivation time constant of N-type channels is between 50 and 80 ms, their kinetics may change greatly from neuron to neuron and this may indicate that a component of the inactivation mechanism is Ca^{2+}-dependent.[77]

N-type channels appear to be sensitive to the peptide toxin from *Conus geographus* (ω-conotoxin GVIA[78]), Cd^{2+}, and various neurochemicals, including norepinephrine, somatostatin, and dynorphin.[79] They are, however, relatively resistant to Ni^{2+} and to the classic organic Ca^{2+} channel blockers.[5] Their pharmacologic properties, ie, their sensitivity to Cd^{2+} and ω-conotoxin GVIA and their insensitivity to DHPs, are consistent with the characteristics of Ca entry pathways underlying transmitter release from sympathetic neurons,[77] motor nerve terminals,[80,81] synaptosomes,[82,83] and brain slices.[84] In line with these observations, ω-conotoxin sites are about 20 times more dense than dihydropyridine sites in brain.[85] Unlike dihydropyridines, ω-conotoxin greatly reduces transmitter release from brain synaptosomes, which is consistent with the theory that N-type channels can play a significant role in mediating transmitter release.

Recently P-type Ca^{2+} channels that appear to be particularly dominant in cerebellar Purkinje cells were described by Llinas and colleagues.[17] P-type channels are a novel class of high-voltage-activated Ca^{2+} channel that are activated over a range of potentials less negative than -50 mV and have a similar single-channel conductance (14 pS) to that of N-type channels. Their inactivation seems to be a slow process with a time constant of approximately 1 s. P-type Ca^{2+} channels are unlike L-type or N-type Ca channels in that they are insensitive to DHPs and ω-conotoxin

GVIA. However, in cerebellar Purkinje cells and squid giant synapses these channels can be potently blocked by FTX, a venom of the funnel web spider, *Agelenopsis aperta*.[17] As described above, a brain Ca^{2+} channel named BI has recently been cloned and its pharmacologic specificity, ie, resistance to DHPs and ω-conotoxin and block by FTX, suggests that it may be related to a P-type channel.[68] In addition, BI channels appear to be expressed at high levels in the cerebellum.[68] Despite these similarities, there are substantial functional differences between P-type and BI-type Ca^{2+} channels, and their relationship remains to be further clarified.[9]

Structure-Function Relationships of VDCCs

Most studies have suggested that the $α_1$ subunit is able to form a functional calcium channel autonomously. Cloning and expression systems have been used to study the participation of different subunits in channel function. Cardiac, smooth muscle (lung), and brain-type calcium channels have been successfully expressed in *Xenopus* oocytes after injection of mRNAs encoding their respective $α_1$ subunits.[60-62] Although the skeletal muscle $α_1$ subunit has not been shown to form a voltage-gated calcium channel in *Xenopus* oocytes, it has been shown to form a DHP-sensitive VDCC when expressed in a murine fibroblast derived cell line.[86] However, the current showed different gating properties from native skeletal muscle channels, which suggests that either the other subunits or the native transverse tubular membrane may be required for full function.[87]

The functional significance of the skeletal muscle $α_1$ subunit has been studied in mice with muscular dysgenesis, which are deficient in this protein.[88] The myocytes of these mice lack significant calcium currents and do not show electrically stimulated excitation-contraction coupling. However, when cDNA encoding the skeletal muscle $α_1$ subunit was injected into their nucleus, both calcium currents and skeletal muscle-type excitation-contraction coupling were restored.[89] The expression of the cardiac muscle $α_1$ subunit by these myotubes produced cardiac-like calcium currents and excitation-contraction coupling[90] that, as in cardiac muscle, was dependent upon extracellular Ca^{2+}.

Studies with chimeric calcium channels have provided information on areas of the channel protein that may be important in conferring the properties of excitation-contraction coupling.[91,92] As described above, excitation-contraction coupling in cardiac, but not skeletal muscle, is dependent upon external calcium ions. Cardiac and skeletal muscle calcium currents are also different, with cardiac calcium channels activating much more rapidly than skeletal channels. When a chimeric calcium channel was expressed in dysgenic myotubes from cDNA-encoding the cardiac α_1 subunit in which the intracellular loop between repeats II and III was replaced by its skeletal muscle counterpart, the cells exhibited excitation-contraction coupling characteristic of skeletal muscle, whereas the calcium current remained to be the cardiac type (fast kinetics).[14] On the other hand, changing the first homologous repeat to that of the skeletal muscle type changed the Ca^{2+} current from the cardiac type (fast kinetics) to the skeletal type (slow kinetics).[92]

Several recent studies have examined the functional role of other protein subunits that, in skeletal muscle, copurify with the α_1 subunit of VDCC. Proteins that are homologous to the disulfide-linked α_2/δ subunits have been identified in cardiac muscle, smooth muscle, and brain.[31,66] Furthermore, antibodies raised against the skeletal muscle α_2 subunit immunoprecipitated DHP-binding proteins from both skeletal muscle and brain, which suggests that α_2 subunits are associated with at least some brain L-type channels.[93] In expression studies, it was found that the magnitude of calcium currents mediated by the cardiac α_1 subunit expressed in oocytes was potentiated by coexpression of the skeletal muscle α_2/δ subunits.[60,94] Furthermore, when the α_2/δ complex was coexpressed with the skeletal muscle α_1 subunit in murine L-cells, the density of DHP-binding sites was increased by about 70%, although there was no change in binding affinity.[95]

A functional role for the β subunit has also been suggested from studies of the coexpression of different combinations of subunits in L cells. Coexpression of the skeletal muscle α_1 and β subunits led to a tenfold increase in DHP-binding sites and an acceleration of the channel activation and inactivation kinetics.[95] In addition, coexpression of these two subunits led to calcium currents that showed apparently normal activation kinetics,[96] in contrast to the abnormal kinetics seen upon the expression of α_1 alone (see above). The skeletal muscle γ subunit also had a modest effect on the current produced by α_1 in L cells, and both enhanced the expression of cardiac α_1 in *Xenopus* oocytes and substantially modified the channel kinetics.[94] The α_2/δ and β subunits show functional synergism with the α_1 subunit since a combination of $\alpha_1/\alpha_2\delta/\beta$ subunits produced a larger peak current compared to either the $\alpha_1/\alpha_2\delta$ or α_1/β combination.[66,68,94]

Regulation of Calcium Channels

The main mechanisms by which calcium channels are thought to be modulated by neurotransmitters and hormones involve second-messenger-mediated phosphorylation/dephosphorylation events and the interaction with activated guanyl nucleotide binding proteins (G proteins).[4,6] The best established mechanism for modulation of the cardiac calcium channel is the phosphorylation by cAMP-dependent protein kinase,[96] which increases the probability of channel opening. In vitro phosphorylation studies have shown that both the α_1 and β subunits of the skeletal muscle channel are substrates for phosphorylation by cAMP-dependent protein kinase.[4] Recently, it has also been reported that the α_1 subunit in primary cultures of skeletal muscle cells[97] and in intact chick muscle[98] could be phosphorylated by stimulation of β-adrenergic receptors and the consequent elevation of cAMP. Other second messengers that have been implicated in calcium channel modulation include cGMP, inositol-1,4,5-trisphosphate, and 1,2-diacylglycerol,[99] which lead to the activation of protein kinase C and Ca/calmodulin-dependent protein kinase, and, depending on the cell and channel type, may either activate or inhibit calcium currents. Regulation by G proteins may be indirect via coupling of hormonal signals to stimulation of protein kinases, but there is also evidence for a stimulatory, direct control of calcium channels in both cardiac[100] and skeletal[101] muscle, via a pathway that does not involve second messengers.

Conclusion

Much progress has been made in the study of the molecular properties of voltage-dependent calcium channels. Despite this progress many essential aspects of channel structure, function, and regulation are, at best, poorly understood. It is anticipated, however, that the continued application of the powerful battery of biophysical, biochemical, and recombinant DNA techniques will bring many answers to fundamental questions about calcium channel proteins. This is essential for our understanding of the mechanisms of electrical excitability and of the role of calcium channels in the regulation of excitation-contraction and excitation-secretion coupling.

Acknowledgments. Work in the authors' laboratory is supported by NIH GM-42375 and the Medical Research Council of Canada. S.M.J.D. is a Scholar of the Alberta Heritage Foundation for Medical Research, and M.B.B. and A.M.Ö. hold AHFMR Studentships.

References

1. Mayer ML, Westbrook GL. Permeation and block of N-methyl-D-aspartic acid channels by divalent cations in mouse cultured central neurons. J Physiol (Lond) 1987; 394:501–527.
2. Benham CD, Tsien RW. A novel receptor-operated Ca^{2+} channel activated by ATP in smooth muscle. Nature 1987; 328:275–278.
3. Catterall W. Structure and function of voltage-sensitive calcium channels. Science 1988; 242:50–61.

4. Hosey MM, Lazdunski M. Calcium channels: Molecular pharmacology, structure and regulation. J Membr Biol 1988; 104:81–105.
5. Bean BP. Classes of calcium channels in vertebrate cells. Annu Rev Physiol 1989; 51:367–384.
6. Hess P. Calcium channels in vertebrate cells. Annu Rev Neurosci 1990; 13:337–356.
7. Dascal N. Analysis and functional characteristics of dihydropyridine-sensitive and -insensitive calcium channel proteins. Biochem Pharmacol 1990; 40:1171–1178.
8. McKenna E, Koch WJ, Slish DF, et al. Toward an understanding of the dihydropyridine-sensitive calcium channel. Biochem Pharmacol 1990; 39:1145–1150.
9. Tsien RW, Ellinor PT, Horne WA. Molecular diversity of voltage-dependent Ca^{2+} channels. Trends Pharmacol Sci 1991; 12:349–354.
10. Slish DF, Schultz D, Shwartz A. Molecular biology of the calcium antagonist receptor. Hypertension 1992; 19:19–24.
11. Miller RJ. Voltage-sensitive calcium channels. J Biol Chem 1992; 267:1403–1406.
12. Barres BM, Chan LLY, Corey DP. Ion channel expression by white matter glial Type-2 astrocytes and oligodendrocytes. Glia 1988; 1:10–30.
13. Fukushima Y, Hagiwara S. Voltage gated Ca^{2+} channel in mouse myeloma cells. Proc Natl Acad Sci USA 1988; 80:2240–2242.
14. Villereal ML, Jamieson GA. Epidermal growth factor stimulates calcium influx via voltage-sensitive calcium channels in cultured human fibroblasts. J Cell Biochem 1988; 159(suppl 12A):C652. Abstract.
15. Nowycky MC, Fox AP, Tsien RW. Three types of calcium channel with different agonist sensitivity. Nature 1985; 316:440–443.
16. Tsien RW, Lipscombe D, Madison DV, et al. Multiple types of neuronal calcium channels and their selective modulators. Trends Neurosci 1988; 11:431–438.
17. Llinas RR, Sugimori M, Lin JW. Blocking and isolation of a calcium channel from neurons in mammals and cephalopods utilizing a toxin fraction (FTX) from funnel web spider poison. Proc Natl Acad Sci USA 1989; 86:1689–1693.
18. Triggle DJ, Janis RA. Calcium channel ligands. Annu Rev Pharmacol Toxicol 1987; 27:347–369.
19. Glossmann H, Ferry DR, Striessnig J, et al. Calcium channels and calcium channel drugs: Recent biochemical and biophysical findings. Drug Res 1985; 35:1917–1935.
20. Triggle DJ, Rampe D. 1,4-Dihydropyridine activators and antagonists: Structural and functional characteristics. Trends Pharmacol Sci 1989; 10:507–511.
21. Fosset M, Jaimovich E, Delpont E, et al. [^{3}H]Nitrendipine receptors in skeletal muscle: Properties and preferential localization in transverse tubules. J Biol Chem 1983; 10:6086–6092.
22. Campbell KP, Leung A, Sharp AH. The biochemistry and molecular biology of the dihydropyridine-sensitive calcium channel. Trends Neurosci 1988; 11:425–430.

23. Curtis BM, Catterall WA. Purification of the calcium antagonist receptor of the voltage-sensitive calcium channel from skeletal muscle transverse tubules. Biochemistry 1984; 23:2113–2118.
24. Borsotto M, Norman RI, Fosset M, et al. Solubilization of the nitrendipine receptor from skeletal muscle transverse tubule membranes: Interactions with specific inhibitors of the voltage-dependent Ca^{2+} channel. Eur J Biochem 1984; 142:449–455.
25. Striessnig J, Moosburger K, Goll A, et al. Stereoselective photoaffinity labelling of the purified 1,4-dihydropyridine receptor of the voltage-dependent calcium channel. Eur J Biochem 1986; 161:603–609.
26. Striessnig J, Knaus H-G, Grabner M, et al. Photoaffinity labelling of the phenylakylamine receptor of the skeletal muscle transverse-tubule calcium channel. FEBS Lett 1987; 212:247–253.
27. Naito K, McKenna E, Schwartz A, et al. Photoaffinity labeling of the purified skeletal muscle calcium channel antagonist receptor by a novel benzothiazepine, [^{3}H]azidobutyryl diltiazem. J Biol Chem 1989; 264:21211–21214.
28. DeJongh KS, Warner C, Catterall WA. Subunits of purified calcium channels: α_2 and δ are encoded by the same gene. J Biol Chem 1990; 265:14738–14741.
29. Jay SD, Sharp AH, Kahl S, et al. Structural characterization of the dihydropyridine-sensitive calcium channel α_2 subunit and the associated δ peptides. J Biol Chem 1991; 3287–3293.
30. Tanabe T, Takeshima H, Mikami A, et al. Primary structure of the receptor for calcium channel blockers from skeletal muscle. Nature 1987; 328:313–318.
31. Ellis SB, Williams ME, Ways NR, et al. Sequence and expression of mRNAs encoding the α_1 and α_2 subunits of a DHP-sensitive calcium channel. Science 1988; 241:1661–1664.
32. Jay SD, Ellis SB, McCue AF, et al. Primary structure of the γ subunit of the DHP-sensitive calcium channel from skeletal muscle. Science 1990; 248:490–492.
33. Ruth P, Röhrkasten A, Biel M, et al. Primary structure of the β subunit of the DHP-sensitive calcium channel from skeletal muscle. Science 1989; 245:1115–1118.
34. Noda M, Shimizu S, Tanabe T, et al. Primary structure of *Electrophorus electricus* sodium channel deduced from cDNA sequence. Nature 1984; 312:121–127.
35. Regulla S, Schnieder T, Nastainczyk W, et al. Identification of the site of interaction of the dihydropyridine calcium channel blockers nitrendipine and azidopine with the calcium-channel α_1 subunit. EMBO J 1991; 10:45–49.
36. Rhodes DG, Sarmiento JG, Herbette LG. Kinetics of binding of membrane-active drugs to receptor sites. Diffusion limited rates for a membrane bilayer approach of 1,4-dihydropyridine calcium channel antagonist to their active site. Mol Pharmacol 1985; 27:612–623.
37. Kass RS, Arena JP, Chin S. Block of L-type calcium channels by charged dihydropyridines: Sensitivity to side of application and calcium. J Gen Physiol 1991; 98:63–75.

38. Nakayama H, Taki M, Striessnig J, et al. Identification of 1,4-dihydropyridine binding regions within the α_1 subunit of skeletal muscle Ca^{2+} channels by photoaffinity labeling with diazepine. Proc Natl Acad Sci USA 1991; 88:9203–9207.

39. Striessnig J, Murphy BJ, Catterall WA. Dihydropyridine receptor of L-type Ca^{2+} channels: Identification of binding domains for [³H](+)-PN200-110 and [³H]azidopine within the α_1 subunit. Proc Natl Acad Sci USA 1991; 88:10769–10773.

40. Affolter H, Coronado R. Sidedness of reconstituted calcium channels from muscle transverse tubules as determined by D600 and D890 blockade. Biophys J 1986; 49:767–771.

41. Striessnig J, Glossmann H, Catterall WA. Identification of a phenylalkylamine binding region within the α_1 subunit of skeletal muscle Ca^{2+} channels. Proc Natl Acad Sci USA 1990; 87:9108–9112.

42. Dunn SMJ, Bladen C. Kinetics of binding of dihydropyridine calcium channel ligands to skeletal muscle membranes: Evidence for low affinity sites and for the involvement of G proteins. Biochemistry 1991; 30:5716–5721.

43. Dunn SMJ, Bladen C. Low affinity sites for 1,4-dihydropyridines in skeletal muscle transverse tubule membranes revealed by changes in the fluorescence of felodipine. Biochemistry 1992; 31:4039–4045.

44. Brown AM, Kunze DL, Yatani A. Dual effects of dihydropyridines on whole cell and unitary calcium currents in single ventricular cells of guinea pig. J Physiol 1986; 379:475–514.

45. Ohkusa T, Carlos AD, Kang J-J, et al. Effects of dihydropyridines on calcium release from the isolated membrane complex consisting of the transverse tubule and sarcoplasmic reticulum. Biochem Biophys Res Commun 1991; 175:271–276.

46. Fleckenstein A, Fleckenstein-Grün G. Effects of and the mechanism of action of calcium antagonists and antianginal agents. In: Sperelakis N, ed. Physiology and Pathophysiology of the Heart. Norwell, Boston, MA: Kluwer Academic; 1989:471–491.

47. Bers DM. Excitation-Contraction Coupling and Cardiac Contractile Force. Norwell, Boston, MA: Kluwer Academic; 1991.

48. Armstrong CM, Bezanilla FM, Horowitcz P. Twitches in the presence of ethylene glycol bis-(aminoethylether)-N,N'-tetraacetic acid. Biochim Biophys Acta 1972; 267:605–608.

49. Frank GB. Roles of extracellular and "trigger" calcium ions in excitation-contraction coupling in skeletal muscle. Can J Physiol Pharmacol 1982; 60:427–439.

50. Lutgau HC, Gottschalk G, Berwe U. The effect of calcium and calcium antagonist on excitation-contraction coupling. Can J Physiol Pharmacol 1986; 60:717–723.

51. Beaty GN, Cota G, Nicola Siri L, et al. Skeletal muscle Ca^{2+} channels. In: Venter JC, Triggle D, eds. Structure and Physiology of the Slow Inward Calcium Channel. New York: Aian R 1987:123–140.

52. Schwartz LM, McCleskey EW, Palade PT. Dihydropyridine receptors in muscle are voltage-dependent but most are not functional calcium channels. Nature 1985; 314:747–751.

53. Lamb GD. Ca^{2+} channels or voltage-sensors? Nature 1991; 352:113.
54. Rios E, Brum G. Involvement of dihydropyridine receptors in excitation-contraction coupling in skeletal muscle. Nature 1987; 325:717–720.
55. DeJongh KS, Merrick DK, Catterall WA. Subunits of purified calcium channels: A 212 kDa form of α_1 and partial amino acid sequence of a phosphorylation site of an independent β subunit. Proc Natl Acad Sci USA 1989; 86:8585–8589.
56. Curtis BM, Catterall WA. Reconstitution of the voltage-sensitive calcium channel purified from skeletal muscle transverse tubules. Biochemistry 1986; 25:3077–3083.
57. Gutierrez LM, Brawley RM, Hosey MM. Dihydropyridine sensitive calcium channels from skeletal muscle. I. Roles of subunits in channel activity. J Biol Chem 1991; 266:3287–3293.
58. Nunoki K, Florio V, Catterall WA. Activation of purified calcium channels by stoichiometric protein phosphorylation. Proc Natl Acad Sci USA 1989; 86:6816–6820.
59. Dunn SMJ. Voltage-dependent calcium channels in skeletal muscle transverse tubules: Measurements of calcium efflux in membrane vesicles. J Biol Chem 1989; 264:11053–11060.
60. Mikami A, Imoto K, Tanabe T, et al. Primary structure and functional expression of the cardiac dihydropyridine-sensitive calcium channel. Nature 1989; 340:230–233.
61. Snutch TP, Leonard JP, Gilbert MM, et al. Rat brain expresses a heterogeneous family of calcium channels. Proc Natl Acad Sci USA 1990; 87:3391–3395.
62. Biel M, Ruth P, Hullin R, et al. Primary structure and functional expression of a high voltage-activated calcium channel from rabbit lung. FEBS Lett 1990; 269:409–412.
63. Snutch TP, Tomlinson WJ, Leonard JP, et al. Distinct calcium channels are generated by alternative splicing and are differentially expressed in mammalian CNS. Neuron 1991; 7:45–57.
64. Koch WJ, Ellinor PT, Schwartz A. cDNA cloning of a dihydropyridine-sensitive calcium channel from rat aorta. J Biol Chem 1990; 265:17786–17791.
65. Hui A, Ellinor PT, Krizanova O, et al. Molecular cloning of multiple subtypes of a novel rat brain isoform of the α_1 subunit of the voltage-dependent calcium channel. Neuron 1991; 7:35–46.
66. Williams ME, Feldman DH, McCue AF, et al. Structure and functional expression of α_1, α_2 and β subunits of a novel human neuronal calcium channel subtype. Neuron 1992; 8:71–84.
67. Seino S, Chen L, Seino M, et al. Cloning of the α_1 subunit of a voltage-dependent calcium channel expressed in pancreatic β cells. Proc Natl Acad Sci USA 1992; 89:584–588.
68. Mori Y, Friedrich T, Kim MS, et al. Primary structure and functional expression from complementary DNA of a brain calcium channel. Nature 1991; 350:398–402.
69. Perez-Reyes E, Wei X, Castellano A, et al. Molecular diversity of L-type calcium channels. J Biol Chem 1990; 265:20430–20436.

70. Swandulla D, Armstrong CM. Fast deactivating calcium channels in chick sensory neurons. J Gen Physiol 1988; 92:197–218.
71. Blaustein MP, Creutzfeldt O, Grunicke H, et al. Reviews of physiology, biochemistry and pharmacology. Heidelberg: Springer-Verlag; 1990:107–207.
72. Gutnick MJ, Lux HD, Swandulla D, et al. Voltage-dependent and calcium-dependent inactivation of calcium channel current in identified snail neurones. J Physiol (Lond) 1989; 412:197–220.
73. Porzig H. Pharmacological modulation of voltage-dependent calcium channels in intact cells. In: Blaustein MP, Creutzfeldt O, Grunicke H, et al., eds. Reviews of Physiology, Biochemistry and Pharmacology. Heidelberg: Springer-Verlag; 1990:209–262.
74. Hagiwara N, Irisawa H, Kameyama M. Contributions of two types of calcium currents to the pacemaker potentials of rabbit sino-atrial node cells. J Physiol (Lond) 1988; 395:233–253.
75. Nilius B, Hess P, Lansman JB, et al. A novel type of cardiac calcium channel in ventricular cells. Nature 1985; 316:443–446.
76. Cota G, Stefani E. A fast-activated inward calcium current in twitch muscle fibres of the frog (Rana montezume). J Physiol (Lond) 1986; 370:151–163.
77. Hirning LD, Fox AP, McCleskey EW, et al. Dominant role of N-type Ca^{2+} channels in evoked release of norepinephrine from sympathetic neurons. Science 1988; 239:57–61.
78. Cruz LJ, Olivera BM. Calcium channel antagonists: ω-Conotoxin defines a new high affinity site. J Biol Chem 1986; 261:6230–6233.
79. Miller RJ. Multiple calcium channels and neuronal function. Science 1987; 235:46–52.
80. Kerr LM, Yashikiami D. A venom peptide with a novel presynaptic blocking action. Nature 1984; 308:282–284.
81. Quastel DM, Saint DA, Guan YY. Does the motor nerve terminal have only one neurotransmitter release system and only one species of Ca^{2+} channel. Soc Neurosci Abst 1986; 12:28.
82. Nachsen DA. The early time course of potassium-stimulated calcium uptake in presynaptic nerve terminals isolated from rat brain. J Physiol 1985; 361:251–268.
83. Reynolds IJ, Wagner JA, Snyder SH, et al. Brain voltage-sensitive calcium channel subtypes differentiated by ω-conotoxin fraction GVIA. Proc Natl Acad Sci USA 1986; 83:8804–8807.
84. Middlemiss DN, Spedding M. A functional correlate for the dihydropyridine binding site in rat brain. Nature 1985; 314:94–96.
85. Wagner JA, Snowman AM, Bismas A. ω-Conotoxin binding to a high-affinity receptor in brain: Characterization, calcium sensitivity and solubilization. J Neurosci 1988; 8:3354–3359.
86. Perez-Reyes E, Kim HS, Lacerda AE, et al. Induction of calcium currents by the expression of the α_1 subunit of the dihydropyridine receptor from skeletal muscle. Nature 1989; 340:233–236.
87. Lacerda AE, Kim HS, Ruth P, et al. Normalization of current kinetics by interaction between the α_1 and β subunits of the skeletal muscle dihydropyridine-sensitive Ca^{2+} channel. Nature 1991; 352:527–530.

88. Knudson CM, Chaudhari N, Sharp AH, et al. Specific absence of the α_1 subunit of the dihydropyridine receptor in mice with muscular dysgenesis. J Biol Chem 1989; 264:1345–1348.

89. Tanabe T, Beam KG, Powell JA, et al. Restoration of excitation-contraction coupling and slow calcium current in dysgenic muscle by dyhydropyridine receptor complementary DNA. Nature 1988; 336:134–139.

90. Tanabe T, Mikami A, Numa S, et al. Cardiac type excitation-contraction coupling in dysgenic skeletal muscle injected with cardiac dihydropyridine receptor cDNA. Nature 1990; 344:451–453.

91. Tanabe T, Beam KG, Adams BA, et al. Regions of the skeletal muscle dihydropyridine receptor critical for excitation-contraction coupling. Nature 1990; 346:567–569.

92. Tanabe T, Adams BA, Numa S, et al. Repeat I of the dihydropyridine receptor is critical in determining calcium channel activation kinetics. Nature 1991; 352:800–803.

93. Ahlijanian MK, Westenbroek RE, Catterall WA. Subunit structure and localization of dihydropyridine-sensitive calcium channels in mammalian brain, spinal cord and retina. Neuron 1990; 4:819–832.

94. Singer D, Biel M, Lotan I, et al. The roles of the subunits in the function of the calcium channel. Science 1991; 253:1499–1500.

95. Varadi G, Lory P, Schultz D, et al. Acceleration of activation and inactivation by the β subunit of the skeletal muscle calcium channel. Nature 1991; 352:159–162.

96. Brum G, Flockerzi V, Hofmann F, et al. Injection of catalytic subunit of cAMP-dependent protein kinase into isolated cardiac myocytes. Pfluegers Arch 1983; 398:147–154.

97. Lai Y, Seagar MJ, Takahashi M, et al. Cyclic AMP-dependent phosphorylation of two size forms of α_1 subunits of L-type calcium channels in rat skeletal muscle cells. J Biol Chem 1991; 265:20839–20848.

98. Mundina-Weilenmann C, Chang CF, Guitterez LM, et al. Demonstration of phosphorylation of dihydropyridine-sensitive calcium channels in chick skeletal muscle and the resultant activation of channels after reconstitution. J Biol Chem 1991; 266:4067–4073.

99. Ferrante J, Triggle DJ. Drug- and disease-induced regulation of voltage-dependent calcium channels. Pharmacol Reviews 1990; 42:29–44.

100. Yatani A, Codina J, Reeves GP, et al. A G protein directly regulates mammalian calcium channels. Science 1987; 238:1288–1292.

101. Yatani A, Imoto Y, Codina J, et al. The stimulatory G protein of adenylate cyclase, G_s, also stimulates dihydropyridine-sensitive Ca^{2+} channels: Evidence for direct regulation independent of phosphorylation by cAMP-dependent protein kinase or stimulation by a dihydropyridine agonist. J Biol Chem 1988; 263:9887–9895.

2
Calcium Signals in Cell Proliferation, Differentiation, and Death

James F. Whitfield

Ca^{2+}, the calcium ion, binding to surface sensor/receptors on the cell membrane, flowing through channels in the cell membrane, and/or surging from internal storage vesicles, "kick starts" and then drives cell cycles, triggers terminal differentiation, and ultimately kills senescent mature cells. This discussion will be divided into four parts: (1) the Ca^{2+} and associated signals, especially the cyclic AMP (cAMP) pulses, that start and then at several key points drive the proliferation of cells in the regenerating rat liver, a physiologically relevant wound-response model; (2) the dramatic liberation from Ca^{2+} control when liver cells start making Ca^{2+}-binding/activated signal proteins in excess and secreting self-stimulating (autocrine) growth factors while on the way to malignancy; (3) the roles of Ca^{2+} and other signals that drive the proliferation, trigger terminal differentiation and finally the death of skin keratinocytes; and the factors that upset the normal delicate balance between Ca^{2+}-driven proliferation, Ca^{2+}-triggered differentiation, and Ca^{2+}-induced death during skin carcinogenesis; and (4) the role of Ca^{2+} in the proliferation, differentiation, and death of colon cells and the possibility that the failure of a Ca^{2+}-sensing mechanism might be the first step on the road to polyps and colon cancer.

Cell Cycle Signals

The mammalian cell cycle commonly has four phases (Figure 2.1). It starts with the prereplicative buildup phase, or G_1 phase, which is characterized by expressions of the so-called immediate-early cell cycle genes, the syntheses of short-lived proteins, and the assembly of ephemeral enzyme complexes. This phase extends from the cell's birth at mitosis, or its emergence from quiescence, to the S phase, when the cell replicates its chromosomes. The S phase is followed by the G_2 phase, when the cell gets rid of its replication machinery and builds a set of protein kinase complexes, the maturation or mitosis-promoting factors

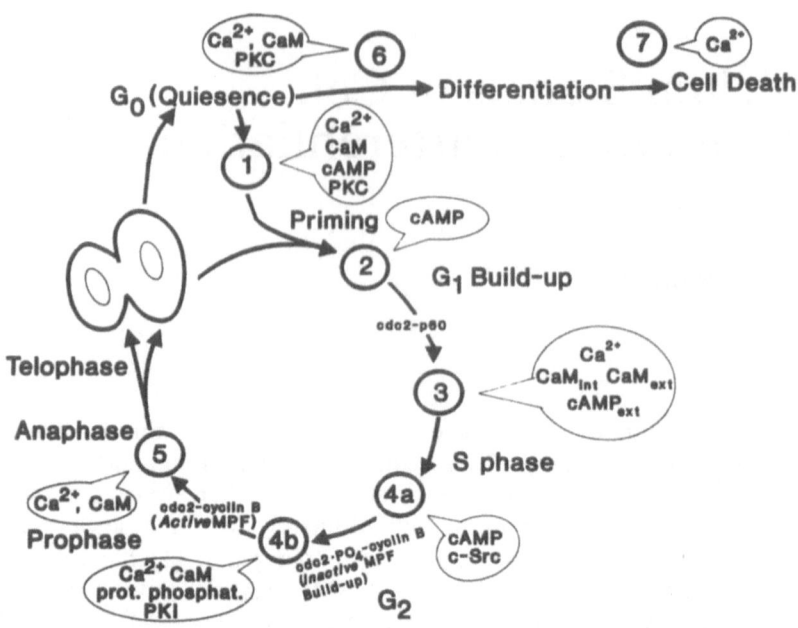

FIGURE 2.1. The 8 signals that trigger the four phases of the cell cycle (G_1, S, G_2, and M), cell differentiation, and cell death. CaM = calmodulin; CaM_{ext} = external calmodulin; CaM_{int} = internal calmodulin; cdc2-cyc = the $p34^{cdc2}$-cyclin protein serine/threonine kinase; $cAMP_{ext}$ = external cyclic AMP; $cAMP_{int}$ = internal cyclic AMP; c-src = $p60^{c-src}$ protein-tyrosine kinase; MPF = mitosis (maturation)-promoting factor; PKC = protein kinase C; PKI = cyclic AMP-dependent protein kinase inhibitor; prot. phosphat. = protein phosphatase.

(MPFs), that start mitosis.[1] Finally, there is mitosis that ends in the destruction of the MPF complexes, nuclear reconstruction, and the birth of two daughter cells.

Continuously cycling cells, such as the precursor cells in the basal layer of the epidermis or the lower two-thirds of the colon crypts, start their cell cycle clocks and the G_1 buildup to DNA replication as soon as they are born at mitosis.[1] Other cells, such as the quiescent or very slowly cycling hepatocytes in a healthy adult liver, need a priming signal (signal 1) from surface receptors to express early cell cycle genes and to get ready to start or accelerate the G_1 buildup proper. The G_1 buildup itself is then started by signal 2 from within the cell and/or from receptors on the cell surface for autocrine or exogenous progression factors. A third signal starts DNA replication (Figure 2.1).[1] Signal 4a is generated at the end of the S phase to start the G_2 buildup of inactive, phosphorylated mitosis-triggering MPFs that are complexes of the $p34^{cdc2}$ protein with mitosis-specific cyclins having serine/threonine protein kinase activity.

Signal 4b dephosphorylates and activates the MPF complexes when they have reached a critical concentration. The last signal, signal 5, is given during mitosis to trigger the anaphase migration of a complete set of replicated chromosomes to each pole of the elongating cell, to inactivate the MPF complexes that started mitosis, and to divide the cell into two daughters. The newborn daughter cells may immediately start a new cycle, but if a key growth factor or nutrient is missing, they will shut themselves down into an initially reversible quiescence (Figure 2.1). Alternatively, one or both daughters may find themselves in differentiation-promoting surroundings and stop proliferating and undergo terminal differentiation in response to signal 6, and perhaps sooner or later "commit suicide" by generating signal 7.

As we shall see, the nature of signal 1 varies according to the kind of cell and whether the cell is mortal, quiescent, and functioning normally in the animal or human, or whether the cell is immortal, cultured, and quiescent because of growth factor deprivation. Signals 2 to 5 are the *basic cell cycle signals* that drive the cycles of all cells.

What are the channels, mediators, and signals that start the different stages of the cell cycle? Are they designed specifically to stimulate proliferation? No, they are multipurpose signals which singly or in various combinations stimulate not only proliferation, but a variety of things such as muscle contraction, migration up the colon crypt, invasion of adjacent tissue, phagocytosis, and secretion of autocrine factors or neurotransmitters.

Liver Regeneration

Signal 1, The Starter

Removing 70% of a young, male, Sprague-Dawley rat's liver sends a signal to the remaining liver cells to express blocks of immediate–early and early cell cycle genes.[1,2] The short-lived products of these first-response genes get the cells growing and, by 8 h, the cells are ready to respond to signal 2 that triggers the G_1 buildup to DNA replication.[2]

The signal consists of surges of Ca^{2+}; bursts of adenylate cyclase, phospholipase A_2, and phospholipase C isozyme activities; a transient surge of nuclear Ca^{2+}-calmodulin-dependent protein kinase activity; the first of two cAMP pulses; a transient surge of ornithine decarboxylase (ODC) activity; a transient surge of prostaglandin E_2; and a transient surge of membrane-associated protein kinase C activity.[1-4] The cAMP pulse increases the impact of Ca^{2+} signals by opening Ca^{2+} channels and stimulating the accumulation of the Ca^{2+}-binding, signal-transducing protein, calmodulin, the level of which plateaus at around three times the basal level during the middle stage of prereplicative development.[1,2] cAMP may increase calmodulin accumulation by stimulating the translation of calmodulin II gene transcripts.[1,2] Alternatively, the cyclic nucleotide may reduce calmodulin degradation by inhibiting the ubiquitination (on lysine 115) that marks the protein for destruction by the ubiquitin-dependent proteolytic pathway.[1,2]

What are the liver's first signaling molecules? Hepatocytes have receptors for calcitonin, glucagon, glucagon-like intestinal peptides (GRP), prolactin, parathyroid hormone (PTH), and vasopressin, all of which surge into the portal blood during the first 15 min to 2 h after partial hepatectomy. The parathyroid gland's chief cells are stimulated to release PTH prolongedly and maximally by the large drop (1 mg/100 ml) in the blood Ca^{2+} concentration that is caused by the operation and is probably triggered by trauma-induced releases of calcitonin from the thyroid gland and interleukin-1 (IL-1) from the Kupffer cells in the liver remnant.[1] Parathyroid hormone is a general wound-response hormone that affects the differentiation, functions, and proliferation of a variety of cells including bone marrow cells, dermal and kidney fibroblasts, keratinocytes, osteoblasts, thymic lymphoblasts, and, most importantly in the present context, hepatocytes.[2] Indeed, the postoperative hypocalcemia-induced PTH release may be responsible, at least in part, for the stimulation of thymic lymphoblast proliferation that follows partial hepatectomy.[5] However, PTH receptor activity is not essential for an effective first signal: other members of the signal chorus can substitute for it when the parathyroid glands are removed before partial hepatectomy. On the other hand, the large surge of blood-borne vasopressin that peaks at 2 hours after partial hepatectomy is essential, because the regeneration response is greatly impaired in Brattleboro rats, which do not make the hormone.[1,2] Another first signaler is the early surge of prostaglandin E_2 mentioned above. Indeed, arachidonic acid, the prostaglandin precursor, which is probably released along with its various metabolites from the liver remnant's Kupffer cells, appears to be a paracrine hepatocyte activator. Indeed, it has been shown that arachiclonic acid is the only thing that can stimulate the proliferation of deeply quiescent mature hepatocytes in primary neonatal rat liver cultures.[1,2] The prompt surge of blood-borne prolactin, probably another response to the surgery-induced drop in the plasma Ca^{2+} concentration, may be another first signal.[1] There is reason to believe that prolactin may actually reach the hepatocyte nucleus, where it stimulates protein kinase C and thus increases gene expression by activating transcription factors and the transport of processed gene transcripts through the nuclear pore complex.[1,7] Finally, there is the hepatocyte growth factor (consisting of a 69 kDa α-subunit and a 34 kDa β-subunit), a paracrine/endocrine factor produced by Kupffer cells and sinusoidal endothelial cells, as well as kidney, lung, and spleen after partial hepatectomy.[6]

There are two models of liver regeneration. According to one model, all hepatocytes are capable of responding to signal 1, but it is the periportal hepatocytes that actually respond, simply because they have the greatest access to oxygen and the blood-borne signals.[1] According to the second model, only the periportal hepatocytes are proliferatively competent progenitor cells and can respond to partial hepatectomy.[1] These cells have a very long (888-h) cycle. They turn off their proliferative machinery and commit all of their resources to functioning when they migrate into the functioning compartment of the liver acinus.[1] The functioning, proliferatively inactive cells migrate slowly into the region around the terminal hepatic vein, where they senesce, "commit suicide" by the Ca^{2+}-mediated "apoptotic" mechanism, and are promptly engulfed by their neighbors.[1] One of the the key Ca^{2+}-induced events in the "apoptosing" hepatocytes is the induction and activation of the Ca^{2+}-dependent protein-glutamine transferase (tissue transglutaminase) which cross-links proteins by forming interchain ε-(γ-glutamyl) lysine linkages.[8-10] Because of this, the dying hepatocytes are converted into wrinkled spherical structures with cross-linked protein shells that resemble the terminally differentiated epidermal keratinous squames (discussed below).[9] According to this model, signal 1 from the partial hepatectomy-induced signalers stimulates the periportal hepatocytes to speed up their cycle and their migration toward the terminal zone.[1]

The blood-borne prolactin and vasopressin, activating prolactin receptors and vasopressin VP1 receptors, cause a phospholipase C-induced breakdown of phosphatidylinositol-4,5-bisphosphate into inositol-1,4,5-trisphosphate (IP_3) and short-lived diacylglycerols.[11] The large surge of blood-borne PTH does the same thing, but it also stimulates hepatocyte adenylate cyclase.[1,2] IP_3 triggers the release of Ca^{2+} from calcium-storage vesicles that have IP_3 receptors while the released Ca^{2+} triggers the release of Ca^{2+} from other calcium-storage vesicles that might, for example, have ryanodine receptors instead of IP_3 receptors.[12] At the same time Ca^{2+} channels in the plasma membrane are opened when phosphorylated by the first surge of cAMP-dependent protein kinase activity or by the activated $GTP.G_{s\alpha}$ protein complexes generated by the binding of glucagon; glucagon-related peptides (GRP), PTH, and/or paracrine (which is produced by Kupffer cells); or prostaglandin E_2 to their receptors.[1,4] The most likely result of this complex interplay of signals and first mediators is a train of Ca^{2+} oscillations and waves of Ca^{2+} that spread out through the cell from the signaling foci on its apical (sinusoidal) surface.[13] The surging Ca^{2+} binds to, and activates, the hepatocyte's distinctive selection from the 170 possible members of the Ca^{2+}-binding protein family.[14] These Ca^{2+}-activated proteins, Ca^{2+}-calmodulin in particular, then activate protein kinases and phosphatases and stimulate cytoskeleton-mediated functions.

In view of the widespread belief in the mitogenicity of Ca^{2+} and the Ca^{2+}-dependent protein kinase C, it is surprising that the phosphoinositide-degrading signals from activated vasopressin VP1 or α_1-adrenergic receptors do not activate hepatocyte proliferation in the rat.[1,2] Indeed, phosphoinositide hydrolysis is either only one part of, or is not involved in, the first signals that start various normal cells cycling in the animal. For example, receptors that activate phospholipase C isozymes (which in turn break down phosphoinositides to trigger Ca^{2+}-surges and bursts of protein kinase C activity in parotid and thyroid glands) stimulate secretion but not proliferation.[1,2] This signal pathway seems to have developed to stimulate cytoskeleton/micromuscle-driven activities such as locomotion, phagocytosis, and secretion.[1,2] However, the Ca^{2+} and protein kinase C signals induced by phospholipid breakdown products are needed by some cells (eg, T-lymphocytes) to proliferate because they stimulate the secretion of signal-2-activating autocrine factors and the synthesis and surface display of the receptors that start the G_1 buildup proper. It appears that the protein kinase C component of this signal pathway somehow becomes associated with proliferogenesis in immortalized cells such as the promiscuously responsive Swiss albino 3T3 mouse fibroblast and carcinogen-initiated colon epithelial cells and skin keratinocytes.

It seems that cAMP is part of the first signals in several kinds of normal cell in the animal. Indeed, cAMP, or an adenylate cyclase stimulator, is a part of the signals that proliferatively activate hepatocytes, parotid gland acinar cells and thyrocytes in the rat while phosphoinositide breakdown products appear only to stimulate product secretion and replacement.[1,2] Adrenocortical parenchymal cells are proliferatively activated by transient Ca^{2+} surges that are triggered by high-affinity ACTH receptors and a cAMP transient that is triggered by adenylate-cyclase-linked, low-affinity ACTH receptors.[1,2] It seems likely that diurnal fluctuations in steroid production are controlled by Ca^{2+} signals from high-affinity ACTH receptors. However, higher levels of circulating ACTH that are caused by a persistent severe steroid shortage stimulate parenchymal cell proliferation and thus increase the gland's production capacity by activating the adenylate-cyclase-coupled, low-affinity ACTH receptors.[1,2] Another example seems to be the parathyroid gland's chief cells, which have a divalent cation (Ca^{2+}, Ba^{2+}, Sr^{2+}) receptor that senses and responds to a lowering of the external Ca^{2+} concentration by generating a burst of adenylate cyclase activity and cAMP production, which then inhibits intracellular parathyroid hormone degradation and stimulates hormone secretion.[1] If the secreted hormone cannot remedy the external Ca^{2+} deficiency, the cAMP signal persists and stimulates the chief cells to proliferate, thus increasing the gland's population of parathyroid hormone producers.[1] Finally, the multi-track signal that activates small T-lymphocytes consists of Ca^{2+} transients and a burst of membrane protein kinase C activity, as well as a cAMP pulse triggered by the co-mitogenic IL-1.[1,2] This signal stimulates the expression of the interleukin-2 (IL-2) and IL-2-receptor genes, whose products establish an autocrine/paracrine loop that generates the signal 2 in the growing, signal 1-"kick-started" lymphocytes to start the G_1 buildup proper.[1,2]

In summary, the key consequence of the Ca^{2+} waves and surges of cAMP-dependent protein kinase and protein kinase C activities, as well as protein-tyrosine kinase and protein-phosphate activities that make up signal 1, are: the activation and release of gene transcription regulators from inhibitors and cytoskeletal attachment sites, stimulation of the transport of these and other proteins into the nucleus, inhibition of the breakdown of the mRNA transcripts of early cell cycle genes and ribosomal RNA, and the stimulation of protein synthesis.[2,15-17] As a result, the stimulated cell starts expressing early cell cycle genes, switches into the proliferation mode, and starts growing and developing into a state of readiness to generate and respond to signal 2.

Signal 2, the G_1 Buildup Trigger

The signal 1-activated liver cells grow into a state of competence to start the G_1 buildup in response to signal 2, which is a second pulse of cAMP.[1,2] This second or middle stage seems not to be Ca^{2+}-dependent, although regenerating rat liver cells (and other cells such as BALB/c 3T3 mouse fibroblasts and human foreskin fibroblasts) start loading their calcium-storage vesicles in preparation for a later round of Ca^{2+} signaling.[2]

The second pulse of cAMP and cAMP-dependent protein kinase activity plays an essential role in the initiation of DNA replication in a wide variety of cells such as the budding yeast *Saccharomyces cerevisiae*, diatoms, *Euglena*, *Tetrahymena*, worms, quail oviduct cells, and various human, mouse, and rat cells.[1,2,18] In quiescent cells, such as hepatocytes after partial hepatectomy, this cAMP pulse follows the pulse generated by signal 1, the master "Zeitgeber" (pacemaker) of the cell cycle clock, and is followed by a third pulse just before mitosis. On the other hand, the "Zeitgeber" in the progeny of continuously cycling cells is something that happens during, or upon the completion of, mitosis. In this case there are only two cAMP pulses, one before DNA replication and one before mitosis.[1,2]

In regenerating rat liver cells, the signal 1 cAMP pulse subsides 6 to 8 h after partial hepatectomy, but the cAMP level soon rises again to a peak by 12 h.[1,2] This second pulse results from adenylate cyclase stimulation by glucagon or, more likely, intestinally-derived GRP (which peaks in the portal blood around 12 h) and reduced cAMP degradation by cyclic nucleotide phosphodiesterase.[1,2] This second cAMP "tick" of the cellular clock has been proven to be necessary for the initiation of DNA replication in budding yeast, B- and T-lymphocytes, *Euglena*, parotid gland acinar cells, skeletoblasts, and our model regenerating rat liver cells.[1,2,18] The crucial role of cAMP and cAMP-dependent protein kinase activity in the G_1 buildup has been demonstrated recently and convincingly by Sheffield,[19] who found that DNA replication in mouse mammary cells induced to cycle by stimulation with fresh fetal bovine serum was inhibited by preventing the rise in cAMP-dependent protein kinase with oligonucleotides that were antisense to the translation-start region of the kinase's catalytic subunit mRNA.

Although the first cAMP pulse activates the type I cAMP-dependent protein kinase at the cell surface and thus participates in surface events such as the activation of Ca^{2+} channels and cytoplasmic protein synthesis, the G_1 cAMP pulse stimulates things in the nucleus.[1,2] To do this, the G_1 surge somehow selectively targets the perinuclear type II cAMP-dependent protein kinases, whose chromatin-binding regulatory RII subunits and the cAMP-induced ternary complexes ($[cAMP]_2.RII_2.C$) of which they are a part are programmed to move into the nucleus,[2,20,21] where they phosphorylate and activate transcription-enhancing proteins bound to the so-called cAMP-responsive element (CRE) octamers (5'-TGACGTCA-3') of specific genes.[1,22] An indication of the importance of cAMP-responsive gene activities at this stage of the G_1 buildup is a three- to fivefold surge of expression of the gene coding for the CRE-BP1 (cAMP-responsive element-binding protein 1) transcription factor, which binds to CREs (as either a homodimer or a heterodimer) with the c-Jun protein transcription factor, and is a target of incoming cAMP-dependent protein kinase complexes.[23-25]

The cAMP pulse may also drive protein kinase C into the nucleus, which would then phosphorylate the lamin B protein on the inner nuclear membrane and gene transcription factors in the chromatin, activate the chromatin-shaping topoisomerases that would change the pattern of chromatin loop displays to make new blocks of genes accessible to transcription factors and the transcription machinery, and promote the transport of mRNAs through the nuclear pore complex into the cytoplasm.[1,2] The particular ability of perinuclear type II cAMP-dependent protein kinase to affect nuclear events when activated by the second cAMP pulse is dramatically illustrated by isoproterenol-stimulated parotid gland acinar cells, in which this pulse stimulates the phosphorylation of 110 and 130 kDa nucleolar proteins that were unaffected by the type I cAMP-dependent protein kinase (in turn stimulated by a large signal 1 cAMP pulse).[1,2] The surge of type II cAMP-dependent protein kinase subunits into the nucleus may also stimulate the expression of the c-H-*ras* and c-K-*ras* proto-oncogenes later in the G_1 buildup of regenerating rat liver cells.[1,2] cAMP signal surges also promote the delivery and attachment of gene transcripts to the ribosomes by activating a messenger RNA transport protein, which delivers RNA messages from the nucleus to cytoplasmic ribosomes, where translation of these messages is promoted by the increased amounts of calmodulin already generated in response to the first (signal 1) cAMP pulse.[1,2]

The G_1 cAMP surge may have yet another job in the buildup to DNA replication. Besides the production of the many enzymes and other components needed to replicate chromosomes, it seems that the products of one or more growth suppressor genes must also be inactivated if the G_1 buildup is to go to completion. One of these suppressors may be the under-phosphorylated form of the 110 kDa RB protein product of the retinoblastoma-susceptibility gene, an ubiquitous nuclear phosphoprotein which affects the activities of proliferation-related genes such as c-*fos* and c-*myc*, and the genes coding for the proliferation-suppressing and differentiation-promoting TGFs-β by binding to the so-called RCE control elements of these genes.[1,26,27] It appears that the ability of the RB protein to bind to its nuclear targets is impaired, and the suppressor protein is thus inactivated, by phosphorylation in middle-to-late G_1 phase.[1,26,28] The fact that this happens along with the G_1 cAMP pulse suggests that RB protein phosphorylation and inactivation are directly or indirectly the results of the nucleus-seeking type II cAMP-dependent protein kinase after its activation by the G_1 cAMP pulse. Inactivation of the RB suppressor in late G_1 phase might be the result of both cAMP-dependent protein kinase activity and the activity of the p33^{cdk2} (G_1-specific) cyclin protein kinase, which is also needed for the initiation of DNA replication and is related to, but may have different substrate specificities from, the p34^{cdc2}-cyclin B protein kinase that triggers prophase.[1,2] The linkage between the equally essential surges of p33^{cdk2}-cyclin and cAMP is indicated by the fact that the initiation of DNA replication in the budding yeast *Saccharomyces cerevisiae*, like that of DNA synthesis in rat liver or mouse mammary cells, requires a burst of cAMP-dependent protein kinase activity that phosphorylates and activates the yeast cell's p34^{cdc2} homolog.[1,2] The cAMP-activated p34^{cdc2} in the yeast cell then induces the expression of unstable (probably because they have a protease-cleaving "destruction box" built into them) G_1-specific cyclins, which combine with the phosphorylated (exclusively on serine-277)/activated p34^{cdc2} to form the p34^{cdc2}-cyclin complexes that are needed to initiate DNA replication.[1,29] The similarity of the cAMP-dependent and p34^{cdc2} homolog-dependent initiation of DNA replication in the budding yeast and mammalian cells is supported by three facts: (1) the cAMP-dependent and p34^{cdc2}-cyclin-dependent duplication of the yeast's equivalent of the mammalian cell's centrosome/centriole complex, the spindle pole body, is an early event that leads to DNA replication; (2) the mammalian cell's type II cAMP-dependent protein kinase, p34^{cdc2}, and centrioles cohabit the centrosome; and (3) centriole duplication, like spindle pole body duplication in the yeast, is coupled to the initiation of DNA replication.[2] Perhaps the most important evidence linking p33^{cdk2}, the G_1 cAMP pulse, and type II cAMP-dependent protein kinase in the G_1 buildup of mammalian cells is the presence of type II cAMP-dependent protein kinase-p34^{cdc2} complexes in human fibroblasts.[30]

p34^{cdc2} may, in turn, be linked to the RB suppressor protein in regenerating liver and other normal mammalian cells. Indeed, the RB protein has several consensus sequences for, and is a substrate of, the p34^{cdc2}cyclin protein kinase in vitro.[1,26,31] Thus one of the important functions of the G_1 cAMP pulse during the G_1 phase may be to stimulate the accumulation of the p33^{cdk2}-cyclin protein kinase, which then hyperphosphorylates and inactivates the RB suppressor, and thus releasing a replication-triggering event.

A possible scenario for the derepression of RB-suppressed genes is suggested by recent experiments of Bandara et al.[32] Active, underphosphorylated RB inhibits the expression of certain genes by complexing with a transcription enhancer, such as the DRTF1 protein bound to their regulatory elements. Cyclin A appears and binds to the gene-bound RB-DRTF1 complexes. The G_1 cAMP pulse phosphorylates the inactive p33^{cdk2} protein kinase, which then binds to the cyclin A in the gene-bound RB-DRTF1-cyclin A complexes. The now-bound p33^{cdk2} is activated and phosphorylates RB, which dissociates from the DRFT1 transcription factor. The now-active DRTF1 transcription factor stimulates late G_1 gene expression.

A 47 kDa protein transcription factor currently being studied in this laboratory appears to be another RB-like suppressor, which in this case binds to the CRE octamer in the regulatory elements of genes such as the c-*fos* proto-oncogene.[1] Phosphorylation or some other change in this factor directly or indirectly induced by a burst of cAMP-dependent protein kinase activity causes it to separate from its chromatin-binding site, thus derepressing target-gene activity.[1] This RB-like behavior contrasts sharply with that of the other cAMP-responsive protein transcription factors, such as CREB and CRE-BP1, which are positive gene regulators that bind more tightly to their target CRE elements and, when phosphorylated by cAMP-dependent protein kinase, stimulate gene transcription.[1,22-25] It should be noted that there ar other transcription inhibitors, such as the α and β isoforms of the 26 kDa cAMP-responsive element modulator (CREM) proteins, that inhibit the stimulation of the transcription of cAMP-responsive genes (such as c-*fos*) by heterodimerizing with the CREB protein or by forming homodimers that displace the CREB homodimers from the gene's cAMP-responsive element.[33,34]

Another suppressor gene encodes the p53 suppressor protein.[35] This protein appears to move into the nucleus where it restrains the initiation of DNA replication.[35] Keeping the wild-type p53 protein from reaching the nucleus by complexing it with a loss-of-function mutant p53 protein prevents it from suppressing cell proliferation.[35] Deletion or loss-of-function mutations of the p53 gene are the most common genetic changes in human cancer.[35] p53 gene expression starts to increase soon after partial hepatectomy and peaks about 14 h later near the end of the G_1

buildup.[2] The protein checks for chromosome damage and prevents a possibly disastrous premature onset of DNA replication. When chromosome damage is repaired and all of the key replication components have accumulated to their threshold levels, the p53 protein is inactivated by signal 3.

Signal 3 and the Climax of the G_1 Buildup

When the G_1 cyclic AMP surge and total messenger RNA content have peaked, the cells are poised to respond to signal 3 and thus raise the curtain on the grand finale of the G_1 buildup (Figure 2.1). All non-neoplastic cells, be they fibroblasts, lymphoblasts, or regenerating rat liver cells, and possibly colon cells and keratinocytes, now need both internal and external Ca^{2+} and, less certainly, functioning Ca^{2+} channels to generate signal 3 and start replicating DNA.[1,2]

A lot now happens both in the cell and, perhaps unexpectedly, on the cell surface. At this point (16 to 18 h after partial hepatectomy) the epithelial growth factor (EGF) level in the portal blood has peaked and the hepatocytes have begun making transforming growth factor-α (TGF-α) and have set up an autocrine loop with these factors using the shared EGF receptor.[1] The cells have also started vigorously endocytosing EGF-EGF receptor complexes and TGF-α-EGF receptor complexes vigorously, and delivering them to the nucleus where they trigger some as yet undefined activity. This is in sharp contrast to earlier stages, when the cells would have delivered such complexes to the lysosomes for destruction of the factor and recycling of the receptors to the cell surface.[1,36,37] One way in which the incoming EGF and TGF-α may directly affect nuclear functions is to stimulate the resident nuclear protein kinase C isozyme(s) and thus stimulate gene transcription and the transport of mRNAs through the nuclear pore complex.[1] Ca^{2+} is expelled from the reloaded calcium-storage vesicles,[2] the G_1 cAMP surge crashes, and the calmodulin level drops to the basal value as they are both pumped out of the cell onto the cell surface, where they stimulate protein kinase activity and affect the activities of other surface components (such as the external domains of receptors and channels).[1,2] Late G_1-specific nuclear transcription factors now appear, which stimulate the expression of the genes coding for DNA-replication enzymes, such as thymidine kinase, thymidylate synthase, and DNA polymerase-α.[1] One of these factors is the c-*myb* protein (or the functionally equivalent a-*myb* and b-*myb* proteins) which stimulates the expression of the DNA polymerase-α gene.[38] Still other factors promote the processing of the stimulated genes' transcripts and thus start the coordinate accumulation of DNA-replicating enzymes.[1]

What now happens to one of the replication enzymes, the all-important ribonucleotide reductase (which, together with thymidylate synthase and the nucleotide kinases, makes the deoxyribonucleotide precursors of DNA), may serve as the model for the other replication enzymes.[1,2] The proliferatively quiescent cells in the intact adult rat liver express the gene coding for the 90 kDa M1 subunit of the enzyme, but they ubiquitinate and promptly destroy the nascent subunits. At the end of the G_1 buildup the liver cells stop ubiquitinating and degrading these nascent M1 subunits, which are then glycosylated and attached to M2 subunits to make holoenzyme complexes. These complexes are then finally inserted into the outer nuclear membrane, where they provide the deoxyribo-nucleoside diphosphates for processing the ultimate deoxyribonucleoside triphosphates for the DNA-replicating machines (or replitases) inside the nucleus.[1]

The exquisite Ca^{2+}-dependence of this still-unexplored stage of the G_1 buildup is vividly demonstrated by the fact that a rather small (about 50%) reduction of the blood Ca^{2+} concentration caused by parathyroidectomy selectively blocks cells at this point in the regenerating liver (as well as in the isoproterenol-stimulated rat parotid acinar cell) without affecting earlier key events. Thus regenerating liver cells (and isoproterenol-stimulated parotid gland acinar cells) still accumulate mRNA and ribosomal RNA, start growing, generate a burst of ODC activity and the two cAMP pulses, and accumulate more cAMP-dependent protein kinase RI and RII subunits. On the other hand, they do not expel Ca^{2+} from their loaded calcium-storage vesicles, and they cannot make more cAMP-dependent protein kinase catalytic subunits to match the accumulating R subunits; nor can they accumulate replication-relevant ribonucleotide reductase in the outer nuclear envelopes, accumulate active thymidylate synthase, make deoxyribonucleotides, accumulate DNA polymerase-α, and therefore cannot initiate DNA replication.[1,2] The message is clear: External Ca^{2+} somehow controls the coordinated expression and accumulation of the various parts of the DNA-replicating machinery.

The lateness of this block in the G_1 phase and the importance of external Ca^{2+} are dramatically demonstrated by the facts that a single 30-min pulse of plasma Ca^{2+} produced at any time between 12 and 15 h by an intraperitoneal injection of 2 mg of $CaCl_2$/100 g of body weight enables hepatocytes to finish the G_1 buildup in the next 3 to 4 h and start replicating DNA on time, while an intravenous infusion of Ca^{2+} during the first 8 h (but no longer) after partial hepatectomy does not.[1,2] Some Ca^{2+}-requiring factors, which, like the divalent cation receptor on the surfaces of parathyroid chief cells, operates optimally only in the presence of external Ca^{2+} concentrations between 0.5 and 1.5 to 2.0 mM, appear on the cell surface near the end of the G_1 buildup.[2] If there is not enough external Ca^{2+} around when it makes its brief appearance, the G_1 buildup of unstable components almost immediately starts dissipating and the cells can no longer be induced to replicate DNA by an injection of Ca^{2+}.

Lowering the external Ca^{2+} concentration to a suboptimal level (eg, from 1.8 mM to 0.01 mM) also arrests cultured BALB/c 3T3 mouse cells, C3H10T1/2 mouse fibroblasts, Swiss albino 3T3 mouse fibroblasts, T51B rat liver cells, and WI-38 human fibroblasts, both at the beginning and at the end of the G_1 phase.[2] Raising the external Ca^{2+} concentration to an optimal level one or two days later causes the cells still arrested in late G_1 phase to initiate DNA replication within an hour or so, while the cells that either were arrested at the start of the G_1 buildup or had reverted from a late to an initial G_1 state surge into the S phase several hours later.[2] The same ephemeral Ca^{2+}-requiring late G_1 event that is seen in regenerating rat liver cells has been seen in BALB/c 3T3 mouse cells, Swiss albino 3T3 mouse fibroblasts, and T51B rat liver cells.[2] Again, if there is not enough external Ca^{2+} around when these cells reach the critical late step, they cannot complete the G_1 buildup and the buildup that has already occurred dissipates at a rate that is determined by the cell and the culture conditions.[2]

The level of $1\alpha,25(OH)_2$vitamin D_3 (Vit.D_3) also controls the completion of the G_1 buildup. Parathyroidectomy causes the plasma Ca^{2+} concentration to drop promptly and the Vit.D_3 concentration to drop more slowly.[1] These drops are associated with a waning responsiveness of liver cells to partial hepatecomy.[1] Feeding the rats a phosphorus-free diet for 5 days before parathyroid removal prevents the fall of plasma Ca^{2+} and Vit.D_3.[1] Under these conditions, the liver cells no longer progressively lose their ability to make active DNA polymerase-α and to replicate DNA after partial hepatectomy.[1] In short, keeping the plasma Ca^{2+} and Vit.D_3 concentrations at their preoperative values maintains the proliferative responsiveness of rat liver cells to partial hepatectomy, despite the lack of the parathyroid hormone.

By the third day after parathyroidectomy, the hepatocytes no longer can be primed for DNA replication by an intraperitoneal Ca^{2+} injection between 12 and 15 h after partial hepatectomy,[1] and an injection of Vit.D_3 (eg, 200 to 400 ng/200 g of body weight), preferably administered at the time of partial hepatectomy, is now necessary.[1] Vit.D_3 does at least two important things: It brings the plasma Ca^{2+} concentration back to its normal level by the time the cells must take the critical Ca^{2+}-dependent step (between 12 and 15 h), and enables the cells to accumulate of the cAMP-dependent protein kinase catalytic subunits needed for the G_1 buildup.[1,2,19]

The G_1 cAMP pulse crashes before the onset of DNA replication, mainly because most of it is pumped out into the medium, possibly by adenylate cyclase itself with its receptor-like, or channel-like, multiple transmembrane domains.[1,2] Although this cAMP pulse and the burst of cAMP-dependent protein kinase activity it causes are needed to trigger key events leading to chromosome replication, they must not persist, otherwise they will block the initiation of DNA replication by phosphorylating and inactivating products of their own actions (such as the accumulating M2 subunit of ribonucletide reductase).[1,2] Failure to understand the inhibitory action of prolonged cAMP surges or cAMP surges that are artificially induced at the wrong time in the cell cycle has given rise to the wrong notion that cAMP normally functions to inhibit the proliferation of various normal cells.[1,2] For example, the addition of an adenylate cyclase stimulator or a cAMP analog, such as dibutyryl-cAMP or 8-Br-cAMP (which can cross the plasma membrane), to a population of proliferating cells (which are already being prodded by programmed pulses of cAMP from their ticking cell cycle clocks) stops proliferation by inhibiting ribonucleotide reductase activity.[1,2]

The secreted cAMP becomes part of another late G_1 autocrine loop, as indicated by evidence from this and other laboratories.[2] Liver cells have cAMP-dependent protein kinases (mainly type I), which accumulate on their surfaces during the G_1 buildup and are then released into the medium at the end of the buildup. BALB/c 3T3 mouse cells and T51B rat liver cells (stalled late in their G_1 buildup by external Ca^{2+} deficiency), as well as Friend murine erythroleukemia cells, can be stimulated to initiate DNA replication by cAMP, type II- (but not type I) cAMP-dependent protein kinase holoenyzme, and the common catalytic subunit of the cAMP-protein kinases, none of which is believed to enter intact cells. This special ability of external type II-cAMP-dependent protein kinase holoenzyme to stimulate DNA replication is probably due to the anti-phosphatase activity of the RII subunit, which would enhance the impact of the enzyme's catalytic subunits on target proteins on the cell's outer surface by inhibiting surface phosphatases.[1,2]

One of the many possible targets of the liver cell's surface cAMP-dependent protein kinases is basic fibroblast growth factor (bFGF) bound to receptors on the cell surface.[29] If phosphorylation increases affinity of bFGF for its high-affinity receptors, or its activity, and/or its bioavail-ability, the release of cAMP from the cell at the end of the G_1 buildup could stimulate some late G_1 event as well as release of active bFGF from the extracellular matrix to stimulate neighboring hepatocytes and stromal cells, as well as the endothelial cells needed to make new blood vessels and sinusoids for the growing remnant.

At the end of the G_1 buildup, the discharge of Ca^{2+} from storage vesicles generates cytoplasmic Ca^{2+}-calmodulin complexes.[1] Ca^{2+} may then be pumped into the nucleus by a Ca^{2+}-calmodulin-activated ATPase in the nuclear envelope,[1] which causes a surge of nuclear Ca^{2+} and then triggers the redistribution of calmodulin from the nucleoli to the nuclear surface.[1] Ca^{2+}-calmodulin binds to several nuclear proteins, some of which, such as α-fodrin and myosin light chain kinase, may be involved in the moving of messages through the nuclear pore complexes, the pulsing of the nuclear skeleton, and the streaming of nuclear contents.[1,2] Other Ca^{2+}-calmodulin complexes are associated with the DNA-replicating machinery. Indeed, this last stage of the G_1 buildup and the triggering of DNA replication in hepatocytes and other cells specifically needs Ca^{2+}-calmodulin activity. Thus the expression of calmodulin antisense mRNA prevents the initiation of DNA replication in mouse C127 cells and Ca^{2+}-calmodulin blockers, such as the naphthalenesulfonamides W7 and W13 "freeze" T51B rat liver cells and CHEF/18 hamster embryo fibroblasts on the threshold of the S phase.[1,40] Removing the inhibitor lifts the block and the cells promptly start replicating DNA.[1,2] The initiation of DNA replication in CHEF/18 hamster cells is associated with the delivery into the nucleus of a 68 kDa Ca^{2+}-calmodulin-binding protein that binds to the multienzyme DNA-replicating replitases.[1] The importance of Ca^{2+}-calmodulin-binding proteins for regulation of the replication complexes is suggested by the presence of several other Ca^{2+}-calmodulin-binding proteins in affinity-purified DNA polymerase-α.[1]

Ca^{2+}-calmodulin may also play a key autocrine role in the climax of the G_1 buildup. Four lines of evidence for this have been reviewed by Whitfield.[1,2] First, external Ca^{2+}-calmodulin stimulates DNA replication in cells such as NRK rat kidney cells and T51B rat liver cells (which have been stalled late in the G_1 buildup by external Ca^{2+} deficiency), B16 mouse melanoma cells, and *Planaria* cells. Second, the calmodulin surge in BALB/c 3T3 mouse cells, isoproterenol-activated rat parotid acinar cells, K562 human leukemia cells, regenerating rat liver and T51B rat liver cells crashes before the onset of DNA replication. In K562 cells this crash is due to the release of the signal protein into the medium. Third, inactivating the released Ca^{2+}-calmodulin with the naphthalenesulfonamide W7 attached to agarose beads (to prevent the inhibitor from getting into the cell) prevents K562 cells from initiating DNA replication. Fourth, polyclonal anti-calmodulin antibodies (which probably do not get into the cell) inhibit DNA replication by K562 cells and block the DNA replication response of late G_1 T51B liver cells to external Ca^{2+}.[1,2]

So the G_1 buildup ends in a swirl of Ca^{2+}-calmodulin-signaled and cAMP-signaled events both on and in the cell. The role of Ca^{2+} channels in these climactic events is not clear. Indeed, a sustained entry of Ca^{2+} through activated, voltage-dependent L-type channels blocks the initiation of DNA replication by GH_4 rat pituitary cells.[41] We also do not know how autocrine events to the cell surface mesh with the events going on inside the cell to trigger chromosome replication. However, it seems that the targets of secreted Ca^{2+}-calmodulin and cAMP are surface (or ecto-) protein kinases. Indeed, the onset of DNA replication in T51B rat liver cells, which have been stalled late in the G_1 buildup by external Ca^{2+} deprivation and then stimulated by Ca^{2+}, Ca^{2+}-calmodulin, cAMP, cAMP-dependent protein kinase catalytic subunit, or type II-cAMP-dependent protein kinase, is preceded by a striking burst of phosphorylation of surface (ecto-) proteins.[1,2] Thus, it seems likely that these ecto-protein kinases may signal internal events by catalyzing the phosphorylation of the external domains of channels, enzymes, receptors, or receptor-bound growth factors.[1,2,39]

Starting DNA Replication—S Phase

When the external and internal Ca^{2+}, external and internal Ca^{2+}-calmodulin (and probably other Ca^{2+}-binding proteins) and external cAMP have done their jobs, the cell is ready to start replicating DNA with replication comlexes assembled and/or activated by the 68 kDa Ca^{2+}-calmodulin-binding protein and the Master of Ceremonies (MC), which is the replication origin-binding/unwinding factor that activates the first battery of replicons.[1,2,42,43] Activated p33[cdk2]-cyclin A protein kinases may participate in this activity by phosphorylating and thus activating MC.[44] Another part of the start-up mechanism may be a partial permeabilization of the nuclear envelope, which allows the entry of triggering agents that have been piling up in the cytoplasm. This putative nuclear permeabilizer might be an EGF/TGF-α-stimulated nuclear protein kinase C that phosphorylates lamin B protein and thus partially disrupts the nuclear lamina; or it could be the late G_1-active p33[cdk2]-cyclin A protein kinase that may briefly destabilize the nuclear envelope by phosphorylating and thus depolymerizing the underlying lamin network, just as the mitosis-specific p34[cdc2]-cyclin B protein kinase does at the start of mitosis. This possibility is suggested by the fact that the fusion of G_1 cells with metaphase cells loaded with mitosis-promoting p34[cdc2]-cyclin B protein kinase triggers a brief premature burst of DNA replication in the G_1 nuclei (which is, of course, promptly terminated by chromatin condensation).[2] At any rate, once the deoxyribonucleotide-processing enzymes and replication complexes are in place on the nuclear skeleton and ribonucleotide reductase holoenzymes are at work in the outer nuclear membrane,[1] the leading replicons are activated by the replication origin-unwinding MC, and the cell starts replicating its chromosomes.

The cell is now in the S phase and on "automatic pilot." No signals from surface receptors are allowed trigger a catastrophic interruption of chromosome replication.

Signals 4 and 5 and the Initiation of Mitosis

Cells do not start mitosis before they have finished replicating their chromosomes and reached the optimal premitotic size for division. It seems that a 45 kDa, DNA-binding nuclear protein, called RCC1, prevents the premature initiation of mitosis.[45] This protein seems to act by stimulating the exchange of GDP for GTP on another nuclear protein, the 25 kDa RAN, that is closely related to the H-*ras* and K-*ras* proteins.[46] It has been suggested that the active GTP.RAN complexes cause some other protein to suppress the generation of the mitosis-initiating $p34^{cdc2}$-cyclin B protein kinase complexes.

When the chromosomes have all been replicated, the cell cycle clock ticks again and generates another cAMP pulse, a part of signal 4 that starts the G_2 buildup to the initiation of mitosis (Figure 2.1). The cause(s) of the G_2 cAMP pulse is unknown. It might be some G_2-specific receptor-activating factor, or it could be a direct receptor-independent activation of adenylate cyclase by the deformations of the cell membrane accompanying the changes in cell shape that occur at this stage of the cycle.[47] The surge could be involved in terminating chromosome replication by phosphorylating, and thus inhibiting, a key replication enzyme such as ribonucleotide reductase.[2] It may be involved in the extensive reorganization of microtubules that happens at this stage;[1] it may also inactivate the RCCl-RAN mitosis-suppressing mechanism.[45,46] The surge might prevent premature nuclear membrane breakdown by phosphorylating certain serine or threonine residues of the lamins A and C that underline and stabilize the nuclear envelope.[48,49] Once again a cAMP pulse that is associated with a $p33^{cdk2}$- or $p34^{cdc2}$-cyclin function and cAMP-dependent protein kinase may indirectly phosphorylate $p34^{cdc2}$ and/or cyclin B and thus promote the association of the two proteins and/or the adoption of a functional configuration by the $p34^{cdc2}$-cyclin B complex.[50]

The association with cyclin B targets $p34^{cdc2}$ for phosphorylation of its tyrosine-15 residue (Y15) by "wee1" kinase.[50-52] The phosphorylation of Y15 inactivates the $p34^{cdc2}$-cyclin B serine/threonine protein kinase.[50-52] This inactivation is meant to delay the initiation of mitosis until the $p34^{cdc2}$-cyclin B complexes and the cell size have reached optimal levels.

When the accumulating inactive, multiply-phosphorylated $p34^{cdc2}$ cyclin B complexes reach the critical level, signal 5 is given to activate the complexes (Figure 2.1). A key part of the signal is a burst of protein-tyrosine phosphatase ("Cdc25") activity that removes the blocking phosphate on Y15 of $p34^{cdc2}$.[50-52] Serine/threonine protein phosphatase(s) (such as protein phosphatase-1[53]) are also stimulated to dephosphorylate other key residues.[50] The cellular Ca^{2+} content rises, possibly because of the activation of L-type Ca^{2+}-channels by $p34^{cdc2}$-cyclin B activity, and there is a Ca^{2+}-calmodulin surge that may do many things including stimulating Ca^{2+}-calmodulin-dependent protein kinase II and protein phosphatase-2B activities.[1,2,40,41,53] The cAMP surge crashes and the cAMP-dependent protein kinase inhibitor (PKI) may be activated.[48,49] This switches off cAMP-dependent protein kinase activity and thus stops the phosphorylation of certain domains of the nuclear lamins A and C that was preventing premature nuclear membrane breakdown.[50,51] This opens the way for the phosphorylation of a different set of serine/ threonine residues by the now-active $p34^{cdc2}$ cyclin B complexes.

The $p34^{cdc2}$-cyclin B complexes coordinately trigger a cluster of reactions that break down the nuclear envelope down into vesicles and condense the chromosomes.[1,2,50,51] Part of the cellular $p34^{cdc2}$ is in the centrosomes along with type II cAMP-dependent protein kinase, the target of the G_2 cAMP pulse.[54,55] Activated $p34^{cdc2}$-cyclin stimulates the accumulation of pericentriolar materials and activates a protein kinase that phosphorylates microtubule-associated proteins. This destabilizes microtubules and activates a factor (EF-1β) that stimulates GTP/GDP exchange and thus promotes the formation of active 51 kDa EF-1α-GTP nucleating factor, which then triggers the assembly of astral and spindle microtubules.[54]

A prerequisite for the transition from metaphase to anaphase is the self destruction of $p34^{cdc2}$-cyclin B complexes.[56] This destruction is caused by the conjugation of the cyclins of the $p34^{cdc2}$-cyclin B complexes with ubiquitin, followed by the proteolysis of the ubiquitinated cyclins by a protease that is activated by the $p34^{cdc2}$-cyclin B complexes themselves.[50,56,57] The remaining $p34^{cdc2}$ may either be reused by the daughter cells if they should initiate a cycle or be destroyed if the daughter cells switch into a functioning mode.

At the end of metaphase Ca^{2+} is released from calcium-storage vesicles clustered in and around the mitotic spindle.[1,2] This and the resulting Ca^{2+}-calmodulin complexes[29] make up the fifth and final signal. Ca^{2+} activates the "Pac-Man"-like chromosome kinetochores that pull their chromosomes toward the spindle poles as they chew up, and thus shorten, attached kinetochore microtubules.[1,2] Ca^{2+} also activates the dynein-like motors that elongate the spindle by causing the overlapping interzonal microtubules of the two half-spindles to slide apart, and it triggers the contraction of the equatorial belt of actomyosin micro-muscles that squeezes the elongating cell into two new daughters.[1,2]

Stromal Cell Activation and the Termination of the Proliferative Response in the Regenerating Liver Remnant

The other regenerating liver cells, the stromal cells, do not seem to respond immediately to the chorus of signals that stimulates hepatocyte proliferation: they do not start replicating their chromosomes until 30 h or more after the hepatocytes.[1,2] This delay suggests that the stromal cells might be activated by a signal 1 from a paracrine growth factor(s) such as phosphorylated bFGF (produced by the hepatocytes' ecto-protein kinases activated by cAMP released in later G_1 stages), TGF-α from the hepatocytes, and/or EGF (which peaks in the portal blood 12 to 18 h after partial hepatectomy).[2,39] The production of such growth factors could be the reason, or part of the reason, why partial hepatectomy enables intraportally injected colon and mammary cancer cells to colonize the liver.[58,59] However, some stromal cells, the sinusoidal cells, start making TGF-β1 around 4 h after partial hepatectomy.[1,2] Hepatocytes do not make TGF-β1, but they have TGF-β receptors on their surfaces and are proliferatively inhibited by the factor.[1,2] It appears that when the flow of TGF-β1 from the sinusoidal cells peaks at around eight times the basal level 72 h after partial hepatectomy, it stops further hepatocyte proliferation by preventing the cells from completing the G_1 buildup.[1,2] Since injecting TGF-β1 into partially hepatectomized rats prevents the initiation of DNA replication in the liver remnant,[1] the TGF-β from the sinusoidal cells is likely to be the principal terminator of hepatocyte proliferation following hepatectomy and other proliferogenic injuries. TGF-β1 may stop the newborn hepatocytes from proliferating by suppressing the expression of the key proliferation-determining c-myc gene and by preventing the further phosphorylation and inactivation of the RB suppressor protein near the end of the G_1 buildup.[1,60]

Liver cells can also be induced to proliferate, and the liver to double its size within 4 to 8 days, by giving rats one injection of lead nitrate or 7 doses of cyproterone (100 ng/kg/day by gavage for the first 3 days; 130 ng/kg/day by gavage for the next 4 days), a synthetic sex steroid.[1,8–10] If no more lead nitrate is injected or the steroid treatment is stopped, cells begin "committing suicide" in order to return the organ to its normal size.[1] Similarly, a burst of proliferation induced by EGF in primary cultures of neonatal rat liver is followed by a wave of cell "suicide," which reduces the cell numbers to the initial value.[1] Hepatocytes "kill themselves" both in vivo and in vitro through a process known as apoptosis (Greek apo, off; pipto, to fall), or active cell death, which is used to shape organs during development and to keep various cell populations at functionally optimal sizes.[1,8–10,61,62] The key players in apoptosis appear to be a Ca^{2+} surge, a Ca^{2+}-activated nuclear endonuclease that kills the cell by cutting the chromatin into nucleosome-sized (175 base pairs) bits, and a Ca^{2+}-dependent transglutaminase that cross-links proteins near the cell surface into a tough shell.[9]

Hepatoma Cells

The proliferation of normal hepatocytes is started, or greatly accelerated, by factors released or generated in response to partial hepatectomy, and it is driven by various signals that transiently stabilize and/or induce the de novo expression of key replication-related components, such as DNA polymerase-α, ribonucleotide reductase M1 subunits, and topoisomerase II.[1,2] As we have seen, normal liver cells need a 1.0 to 2.0 mM concentration of external Ca^{2+} to activate surface divalent cation receptors and feed the cell-cycle-driving signal mechanisms (especially signal 3), and their cycling is terminated by signals from TGF-β receptors.[1,2] By contrast, hepatoma cells maintain pools of 90 kDa ribonucleotide reductase M1 subunits and topoisomerase II, proliferate spontaneously without signals from exogenous hormones in very Ca^{2+}-deficient (eg, 0.02 mM) medium or when suspended in soft agar, and they cannot be stopped by TGF-β1, perhaps because they may no longer make the TGF-β1-activatable RB-growth-suppressor protein.[1,2,63]

As expected from the importance of cAMP pulses in driving the cell cycle, there is an increased responsiveness of adenylate cyclase and an increased cAMP content in the early stages of chemically induced carcinogenes.[1,20] Indeed, cAMP may be one of the prime initiators of carcinogenesis.[1,2] The hepatoma cell's cAMP-dependent controls also seem to break down, as indicated by the loss of the 47 kDa transcription suppressor, which binds to the CRE of cAMP-responsive genes (such as the proliferation-related c-*fos* proto-oncogene.[1] The loss of this transcription suppressor may be responsible for the overexpression of the c-*fos* gene in precancerous liver and liver tumors.[1] Unfortunately there are no reports of the marked shift in the balance of the expression of the RI and RII subunits in favor of RI, which is a striking feature of colon and other cancer cells.[20]

Hepatoma cells do not need exogenous hormones to be switched into the proliferative mode or a fast-cycling, non-senescing mode. Unlike the proliferatively inactive normal hepatocytes, hepatoma cells constitutively express the c-*fos* and c-*myc* early cell cycle genes, and they make and secrete TGFα.[1,2,64] Consequently they are permanently competent and able to respond immediately to the secreted TGF-α.[1,2] Hepatoma cells also make and respond to other autocrine factors such as a bFGF-like factor and a 60 kDa hepatocyte growth factor.[1,2] Since EGF drastically reduces the amount of external Ca^{2+} needed for proliferation as dramatically as does neoplastic transformation,[2] the stimulation of EGF receptors by large amounts of autocrine TGF-α could explain the greatly reduced external Ca^{2+} requirement of hepatoma cells.[1,2] The bFGF-like factor may be responsible for the anchorage-independent proliferation of its producer cells. It may also stimulate the angiogenesis needed for sustained tumor growth as well as the proliferation of the tumor's stromal cells.[1,2]

Hepatoma cells overproduce calmodulin, which could be regarded as an autocrine growth factor since external Ca^{2+}-calmodulin stimulates the proliferation of rat liver cells and other cells stalled in late G_1 phase by external Ca^{2+} deprivation. Indeed, human K562 myeloid leukemia cells release calmodulin onto their surfaces, where it activates a process required for the initiation of DNA replication and proliferation.[1,2] Since normal cells adjust the production of calmodulin to avoid an excess of calmodulin over its target proteins, the overproduction of calmodulin by hepatoma cells must greatly affect Ca^{2+} signaling and upset the regulation of a variety of cell functions, which range from cytoskeleton activities to protein kinase and phosphatase activities.[1,40] Indeed, the constitutive overexpression of calmodulin stimulates the proliferation of mouse fibroblasts.[1,40] Hepatoma cells also make another Ca^{2+}-binding signaling protein, the oncotrophoblastic 11.7 kDa oncomodulin, which was discovered and named in this laboratory.[1,2] Although we know almost everything about oncomodulin's structure, we know nothing about its function in placenta and cancer cells. However, we know that under oxidizing conditions it can dimerize into a calmodulin-like dumbbell dimer that binds to and activates all calmodulin targets as effectively as calmodulin itself; while in its monomeric form it can inhibit glutathione reductase, an enzyme that is unaffected by calmodulin.[1,2] It is a member of the parvalbumin family of Ca^{2+}-binding proteins.[2] But it is a parvalbumin with a punch!

Parvalbumins are usually Ca^{2+}-binding/Ca^{2+}-buffering, but not signaling, proteins that do not bind to target proteins and thus simply terminate Ca^{2+} signals.[40] Excessive levels of the Ca^{2+}-buffering parvalbumins slow the progression through the G_1 and mitotic phases of the cell cycle.[1,40] However, oncomodulin does bind to target proteins and is thus a signaling parvalbumin that can stimulate DNA replication.[1,2] Oncomodulin is normally made and functions in placental cytotrophoblast cells, but not in proliferating or quiescent cells of either the embryo proper or the adult.[1,2] However, the oncomodulin gene is switched on, and oncomodulin appears during carcinogenesis in various tissues, including colon, liver, and skin.[1,2] The inhibition of glutathione reductase by the reappearing oncomodulin could promote proliferation and carcinogenesis by reducing the cellular pool of reduced glutathione (GSH), which would result in increased accumulations of peroxides and peroxide-induced arachidonic acid derivatives (which are normally restrained by the GSH-dependent activity of glutathione peroxidase). Clearly the changes in the production of these two Ca^{2+}-binding proteins (and maybe others) and the related disturbances of Ca^{2+}-signal transduction and Ca^{2+}-dependent functions must be important events in carcinogenesis.

The malignant transformation of liver cells appears to be associated with the inactivation or deletion of the chromosome-monitoring, mutation-eliminating p53 gene.[65-67] It is also accompanied by invasiveness, which requires that the cells no longer have the normally tight adhesion to neighboring cells that is partly due to another Ca^{2+}-binding protein, type E cadherin.[68,69] Indeed, malignant hepatoma cells do not make type E cadherin. Thus, in a survey of the cadherins of 24 primary human hepatomas, Shimoyama and Hirohashi[70] found that the cells of the one grade IV hepatoma in the group did not express type E cadherin and the cell-cell contacts were loose.

The growth of foci of transforming liver cells is a function of the rate of cell proliferation and the frequency of cell "suicide" by Ca^{2+}-mediated apoptosis, a homeostatic mechanism designed to eliminate excess liver tissue. Protein kinase C-activating phorbol esters promote the proliferation and survival of premalignant cells and the ultimate emergence of cancer cells in two ways. Although phorbol esters do not directly stimulate normal hepatocytes to proliferate, they do directly stimulate carcinogen-initiated hepatocytes to proliferate and at the same time they enhance tumor growth by inhibiting the triggering of tissue-homeostatic apoptotic cell death.[1]

Keratinocytes

The skin is maintained by the continuous multiplication of basal lamina-bound precursor cells which are themselves the progeny of more slowly cycling, self-renewing or immortal, pluripotent stem cells. This process is balanced by differentiation and an extraordinary kind of programmed "suicide" of the precursor cells' progeny after they lose contact with the basal lamina.[1,71] Signals from cell surface receptors for basal lamina components (eg, collagen IV, laminin), particularly the specific kinds of Ca^{2+}-binding integrin receptor that connect these external ligands to the cytoskeleton and signal-generating and transmission mechanisms,[72] keep the cell cycle clock ticking, the proliferation-related genes such as c-*myc* active, and the autocrine proliferogens flowing. In other words, the basal lamina-bound precursor cells start building up to DNA replication as soon as they are born at mitosis.

We do not know what signals drive the keratinocyte cycle, but they are likely to be the same basic ones that drive the G_1 buildup, the initiation of DNA replication, and the initiation of mitosis in regenerating rat liver cells primed by the signal-1-triggered cascade of competence-establishing events (Figure 2.1). However, we do know that cultured skin keratinocytes need a small amount (0.01 to 0.1 mM) of Ca^{2+} and cAMP to proliferate indefinitely and maintain their morphological identity.[1] We also know that signals from activated receptors for β-adrenergic agonists—EGF, bFGF, paracrine FGF-related KGF (keratinocyte growth factor), autocrine IGF-I, and probably most importantly, autocrine TGF-α—drive keratinocyte proliferation by stimulating key cell cycle events, promoting attachment to the basal lamina, and blocking differentiation.[1,2] The continuous cycling of the keratinocyte precursors is maintained by a very low external Ca^{2+} concentration, but proliferation stops and differentiation is initiated by a Ca^{2+} signal (signal 7), which stimulates phosphatidylinositol breakdown and a surge of membrane protein kinase C activity.[1] Keratinocytes cycling on the basal lamina make a skin-specific Ca^{2+}-binding/buffering protein (ScaBP) that would reduce the size of any unscheduled Ca^{2+} surge which, without this protein, could prematurely stop their proliferation.[1]

The Ca^{2+}-driven replication and differentiation mechanism must obviously be balanced to maintain a proliferating population of precursor cells and, at the same time, a steady supply of terminally differentiating progeny. One way of doing this is to keep the Ca^{2+} concentration in the basal proliferation zone below the threshold level for triggering differentiation. Indeed, the intercellular Ca^{2+} concentration in the epidermis is lowest in the basal layer and highest at the top of the granulosal layer.[1,2] The low Ca^{2+} level in the basal zone is probably suboptimal for unassisted keratinocyte cycling, but is adequate in the presence of EGF, or more likely the autocrine TGF-α, which enables the cells to cycle in the presence of a very low external Ca^{2+} concentration by activating EGF-receptors.[1,2]

We do not know what triggers differentiation in the skin. In cultured keratinocytes it can be triggered by reducing the cells' contact with their substrate.[71] It can also be triggered simply by raising the Ca^{2+} concentration in the medium from 0.05 mM to 1.4 or 1.8 mM,[2,73] which triggers a sustained 30-min to 4-h intracellular Ca^{2+} surge and increases membrane-protein kinase C activity and production of TGF-β2.[1] The same responses are triggered by the phorbol ester TPA (12-O-tetradecanoyl phorbol-13-acetate), which directly stimulates membrane-associated protein kinase C activity.[1,2] Vit.D$_3$ also promotes differentiation, probably by stimulating an intracellular Ca^{2+} surge, phosphoinositide breakdown, and a surge of membrane-associated protein kinase C activity.[1]

It might be expected from these in vitro observations that at some point after severing its ties with the basal lamina the keratinocyte stops cycling and starts its terminal differentiation program with a Ca^{2+}/protein kinase C signal. In keratinocyte cultures the synthesis of involucrin, the cornified envelope precursor, starts immediately above the basal layer, but in the skin it usually begins in the upper spinous layer well after the cell has stopped cycling.[71,74] The prime signaler might be autocrine PTHrP (parathyroid hormone-related protein), which is made and secreted only by the spinous (or prickle) cells.[1,76,77] Indeed, PTHrP (and PTH) have been shown to trigger internal Ca^{2+} surges, stop proliferation, and stimulate involucrin production and cornified envelope formation in human keratinocyte cultures.[1,76] Moreover, we have recently found that exposing mouse BALB/MK2 keratinocytes to 100 to 500 pM PTHrP[1–40] in low (0.05 mM) Ca^{2+} medium nearly doubled their membrane-associated protein kinase C activity. Since human and murine keratinocytes secrete PTHrP, and since these cells have the PTH receptors that bind this protein,[1] a PTHrP/PTH receptor autocrine loop must operate in the spinous cells. The murine skin keratinocyte PTH receptor is an unconventional one that stimulates membrane-associated protein kinase C activity, without activating adenylate cyclase.[1] Since protein kinase C can stimulate TGF-β gene expression, the membrane-associated protein kinase C signal from the PTHrP/PTH receptor autocrine loop might be the initiator of the TGF-β/TGF-β receptor autocrine loop that suppresses the transcription of the key c-*myc* cell-cycle gene (probably via the RB protein that binds to the c-*myc* gene's RCE 5'-control elements) and thus keeps the suprabasal cells proliferatively quiescent but still growing.[1,60,78-80] However, the TGF-β/TGF-β receptor autocrine loop might in turn further stimulate the transcription of the PTHrP gene and the production of PTHrP.[77] The continuing cell growth is essential for differentiation, and the cells may grow to as much as 10 times their initial size.[1,71]

The PTHrP/PTH receptor autocrine loop may stimulate the 1α-hydroxylase that catalyzes the formation of Vit.D_3 from 25(OH)vitamin D_3.[1] This, in turn, could further enhance keratinocyte differentiation by contributing to the increase of cytosolic Ca^{2+} concentration and membrane-protein kinase C activity through an action at the cell surface.[1,2]

PTHrP, like other factors secreted by the spinous cells, may also establish a paracrine loop by binding to the adenylate cyclase-linked PTH receptors on the underlying dermal fibroblasts.[1,2] The PTHrP seeping down from the spinous layer may stimulate the dermal fibroblasts to make and secrete cytokines and other factors that directly regulate the proliferation of the basal keratinocytes and thus ultimately regulate the generation of spinous cells. Raising the external Ca^{2+} concentration from 0.5 to 2.0 nM doubles the gap-junctional communication between the cells of primary mouse skin keratinocytes within 12 h.[81] In skin, the rising, growing keratinocytes continue talking to the basal cells through their Ca^{2+}-modulated gap junctions.[1] As expected from the response of cultured keratinocytes to a differentiation-triggering Ca^{2+} signal, they increase their gap-junctional communication with spinous and granulosa cells.[1]

The rising keratinocyte stops making the light, proliferation-specific K5 and K14 keratins and starts making the heavier type II K1 keratin, which, after reaching a certain level, stimulates the expression of its type I partner, the K10 keratin.[1,2] This shift to the heavier keratins may contribute to the irreversibility of the differentiated state because these keratins form large aggregates that could interfere with cell division.[1] The cell also stops making the Ca^{2+}-binding/buffering SCaBP and thus becomes vulnerable to Vit.D_3-enhanced Ca^{2+} surges.[1]

When the rising, growing keratinocyte reaches a critical size it starts accumulating involucrin, plasma membrane-anchored, epidermis-specific type I transglutaminase, and keratohyalin granules containing heavily phosphorylated 350 kDa profilaggrin.[1] Upon reaching the top of the granular layer with its high intercellular Ca^{2+} concentration, the granulosal cell, now no longer buffered by the SCaBP, somehow generates a massive influx of Ca^{2+} (signal 7) from the large intercellular reservoir that severs all intercellular gap-junctional communication and triggers the final events that kill the cell.[1] The granulosa cell seems to commit a special kind of "suicide" based on the Ca^{2+}-mediated apoptosis program.[1] However, because the Ca^{2+} surge induces the formation of a cornified envelope and network of supporting keratin struts, the dead cell does not disintegrate. The "killer" Ca^{2+} signal switches on the plasma membrane-bound type I transglutaminase, possibly in part by stimulating calcium-dependent neutral protease(s) (CANP(s) or calpain(s)), which is known to activate the transglutaminase by limited proteolysis,[82] and in part by directly stimulating this Ca^{2+}-dependent enzyme. The activated transglutaminase cross-links the accumulated involucrin molecules into a tough, cornified submembranous envelope. The Ca^{2+} signal also activates the breakdown of the 350 kDa profilaggrins of the keratohyalin granules into the 37 kDa filaggrins that bundle the heavy keratins into macrofilaments or cables.[1] The Ca^{2+} surge generates Ca^{2+}-calmodulin complexes that activate the Ca^{2+}-ATPase pumps in the cell's membrane. The Ca^{2+}-activated CANP(s) may also bind via its calmodulin-like domains to the calmodulin-binding regulatory sites of the Ca^{2+} pumps and then maximally stimulate these pumps by limited proteolysis.[83] This double stimulation ensures that once it has done its job most of the exchangeable Ca^{2+} is pumped back into the intercellular space, preventing a continuous loss of calcium through the skin.[1] The activated Ca^{2+}/Mg^{2+}-dependent endonucleases, phosphatases, phospholipases, and proteases now destroy all of the cell organelles, leaving the final functional form—a dead, cornified envelope purged of its Ca^{2+} and held together by a latticework of keratin cables.

The hypothesis that squame formation is a special case of apoptosis is strongly supported by the similarities between normal senescing hepatocytes and the excess hepatocytes in an involuting lead nitrate-induced hyperplastic liver that are known to be killed by apoptosis.[1,8-10] In both cases there is a burst of Ca^{2+}-dependent transglutaminase activity that encases the dying hepatocyte in a cross-linked protein shell very much like the keratinous squame's cornified envelope.[9]

Cancerous Keratinocytes

During carcinogenesis, basal keratinocytes no longer need attachment to the basal lamina, exogenous growth factors, and a low external Ca^{2+} concentration to proliferate, probably partly because they step up their production of autocrine growth factors, such as TGF-α, and display more receptors, such as EGF receptors, on their surface.[1,2] The cells also may increase their cAMP content, which drives their proliferation and would hold them in the proliferation mode.[1,20] Perhaps most importantly they become unable to respond to Ca^{2+} signals that in normal keratinocytes would stop the proliferation and trigger the complete differentiation program leading to cornified envelope formation.[1,84] They also become insensitive to the cell cycle-stopping action of the Ca^{2+}- and protein-kinase-C-induced TGF-β because: (1) they cannot activate the inactive secreted form of the factor; or (2) they can no longer display TGF-β receptors; or (3) they cannot respond to the TGF-β receptor signal perhaps because they no longer make active RB protein that seems to mediate suppression of the c-*myc* cell cycle gene.[78,79]

The keratinocyte's response to the protein-kinase-C-stimulating phorbol esters radically changes. Normally TPA triggers the differentiation program that may include an initial brief burst of proliferative activity, which gives way to proliferative inhibition and terminal differentiation into keratinous squames.[1,2] But TPA-treated carcinogen-initiated cells do not stop proliferating, nor do they readily differentiate into dead cornified envelopes.[1] Thus TPA-treated premalignant cells continue to proliferate while their TPA-treated neighbors stop proliferating, differentiate, and ultimately become dead squames.[1] This suggests that the continuous overproduction and secretion of the protein-kinase-C-stimulating (hence TPA-like) PTHrP by neoplastic keratinocytes might be an autocrine proliferation enhancer for the producer cells and a paracrine-proliferation-suppressing, differentiation-stimulating factor for their normal neighbors.[1,76] However, PTHrP is expressed in squamous cell carcinomas that contain differentiated keratinocytes, but not in basal cell carcinomas that do not contain differentiated cells.[76] Thus PTHrP expression remains strictly coupled to differentiation and PTHrP is made only if neoplastic keratinocytes start differentiating. Besides selectively stimulating the expansion of initiated clones at the expense of their normal neighbors, the protein kinase C stimulators set off a shower of active oxygen radicals by activating plasma membrane NAD(P)H oxidase. These radicals generate clastogenic agents and Ca^{2+}-dependent endonuclease activity that break chromosomes and cause unequal chromosome exchanges that result in gene deletions and rearrangements, which then result in the loss of tumor suppressor genes and activation of other genes with the ultimate expression of the malignant phenotype.[1,2] The protein-kinase-C-activating promoters could also stimulate the phosphorylation of the mRNA transport receptors in the nuclear pore complexes.[1] This would contribute to the expression of the emerging malignant phenotype by increasing the affinities of low-abundance messages for these nuclear pore receptors and thus "scramble" the abundance class distribution of mRNAs being transported through the nuclear pores and into the cytoplasm.[1] Eventually successful senescence-proof malignant variants appear, which are unresponsive to Ca^{2+} signals and produce the activated RAS proteins, angiogenic factors, autocrine proliferogens, hydrolases, and other things that when together enable them to invade and metastasize.[1]

Colon Cells

An immortal, self-renewing, pluripotent stem cell cycles slowly in the bottom of the colon crypt and generates a large population of rapidly cycling precursors of the different terminally differentiated colon cells.[1] These precursor cells, which express such proliferation-related genes as c-*myc*, glide along the basal lamina and up the crypt wall in step with gap-junctionally coupled, fibroblast-myocyte pericryptal cells of the underlying lamina propria.[1] The proliferation of these precursor cells is driven primarily by signals from receptors for basal lamina components of lower crypt and by a TGF-α/EGF receptor autocrine loop.[1,85] This core mechanism is modulated by the ebb and flow of other factors such as EGF and gastrin from the upper gastrointestinal tract and noradrenergic agonists from nerve nets that envelope the lower crypts.[1] These factors establish a proliferative circadian rhythm and provide input from the diet and other external stimuli.

As limited as it is, the available information[1] strongly indicates that Ca^{2+} signals drive the cycles of colon crypt cells just as they drive the cycles of activated hepatocytes, basal skin keratinocytes, and other normal cells. Cells freshly isolated from the normal human colon need about 0.1 mM Ca^{2+} in their medium to proliferate optimally, but their proliferation and the proliferation of normal rat colon cells, again like that of normal basal skin keratinocytes, are stopped by the much higher (0.8 to 2.2 mM) Ca^{2+} concentrations in the blood and fecal water. It seems that colon cells have a Ca^{2+}-sensing receptor on their surfaces (like those that control PTH secretion by parathyroid chief cells and the differentiation of myoblasts), which can give a signal to stop proliferation when the external Ca^{2+} concentration reaches a critical level.[1,2]

Evidently something must shield the cycling cells in the lower crypt from contact with the large amounts of Ca^{2+} streaming by in the feces, which otherwise would sooner or later seep down into the lowest reaches of the crypt and stop cell proliferation. The likely candidates for the anti-Ca^{2+} barrier are the Ca^{2+}-binding chrondroitin sulfates, heparan sulfates, and particularly the sialo- and sulfomucins secreted by the goblet cells concentrated in the lower regions of the crypt. A decreasing content of Ca^{2+}-binding mucins from the base to the mouth of the crypt would establish an external Ca^{2+}-gradient like that in the epidermis.[1]

The proliferation of precursor cells in the colon also requires cAMP signals and a predominance of the proliferation-promoting type I isozymes over the differentiation-promoting type II isozymes.[1] It appears that the proliferation-promoting EGF receptor signals triggered by EGF or autocrine/paracrine TGF-α increase the expression of the type I regulatory subunit gene and reduce that of the type II regulatory subunit gene.[20]

At a certain point the upwardly gliding, cycling cell emerges from behind the anti-Ca^{2+} barrier and starts receiving different signals from receptors for components of the basal lamina, extracellular matrix, and cell adhesion molecules.[1] This would be analogous to the lifting of skin keratinocytes from the basal lamina. Signals from the cell surface Ca^{2+}

sensor stop the cell from starting a new cycle in the upper crypt by stimulating the expression of a TGF-β isoform and its receptor. This would establish an autocrine loop that changes the expression of the cell's Ca^{2+}-dependent integrin and other receptors for extracellular components, down-regulates TGF-α production, and shifts the balance of the cAMP-dependent protein kinases from the proliferation-promoting type I to the differentiation promoting type II isozymes.[1,20] The differentiating cell shuts down its proliferation-related c-*myc* and c-*myb* genes and increases by 3 to 5 times the expression of its normal H-*ras* gene and to a lesser extent its normal K-*ras* gene.[1,86] The products of these normal *ras* genes may stimulate functions such as mucin secretion by the mature goblet cells. At this point one must remember that protein kinase C activity in normal colon cells is involved in functions such as secretion rather than proliferation. Indeed, protein kinase C stimulation by phorbol ester tumor promoters does not stimulate normal colon cells to proliferate as it does transforming colon cells.[1]

To quote Pienta et al.: "What a cell touches determines what a cell does."[87] The rising colon epithelial cell starts touching a changing extracellular matrix differently. The Ca^{2+}-TGF-β signals have stopped the cell from cycling, and they have also stimulated the expression of a new set of surface receptors such as different types of the Ca^{2+}-dependent integrins that couple the cytoskeleton to the exptracellular matrix.[87–89] These receptors are activated when the cells binds to extracellular matrix components (eg, collagen I); they send signals to the nucleus that stop the expression of genes encoding replication-related components, and they trigger the coordinate expression of genes whose products determine specific cell shape and functions.[88–90] The composition of the extracellular matrix is also determined by products of the fibroblast-myocytes of the pericryptal lamina propria, which respond to paracrine factors such as TGF-β from the epithelial cells.[88–90]

An indication of how signals from cell surface components may dampen proliferation and promote differentiation is provided by neural cell-adhesion molecules (N-CAMs). Thus, one N-CAM, the placental LAR receptor, appears to have two protein-tyrosine phosphatase domains in its cytoplasmic portion.[53] When expressed on the cell surface and activated by cell–cell adhesion, such receptors can stop cell cycling and promote differentiation by dephosphorylating, and consequently deactivating, autophosphorylated active mitogenic protein-tyrosine kinase receptors and their substrates.[53]

Some cells become absorptive enterocytes with apical brush borders that face the colon lumen armed with hydrolytic enzymes such as disaccharidases, peptidases, and Na-dependent glucose co-transporters that collect water and nutrients from the passing fecal stream. Their basolateral borders contain $α_2$- and β-adrenergic receptors, VIP (vasoactive intestinal peptide) receptors, and outwardly directed Na-independent GluT2 transporters to release the apically absorbed glucose into the blood.[1] Other cells become goblet cells, most of which stay in the crypt to produce the mucins of the Ca^{2+} shield.

Upon emerging from the crypt, the now mature, functioning cell becomes part of the colon's flat mucosal surface. As long as the cell is not neoplastically transformed, it will not, indeed it cannot, proliferate because of its functioning TGF-β receptor autocrine loop and the high Ca^{2+} content of the fecal stream.[1] Eventually the cell is sloughed off the basal lamina at an extrusion zone and swept away in the feces. Actually, it is not known whether the terminal cell is simply pushed off the basal lamina by the pressure of cells coming up behind it and cells coming from neighboring crypts, or whether, like keratinocytes, it also "kills" itself at the appropriate moment by taking up Ca^{2+} and thus activating Ca^{2+}-dependent endonucleases, phospholipases, and proteases.[1] However, Iyengar et al.[91] have reported that human feces contain a substantial number of extruded, viable (ie, trypan blue-extruding), squame-like colon epithelial cells.

Colon Cancer

Lurking in the fecal stream are carcinogens and tumor promoters such as secondary bile acids and, maybe most importantly, diacylglycerols and free fatty acids from the hydrolysis of dietary fats.[1] These initiate and promote carcinogenesis. It would seem that the factor limiting the appearance of colon cancers is the fecal load of tumor promoters.

The carcinogen-initiated colon cells increase their proliferative activity by becoming more responsive to various combinations of growth factors and by starting to produce or overproduce autocrine proliferogens such as PDGF subunits, IGF-I and IGF-I variants, as well as the prime proliferogen, TGF-α.[1] They may also increase the impact of their autocrine TGF-α and exogenous EGF by increasing their complement of EGF receptors.[1,92] Moreover, at some point transforming cells appear, which are insensitive to TGF-β and thus cannot down-regulate TGF-α production, a key part of the differentiation program.[1] The cells might also start producing or perhaps overproducing PTHrP, which is known to stimulate adenylate cyclase and proliferation of cells of the human LoVo colon carcinoma cell line (which therefore must have PTH/PTHrP receptors on their surfaces).[93] The upshot is a persistent expression or overexpression of cell cycle-driving genes such as c-*myc* and c-*myb*.

It has been suggested that the tumor-promoting proliferative hyperactivity might be a compensatory response to the heightened killing or damaging of mucosal cells by secondary bile acids or free fatty acids.[1] The crypt cells would be responding to positive signals, or to a reduced flow of inhibitory signals, coming down to them along some intercellular communication pathway(s), such as the gap junctions.[1] Reducing the soluble secondary bile acid and free fatty acid contents of the fecal stream by increasing the dietary Ca^{2+} load (Ca^{2+} binds avidly to these acids and renders them insoluble) does seem to reduce crypt cell proliferation and the development of colon cancer.[1] Indeed, dietary Ca^{2+}-loading, together with certain site-specific cAMP analogs to be discussed below, is a primary tool for the prevention and "differentiation-therapy" of colon cancer.[1,2,20]

The increasingly self-driven initiated colon cells, like initiated keratinocytes, respond only partly or not at all to differentiation signals and thus tend to continue proliferating when they reach the Ca^{2+}-rich upper levels of the crypt or the mucosal surface. This is the crucial change in cellular behavior that opens the way to the formation of polyps, adenomas, and finally carcinomas.[1] To take this step, the cells must lose, or become much less responsive or even unresponsive to signals from the mechanism that normally senses an increase in the external Ca^{2+} concentration above a critical threshold and signals normal cells to stop cycling and start differentiating.[1] A clue to how this change might come about has come from the study of patients with familial adenomatous polyposis (FAP) (or adenomatous polyposis coli [APC]), whose colons become carpeted with thousands of polyps, some of which ultimately spawn cancers.[1] According to the currently favored hypothesis, FAP is the result of the deletion or mutation of a "tumor suppressor" (or differentiation-promoting) gene.[94] FAP is an autosomal dominant disorder, but it is not obvious how this can be because the heterozyous patient's polyp cells still have one normal chromosome 5 with one functioning "suppressor" gene.[94] The most likely solution to this problem is that one functional of the FAP gene cannot produce enough suppressor product. It is believed that FAP results from the deletion or a loss-of-function mutation (usually the creation of "stop" codons or altered splice consensus sequences) of a gene, the APC (*A*denomatous *P*olyposis *C*oli) gene, in the 21 band of the long (or q) arm of chromosome 5.[1] A part of both the APC gene and the contiguous MCC (*M*issing in *C*olon *C*ancer) gene, whose deletion or loss-of-function mutation is associated with sporadic colon cancers, is homologous to the G-protein-binding domain of the m3 muscarinic cholinergic receptor involved in the generation of Ca^{2+} surges and bursts of membrane-protein kinase C activity.[1] Another part of the APC protein is homologous to the yeast Ral-2 protein that controls the activity of RAS proteins, which are also involved in Ca^{2+} and protein kinase C signaling.[1,2] All of this suggests that colon cells of FAP patients cannot monitor and/or respond to changes in the external Ca^{2+} concentration.[1] Indeed, freshly isolated FAP (APC) cells, as well as colon adenoma and adenocarcinoma cells, do not stop proliferating (or may even be stimulated) when exposed to 2.2 mM Ca^{2+} in vitro, as do normal colon cells.[1] It has been claimed, however, that the proliferation of the cells of some cultured colorectal lines (eg, the CACO-2 line) can still be inhibited by exposure to higher Ca^{2+} concentrations such as 1.8 mM.[95]

However, this is not really true because the proliferative activity of these cancer cells in the presence of 1.8 mM Ca^{2+} was still 70% of the activity of cells in the presence of 0.25 mM Ca^{2+}.[95] This greatly reduced sensitivity of proliferative activity to the external Ca^{2+} concentration could be the reason why colon cells from the early to the most advanced stages of carcinogenesis can proliferate at all levels of the crypt and even on the mucosal surface.

Whatever the deletion or loss-of-function mutation of the genes on the long arm of chromosomes 5 does to upset the Ca^{2+} control of differentiation and proliferation, it currently seems to apply almost exclusively to colon epithelial cells, because there is no clear indication of involvement of other tissues. Thus a change in something specific for colon cells leads ultimately to the disturbance of a basic proliferation/differentiation control mechanism that also operates in other epithelial cells.

The average intracellular Ca^{2+} concentration of colon cancer cells shortly after isolation (130 ± 11 nmol/l in the presence of 0 mM external Ca^{2+} or 186 ± 17 nmol/l in the presence of 1 mM external Ca^{2+}) is slightly, but significantly, lower than the average concentration in cells freshly isolated from normal mucosa (151 ± 10 nmol/l in the presence of 0 mM external Ca^{2+} or 256 nmol/l in the presence of 1 mM external Ca^{2+}).[96] In contrast, cells from "normal" colon mucosa adjacent to colon carcinomas have 2.4 to 9.6 times higher concentrations of internal Ca^{2+},[96] which suggests a higher level of Ca^{2+} signaling. Thus a larger fraction of these cells may be induced to cycle by paracrine growth factors secreted into the local region by the cancer cells. Alternatively, it might reflect the hyperresponsiveness of the cells in a cancer-prone field to fecally-borne, protein-kinase-C-activating, Ca^{2+}-mobilizing tumor promoters. Finally, it might be the result of Ca^{2+}-loading due to cell injury by factors coming from the cancer or by fecal secondary bile acids or free fatty acids.[97] At any rate, once they have achieved the cancerous state, cells apparently no longer respond to these Ca^{2+}-elevating factors, whatever they might be.

A commonly observed premalignant change is a 40 to 50% drop in the cellular diacylglycerol content and the total protein kinase C activity, without a change in the fraction protein kinase C isozymes (about 30%) normally with the cell membrane.[1,98,99] However, the membrane-associated fraction rises to 45% in colon cancer cells.[1] Since protein kinase C inhibits proliferation of normal colon cells and is part of the differentiation signal,[1] the depletion of the total enzyme(s) pool could contribute to the resistance of neoplastic colon cells to the signal that stops proliferation and starts differentiation.

Protein kinase C stimulators such as long-chain diacylglycerols (especially those with $C_{18.1}$ oleic acid residues), TPA, and secondary bile acids do not directly induce or stimulate the proliferation of normal colon cells, but they do stimulate the poliferation of transforming mucosal cells, starting with the early premalignant FAP cells.[1] (However, it must be noted that overexpression of the β1 isoform of protein kinase C inhibits the proliferation of some colon cancer cells.[98]) A clue to the reason for this dramatic conversion of membrane-associated protein kinase C into a mitogen during carcinogenesis is provided by the observation that a burst of membrane-associated protein kinase C activity, which is induced by TPA or the abundant fecal diacylglyceride, diolein, triggers proliferation and sustains the activity of a protein-tyrosine kinase that then catalyzes the phosphorylation of tyrosine residues of a 63 kDa protein in premalignant and malignant colon cells, but does not do so in normal colon cells or in colon carcinoma cells induced to differentiate.[1,100] This protein-tyrosine kinase is probably pp60^{c-src}, which is known to be activated early in colon carcinogenesis, and which can indeed be stimulated by TPA-activated protein kinase C.[1,94] This protein-tyrosine kinase and its 63 kDa substrate may be parts of a protein kinase C-activated mitogenic mechanism, which is suppressed in normal cells and in differentiating colon carcinoma cells,[1,100] possibly via TGFs-β that stop proliferation by preventing inactivation of some RB-like growth suppressor protein(s). Thus TPA-induced protein kinase C might stop normal cells cycling and promote their differentiation by stimulating the production of TGF-β which, in turn, prevents the inactivation of the cycle-blocking suppressor protein(s),[1] but this system would not function in transforming cells that no longer make effective suppressor protein(s). The key consequence of this conversion of protein kinase C from a promoter of differentiation to a promoter of proliferation is that protein kinase C-stimulating colon tumor promoters such as TPA, diolein, and free secondary bile acids promote the clonal expansion of premalignant colon epithelial cells at the expense of their unstimulable or proliferatively suppressed normal neighbors.[1]

Protein-kinase-C-activating bile acids or fecal diacylglycerols could promote the neoplastic transformation of colon cells in several different ways. Membrane-associated protein kinase C is a key player in secretion.[1] Therefore tumor promoters may stimulate the secretion of the autocrine mitogens that drive the transforming cells' proliferation. They also may stimulate the secretion of proteases, such as urokinase, which are expressed by colon cells in the later stages of neoplastic transformation; these contribute to the selective proliferation of the cells by shutting off gap-junctional cross-communication with neighboring normal cells, and may actually destroy neighboring normal cells and less advanced premalignant cells.[1] As discussed below, a common later step in colon carcinogenesis is the mutation of a H-*ras*, or, more often, a K-*ras*, gene whose wild-type product normally functions in the proliferatively shut-down mature cells on the mucosal surface.[1] The mutant or activated RAS proteins block differentiation, are mitogenic, and are further activated when phosphorylated by tumor promoter-induced, membrane-associated protein kinase C activity.[1,2] Perhaps one of the most important effects of the membrane-protein-kinase-C activity stimulated by tumor promoters is a surge of oxygen radicals generated by protein-kinase C-activated NAD(P)H oxidase in the plasma membrane.[1,2] These radicals may generate clastogenic agents that operate through Ca^{2+}-dependent nucleases to promote the chromosome changes and gene deletions that mark the later stages of carcinogenesis.

An important factor in the failure of colon cancer cells to respond to differentiation signals and proliferate inappropriately may be the overexpression of the proliferation-promoting type I cAMP-dependent protein kinase and the suppression of the differentiation-promoting type II cAMP-dependent protein kinase.[13] Indeed, Bradbury et al.[101] found that all of 50 colorectal cancer specimens had no detectable RII subunits of type II cAMP-dependent protein kinase while "normal" cells of carcinoma-adjacent and of distant mucosa had detectable amounts of both RII and RI subunits. The importance of this abnormal ratio of RI to RII subunit levels for colon cancer cell behavior is indicated by the fact that the proliferation of colon cancer cells is inhibited by 8-Cl-cAMP, a cAMP analog that selectively up-regulates the expression of the type II isozyme and down-regulates the expression of the type I isozyme.[20] Furthermore, the reduction of RI subunit expression with antisense oligodeoxynucleotides targeted against the first 21 bases of the human RIα mRNA arrests the proliferation of LS-147T colon carcinoma cells without cytotoxicity, while the subcutaneous injection of the antisense deoxyribonucleotides twice per week reduces (by 50%) the growth of these LS-147T tumor cells in athymic nude mice.[102] Thus restoring the normal balance of the expression of cAMP-dependent protein kinase isozymes with site-specific cAMP analogs may someday prove to be an effective "differentiation therapy" for colon cancer. Indeed, combining such a therapy with the use of other differentiation-triggering agents such as calcium, 1α,25(OH)$_2$vitamin D$_3$, retinoic acid, or butyric acid might provide an even better chemoprevention and/or chemotherapy for colorectal cancer.

As mentioned above, variants appear in the transforming colon cell population that has mutant *ras* genes coding for activated RAS proteins. These activated Ras proteins, unlike their normal counterparts, are proliferogenic, block differentiation, and are probably responsible for the loss of the dependence on external Ca^{2+} for optimal proliferate that is the hallmark of the most advanced adenocarcinoma cells.[1,2] These activated Ras proteins also stimulate the expression of the c-*fos* proto-oncogene, of the genes coding for autocrine motility factor(s), and of collagenase IV, which are key parts of the invasiveness/metastasis phenotype.[1] A role for *ras* gene mutations and activated Ras proteins in the optimal progression of colon carcinogenesis is indicated by the ability of fecal Ca^{2+} loading to prevent mutations of the K-*ras* gene and reduce the number and size of colon cancers induced in male Sprague-Dawley rats by 1,2-dimethylhydrazine.[103]

As the protein-kinase-C-stimulating tumor promoters relentlessly drive the carcinogenic process forward from one generation of cells to another, there is an increased likelihood that malignant carcinoma cells with deleted or mutated tumor suppressor (differentiation-promoting, proliferation-suppressing) genes such as the p53 gene on chromosome 17q may appear among the benign ademona cells.[1,35,94,104,105] This final malignant transformation may begin with a selection for cells with "dominant" loss-of-function mutations of the p53 gene and overproduction of inactive p53 proteins capable of forming complexes with the wild-type p53 protein and thus preventing it from entering the nucleus and monitoring chromosomal damage.[35] This is often followed by a second selection for the deletion of the remaining wild-type gene and further loss of its restraining function.[104,105] The elimination of this gene permits the generation of malignant variants of adenoma cells, but is not involved in the initial loss of Ca^{2+} control mechanisms, the resulting expanded proliferative zone, and the increased proliferative activity in the pre-malignant polyps and adenomas, whose cells are either not hemizygous for the wild-type p53 gene or have not suffered loss-of-function mutations of the p53 gene.[35,94,104,105]

As the carcinogenic process advances, variant cells appear that are increasingly incapable of adopting the shapes and functions of mature colon cells in response to the upper crypt's extracellular matrix as normal colon cells do, possibly because they fail to respond to their own autocrine TGF-β signals and thus do not express matrix-binding, differentiation-triggering receptors such as specific types of integrin receptor; consequently they cannot transmit information from the extracellular matrix to the cytoskeleton, the nucleus, and the genes of the differentiation programs.[87,89,90] Malignant transformation also includes the severing of the ties that bind cells to their neighbors[1] through failure to express cell adhesion molecules (CAMs), such as the Ca^{2+}-binding E-cadherins and the Ig-related, L-CAM-like, glycoprotein product of the DCC (Deleted in Colon Cancer) gene on chromosome 18q.[1,69]

Carcinogenesis is finally complete with the appearance of malignant colon cells, which overexpress proliferation-related type I cAMP-dependent protein kinases; which produce activated Ras proteins; which drive their own proliferation regardless of the extracellular Ca^{2+} concentration and the availability of exogenous growth factors; which cannot differentiate and do not senesce; which make and secrete angiogenic and autocrine motility factor(s); and which, like embryonic trophoblast cells and migrating monocytes, have surfaces loaded with the various hydrolases and receptors needed to bind to, and chew through, basal laminae, pry apart vascular endothelial cells and enter blood vessels, embolize with platelets and other malignant cells, and finally leave the blood vessels and invade and colonize distant organs, especially the liver.[1]

Acknowledgment. I thank Lynda Boucher and Kathy Hamelin for processing the words and Geoff Mealing for preparing Figure 2.1.

References

1. Whitfield JF. Calcium and cancer. Crit Rev Oncogenesis 1992; 3:55–90.
2. Whitfield JF. Calcium, Cell Cycles, and Cancer. Boca Raton, Fla.: CRC Press; 1990:263.
3. Rhee SG. Inositol phospholipid-specific phospholipase C: Interaction of the $gamma_1$ isoform with tyrosine kinase. Trends Biochem Sci 1991; 16:297–301.
4. Brown AM. A cellular logic for G-protein-coupled ion channel pathways. FASEB J 1991; 5:2175–2179.
5. Desser-Wiest L, Desser H. The proliferation, activity of the rat thymocytes after partial hepatectomy. Exp Path 1990; 18:167–169.
6. Matsumoto K, Nakamura T. Hepatocyte growth factor: Molecular structure, roles in liver regeneration, and other biological functions. Crit Rev Oncogenesis 1992; 3:27–54.
7. Giss BJ, Walker AM. Mammotroph autoregulation: Intracellular fate of internalized prolaction. Mol Cell Endocrinol 1985; 42:259–267.
8. Fesus L, Tomazy V. Searching for the function of tissue transglutaminase: Its possible involvement in the biochemical pathway of programmed cell death. Adv Exp Med Biol 1988; 231:119–134.
9. Fesus L, Tomazy V, Autouri F, Taresa E, Piacentini M. Apoptotic hepatocytes become insoluble in detergents and chaotropic agents as a result of transglutaminase actin. FEBS Lett 1989; 245:150–154.
10. Fesus L, Tomazy V, Falus A. Induction and activation of tissue transglutaminase during programmed cell death. FEBS Lett 1987; 224:104–108.
11. Martin TFJ. Receptor regulation of phosphoinositidase C. Pharmacol Ther 1991; 49:329–345.
12. Taylor CW, Richardon A. Structure and function of inositol triphosphate receptors. Pharmacol Ther 1991; 51:97–137.
13. Roony TA, Thomas AP. Organizations of intracellular calcium signals generated by inositol lipid-dependent hormones. Pharmacol Ther 1991; 49:223–237.
14. Heizmann CW, Schäfer BW. International calcium-binding proteins. Seminars Cell Biol 1990; 1:277–282.
15. Ben-Ze'ev A. Animal cell shape changes and gene expression. Bic Essays 1991; 13:207–212.
16. Morley SJ, Thomas G. Intracellular messengers and the control of protein synthesis. Pharmacol Ther 1991; 50:291–319.
17. Shalloway D, Shenoy S. Oncoprotein kinases in mitosis. Adv Cancer Res 1991; 57:185–225.

18. Petrovic AG, Oudet CL, Stutzmann JL. Temporal organization of rat and human skeletal cells. In: Edmunds LN, ed. Cell Cycle Clocks. New York: Marcel Dekker; 1984:325–347.

19. Sheffield LG. Oligonucleotides antisense to catalytic subunit of cyclic AMP-dependent protein kinase inhibit mouse mammary epithelial cell DNA synthesis. Exp Cell Res 1991; 192:307–310.

20. Cho-Chung YS, Clair T, Tortora G, Yokonaki H. Role of site-selective cAMP analogs in the control and reversal of malignancy. Pharmacol Ther 1991; 50:1–33.

21. Scott JD. Cyclic nucleotide-dependent protein kinases. Pharmacol Ther 1991; 50:123–145.

22. Montminy MR, Gonzalez GA, Yamamoto KK. Characteristics of the cAMP response unit. Recent Progr Hormone Res 1990; 46:21–230.

23. Benbrook DM, Jones NC. Heterodimerformation between CREB and JUN proteins. Oncogene 1990; 5:295–302.

24. Macgregor PF, Abate C, Curran T. Direct cloning of leucine zipper proteins: Jun binds cooperatively to the CRE with CRE-BP1. Oncogene 1990; 5:451–458.

25. Takeda J, Mackawa T, Sudo T, Seino Y, Imura H, Naoaki S, Tanaka C, Ishii S. Expression of the CRE-BP1 transcriptional regulator binding to the cyclic AMP response element in central nervous system, regenerating liver, and human tumors. Oncogene 1991; 6:1009–1014.

26. Bookstein R, Lee WH. Molecular genetics of the retinoblastoma suppressor gene. Crit Rev Oncogenesis 1991; 2:211–227.

27. Kim SJ, Lee HD, Robbins PD, Busam K, Sporn MB, Roberts AB. Regulation of transforming growth factor B1 gene expression by the product of the retinoblastoma-susceptibility gene. Proc Natl Acad Sci USA 1991; 88: 3052–3056.

28. Mittnacht S, Weinberg RA. G1/S phosphorylation of the retinoblastoma protein is associated with an altered affinity for the nuclear compartment. Cell 1991; 65:381–393.

29. Krek W, Nigg EA. Differential phosphorylation of vertebrate p34^{cdc2} kinase at the G_1/S and G_2/M transitions of the cell cycle: identification of major phosphorylation sites. EMBO J 1991; 10:305–316.

30. Tournier S, Raynaud F, Gerbaud P, Lohmann SM, Dorée M, Evain-Brion D. Association of type II cAMP-dependent protein kinase with p34^{cdc2} protein kinase in human fibroblasts. J Biol Chem 1991; 266:19018–19022.

31. Lin BT-Y, Gruenwald S, Morla AO, Lee W-H, Wang JYJ. Retinoblastoma cancer suppressor gene product is a substrate of the cell cycle regulator cdc2 kinase. EMBO J 1991; 10:857–864.

32. Bandara LR, Adamczewski JP, Hunt T, Sassone-Corsi P. Cyclin A and the retinoblastoma gene product complex with a common transcription factor. Nature 1991; 352:249–251.

33. Foulkes NS, Laoide BM, Schlotter F, et al. Transcriptional antagonist cAMP-responsive element modulator (CREM) down-regulates c-*fos* cAMP-induced expression. Proc Natl Acad Sci USA 1991; 88:5448–5452.

34. Foulkes NS, Borrelli E, Sassone-Corsi P. CREM gene: Use of alternative DNA-binding domains generates multiple antagonists of cAMP-induced transcription. Cell 1991; 64:739–749.

35. Levine AJ, Momand J, Finlay CA. The p53 tumor suppressor gene. Nature 1991; 351:453–456.
36. Miller M, Park MK, Hanover JA. Nuclear pore complex: Structure, function, and regulation. Physiol Rev 1991; 71:909–949.
37. Raper SE, Buswen SJ, Basker ME, Jones AL. Translocation of epidermal growth factor to the hepatocyte nucleus during rat liver regeneration. Gastroenterology 1987; 92:1243–1260.
38. Calabretta B, Gewirtz AM. Functional requirements of c-myb during normal and leukemic hematopoiesis. Crit Rev Oncogenesis 1991; 2:187–194.
39. Vilgrain I, Baird A. Phosphorylation of basic fibroblast growth factor by a protein kinase associated with the outer surface of a target cell. Mol Endocrinol 1991; 5:1003–1012.
40. Means AR, Van Berkum MFA, Bagchi I, Lu KP, Rasmussen CD. Regulatory functions of calmodulin. Pharmacol Ther 1991; 50:255–270.
41. Ramsdell JS. Voltage-dependent calcium channels regulate GH_4 pituitory cell proliferation at two stages of the cell cycle. J Cell Physiol 1991; 146: 197–206.
42. Reddy GPV. Allosteric interactions between the enzymes of DNA biosynthesis in mammalian. In: Campisi J, Cunningham DD, Inouye M, Riley M, eds. Pespectives on Cellular Regulation: From bacteria to Cancer. New York: Wiley; 1991:315–329.
43. Roberts JM, d'Urso G. An origin unwinding activity regulates initiation of DNA replication during mammalian cell cycle. Science 1988; 241:1486–1489.
44. McVey D, Brizuela L, Mohr I, Marshak DR, Gluzman Y, Beach D. Phosphorylation of large tumour antigen by cdc 2 stimulates SV40 DNA replication. Nature 1989; 341:503–507.
45. Nishitani H, Ohtsubo M, Yamashita K, Iida H, Pines J, Yasudo H, Shibata Y, Hunter T, Nishimoto T. Loss of RCC1, a nuclear DNA-binding protein, uncouples the completion of DNA replication from the activation of cdc2 protein kinase and mitosis. EMBO J 1991; 10:1555–1564.
46. Bischoff FR, Ponstingl H. Catalysis of guanine nucleotide exchange on Ran by the mitotic regulator RCC1. Nature 1991; 354:80–82.
47. Watson PA. Function follows form: Generation of intracellular signals by cell deformation. FASEB J 1991; 5:2013–2019.
48. Fernandez A, Lamb N. Mitotic control in mammalian cells, positive and negative regulation by protein phosphorylation. Berlin: Springer-Verlag; 1991:397–409.
49. Lamb NJC, Cavadore J-C, Labbe J-C, Maurer RA, Fernandez A. Inhibition of cAMP-dependent protein kinase plays a key role in the induction of mitosis and nuclear envelope breakdown in mammalian cells. EMBO J 1991:1523–1533.
50. Murray A, Hunt T. The Cell Cycle. New York: WH Freeman and Company; 1993:251.
51. Meijer L, Azzi L, Wang JYJ. Cyclin B targets $p34^{cdc2}$ for tyrosine phosphorylation. EMBO J 1991; 10:1545–1554.
52. Parker LL, Atherton-Fessler S, Lee MS. Cyclin promotes the tyrosine phosphorylation of $p34^{cdc2}$ in a wee 1+ dependent manner. EMBO J 1991; 10:1255–1263.

53. Shenolikar S, Nairn AC. Protein phosphatases: Recent progress. Adv Second Messenger Phosphoprot Res 1991; 23:1–121.
54. Sakai H, Ohta K. Centrosome signalling at mitosis. Cell Signal 1991; 3:267–272.
55. Pavelka M. Functional morphology of the Golgi apparatus. Adv Anal Embryol Cell Biol 1987; 106:1–94.
56. Goldbeter A. A minimal cascade model for the mitotic oscillator involving cyclin and cdc kinase. Proc Natl Acad Sci USA 1991; 88:9107–9111.
57. Glotzer M, Murray AW, Kirschner MW. Cyclin is degraded by the ubiquitin pathway. Nature 1991; 349:132–138.
58. Fischer B, Fischer ER. Experimental studies of factors influencing hepatic metastases. III. Effect of surgical trauma with special reference to liver injury. Ann Surg 1959; 150:731–744.
59. van Dale P, Galand P. Effect of partial hepatectomy on experimental liver invasion by intraportally injected colon carcinoma cells in rats. Invasion Metastasis 1988; 8:217–227.
60. Pietenpol JA, Holt JT, Stein RW, Moses HL. Transforming growth factor B1 suppression of c-*myc* gene transcription: Role in inhibition of keratinocyte proliferation. Proc Natl Acad Sci USA 1990; 87:3758–3762.
61. Arends MJ, Wyllie AH. Apoptosis: Mechanism and roles in pathology. Int Rev Exp Pathol 1991; 32:223–254.
62. Goldstein P, Ojcius DM, Young DE. Cell death mechanisms and the immune system. Immunol Revs 1991; 121:30–65.
63. Walker GJ, Hayward NK, Falvey S, Cooksley WGE. Loss of somatic heterozygosity in hepatocellular carcinoma. Cancer Res 1991; 51:4367–4370.
64. Arbuthnot P, Kew M, Fitschen W. c-*fos* and c-*myc* oncoprotein expression in human hepatocellular carcinomas. Anticancer Res 1991; 11:921–924.
65. Hayward NK, Walker GJ, Graham W, Cooksley WGE. Hepatocellular carcinoma mutation. Nature 1991; 352:764.
66. Hsu IC, Metcalf RA, Sun T, Welsh JA, Wang NJ, Harris CC. Mutational hotspot in p53 gene in human hepatocellular carcinomas. Nature 1991; 350:427–428.
67. Bressac B, Kew M, Wands J, et al. Selective G to T mutations of p53 gene in hepatocellular carcinoma from southern Africa. Nature 1991; 350:429–431.
68. Behrens J, Weidner KM, Frixen UH, Schipper JH, Sachs M, Arakaki N, Daikura Y, Birchmeier W. The role of E-cadherin and scatter factor in tumor invasion and cell motility. In: Goldberg ID, ed. Cell Motility Factors. Basel: Birkhäuser; 1991:109–126.
69. Takeichi M. Cadherin cell adhesion receptors as a morphogenetic regulator. Science 1991; 251:1451–1454.
70. Shimoyama Y, Hirohashi S. Cadherin intercellular adhesion molecule in hepatocellular carcinomas: Loss of E-cadherin expression in an undifferentiated carcinoma. Cancer Letters 1991; 57:131–135.
71. Watt FM, Jordan PW, O'Neill CH. Cell shape controls terminal differentiation of human epidermal keratinocytes. Proc Natl Acad Sci USA 1988; 85:5576–5580.
72. Ruoslahti E. Integrins. J Clin Invest 1991; 87:1–5.
73. Manciant ML, Su MJ, Pillai S, Bikle BB. Calcium regulation of involucrin expression in human keratinocytes. J Invest Dermatol 1991; 96:Abstract 392.

74. Cook PW, Pittelkow MR, Shipley GD. Growth factor-independent proliferation of normal human neonatal keratinocytes: Production of autocrine- and paracrine-acting mitogenic factors. J Cell Physiol 1991; 146:277–289.
75. Watt FM, Boukamp P, Hornung J. Effect of growth environment on spatial expression of involucrin by human epidermal keratinocytes. Arch Dermatol Res 1987; 279:335–340.
76. Hayman JA, Danks JA, Ebeling PR, Moseley JM, Kemp BE, Martin TJ. Expression of parathyroid hormone related protein in normal skin and in tumors of skin and skin appendages. J Pathol 1989; 158:293–296.
77. Martin TJ, Moseley JM, Gillespie MT. Parathyroid hormone-related protein: Biochemistry and molecular biology. Crit Rev Biochem Mol Biol 1991; 26:377–395.
78. Pietenpol JA, Holt JT, Stein RW, Moses HL. Transforming growth factor β1 suppression of c-myc gene transcription: Role in inhibition of keratinocyte proliferation. Proc Natl Acad Sci USA 1990; 87:3758–3762.
79. Pietenpol J, Stein RW, Moran E, Yaciuk R, Schlegel R, Lyons RM, Pittelkow MR, Münger K, Howley PM, Moses HL. TGF-β1 inhibition of c-myc transcription and growth in keratinocytes is abrogated by viral transforming proteins with pRB binding domains. Cell 1990; 61:777–785.
80. Mallette LE. The parathyroid polyhormones: New concepts in the spectrum of peptide hormone action. Endocrine Revs 1991; 12:110–117.
81. Dotto GP, El-Fouly MH, Nelson C, Trosko JE. Similar and synergistic inhibition of gap-junctional communication by ras transformation and tumor promoter treatment of mouse primary keratinocytes. Oncogene 1989; 4: 637–641.
82. Ando Y, Imamura S, Murachi T, Kennagi R. Calpain activates two transglutaminases from porcine skin. Arch Dermatol Res 1988; 280:380–384.
83. Croall DE, DeMartino GN. Calcium-activated neutral protease (calpain) system: Structure, function, and regulation. Physiol Rev 1991; 71:813–847.
84. Wyatt GP, Steele VE, Elmore E. Differential growth response to exogenous calcium in normal and carcinogen-exposed primary human keratinocyte cell cultures. In: Kelloff G, Steele VE, Stoner GD, eds. Proceedings of workshop on cellular and molecular targets for chemoprevention. National Cancer Institute, Chemoprevention Branch, Bethesda, 1991. p. 35.
85. Suemori S, Ciacci C, Podolsky DK. Regulation of transforming growth factor expression in rat intestinal epithelial cell lines. J Clin Invest 1991; 87:2216–2221.
86. Torelli G, Venturelli D, Colo A, et al. Expression of c-myb protooncogene and other cell cycle-related genes in normal and neoplastic human colonic mucosa. Cancer Res 1987; 47:5266–5269.
87. Pienta KJ, Getzenberg RH, Coffey S. Cell structure and DNA organization. Crit Rev Eukaryotic Gene Expression 1991; 1:355–385.
88. Pignatelli M, Bodmer WF. Genetics and biochemistry of collagen binding-triggered glandular differentiation in a human colon carcinoma cell line. Proc Natl Acad Sci USA 1988; 85:5561–5565.
89. Pignatelli M, Bodmer WF. Integrin-receptor-mediated differentiation and growth inhibition are enhanced by transforming growth factor-β in colorectal tumour cells grown in collagen gel. Int J Cancer 1989; 44:518–523.
90. Richman PJ, Bodmer WF. Control of differentiation in human colorectal carcinoma cell lines: Epithelial-mesenchymal interactions. J Pathol 1988; 156:197–211.

91. Iyengar V, Albaugh GP, Lohani A, et al. Human stools as a source of viable colonic epithelial cells. FASEB J 1991; 5:2856–2859.
92. Gross ME, Zorbas MA, Danels YJ, et al. Cellular growth response to epidermal growth factor in colon carcinoma cells with an amplified epidermal growth factor receptor derived from a familial adenomatous polyposis patient. Cancer Res 1991; 51:1452–1459.
93. Yu D, Seitz PK, Rajaraman S, et al. Stimulation of intracellular cAMP and ornithine decarboxylase activity in a human colonic cell line by parathyroid hormone-related protein. J Bone Min Res 1991; 6:S116.
94. Bishop DT, Thomas HJW. The genetics of colon cancer. Cancer Surveys 1990; 9:585–604.
95. Cross HS, Huber C, Peterlik M. Antiproliferative effect of 1,25-dihydroxyvitamin D_3 and its analogs on human colon adenocarcinoma cells (CACO-2): Influence of extracellular calcium. Biochem Biophys Res Commun 1991; 179:57–62.
96. Edelstein PS, Thompson SM, Davies, RJ. Altered intracellular calcium regulation in human colorectal cancers and in "normal" adjacent mucosa. Cancer Res 1991; 51:4492–4494.
97. Whitfield JF. Calcium switches, cell cycles, differentiation, and death. In: Lipkin M, Newmark HL, Kelloff G, eds. Calcium, Vitamin D, and Prevention of Colon Cancer. Boca Raton, Fla.: CRC Press; 1991:31–77.
98. Phan SC, Morotomi M, Guillem JG, et al. Decreased levels of 1,2-*sn*-diacylglycerol in human colon tumors. Cancer Res 1991; 51:1571–1573.
99. Sauter G, Nerlick A, Spengler U, et al. Low diacylglycerol values in colonic adenomas and colorectal cancer. Gut 1990; 31:1041–1045.
100. Lee H, Winawer S, Friedman E. Signal-transduction through pp63 tyrosine phosphorylation in human colon carcinoma cell lines. Proc Amer Assoc Cancer Res 1991; 51:Abstract 450.
101. Bradbury AW, Miller WR, Clair T, et al. Overexpressed type I regulatory subunit (RI) of cAMP-dependent protein kinase (PKA) as tumor marker in colorectal cancer. Proc Amer Assoc Cancer Res 1990; 31:172.
102. Clair T, Yokozaki H, Tortora G, et al. An antisense oligodeoxynucleotide targeted against the type I regulatory subunit (RIα) mRNA of cAMP-dependent protein kinase (PKA) inhibits the growth of LS-147T human colon carcinoma in athymic nude mice. Proc Amer Assoc Cancer Res 1991; 51:Abstract 1645.
103. Lior X, Jacoby RF, Teng BB, et al. K-*ras* mutations in 1,2-dimethylhydrazine-colonic tumors: Effects of supplemental dietary calcium and vitamin D deficiency. Cancer Res 1991; 51:4305–4309.
104. Rodrigues NR, Rowan A, Smith MEF, et al. p53 mutations in colorectal cancer. Proc Natl Acad Sci USA 1990; 87:7555–7559.
105. Campo F, Calle-Martin O, Miquel R, et al. Loss of heterozygosity of p53 gene and p53 protein expression in human colorectal carcinomas. Cancer Res 1991; 51:4436–4442.

3
Role of Calcium in Stimulus-Secretion Coupling in Exocrine Glands

CATHERINE S. CHEW

Calcium plays a key role not only in the control of exocrine secretion but also in an extraordinarily diverse group of cellular activities including endocrine secretion, neural transmission, cellular differentiation and growth, intermediary metabolism, protein synthesis, and muscle contraction.[1] The overall mechanisms by which intracellular calcium concentrations are regulated in non-excitable cells include the release of calcium from intracellular storage sites and the influx of calcium into the cell from the extracellular environment. In general, elevation of intracellular calcium is thought to result in initial activation of calcium and phospholipid-dependent protein kinases. Through cell specific phosphorylation mechanisms these kinases, in conjunction with opposing phosphatases, may modulate the activities of membrane-bound ion channels, agonist receptors and receptor-associated regulatory enzymes, and many other enzymes present in different cellular compartments.

The specific functions of calcium in the regulation of exocrine secretion has been the subject of scientific investigation for over a half a century.[2–5] In 1968 Douglas[6] coined the term "stimulus-secretion coupling" to describe perceived similarities between calcium-dependent activation events associated with exocrine secretion and excitation-contraction coupling in muscle. Only recently, however, has it been demonstrated that the similarities envisioned by Douglas extend to at least some of the molecular mechanisms associated with these events. Recent advances in our understanding of calcium-dependent secretion can be attributed to several important technologic advances. These include the development of cell peermeant, fluorescent calcium-sensitive probes and non-fluorescent intracellular calcium chelators in parallel with complementary measurement techniques, including digitized video image analysis of intracellular free calcium concentrations ($[Ca^{2+}]_i$) in single living cells and spectrofluorometric analyses of $[Ca^{2+}]_i$ in cell populations. Patch clamp techniques have allowed rapid advances in the electrochemical characterization of a variety of membrane-associated ionic channels. In addition, cell isolation and primary culture techniques have been improved to the extent that it is now possible, in some cell types, to measure agonist-stimulated calcium signaling and secretion simultaneously within the same cell. Confocal microscopy has significantly enhanced the ability to localize proteins within specific cellular compartments. This technique has also been utilized to detect localized increases in calcium in single living cells. The development of powerful new methods of molecular biology, including the polymerase chain reaction, has led to rapid advances in recognition and characterization of specific classes of calcium channels, calcium binding proteins, and enzymes involved in calcium-dependent processes. This chapter will consider recent experimental data acquired using these techniques, focusing mainly on the requirement for calcium in stimulus-secretion coupling and the molecular mechanisms by which agonists modulate intracellular calcium concentrations.

Requirement for Calcium in Exocrine Secretion

The earliest evidence suggesting that calcium is necessary for exocrine secretion was based on rather crude in vitro experiments in which calcium concentrations in extracellular media were reduced to negligible levels by chelation with EGTA. Because calcium chelation leads to disintegration of cell–cell junctions in tissue sheets and can affect the integrity of the plasma membrane, many of these experiments were of doubtful significance.[2,7,8] Nonetheless, the data generated were important because they supported a role for calcium in the maintenance of secretion stimulated by specific classes of agonists. In other experiments, calcium ionophores, which generate non-specific increases in $[Ca^{2+}]_i$ by forming calcium permeant pores in the plasma membrane, were tested to determine whether or not a simple rise in $[Ca^{2+}]_i$ is sufficient to stimulate secretion. The ionophores stimulated secretion in exocrine cells that exocytose packaged secretory products such as pancreatic acini and gastric chief cells, but were less effective than calcium-dependent neurotransmitters and hormones such as acetylcholine and cholecystokinin.[5,9,10] In exocrine cells that secrete fluids and electrolytes such as the gastric parietal cell and eccrine sweat glands, the calcium ionophores A23187 and ionomycin did not stimulate secretion when used at concentrations that cause increases in $[Ca^{2+}]_i$ to levels equal to or greater than those obtained with calcium-dependent agonists.[2,11] In rat parotid tissue, however, A23187 was found to mimic the effect of muscarinic or α-adrenergic agonists in the stimulation of K loss from the cells.[3] The mixed results with calcium ionophores raised the question as to whether or not calcium is sufficient to establish and maintain secretion.

Recently two classes of agents with very different actions have been used to provide more specific information on the role of calcium in stimulus-secretion coupling in non-excitable cells. The first class is best represented by the cell permeant calcium chelator, BAPTA/AM, which has a structure similar to EGTA. Like the cell-permeant fluorescent calcium indicators, fura-2 and indo-1, BAPTA is trapped within cells by the action of cellular esterases which cleave off the acetoxymethylester (AM) portion of the molecule rendering it membrane impermeant.[12] When cells sufficiently loaded with BAPTA are then stimulated with calcium-dependent agonists, there is no increase in intracellular free calcium concentrations and secretory responses are inhibited. In contrast, BAPTA has little effect on secretion stimulated by cAMP-dependent agonists.[13-15] The second class is best represented by thapsigargin, an agent that elevates $[Ca^{2+}]_i$ by inhibiting ATP-dependent inwardly-directed calcium pumps located on intracellular calcium-containing organelles, but not on plasma membranes.[16] Another agent, tBuBHQ (2,5-Di-(*tert*-butyl)-1,4-benzohydroquinone) appears to have an action similar to that of thapsigargin.[17] The plant, *Thapsia garganica*, is the source of thapsigargin, which has been used for many years as a counterirritant in the treatment of rheumatic pain,[18] a possibly hazardous practice since thapsigargin is a tumor promoter in mouse skin when applied with a subthreshold dose of a primary carcinogen.[19] Since thapsigargin does not activate protein kinase C (a calcium-phospholipid-dependent enzyme that is activated by the phorbol ester tumor promoter, TPA), it is referred to as a non-phorbol tumor promoter. The mechanism by which thapsigargin promotes tumorigenesis is unknown, but is thought to be related to the ability to elevate intracellular calcium concentrations for a prolonged period of time.

Although there is a growing body of information on the effects of thapsigargin on calcium entry and intracellular release mechanisms, there is presently limited information as to the effects of thapsigargin on exocrine cell secretion. In some non-exocrine cell types, such as mast cells and adrenal chromaffin cells, thapsigargin causes a sustained elevation in $[Ca^{2+}]_i$ and stimulates secretion.[16] In other cell types, secretion is not stimulated by thapsigargin. For example, the gastric parietal cell secretes little or no HCl when exposed to thapsigargin under conditions where $[Ca^{2+}]_i$ is clearly elevated, and thapsigargin appears to act on the same calcium pool(s) as the cholinergic agonist, carbachol. In contrast, when thapsigargin is added in combination with agents that act via cAMP-dependent pathways, the acid secretory response is potentiated.[20] The variability in the ability of thapsigargin to stimulate secretion in different cell types may reflect differences in the stimulatory action of calcium itself and its participation in more specialized and complex secretory mechanisms. Thus a simple rise in calcium in salivary acinar cells may be sufficient to activate K channels in the basolateral membrane.[21] At the opposite extreme is the gastric parietal cell, which is strongly dependent on cAMP-modulated pathways for initiation of HCl secretion and in which increased calcium is not sufficient to stimulate the complex membrane rearrangements that are required for activation of HCl secretion (ie, fusion of intracellular tubulovesicles containing the proton pump (H,K-ATPase) with the apical, canalicular membrane that leads to pump activation.[2,22]

Calcium Entry, Release, and Efflux Mechanisms

Much attention has been focused not only on the mechanisms by which calcium-dependent agonists elevate $[Ca^{2+}]_i$, but also on the mechanisms that cells utilize to maintain the concentration of calcium of extracellular calcium in the millimolar range and that of intracellular free calcium in the nanomolar-to-micromolar range. If intracellular free calcium is increased above physiologic levels for more than a brief period of time, cell death results;[23] however, elevation of $[Ca^{2+}]_i$ within physiologic limits serves to control many cellular activities including exocrine secretion. Cells must therefore have the capacity not only to maintain but also to elevate intracellular calcium within narrowly-defined limits. This function requires the coordination of calcium influx, efflux, and intracellular storage-release mechanisms. Since these mechanisms have been best described in excitable tissues, it is useful to consider briefly what is presently known about calcium homeostatic mechanisms in excitable cells in order to place current knowledge of exocrine mechanisms in the proper perspective.

Mechanisms of Calcium Influx and Efflux Across the Plasma Membrane in Excitable Cells

The calcium channels within the plasma membrane in excitable cells are considered to be subclasses of ionic channels first recognized in the 1950s for their involvement in the generation of propagated action potentials and end plate potentials in neuromuscular junctions. There are thought to be two major classes of calcium channels—voltage-gated, and agonist-gated or receptor-operated. From these two major classes of calcium channels, more specific subclasses appear to have evolved.[24]

Voltage-Gated and Agonist-Gated Calcium Channels

A major pathway through which calcium enters muscle, nerve, and various endocrine cells is represented by the voltage-gated or voltage-operated calcium channels (VOCCs), which are proteins that span the plasma membrane. Several categories of VOCCs have been described according to the degree of membrane depolarization required for activation, the effects of pharmacologic antagonists such as dihydropyridines and ω-conotoxin, and the length of time the channels remain open under conditions of continuous depolarization. The best-characterized VOCCs are the L-type calcium channels. These channels are blocked by the organic cations, Mn, Ni, Co, Cd, and La, and by low concentrations (1–100 nM) of dihydropyridines (as nifedipine, nitrendipine), phenyl-alkylamines (verapamil, D600), and benzothiazepines (diltiazem).[25–29] Dihydropyridine-sensitive L-type channels, which are also called di-hydropyridine receptors, are found in high concentrations in skeletal muscle transverse (T) tubule membranes. The L-type channel is considered to be an example of the evolution of an ancient, voltage-sensitive channel to perform a specific function in excitation-contraction coupling.[24] The skeletal muscle dihydropyridine receptor (DHPR) protein is composed of five different subunits, α_1, α_2, β, γ, and δ. Although the specific function of the DHPR in skeletal muscle is controversial, there is good evidence that the α_1 subunit is the calcium channel component of the protein. Thus L-type calcium currents are not detected in the skeletal muscles of mice that are homozygous for muscular dysgenesis (*mdg*); although these mice lack this subunit, they express the α_2, β, δ, γ subunits, and the calcium current is restored by expression of the DHPR α_1 subunit in myotubes cultured from these mice.[30]

Both the α_1 and β subunits of the DHPR can be phosphorylated by cAMP-dependent protein kinase, which suggests that the DHPR calcium channel may be regulated by receptor-mediated second messenger mechanisms. Heterotrimeric GTP-binding proteins (G proteins) also regulate DHPR calcium channels.[28,31] These G proteins are composed of three subunits, α, β, γ, and are classified according to the characteristics of the α subunit. Molecular cloning experiments indicate that there are at least 16 α subunits encoded by separate genes and that still others arise from alternate gene splicing.[32] G proteins transduce signals from plasma membrane-bound receptors to effectors such as adenylyl cyclase, the enzyme responsible for cAMP production, and to other ion channels such as voltage-sensitive potassium channels.[32] Plasma membrane receptors that are coupled to G proteins (G protein-coupled receptors) form a separate gene superfamily characterized by slower intracellular response compared to agonist-gated (receptor-operated) channels.[24]

In addition to the major agonist-gated calcium channels, the plasma membrane of excitable and non-excitable cells contains other voltage-insensitive calcium channels. Agonist-gated or receptor-operated channels (ROCs) open directly when agonists bind to their specific receptors. Examples of such channels include the nicotinic acetylcholine receptor, glutamate, GABA$_A$, and N-methyl-D-aspartate (NMDA) receptors. These receptors have now been cloned and shown to form calcium channels. Another class of voltage-insensitive calcium channels is activated following agonist-receptor binding and generation of second messengers and are called second-messenger-operated channels or SMOCs.[27] The classification of ROCs and SMOCs varies in the literature with SMOCs being categorized either as a separate entity or as a ROC subclass. Unfortunately, many publications loosely refer to calcium entry mechanisms as ROCs even in the absence of any good biochemical evidence to support this mechanism. Another calcium channel type, the calcium-activated nonspecific (CAN) channel, was first demonstrated in cardiac Purkinje cells.[33] This putative channel exhibits calcium sensitivity and allows the passage of other cations such as sodium, but is not yet well defined.

Plasma Membrane Calcium Pumps and Exchangers

In most cell types there are to major calcium efflux pathways in the plasma membrane: calcium "pumps" (Ca^+-ATPases) and Na/Ca exchangers. The calcium pumps, which have been studied the most in red blood cells and are thought to be the main calcium extrusion pathway, have a higher affinity but a lower capacity for calcium transport than the exchangers. The complete amino acid sequences of plasma membrane calcium pumps from humans and rats have been determined and it is now clear that there are a number of pump isoforms encoded by a multigene family.[34,35] There is reasonably good evidence that calcium pumps in many different cell types are regulated by calmodulin, protein kinase C, and cyclic nucleotides.[23,36,37] A recent finding of potential clinical interest is that a decrease in the number of calcium pumps in brain has been associated with senescence in Alzheimer's disease.[38]

The existence of an electroneutral Na/Ca exchange mechanism was initially offered as an explanation for the changes in ^{45}Ca efflux in isolated cardiac muscle following removal of calcium or sodium from the bath solution.[39] It has since been demonstrated in both excitable and non-excitable cells but is thought to be more important in regulating calcium homeostasis in excitable tissues. In cardiac tissue, for example, the exchanger may operate in a bidirectional manner, transporting calcium into the cell when the plasma membrane is depolarized and out of the cell when the membrane is repolarized.[36] Little is known about the role of the exchanger in non-excitable cells, except that it appears to be associated only with calcium efflux. Thus far the Na/Ca exchanger has proven difficult to isolate and characterize, mainly because of the lack of a specific ligands and inhibitors. In addition, non-specific effects of extracellular sodium removal and choline replacement have made it difficult to study the physiologic function of the exchanger in isolated cells.[40,41] Putative Na/Ca exchanger proteins ranging in molecular weight from 33,000 to 220,000 have been purified from heart, brain, kidney, and rod outer segment membranes.[42]

Intracellular Calcium Release and Uptake Mechanisms

Intracellular Calcium Release: Ryanodine-Sensitive Calcium Channels

There is another class of calcium channels located within the cell interior that allows the efflux of calcium from intracellular calcium stores into the cytosol. These channels may have evolved independently of the voltage- and agonist-gated plasma membrane-assocated channel superfamilies.[24] Thus a calcium channel protein that allows efflux of calcium from the skeletal muscle sarcoplasmic reticulum (SR) has recently been identified and purified. A similar, but not identical protein has been isolated from brain.[43] The SR calcium channel is frequently called the ryanodine receptor because it binds ^{3}H-ryanodine with high affinity and low concentrations of ryanodine release calcium from the SR. The SR channel is a tetrameric protein ($M_r \sim 450\,kDa$) that possesses intrinsic calcium channel activity and a "foot" structure morphologically identical to a structure seen in electron micrographs that is located within the junctional area between SR and T-tubule membranes in skeletal muscle.[29,44,45] Several differences between ryanodine receptors in brain and skeletal muscle have been documented. For instance, in experiments with microsomal fractions isolated from canine cerebrum and cerebellum, caffeine-stimulated calcium efflux was only weakly inhibited with ruthenium red and magnesium, whereas low concentrations of ryanodine had no calcium-releasing effect.[46] In contrast, ruthenium red and magnesium are potent inhibitors of caffeine-stimulated calcium efflux in skeletal muscle and, as discussed above, low concentrations of ryanodine activate these calcium channels in skeletal muscle.

Calcium Pumps Within Intracellular Vesicular Compartments

ATP-dependent calcium uptake into intracellular compartments occurs not only in skeletal muscle sarcoplasmic reticulum, but also in unidentified organelles in other non-muscle cell types. The skeletal muscle SR enzyme has a molecular mass of 110 kDa, whereas the red blood cell plasma membrane Ca^{2+} pump has a mass of 140 kDa. Unlike the plasma membrane ATP-dependent calcium pump(s) in most cell types (liver being a possible exception), the SR enzyme is not stimulated by calmodulin.[47] Interestingly, two immunologically distinct Ca^{2+} ATPases with different subcellular distribution have been detected in adrenal chromaffin cells.[48] In skeletal muscle the uptake and release of calcium appear to occur in different regions of the SR as the inwardly-directed, ATP-dependent calcium pumps responsible for calcium reuptake are located at some distance from the calcium-releasing ryanodine receptor.[40] The relationship of these functions in exocrine and other non-muscle tissues is discussed in the following section.

Cellular Mechanisms Involved in Agonist-Stimulated Increases in Intracellular Calcium in Exocrine Cells

In exocrine cells as well as other non-excitable cells, agonist-stimulated increases in $[Ca^{2+}]_i$ has been divided into two major phases: initial release of calcium from intracellular store(s) followed by a sustained influx of calcium across the plasma membrane. Calcium movement into cells is balanced by ATP-dependent extrusion from the cell and ATP-dependent reuptake into intracellular stores. There is increasing evidence that calcium metabolism in exocrine cells is modulated not by voltage or agonist-gated mechanisms but by indirect activities involving the super-family of G protein-coupled receptors, including the transduction of extracellular signals to the cell interior by means of heterotrimeric GTP-binding (G) proteins. The nature of the regulatory G proteins involved in these mechanisms is not as well understood as the molecular events associated with the release of calcium from intracellular stores, which is known to be mediated, at least in part, by inositol-1,4,5-trisphosphate (IP_3). The events that lead to generation of IP_3 include receptor-mediated activation of phospholipase C (PLC), which may involve a putative G protein, with resultant breakdown PIP_2 to form IP_3 and diacylglycerol (DAG). Elevation of DAG, sometimes in concert with elevated calcium levels, activates one or more protein kinase C isozymes. Elevated calcium levels also lead to activation of other calcium-dependent enzymes, such as calcium-calmodulin-dependent protein kinases. It is not yet clear whether agonist-induced calcium release precedes or follows calcium influx.

Measurement of $[Ca^{2+}]_i$ in single cells suggest that agonist-induced oscillations are not uncommon and that they may represent the standard response in some cell types. Although calcium oscillations are not detected when cell population responses are measured using spectrofluorimetric techniques, earlier electrophysiologic studies on blowfly salivary glands suggested that calcium levels oscillate upon stimulation with cholinergic agonists.[50] In hepatocytes, regular oscillations in $[Ca^{2+}]_i$ occur upon stimulation with vasopressin and α adrenergic agonists, and the frequency of oscillations increases with increasing doses of these agonists.[51–53] The pattern of oscillations may vary among individual cells.[54,55] In pancreatic acinar cells, calcium oscillations are more sinusoidal than in hepatocytes and appear to be more pronounced at lower temperatures.[56] There is some evidence suggesting that cholecystokinin-induced calcium oscilla-tions in acinar cells are initiated by high but not by low-affinity receptors.[57] Such observations have stimulated a considerable amount of research and discussion on the mechanisms and physiologic relevance of calcium oscillations. These issues are discussed below. More detailed discussions can be found in recent reviews.[52,58,59]

Intracellular Calcium Release in Exocrine Cells

Role of Inositol-1,4,5-Trisphosphate in Intracellular Calcium Mobilization

Whether or not all exocrine cell types responded to calcium-dependent agonists by releasing calcium from intracellular stores has been a matter of debate. Indeed, early investigations suggested that calcium influx alone was responsible for agonist-induced increases in $[Ca^{2+}]_i$.[60,61] Much of this controversy may be attributed to the use of a first generation fluorescent intracellular calcium probe, quin-2, which has a low fluorescent quantum yield and must be used in high concentrations to produce a measureable signal. Since all of the fluorescent calcium-sensitive probes presently available are calcium chelators, high concentrations will buffer changes in intracellular calcium.[12] With the use of more sensitive second-generation probes, such as fura-2 and indo-1, it became apparent that calcium-dependent agonists stimulate both release and influx of $[Ca^{2+}]_i$ in exocrine cells.[2,4,5,8,9,62]

The work leading to the concept that InsP$_3$ can release calcium from intracellular storage sites began in 1953 when Hokin and Hokin[63] demonstrated that stimulation with acetylcholine increases phosphatidylinositol turnover in pancreatic fragments. It was not until 1975 that Michelle[64] proposed that PI turnover is linked to calcium fluxes. A few years later Berridge[65] demonstrated that enzymatic breakdown of phosphoinositol-4,5-bisphosphate (PIP$_2$) produces inositol-1,4,5-trisphosphate (IP$_3$) and 1,2-diacylglycerol. The same year Streb and colleagues[66] demonstrated that both IP$_3$ and hormones can release calcium from non-mitochondrial calcium pools, which appeared to be located in the endoplasmic reticulum.[67] Since these seminal discoveries, the literature on the role of IP$_3$ in modulating release of intracellular calcium has grown exponentially. It is now generally accepted that IP$_3$ plays an important role in the release of intracellular calcium in smooth muscle and in a variety of non-muscle cells including many exocrine cells, although probably not in the release of calcium associated with excitation-contraction coupling in cardiac and skeletal muscle.[2,4,5,9,29,68–70]

Evidence for G Protein Coupling of Receptors to Phospholipase C

Much has been learned recently about mechanisms by which calcium-dependent agonists stimulate PIP$_2$ breakdown to increase IP$_3$ and di-acylglycerol (DAG) production. Five different isoforms of phospholipase C (PLC), designated α, β, β_1, γ, and δ, have been detected in molecular cloning experiments,[71,72] and there is evidence suggesting that the mode of G protein coupling may depend on the PLC isozyme involved.[73] In general, agonists that bind to receptors with seven membrane-spanning regions are thought to promote coupling of the receptors to the phospholipase C_β isoform through heterotrimeric G proteins. In contrast, growth factors that elevate $[Ca^{2+}]_i$ by binding to receptors with tyrosine kinase activity are thought to promote association of the receptor with PLC-γ as well as other enzymes such as PI-3-kinase.[69,74,75] Several different G proteins may regulate PLC activation, including both pertussis toxin-sensitive and insensitive forms.[32,69,76]

The first suggestion that G proteins might be involved in calcium-mediated responses in exocrine cells came from studies of isolated cell membrane fractions and cells permeabilized with the mild detergents, digitonin and saponin. In these studies it was observed that both GTP and GTPγS (a poorly hydrolyzable GTP analog) facilitates agonist-induced formation of IP_3 and DAG.[4,75,77,78] Both nucleotides are known to promote dissociation of the α subunit of heterotrimeric G proteins, an event that leads to activation of enzymes associated with receptor-G protein complexes.[74] Other less direct experiments have shown that the combination of NaF and $AlCl_3$, which activates heterotrimeric G proteins in vitro, also stimulates PIP_2 breakdown and elevates $[Ca^{2+}]_i$ in several exocrine cell types, including parotid and pancreatic acini, avian nasal glands, and hepatocytes.[79-85] Although suggestive, these experiments are not conclusive because fluoride is a universal activator of several known heterotrimeric G proteins, including those coupled to adenylyl cyclase, and because fluoride inhibits a variety of cellular phosphatases and may also inhibit the enzyme PIP kinase that converts PI to PIP_2 and PIP.[86,87] Thus the possibility that fluoride may have an effect on G proteins coupled to phospholipase C exists, but does not unequivocally demonstrate such a mechanism.

Initially G_i (a G protein specifically inhibited by *Bordetella Pertussis* toxin through a specific ADP-ribosylation mechanism) was considered as a possible mediator of agonist-induced activation of phospholipase C. However, it now appears that G_i modulates PLC activation in cells of bone marrow origin but not in isolated exocrine cells. Thus, as mentioned above, there is evidence suggesting that there are at least two classes of G proteins capable of activating PLC: those that are sensitive to pertussis toxin and those that are not. Among the pertussis toxin-insensitive G protein is G_q. (Before G_q was identified, a putative G protein was postulated to exist and this protein was labeled G_q with "p" referring to PLC).[69] In the G_q family of proteins, there are five known G_α subunits: $G_{\alpha11}$, $G_{\alpha14-16}$ and G_{q-32} $G_{\alpha q}$ and $G_{\alpha11}$ appear to activate specifically the β isoform of PLC.[88,89] In other experiments, preincubation of liver plasma membranes with the calcium-dependent agonists, vasopressin, angiotensin II, or epinephrine, induced labeling of two proteins with a photoactivatable GTP analog. Since both proteins could be immunoprecipitated with antisera raised against $G_{\alpha q}$ and $G_{\alpha11}$, these data also suggested a regulatory function of $G_{\alpha q}$ proteins.[90]

Other investigators have proposed that PLC might be regulated by member(s) of a recently discovered new class of small, momomeric GTP-binding proteins.[69] These proteins, which have been grouped into four subfamilies on the basis of structure, include RAS-, RAB-, RHO-, and ARF-related proteins. There are at least 30 such proteins in search of a function.[91,92] Since fluoride does not appear to activate small GTP-binding proteins,[93] there is presently no strong evidence that small GTP-binding proteins are involved in the control of PIP_2 hydrolysis, particularly in cell types in which AIF_3 stimulates PIP_2 breakdown and calcium mobilization. Small GTP-binding proteins may, however, be involved in membrane fusion events associated with exocytosis and insertion of transporters into the plasma membrane.[91,92] Recently 20 such proteins were detected in pancreatic acinar cells with high resolution two-dimensional gel techniques, and two of these proteins were found to be phosphorylated following stimulation with the cholinergic agonist carbachol.[94] Thus, although it is unlikely that the small GTP-binding proteins regulate PLC directly, there is increasing evidence that these proteins may play an important role in regulating stimulus-secretion coupling.

The heterotrimeric G-protein-coupling mechanisms are of clinical interest because there is evidence that mutations resulting in chronic activation of G_α subunits create oncogenes. It is now known, for example, that prolonged activation of receptors coupled to the PI-PLC signaling system stimulates mitogenesis and tumorigenesis. Thus, mutations in $G_{\alpha q}$ might create oncogenes similar to the α_s and α_{i2} oncogenes already detected.[95,96]

Characteristics of the IP₃ Receptor

Well before the IP_3 receptor was isolated, its existence was postulated on the basis of the observation that the addition of IP_3 to permeabilized cells caused release of calcium from intracellular stores. The IP_3 receptor was first characterized by $[^3H]$ or $[^{32}P]$-labeled IP_3 binding studies. Specific binding was localized to particulate fractions of homogenates from several different tissue types including liver and salivary glands, and was associated with calcium-releasing activity in vesicular structures that were enriched in plasma membrane markers. An IP_3 receptor protein was first isolated from bovine and rat cerebella and functionally reconstituted into liposomes in 1988–1989.[97,98] A similar protein with high affinity for IP_3 and with properties of a calcium-mobilizing receptor has now been detected in liver.[99–106] The IP_3 receptor is identical to a protein (P_{400}) that was detected in abundance in cerebellar Purkinje cells some years ago.[107–108] The receptor has now been cloned and shown to function as a calcium channel. It shares almost 50% homology with the skeletal muscle ryanodine receptor in one putative transmembrane region (which possesses calcium channel activity) and forms homotetramers like the ryanodine receptor.[104,109-112] The similarities between the IP_3 receptor and the ryanodine receptor suggest that these proteins may have analogous functions. Mutagenesis studies with COS cells (ie, fibroblast-like cell line from SV40 transformed African Green monkey Kidney) indicate that each of the N-terminal sequences of the IP_3 homotetramer contains an IP_3 binding site that is separate from the channel-forming transmembrane region of the molecule. These studies also suggest that the receptor undergoes a significant conformational change upon binding IP_3.[112] A cAMP-dependent phosphorylation site and a possible modulatory ATP-binding site appear to be juxtaposed between the IP_3 binding site and the transmembrane calcium channel region.[111,112] Recent work suggests the presence of different IP_3 receptor subtypes which are expressed in a tissue specific and developmentally regulated manner. These isoforms may arise from alternate splicing of the same gene. The variations in the subtypes appear to be localized to the IP_3 binding site and to putative ATP-binding and phosphorylation sites.[113] Variations in IP_3 expression may explain the different binding affinities of IP_3 reported for cerebellum (dissociation constant (K_d) 80–100 nM[97,102] and peripheral tissues, K_d 1.7–10.4 nM).[103-105] Variations in modulatory sequences may also affect the ability of ATP and of phosphorylation to regulate IP_3-induced calcium release. However, since assay conditions have not been standardized, the apparent variations in binding affinities might also reflect different in vitro conditions such as pH.

Localization of IP_3-Sensitive and -Insensitive Intracellular Calcium Storage Sites

The precise localization of calcium storage sites in non-muscle tissues (including exocrine cells) is controversial. It has been established that in skeletal and cardiac muscle the SR is the major calcium storage site, whereas in smooth muscle there may be two classes of intracellular calcium stores, one that is sensitive to IP_3 and one that is sensitive to both IP_3 and caffeine.[114] As discussed above, the ER is believed to be an important calcium storage organelle in exocrine and other non-excitable cells. More recent studies have emphasized similarities between the ER and SR. For example, calcium uptake into ER-enriched vesicles from pancreatic acini was found to be mediated by a Ca-ATPase with structural similarities to the Ca-ATPase found in cardiac muscle SR.[115] However, since this information was obtained using cell fractions, the ER has not yet been established unambiguously as a calcium storage organelle. Indeed, several immunologic and cell fractionation studies have challenged the role of the ER in calcium signaling and there is also recent evidence suggesting the presence of IP_3 receptors capable of releasing calcium in the nucleus of hepatocytes.[103,116-119]

In 1988 Volpe and colleagues[120] postulated the existence of a separate organelle, which they named the "calcisome." This hypothesis, which has received much attention, was based originally on immunocytochemical studies in which it was found that antibodies raised against skeletal muscle SR calcium-ATPase and a calcium-binding protein, calsequestrin (which is abundant within the junctional ryanodine receptor-rich region of the SR) cross-reacted with small, distinct organelles with smooth membranes that were scattered throughout the cytosol in several different cell types, including pancreas and liver.[119,120] Recent evidence, however, suggests that the calcisome may be an artifact caused by formaldehyde fixation that can produce ER vesiculation.[121,122] Furthermore, it has been suggested that non-muscle cells contain not calsequestrin, but another calcium-binding protein, calreticulin, present in low concentrations also in the lumen of skeletal muscle SR.[121-124] The calsequestrin–calreticulin controversy was partially resolved when it was found that although the sequences of calsequestrin and calreticulin are distinct, there is sufficient sequence similarity to allow cross-reaction with some monoclonal antibodies.[125,126]

The amino acid sequence of calreticulin has been obtained from cDNA and protein sequencing analyses,[127] and found to be identical to two previously identified non-muscle proteins, calregulin and CRP55.[128-130] Calreticulin immunoreactivity has now been detected in ER-like membranes of a variety of cultured non-muscle cell types, in nucleoli-like structures in proliferating muscle cells and in the nuclear envelope of MDCK cells (a cultured cell line in which the ER appears to be continuous with the nuclear membrane).[131] These and similar observations suggest that calreticulin is the major calcium-sequestering protein of non-muscle cells, including pancreatic acinar cells. The matter is not entirely settled, however, because in at least one cell type, the chicken Purkinje cell, both calreticulin and calsequestrin may be present.[132,133]

Theoretical Models Describing Intracellular Location of Calcium Release Sites

An analysis of several recent immunohistochemical studies performed mainly in chicken Purkinje cells, led Burgoyne and Cheek[134] to propose a model for the distribution of the IP_3 and ryanodine receptors and calsequestrin (calreticulin?)-containing organelles in neurons. In this model the IP_3 receptors are localized in the rough ER and in the cisternal stacks, two possible subdomains of the rough ER, as well as in the nuclear envelope and in other unidentified organelles. The distribution of the ryanodine receptor overlaps that of the IP_3 receptor. Each receptor is also uniquely present in some organelles. The distribution of calsequestrin is best correlated with the location of the ryanodine receptor. Whether or not a similar model can be applied to exocrine and other non-excitable cells is presently unclear. As these authors point out, the neuronal model should be interpreted cautiously because receptors detected using immunohistochemical techniques may be present not only in functional locations but also in sites of synthesis and/or transport pathways. In addition, there is less immunohistochemical information available and no firm biochemical data to support the presence of ryanodine receptors in exocrine cells. Nonetheless, a number of interesting models have been proposed to explain the morphologic relationships associated with intracellular calcium release and influx. One model, similar to the one proposed for skeletal muscle, assumes that the IP_3-sensitive calcium release sites and intracellular calcium pumps are contained within the same organelle but are spatially separated from one another.[135] A similar, more elaborate proposal was suggested by Rossier and colleagues[136] on the basis of cell fractionation studies. These authors found that the distribution of IP_3 receptors varied with the method of homogenization. Under standard conditions, binding was associated with plasma membrane markers; however, when membranes were freeze-thawed or exposed to disrupters of actin microfilaments such as cytochalasin B, this association disappeared. To explain the data, the authors postulated that specific calcium-containing organelles that possess IP_3 receptors are linked to the plasma membrane by a mesh of actin microfilaments. This model is attractive because it appears to solve the conflicting observations in which IP_3 binding and calcium-releasing activity were associated with either the ER or the plasma membrane.[67,136,137]

More recently Putney and colleagues have provided evidence that the calcium uptake and the calcium release sites are morphologically separated. In these experiments saponin-permeabilized rat parotid acinar cells were exposed to poorly metabolized IP_3 analogs and thapsigargin.[138] It will be interesting to know whether or not these putative calcium pools possess junctional complexes similar to those seen in smooth muscle cells,[49] or whether they are contained in separate, distinct organelles. Another model suggests a completely different morphologic arrangement in which the IP_3-sensitive organelle is localized in secretory granules near the apical membrane of the cell rather than in the proximity of the agonist receptor-containing basolateral membrane.[139] This location was suggested by digitized video image analysis studies of single pancreatic acinar cells, which purported to show a localized rise in $[Ca^{2+}]_i$ near the apical pole of the cell.[140] The model predicts that following apical release, a generalized rise occurs as a result of a wavelike propagation through the ER via a calcium-induced calcium release (CICR) mechanism (described below). The apical model is interesting but requires the difficult assumption that IP_3 diffuses all the way across the cell to the apical membrane to exert an effect. In addition, other studies have suggested a basal rather than an apical localization for initial calcium release.[141,142] Since the ability to localize accurately intracellular calcium release sites using currently available low-resolution video cameras is limited, and approaches using laser-based confocal microscopic techniques in living cells are not well established,[59] it may be some time before this question is adequately resolved.

IP$_3$-Insensitive Intracellular Calcium Pools: Potential Interrelationships Between Calcium-Induced Calcium Release (CICR) and Calcium Oscillations

The existence of IP$_3$-insensitive calcium pools was first proposed to explain the observation that only 30–50% of calcium taken up by non-mitochondrial pools is released by IP$_3$.[143] The concept was later used to explain the observation that, in single cells, agonist-induced changes in $[Ca^{2+}]_i$ are not biphasic, but rather oscillate at different frequencies depending on the dose of agonist used. In 1988 Berridge and Gallione[58] considered several models to account for the oscillations in $[Ca^{2+}]_i$ observed in cell types such as blowfly salivary gland cells, hepatocytes, and parotid acinar cells. In the receptor-controlled models, cytosolic calcium oscillations were generated via oscillations in IP$_3$ levels as originally proposed by Cobbold and colleagues.[51,144] A second messenger-controlled group of models assumed that oscillations occurred as a result of second messenger actions on the ER. One of these second messenger models proposed the existence of both IP$_3$-sensitive and -insensitive calcium pools, which interacted to generate oscillations through a calcium-induced calcium release (CICR) mechanism. Indeed, a release of calcium from intracellular stores following a sudden increase in free calcium concentration was observed some years ago in skeletal and cardiac muscle fibers and SR vesicles and termed "calcium-induced calcium release."[145,146] In muscle, CICR appears to be mediated by the ryanodine receptor because it is enhanced by low concentrations of this plant alkaloid. Both the rate of change and the concentration of calcium appear to be important in the activation of CICR. At extravesicular calcium concentrations below approximately 10 µM, CICR is stimulated; however, when calcium levels are increased above this level, CICR is inhibited. CICR is enhanced by caffeine, which activates a calcium channel in cardiac muscle SR,[149] and CICR is inhibited by ruthenium red and by the local anaesthetic, procaine.[29,146]

Although the physiologic significance of CICR in muscle (particularly in skeletal muscle) has been questioned,[70,146] some investigators believe CICR plays an important role in cardiac contractility; there are data, albeit obtained under non-physiologic conditions, that suggest that CICR may occur in exocrine cells. For example, in patch-clamped pancreatic acinar cells, infusion of calcium elicits repetitive spikes in calcium-dependent Cl^- current, an event that parallels changes in $[Ca^{2+}]_i$. These oscillations appear to mimic the effects of acetylcholine and of internal application of IP_3. Since the IP_3 antagonist, heparin, blocked the IP_3 but not the calcium-induced oscillations and caffeine enhances the effects of subthreshold calcium infusions, Petersen and colleagues proposed that acetylcholine-induced calcium oscillations are caused by pulsatile release of calcium through a caffeine-sensitive calcium channel present in intracellular membranes. Previous experimental data by these authors indicated that a steady infusion of IP_3 into acinar cells can elicit an oscillatory calcium response. Thus it was concluded that CICR is initially elicited by an agonist-induced steady-state elevation in IP_3.[148,150,151]

There is also some indirect evidence for CICR in hepatocytes, and lacrimal and pancreatic acinar cells.[147,152,153] Experiments in brain suggest a potential modulatory mechanism for CICR based on apparent differing sensitivities of IP_3 and ryanodine-sensitive calcium channels. Indeed, in lipid bilayers, reconstituted calcium channels from ER-enriched membranes from cerebellum exhibit optimal IP_3-sensitive channel opening at $0.2 \mu M$ calcium. In contrast, the caffeine-sensitive ryanodine receptor channel opens at calcium concentrations between 1 and $100 \mu M$.[154]

Other observations on pancreatic acinar cells provided indirect evidence suggesting the presence of two or there functionally distinct non-mitochondrial calcium pools. Calcium pools with different sensitivities to caffeine and IP_3 have been described in smooth muscle SR and in pancreatic acinar cells.[155,156] IP_3-sensitive and -insensitive calcium pools have also been detected in isolated liver nuclei.[118] With the exception of experiments performed by Foskett and Wong,[153] who showed that caffeine, when added after thapsigargin, increases calcium oscillations in intact rat parotid acini, most evidence for the existence of different calcium pools has been obtained using isolated vesicle preparations or permeabilized cells rather intact exocrine cells. Since neither caffeine nor ryanodine, when administered alone, has been shown to affect calcium signaling patterns or secretion, there is presently little evidence to support a role for the type of CICR mechanism observed in muscle. In addition, since cell fractionation is likely to disrupt contiguous pools and detergent permeabilization may release IP_3-sensitive structures,[157] it is not clear whether IP_3-sensitive and -insensitive calcium pools are physically separate entities or whether they reside in the same vesicular compartment. Moreover, it should be emphasized that neither a CICR- nor an IP_3-induced increase in nuclear calcium concentration has yet been shown to have a physiologically relevant role in exocrine cell function.

Does GTP Allow Communication Between Calcium Pools?

A few studies have suggested a potential role for GTP in the regulation of IP_3-sensitive and -insensitive calcium pools apparently unrelated to the GTP-dependent activation of regulatory G proteins. In 1985 Dawson[158] first observed that GTP induced a release of calcium from isolated liver microsomes and that this effect was enhanced by polyethylene glycol. Initially it was thought that GTP was activating an IP_3-sensitive calcium channel. However, other data suggested that GTP might promote membrane fusion, thus facilitating the association of IP_3-sensitive and -insensitive pools.[135,159] GTP appears to induce calcium release from permeabilized hepatocytes, but not from permeabilized exocrine cell types.[138] Thus, even though the idea of a GTP-dependent regulation of different intracellular calcium pools is attractive in light of the demonstrated role of GTP-binding proteins in intracellular membrane trafficking,[91,92] supporting data are not strong.

Potential Role for Cyclic ADP Ribose in Intracellular Calcium Release

A metabolite of NAD^+, as potent as IP_3 in releasing calcium from intracellular stores in sea urchin eggs, has been identified as cyclic ADP-ribose (cADPR).[160] Although the effects of this compound have not yet been characterized in exocrine cells, there is some evidence that the enzyme that cyclizes NAD to cADPR is present in mammalian cells extracts,[161] and that cADPR releases calcium from permeabilized pituitary cells,[162] through an action on the IP_3-sensitive calcium pools.[163] Recent experiments showing that, in sea urchin egg homogenates, cADPR-induced calcium release was blocked by ruthenium red and procaine and that cADPR itself prevented caffeine- but not IP_3-induced calcium release led to the suggestion that this cyclic compound may act as an endogenous modulator of CICR,[164] especially since cADPR is synthesized from NAD and NADH (the concentrations, of which may in turn change upon agonist stimulation).

Calcium Entry Mechanisms in Exocrine Cells

The mechanisms that regulate the entry of calcium into cells required to sustain calcium-dependent secretion in exocrine cells are not clear. Early reports that calcium antagonists that inhibit L-type channels in excitable cells also inhibited secretion and calcium fluxes in some exocrine cell types proved to be misleading and due to non-specific effects produced by using these compounds at concentrations 10–100× higher than those required for inhibition in excitable cells. Moreover, inhibition was found with some antagonists but not others, apparently because some calcium channel antagonists can also antagonize agonist-receptor binding. Thus there is presently no good physiologic or pharmacologic evidence for diydropyridine-sensitive L-type channels in exocrine tissue.[2,3,80,165–167] Nevertheless, a pancreatic cell line, AR4-2J, does express DHP-inhibitable calcium channels even though normal pancreatic acinar cells do not.[168] The meaning of this phenomenon is unclear, except to suggest that data obtained with cancer cell lines may not necessarily reflect normal cell functions.

Since agonist-induced elevation of $[Ca^{2+}]_i$ in exocrine cells is slow in comparison to agonist-gated calcium movement in excitable cells, it is unlikely that classic receptor-operated mechanisms are involved. There are presently two opposing views concerning calcium entry into non-excitable cells in general. The first is that calcium entry follows an initial release of calcium from intracellular stores. The second is that calcium entry is independent of intracellular calcium release.[75] There are many interpretations of these central themes. In support of the first hypothesis, Putney[169] initially proposed a "capacitive" model in which depletion of intracellular calcium pools initiated calcium entry. In this model, calcium enters directly into plasma membrane-associated organelles, bypassing the cytosol. This model was recently revised to account for data showing a rise in cytosolic calcium during refilling of intracellular calcium pools.[170–173] The role of IP_3 and/or IP_4 in mediation of calcium entry is also uncertain since thapsigargin, which does not increase the concentration of either metabolite, increases calcium entry.

However, Irvine has proposed a second messenger model as an extension of the second hypothesis to explain how calcium entry and release from intracellular stores may be coordinated. In this model, which mimics the configuration in skeletal muscle where the ryanodine receptor bridges the gap between the SR membrane and the sarcolemma (T-tubule membrane), IP_4 induces conformational changes in an receptor protein located in the plasma membrane that is positioned near an IP_3 receptor situated on a nearby intracellular calcium storage site. In the absence of agonists, the IP_3 and IP_4 receptor proteins associate with one another. Following agonist-induced elevation of IP_3 and IP_4, the receptor proteins dissociate, allowing calcium influx and release.[177] As yet there is no direct evidence for an IP_4 receptor in the plasma membrane of exocrine cells; therefore this interesting model remains speculative. It may well be that more than one mechanism exists for controlling calcium entry. In lacrimal acinar cells, for example, microinjection of IP_3/IP_4 increases calcium influx.[174,175] A small conductance calcium channel has also been detected in endothelial cells. This channel could be activated by calcium alone or by IP_4 in the presence of relatively high calcium concentrations.[176]

Potential Role of Non-Selective Cation Channels in Calcium Entry

Non-selective cation channels (plasma membrane-associated proteins that allow passage of cations in addition to calcium) have been detected in pancreatic acini, cultured mandibular cells, and hepatoma cells,[178–180] and may represent a significant pathway of basal calcium influx.[180] Whether or not these or similar channels also function as agonist-sensitive calcium entry pathways is debatable. It has been reported that extracellular ATP promotes calcium influx via a non-selective, 10 pS calcium channel,[181] and that manganese (Mn) competes with calcium for agonist-stimulated entry in certain cell types[182] such as pancreatic acinar[183] and parotid acinar cells,[184] hepatocytes,[80] and rat sublingual mucous acinar cells.[166] However, negative results have also been reported for parotid acinar[185] and lacrimal acinar cells,[171] and avian salt glands.[167] Part of the problem may be due to the presence of basal-, but not agonist-stimulated, Mn entry pathways, and/or differences in experimental protocols, as well as technical problems associated with measurement procedures.

Summary and Conclusions

Figure 3.1 is a diagram of known and postulated mechanisms controlling calcium fluxes in non-excitable cells. It should be emphasized that many of the represented pathways have not yet been demonstrated in exocrine gland cells. It has been established that calcium-dependent agonists stimulate intracellular release and extracellular influx of calcium. Intracellular release is mediated, at least partially, by agonist-induced generation of IP_3. There is reasonably good evidence to suggest that IP_3 production involves the activation of heterotrimeric G proteins of the Gq family and the PLC_β isozymes. The IP_3 receptor has characteristics similar to those of the skeletal muscle ryanodine receptor, which is located on the SR membrane. Whether or not IP_3 receptors are localized in the ER or on some other specialized intracellular membrane is uncertain; however, it appears that the membranous compartment that contains the IP_3 receptor is closely, albeit indirectly, associated with the plasma membrane, perhaps through the actin microfilament network. Current evidence does not allow definite conclusions about the relative involvement of IP_3-sensitive and -insensitive calcium stores in generalized increases in $[Ca^{2+}]_i$ or in the production of calcium oscillations. The physiologic relevance of calcium oscillations is uncertain; however, it has been suggested that different patterns of oscillations may control different intracellular activation events. Calcium oscillations may also allow frequency-dependent coding of cellular activities. In this case, the amplitude of the calcium signal is invariant and the frequency of spikes controls the magnitude of the response.[51,58,186,187] The mechanism of calcium entry into exocrine cells remains poorly described, although it is tempting to speculate that agonist-sensitive calcium entry channels may function in a manner analogous to that of the skeletal muscle DHP

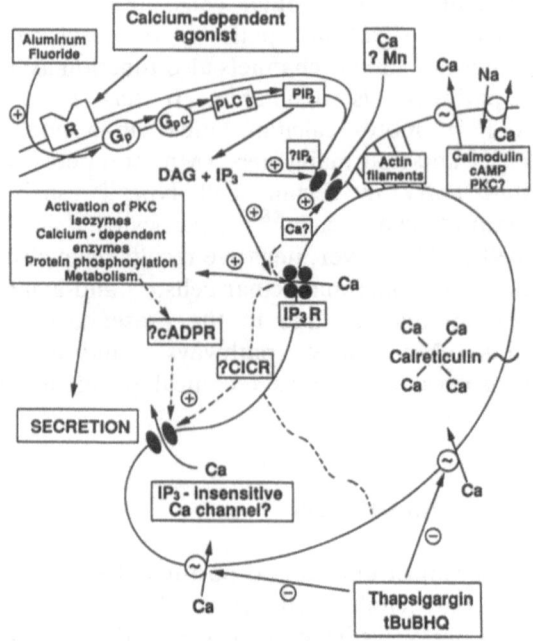

FIGURE 3.1. Hypothetical model of calcium influx and intracellular release pathways in exocrine and other non-excitable cells. Calcium-dependent agonists bind to their specific membrane-bound receptors. Agonist-receptor binding leads to activation of heterotrimeric G proteins, possibly G_p. The α subunit of the G protein activates phospholipase C_β isozyme(s), which break down PIP_2 to form IP_3 and diacylglycerol (DAG). IP_3 binds to specific receptors located on intracellular calcium storage site(s), which allows calcium influx into the cytosol via the receptor channel. Calreticulin serves as a calcium binding protein within calcium storage organelles. Actin microfilaments may serve to "dock" calcium storage organelles near the plasma membrane. Calcium oscillations observed in some cell types may be produced by a calcium-induced calcium release (CICR) mechanism similar to that observed in excitable tissue; however, evidence for such a mechanism is not strong. There are also preliminary data suggesting that cyclic ADP-ribose (cADPR) may regulate the release of intracellular calcium. Both IP_3-sensitive and -insensitive calcium stores may be present in adjacent or separate pools. Intracellular energy-dependent calcium "pumps" (ATPases), but not plasma membrane-bound calcium pumps, are sensitive to inhibition by thapsigargin and tBuBHQ. The influx of calcium into the cell may be activated by IP_4, IP_3, and/or intracellular calcium. The relative amount of calcium released from intracellular storage sites may also influence calcium influx (capacitive entry model[173]). The model suggests possible homologies between the DHPR receptors and ryanodine-sensitive SR calcium channels in skeletal muscle and the putative plasma membrane-bound and intracellular InsP3-sensitive calcium channels in non-excitable cells. Elevation of intracellular free calcium concentrations leads to activation of calcium-dependent enzymes, metabolism, and protein phosphorylation. These pathways are not well understood.

receptor. It is also possible that the release of calcium from intracellular organelles precedes calcium influx.[173,177] There is much to be learned about the role of second messengers in the regulation of calcium entry and release mechanisms.

Beyond the mechanisms of calcium release, reuptake, and influx lies a complex series of calcium-dependent intracellular events. In addition to the intracellular calcium binding proteins, calreticulin and calregulin, there are a number of other calcium-binding proteins and calcium-activated enzymes including, for example, calmodulin and other protein members of the EF-hand family, calmodulin-dependent protein kinases, protein kinase C, the calpains, and the annexin family.[188] The physiologic functions of these proteins are poorly understood.[189,190] To date only a few agonist-responsive phosphoproteins have been detected and little is known of their intracellular functions or the specific protein kinases that are involved in their phosphorylation. It has been observed, however, that the concentrations of some calcium-binding are altered in Parkinson's disease, Alzheimer's disease, epilepsy, hypertension, cystic fibrosis, acute and inflammatory lesions, and tumorigenesis.[190]

The discovery of the IP$_3$ calcium release pathway has stimulated much research on calcium signaling mechanisms. The importance of calcium influx pathways becomes apparent when one considers that calcium release without influx does not sustain calcium-dependent cellular activities. The use of specific blockers of agonist-stimulated calcium influx into exocrine cells would greatly enhance our understanding of exocrine and other cell functions, including growth, may prove to have important clinical applications. Although such blockers are not presently available, rapid developments in the area are expected to lead to their identification within the next few years.

References

1. Campbell AK. Intracellular Calcium—Its Universal Role as Regulator. New York: Wiley; 1983:1–556.
2. Chew CS. Intracellular activation events for parietal cell HCl secretion. In: Forte JG, ed. Handbook of Physiology, Section 6: The Gastrointestinal System. New York: Oxford University Press; 1989:255–266.
3. Putney JW, Jr. Calcium signaling system in salivary glands. In: Forte JG, ed. Handbook of Physiology, Section 6: The Gastrointestinal System. New York: Oxford University Press; 1989:51–61.
4. Schulz I. Signaling transduction in hormone- and neurotransmitter-induced enzyme secretion from the exocrine pancreas. In: Forte JG, ed. Handbook of Physiology, Section 6: The Gastrointestinal System. New York: Oxford University Press; 1989:443–463.
5. Williams JA, Burnham DB, Hootman SR. Cellular regulation of pancreatic secretion. In: Forte JG, ed. Handbook of Physiology, Section 6: The Gastrointestinal System. New York: Oxford University Press; 1989:419–441.

6. Douglas WW. Stimulus-secretion coupling: the concept and clues from chromaffin and other cells. Br J Pharmacol 1968; 34:451–474.

7. Joseph SK, Coll KE, Thomas AP, Rubin R, Williamson JR. The role of extracellular Ca^{2+} in the response of the hepatocyte to Ca^{2+}-dependent hormones. J Biol Chem 1985; 260:12508–12515.

8. Shuttleworth TJ, Thompson JL. Intracellular $[Ca^{2+}]$ and inositol phosphates in avian nasal gland cells. Am J Physiol 1989; C1020–C1029.

9. Muallem S. Calcium transport pathways of pancreatic acinar cells. Annu Rev Physiol 1989; 51:83–105.

10. Hersey SJ. Cellular basis of pepsinogen secretion. In: Forte JG, ed. Handbook of Physiology, section 6: The Gastrointestinal System. New York: Oxford University Press; 1989:267–278.

11. Sato K, Sato F. Relationship between quin2-determined cytosolic $[Ca^{2+}]$ and sweat secretion. Am J Physiol 1988; 254(Cell Physiol 23):C310–C317.

12. Tsien RY. Fluorescent indicators of ion concentrations. Meth Cell Biol 1989; 30:127–156.

13. Foskett JK, Melvin JE. Activation of salivary secretion: coupling of cell volume and $[Ca^{2+}]_i$ in single cells. Science 1989; 244:1582–1585.

14. Negulescu PA, Reenstra WW, Machen TE. Intracellular Ca requirements for stimulus-secretion coupling in the parietal cell. Am J Physiol 1989; 256:C241–C251.

15. Brown MR, Chew CS. Carbachol-induced protein phosphorylation in parietal cells: Regulation by $[Ca^{2+}]_i$. Amer J Physiol 1989; 257(Gastrointest Liver Physiol 20):G99–G110.

16. Thastrup O, Dawson AP, Scharff O, Foder B, Cullen PJ, Drobak BK, Bjerrum PJ, Christensen SB, Hanley MR. Thapsigargin, a novel molecular probe for studying intracellular calcium release and storage. Agents and Actions 1989; 27:17–23.

17. Llopis J, Chow SB, Kass GEN, Gahm A, Orrenius S. Comparison between the effects of the microsomal Ca^{2+}-translocase inhibitors thapsigargin and 2,5-di-(t-butyl)-1,4-benzohydroquinone on cellular calcium fluxes. Biochem J 1991; 277:553–556.

18. Thastrup O, Foder B, Scharff O. The calcium mobilizing and tumor promoting agent, thapsigargin, elevates the platelet cytoplasmic free calcium concentration to a higher steady state level: A possible mechanism of action for the tumor promotion. Biochem Biophys Res Commun 1987; 142:654–660.

19. Hakil H, Fujiki H, Suganuma M, Nakayasu M. Thapsigargin, a histamine secretagogue is a non-12-O-tetradecanoylphorbol-13-acetate type tumor promoter, in two-stage mouse skin carcinogenesis. J Cancer Res Clin Oncol 1986; 111:177–181.

20. Chew CS, Petropoulos AC. Thapsigargin potentiates histamine-stimulated HCl secretion, in gastric parietal cells but does not mimic cholinergic agonists. Cell Regul 1991; 2:27–39.

21. Petersen OH, Findlay I. Electrophysiology of the pancreas. Physiol Rev 1987; 67:1054–1116.

22. Forte JG, Soll AH. Cell biology of hydrochloric acid secretion. In: Forte JG, ed. Handbook of Physiology Section 6: The Gastrointestinal System New York: Oxford University Press; 1989:207–228.

23. Rasmussen H, Rasmussen JE. Calcium as intracellular messenger: From simplicity to complexity. Curr Topics Cell Regul 1990; 31:1–127
24. Hille B. Ionic channels: Evolutionary origin and modern roles. Quarterly Jour Exper Physiol 1989; 74:785–804.
25. Bean BP. Classes of calcium channels in vertebrate cells. Annu Rev Physiol 1989; 51:387–384.
26. Hosey MM, Lazdunski M. Calcium channels: Molecular pharmacology, structure and regulation. J Membr Biol 1988; 104:81–105.
27. Tsien RW, Tsien RY. Calcium channels, stores and oscillators. Ann Rev Cell Biol 1990; 6:715–760.
28. Hofmann F, Flockerzi V, Nastainczyk W, Ruih P, Schnelder T. The molecular structure and regulation of muscular calcium channels. Current Topics in Cellular Regul 1990; 31:233–239.
29. Fleischer S, Inui M. Biochemistry and biophysics of excitation—contraction coupling. Annu Rev Biophys Chem. 1989; 18:333–364.
30. Tanabe T, Beam KG, Powell JA, Numa S. Restoration of excitation-contraction coupling and slow calcium current in dysgenic muscle by dihydropyridine receptor complementary DNA. Nature 1988; 336:134–139.
31. Caterall WA, Genetic analysis of ion channels in vertebrates. Annu Rev Physiol 1988; 50:395–408.
32. Simon MI, Strathmann MP, Gaulam N. Diversity of G proteins in signal transduction. Science 1991; 252:802–808.
33. Kass RS, Lederer WJ, Tsien RW, Weingart R. Role of calcium ions in transient inward currents and latter contractions induced by strophanthidin in cardiac Purkinje fibres. J Physiol 1978; 281:187–208.
34. Shull GE, Greeb J. Molecular cloning of two isoforms of the Ca^{2+},-transporting ATPase from rat brain: Structural and functional domains exhibit similarlty to Na^+, K^+ and other cation transport ATPases. J Biol Chem 1988; 263:8646–8657.
35. Verma AK, Filoteo AG, Stanford DR et al. Complete primary structure of a human plasma membrane Ca^{2+} pump. J Biol Chem, 1988; 263:14152–14159.
36. Carafoli E. The calcium pumping ATPase of the plasma membrane. Annu Rev Physiol 1991; 53:531–547.
37. Grover AK, Kahn I. Calcium pump Isoforms: Diversity, selectivity and plasticity. Cell Calcium 1992; 13:9–17.
38. Ross GS. Hormone/neurotransmitter action during aging: The calcium hypothesis of impaired signal transduction. Rev Biol Res Aging 1990; 4:243–252.
39. Reuter H, Seitz N. The dependence of Ca^{2+} efflux from cardiac muscle on temperature and external ion composition. J Physiol 1968; 195:451–470.
40. Muallem S, Beeker T, Pandol SJ. Role of Na/Ca exchange and plasma membrane Ca pump in hormone-related Ca efflux from pancreatic acini. J Membr Biol 1988; 102:155–162.
41. Negulescu PA, Machen TE. Lowering extracellular sodium of pH raises intracellular calcium in gastric cells. J Membr Biol 1990; 116:239–248.
42. Rahamimoff H. Na^+-Ca^{2+} exchanger: The elusive protein. Curr Topics Cell Regul 1990; 31:241–271.
43. McPherson PS, Campbell KP. Solubilization of biochemical characterization of the high affinity [^{3}H]ryanodine receptor from rabbit brain membranes. J Biol Chem 1990; 265:18454–18460.

44. Lai F, Erickson HP, Rousseau E, Liu Q, Meissner G. Purification and reconstitution of the calcium release channel from skeletal muscle. Nature 1988; 331:315–319.

45. Block BA, Imagawa T, Campbell KP, Franzi-Armstrong C. Structural evidence for direct interaction between the molecular components of the transverse tubule/sarcoplasmic reticulum junction in skeletal muscle. J Cell Biol 1988; 107:2587–2600.

46. Mészaros LG, Volpe P. Caffeine- and ryanodine-sensitive Ca^{2+} stores of canine cerebrum and cerebellum neurons. Am J Physiol 1991; 261:C1048–C1054.

47. Schatzmann HJ. The calcium pump of the surface membrane and of the sarcoplasmic reticulum. Annu Rev Physiol 1989; 51:473–495.

48. Burgoyne RD, Cheek TR, Morgan A, O'Sullivan AJ, Moreton RB, Berridge MJ, Mata AM, Colyer J, Lee AG. Distribution of two distinct Ca^{2+}-ATPase-like proteins and their relationships to the agonist-sensitive calcium store in adrenal chromaffin cells. Nature 1989; 342:72–74.

49. Somlyo AP. Excitation-contraction coupling and the structure of smooth muscle. Circ Res 1985; 57:497–507.

50. Berridge MJ, Rapp PE. A comparative survey of the function, mechanism and control of cellular oscillators. J Exp Biol 1979; 81:217–280.

51. Woods NM, Cuthbertson KSR, Cobbold PH. Agonist-induced oscillations in cytoplasmic free calcium concentration in single rat hepatocytes. Cell Calcium 1987; 8:79–100.

52. Jacob R. Calcium oscillations in electrically non-excitable cells. Biochim Biophys Acta 1990; 1052:427–438.

53. Berridge MJ. Calcium oscillations. J Biol Chem 1990; 265:9583–9586.

54. Rooney TA, Sass EJ, Thomas AP. Characterisation of cytosolic calcium oscillations induced by phenylephrine and vasopressin in single fura-2-loaded hepatocytes. J Biol Chem 1989; 264:17131–17141.

55. Ljungström M, Chew CS. Calcium oscillations and morphological transformations in single cultured gastric parietal cells. Am J Physiol 1991; 260:C67–C78.

56. Gray PTA. Oscillations of free cytosolic calcium evoked by cholinergic and catecholaminergic agonists in rat parotid acinar cells. J Physiol (Lond) 1988; 406:35–53.

57. Matozaki T, Göke B, Tsunoda Y, Rodriguez M, Martinez Jr, Williams JA. Two functionally distinct cholecystokinin receptors show different modes of actions on Ca^{2+} mobilization and phospholipid hydrolysis in isolated rat pancreatic acini. J Biol Chem 1990; 265:6247–6254.

58. Berridge MJ, Gallione A, Cytosolic calcium oscillators. FASEB J 1988; 2:3074–3082.

59. Chew CS, Ljungström M. Measurement and manipulation of oscillations in cytoplasmic calcium. In Herman B, Lemasters J, New Methods of Microscopy, San Diego: Academic Press; 1993:133–175.

60. Snider RM, Roland RM, Lowy RJ, Agranoff BW, Ernst SA. Muscarinic receptor-stimulated Ca^{2+} signaling and inositol lipid metabolism in avian salt gland cells. Biochim Biophys Acta 1986; 889:216–224.

61. Muallem S. Sachs G. Ca^{2+} metabolism during cholinergic stimulation of acid secretion. Am J Physiol 1985; 248:G216–G228.

62. Chew CS, Brown MR. Release of intracellular Ca^{2+} and elevation of inositol trisphosphate by secretagogues in parietal and chief cells isolated from rabbit gastric mucosa. Biochim Biophys Acta 1986; 888:116–125.
63. Hokin MR, Hokin LE. Enzyme secretion and the incorporation of ^{32}P into phospholipids of pancreas slices. J Biol Chem 1953; 203:967–977.
64. Michelle RH. Inositol phospholipids and cell surface receptor function. Biochim Biophys Acta 1975; 415:81–147.
65. Berridge MJ. Rapld accumulation of inositol trisphosphate reveals that agonists hydrolyze polyphospholnositides instead of phosphatidylinositol. Biochem J 1983; 212:849–858.
66. Streb H, Irvine RF, Berridge MJ, Schulz I. Release of Ca^{2+} from a non-mitochondrial intracellular store in pancreatic acinar cells by inositol-1,4,5-trisphosphate. Nature 1983; 306:67–69.
67. Sayerdorffer E, Streb H, Eckhardt L, Haase W, Schulz I. Characterization of calcium uptake into rough endoplasmic reticulum of rat pancreas. J Membr Biol 1984; 81:69–82.
68. Berridge MJ, Irvine RF. Inositol phosphates and cell signalling. Nature 1989; 341:197–205.
69. Catt KJ, Hunyady L, Balla T. Second messengers derived from inositol lipids. J Bioenergetics and Biomembranes 1991; 23:7–27.
70. Rios EJ. The mechanical hypothesis of excitation-contraction (EC) coupling in skeletal muscle. Muscle Res Cell Motil 1991; 12:127–135.
71. Rhee SG, Suh PG, Hyu SH, Lee SY, Studies of inositol phospholiplid-specific phospholipase C. Science 1989:546–550.
72. Suh PG, Ryu SH, Choi WC, Lee KY, Rhee SG, Monoclonal antibodies to three phopholipase C isozymes from bovine brain. J Biol Chem. 1988; 263:14497–14504.
73. Martin TFJ, Lewis JF, Kowalchk JA. Phospholipase C-β1 is regulated by a pertussis toxin-insensitive G protein. Biochem J 1991; 280:753–760.
74. Gilman AG. G proteins: Transducers of receptor-generated signals. Annu Rev Biochem 1987; 56:615–649.
75. Exion JH. Mechanisms of action of calcium-mobilizing agonists: Some variations on a young theme. FASEB J. 1988; 2:2670–2676.
76. Schnefel S, Bantic H, Eckhardt L, Schultz G, Schulz I. Acetylcholine and cholecystokinin receptors functionally coupled by different G-proteins to phospholipase C in pancreatic acinar cells. FEBS Lett 1988; 230:125–130.
77. Berridge MJ. Inositol trisphosphate and diacyl glycerol: Two interacting second messengers. Annu Rev Biochem 1987; 56:159–193.
78. Rana RS, Hokin LE. Role of phosphoinositides in transmembrane signaling. Physiol Rev 1990; 70:115–163.
79. Duddy SK, Kass GEN, Orrenius S. Ca^{2+}-mobilizing hormones stimulate Ca^{2+} efflux from hepatocytes. J Biol Chem 1989; 264:20863–20866.
80. Barritt GJ, Hughes BP. The nature and mechanism of activation of the hepatocyte receptor-activated Ca^{2+} inflow system. Cellular Signalling 1991; 3:283–292.
81. Shuttleworth TJ. Fluoroaluminate activation of different components of the calcium signal in an exocrine cell. Biochem J 1990; 269:417–422.
82. Blackmore PF, Bocekino SR, Waynick CE, Exton JH. Role of a guanine-nucleotide-binding regulatory protein in the hydrolysis of hepatocyte

phosphatidylinositol 4,5-bisphosphate by calcium-mobilizing hormones and the control of cell calcium studies utilizing aluminum fluoride. J Biol Chem 1985; 260:14477–14483.

83. Taylor CW, Merritt JE, Putney JW Jr, Rubin RP. A guanine nucleotide-dependent regulatory protein couples substance P receptors to phospholipase C in rat parotid glands. Biochem Biophys Res Commun 1986; 136:362–368.

84. Merritt JE, Rink TJ. Regulation of cytosolic free calcium in Fura-2 loaded rat parotid acinar cells. J Biol Chem 1987; 262:17362–17369.

85. Matozaki T, Sakamoto C, Nagao M, Nishizaki H, Baba S. G protein in stimulation of PI hydrolysis by CCK in isolated rat pancreatic acinar cells. Am J Physiol 1988; 255:E652–E659.

86. Claro E, Wallace MA, Fain JN. Dual effect of fluoride on phosphoinositide metabolism in rat brain cortex. Biochem J 1990; 268:733–737.

87. Lange AJ, Arion WJ, Burcell A, Burcell B. Aluminum ions are required for stabilization and inhibition of hepatic microsomal glucose-6-phosphatase by sodium fluoride. J Biol Chem 1986; 261:101–107.

88. Smrcka AV, Hepler JR, Brown KO, Sternweis PC. Regulation of polyphosphoinositide-specific phospholipase C activity by purified G_q. Science 1991; 251:804–807.

89. Taylor SJ, Choe HZ, Rhee SG, Exton JH. Activation of the B_1 isozyme of phospholipase C by α subunits of the G_q class of G proteins. Nature 1991; 350:516–518.

90. Wange RL, Smrcka AV, Sternweis PC, Exton JH. Photoaffinity labeling of two rat liver plasma membrane proteins with [^{32}P]γ-azidoaniliado GTP in response to vasopressin. J Biol Chem 1991; 266:11409–11412.

91. Bourne HR, Saunders DA, McCormick F. The GTPase superfamily: A conserved switch for diverse cell functions. Nature 1990; 348:125–132.

92. Hall A. The cellular Functions of small GTP-binding proteins. Science (Wash DC) 1990; 249:635–640.

93. Kahn RA. Fluoride is not an activator of the smaller (20–25 kDa) GTP-binding proteins. J Biol Chem 1991; 266:15595–15597.

94. Göke B, Williams JA, Wishart MJ, De Lisle RC. Low molecular mass GTP-binding proteins in subcellular fractions of the pancreas: Regulated phosphoryl G proteins. Am J Physiol 1992; 262:C493–C500.

95. Landis CA, Masters SB, Spada A, Pace AM, Bourne HR, Vallar L. GTPase inhibiting mutations activate the α chain of G_s and stimulate adenylyl cyclase. Nature 1989; 340:692–696.

96. Gutkind JS, Novotny EA, Brann MR, Robbins KC. Muscarinic acetylcholine receptor subtypes as agonist-dependent oncogenes. Proc Natl Acad Sci USA 1991; 88:4703–4707.

97. Suppatapone S, Worley PF, Baraban JM, Snyder SH. Solubilization, purification, and characterization of an inositol trisphosphate receptor. J Biol Chem 1988; 263:1530–1534.

98. Ferris CD, Huganir RL, Supattapone S, Snyder SH. Purified inositol 1,4,5-trisphosphate receptor mdiate calcium flux in reconstituted lipid vesicles. Nature 1989; 342:87–89.

99. Nunn DL, Taylor CW. Liver inositol 1,4,5-trisphosphate binding sites are the Ca^{2+}-mobilizing receptors. Biochem J 1990; 270:227–232.

100. Nunn DL, Potter BVL, Taylor CW. Molecular target sites of inositol 1,4,5-trisphosphate receptors in liver and cerebellum. Biochem J 1990; 265:393–398.
101. Furuichi T, Shiota C, Mikoshiba K. Distribution of inositol 1,4,5-trisphosphate receptor mRNA in mouse tissues. FEBS Lett 1990; 267:85–88.
102. Maeda N, Niinobe M, Nakahira K, Mikoshiba KJ. Purification and characterization P_{400} protein, a glycoprotein characteristic of the Purkinje cell from mouse cerebellum. Neurochemistry 1988; 51:1724–1730.
103. Guillemette G, Balla T, Baukal AJ, Spat A, Catt KJ. Inositol 1,4,5-trisphosphate binds to a specific receptor and releases microsomal calcium in the anterior pituitary gland. J Biol Chem 1987; 262:1010–1015.
104. Chadwick CC, Saito A, Fleischer S. Isolation and characterization of the inositol trisphosphate receptor from smooth muscle. Proc Natl Acad Sci USA 1990; 87:2132–2136.
105. Mourey RJ, Verma A, Supattapone S, Snyder SH. Purification and characterization of the inositol 1,4,5-trisphosphate receptor protein from rat vas deferens. Biochem J 1990; 272:383–389.
106. Mignery GA, Sudhof TC, Takei K, De Camilli P. Putative receptor for inositol 1,4,5-trisphosphate similar to ryanodine receptor. Nature 1989; 342:192–195.
107. Mikoshiba K, Huchet M, Changeux JP. Biochemical and immunological studies on the P_{450} protein, a protein characteristic of the Purkinje cell from mouse and rat cerebellum. Dev Neurosci 1979; 2:254–275.
108. Maeda N, Niinobe M, Mikoshiba K. A cerebellar Purkinje marker P_{400} protein is an inositol 1,4,5-trisphosphate (Ins P_3) receptor protein. Purification and characterization of Ins P_3 receptor complex. EMBO J 1990; 9:61–67.
109. Furuichi T, Yoshikawa S, Miyawaki A, Wada K, Maeda N, Mikoshiba K. Primary structure and functional expression of the inositol 1,4,5-trisphosphate-binding protein P_{400}. Nature 1989; 342:32–38.
110. Takeshima H, Nishimura S, Matsumoto T, Ishida H, Kangawa K, Minamino N, Matsuo H, Ueda M, Hanaoka M, Hirose T, Numa S. Primary structure and expression from complementary DNA of skeletal muscle ryanodine receptor. Nature 1989; 339:439–445.
111. Mignery GA, Newton CL, Archer BT, Sudhof TC. Structure and expression of the rat inositol 1,4,5-trisphosphate receptor. J Biol Chem 1990; 265:12679–12685.
112. Mignery GA, Sudhof TC. The ligand binding site and transduction mechanism in the inositol-1,4,5-triphosphate receptor. EMBO J 1990; 9:3893–3898.
113. Nakagawa T, Okano H, Furuichi T, Aruga J, Mikoshiba K. The subtypes of the mouse inositol 1,4,5-trisphosphate receptor are expressed in a tissue-specific and developmentally specific manner. Proc Natl Acad Sci USA 1991; 88:6244–6248.
114. Iino M, Kobayashi T, Endo M. Use of ryanodine for functional removal of the calcium store in smooth muscle cells of the guinea pig. Biochem Biophys Res Commun 1988; 152:417–422.
115. Burk SE, Lytton J, MacLennan DH, Shull GE. cDNA cloning, functional expression, and mRNA tissue distribution of a third organellar Ca^{2+} pump. J Biol Chem 1989; 264:18,561–568.

116. Rossier MF, Capponi AM, Vallotton MB. The inositol 1,4,5-trisphosphate-binding site in adrenal cortical cells is distinct from the endoplasmic reticulum. J Biol Chem 1989; 264:14078–14084.

117. Malviya AN, Rogue P, Vincendon G. Stereospecific inositol 1,4,5-[^{32}P] trisphosphate binding to isolated rat liver nuclei: Evidence for inositol trisphosphate receptor-mediated calcium release from the nucleus. Proc Natl Acad Sci USA 1990; 87:9270–9274.

118. Nicotera P, Orrenius S, Nilsson T, Berggren P-O. An inositol 1,4,5-trisphosphate-sensitive Ca^{2+} pool in liver nuclei. Proc Natl Acad Sci USA 1990; 87:6858–6862.

119. Hashimoto S, Bruno G, Lew DP, Pozzan T, Volpe P, Meldolesi J. Immunocytochemistry of calcisomes in liver and pancreas. J Cel Biol 1988; 107:2523–2531.

120. Volpe P, Krause K-H, Hashimoto S, Zorzato F, Pozzan T, Meldolesi J, Lew DP. "Calciosome," a cytoplasmic organelle: The inositol 1,4,5-trisphosphate-sensitive Ca^{2+} store of nonmuscle cells? Proc Natl Acad Sci USA 1988; 85:1091–1095.

121. Opas M, Dziak E, Fliegel L, Michalak M. Regulation of expression and intracellular distribution of calreticulin, a major calcium binding protein of nonmuscle cells. J Cell Physio 1991; 49:160–171.

122. Michalak M, Baksh S, Opas M. Identification and immunolocalization of calreticulin in pancreatic cells: No evidence for "Calciosomes." Exp Cell Res 1991; 197:91–99.

123. Van PN, Peter F, Söling HD. Four intracisternal calcium-binding glyco-proteins from rat liver microsomes with high affinity for calcium. J Biol Chem 1989; 264:17494–17501.

124. Milner RE, Baksh S, Shemanko C, Carpenter MR, Smillie L, Vance JE, Opas M, Michalak M. Calreticulin, and not calsequestrin, is the major calcium binding protein in smooth muscle sarcoplasmic reticulum and liver endoplasmic reticulum. J Biol Chem 1991; 266:7155–7165.

125. Treves S, de Mattei M, Lanfredi M, Villa A, Green NM, MacLennan DH, Meldolesi J, Pozzan T. Calreticulin is a candidate for a calsequestrin-like function Ca^{2+}-storage compartments (calciosomes) of liver and brain. Biochem J 1990; 271:473–480.

126. Krause KH, Simmerman HKB, Jones LR, Campbell KP. Sequence similarity of calreticulin with a Ca^{2+}-binding protein that co-purifies with an Ins(1,4,5)P$_3$-sensitive Ca^{2+} store in HL-60 cells. Biochem J 1990; 270:545–548.

127. Fliegel L, Burns K, MacLennan DH, Reithmeier RAF, Michalak M. Molecular cloning of the high affinity calcium binding protein (calreticulin) of skeletal muscle sarcoplasmic reticulum. J Biol Chem 1989; 264:21522–21528.

128. Waisman D, Salimath BP, Anderson MI. Isolation and characterization of CAB-63, a novel calcium-binding protein. J Biol Chem 1985; 260:1652–1660.

129. Macer DRJ, Koch GLE. Identification of a set of calcium-binding proteins in reticuloplasm, the luminal content of the endoplasmic reticulum. J Cell Sci 1988; 91:61–70.

130. Smith M, Koch GLE. Multiple zones in the sequence of calreticulin (CRP55, calregulin, HACBP), a major calcium binding protein. EMBO J 1989; 8:3581–3586.
131. Ross CA, Meldolesi J, Milner TA, Satoh T, Supattapone S, Snyder SH. Inositol 1,4,5-trisphosphate receptor localized to endoplasmic reticulum in cerebellar Purkinje neurons. Nature 1989; 339:468–470.
132. Volpe P, Anderson-Lang BH, Madedder L, Damiani E, et al. Calsequestrin, a component of the inositol 1,4,5-trisphate-sensitive Ca^{2+} store of chicken cerebellum. Neuron 1990; 5:713–721.
133. Villa A, Podini P, Clegg Do, Pozzan T, Meldolesi J. Intracellular Ca^{2+} stores in chicken Purkinije neurons: Differential distribution of the low affinity–high capacity Ca^{2+} binding protein, calsequestrin, of Ca^{2+} ATPase and the ER lumenal protein, BiP_1. J Cell Biol 1991; 113:779–791.
134. Burgoyne RD, Cheek TR. Locating intracellular calcium stores. Trends Biochem Sci 1991; 16:319–320.
135. Dawson AP, Comerford JG. Effects of GTP on Ca^{2+} movements across endoplasmic reticulum membranes. Cell Calcium 1989; 10:343–350.
136. Rossier MF, Bird GS, Putney JW Jr. Subcellular distribution of the calcium-storing inositol 1,4,5-trisphosphate-sensitive organelle in rat liver. Biochem J 1991; 274:643–650.
137. Henne V, Piiper A, Soling H-D. Inositol 1,4,5-trisphosphate and 5'γ-GTP induce calcium release from different intracellular pools. FEBS Let 1987; 218:153–158.
138. Menniti FS, Bird GS, Takemura H, Thastrup O, Potter BVL, Putney JW Jr. Mobilization of calcium by inositol trisphosphate from permeabilized rat parotid acinar cells. J Biol Chem 1991; 266:13646–13653.
139. Marty A. Calcium release and internal calcium regulation in acinar cells of exocrine glands. J Membr Biol 1991; 124:189–197.
140. Kasai H, Augustine GJ. Cytosolic Ca^{2+} gradients triggering unidirectional fluid secretion from exocrine pancreas. Nature 1990; 348:735–738.
141. Dissing S, Nauntofte B, Sten-Knudsen O. Spatial distribution of intra-cellular, free Ca^{2+} in isolated rat parotid acini. Pfluegers Arch 1990; 417:1–12.
142. Foskett JK, Gunter-Smith PJ, Melvin JE, Turner RJ. Physiological localization of an agonist-sensitive pool of Ca^{2+} in parotid acinar cells. Proc Natl Acad Sci USA 1989; 86:167–171.
143. Berridge MJ, Irvine RF. Inositol trisphosphate, a novel second messenger in cellular signal transduction. Nature 1984; 312:315–321.
144. Cobbold PH. Oscillatory calcium signals in hormone-stimulated cells. News in Physiological Sciences 1989; 4:211–215.
145. Fabiato A. Calcium-induced calcium release from canine sarcoplasmic reticulum. Am J Physiol 1983; 245:C1–C14.
146. Endo M. Calcium release from the sarcoplasmic reticulum. Curr Top Membr Transp 1985; 25:181–230.
147. Rooney TA, Renard DC, Sass EJ, Thomas AP. Oscillatory cytosolic calcium waves independent of stimulated inositol 1,4,5-trisphosphate formation in hepatocytes. J Biol Chem 1991; 266:12272–12282.

148. Wakui M, Osipchuk YV, Petersen OH. Receptor-activated cytoplasmic Ca^{2+} spiking mediated by inositol trisphosphate is due to Ca^{2+}-induced Ca^{2+} release. Cell 1990; 63:1025–1032.

149. Rousseau E, Meissner G. Single cardiac sarcoplasmic reticulum Ca^{2+}-release channel: Activation by caffeine. Am J Physiol 1989; 256:H328–H333.

150. Osipchuk YV, Wakui M, Yule DI, Gallacher DV, Petersen OH. Cytoplasmic Ca^{2+} oscillations evoked by receptor stimulation G-protein, internal application of inositol trisphosphate or Ca^{2+}: Simultaneous microfluorimetry and Ca^{2+} dependent Cl^- current recording in single pancreatic acinar cells. EMBO J 1990; 9:697–704.

151. Wakui M, Potter BVL, Petersen OH. Pulsatile intracellular calcium release does not depend on fluctuations in inositol trisphosphate concentration. Nature 1989; 339:317–320.

152. Marty A, Tan YP. The initiation of calcium release following muscarinic stimulation in rat lacrimal glands. J Physiol 1989; 419:665–687.

153. Foskett JK, Wong D. Free cytoplasmic Ca^{2+} concentration oscillations in thapsigargin-treated parotid acinar cells are caffeine- and ryanodine-sensitive. J Biol Chem 1991; 266:14535–14538.

154. Bezprozvanny I, Watras J, Ehrlich BE. Bell-shaped calcium responses of inositol 1,4,5-trisphosphate-gated and calcium-gated channels from endoplasmic reticulum of cerebellum. Nature 1991; 351:751–754.

155. Ehrlich BE, Watras J. Inositol 1,4,5-trisphosphate activates a channel from smooth muscle sarcoplasmic reticulum. Nature (London) 1988; 336:583–586.

156. Thevenod F, Dehlinger-Kremer M, Kemmer TP, Christian A-L, Potter BVL, Schulz I. Characterization of inositol 1,4,5-trisphosphate-sensitive (IsCaP) and -insensitive (IisCaP) nonmitochondrial Ca^{2+} pools in rat pancreatic acinar cells. J Membr Biol 1989; 109:173–186.

157. Champeil P, Combettes L, Berthon B, Doucet E, Orlowski S, Claret M. Fast kinetics of calcium release induced by myo-inositol trisphosphate in permeabilized rat hepatocytes. J Biol Chem 1989; 264:17665–17673.

158. Dawson AP. GTP enhances inositol trisphosphate-stimulated Ca^{2+} release from rat liver microsomes. FEBS Lett 1985; 185;147–150.

159. Mullaney JM, Yu M, Ghosh TK, Gill DL. Calcium entry into the inositol 1,4,5-trisphosphate-releasable calcium pool is mediated by a GTP-regulatory mechanism. Proc Natl Acad Sci USA 1988; 85:2499–2503.

160. Lee HC, Wadseth TF, Bratt GT, Hayes RN, Clapper DC. Structural determination of a cyclic metabolite of NAD^+ with intracellular Ca^{2+}-mobilizing activity. J Biol Chem 1989; 264:1608–1615.

161. Rusinko N, Lee HC. Widespread occurrence in animal tissues of an enzyme catalyzing the conversion of NAD^+ into a cytosolic metabolite with intracellular Ca^{2+}-mobilizing activity. J Biol Chem 1989; 264:11725–11731.

162. Koshiyama H, Lee HC, Tashjian AH Jr. Novel mechanism of intracellular calcium release in pituitary cells. J Biol Chem 1991; 266:16985–16988.

163. Dargi PL, Agre MC, Lee HC. Comparison of Ca^{2+}-mobilizing activities of cyclic ADP-ribose and inositol trisphosphate. Cell Regul 1990; 1:279–290.

164. Galione A, Lee HC, Busa WB. Ca^{2+}-induced Ca^{2+} release in sea urchin egg homogenates: Modulation by cyclic ADP-ribose. Science 1991; 253:1143–1146.

165. Chew CS. Differential effects of extracellular calcium removal and non-specific effects of Ca^{2+} antagonists on acid secretory activity in isolated gastric glands. Biochim Biophys Acta 1985; 846:370–378.

166. Melvin JE, Koek L, Zhang GH. A capacitative Ca^{2+} influx is required for sustained fluid secretion in sublingual mucous acini. Amer J Physiol 1991; 261:G1043–G1050.

167. Shuttleworth TJ. Receptor-activated calcium entry in exocrine cells does not occur via agonist-sensitive intracellular pools. Biochem J 1990; 266:719–726.

168. Bird GS, Takemura H, Thastrup O, Putney JW Jr, Menniti F. Mechanisms of activated Ca^{2+} entry in the rat pancreatoma cell line, AR4-2. Cell Calcium 1992; 13:41–48.

169. Putney JW. A model for receptor-regulated calcium entry. Cell Calcium 1986; 7:1–12.

170. Pandol SJ, Schoeffield MS, Fimmel JC, Muallem S. The agonist-sensitive calcium pool in the pancreatic acinar cell: Activation of plasma membrane Ca^{2+} influx mechanism. J Biol Chem 1987; 262:16963–16968.

171. Kwan C-Y, Putney JW Jr. Uptake and intracellular sequestration of divalent cations in resting and methacholine-stimulated mouse lacrimal acinar cells. J Biol Chem 1990; 265:678–684.

172. Negulescu PA, Machen TE. Release and reloading of intracellular Ca stores after cholinergic stimulation of the parietal cell. Am J Physiol 1988; 254:C498–C504.

173. Putney JW Jr. Capacitative calcium entry revisited. Cell Calcium 1990; 11:611–624.

174. Morris AP, Gallacher DV, Irvine RF, Petersen OH. Synergism of inositol trisphosphate and tetrakisphosphate in activating Ca^{2+}-dependent K^+ channels. Nature 1987; 330:653–655.

175. Changya L, Gallacher DV, Irvine RF, Potter BVL, Petersen OH. Inositol 1,3,4,5-tetrakisphosphate is essential for sustained activation of the Ca^{2+}-dependent K^+ current in single internally perfused mouse lacrimal acinar cells. J Membr Biol 1989; 109:85–93.

176. Luckhoff A, Clapham DE. Inositol 1,3,4,5-tetrakisphosphate activates an endothelial Ca^{2+}-permeable channel. Nature 1992; 355:356–358.

177. Irvine RF. "Quantal" release and the control of Ca^{2+} entry by inositol phosphates—a possible mechanism. FEBS Lett 1990; 263:5–9.

178. Petersen OH, Maruyama Y. What is the mechanism of the calcium influx to pancreatic acinar cells evoked by secretagogues. Pfluegers Arch 1983; 396:82–84.

179. Bear CE, Li C. Calcium-permeable channels in rat hepatoma cells are activated by extracellular nucleotides. Am J Physiol 1991; 261:C1018–C1024.

180. Poronnik P, Cook DI, Allen DG, Young JA. Diphenylamine-2-carboxylate (DPC) reduces calcium influx in a mouse mandibular cell line (ST_{385}). Cell Calcium 1991; 12:441–447.

181. Sasaki T, Gallacher DV. Extracellular ATP activates receptor-operated cation channels in mouse lacrimal acinar cells to promote calcium influx in the absence of phosphoinositide metabolism. FEBS Lett 1990; 264:130–134.
182. Hallam TJ, Rink TJ. Receptor-mediated Ca^{2+} entry: diversity of function and mechanisms. Trends Pharmacol Sci 1989; 10:8–10.
183. Muallem S, Khademazad M, Sachs G. The route of Ca^{2+} entry during reloading of the intracellular Ca^{2+} pool in pancreatic acini. J Biol Chem 1990; 265:2011–2016.
184. Mertz LM, Baum BJ, Ambudkar IS. Refill status of the agonist-sensitive Ca^{2+} pool regulates Mn^{2+} influx into parotid acini. J Biol Chem 1990; 265:15010–15014.
185. Merritt JE, Hallam TJ. Platelets and parotid acinar cells have different mechanisms for agonist-stimulated divalent cation entry. J Biol Chem 1988; 263:6161–6164.
186. Harootunian AT, Kao JPY, Tsien RY. Agonist-induced calcium oscillations in depolarized fibroblasts and their manipulation by photoreleased $Ins(1,4,5)P_3$, and Ca^{2+} buffer. Cold Spring Harbor Symp Quant Biol 1988; 53:935–943.
187. Goldbeter A, Dupont G, Berridge MJ. Minimal model for signal-induced Ca^{2+} oscillations and for their frequency encoding through protein phosphorylation. Proc Natl Acad Sci USA 1990; 87:1461–1465.
188. Crumpton MJ, Dedman JR. Protein terminology tangle. Nature 1990; 345:212.
189. Smith VL, Kaetzel MA, Dedman JR. Stimulus-response coupling: The search for intracellular calcium mediator proteins. Cell Regul 1990; 1:165–172.
190. Heizmann CW, Hunziker W. Intracellular calcium-binding proteins: More sites than insights. Trends Biochem Sci 1991; 16:98–103.

4
Calcium Channels, the Pancreatic Islet, and Endocrine Secretion

Joseph C. Dunbar

Insulin release from the pancreatic islet is stimulated and/or modulated by a variety of metabolic and hormonal factors, paramount among them the concentration of plasma glucose. As is the case for other endocrine and exocrine secretions, Ca^{2+} appears to be a major mediator of these secretory stimuli. Grodsky and Bennett[1] and Milner and Hales[2] were the first to show that insulin secretion was markedly impaired in the absence of calcium, an observation later confirmed by other investigators.[3,4] Subsequent studies demonstrated that glucose itself increases the concentration of calcium in the cytosol of the β cell, both by stimulating calcium uptake and by inhibiting calcium efflux through the cell membrane.[5-10]

Glucose-induced insulin release is also significantly correlated with an increase in electrical activity of the β cell and occurs in pulses whose frequency corresponds to rhythmic oscillations of the membrane potential.[11] When β cells are exposed to glucose, the membrane potential increases, reaching a plateau upon which bursts of high frequency spikes are superimposed.[12,13] This association between insulin secretion, electrical activity, and the ionic environment led to the suggestion that metabolically-induced changes in membrane potential may represent the initial step in the transduction of the secretory stimulus.[14] Briefly, it is believed that glucose, after being transported into the cell, is phosphorylated and metabolized producing ATP. The ATP-sensitive K channels then close, depolarizing the cell membrane and opening the voltage-dependent calcium channels to the influx of Ca^{2+}.[4,14,15]

Ion Channels in β Cells

The existence of a membrane potential in the pancreatic β cells, its decrease upon exposure to glucose, and the fact that, like that of other excitable tissues, the membrane of the β cell contains specific channels that control ion flow and thus regulate the membrane potential have been repeatedly demonstrated.[11,16,17] In particular, the β cell contains three types of potassium channels: a low conductance type about which little is known, an ATP-sensitive channel with a conductance of 50–60 pS and a high conductance (250 pS) Ca^{2+} and voltage-dependent type.[14,18-20] In addition there are two types of calcium channels: an L or long-lasting channel, and a T or transient type.[21-28]

Studies using patch-clamp techniques indicate that in normal β cells, as well as in SV40 transformed (HIT) and RINm5F insulinoma cell lines, the resting membrane potential depends upon the activity of the ATP-sensitive potassium channels.[23,25,29,30] Secretagogues, like glucose or amino acids, would decrease K conductance through these ATP-sensitive channels, which would lead to β cell depolarization and, thus, an increased calcium influx through voltage-dependent calcium channels. Coupling of the stimulus to the secretory process by calcium ensues. Repolarization of the cell is achieved by potassium egress through the calcium-dependent potassium channels.[31-36]

The characteristics of the potassium channels and their regulation have been reviewed elsewhere,[14] and this chapter will limit itself to a brief discussion of the L and T forms of calcium channels. The L channels are activated at a membrane potential of -30 to -40 mV, their average conductance is 30 pS, and their inactivation is calcium-dependent. On the other hand, the T channels are activated at negative voltages greater than -50 mV, and their average conductance is 15 pS. The L channels of β cells, like those of other tissue, bind dihydropyridine (DHP), are activated by DHP-receptor agonists such as Bay K8644, and are blocked by DHP-receptor antagonists such as nitrendipine, nimodipine, and nifedipine, with the expected changes in calcium flow, intracellular calcium concentration ($[Ca^{2+}]_i$), and insulin secretion. Other drugs, such as verapamil and diltiazem, also bind to and effectively block these calcium channels, reducing calcium influx and insulin secretion.[29,37,38]

T-type calcium channels that are activated at -50 to -40 mV and remain open only for 10–100 ms have also been found in the β cells by some, but not all, investigators, and there is no general agreement on their significance.[22,27]

Insulinotropic agents may also regulate the activity of the Ca^{2+} channels by stimulating the formation of cyclic AMP (cAMP) or the activity of the phosphoinositide transduction system. Thus the secretion of insulin brought about by agents that activate adenylatecyclase (such as glucagon, gastric inhibitory peptide, β adrenergic agonists, or forskolin) or that inhibit cAMP diesterase (such as methyl xanthine) is associated with a decreased conductance of the K^+ channels and membrane depolarization. Indeed, there is good evidence that cAMP may enhance Ca^{2+} influx through the voltage-dependent channels through the phosphorylation of the channel proteins by cAMP-dependent kinases.

The G proteins of cell membranes may also play a modulatory role on the calcium channels in β cells. Thus, somatostatin and α 2 adrenergic agonists, such as norepinephrine, epinephrine, and clonidine, may inhibit insulin secretion via a G protein, as suggested by the observation that pretreatment of β cells with pertussis toxin abolishes the effect of somatostatin on calcium currents and insulin secretion.[30,39–47]

Insulin secretion may also be increased through activation of the phosphoinositol pathway.[41,42] Thus a stimulation of phospholipase C by cAMP or by the direct effect of other secretagogues, such as cholecystokinin would lead to the hydrolysis of PIP_2 and to an increased production of inositol trisphosphate (IP_3) and diacylglycerol (DAG) in the β cells. A combination of increased release of Ca^{2+} from the endoplasmic reticulum by IP_3 and stimulation of protein kinase C, and of Ca^{2+} influx by DAG would lead to an increase in $[Ca^{2+}]_i$ and in the activity of the cytosolic Ca^{2+}-calmodulin-dependent kinases. It is believed that somatostatin may inhibit insulin secretion in HIT cells also by decreasing this pathway, and that a defective phosphoinositol hydrolysis may contribute to the impairment of glucose-induced insulin secretion found in chronic hyperglycemia or after exposure to high concentrations of glucose, the so-called toxic effect of glucose on the β cell.[42,48–51]

Calcium Channels in α Cells

Two types of calcium channels have been described in the α cells: an L-type characterized by rapid activation and slow inactivation, and a T-type activated at more negative potentials and characterized by rapid inactivation. The L-type Ca^{2+} channel, which can be blocked by antagonists such as nifedipine, appears to be the main conduit for the rapid entry of calcium into the cell and for the increase in $[Ca^{2+}]_i$ that initiates the secretion of glucagon. Contrary to the β cells, the α cells are characterized by a high rate of action potentials in the resting state. The frequency of these potentials is increased by arginine in concentrations known to stimulate glucagon secretion. Although glucose does not affect the electrical activity, it has been suggested that its metabolism by α cells may reduce $[Ca^{2+}]_i$ by increasing the intracellular sequestration of Ca^{2+} or increasing Ca^{2+} efflux.[52,53]

Other activators of glucagon secretion such as the catecholamines may act by activating adenyl cyclase. Indeed, cAMP increases calcium currents in the α cells and thus increases calcium influx and $[Ca^{2+}]_i$.

The T channels of the α cells are not sensitive to DHP blockers and, although their function has not been clarified, it has been suggested that since they are activated at a high voltage, they may serve as pacemakers for the low threshold Ca^{2+} currents.

Conclusions

The insulin secretory response to glucose starts with the transport of glucose through the membrane of the β cell and with an increase in ATP production subsequent to its metabolism. The resulting increase in the ATP/ADP ratio decreases potassium conduction in the ATP-dependent potassium channels and leads to depolarization of the cell membrane. When depolarization reaches the range of -70 to $-30\,mV$, the conduction in the voltage-dependent calcium channels increases, increasing the influx of calcium in the cytoplasm and setting in motion calcium-dependent cellular mechanisms, among them the secretion of insulin. Indeed, a strict correlation between intracellular Ca^{2+} and glucose-induced insulin secretion has been established and confirmed. The crucial role of calcium in stimulus-secretion coupling is underscored by the observation that calcium ionophores can trigger insulin secretion in the presence of permissive, but not stimulating, concentrations of glucose. Upon cessation of the stimulus, calcium itself deactivates the calcium channels and activates the calcium-dependent potassium channels, decreasing the secretory response and leading to repolarization of the β cell membrane. The same series of events may be set in motion by sugars other than glucose and by amino acids that are metabolized and increase ATP production.

Other modulators of insulin secretion increase cytoplasmic cAMP or IP_3 that lead to an increase or decrease in the activity of calcium channels. It appears the action of these modulators is transduced via G proteins and by the phosphorylation of the calcium channel proteins.

References

1. Grodsky GM, Bennett LL. Cation requirement for insulin secretion in isolated perfused pancreas. Diabetes 1966; 15:910–913.
2. Milner RDG, Hales CN. The role of calcium and magnesium in insulin secretion from rabbit pancreas studied in vitro. Diabetologia 1967; 3:47–49.
3. Milner RDG, Hales CN. Cations and the secretion of insulin. Biochim Biophys Acta 1968; 150:165–167.
4. Wollheim CB, Sharp GWG. Regulation of insulin release by calcium. Physiol Rev 1981; 61:914–972.
5. Malaisse-Lagae F, Malaisse WL. Stimulus-secretion coupling of glucose-induced insulin release. III. Uptake of [45]calcium by isolated islets of Langerhans. Endocrinology 1971; 88:72–80.
6. Malaisse WL, Malaisse-Lagae F, Brisson JR. The stimulus-secretion coupling of glucose-induced insulin release. II. Interactions of alkai and alkaline earth cations. Horm Metab Res 1971; 3:65–70.
7. Malaisse WJ, Brisson GR, Baird LE. Stimulus-secretion coupling of glucose-induced insulin release. X. Effect of glucose on [45]Ca efflux from perifused islets. Am J Physiol 1973; 224:389–394.

8. Malaisse WJ. Insulin secretion: Multifactorial regulation for a single process of release. Diabetologia 1973; 9:167–173.

9. Hellman B, Sehlin J, Taljedal IB. Calcium uptake by pancreatic B-cells as measured with the aid of ^{45}Ca and mannitol-^{4}H. Am J Physiol 1971; 221:1795–1801.

10. Kikuchi M, Wollheim CB, Cuendet GS, Renold AE, Sharp GWG. Studies on the dual effect of glucose on ^{45}Ca^{2+} efflux from isolated rat islets. Endocrinology 1978; 102:1339–1349.

11. Matthews EK, Dean PM, Sakamoto Y. The bioelectrical activity of the islet cell membrane. In: Hasselblatt A, Bruchhausen FV, eds. Handbook of Experimental Pharmacology. XXXII/2. Insulin. New York: Springer-Verlag; 1975:157–173.

12. Sherman A, Rinzel J, Keizer J. Emergence of organized bursting in clusters of pancreatic B-cells by channel sharing. Biophysic J 1988; 54:411–425.

13. Chay TR, Kang HS. Role of single-channel stoichastic noise on bursting clusters of pancreatic B-cells. Biophysic J 1988; 54:427–435.

14. Rajan AS, Aguilar-Bryan L, Nelson DA, Yaney GC, Hsu WH, Konze DL, Boyd AE. Ion channels and insulin secretion. Diabetes Care 1990; 13:340–363.

15. Hedeskov CJ. Mechanism of glucose-induced insulin secretion. Physiol Rev 1990; 60:442–509.

16. Dean PM, Matthews EK. Glucose-induced electrical activity in pancreatic islet cells. J Physiol 1970; 210:255–264.

17. De Weille J, Schmid-Antomarchi H, Fosset M, Lazdunski M. ATP-sensitive K channels that are blocked by hypoglycemia inducing sulfonylureas in insulin secreting cells are activated by galanin 4 hyperglycemia inducing hormone. Proc Natl Acad Sci USA 1988; 85:1313–1316.

18. Cook DL, Hales CH. Intracellular ATP directly blocks K$^+$ channels in pancreatic B cells. Nature 1984; 311:271–273.

19. Cook DL, Ikeuchi M, Fijimoto WY. Lowering of pH inhibits Ca^{++}-activated K$^+$ channels in pancreatic B cells. Nature 1984; 311:269–271.

20. Findlay I, Dunne MJ, Petersen OH. High conductance K$^+$ channels in pancreatic islet cells can be activated and inactivated by internal calcium. J Membr Biol 1985; 83:169–175.

21. Hiriart M, Matteson DR. Na channels and two types of calcium channels in rat pancreatic B cells identified with the reverse hemolytic plaque assay. J Gen Physiol 1988; 91:617–639.

22. Aschroft FM, Kelly RP, Smith PA. Two types of Ca channels in rat pancreatic B cells. Eur J Physiol 1990; 415:504–506.

23. Findlay I, Aschroft FM, Kelly RP, Rorsman P, Petersen OH, Trube G. Calcium currents in insulin-secreting B-cells. Ann NY Acad Sci 1989; 560:403–409.

24. Satin LS, Cook DL. Voltage gated calcium current in pancreatic B cells. Pfluegers Arch 1985; 404:385–387.

25. Satin LS, Cook DL. Evidence for two calcium currents in insulin-secreting cells. Pfluegers Arch 1988; 411:401–409.

26. Plant TD. Properties and calcium-dependent inactivation of calcium currents in cultured mouse pancreatic B-cells. J Physiol 1988; 404:731–747.

27. Rorsman P, Ashcroft FM, Trube G. Single Ca channel currents in mouse pancreatic B-cells. Eur J Physiol 1988; 412:597–603.

28. Ashcroft FM, Rorsman P, Trube G. Single calcium channel activity in mouse pancreatic B-cells. Ann NY Acad Sci 1989; 560:410–412.
29. Keahey HH, Rajan AS, Boyd AE III, Kunze DL. Characterization of voltage-dependent Ca^{2+} channels in B-cell line. Diabetes 1989; 38:188–193.
30. Keahey HH, Boyd AE III, Kunze DL. Catecholamine modulation of calcium currents in clonal pancreatic B-cells. Am J Physiol 1989; 257:C1171–C1176.
31. Ashcroft FM, Harrison DE, Ashcroft SJH. Glucose induces closure of single potassium channels in isolated rat pancreatic B-cells. Nature 1984; 312:446–448.
32. Misler S, Falke LC, Gillis K, McDaniel ML. Metabolite regulated potassium channels in rat pancreatic B-cells. Proc Natl Acad Sci USA 1986; 83:7119–7123.
33. Grynkiewicz G, Poenic M, Tsien RY. A new generation of calcium indicators with greatly improved fluorescence properties. J Biol Chem 1985; 260:3440–3450.
34. Tsien RY, Pozzan T, Rink RY. Calcium homeostasis in intact lymphocytes: Cytoplasmic free calcium monitoring with a new intracellular trapped fluorescent indicator. J Cell Biol 1982; 94:325–334.
35. Ribes G, Siegel EG, Wollheim CB, Renold AE, Sharp GWG. Rapid changes in calcium content of rat pancreatic islets in response to glucose. Diabetes 1981; 30:52–55.
36. Naber SP, McDaniel ML, Lacy PE. The effect of glucose on the acute uptake and efflux of calcium-45 in isolated rat islets. Endocrinology 1977; 101:686–693.
37. Smith PA, Rorsman P, Ashcroft FM. Modulation of dihydropyridine-sensitive Ca^{2+} channels by glucose metabolism in mouse pancreatic B-cells. Nature 1989; 342:550–553.
38. Triggle DJ, Janis RA. Calcium channel ligands. Ann Rev Pharmacol Toxicol 1987; 27:347–369.
39. Hill RS, Oberwetter JM, Boyd AE III. Increase in cAMP levels in B-cell line potentiates insulin secretion without altering cytosolic free-calcium concentration. Diabetes 1987; 36:440–446.
40. Rajan AS, Hill RS, Boyd AE III. Effect of rise in cAMP levels on Ca^{++} influx through voltage-dependent Ca^{++} channels in HIT cells. Second-messenger synarchy in beta cells. Diabetes 1989; 38:874–880.
41. Zawalich WS. Modulation of insulin secretion from B-cells by phosphoinositol-derived second-messenger molecules. Diabetes 1988; 37:137–141.
42. Hsu WH, Xiang H, Rajan AS, Kunze DL, Boyd AE. Somatostatin inhibits insulin secretion by a G-protein-mediated decrease in Ca^{2+} entry through voltage-dependent Ca^{2+} channels in the beta cell. J Biol Chem 1991; 266:837–843.
43. Di Virgilio F, Pozzan T, Wolheim CB, Vicentini LM, Meldolesi J. Tumor promoter phorbol myristate acetate inhibits Ca^{++} influx through voltage-dependent calcium channels in two secretory cell lines, PC12 and RIN_m5F. J Biol Chem 1986; 261:32–35.
44. Yada T, Russo LL, Sharp GWG. Phorbol ester-stimulated insulin secretion by RIN_m5f insulinoma cells is linked with membrane depolarization and an

increase in cytosolic free Ca^{++} concentration. J Biol Chem 1989; 264:2455–2462.

45. Sharp GWG, LeMarchand-Brustel Y, Yada T, et al. Galanin can inhibit insulin release by a mechanism other than membrane hyperpolarization or inhibition of adenylate cyclase. J Biol Chem 1989; 264:7302–7309.

46. Dunne MJ, Bullet MJ, Li G, Wolheim CB, Petersen OH. Galanin activates nucleotide-dependent K channels in insulin-secreting cells via a pertussis toxin-sensitive G-protein. EMBO J 1989; 8:413–420.

47. Nilsson T, Arkhammer P, Rorsman P, Berggren P-O. Suppression of insulin release by galanin and somatostatin is mediated by a G protein: An effect involving repolarization and reduction in cytoplasmic free Ca^{++} concentration. J Biol Chem 1989; 264:973–980.

48. Dunbar JC, Houser F, Levy J. Beta cell desensitization to glucose induced by hyperglycemia is augmented by increased calcium. Diab Res Clin Pract 1989; 7:187–196.

49. Hoenig M, Culberson LH, Furguson DC. Calcium transport by plasma membranes from a glucose-responsive rat insulinoma. Endocrinology 1991; 128:1381–1384.

50. Bolaffi JL, Rodd GG, Ma YH, Bright D, Grodsky GM. The role of Ca^{2+}-related events in glucose-stimulated desensitization of insulin secretion. Endocrinology 1991; 129:2131–2138.

51. Leahy JL, Bonner-Weir S, Weir GC. B-cell dysfunction induced by chronic hyperglycemia. Diabetes Care 1992; 15:442–455.

52. Rorsman P. Two types of Ca^{++} currents with different sensitivities to organic Ca^{++} channel antagonists in guinea pig pancreatic α_2 cells. J Gen Physiol 1988; 91:243–254.

53. Rorsman P, Hellman B. Voltage-activated currents in guinea big pancreatic α_2 cells: Evidence for Ca^{++}-dependent action potentials. J Gen Physiol 1988; 91:223–242.

5
Calcium Channels in Cells of the Anterior Pituitary

David M. Lawson

The mechanisms by which hypothalamic releasing and inhibiting peptides and amines control the release of anterior pituitary hormones have been the object of intense investigation for many years. They involve membrane-bound enzymes such as adenylate cyclase, and phospholipases C and A_2, which liberate second messengers such as cAMP, inositol phosphates, free fatty acids, and diacylglycerides, as well as membrane-resident regulatory proteins (G proteins), which couple these enzymes to the hypothalamic hormone receptors. In addition, intracellular kinases and phosphatases that control the degree of phosphorylation and dephosphorylation of important intracellular proteins participate in the control of pituitary hormone secretion. The actions of these intracellular substances are intimately related to those of the central regulator of secretion, the calcium ion.

Compelling evidence accumulated since the pivotal work of Douglas[1] has confirmed and extended the concept of calcium-regulated "stimulus-secretion coupling" in the pituitary, and has recognized the dominant role of free intracellular calcium in the release of pituitary hormones. The concentration of cytoplasmic calcium ($[Ca^{2+}]_i$) depends on the release of calcium ions from intracelluar sequestration sites (endoplasmic reticulum and mitochondria), which is in turn linked to the stimulus via a cascade of membrane reactions, terminating with the production of inostiol-trisphosphate (IP_3) from membrane phosphoinostol by the actions of phospholipase C. In addition, $(Ca^{2+})_i$ depends on the entry of calcium ions from the extracellular space, through calcium channels whose activity is regulated by hypothalamic hormones or by changes in membrane potential.

The objective of this chapter is to review recent information concerning the role of ion channels in regulating anterior pituitary cells. Most of this recent work has examined voltage-dependent channels, the opening and closing of which can be directly monitored using voltage and current patch-clamp techniques. Pharmacologic agonists and antagonists that induce or prevent influx of calcium through membrane channels have also been used extensively.

General Properties of Calcium Channels in Pituitary Cells

Previously reviewed[2] work by Douglas has shown that pituitary tumor cells and normal rat pituitary cells are electrically active and that action potentials induced by current pulses or by hypothalamic hormones, like TRH (Thyrotropin Releasing Hormone), are calcium- and not sodium-dependent since they can be blocked by La^{3+} or by methoxyverapamil, but not by the sodium channel blocker tetrodotoxin or by the replacement of Na^+ with Tris.

Other investigators, using pituitary tumor cell lines, confirmed and extended these initial observations[3-8]. Thus, Hagiwara and Ohmori[3] observed that calcium currents could be detected by patch-clamp techniques, if large inward sodium and outward potassium currents were blocked. Adler et al.[4] observed both calcium- and sodium-dependent potentials in AtT20 mouse pituitary cells, but although the sodium "spikes" were necessary to maintain normal electrical activity, their role in hormone secretion was not clear. Tan and Tashjian[5] showed that calcium uptake by cells of the GH_4C_1 pituitary tumor cell line could be induced by depolarizing concentrations of potassium, and that this uptake could be blocked by La^{3+}, by Co^{2+}, and by the calcium channel blockers verapamil and nifedipine, but not by the sodium-channel blocker tetrodotoxin. Dubinsky and Oxford[6] observed voltage-dependent sodium, calcium, and potassium currents in this same cell line, but, in contrast to the conclusions of Adler et al.,[4] these authors concluded that sodium-channel activity did not make a major contribution to the overall electrical activity of the cells.

Matteson and Armstrong[7] described the activation and inactivation properties of sodium and calcium channels in a rat pituitary tumor cell line (GH_3) and found that the sodium conductance had fast activation and inactivation kinetics, whereas calcium currents activated more slowly and inactivated in a biphasic manner.

Subsequently these authors described two subsets of calcium channels in GH_3 cells based on their closing (deactivation) kinetics.[8] One of these subtypes deactivated rapidly, the other more slowly. Matteson and Armstrong[9] further characterized the fast-deactivating (FD) and slow-deactivating (SD) channels on the basis of their activation (opening) and the inactivation properties. The FD channels activated more rapidly than the SD channels at positive potentials, and did not inactivate. In contrast the SD channels activated at more negative potentials and inactivated almost completely. The authors concluded that SD channels are probably involved in electrical events near threshold for action potentials, whereas FD channels are probably more important in increasing calcium conductance once a spike in membrane potential has been initiated. The terminology currently used to describe ion channels in other tissues is also applied for these channels: the FD channels in pituitary cells are also called L-type (or long-lasting) channels and the SD channels are also referred to as T-type (or transient) channels. Since the L-type channel is blocked by dihydropyridines, such as nifedipine, it is also known as the dihydropyridine-sensitive channel. L-type channels are also activated by Bay K8644. It is generally accepted that T-type channels are not affected by dihydropyridines but can be blocked with such agents as Ni^{2+}.

L-type channels are controlled by a number of intracellular mediators, such as cAMP and products in the diacylglyceride-mediated pathway,[10] and by hypothalamic hormones, such as TRH.[11] Thus cAMP and phorbol esters stimulate calcium entry through the L-type channel,[11] while somatotstatin blocks Ca^{2+} currents in AtT20 cells via a pertussis-toxin-sensitive pathway,[12] probably involving a G protein. On the other hand, TRH inhibits gating of the L-type channel.[11] TRH increases intracellular calcium in pituitary cells in two phases, a first and most dramatic phase apparently caused by release of calcium from intracellular sequestration sites; and a second, more prolonged phase of lower amplitude, apparently due to extracellular calcium entry via T-type channels.[13] Indeed, the second phase is blocked by Ni^{2+}, but not by ω-conotoxin, another L-type channel blocker.

Calcium Channels in Pituitary Lactotrophs

Many groups of investigators have studied the link between prolactin secretion and the activity of calcium channels (whose general properties have been outlined above). Ozawa and Kimura[14] have shown that action potentials in GH_3 cells in Na^+-free solution and prolactin release induced by TRH, Ba^{2+}, or K^+ could be blocked by verapamil. Tan and Tashjian[15] reported that although TRH evoked prolactin release in GH_4C_1 cells by two calcium-dependent mechanisms, the predominant mechanism was dependent on extracellular calcium and could be blocked by verapamil, nifedipine, and Co^{2+}, an observation in direct contrast to the findings of Suzuki et al.[13] Thaw, Raaka, and Gershengorn[16] also observed that Co^{2+} inhibited TRH-induced prolactin release but found that TRH was also active in the presence of low extracellular calcium, which suggests that Co^{2+} may have actions other than blocking calcium channels.

It has been shown that the calcium-channel agonist Bay K8644 increases $^{45}Ca^{2+}$ uptake and the simultaneous secretion of prolactin from both GH_4C_1 and normal pituitary cells, and that these effects could be blocked by nisoldipine, a calcium-channel antagonist.[17] Bay K8644 also increases long-term production of prolactin, whereas nomodopine suppresses it.[18] Apparently the action of Bay K8644 depends upon a stimulation of prolactin gene expression in GH_3 cells, an effect that can be blocked by a calcium-channel antagonist PN 200-110,[19] and by the blockade of L- and T-type channels with diphenylbutylperidine antipsychotics.[20] The stimulatory effect of Bay K8644 is potentiated by estradiol and blocked by the hypothalamic inhibitor of prolactin release, dopamine.[21]

Calcium influx into normal pituitary cells and prolactin release are also increased by neurotensin.[22,23] Both effects can be blocked by dopamine[22] and by the removal of extracellular Ca^{2+},[23] but not by verapamil or nifedipine,[22] which suggests that neurotensin may operate via dihydropyridine-insensitive (ie, T-type) channels.

Dopamine did not inhibit calcium influx induced by 25 mM K^+ or by the sodium-channel agonist veratridine, but dopamine did inhibit the prolactin-releasing effects of vasoactive intestinal peptide (VIP). However, VIP had no effect on calcium influx and apparently its action is mediated by cAMP.[22] Dopamine has been shown to depress all calcium currents in normal rat lactotrophs, albeit the effects vary with the type of channel examined.[24] For example, dopamine increases the threshold for activation for L-type channels while it enhances the inactivation of T-type channels. A recent report[25] by Lledo et al. indicates that calcium currents through both L- and T-type channels increase with time in culture, are protein synthesis-dependent, and can be blocked by chronic exposure to a dopamine agonist RU 24213, which has no effect on K^+ currents.

Somatostatin has been shown to depress spontaneous and evoked action potentials and voltage-dependent Ca^{2+} currents measured by patch-clamp methods in GH_3 and human prolactinoma cells.[26] These effects could be prevented by pertussis toxin, which suggests that somatostatin acts through a G protein. A similar finding was reported for dopamine.[24]

In addition to the multiplicity of substances that control prolactin secretion via calcium channels, the heterogeneity of lactotrophs themselves further complicates the interpretation of electrophysiologic and secretory data. Thus Lewis et al.,[27] while simultaneously measuring calcium currents and intracellular calcium concentrations in pituitary cells separated by fluoresence-activated cell sorting, observed two groups of lactotrophs, one silent and one spontaneously active. In a very comprehensive study, Lledo et al.[28] isolated cells of two different densities (defined as "light" and "heavy") from lactating rat pituitaries and measured calcium currents by whole cell patch clamp, intracellular calcium concentration by indo-1 fluorescence, and the release of prolactin from perifused cells or individual cells in a reverse hemolytic plaque assay. In the presence of dopamine or TRH, they found significant differences in the electrical and secretory responses in the behavior of the calcium channels between the two types of cells. For example, these investigators showed that dopamine hyperpolarized "light" cells but not "heavy" cells, while TRH depolarized "heavy" but not "light" cells. "Light" cells had a higher basal cytoplasmic level of calcium that could be reduced by dopamine, whereas TRH increased cytoplasmic calcium in "heavy" but not in "light" cells. Interestingly, prolactin release from "light" cells was inhibited by dopamine to a greater extent than was the release from "heavy" cells, while TRH was twice as effective in releasing prolactin from "heavy" rather than from "light" cells. Finally, "light" cells had twice the amount of calcium current through T-type channels than through L-type channels, whereas in "heavy" cells the contribution of the two was the same. In contrast with the work of Lledo et al.,[28] Cota et al.[29] reported that lactotrophs that secreted large amounts of prolactin as measured by reverse hemolytic plaque showed more L-type than T-type channel activity. However, these investigators used cells obtained from male rats.

Calcium Channels in Somatotrophs

An early report by Bicknell and Schofield[30] showed that veratridine, the sodium-channel activator, increased growth hormone secretion in dispersed bovine pituitary cells, and that these responses were abolished by verapamil, which suggests that sodium influx depolarized the cells and opened voltage-dependent calcium channels, which in turn resulted in growth hormone secretion. Somatostain blocked veratridine-induced growth hormone release but not the agonist-induced increases in sodium or calcium fluxes. Later, Mason and Rawlings,[31] using whole-cell patch-clamp techniques, identified sodium, calcium, and potassium currents in bovine somatotrophs, and demonstrated that the sodium current was not critical for growth hormone release under basal or growth-hormone-releasing hormone (GHRH)-stimulated conditions. In either case, secretion was blocked by Co^{2+}. Still more recently, Chen et al.[32] reported that the development of action potentials in rat somatotrophs required both sodium and calcium fluxes and that tetrodotoxin, a sodium-channel antagonist, slightly decreased basal but not GHRH-induced growth hormone release, whereas Co^{2+} totally blocked GHRH-induced release. Bresson et al.[33] demonstrated that GHRH induced an immediate and transient increase in cytoplasmic free calcium in GH_3 cells.

Drouva et al.[21] have shown that Bay K8644, an L-type calcium-channel agonist, evokes GH release from rat pituitary cells, and that this effect can be reversed by somatostatin. Indeed, somatostatin appears to directly inhibit voltage-activated calcium channels in rat somatotrophs,[34] while its inhibitory effect on the action of Bay K8644 appears to be mediated by a G protein, since it is abolished by petussis toxin.[35]

Like lactotrophs, somatotrophs appear to be somewhat heterogeneous, as some of them are electrically silent and have stable intracellular calcium concentrations, while others are spontaneously active and show wide calcium fluctuations.[27] In contrast to lactotrophs, most calcium channels of somatotrophs are of the T-type, although L-type channels are also present.[27]

Calcium Channels in Corticotrophs and Thyrotrophs

Corticotrophin-releasing hormone (CRH)-stimulated ACTH secretion is abolished by verapamil, Co^{2+}, and nimodipine, which suggests that L-type calcium channels are functional in corticotrophs.[36] Both L- and T-type currents have been shown to exist in human ACTH-secreting tumor cells and normal rat corticotrophs.[37] Antidiuretic hormone (ADH), another ACTH secretagogue, increases the amplitude of calcium current through the L-type, but not the T-type, channel.[38]

Low extracellular calcium concentrations block the stimulatory effect of TRH, K^+, or the calcium ionophore A 23187 on TSH secretion from normal pituitary cells, but not the release of TSH induced by agents that increase intracellular cAMP (theophilline and IBMX).[39] On the other hand, methoxyverapamil blocks the responses to TRH, K^+, and IBMX, but not the response to A 23187. The stimulatory effect of TRH on TSH release is potentiated by GABA, an effect that is apparently calcium-dependent since it is blocked by Co^{2+} and nifedipine.[40]

Calcium Channels in Gonadotrophs

Conn and Rogers have demonstrated that veratridine, a sodium-channel agonist, increases LH release from primary cultures of rat pituitary cells in a sodium- and calcium-dependent manner, and that LH release induced by gonadotropin-releasing hormone (GnRH) required calcium, but not sodium.[41] Other investigators[42-45] have shown that ovine gonadotrophs are electrically quiet in the absence of GnRH, that electrical depolarization opens both sodium and calcium channels, and that GnRH induces the release of LH and the induction of an inward calcium current, as well as two outward potassium currents. The calcium currents appeared to be of the L- and T-type.

Calcium channels with L- and T-type kinetics have also been identified in rat gonadotrophs by Stutzin et al.[46] and Marchetti et al.[47] These cells differ from lactotrophs and somatotrophs because, in contrast to prolactin and growth hormone, basal LH secretion is not promoted by the calcium channel agonist Bay K8644.[21,48] However, Bay K8644 does enhance the LH response to GnRH.[21,48] Prolonged or repeated GnRH exposure impairs the activity of voltage-dependent calcium channels and the release of LH.[49] Thus this desensitization to GnRH, a well-recognized phenomenon with regard to LH release, may be the result calcium-channel inactivation.

Protein kinase C (PKC) has been implicated in LH release. Activators of PKC, such as phorbol esters and diacylglycerides, have been shown to affect calcium fluxes in gonadotrophs.[50,51] Indeed, Izumi et al.[50] have shown that the increase in intracellular calcium in rat gonadotrophs induced by phorbol esters and by diacylglycerols depend on extracellular calcium and are blocked by nifidipine. Stojilkovic et al.[51] have also shown that phorbol esters increase intracellular calcium in quiescent gonadotrophs, although they inhibit this increase when voltage-sensitive calcium channels are already activated. However, even when calcium influx is inhibited, phorbol esters still stimulate the release of LH.

Agents other than GnRH have been shown to increase LH release in vitro. Among them is GABA.[52] Although GABA is much less potent than GnRH, its effect is blocked by nifedipine, indicating that GABA acts, in part, through calcium channels. Endothelin, a vasoactive peptide liberated by endothelial cells, increases intracellular calcium in a manner very similar to GnRH and also produces a moderate release of LH.[53] Finally, it has been demonstrated that neuropeptide Y enhances the GnRH-induced LH release and the rise in intracellular calcium.[54] The potentiation of LH release could be blocked by nitrendipine, and the enhancement of intracellular calcium was not seen in the presence of extracellular EGTA. These results strongly suggest that neuropeptide Y also has an action on voltage-dependent calcium channels.

Conclusion

From the preceeding discussion, it is clear that voltage-dependent channels are involved in the release of anterior pituitary hormones. Both L- and T-type channels are present in most cells of the anterior pituitary, although their relative distributions vary considerably between cell types and sometimes between subpopulations of the same cell type (eg, lactotrophs). Hypothalamic-releasing and -inhibiting hormones often regulate the activity of the L-type channel, although in some cases (eg, TRH and neurotensin) they act on T-type channels. The T-type channels seem to be active primarily in spontaneously secreting cells and may subserve a secretagogue-independent or continuous hormone release. Voltage-dependent calcium channels seem to be intimately related to other membrane-bound proteins, such as G proteins or protein kinase C. This association is probably important for the integrated regulation of pituitary cell activity and needs to be examined more closely.

References

1. Douglas WW. Stimulus-secretion coupling: The concept and clues from chromaffin and other cells. Br J Pharmacol 1968; 34:451–474.
2. Douglas WW. Aspects of the calcium hypothesis of stimulus-secretion coupling: Electrical activity in adenohypophyseal cells, and membrane retrieval after exocytosis. In: Hand AR, Oliver C, eds. Methods of Cell Biology. New York: Academic Press; 1981:483–501.
3. Hagiwara S, Ohmori H. Studies of calcium channels in rat clonal pituitary cells with patch electrode voltage clamp. J Physiol 1982; 331:231–252.
4. Adler M, Wong BS, Sabol SL, Busis N, Jackson MB, Weight FF. Action potentials and membrane ion channels in clonal anterior pituitary cells. Proc Natl Acad Sci USA 1983; 80:2086–2090.
5. Tan K-N, Tashjian AH. Voltage-dependent calcium channels in pituitary cells in culture. 1. Characterization by $^{45}Ca^{2+}$ fluxes. J Biol Chem 1984; 259:418–426.
6. Dubinsky JM, Oxford GS. Ionic currents in two strains of rat anterior pituitary tumor cells. J Gen Physiol 1984; 83:309–339.
7. Matteson DR, Armstrong CM. Na and Ca channels in a transformed line of anterior pituitary cells. J Gen Physiol 1984; 83:371–394.
8. Armstrong CM, Matteson DR. Two distinct populations of calcium channels in a clonal line of pituitary cells. Science 1985; 227:65–67.
9. Matteson DR, Armstrong CM. Properties of two types of calcium channels in clonal pituitary cells. J Gen Physiol 1986; 87:161–182.
10. Schofl C, Meier K, Gotz DM, Knepel W. cAMP- and diacylglycerol-mediated pathways elevate cytosolic free calcium concentrations via dihydropyridine-sensitive, ω-conotoxin insensitive calcium channels in normal rat anterior pituitary cells. Nauyn-Schmiedeberg's Arch Pharmacol 1989; 339:1–7.
11. Levitan ES, Kramer RH. Neuropeptide modulation of single calcium and potassium channels detected with a new patch clamp configuration. Nature 1990; 348:545–547.
12. Lewis DL, Weight FF, Luini A. A guanine nucleotide-binding protein mediates the inhibition of voltage-dependent calcium current by somatostatin in a pituitary cell line. Proc Natl Acad Sci USA 1986; 83:9035–9039.
13. Suzuki N, Kudo Y, Takagi H, Yoshioka T, Tanakadate A, Kano M. Participation of transient-type Ca^{2+} channels in the sustained increase of Ca^{2+} level in GH_3 cells. J Cell Physiol 1990; 144:62–68.
14. Ozawa S, Kimura N. Calcium channel and prolactin release in rat clonal pituitary cells: Effects of verapamil. Am J Physiol 1982; 243:E68–E73.
15. Tan K-N, Tashjian AH. Voltage-dependent calcium channels in pituitary cells in culture. II. Participation in thyrotropin-releasing hormone action on prolactin release. J Biol Chem 1983; 258:418–426.
16. Thaw CN, Raaka EG, Gershengorn MC. Evidence that cobalt ion inhibition of prolactin secretion occurs at an intracellular locus. Am J Physiol 1984; 247:C150–C155.
17. Enyeart JJ, Aizawa T, Hinkle PM. Interaction of dihydropyridine Ca^{2+} agonist Bay K8644 with normal and transformed pituitary cells. Am J Physiol 1986; 250:C95–C102.

18. Enyeart JJ, Sheu S-S, Hinkle PM. Dihydropyridine modulators of voltage-sensitive Ca^{2+} channels specifically regulate prolaction production by GH_4C_1 pituitary tumor cells. J Biol Chem 1987; 262:3154–3159.

19. Laverriere J-N, Richard J-L, Buisson N, Martial JA, Tixier-Vidal A, Gourdji D. Thyroliberin and dihydropyridines modulate prolactin gene expression through interacting pathways in GH_3 cells. Neuroendocrinology 1989; 50:693–701.

20. Enyeart JJ, Biagi BA, Day RN, Sheu S-S, Maurer RA. Blockade of low and high threshold Ca^{2+} channels by diphenylbutylpiperidine antipsychotics linked to inhibition of prolactin gene expression. J Biol Chem 1990; 265:16373–16379.

21. Drouva SV, Rerat E, Bihoreau C, Laplante E, Rosolonjanahary R, Clauser H, Kordon C. Dihydropyridine-sensitive calcium channel activity related to prolactin, growth hormone, and luteinizing hormone release from anterior pituitary cells in culture: Interactions with somatostatin, dopamine and estrogens. Endocrinology 1988; 123:2762–2773.

22. Memo M, Castelletti L, Missale C, Valerio A, Carruba M, Spano PF. Dopaminergic inhibition of prolactin release and calcium influx induced by neurotensin in anterior pituitary is independent of cyclic AMP system. J Neurochem 1986; 47:1689–1695.

23. Memo M, Castelletti L, Valerio A, Missale C, Spano PF. Identification of neurotensin receptors associated with calcium channels and prolactin release in rat pituitary. J Neurochem 1986; 47:1682–1688.

24. Lledo P-M, Legendre P, Israel J-M, Vincent J-D. Dopamine inhibits two characterized voltage-dependent calcium currents in identified rat lactotroph cells. Endocrinology 1990; 127:990–1001.

25. Lledo P-M, Israel J-M, Vincent J-D. Chronic stimulation of D_2 dopamine receptors specifically inhibits calcium but not potassium currents in rat lactotrophs. Brain Res 1991; 558:231–238.

26. Mollard P, Vacher P, Dufy B, Barker JL. Somatostatin blocks Ca^{2+} action potential activity in prolactin-secreting tumor cells through coordinate actions on K^+ and Ca^{2+} conductances. Endocrinology 1988; 123:721–732.

27. Lewis DL, Goodman MB, St. John PA, Barker JL. Calcium currents and Fura-2 signals in fluorescence-activated cell sorted lactotrophs and somatotrophs of rat anterior pituitary. Endocrinology 1988; 123:611–612.

28. Lledo P-M, Guerineau N, Mollard P, Vincent J-D, Israel J-M. Physiological characterizatin of two functional states in subpopulations of prolactin cells from lactating rats. J Physiol 1991; 437:477–494.

29. Cota G, Hiriart M, Horta J, Torres-Escalante JL. Calcium channels and basal prolactin secretion in single male rat lactotrophs. Am J Physiol 1990; 259:C949–C959.

30. Bicknell RJ, Schofield JG. Inhibition by somatostatin of bovine growth hormone secretion following sodium channel activation. J Physiol 1981; 316:85–96.

31. Mason WT, Rawlings SR. Whole cell recordings of ionic currents in bovine somatotrophs and their involvement in growth hormone secretion. J Physiol 1988; 405:577–593.

32. Chen C, Zhang J, Vincent J-D, Israel J-M. Sodium and calcium currents in action potentials of rat somatotrophs: Their possible functions in growth hormone secretions. Life Sci 1990; 46:983–989.

33. Bresson L, Fahmi M, Sartor P, Dufy B, Dufy-Barbe L. Growth hormone releasing factor stimulates calcium entry in the GH_3 pituitary cell line. Endocrinology 1991; 129:2126–2130.

34. Nussinovitch I. Somatostatin inhibits two types of voltage-activated calcium currents in rat growth hormone secreting cells. Brain Res 1989; 504:136–138.

35. Reisine T. Cellular mechanisms of somatostatin inhibition of calcium influx in anterior pituitary cell line AtT-20. J Pharmacol Exp Ther 1990; 254:646–651.

36. Richardson UI. Multiple classes of calcium channels in mouse pituitary tumor cells. Life Sci 1986; 38:41–50.

37. Guerineau N, Corcuff J-B, Tabarin A, Mollard P. Spontaneous and corticotropin-releasing factor induced cytosolic calcium transients in corticotrophs. Endocrinology 1991; 129:409–420.

38. Mollard P, Vacher P, Rogawski MA, Dufy B. Vasopressin enhances a calcium current in human ACTH-secreting pituitary adenoma cells. FASEB J 1988; 2:2907–2912.

39. Schrey MP, Brown BL, Ekins PR. Studies on the role of calcium and cyclic nucleotides in the control of TSH secretion. Mol Cell Endocrinol 1978; 11:249–264.

40. Roussel J-P, Astier H. Involvement of dihydropyridine-sensitive calcium channels in the $GABA_A$ potentiation of TRH-induced TSH release. Eur J Pharmacol 1990; 190:135–145.

41. Conn PM, Rogers DC. Gonadotropin release from pituitary cultures following activation of endogenous ion channels. Endocrinology 1980; 107:2133–2134.

42. Mason WT, Waring DW. Electrophysiological recordings from gonadotrophs. Neuroendocrinology 1985; 41:258–268.

43. Sikdar SK, Waring DW, Mason WT. Voltage activated ionic currents in gonadotrophs of the ovine pars tuberalis. Neurosci Lett 1986; 71:95–100.

44. Mason WT, Waring DW. Patch clamp recordings of single ion channel activation by gonadotropin releasing hormone in ovine pituitary gonadotrophs. Neuroendocrinology 1986; 43:205–219.

45. Mason WT, Sikdar SK. Characteristics of voltage gated Ca^{2+} currents in ovine gonadotrophs. J Physiol 1989; 415:367–391.

46. Stutzin A, Stojilkovic SS, Catt KJ, Rojas E. Characteristics of two types of calcium channels in rat pituitary gonadotrophs. Am J Physiol 1989; 257:C865–C874.

47. Marchetti C, Childs GV, Brown AM. Voltage-dependent calcium currents in rat gonadotropes separated by centrifugal elutriation. Am J Physiol 1990; 258:E589–E596.

48. Chang JP, McCoy EE, Graeter J, Tasaka K, Catt KJ. Participation of voltage-dependent calcium channels in the action of gonadotropin-releasing hormone. J Biol Chem 1986; 261:9105–9108.

49. Stojilkovic SS, Rojas E, Stutzin A, Izumi S-I, Catt KJ. Desensitization of pituitary gonadotropin secretion by agonist-induced inactivation of voltage-sensitive calcium channels. J Biol Chem 1989; 264:10939–10942.

50. Izumi S-I, Stojilkovic SS, Iida T, Krsmanovic LZ, Omeljaniuk RJ, Catt KJ. Role of voltage sensitive calcium channels in $[Ca^{2+}]$ and secretory responses to activators of protein kinase C in pituitary gonadotrophs. Biochem Biophys Res Commun 1990; 170:359–367.

51. Stojilkovic SS, Iida T, Merelli F, Torsello A, Krsmanovic LZ, Catt KJ. Interactions between calcium and protein kinase C in the control of signaling and secretion in pituitary gonadotrophs. J Biol Chem 1991; 266:10377–10384.
52. Virmani MA, Stojilkovic SS, Catt KJ. Stimulation of luteinizing hormone release by gamma aminobutyric acid (GABA) agonists: Mediation by $GABA_A$-type receptors and activation of chloride and voltage-sensitive calcium channels. Endocrinology 1990; 126:2199–2505.
53. Stojilkovic SS, Iida T, Merelli F, Catt KJ. Calcium signaling and secretory responses in endothelin-stimulated anterior pituitary cells. Mol Pharmacol 1991; 39:762–770.
54. Crowley WR, Shah GV, Carroll BL, Kennedy D, Dockter ME, Kalra SP. Neuropeptide Y enhances luteinizing hormone (LH)-releasing hormone induced LH release and elevations in cytosolic Ca^{2+} in rat anterior pituitary cells: Evidence for involvement of extracellular Ca^{2+} influx through voltage sensitive channels. Endocrinology 1990; 127:1487–1494.

6
Role of Calcium in the Secretion of Atrial Natriuretic Peptide

Mary F. Walsh

Atrial natriuretic peptide (ANP), a potent natriuretic, diuretic, and smooth muscle relaxant, is released from atrial cardiomyocytes in response to a variety of stimuli, such as atrial distension and tachycardia, and in response to various vasoconstrictors.[1,2] Although ANP appears to be stored in secretory granules in the form of a pre-hormone and to be proteolytically cleaved only upon secretion,[3] the functional morphology of these granules is analogous to that of other endocrine secretory tissues, which suggests that calcium may play a role in their release.[4] Indeed, the heart possesses a number of complex mechanisms for regulating the concentration of intracellular calcium ($[Ca]_i$). Most of them involve the release of Ca^{2+} from a large intracellular pool in the sarcoplasmic reticulum (SR), although extracellular calcium $[Ca]_o$ may also play a role in contractile function.[5,6]

Many known stimulators of ANP secretion, such as α_1 agonists and endothelin, activate phospholipase C with resulting generation of IP_3 and diacylglycerol, and although IP_3 mobilizes $[Ca]_i$ from the intracellular pool, its action is slow (seconds to minutes), while the Ca^{2+} transients that drive contraction are much more rapid (0.2 s). In addition, these transients are thought to be independent of the formation of inositol phosphates.[7] The difficulty then has been to separate ion transport processes involved in ANP secretion from other concomitant cyclical ion fluxes, such as transplasmalemmal fluxes of Ca^{2+}, K^+, and Na^+, associated with the action potential and the resulting Ca^{2+}-induced Ca^{2+} release from the SR. In addition, atrial stretch in itself appears to be the major stimulus for ANP secretion, although nothing is known about the coupling mechanism(s) between mechanical stretch and hormone release. It has been shown that stretch increases $[Ca]_i$ in both vascular smooth muscle and skeletal muscle,[8,9] but whether it plays a similar role in the heart remains to be determined.

Although it is generally accepted that activation of protein kinase C (PKC) is instrumental in ANP secretion, the role of $[Ca]_i$ in this activation is being questioned. This chapter will discuss the controversial results obtained with three in vitro models.

The Langendorff Preparation

In this preparation, the entire heart is perfused in a retrograde fashion through the aorta and the effluent collected. Heart rate may change spontaneously or be controlled with an electrode. As the atria do not fill, there is minimal stretch and the effect of pacing alone may be studied. Alternately, the preparation may be modified to allow studies of the effect of atrial stretch.

Saito et al.[10] reported that the voltage-sensitive Ca^{2+} channel agonist Bay K8644 increased ANP secretion in the perfused rat heart. Since this effect was blocked by nifedipine, these authors concluded that this was the result of an increase in probability and duration of Ca^{2+} channel opening. However, Bay K8644 also increased heart rate, while A23187 induced a sharp increase in ANP secretion and superfusion with TPA (a phorbol ester activator of PKC) led to a progressive release of the hormone. The combination was synergistic, implying that Ca^{2+} mobilization and PKC activation were both necessary to enhance ANP release.[11,12] Forskolin also increased ANP release, perhaps because cAMP leads to the recruitment of depolarization-activated Ca^{2+} channels and thus enhances Ca^{2+} influx. However, when stretch was added to the experimental condition, both forskolin and Bay K8644 inhibited ANP, which led the author to conclude that increasing $[Ca]_i$ had a negative effect on stretch-induced ANP release.[13] Stretch-induced, but not basal, ANP release was also blocked by a decrease of $[Ca]_o$.

Doubell reported that ANP release in response to rapid pacing (500 beats/min) was abolished by both verapamil and nifedipine.[14] Nisoldipine decreased ANP secretion while W-7, a calmodulin inhibitor, did not. However, the nisoldipine decreased the heart rate as well. The author concluded that lowered $[Ca]_i$ was the common mechanism for the decreased inotropy and ANP secretion.[15]

Isolated Atrial Preparations

There are several types of atrial preparations: the right (beating) or sometimes the left (non-beating) atria may be used; the atria may be stretched and/or paced, and/or superfused with various agents. In isolated non-paced left atria, Ishida et al. reported that both arginine vasopressin (AVP) and phenylephrine (PE) increased ANP release. The effect was blocked by the calmodulin inhibitor W-7 and the PKC inhibitor H-7. However, neither agent inhibited stretch-induced ANP release, which suggests that stretch and hormonal agonists stimulate ANP secretion by different mechanisms.[16]

A number of studies aimed at delineating the role of Ca^{2+} in ANP release have been carried out by Schiebinger and his colleagues. β agonists influence Ca^{2+} flux through L-type channels and consequently raise the amount of Ca^{2+} in the SR storage pool. As expected, isoproterenol increased ANP secretion in paced left atria; this effect was blocked by lowering $[Ca]_o$ to 0.2 mM, using ryanodine and nitrendipene. Thus isoproterenol-stimulated ANP secretion is dependent both on influx of extracellular Ca^{2+} and Ca^{2+} release from the SR. Bay K8644 mimicked the action of isoproterenol, which suggests that Ca^{2+} entry occurred through voltage-sensitive Ca^{2+} channels. These investigators further demonstrated that stimulation by isoproterenol is dependent on the electrical activity of the membrane, since it does not occur in quiescent atria.[17]

Endothelin (ET) and α_1 agonists activate phospholipase C; they also stimulate ANP release from stretched, paced atria, but Ca^{2+} may play a slightly different role in this stimulation. Nitrendipine blocked both ET- and PE-induced ANP secretion by 50%. Ryanodine had no effect. Thus both agonists are independent of SR Ca^{2+}, but somewhat dependent on Ca^{2+} flux through voltage-dependent channels. Initiation of secretion in response to PE required Ca^{2+} and was abolished by lowering medium $[Ca]_o$ to 0.2 mM, although maintenance of secretion was not dependent on influx.[18] This was not the case for ET; some secretion occurred even at low $[Ca]_o$.[19] This differs from β-agonist-induced ANP stimulation, which is totally dependent on Ca^{2+} influx and subsequent storage in the SR.[17]

The above results are at odds with those obtained by de Bold, whose experiment allowed superfused right atria to beat spontaneously. Both basal and stretch-induced ANP secretion were enhanced when $[Ca]_o$ was decreased to 0.2 mM or EGTA was added.[20] In other studies extracellular Ca^{2+} was replaced with Ba^{2+}, Sr^{2+}, and La^{2+}, and intracellular [Ca] was altered with the use of caffeine, ryanodine, and depolarizing K^+ concentrations. Both Ba^{2+} and Sr^{2+} abolished contractile activity but did not change ANP secretion in response to stretch. K^+ affected basal and stretch-induced ANP secretion at low, but not at high (1.25–2.5 mM), Ca^{2+}. Caffeine had no effect. Ryanodine decreased ANP secretion as well as heart rate and resting tension.[21] The authors concluded that basal ANP secretion is not dependent on either contractile activity, or Ca^{2+} transients from either outside the cell or from intracellular stores; they also concluded that Ca^{2+} may tonically inhibit ANP release. However, a ryanodine-sensitive compartment may be partly responsible for the response to stretch. Some of the differences in the results obtained may be due to the fact that in the latter experiments the heart rate was not controlled.

Because the study of the role of Ca^{2+} is complicated by action potentials and contractions, Page et al. induced quiescence in an isolated combined right and left atrial preparation with tetrodotoxin at doses high enough to block the "fast" Na channel. They also lowered $[Ca]_o$, blocked Ca^{2+} influx with nitrendipine, and inhibited SR release with ryanodine.

The results obtained under these conditions indicated that stretch-activated ANP secretion requires Ca^{2+} influx but is independent of the SR.[22,23] Thus, in contrast to the results obtained by de Bold in beating atria, those of Page et al. show that, in the absence of contraction, both basal and stretch-induced ANP release are positively modulated by the presence of Ca^{2+} in the medium.

Cardiomyocytes in Culture

Adult cardiomyocytes are very difficult to maintain in culture. Using primary cultures of neonatal cardiomyocytes, Matsubara et al. found that both α_1 and muscarinic cholinergic agonists stimulated both ANP secretion via receptor-mediated mobilization of Ca^{2+} and activation of PKC, since the actions of TPA and ionomycin were synergistic.[24] Further work corroborated the conclusion that voltage-sensitive Ca^{2+} channels were involved, since nitrendipine could block the ANP stimulatory effect of Bay K8644.[25]

Hypotonic swelling (300 to 200 mOsm) is believed to mimic stretch by opening selective Ca-permeable channels. Neonatal cardiomyocytes exposed to medium or low osmolality respond by increasing ANP release. When loaded with the Ca^{2+} indicator indo-1, $[Ca]_i$ increased in these cells with decreasing osmolality as well. However, ANP secretion in response to hyposmolar stretch was enhanced in the absence of $[Ca]_o$. The addition of ionomycin moderately inhibited basal ANP release and abolished the stimulation seen with osmotic stretch. On the other hand, when $[Ca]_i$ was chelated with BAPTA, both basal and osmotic stretch-induced ANP release were enhanced.[26] Thus ANP secretion was temporally related to, but not dependent upon, a rise in $[Ca]_i$.

Compatible with these results are those of Uusimaa et al., who used neonatal rat myocytes that were grown on microcarrier beads and loaded with fura-2 to monitor $[Ca]_i$. Exposure of this preparation to K^+, Bay K8644, and ionomycin elevated $[Ca]_i$, but not ANP.[27] Endothelin increased both $[Ca]_i$ and ANP secretion, but although diltiazem blocked the rise in $[Ca]_i$, it had no effect on ANP.[28] The authors concluded that ET-induced ANP release did not require Ca^{2+} influx.

However, results of other studies tend to support a role for Ca^{2+} in mediating both agonist- and stretch-induced ANP secretion. In unstretched, primary neonatal myocytes in culture, Sei and Glembotski studied ANP secretory response to activation of voltage-dependent Ca^{2+} channels by KCl, and to PE and ET. In low $[Ca]_o$ (0.2 mM), or in the presence of nifedipine and verapamil there was no secretion with KCl, although responses to PE and ET were partially preserved. All secretion was inhibited when $[Ca]_i$ pools were depleted by ionomycin.[29] These results suggest that both intracellular and extracellular Ca^{2+} pools are important for ANP secretion.

Although exposure to hyposmolar medium enhances ANP secretion, this may be the result of changes in intracellular osmolality or of stretch. To clarify this point, myocytes were grown on flexible plates and stretch was induced by deformation of the plates. Although verapamil had no effect, calmidazolium (a calmodulin inhibitor) decreased both basal and stretch-induced ANP release, which implied that $[Ca]_i$ is important while voltage-gated influx is not.[30] Finally, 24-hr exposure to high $[Ca]_o$ or A23187 enhanced both ANP release and mRNA in myocyte cultures, an effect that was blocked by diltiazem, nifedipine, and verapamil.[31]

In conclusion, although the results of most studies reviewed above suggest that ANP secretion is Ca^{2+}-dependent, the exact role of Ca^{2+} remains in doubt. The fact that electrical activity, wall tension, and frequency of contraction affect both Ca^{2+} homeostasis and ANP secretion in cardiac tissue increases the difficulty already inherent in dissecting the role of Ca^{2+} in the secretion of this hormone.

References

1. deBold AJ. Atrial natriuretic factor: A hormone produced by the heart. Science 1985; 230:767–770.
2. Schiebinger RJ, Baker MZ, Linden J. Effect of adrenergic and muscarinic cholinergic agonists on atrial natriuretic peptide secretion by isolated rat atria. J Clin Invest 1987; 80:1687–1691.
3. Vulteenako O, Arjamaa O, Ling N. Atrial natriuretic polypeptides (ANP): Rat atria store high molecular weight precursor but secrete processed peptides of 25–35 amino acids. Biochem Biophys Res Comm 1985; 129:82–88.
4. Rasmussen H. The calcium messenger system. N Engl J Med 1986; 314:1094–1101.
5. Morgan JP, Morgan KG. Calcium and cardiovascular function: Intracellular calcium levels during contraction and relaxation of mammalian cardiac and vascular smooth muscle as detected with aequorin. Am J Med 1984; 77:33–46.
6. Bers DM. Early transient depletion of extracellular Ca during individual cardiac muscle contraction. Am J Physiol 1986; 244:H462–H468.
7. Movsesian MA, Thomas AP, Selak M, Williamson JR. Inositol triphosphate does not release Ca^{++} from permeabilized cardiac myocytes and sarcoplasmic reticulum. FEBS Lett 1985; 185:328–332.
8. Laher I, Beban JA. Protein kinase C activation selectively augments a stretch-induced calcium dependent tone in vascular smooth muscle. J Pharmacol Exp Ther 1987; 242:566–572.
9. Guharay F, Sachs F. Stretch-activated single ion channel currents in tissue-cultured embryonic chick skeletal muscle. J Physiol 1984; 352:685–701.
10. Saito Y, Nakao K, Morii N, Sugawara A, Shiono S, Yamada T, Itoh H, Sakamoto M, Kurahashi K, Fujiwara M, Imura H. Bay K8644, a voltage-sensitive calcium channel agonist, facilitates secretion of atrial natriuretic polypeptide from isolated perfused rat hearts. Biochem Biophys Res Comm 1986; 138:1170–1176.

11. Ruskoaho H, Toth M, Lang RE. Atrial natriuretic peptide secretion: Synergistic effect of phorbol ester and A23187. Biochem Biophys Res Comm 1985; 133:581–588.

12. Ruskoaho H, Toth M, Ganten D, Unger T, Lang RE. The phorbol ester induced atrial natriuretic peptide secretion is stimulated by forskolin and Bay K8644 and inhibited by 8-bromo-cyclic GMP. Biochem Biophys Res Comm 1986; 139:266–274.

13. Ruskoaho H, Vuolteenaho O, Leppalluoto J. Phorbol esters enhance stretch-induced atrial natriuretic peptide secretion. Endocrinology 1990; 127:2445–2455.

14. Doubell AF. The effect of calcium antagonists on atrial natriuretic peptide (ANP) release from the rat heart during rapid cardiac pacing. J Mol Cell Cardiol 1989; 21:437–440.

15. Doubell AF. The second messenger system(s) mediating the secretion of atrial natriuretic peptide (ANP) from the isolated rat heart during rapid cardiac pacing. Life Sci 1989; 45:2193–2200.

16. Ishida A, Tanakashi T, Okumura K, Hashimoto H, Ito T, Ogawa K, Satake T. A calmodulin antagonist (W-7) and a protein kinase C inhibitor (H-7) have no effect on atrial natriuretic peptide release induced by atrial stretch. Life Sci 1988; 42:1659–1667.

17. Schiebinger RJ. Calcium, its role in isoproterenol-stimulated atrial natriuretic peptide secretion by superfused rat atria. Circ Res 1989; 65:600–606.

18. Schiebinger RJ, Parr HG, Cragoe EJ Jr. Calcium: Its role in α_1-adrenergic stimulation of atrial natriuretic peptide secretion. Endocrinology 1992; 130:1017–1023.

19. Schiebenger RJ, Gomez-Sanchez CE. Endothelin: A potent stimulus of atrial natriuretic peptide secretion by superfused rat atria and its dependency on calcium. Endocrinology 1990; 127:119–125.

20. deBold ML, deBold AJ. Effect of manipulations of Ca^{++} environment on atrial natriuretic factor release. Am J Physiol 1989; 256:H1588–H1594.

21. Kuroski-deBold ML, deBold AJ. Stretch-secretion coupling in atrial cardiocytes: Dissociation between atrial natriuretic factor release and mechanical activity. Hypertension 1991; 18:III-169–III-178.

22. Page E, Goings GE, Power B, Upshaw-Earley J. Basal and stretch-augmented natriuretic peptide secretion by quiescent rat atria. Am J Physiol 1990; 259:C801–C818.

23. Page E, Upshaw-Earley J, Goings GE, Hanek DA. Effect of external Ca^{++} concentration on stretch-augmented natriuretic peptide secretion by rat atria. Am J Physiol 1991; 260:C756–C762.

24. Matsubara H, Hirata Y, Yoshimi H, Takata S, Takagi Y, Umeda Y, Yamane Y, Inada M. Role of calcium and protein kinase C in ANP secretion by cultured rat cardiocytes. Am J Physiol 1988; 255:H405–H409.

25. Hirata Y, Matsubara H, Fukuda Y, Yoshimi H. Cellular mechanisms of atrial natriuretic factor secretion by cultured rat cardiocytes. J Hypertension 1988; 6:S295–S296.

26. Greenwald JE, Apkon M, Hruska KA, Needleman P. Stretch-induced atriopeptin secretion in the isolated rat myocyte and its negative modulation by calcium. J Clin Invest 1989; 83:1061–1065.

27. Uusimaa PA, Ruskoako H, Leppaluoto J, Hassinen IE. Cytosolic Ca^{++} during atrial natriuretic peptide secretion from cultured neonatal cardiomyocytes. Mol Cell Endocrinol 1990; 73:153–163.
28. Uusimaa PA, Hassinen IE, Vuolseenako O, Ruskoaho H. Endothelin-induced atrial natriuretic peptide release from cultured neonatal cardiac myocytes: the role of extracellular calcium and protein kinase C. Endocrinology 1992; 130:2455–2464.
29. Sei CA, Glembotski CC. Calcium dependence of phenylephrine-, endothelin-, and potassium chloride-stimulated atrial natriuretic factor secretion from long term primary neonatal rat atrial cardiocytes. J Biol Chem 1990; 265:7166 27172.
30. Gardner DG, Wirtz H, Dobbs LG. Stretch-dependent regulation of atrial peptide synthesis and secretion in cultured atrial cardiocytes. Am J Physiol 1992; 263:E239–E244.
31. LaPointe MC, Deschepper CF, Wu J, Gardner DG. Extracellular calcium regulates expression of the gene for atrial natriuretic factor. Hypertension 1990; 15:20–28.

7
Intracellular Ca^{2+} and Insulin Action: Possible Role in the Pathogenesis of Syndrome X

Joseph Levy and James R. Sowers

Overwhelming evidence suggests that binding to the receptor followed by receptor autophosphorylation are the initial steps by which insulin exerts its biologic effects.[1-5] However, the subsequent steps in the propagation of the signal are not clear. Indeed, since the biologic effects of insulin are many, it has been suggested that more than one second messenger may be involved in the propagation of the signal.[6] Intracellular Ca^{2+} ($[Ca^{2+}]_i$) may have a significant role in this regard,[5] thus explaining why abnormal $[Ca^{2+}]_i$ homeostasis may be associated with impaired insulin action[7,8] or with insulin resistance.[9,10]

Insulin resistance, on the other hand, may play an essential central role in a variety of disease states and may be the critical factor in Syndrome X,[11,12] a new clinical entity consisting of android obesity, hypertension, non-insulin-dependent diabetes mellitus (NIDDM), high plasma levels of low density lipoproteins (LDL), and low plasma levels of high density lipoproteins (HDL). Since abnormal $[Ca^{2+}]_i$ values and insulin resistance are characteristic of all three clinical conditions, it has been proposed that an altered $[Ca^{2+}]_i$ homeostasis may be a potential common denominator in the pathogenesis of Syndrome X.[7] This chapter will review the evidence suggesting that impaired $[Ca^{2+}]_i$ homeostasis has a fundamental role in the pathogenesis of Syndrome X. New information, which suggests a potential role of increased $[Ca^{2+}]_i$ in the pathogenesis of atherosclerosis and in the increased incidence of atherosclerosis observed in diabetes, hypertension, and obesity, will also be reviewed.

Physiologic Regulation of the Intracellular Calcium Concentration

Many vital cell functions depend upon the regulation of $[Ca^{2+}]_i$, which normally is 10^{-4}-fold lower than that of extracellular Ca^{2+}.[13-15] Persistent increases in $[Ca^{2+}]_i$ are deleterious to cell metabolism and may lead to cell death.[16,17] Maintenance of the physiologic low $[Ca^{2+}]_i$ is a dynamic

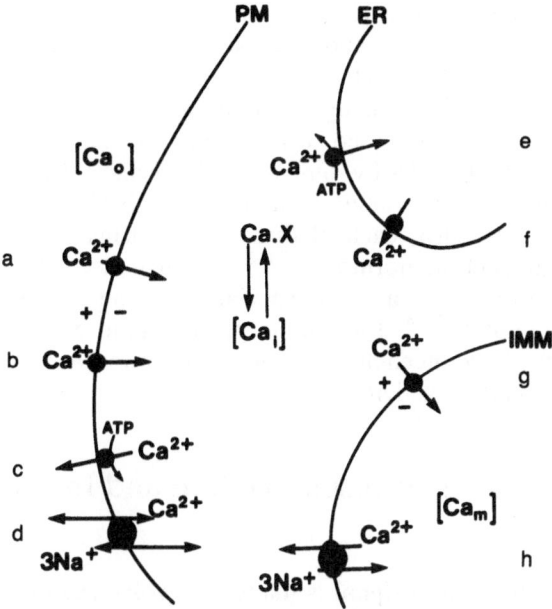

FIGURE 7.1. Major pathways for the transport of Ca^{2+} across cellular membranes. PM = plasma membrane; a = steady Ca^{2+} influx; b = voltage-dependent Ca^{2+} influx; c = plasma membrane Ca^{2+}-ATPase; d = plasma membrane Na^+/Ca^{2+} exchanger (direction depends on thermodynamic balance between Na^+ and Ca^{2+}; electrochemical potential gradients and can vary with tissue and condition). ER = endoplasmic reticulum; e = endoplasmic reticulum Ca^{2+}-ATPase; f = second-messenger-activated Ca^{2+}-efflux; IMM = inner mitochondrial membrane; g = mitochondrial Ca^{2+}-uniporter; h = mitochondrial Na^+/Ca^{2+} exchange; Ca.X = CaX-binding of Ca^{2+} to cytosolic molecules. (Data from Nicholls.[15])

process and is achieved by the concerted effects of a variety of mechanisms operating at various subcellular levels[15] (Figure 7.1). Ca^{2+} enters the cell via voltage-dependent and hormone-dependent channels; and leaves the cell through the operation of a Mg^{2+}-dependent Ca^{2+}-ATPase [$(Ca^{2+} + Mg^{2+})$-ATPase] and a Na^+-Ca^{2+} exchanger. The activity of these mechanisms varies from cell type to cell type.[14,18–20] Uptake of Ca^{2+} by intracellular organelles such as the mitochondria and the endoplasmic reticulum also helps maintain a low $[Ca^{2+}]_i$. Ca^{2+} enters the mitochondria via a Ca^{2+} uniporter and the endoplasmic reticulum via a distinct Ca^{2+}-ATPase, which is different from the $(Ca^{2+} + Mg^{2+})$-ATPase located on the plasma membrane. Ca^{2+} leaves the mitochondria via a Na^+-Ca^{2+} exchanger and the endoplasmic reticulum via a constant efflux, which increases in the presence of inositol-1,4,5,-trisphosphate (IP_3).[15]

Other mechanisms regulate $[Ca^{2+}]_i$ indirectly, among them the cation pump $(Na^+ + K^+)$-ATPase and the Na^+/H^+ exchanger. Decreased activity of the $(Na^+ + K^+)$-ATPase increases intracellular Na^+ ($[Na^+]_i$), inhibits the Na^+-Ca^{2+} exchanger, reduces Ca^{2+} efflux, and thus increases $[Ca^{2+}]_i$.[21-23] Increased activity of the Na^+/H^+ exchanger increases $[Na^+]_i$ and $[Ca^{2+}]_i$ by similar mechanisms.[22] Increased amino acid uptake, which is associated with increased Na^+ influx, may also increase $[Ca^{2+}]_i$.[22] In turn, each of these mechanisms is regulated by a variety of factors, such as hormones,[7,8,24-35] nutrients,[36-38] kinases,[39-41] peptidases, calmodulin,[6,41] and, of special importance, the membrane phospholipid content.[36,42-47] Membrane phospholipid content influences Ca^{2+} influx and efflux mechanisms, and acidic phospholipids stimulate $(Ca^{2+} + Mg^{2+})$-ATPase activity.[42]

The Role of $[Ca^{2+}]_i$ in Insulin Action and Insulin Resistance

Transduction of the insulin signal is likely to involve several mediators.[5] Although the direct role of Ca^{2+} in the mechanism of insulin action is controversial,[48,49] a large body of evidence corroborates its importance.[6,7] Thus, insulin-sensitive enzymes are Ca^{2+}-dependent,[48,50-55] while a number of insulin effects on cellular metabolism[55,56,60] require extracellular Ca^{2+} and may be mimicked by compounds that alter the

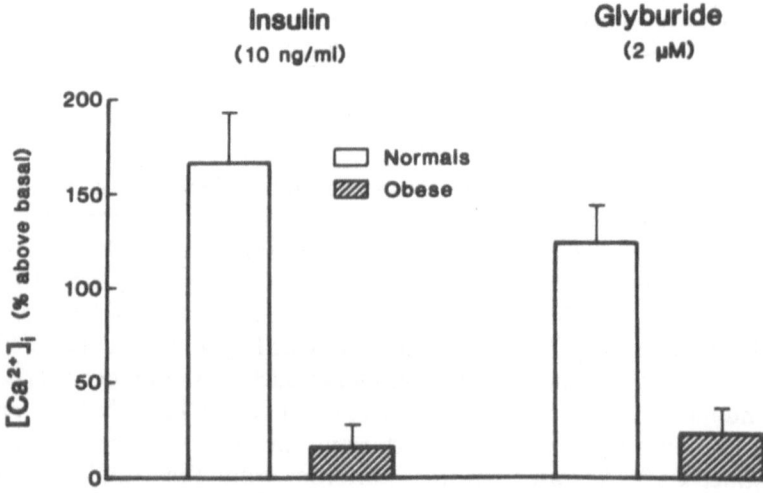

FIGURE 7.2. Effect of insulin and glyburide upon $[Ca^{2+}]_i$ in adipocytes obtained from six normal weight controls and six obese women after overnight fast. $p < 0.01$ (obese vs normals). (Data from Draznin et al.[24])

concentration of $[Ca^{2+}]_i$.[57,58] Among the pertinent effects of insulin are changes in the Ca^{2+} homeostasis of plasma membranes and in the Ca^{2+} fluxes of intact cells,[8,59,61-63,69] the phosphorylation of calmodulin in subcellular preparations and intact cells,[6,62] and the resulting Ca^{2+} calmodulin-mediated stimulation of the insulin receptor tyrosine kinase activity.[6] Chelation of $[Ca^{2+}]_i$ prevents stimulation of glucose transport by insulin and insulinomimetic agents in adipocytes[64] and, while this effect is still controversial, insulin appears to increase $[Ca^{2+}]_i$ concentration in adipocytes[24] and platelets[32] (Figure 7.2).

Of significance are the direct effects of insulin on enzymes that regulate $[Ca^{2+}]_i$. Although there are reports to the contrary,[61] overwhelming evidence suggests that insulin stimulates the plasma membrane (Ca^{2+} + Mg^{2+})-ATPase[8,65-68] (Figure 7.3). This effect is achieved at physiologic concentrations of insulin, is dose-dependent, and is specific for the active insulin peptide.[8] Insulin also stimulates the Na$^+$/Ca^{2+} exchanger in heart sarcolemma.[67] The effect of insulin on the (Ca^{2+} + Mg^{2+})-ATPase may be mediated, in part, by direct phosphorylation of the ATPase, and in part by insulin-induced acute changes in the phospholipid content of the plasma membrane.[71,72] This is suggested by the observation that insulin rapidly increases the phosphatidic acid content of the plasma membrane (possibly via an activation of phospholipase D[71]), and that acidic phospholipids stimulate the (Ca^{2+} + Mg^{2+})-ATPase[71] (Figure 7.4). Inhibition of the effect on ATP-ase inhibits insulin stimulation of glucose oxidation and lipogenesis in adipocytes.[73] Insulin can also modulate cellular Ca^{2+}

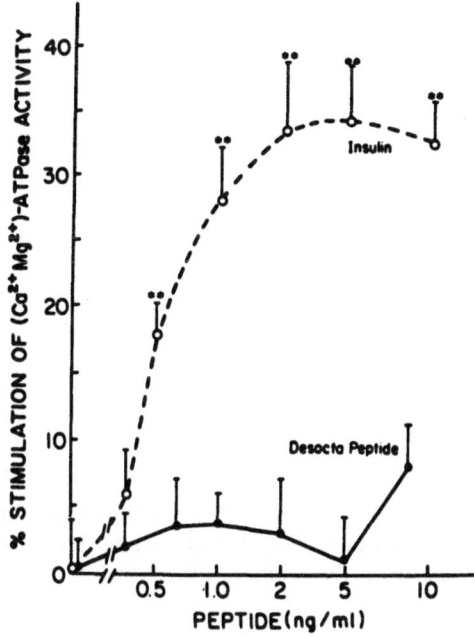

FIGURE 7.3. Effect of insulin and desoctapeptide porcine insulin on activity of (Ca^{2+} + Mg^{2+})-ATPase in kidney basolateral membranes (BLM) from nondiabetic rats. **$p < 0.01$ (insulin vs desoctapeptide insulin). (Data from Levy et al.[8])

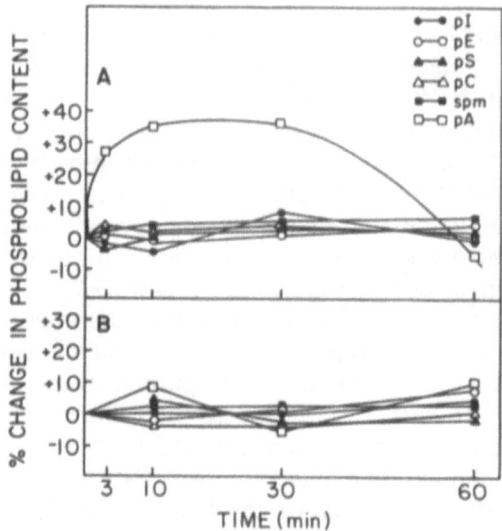

FIGURE 7.4. Time-dependent effect of insulin in vitro (100 μU/ml) on the phospholipid content of kidney basolateral membranes (BLM) of control rats (A) and rats with NIDDM (B). Results represent mean of duplicate measurements. Plasma membrane content of each phospholipid was calculated as percent of total phospholipid. pA = phosphatidic acid; pS = phosphatidylserine; spm = sphingomyelin; pE = phosphatidylethanolamine; pI = phosphatidylinositol; pC = phosphatidylcholine. (Data from Levy.[167])

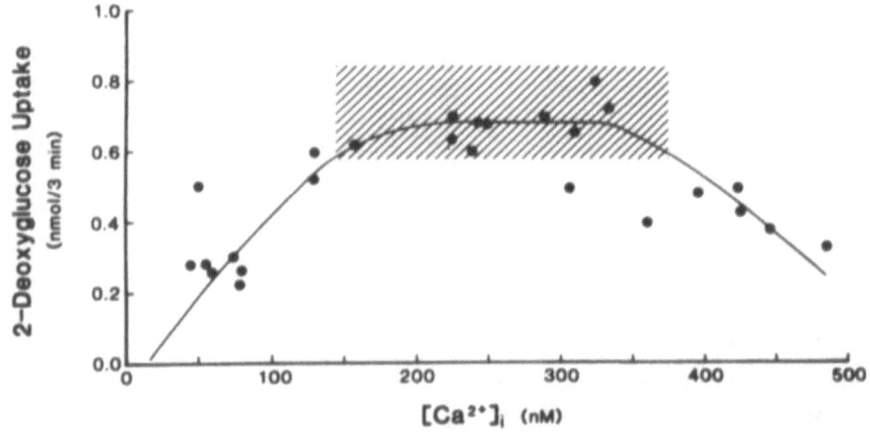

FIGURE 7.5. Rates of maximally stimulated 2-deoxyglucose uptake (at insulin concentration 25 ng/ml) at various levels of cytosolic free Ca^{2+}. Hatched area represents the window of optimal response to insulin. (Data from Draznin et al.[77])

homeostasis by affecting the movement of other cations. For example, insulin can stimulate the Na^+/H^+ exchange in adipocytes[70] and the $(Na^+ + K^+)$-ATPase activity in a variety of tissues,[74-76] altering $[Na^+]_i$ and affecting cell $[Ca^{2+}]_i$ homeostasis via the Na^+/Ca^{2+} exchanger.[21-23]

Finally, it has been shown that there is an optimal range of $[Ca^{2+}]_i$ for mediating insulin action. $[Ca^{2+}]_i$ levels higher or lower than the optimal range are associated with impaired insulin action[8,77-79] (Figure 7.5). It may be concluded that conditions in which $[Ca^{2+}]_i$ homeostasis is impaired and insulin fails to regulate $[Ca^{2+}]_i$ may be associated with insulin resistance.[8,78,79]

Experimental evidence supports this conclusion. $(Ca^{2+} + Mg^{2+})$-ATPase activity is increased in erythrocyte and kidney membranes from NIDDM rats and is decreased in membranes from obese rats, and insulin fails to stimulate the $(Ca^{2+} + Mg^{2+})$-ATPase in both types of membranes.[8,78] As expected, both rat models show marked insulin resistance.[80-82] Improvement in the insulin resistance in the NIDDM

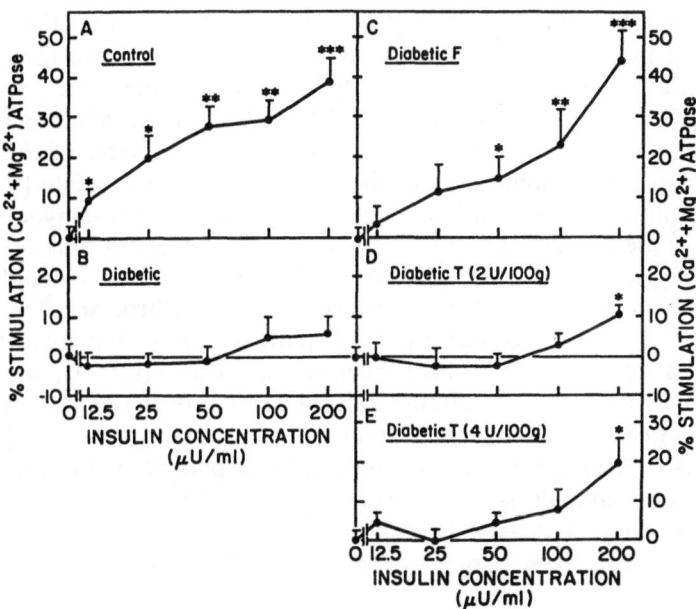

FIGURE 7.6. Insulin effect on the $(Ca^{2+} + Mg^{2+})$-ATPase activity in rat kidney basolateral membranes (BLM): (A) control (B) diabetic, (C) diabetic F (fasting, in which insulin resistance improved), (D) diabetic treated with insulin (2 U/100 g), and (E) diabetic treatment with insulin (4 U/100 g). Results are given as mean ± SE of 6 experiments and presented as percentage increase over basal activity (in the absence of insulin). Experiments were performed at free Ca^{2+} concentration of 0.1 μmol/l. *$p < 0.05$; **$p < 0.02$, ***$p < 0.01$ vs basal activity of each membrane preparation. (Data from Levy et al.[80])

rats is associated with regained ability of insulin to stimulate the $(Ca^{2+} + Mg^{2+})$-ATPase[80] (Figure 7.6). In addition, although insulin increases $[Ca^{2+}]_i$ in normal adipocytes, this increase is blunted in adipocytes obtained from obese (Figure 7.2) or aged individuals,[79,83] conditions characterized by high basal $[Ca^{2+}]_i$ and significant degrees of insulin resistance.[79,83]

The exact mechanism by which impaired $[Ca^{2+}]_i$ homeostasis impairs insulin action is not known. However, it has been demonstrated that high $[Ca^{2+}]_i$ impairs the dephosphorylation of the insulin receptor and of glycogen synthase,[84] a major target for insulin effect.

Abnormal $[Ca^{2+}]_i$ Homeostasis in Diabetes

Studies of rats with streptozotocin-induced diabetes, patients with insulin-dependent diabetes mellitus (IDDM), and patients with non-insulin-dependent diabetes mellitus (NIDDM) suggest that abnormal $[Ca^{2+}]_i$ homeostasis is a generalized defect common to all forms of the disease. It can manifest itself by changes in $[Ca^{2+}]_i$ level, and/or malfunction of one or more of the mechanisms that regulate $[Ca^{2+}]_i$, mainly the function of the plasma membrane $(Ca^{2+} + Mg^{2+})$-ATPase, the $(Na^+ + K^+)$-ATPase, and the Ca^{2+} entry channels. The nature of the defect in $[Ca^{2+}]_i$ homeostasis is tissue-specific. This is not surprising in view of the tissue specificity in the predominance of the mechanisms that regulate $[Ca^{2+}]_i$.

Abnormal $[Ca^{2+}]_i$ regulation has been described in cardiomyocytes,[85–96] aorta,[97] mesenteric and coronary arteries,[98,99] skeletal muscle,[100–104] kidney,[80,105] liver,[106,107] erythrocytes,[108,109] lens,[110,111] and osteoblasts[112] of diabetic animals, as well as in arteries,[113–115] erythrocytes,[36,37,116–119] platelets,[120–123] and adipocytes[124] of patients with diabetes. Occult increases in plasma calcium have been observed in NIDDM patients[125–127] and decreased levels of plasma ionized calcium have been observed in IDDM patients.[128] The most common abnormality in cell $[Ca^{2+}]_i$ homeostasis observed in tissues obtained from patients with diabetes or from diabetic animals is Ca^{2+} overload.[86,106,108,112,117,119,124] However, normal or even decreased $[Ca^{2+}]_i$ have also been found in platelets, osteoblasts, and liver tissue of both NIDDM patients and diabetic rats.[80,112,116,129]

It cannot be stated at this time which of the multiple mechanisms that regulate $[Ca^{2+}]_i$ may represent the primary defect in diabetes. One mechanism for Ca^{2+} overload in platelets taken from insulin-treated patients with diabetes and the arteries and skeletal muscle of diabetic rats may be an increased cation concentration-dependent influx of Ca^{2+}.[98,99,116–119,130] A decrease in Ca^{2+}-ATPase activity that was observed in tissues of diabetic patients[117] (possibly resulting from glycosylation)[37,132] may decrease Ca^{2+} efflux, thus contributing to the increased $[Ca^{2+}]_i$. On the other hand, changes in membrane phospholipid content[44,131] may explain the increased $(Ca^{2+} + Mg^{2+})$-ATPase activity observed under some diabetic conditions.[8,44,87,102,103,109] In addition, the $(Na^+ + K^+)$-ATPase activity, which tends to be decreased in tissues from patients and animals with diabetes,[97,119,121,133] may contribute to the increased $[Ca^{2+}]_i$ by inhibiting the Na^+/Ca^{2+} exchanger.[21]

Finally, since diabetes is characterized by low tissue levels of Mg^{2+},[134–136] and since the activity of the $(Na^+ + K^+)$-ATPase and of the membrane Ca^{2+}-ATPase are Mg^{2+}-dependent,[8,137] Mg^{2+} deficiency may also contribute to the decreased activity of these cation pump enzymes and thus increase $[Ca^{2+}]_i$. It is not clear whether the defects in $[Ca^{2+}]_i$ observed in diabetes are primary genetically-determined defects, or if they are secondary to the metabolic abnormalities of diabetes. It is possible that longitudinal studies assessing $[Ca^{2+}]_i$ metabolism in family members of diabetic patients, before and after the development of overt diabetes, may provide the answer to this question.

Abnormal $[Ca^{2+}]_i$ Homeostasis in Hypertension and Obesity

Abnormalities in $[Ca^{2+}]_i$ as well as defects in membranes binding and transport kinetics of Ca^{2+} have been identified in erythrocytes, platelets, and adipocytes of patients with essential hypertension.[138–143] There is evidence of decreased ATP-dependent calcium transport by isolated vesicles prepared from vascular tissue of spontaneously hypertensive rats and of a decrease in Ca^{2+}-ATPase activity in red blood cell membranes of patients with essential hypertension.[144,145] These changes may explain why erythrocyte $[Ca^{2+}]_i$ and total skeletal muscle calcium are increased in patients with essential hypertension.[143,146] $[Ca^{2+}]_i$ is increased in platelets of patients with essential hypertension. Successful antihypertensive therapy that decreases the peripheral vascular resistance also reduces the elevated platelet $[Ca^{2+}]_i$.[142] $[Ca^{2+}]_i$ is also increased in lymphocytes, but not in granulocytes, of hypertensive patients.[144,148] Thus, while a certain degree of tissue specificity with regard to changes in cell Ca^{2+} homeostasis also exist in hypertension, high $[Ca^{2+}]_i$ appears to be the predominant abnormality. However, whether the altered Ca^{2+} metabolism in the red blood cells of hypertensive patients is a marker or a consequence of essential hypertension is still a matter of controversy.[143]

High $[Ca^{2+}]_i$ and increased Ca^{2+} influx have been observed also in vascular smooth muscle cells from spontaneously hypertensive rats,[144,147] and it has been suggested that increased $[Ca^{2+}]_i$ in vascular smooth muscle cells of patients with essential hypertension may be of primary importance for the maintenance of increased peripheral vascular resistance and decreased blood flow, which are characteristics of the disease.[149,150] The origin of the cellular Ca^{2+} shifts in essential hypertension remains unclear. The content and distribution of calmodulin in erythrocytes of patients with essential hypertension are normal[151] and therefore cannot be the direct cause of observed changes in $[Ca^{2+}]_i$ homeostasis. However, the ability of calmodulin to activate the Ca^{2+}-ATPase in platelets from patients with essential hypertension is impaired.[145] Since the physiologic activities of $(Ca^{2+} + Mg^{2+})$-ATPase and of $(Na^{+} + K^{+})$-ATPase depend on their phospholipid milieu,[85] changes in membrane lipid composition may also be involved.[144,152]

Changes in cell Ca^{2+} regulation have also been described in obesity. High $[Ca^{2+}]_i$ levels were found in adipocytes of obese patients and in bone cells of genetically obese rats.[112,124] The mechanisms for this increase are largely unknown. Low $(Ca^{2+} + Mg^{2+})$-ATPase activity has been observed in kidney and in erythrocyte membranes from obese rats,[78,108] while red-cell membrane ouabain binding (an indirect measure of $(Na^{+} + K^{+})$-ATPase activity) is decreased in human obesity.[153] However, since the level of $(Na^{+} + K^{+})$-ATPase activity is genetically determined and may change in different ethnic groups, the implication of this observation is uncertain.[154] Pasquali et al.[155] found heterogeneity of erythrocyte $(Na^{+} + K^{+})$-ATPase activity and also in the number of pump units in human obesity, and suggested that it may be evidence of the abnormal endocrine milieu.

Abnormal $[Ca^{2+}]_i$ and Syndrome X

Insulin resistance appears to be the common denominator of the three clinical components of Syndrome X (NIDDM, obesity, and hypertension), and although its most common and recognized manifestations are glucose intolerance and hyperglycemia, insulin resistance may have other metabolic consequences, and may help elevate blood pressure.[7,148] In turn it could be argued that, since an abnormal $[Ca^{2+}]_i$ homeostasis and high $[Ca^{2+}]_i$ are hallmarks of impaired insulin action[7,8,78,79] and are commonly observed in obesity, diabetes, and hypertension,[7] a high $[Ca^{2+}]_i$ may be the basic pathogenetic factor of the syndrome.

High $[Ca^{2+}]_i$ is the most common abnormality of $[Ca^{2+}]_i$ homeostasis seen in diabetes, obesity, and hypertension. Increased $[Ca^{2+}]_i$ in vascular smooth muscle cells, as is seen in spontaneously hypertensive rats,[144,147,156] increases vascular contraction, decreases the vessel luminal size and thus increases vascular resistance and blood pressure, and decreases the blood flow.

In turn, decreased muscle blood flow has been described in hypertension, obesity, and diabetes,[149,157–159] due in part to a loss of the physiologic vasodilator effect of insulin.[157–159] This may lead to decreased utilization of glucose by the tissue in response to insulin,[160] possibly by decreasing availability of hormone and substrates. The decreased blood flow may be a consequence of increased contraction of the vessels secondary to high $[Ca^{2+}]_i$. However, non-calcium related changes in skeletal muscle morphology and vascular rarefaction can also contribute to insulin resistance.[149,160] The loss of the physiologic vasodilatory effect of insulin[157-159] in the insulin-resistant conditions may result in part from the inability of insulin to regulate $[Ca^{2+}]_i$ in these conditions.[7,8,78,79,83]

Such a scenario suggests that obesity, hypertension, diabetes, and possibly the plasma lipid abnormalities are different manifestations of a common basic mechanism. The severity of each manifestation may depend upon additional factors. For example, the co-existence of obesity and diabetes may be determined by the residual pancreatic β cell mass and the ability of the β cells to increase insulin secretion, a function also negatively affected by persistent high $[Ca^{2+}]_i$.[161]

High $[Ca^{2+}]_i$ and Atherosclerosis

Atherosclerosis is characterized by increased endothelial permeability, monocyte infiltration, intimal smooth muscle cell proliferation, increased platelet aggregation, and accumulation of lipids, calcium, and extracellular matrix components in the vessel wall,[162,163] processes that can be accelerated by elevations in $[Ca^{2+}]_i$.[162-164] Thus the atherosclerotic changes observed in rabbits fed a high-cholesterol diet are associated with increased Ca^{2+} influx and increased $[Ca^{2+}]_i$ of the aortic smooth muscle cells.[162] Increased Ca^{2+} influx increases the influx of low density lipoproteins across the endothelial cell layer,[164] an effect that can be inhibited by Ca^{2+}-channel blockers.[164] The increase in DNA synthesis in smooth muscle cells, an indicator of cell proliferation, is also associated with an increase in cell $[Ca^{2+}]_i$.[164] Additional support for the view that Ca^{2+} has a role in the atherosclerotic process can be derived from epidemiologic studies. For example, the age-related increase in the incidence of atherosclerosis is associated with a parallel accumulation of cholesterol and calcium in the arterial wall,[16] a process markedly increased in diabetic patients.[16] Although the Ca^{2+} deposition has been considered a result of atherosclerosis, there is evidence that it may be involved in its development.[164] Indeed, it has been demonstrated that human aortic medial calcification may precede the development of the atherosclerotic plaque,[165] a notion supported by experimental evidence. In normolipidemic vitamin D-treated rabbits and swine, atherosclerotic lesions appear only after fibromuscular changes, cell necrosis, and calcifications have occurred.[166] Increased cell Ca^{2+} uptake and Ca^{2+} overload in the mitochondria may cause cell necrosis, and calcium-loaded cell remnants may initiate the arterial wall calcifications.[166] Ca^{2+} channel blockers inhibit smooth muscle cell proliferation and migration, improve cellular lipoprotein metabolism in vascular cells, alter phospholipid turnover, decrease platelet adhesion to the vessel wall, reduce extracellular matrix synthesis, protect against radical cell damage,[19,162–164] and consequently inhibit the development of the atherosclerotic lesion. Although the extent to which these drugs induce their effect via inhibition of the voltage-dependent Ca^{2+} channels is controversial, there is evidence that they may inhibit the increased $[Ca^{2+}]_i$ by other pathways.[164]

Conclusions

Studies on the pathogenesis of Syndrome X (hypertension, obesity, and non-insulin-dependent diabetes) carried out in the last decade suggest that abnormalities of intracellular Ca^{2+} metabolism may have a central role, possibly by being a factor in the development of insulin resistance. These studies also suggest that since abnormal $[Ca^{2+}]_i$ homeostasis may accelerate the atherosclerotic process, the impaired $[Ca^{2+}]_i$ regulation seen in obesity, hypertension, and diabetes may contribute to the increased incidence and progression of atherosclerosis characteristic of these conditions.

Acknowledgment. The excellent secretarial help of Lois Wiggins is gratefully acknowledged.

References

1. Rosen OM. After insulin binds. Science 1987; 237:1452–1458.
2. Zick Y. The insulin receptor: Structure and function. Crit Rev Biochem Mol Biol 1989; 24:217–269.
3. Denton RM. Search for the missing links. Nature 1990; 348:286–287.
4. Khan RC, White MF, Grigorescu F, Takayama S, Häring HU, Crettaz M. The insulin receptor protein kinase. In: Czech MP, ed. Molecular Basis of Insulin Action. New York: Plenum; 1985:67–93.
5. Gherzi R, Russell DS, Taylor SI, Rosen OM. Reevaluation of the evidence that an antibody to the insulin receptor is insulinminetic without activating the protein tyrosine kinase activity of the receptor. J Biol Chem 1987; 262:16900–16905.
6. McDonald JM, Pershadsingh HA, Colca J. The role of calcium and calmodulin in insulin receptor function in the adipocyte. Ann NY Acad Sci 1986; 488:406–418.
7. Levy J, Zemel MB, Sowers JR. Role of cellular calcium metabolism in abnormal glucose metabolism and diabetic hypertension. Am J Med 1989; 87(suppl 6A):7S–15S.

8. Levy J, Gavin JR III, Hammerman MR, Avioli LV. Ca^{2+} + Mg^{2+}-ATPase activity in kidney basolateral membrane in non-insulin-dependent diabetic rats: Effect of insulin. Diabetes 1986; 35:899–905.
9. Draznin B. Cytosolic calcium: A new factor in insulin resistance? Diab Res Clin Pract 1991; 11:141–146.
10. Levy J, Grunberger G, Karl I, Gavin JR III. Effects of food restriction and insulin treatment on (Ca^{2+} + Mg^{2+})-ATPase response to insulin in kidney basolateral membranes of non-insulin-dependent diabetic rats. Metabolism 1990; 39:25–33.
11. Reaven GM. Role of insulin resistance in human disease. Diabetes 1988; 37:1595–1607.
12. Zimmet P. Non-insulin-dependent (type 2) diabetes mellitus: Does it really exist? Diab Med 1989; 6:728–735.
13. Carafoli E. Intracellular calcium homeostasis. Annu Rev Biochem 1987; 56:395–433.
14. Carafoli E. Calcium-transporting systems of plasma membranes, with special attention to their regulation. In: Greengard P, Robison GA, Paoletti RP, eds. Advances in Cyclic Nucleotide and Protein Phosphorylation Research, Vol. 17. New York: Raven; 1984:534–549.
15. Nicholls DG. Intracellular calcium homeostasis. Br Med Bull 1986; 42:353–358.
16. Fleckenstein A, Frey M, Zorn J, Fleckenstein-Grum G. The role of calcium in the pathogenesis of experimental arteriosclerosis. TIPS 1987; 8:496–501.
17. Avioli LV. Calcium, cell function and cell death. Adv Exp Med Biol 1986; 209:9–15.
18. Lidofsky SD, Xie MH, Scharschmidt BF. Na$^+$–Ca^{2+} exchange in cultured rat hepatocytes: Evidence against a role in cytosolic Ca^{2+} regulation or signaling. Am J Physiol 1990; 259:G56–G61.
19. Schmitz G, Hankowitz J, Kovacs EM. Cellular processes in atherogenesis: Potential targets of Ca^{2+} channel blockers. Atherosclerosis 1991; 88:109–132.
20. Zidek W, Vetter H. Cellular calcium metabolism in primary hypertension. Klin Wochenschr 1987; 65:155–160.
21. Blaustein MP. Sodium ions, blood pressure regulation and hypertension: A reassessment of a hypothesis. Am J Physiol 1977; 232:C165–C173.
22. Moore RD. The case of intracellular pH in insulin action. In: Czech MP, ed. Molecular Basis of Insulin Action. New York: Plenum; 1985:145–170.
23. El-Mallakh RS. Hypertension and diabetes in obesity: A review and new ideas on the contributing role of ions. Med Hypotheses 1986; 19:44–45.
24. Draznin B, Kao M, Sussman KE. Insulin and glyburide increase cytosolic free-Ca^{2+} concentration in isolated rat adipocytes. Diabetes 1986; 36:174–178.
25. Standley PR, Zhang F, Ram JL, Zemel MB, Sowers JR. Insulin attenuates vasopressin-induced calcium transients and a voltage-dependent calcium response in rat vascular smooth muscle cells. J Clin Invest 1991; 88:1230–1236.
26. Lin SH, Wallace MA, Fain JN. Regulation of Ca^{2+}–Mg^{2+}-ATPase activity in hepatocyte plasma membranes by vasopressin and phenylephrine. Endocrinology 1983; 113:2268–2275.

27. Mine T, Kojima I, Ogato E. Calcium rather than cyclic AMP is an intracellular messenger of parathyroid hormone action on glycogen metabolism in isolated rat hepatocytes. Biochem J 1989; 258:889–894.

28. Rasmussen H. The messenger function of Ca^{2+}: From PTH action to smooth muscle contraction. Bone and Mineral 1989; 5:233–248.

29. Rosci NK, Stanchaert ML, Pollet RJ. The mechanism of insulin stimulation of $(Na^+ + K^+)$-ATPase transport activity in muscle. J Biol Chem 1985; 260:6206–6212.

30. Lynch CJ, Wilson PB, Blackmore PF, Exton JH. The hormone-sensitive hepatic Na^+-pump. J Biol Chem 1986; 261:14551–14556.

31. Muallem S, Pandol SJ, Becker TG. Calcium mobilizing hormones activate the plasma membrane Ca^{2+} pump of pancreatic acinar cells. J Membr Biol 1988; 106:57–69.

32. Blackmore PF, Augert G. The effect of hormones on cytosolic free calcium in adipocytes. Cell Calcium 1989; 10:561–567.

33. Levy J, Gavin JR III, Morimoto S, Hammerman MR, Avioli LV. Hormonal regulation of $(Ca^{2+} + Mg^{2+})$-ATPase activity in canine renal basolateral membrane. Endocrinology 1986; 119:2405–2410.

34. Hernandez H, Spencer BA, Arsenis G. 1986; Stimulation of Na^+/H^+ exchange by insulin (Ins) and isoproterenol (Iso) in rat adipocytes. Diabetes 1989; 38(suppl 2):182A. Abstract.

35. Gupta MP, Makino N, Khatter K, Dhalla NS. Stimulation of Na^+-Ca^{2+} exchange in heart sarcolemma by insulin. Life Sci 1986;39:1077–1083.

36. Davis FB, Davis PJ, Blas SD, Schoenl M. Action of long-chain fatty acids in vitro on Ca^{2+}-stimulatable, Mg^{2+}-dependent ATPase activity in human red cell membranes. Biochem J 1987; 248:511–516.

37. Hoeing M, Lee RJ, Ferguson DC. Glucose inhibits the high-affinity $(Ca^{2+} + Mg^{2+})$-ATPase in the plasma membrane of a glucose-responsive insulinoma. Biochim Biophys Acta 1990; 1022:333–338.

38. Obando MA, Marin R, Proverbio T, Proverbio F. High sodium diet and Na^+ stimulated ATPase activities in basolateral plasma membranes from rat kidney proximal tubular cells. Biochem Pharmacol 1987; 36:7–11.

39. Smallwood JI, Gugi B, Rasmussen H. Modulation of erythrocyte Ca^{2+} pump activity by protein kinase C. J Biol Chem 1988; 263:2195–2202.

40. Neyses L, Reinlib L, Carafoli E. Phosphorylation of the Ca^{2+}-pumping ATPase of heart sarcolemma and erythrocyte plasma membrane by the cAMP-dependent protein kinase. J Biol Chem 1985; 260:10283–10287.

41. Carafoli E. The calcium pumping ATPase of the plasma membrane. Annu Rev Physiol 1991; 53:531–547.

42. Niggli V, Adunyah ES, Carafoli E. Acidic phospholipids, unsaturated fatty acids, and limited proteolysis mimic the effect of calmodulin on the purified erythrocyte Ca^{2+}-ATPase. J Biol Chem 1981; 256:395–401.

43. Froud RJ, East JM, Jones OT, Lee AG. Effects of lipids and long-chain alkyl derivatives on the activity of $(Ca^{2+} + Mg^{2+})$-ATPase. Biochemistry 1986; 25:7544–7552.

44. Levy J, Suzuki Y, Avioli LV, Grunberger G, Gavin JR III. Plasma membrane phospholipid content in non-insulin-dependent streptozotocin-diabetic rats-effect of insulin. Diabetologia 1988; 31:315–321.

45. Warren GB, Hously MD, Metcalfe JC, Birdsall NJM. Cholestrol is excluded from the phospholipid annulus surrounding an active calcium transport protein. Nature 1975; 255:684–687.
46. Bennett JP, Smith GA, Hously MD, Hesketh TR, Metcalfe JC, Warren GB. The phospholipid headgroup specificity of an ATP-dependent calcium pump. Biochim Biophys Acta 1978; 513:310–320.
47. Ortego A, Mas-Oliva J. Direct regulatory effect of cholestrol on the calmodulin stimulated calcium pump of cardiac sarcolemma. Biochem Biophys Res Commun 1986; 139:868–874.
48. Pershadsingh HA, McDonald JM. Hormone receptor coupling and the molecular mechanism of insulin action in the adipocyte: A paradign for Ca^{2+} homeostasis in the initiation of the insulin induced metabolic cascade. Cell Calcium 1984; 5:111–130.
49. Klip A, Li G, Logan WJ. Role of calcium ions in insulin action on hexose transport in L$_6$ muscle cells. Am J Physiol 1984; 24:E297–E304.
50. Vydelingum N, Kissebah AH, Wynn V. The role of calcium in insulin action. V. Importance of cyclic guanosine 3'5' monophosphate and calcium ions in insulin stimulation of lipoprotein lipase activity and portein synthesis in adipose tissue. Horm Metab Res 1978; 10:38–46.
51. Hope-Gill HR. The effect of calcium on purified human adipose tissue "I" form glycogen synthase: A possible mechanism of hormonal regulation. Horm Metab Res 1976; 8:321–322.
52. Van de Werve G. Ca^{2+}, a newly discovered regulatory of liver microsomal glucose 6-phosphatase activity. Diabetes 1989; 38(suppl 2):42A. Abstract.
53. Ochs R. Does calcium regulate pyruvate kinase? TIBS 1988; 13:2–3.
54. Lawrence JC Jr, Larner J. Effect of insulin, methoxamine, and calcium on glycogen synthase in rat adipocytes. Mol Pharmacol 1978; 14:1079–1091.
55. Bonne D, Belhadu O, Cohen P. Modulation by calcium on the insulin action and of the insulin-like effect of ocytocin on isolated rat adipocytes. Eur J Biochem 1977; 75:101–105.
56. Clausen T, Elbrink J, Martin BR. Insulin controlling calcium distribution in muscle and fat cell. Acta Endocrinol (Copenhagen) 1974; 77(suppl 191):137–143.
57. Saggerson ED, Sooranna SR, Evans CJ. Insulin-like actions of nickel and other transtion-metal ions in rat fat-cells. Biochem J 1976; 154:349–357.
58. Kissebah AH, Tulloch BR, Vydelingum N, Hope-Gill H, Clarke P, Fraser TR. The role of calcium in insulin action. II. Effects of insulin and procaine hydrochloride on lipolysis. Horm Metab Res 1974; 6:357–364.
59. Pierce GN, Kutryk MJB, Dhalla NS. Alterations in Ca^{2+} binding by and composition of the cardiac sarcolemmal membrane in chronic diabetes. Pro Natl Acad Sci USA 1983; 80:5412–5416.
60. Clausen T. The role of calcium in the activation of the glucose transport system. Cell Calcium 1980; 1:311–325.
61. Pershadsingh HA, McDonald JM. Direct addition of insulin inhibits a high affinity Ca^{2+}-ATPase in isolated adipocyte plasma membranes. Nature 1979; 281:495–497.
62. Wong ECC, Sacks DB, Laurino JP, McDonnald JM. Characteristics of calmodulin phosphorylation by the insulin receptor kinase. Endocronology 1988; 123:1830–1836.

63. Sandra A, Fyler DJ. Effects of liposome-adipocyte interaction on calcium binding and insulin action. Endoc Res Commun 1982; 9:107–120.
64. Pershadsigh HA, Shade DL, Delfert DM, Macdonald JM. Chelation of intracellular calcium blocks insulin action in the adipocyte. Proc Natl Acad Sci USA 1987; 84:1025–1029.
65. Levy J, Gavin JR III, Morimoto S, Hammerman MR, Avioli LV. Hormonal regulation of $(Ca^{2+} + Mg^{2+})$-ATPase activity in canine renal basolateral membrane. Endocrinology 1986; 119:2405–2410.
66. Gavin JR III, Lowry M, Levy J. $(Ca^{2+} + Mg^{2+})$-ATPase activity is stimulated by insulin in canine and rat fat membranes. Diabetes 1987; 36(suppl 1):50A. Abstract.
67. Gupta MP, Makino N, Khatter K, Dahalla NS. Stimulation of Na^+–Ca^{2+} exchange in heart sarcolemma by insulin. Life Sci 1986; 39:1077–1083.
68. Hope-Gill HF, Nanada V. Stimulation of calcium ATPase by insulin, glucagon, cyclic AMP and cyclic GMP in triton-X 100 extracts of purified rat liver plasma membrane. Horm Metab Res 1979; 11:698–700.
69. Zierler K. Insulin's block of slow Ca^{2+} conduction requires hydrolyzable GTP. Clin Res 1989; 37:610A. Abstract.
70. Hernandez H, Spencer BA, Arsenis G. Stimulation Na^+/H^+ exchange by insulin (Ins) and isoproterenol (Iso) in rat adipocytes. Diabetes 1989; 38(suppl 2):182A. Abstract.
71. Levy J, Suzuki Y, Avioli LV, Grunberger G, Gavin JR III. Plasma membrane phospholipid content in non-insulin-dependent streptozotocin-diabetic rats-effect of insulin. Diabetologia 1988; 31:315–321.
72. Levy J, Rempinski D. Insulin phosphorylates the membrane $(Ca^{2+} + Mg^{2+})$-ATPase in kidney basolateral membranes. Clin Res 1991; 39(3):698A. Abstract.
73. Levy J, Rempinski D. Inhibition of insulin bioeffects in adipocytes by monoclonal antibodies to the membrane $(Ca^{2+} + Mg^{2+})$-adenosine triphosphatase. Endocrinology Proceeding 74th Annual Meeting. San Antonio, TX: Endocrinology Society 1992; 22. Abstract.
74. Levy J, Avioli LV, Roberts ML, Gavin JR III. $(Na^+ + K^+)$-ATPase activity in kidney basolateral membranes of noninsulin dependent diabetic rats. Biochem Biophys Res Commun 1986; 139:1313–1319.
75. Rosic NK, Standaert ML, Pollst RJ. The mechanism of insulin stimulation of $(Na^+ + K^+)$-ATPase transport activity in muscle. J Biol Chem 1985; 260:6206–6212.
76. DeLuise MA, Harker M. Insulin stimulation of $Na^+ + K^+$ pump in clonal rat osteosarcoma cells. Diabetes 1988; 37:33–37.
77. Draznin B, Sussman K, Kao M, Lewis D, Sherman N. The existence of an optimal range of cytosolic free calcium for insulin-stimulated glucose transport in rat adipocytes. J Biol Chem 1987; 262:14385–14388.
78. Levy J, Rempinski D. A hormone specific defect in insulin regulation of the membrane $(Ca^{2+} + Mg^{2+})$-ATPase in obesity. Clin Res 1989; 37:455. Abstract.
79. Draznin B, Sussman KE, Ekel RH, Kao M, Yost T, Shermann NA. Possible role of cytosolic free calcium concentrations in mediating insulin resistance of obesity and hyperinsulinemia. J Clin Invest 1988; 28:1848–1852.

80. Levy J, Grunberger G, Karl I, Gavin JR III. Effects of food restriction and insulin treatment on (Ca^{2+} + Mg^{2+})-ATPase response to insulin in kidney basolateral membranes of noninsulin-dependent-diabetic rats. Metabolism 1990; 39:25–33.
81. Karl IE, Gavin JR III, Levy J. Effect of insulin on glucose utilization in epitrochleasis muscle of rats with streptozotocin-induced NIDDM. Diabetes 1990; 39:1106–1115.
82. Carnie JA, Smith DG, Marvis-Vavayannis M. Effects of insulin on lipolysis and lipogenesis in adipocytes from genetically obese (ob/ob) mice. Biochem J 1979; 184:107–112.
83. Draznin B. Intracellular calcium, insulin secretion and action. Am J Med 1988; 85(suppl 5A):44–58.
84. Begum N, Sussman KE, Draznin B. High level of cytosolic free calcium inhibit dephosphorylation of insulin receptor and glycogen synthase. Cell Calcium 1991; 12:423–430.
85. Allo SN, Lincoln TM, Wilson GL, Green FJ, Watanabe AM, Schaffer SW. Non-insulin-dependent diabetes-induced defects in cardiac cellular calcium regulation. Am J Physiol 1991; 260:C1165–C1171.
86. Heijnis JB, Mathy MJ, Van Zwieten PA. Effects of various calcium antagonists in isolated perfused hearts from diabetic and age-matched control rats. J Cardiovasc Pharmacol 1991; 17:983–989.
87. Borda E, Pascual J, Wald M, Sterin-Borda L. Hypersensitivity to calcium associated with an increased sarcolemmal Ca^{2+}-ATPase activity in diabetic rat heart. Can J Cardiol 1988; 4:97–101.
88. Ganguly PK, Pierce GN, Dhalla KS, Dhalla SN. Defective sarcoplasmic reticular calcium transport in diabetic cardiomyopathy. Am J Physiol 1983; 244:E528–E535.
89. Makino N, Dhalla KS, Elimban V, Dhalla NS. Sarcolemmal Ca^{2+} transport in streptozotocin-induced diabetic cardiomyopathy in rats. Am J Physiol 1987; 253:E202–E207.
90. Russ M, Reinaver H, Eckel J. Diabetes-induced decrease in the mRNA coding for sarcoplasmic reticulum Ca^{2+}-ATPase in adult rat cardiomyocytes. Biochem Biophys Res Comm 1991; 178:905–912.
91. Black SC, Katz S, NcNeill H. Cardiac performance and plasma lipids of omega-3 fatty acid-treated streptozocin-induced diabetic rats. Diabetes 1989; 38:969–974.
92. Nobe S, Aomine M, Artia M, Ito S, Takaki R. Chronic diabetes mellitus prolongs potential duration of rat ventricular muscles: Circumstantial evidence for impaired Ca^{2+} channel. Cardiovasc Res 1990; 24:381–389.
93. Pierce GN, Dhalla NS. Cardiac myofibrillar ATPase activity in diabetic rats. J Mol Cell Cardiol 1981; 13:1063–1069.
94. Afzal N, Ganguly PK, Dhalla KS, Pierce GN, Singal PK, Dhalla NS. Beneficial effects of verapamil in diabetic cardiomyopathy. Diabetes 1988; 37:936–942.
95. Nagase N, Tamura Y, Kobayashi S, Saito K, Saito M, Niki T, Chikamori K, Mori H. Myocardial disorders of hereditarily diabetic KK mice. J Mol Cell Cardiol 1981; 13(suppl 2):70. Abstract.
96. Dhalla NS, Pierce GN, Innes IR, Beamish RE. Pathogenesis of cardiac dysfunction in diabetes mellitus. Can J Cardiol 1985; 1:263–281.

97. Ohara T, Sussman KE, Draznin B. Effect of diabetes on cytosolic free Ca^{2+} and Na^{+}-K^{+}-ATPase in rat aorta. Diabetes 1991; 40:1560–1563.
98. Agrawal KD, McNeill JH. Vascular responses to agonists in rat mesenteric artery from diabetic rats. Can J Physiol Pharmacol 1987; 65:1484–1490.
99. Piper GM, Gross GJ. Diabetes enhances vasoreactivity to calcium entry blockers. Artery 1989; 16:263–271.
100. Nakagawa M, Kobayashi S, Kimura I, Kimura M. Diabetic state-induced modification of Ca, Mg, Fe, and Zn content of skeletal, cardiac and smooth muscles. Endocrinol Jpn 1989; 36:795–807.
101. Kobayashi S, Fujihara M, Hoshino N, Kimura I, Kimura M. Diabetic state-induced activation of calcium-activated neutral proteinase in mouse skeletal muscle. Endocrinol Jpn 1989; 36:833–844.
102. Taira Y, Hata T, Ganguly PK, Elimban V, Dhalla S. Increased sarcolemmal Ca^{2+} transport activity in skeletal muscle of diabetic rats. Am J Physiol 1991; 260:E626–E632.
103. Ganguly PK, Mathur S, Gupta MP, Beamish RE, Dhalla NS. Calcium pump activity of sarcoplasmic reticulum in diabetic rat skeletal muscle. Am J Physiol 1986; 251:E515–E523.
104. Nishida K, Ohara T, Johnson J, Wallner JS, Wilk J, Sherman N, Kawakami K, Sussman KE, Draznin B. Na^{+}/K^{+}-ATPase activity and its αII subunit gene expression in rat skeletal muscle: Influence of diabetes, fasting and refeeding. Metabolism 1992; 41:56–63.
105. Sahai A, Ganguly PK. Lack of response of (Ca^{2+} + Mg^{2+})-ATPase to atrial natriuretic peptide in basolateral membranes from kidney cortex of chronic diabetic rats. Biochem Biophys Res Commun 1990; 169:537–544.
106. Studer RK, Ganas L. Effect of diabetes on hormone-stimulated and basal hepatocyte calcium metabolism. Endocrinology 1989; 125:2421–2433.
107. Chan KM, Junger KD. The effect of streptozocin-induced diabetes on the plasma membrane calcium uptake activity of rat liver. Diabetes 1984; 33:1072–1077.
108. Zemel MB, Sowers JR, Shehin S, Walsh MF, Levy J. Impaired calcium metabolism associated with hypertension in Zucker obese rats. Metabolism 1990; 39:704–708.
109. Levy J, Sowers JR, Zemel MB. Abnormal Ca^{2+}-ATPase activity in erythrocytes of non-insulin-dependent diabetic rats. Horm Metab Res 1990; 22:136–140.
110. Pierce GN, Afzal N, Kroeger EA, Lockwood MK, Kutryk JB, Eckhert CD, Dhalla NS. Cataract formation is prevented by administration of verapamil to diabetic rats. Endocrinology 1989; 125:730–735.
111. Cowan T, Levy J, Dunbar J. Role of cell calcium metabolism in cataract formation of diabetic rats. Clin Res 1992; 40:239A. Abstract.
112. Levy J, Reid I, Halstad L, Gavin JR III, Avioli LV. Abnormal cell calcium concentration in cultured bone cells obtained from femurs of obese and noninsulin-dependent diabetic rats. Calcif Tissue Int 1989; 44:131–137.
113. Maser RE, Wolfson SK, Ellis D, Stein EA, Drash AL, Becker DJ, Dorman JS, Orchard TJ. Cardiovascular disease and arterial calcification in insulin-dependent diabetes mellitus: Interrelations and risk factor profiles. Pittsburgh Epidemiology of Diabetes Complications Study—V. Arterioscler Thromb 1991; 11:958–965.

114. Fleckenstein A, Frey M, Fleckenstein-Grum G. Antihypertensive and arterial anticalcinotic effects of calcium antagonists. Am J Cardiol 1986; 57:1D–10D.

115. Katz MA, McNeill G. Defective vasodilation response to exercise in cutaneous precapillary vessels in diabetic humans. Diabetes 1987; 36:1386–1396.

116. Baldini P, Incerpi S, Lambert-Gardini S, Spinedi A, Luly P. Membrane lipid alterations and Na$^+$-pumping activity in erythrocytes from IDDM and NIDDM subjects. Diabetes 1989; 38:825–831.

117. Schaefer W, Priben J, Mannhold R, Gries AF. (Ca^{2+} + Mg^{2+})-ATPase activity of human red blood cells in healthy and diabetic volunteers. Klin Wochenschr 1987; 65:17–21.

118. Zemel MB, Bedford BA, Zemel PC, Marwah O, Sowers JR. Altered cation transport in non-insulin dependent diabetic hypertension: Effect of dietary calcium. J Hypertension 1988; 6(suppl 4):S228–S230.

119. Rahmini-Jourdheuil, Mourayre Y, Vague P, Boyer J, Juhan-Vague I. In vivo insulin effect on ATPase activities in erythrocyte membrane from insulin-dependent diabetics. Diabetes 1987; 36:991–995.

120. Ishi H, Umeda F, Hashimoto T, Hawata H. Changes in phosphoinositide turnover, Ca^{2+} mobilization, and protein phosphorylation in platelets from NIDDM patients. Diabetes 1990; 39:1561–1568.

121. Mazzanti L, Rabini RA, Faloia E, Fumelli P, Bertoli E, De-Pirro R. Altered cellular Ca^{2+} and Na$^+$ transport in diabetes mellitus. Diabetes 1990; 39:850–854.

122. Ishii H, Umeda F, Hashimoto T, Nawata H. Increased intracellular calcium mobilization in platelets from patients with type 2 (non-insulin-dependent) diabetes mellitus. Diabetologia 1991; 34:332–336.

123. Tschöpe D, Rösen P, Gries FA. Increase in cytosolic concentration of calcium in platelets of diabetes type II. Thromb Res 1991; 62:421–428.

124. Segal S, Lloyd S, Sherman N, Sussman KE, Draznin B. Postprandial changes in cytosolic free calcium and glucose uptake in adipocytes in obesity and non-insulin dependent diabetes mellitus. Horm Res 1990; 34:39–44.

125. Levy J, Stern Z, Gutman A, Naparstek Y, Gavin JR III, Avioli LV. Plasma calcium and phosphate levels in an adult noninsulin-dependent diabetic population. Calcif Tissue Int 1986; 39:316–318.

126. Sorva A, Tilvis RS. Low serum ionized to total calcium ratio: Association with geriatric diabetes mellitus and with other cardiovascular risk factors? Gerontology 1990; 36:212–216.

127. Pedrazzoni M, Ciotti G, Pioli G, Girasole G, Davioli L, Palummeri E, Passeri M. Osteocalcin levels in diabetic subjects. Calcif Tissue Int 1989; 45:331–336.

128. Levy J, Teitelbaum SL, Gavin JR III, Fausto A, Kurose H, Avioli LV. Bone calcification and calcium homeostasis in rats with non-insulin-dependent diabetes induced by streptozotocin. Diabetes 1985; 34:365–372.

129. Khandelwal RL, Zinman SM, Zebrowski EJ. The effect of streptozotocin-induced diabetes and insulin supplementation on glycogen metabolism in rat liver. Biochem J 1977; 168:541–548.

130. Lee SL, Dhalla NS. Ca^{2+}-channels and adrenoceptors in diabetic skeletal muscle. Biochem Biophys Res Commun 1992; 184:353–358.

131. Kalofoutis A, Lekakis J. Changes of platelet phospholipids in diabetes mellitus. Diabetologia 1981; 21:540–543.
132. Flecha FLG, Bermudez MC, Cedola NV, Gagliardino JJ, Rossi JPFC. Decreased Ca^{2+}-ATPase activity after glycosylation of erythrocyte membranes in vivo and in vitro. Diabetes 1990; 39:707–711.
133. Greene DA, Lattimer SA. Impaired rat sciatic nerve sodium-potassium adenosine triphosphatase in acute streptozotocin diabetes and its correction by dietary myo-inositol supplementation. J Clin Invest 1983; 72:1058–1063.
134. Resnick LM, Gupta RK, Bhargava KK, Gruenspan H, Alderman MH, Laragh JH. Cellular ions in hypertension, diabetes and obesity: A nuclear magnetic resonance spectroscopic study. Hypertension 1991; 17:951–957.
135. Fuijii S, Takemura T, Wada M, Akia T, Okuda K. Magnesium levels in plasma, erythrocyte and urine in patients with diabetes mellitus. Horm Metab Res 1982; 14:161–62.
136. Grafton G, Bunce CM, Sheppard MC, Brown G, Baxter MA. Effect of Mg^{2+} on Na^+-dependent inositol transport. Role for Mg^{2+} in etiology of diabetic complications. Diabetes 1992; 41:35–39.
137. Levy J, Avioli LV, Roberts ML, Gavin JR III. ($Na^+ + K^+$)-ATPase activity in kidney basolateral membranes of non insulin dependent diabetic rats. Biochem Biophys Res Commun 1986; 139:1313–1319.
138. Strazzullo P, Nunziata V, Cirillo M, Biannattasio R, Ferrara LA, Mattioli PL, Mancini M. Abnormalities of Calcium metabolism in essential hypertension. Clin Sci 1983; 65:137–141.
139. Resnick LM, Laragh JH, Sealey JE, Alderman MH. Divalent cations in essential hypertension. Relations between serum Ionized calcium, magnesium and plasma renin activity. N Engl J Med 1983; 309:888–891.
140. Orlov SN, Postnov YV. Ca^{2+} binding and membrane fluidity in essential and renal hypertension. Clin Sci 1982; 63:281–284.
141. Vincenzi FF, Morris CD, Kinnel LB, Kenny M, McCarron DA. Decreased calcium pump adenosine triphosphatase in red cells of hypertensive subjects. Hypertension 1986; 8:1058–1066.
142. Erne P, Bolli P, Burgisser E, Buhler FR. Correlation of platelet calcium with blood pressure. Effect of antihypertensive therapy. N Engl J Med 1984; 310:1084–1088.
143. Wheling M, Theisen K. Altered calcium metabolism in red blood cells of hypertnesives: Persistent marker or sequel of essential hypertension? Klin Wochenschr 1987; 65:769–772.
144. Dominiczak AF, Bohr DF. Cell membrane abnormalities and the regulation of intracellular calcium concentration in hypertension. Clin Sci 1990; 79:415–423.
145. Resnik TJ, Tkachuk VA, Erne P, Buhler FR. Platelet membrane calmodulin-stimulated calcium-adenosine triphosphatase. Altered activity in essential hypertension. Hypertension 1986; 8:159–166.
146. Zemel MB, Kraniak J, Standley PR, Sowers JR. Erythrocyte cation metabolism in salt-sensitive blacks as affected by dietary sodium and calcium. Am J Hypertens 1988; 1:386–392.
147. Sugiyama T, Yoshizumi M, Takaku F, Urabe H, Tsukakoschi M, Kasuya T, Yazaki Y. The elevation of the cytoplasmic calcium ions in vascular smooth muscle cells in SHR. Measurement of the free calcium ions in single living

cells by laser micro fluorospectroscopy. Biochem Biophys Res Commun 1986; 141:340–345.

148. Bruschi G, Bruschi ME, Caroppo M, Orlandini G, Pavarani C, Cavatorta A. Intracellular free [Ca^{2+}]$_i$ in circulating lymphocytes of spontaneously hypertensive rats. Life Sci 1984; 35:535–542.

149. Epstein M, Sowers JR. Diabetes mellitus and hypertension. Hypertension 1992; 19:403–418.

150. Gilbert D'Angelo EK, Singer HA, Rembold CM. Magnesium relaxes arterial smooth muscle by decreasing intracellular Ca^{2+} without changing intracellular Mg^{2+}. J Clin Invest 1992; 89:1988–1994.

151. Postnov YV, Orlov SN, Reznikova MB, Rjazhsky GG, Podukin NI. Calmodulin distribution and Ca^{2+} transport in the erythrocytes of patients with essential hypertension. Clin Sci 1984; 66:459–463.

152. Robinson BF. Altered calcium handling as a couse of primary hypertension. J Hypertension 1984; 2:453–460.

153. De Luise M, Blackburn GL, Flier JS. Reduced activity of the red-cell sodium-potassium pump in human obesity. N Engl J Med 1980; 303:1017–1022.

154. Beutler E, Kuhl W, Sacks P. Sodium-potassium-ATPase activity is influenced by ethnic origin and not by obesity. N Engl J Med 1983; 309:756–760.

155. Pasquali R, Strocchi E, Malini P, Casimirri F. Heterogeneity of the erythrocyte N–K pump status in human obesity. Metabolism 1985; 34:802–807.

156. Erne P, Hermsmeyer K. Intracellular vascular muscle Ca^{2+} modulation in genetic hypertension. Hypertension 1989; 14:145–151.

157. Laasko M, Edelman SV, Brechtel G, Baron AD. Decreased effect of insulin to stimulate skeletal muscle blood flow in obese man: A novel mechanism for insulin resistance. J Clin Invest 1990; 85:1844–1852.

158. Baron AS, Laakso M, Brechtel G, Edelman SV. Mechanism of insulin resistance in insulin-dependent diabetes mellitus: A major role for reduced skeletal muscle blood flow. J Clin Endocrinol Metab 1991; 73:637–643.

159. Baron AD, Laakso M, Brechtel G, Holt B, Watt C, Edelman SV. Reduced postprandial skeletal muscle blood flow contributes to glucose intolerance in human obesity. J Clin Endocrinol Metab 1990; 70:1525–1533.

160. Lillioja S, Young AA, Culter CL. Skeletal muscle capillary density and fiber type are possible determinants of in vivo imsulin resistance in man. J Clin Invest 1987; 80:415–424.

161. Hellman B, Berne C, Grapengiesser E, Grill V, Gylfe E, Lund PE. The cytoplasmic Ca^{2+} response to glucose as an indicator of impairment of the pancreatic beta-cell function. Eur J Clin Invest 1990; (suppl 1):S10–S17.

162. Stickberger AS, Russek LN, Phair RD. Evidence for increased aortic plasma membrane calcium transport caused by experimental atherosclerosis in rabbits. Circ Res 1988; 62:75–80.

163. Ross R, Glomset JA. The pathogenesis of atherosclerosis. N Engl J Med 1976; 295:420–425.

164. Orimo H, Ouchi Y. The role of calcium and magnesium in the development of atherosclerosis. Experimental and clinical evidence. Ann NY Acad Sci 1990; 598:444–457.

165. Blumenthal HT, Lansing AI, Wheeler PA. Calcification of the media of the human aorta and its relation to intimal atherosclerosis, aging, and disease. Am J Pathol 1944; 20:695–679.
166. Toda T, Leszczynski DE, Kummerow FA. The role of 25-hydroxy-vitamin D_3 in the induction of atherosclerosis in swine and rabbit by hypervitaminosis D. Acta Pathol Jpn 1983; 33:37–44.
167. Levy J. Insulin resistance in neonatal NIDDM: Role of abnormal cell Ca^{2+} homeostasis. In: Shafrir E, ed. Frontiers in Diabetes Research: Lessons from Animal Diabetes III. London: Smith-Gordon Press; 1991:567–573.

8
$[Ca^{2+}]_i$ and Contraction of Arterial Smooth Muscle

CHRISTOPHER M. REMBOLD

Most contractile stimuli induce arterial smooth muscle contraction by increasing myoplasmic $[Ca^{2+}]$ ($[Ca^{2+}]_i$). Ca^{2+} binds to calmodulin and activates myosin light chain kinase.[1] The activated myosin light chain kinase phosphorylates the 20 kDa light chain of myosin at serine 19 and activates the myosin's ATPase.[2] The phosphorylated myosin binds to actin filaments, cycles, and produces shortening and/or force.[3] This is the most widely accepted mechanism for the regulation of smooth muscle contraction. In this chapter, I will discuss in detail the mechanisms regulating $[Ca^{2+}]_i$ and the dependence of contractile force of myosin light chain phosphorylation. I will also briefly discuss the mechanisms altering the dependence of myosin phosphorylation on $[Ca^{2+}]_i$ (ie, the $[Ca^{2+}]_i$ sensitivity of phosphorylation).

Regulation of Myoplasmic $[Ca^{2+}]$

Stimuli can regulate $[Ca^{2+}]_i$ either by changes in membrane potential (ie, electromechanical coupling) or independently of changes in membrane potential (ie, pharmacomechanical coupling).[4] The latter includes: (1) receptor-dependent release of intracellular Ca^{2+} by inositol-1,4,5-trisphosphate (IP$_3$); (2) receptor-dependent modulation of Ca^{2+} influx; and (3) modulation of Ca^{2+} efflux. These three mechanisms alter contraction by changing $[Ca^{2+}]_i$ in smooth muscle. In addition, (4) some stimuli can alter the $[Ca^{2+}]_i$ sensitivity of the contractile apparatus independently of changes in $[Ca^{2+}]_i$ (a fourth form of pharmacomechanical coupling).

Potential-Dependent Ca^{2+} Influx

Smooth muscle contracts in response to action potentials or depolarization through activation of L-type Ca^{2+} channels.[5] In the swine carotid, depolarization induced contractions are completely dependent on extracellular Ca^{2+} and are blocked by L-channel blockers.[6,7] Some agents induce relaxation by activating K^+ channels, inducing hyperpolarization, and decreasing Ca^{2+} influx through L channels.[8,9] K^+ channel activity can be directly activated by a number of pharmacologic agents.[8] K^+ channel activity may also be regulated by second messengers such as cGMP.[9]

Receptor-Mediated Intracellular Ca^{2+} Release

In vascular smooth muscle, binding of various contractile agonists (eg, norepinephrine, angiotensin II, histamine) to their specific receptors activates phospholipase C on the inner surface of the plasma membrane through the mediation of a G protein.[10-12] Activated phospholipase C hydrolyses membrane phosphatidylinositol-4,5-bisphosphate (PIP_2), which produces IP_3 and 1,2-diacylglycerol (DAG). IP_3 appears to be the major stimulus for release of Ca^{2+} from non-mitochondrial intracellular stores.[13] In skinned smooth muscle, physiologic concentrations of IP_3 have been shown to cause the release of sufficient Ca^{2+} rapidly enough to account for transient contractions.[14-18] This response can be blocked by intracellular heparin,[19] which, at least in the cerebellum, acts as a competitive blocker of IP_3 receptors.[20] Agonist stimulation increases total IP_3 in smooth muscle cells[21] and tissues.[22-26] Agonists also increased [IP_3] (measured with a receptor-binding assay).[27] However, results from the receptor binding assay are questionable because the estimated [IP_3] levels in tissue homogenates are much higher (10 to 50 µM) than the concentration necessary to release Ca^{2+} from skinned tissues or isolated vesicles (<1 µM).[14] IP_3 does not typically release the entire intracellular Ca^{2+} store (as measured by caffeine-induced intracellular release). In skinned smooth muscle, GTP can release Ca^{2+} from a non-IP_3 sensitive compartment,[28,29] which presumably reflects G protein mediated transfer of Ca^{2+} from a non-IP_3-dependent store to an IP_3-responsive store.[30,31]

Receptor-Mediated Ca^{2+} Influx

Sustained smooth muscle contraction is dependent on extracellular Ca^{2+}.[32-34] In the absence of extracellular Ca^{2+}, agonist-induced contractions quickly relax as the intracellular Ca^{2+} pool is depleted.[35] Thus sustained agonist-induced contractions are associated with sustained increases in Ca^{2+} influx (as measured with $^{45}Ca^{2+}$)[36,37] and with sustained elevations of myoplasmic [Ca^{2+}],[38-41] although this is not a consistent finding.[42,43] There are three potential mechanisms to explain the agonist induced increase in Ca^{2+} influx: (1) the agonist could open "receptor-operated" Ca^{2+} channels and directly increase Ca^{2+} influx;[44] (2) The agonists could activate a nonspecific ion channel, which would induce depolarization and activation of L channels; and (3) the agonists could directly activate L channels without causing changes in membrane potential. There are data supporting all three mechanisms. For example, ATP and carbachol appear to activate nonspecific ion channels in some smooth muscle.[45,46] Histamine increases Ca^{2+} permeability in plasmalemmal membrane fractions of swine aorta (although the proposed ion channel has yet to be characterized).[47] Norepinephrine opens a cation channel in rabbit portal vein, even though the Ca^{2+} conductance is very low.[48] Norepinephrine and angiotensin II increase conductance through L-type Ca^{2+} channels without causing changes in membrane potential.[49,50]

Ca^{2+} Extrusion and Sequestration

Myoplasmic $[Ca^{2+}]$ can also be regulated by mechanisms that remove Ca^{2+} from the cytoplasm, such as the plasma membrane and sarcoplasmic reticulum Ca^{2+} pumps regulated by cAMP and/or cGMP.[51-53] Additionally, $[Ca^{2+}]_i$ may be regulated by the plasma membrane Na^+ gradient, via Na^+/Ca^{2+} exchange.[54] At membrane potential and $[Na^+]_i$ in the physiologic range, small changes in $[Na^+]_i$ may have only a small effect on $[Ca^{2+}]_i$.[55] However, since there is a steep relationship between $[Ca^{2+}]_i$ and force, small changes in $[Ca^{2+}]_i$ could produce large changes in force.[56]

Regulation of Myosin Light Chain Phosphorylation ($[Ca^{2+}]_i$ Sensitivity)

Stimuli can also affect smooth muscle contractile behavior independently of any changes in $[Ca^{2+}]_i$ by modifying the relative dependence of myosin light chain phosphorylation on $[Ca^{2+}]_i$ (ie, the $[Ca^{2+}]_i$ sensitivity of phosphorylation).[38] A detailed discussion of $[Ca^{2+}]_i$ sensitivity is beyond the scope of this chapter. Briefly, it appears to be regulated by two mechanisms: (1) a Ca^{2+}-dependent decrease in the $[Ca^{2+}]_i$ sensitivity of phosphorylation induced by phosphorylation of myosin light chain kinase,[57,58] and (2) an agonist-dependent increase in the $[Ca^{2+}]_i$ sensitivity of phosphorylation that is potentially induced by inhibition of myosin light chain phosphatase.[59]

Regulation of Contractile Force (The Latch Phenomenon)

Activated MLCK phosphorylates the 20 kDa light chain of myosin on serine 19,[2] and this phosphorylation is associated with an increase in the actin-activated myosin ATPase activity. This finding suggested the phosphorylation "switch" hypothesis, in which phosphorylation "turns on" myosin, allowing cross bridge interactions. Such a mechanism is equivalent to Ca^{2+}-dependent troponin C activation of thin filaments in skeletal muscle. Supporting this hypothesis are: (1) the near linear dependence of force on $[Ca^{2+}]$ or phosphorylation in many skinned smooth muscle preparations;[60,61] (2) the correlation between aequorin estimated myoplasmic $[Ca^{2+}]$ and phosphorylation in depolarized intact smooth muscle;[62] (3) the correlation between peak myoplasmic $[Ca^{2+}]$ and the rate of stress development;[63] (4) the finding that isolated smooth muscle cells contract when injected with preactivated, Ca^{2+}-independent MLCK;[64] and (5) the finding that incubation of skinned *taenia coli* with antibodies to MLCK decreased both force and myosin phosphorylation values.[65]

The phosphorylation "switch" hypothesis may explain the initial phase of smooth muscle contraction (this has been called the phasic part of contraction). However, the switch hypothesis suggests that contractile force should always be proportional to the level of myosin light chain phosphorylation. This is not the case. Maximal stimulation of intact smooth muscle is associated with sustained maximal force generation. Despite maximal force, $[Ca^{2+}]_i$,[38,42] myosin phosphorylation,[66] unloaded shortening velocity (V_0, an estimate of cross bridge cycling rates),[67,68] and energy consumption[69] all decrease from their peak to intermediate (yet still suprabasal) values. This has been termed the "latch" phenomenon (high stress with intermediate activation). An acceptable explanation of latch must be consistent with several key observations:

1. There appears to be a hyperbolic relationship between myosin phosphorylation and stress during steady-state conditions with most stimuli. Recent data show that in the swine carotid there is a steep relationship between phosphorylation and stress as phosphorylation increased from resting values of 5–10% to approximately 30%.[38,56,70] Higher levels of phosphorylation do not result in a further increase in force. Earlier studies had reported either a linear relationship[71,72] or high levels of stress with no change in phosphorylation.[68,73-75] More recent data suggest that, in these earlier experiments, the resting level of phosphorylation may have been artifactually high, thus obscuring stimulus-induced changes.[70,76] Several exceptions to this relationship between phosphorylation and steady-state stress have been reported, among them: nitrovasodilators, high extracellular $[Mg^{2+}]$, Mn^{2+}, Ca^{2+} repletion, okadaic acid, or phorbol diesters.[77-83]

2. The unloaded shortening velocity (V_0, a measure of mean cross bridge cycling rates, the biophysical correlate of ATPase activity) is a regulated parameter and is linearly dependent on phosphorylation.[38,67,68] This is substantially different from skeletal muscle in which V_0 is not

regulated. Similar dependence of velocity on phosphorylation has been found with motility assays using smooth muscle proteins.[84]

3. The mechanical characteristics of smooth muscle are similar during high force/high activation and high force/low activation (latch) states.[85] These findings suggest that the force maintaining structure(s) during "latch" are probably cross bridges (or at least mechanically very similar to cross bridges).

4. The release of ADP decreases during sustained contractions.[86]

Several mechanisms have been proposed to explain the latch phenomenon:

1. The thin filament proteins leiotonin, caldesmon, and calponin may regulate the thin filament, potentially in a [Ca^{2+}]$_i$-dependent manner similar to that of skeletal muscle troponin C.[87] The binding of caldesmon to actin inhibits myosin's ATPase[88] and its binding to actin can be inhibited by either Ca^{2+}-calmodulin[89] or Ca^{2+}-dependent phosphorylation.[90,91] However, since these effects appear to require concentrations of [Ca^{2+}]$_i$ higher than that required by myosin light chain kinase, the physiologic significance of these observations is unclear. Calponin also binds to actin and inhibits myosin's ATPase[92,93] and its action can also be inhibited by either Ca^{2+}-calmodulin[94] or Ca^{2+}-dependent phosphorylation.[92] Just like caldesmon, inhibition of calponin by [Ca^{2+}] requires concentrations of [Ca^{2+}]$_i$ that are higher than that required by myosin light chain kinase. One group of investigators found that calponin was not phosphorylated during contractions,[95] while another group found that calponin was phosphorylated only during the initial phase of contraction, not during the sustained phase of contraction.[96] Therefore it seems unlikely that either caldesmon or calponin can regulate "latch."

2. [Ca^{2+}]$_i$-dependent thick filament regulation, distinct from phosphorylation of the myosin light chain, has also been proposed as a possible explanation of the latch phenomenon.[87,97] There is a Ca^{2+}-binding site on the 20 kDa light chain of myosin, which is regulatory in some invertebrate myosins. However, myosin's Ca^{2+}-binding affinity appears to be too low for any physiologic significance.[97] The myosin heavy chain is phosphorylated during contraction. However, the myosin heavy chains are dephosphorylated much too slowly during relaxation to explain the latch phenomenon.[87]

3. Another possibility is that protein kinase C may regulate force.[43] However, direct phosphorylation of myosin at protein kinase C sites has not been observed in physiologically stimulated intact smooth muscle.[98] Furthermore, low-dose phorbol diesters induce contraction by [Ca^{2+}]$_i$-dependent increases in myosin phosphorylation at myosin light chain kinase phosphorylation sites.[99,100] Higher concentrations of phorbol diesters induce myosin phosphorylation at both myosin light chain kinase and protein kinase C sites. The high-dose phorbol diesters also induce

contractions independent of changes in $[Ca^{2+}]_i$.[99,101,103] However, the significance of such supraphysiologic protein kinase C stimulation is unclear.

4. Sustained contractile stress could also be maintained by structures other than cross bridges.[43,103] Small et al.[103] reported the existence of two domains in smooth muscle: a contractile domain containing myosin, actin, and caldesmon, and a cytoskeletal domain containing actin and filamen. Rasmussen et al.[43] suggested that non-myosin cross-links in the cytoskeletal domain could explain latch. However, stress is mechanically similar at both high and low phosphorylation levels (observation 3, above).[85] Therefore any hypothesized cytoskeletal cross-links should be mechanically similar to cross bridges. Such cross-links have yet to be described.

5. A variable myosin ATPase activity has also been proposed to explain the variable V_0 in smooth muscle.[104] However, no specific

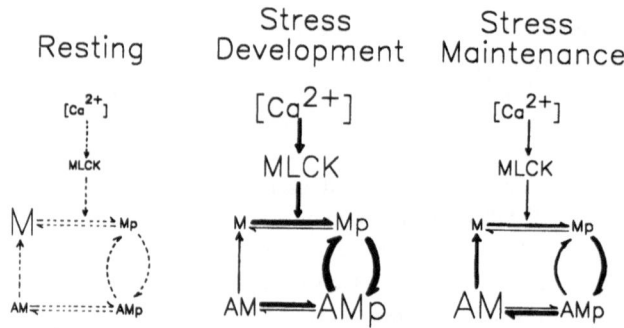

FIGURE 8.1. The latch-bridge model of smooth muscle contraction. The size of the symbols and arrows represent the relative concentration and the relative flux between each cross-bridge species. M = unphosphorylated myosin, Mp = phosphorylated myosin, AMp = phosphorylated myosin attached to actin, AM = unphosphorylated myosin attached to actin (ie, a latch bridge). In an unstimulated muscle, $[Ca^{2+}]_i$ and myosin light chain kinase (MLCK) activity are low, the myosin remains unphosphorylated, and little force is generated because few cross bridges are attached. When $[Ca^{2+}]_i$ is high during the initial phase of contraction, myosin light chain kinase activity is high, most myosin is phosphorylated, and there is rapid cross-bridge cycling (between Mp and AMp) and rapid force development. During the sustained phase of contraction, $[Ca^{2+}]_i$ and myosin phosphorylation decrease. Many of the cross bridges that are being dephosphorylated remain attached (AM) and maintain force at high levels despite the decreasing $[Ca^{2+}]_i$ and myosin phosphorylation. The accumulation of these latch bridges (AM) is responsible for maintained high force with lower $[Ca^{2+}]_i$, myosin phosphorylation, shortening velocity (V_0), and ATP consumption. (From: Murphy RA, Rembold CM, and Hai C-M. Contraction in Smooth Muscle. Alan R. Liss, New York, 1990, with permission.)

regulatory scheme has been suggested. Without multiple regulatory sites, it is difficult to imagine a molecular mechanism for continuously varying the kinetics of individual cross bridges.

None of these five mechanisms offers a satisfactory mechanistic explanation or a quantitative prediction of the contractile behavior of smooth muscle. However, the latch bridge hypothesis of Murphy et al. is consistent with all of the above observations (Figure 8.1). This latch bridge model[67,105] proposes the existence of a latch bridge, that is, an attached dephosphorylated cross bridge formed by dephosphorylation of an attached phosphorylated cross bridge. Dephosphorylation would not result in the loss of force because the latch bridge is proposed to maintain force like a phosphorylated cross bridge. Therefore force can be maintained at higher levels than what would be expected based on the number of phosphorylated cross bridges, hence the hyperbolic relationship between phosphorylation and force. Latch bridges are assumed to be identical to phosphorylated cross bridges, except that they have a fivefold slower detachment rate, accounting for the reduction in V_0 and ATP consumption. We have showed that a four-state cross-bridge model, dependent only on changes in measured [Ca^{2+}]$_i$ with resulting changes in MLCK activity, was sufficient to quantitatively predict most of the resultant contractile response when the stimuli was limited to agonist-induced contractions and cAMP-induced relaxations.[106] This result strongly supports the hypothesis that crossbridge phosphorylation is the primary determinant of force in vascular smooth muscle. However, as noted above, some stimuli change the relationship between myosin phosphorylation and force, which suggests that other mechanisms may also regulate force.

Conclusion

The data discussed in this chapter suggest that contraction and relaxation of arterial smooth muscle are regulated by multiple mechanisms that alter [Ca^{2+}]$_i$. The mechanical response can also be altered by changes in [Ca^{2+}]$_i$ sensitivity of phosphorylation. Myosin phosphorylation appears to be the primary regulator of contractile force.

Acknowledgments. Christopher M. Rembold is a Lucille P. Markey scholar; this work was supported by a grant from the Lucille P. Markey Charitable Trust. In addition our work was supported by the National Institutes of Health Grant RO1-HL38918 and the Virginia Affiliate of the American Heart Association.

144   C.M. Rembold

References

1. Means AR, VanBerkum MFA, Bagchi I, Lu KP, Rasmussen CD. Regulatory functions of calmodulin. Pharmacol Ther 1991; 50:255–270.
2. Ikebe M, Koretz J, Hartshorne DJ. Effects of phosphorylation of light chain residues threonine 18 and serine 19 on the properties and conformation of smooth muscle myosin. J Biol Chem 1988; 263:6432–6437.
3. Hai C-M, Murphy RA. Ca^{2+}, crossbridge phosphorylation, and contraction. Annu Rev Physiol 1989; 51:285–298.
4. Somlyo AV, Somlyo AP. Electromechanical and pharmacomechanical coupling in vascular smooth muscle. J Pharmacol Exp Ther 1968; 159:129–145.
5. Hermsmeyer K, Sturek M, Rusch NJ. Calcium channel modulation by dihydropyridines in vascular smooth muscle. Ann NY Acad Sci 1988; 522:25–31.
6. Rembold CM. Desensitization of swine arterial smooth muscle to transplasmalemmal Ca^{2+} influx. J Physiol (Lond) 1989; 416:273–290.
7. Rembold CM. Relaxation, $[Ca^{2+}]_i$, and the latch-bridge hypothesis in swine arterial smooth muscle. Am J Physiol 1991; 261(Cell Physiol):C41–C50.
8. Nelson MT, Patlak JB, Worley JF, Standen NB. Calcium channels, potassium channels, and voltage dependence of arterial smooth muscle tone. Am J Physiol 1990; 259(Cell Physiol):C3–C18.
9. Thornbury KD, Ward SM, Dalziel HH, Carl A, Westfall DP, Sanders KM. Nitric oxide and nitrosocysteine mimic nonadrenergic, noncholinergic hyperpolarization in canine proximal colon. Am J Physiol 1991; 261(Gastrointest Liver Physiol):G553–G557.
10. Gaul G, Gierschik P, Marme D. Pertussis toxin inhibits angiotensin II-mediated phosphatidyl-inositol breakdown and ADP-ribosylates A 40 KD protein in cultured smooth muscle cells. Biochem Biophys Res Commun 1988; 150(2):841–847.
11. Litosch I, Fain JN. Minireview: Regulation of phosphoinositide breakdown by guanine nucleotides. Life Sci 1986; 39:187–194.
12. Zeng YY, Benishin CG, Pang PKT. Guanine nucleotide binding proteins may modulate gating of calcium channels in vascular smooth muscle. I. Studies with fluoride. J Pharmacol Exp Ther 1989; 250:343–351.
13. Berridge MJ. Inositol trisphosphate and diacylglcerol: Two interacting second messengers. Annu Rev Biochem 1987; 56:159–193.
14. Somlyo AV, Bond M, Somlyo AP, Scarpa A. Inositol trisphosphate-induced calcium release and contraction in vascular smooth muscle. Proc Natl Acad Sci USA 1985; 82:5231–5235.
15. Suematsu E, Hirata M, Hashimoto T, Kuriyama H. Inositol 1,4,5-trisphosphate releases Ca^{2+} from intracellular store sites of skinned single cells of porcine coronary artery. Biochem Biophys Res Commun 1984; 120:481–485.
16. Walker JW, Somlyo AV, Goldman YE, Somlyo AP, Trentham DR. Kinetics of smooth and skeletal muscle activation by laser pulse photolysis of caged inositol 1,4,5-trisphosphate. Nature 1987; 327:249–252.
17. Baron CB, Pring M, Coburn RF. Inositol lipid turnover and compartmentation in canine trachealis smooth muscle. Am J Physiol 1989; 256:C375–C383.

18. Abdel-Latif AA. Calcium-mobilizing receptors, polyphosphoinositides, generation of second messengers and contraction in the mammalian iris smooth muscle: Historical perspectives and current status. Life Sci 1989; 45:757–786.

19. Kobayashi S, Somlyo AV, Somlyo AP. Heparin inhibits the inositol 1,4,5-trisphosphate-dependent, but not the independent, calcium release induced by guanine nucleotide in vascular smooth muscle. Biochem Biophys Res Commun 1988; 153:625–631.

20. Worley PF, Baraban JM, Supattapone S, Wilson VS, Snyder SH. Characterization of inositol triphosphate receptor in brain. Regulation by pH and calcium. J Biol Chem 1987; 262:12132–12136.

21. Griendling KK, Rittenhouse SE, Brock TA, Ekstein LS, Gimbrone MA Jr, Alexander RW. Sustained diacylglycerol formation from inosital phospholipids in angiotensin II-stimulated vascular smooth muscle cells. J Biol Chem 1986; 261(13):5901–5906.

22. Howe PH, Akhtar RA, Naderi S, Abdel-Latif AA. Correlative studies on the effect of carbachol on myo-inositol triphosphate accumulation, myosin light chain phosphorylation and contraction in sphincter smooth muscle of rabbit iris. J Pharmacol Exp Ther 1986; 239:574–583.

23. Long CJ, Stone IW. Adenosine reduces agonist-induced production of inositol phosphates in rat aorta. J Pharm Pharmacol 1987; 39:1010–1014.

24. Takuwa Y, Takuwa N, Rasmussen H. Carbachol induces a rapid and sustained hydrolysis of polyphosphoinositide in bovine tracheal smooth muscle measurements of the mass of polyphosphoinositides, 1,4-diacylglycerol, and phosphatidic acid. J Biol Chem 1986; 261(31):14670–14675.

25. Miller-Hance WC, Miller JR, Wells JN, Stull JT, Kamm KE. Biochemical events associated with activation of smooth muscle contraction. J Biol Chem 1988; 263:13979–13982.

26. Salmon DMW, Bolton TB. Early events in inositol phosphate metabolism in longitudinal smooth muscle from guinea-pig intestine stimulated with carbachol. Biochem J 1988; 254:553–557.

27. Chilvers ER, Challiss RAJ, Barnes PJ, Nahorski SR. Mass changes of inositol(1,4,5)trisphosphate in trachealis muscle following agonist stimulation. Eur J Pharmacol 1989; 164:587–590.

28. Chuch SH, Mullaney JM, Ghosh TK, Zachary AL, Gill DL. GTP and inositol 1,4,5-trisphosphate-activated intracellular calcium movements in neuronal and smooth muscle cell lines. J Biol Chem 1987; 262(No. 28):13857–13864.

29. Kobayashi S, Somlyo AP, Somlyo AV. Guanine nucleotide- and inositol 1,4,5-trisphosphate-induced calcium release in rabbit main pulmonary artery. J Physiol 1988; 403:601–619.

30. Gill DL, Ghosh TK, Mullaney JM. Calcium signalling mechanisms in endoplasmic reticulum activated by inositol 1,4,5-trisphosphate and GTP. Cell Calcium 1989; 10:363–374.

31. Burgoyne RD, Cheek TR, Morgan A, et al. Distribution of two distinct Ca^{2+}-ATPase-like proteins and their relationships to the agonist-sensitive calcium store in adrenal chromaffin cells. Nature 1989; 342:72–74.

32. Deth R, Van Breemen C. Agonist induced release of intracellular Ca^{2+} in the rabbit aorta. J Membr Biol 1977; 30:363–380.

33. Casteels R, Droogmans G. Exchange characteristics of the noradrenaline sensitive calcium store in vascular smooth muscle cells of rabbit ear artery. J Physiol (Lond) 1981; 317:263–279.

34. Ratz P, Murphy RA. Contributions of intracellular and extracellular Ca^{2+} pools to activation of myosin phosphorylation and stress in swine carotid media. Circ Res 1987; 60:410–421.

35. Bozler E. Role of calcium in initiation of activity of smooth muscle. Am J Physiol 1969; 216:671–673.

36. Van Breemen C, Lukeman S, Leijten P, Yamamoto H, Loutzenhiser R. The role of superficial SR in modulating force development induced by Ca entry into arterial smooth muscle. J Cardiovasc Pharmacol 1986; 8:S111–S116.

37. Forder J, Scriabine A, Rasmussen H. Plasma membrane calcium flux, protein kinase C activation and smooth muscle contraction. J Pharmacol Exp Ther 1985; 235:267–273.

38. Rembold CM, Murphy RA. Myoplasmic $[Ca^{2+}]$ determines myosin phosphorylation in agonist-stimulated swine arterial smooth muscle. Circ Res 1988; 63:593–603.

39. Gunst SJ, Bandyopadhyay S. Contractile force and intracellular Ca^{2+} during relaxation of canine tracheal smooth muscle. Am J Physiol 1989; 257:C355–C364.

40. Himpens B, Somlyo AP. Free calcium and force transients during depolarization and pharmacomechanical coupling in guinea-pig smooth muscle. J Physiol 1988; 395:507–530.

41. Himpens B, Matthijs G, Somlyo AV, Butler TM, Somlyo AP. Cytoplasmic free calcium, myosin light chain phosphorylation, and force in phasic and tonic smooth muscle. J Gen Physiol 1988; 92:713–729.

42. Morgan JP, Morgan KG. Stimulus-specific patterns of intracellular calcium levels in smooth muscle of ferret portal vein. J Physiol (Lond) 1984; 351:155–167.

43. Rasmussen H, Takuwa Y, Park S. Protein kinase C in the regulation of smooth muscle contraction. FASEB J 1987; 1:177–185.

44. Bolton TB. Mechanism of action of transmitters and other substances on smooth muscle. Physiol Rev 1979; 59:606–718.

45. Benham CD, Tsien RW. A novel receptor-operated Ca^{2+}-permiable channel activated by ATP in smooth muscle. Nature 1987; 328:275–278.

46. Ganitkevich V, Isenberg G. Isolated guinea pig coronary smooth muscle cells: Acetylcholine induces hyperpolarization due to sarcoplasmic reticulum calcium release activating potassium channels. Circ Res 1990; 67:525–528.

47. Blayney LM, Newby AC. Histamine and a guanine nucleotide increase calcium permeability in pig aortic microsomal fractions. Biochem J 1990; 267:105–109.

48. Byrne NG, Large WA. Membrane ionic mechanisms activated by noradrenaline in cells isolated from rabbit portal vein. J Physiol (Lond) 1988; 404:557–573.

49. Nelson MT, Standen NB, Brayden JE, Worley JF III. Noradrenaline contracts arteries by activating voltage-dependent calcium channels. Nature 1988; 336:382–385.

50. Ohya Y, Sperelakis N. Involvement of a GTP-binding protein in stimulating action of angiotensin II on calcium channels in vascular smooth muscle cells. Circ Res 1991; 68:763–771.
51. Kimura M, Kimura I, Kobayashi S. Relationship between cyclic AMP-dependent protein kinase activation and calcium uptake increase of sarcoplasmic reticulum fraction of hog biliary muscles relaxed by cholecystokinin-C-terminal peptides. Biochem Pharmacol 1982; 31(19):3077–3083.
52. Furakawa K, Tawada Y, Shigekawa M. Regulation of the plasma membrane Ca^{2+} pump by cyclic nucleotides in cultured vascular smooth muscle cells. J Biol Chem 1988; 263:8058–8065.
53. Francis SH, Noblett BD, Todd BW, Wells JN, Corbin JD. Relaxation of vascular and tracheal smooth muscle by cyclic nucleotide analogs that preferentially activate purified cGMP-dependent protein kinase. Mol Pharmacol 1988; 34:506–517.
54. Blaustein MP. Sodium/calcium exchange and the control of contractility in cardiac muscle and vascular smooth muscle. J Cardiovasc Pharmacol 1988; 12(suppl 5):S56–S68.
55. Aaronson PI, Benham CD. Alterations in [Ca^{2+}]$_i$ mediated by sodium-calcium exchange in smooth muscle cells isolated from the guinea-pig ureter. J Physiol (Lond) 1989; 416:1–18.
56. Rembold CM, Murphy RA. Myoplasmic [Ca^{2+}] determines myosin phosphorylation and isometric stress in agonist-stimulated swine arterial smooth muscle. J Cardiovasc Pharmacol 1988; 12(suppl 5):S38–S42.
57. Stull JT, Hsu L-C, Tansey MG, Kamm KE. Myosin light chain kinase phosphorylation in tracheal smooth muscle. J Biol Chem 1990; 265:16683–16690.
58. Gilbert EK, Weaver BA, Rembold CM. Depolarization decreases the [Ca^{2+}]$_i$-sensitivity of myosin light chain kinase in arterial smooth muscle: A comparison of aequorin and Fura-2 [Ca^{2+}] estimates. FASEB J 1991; 5:2593–2599.
59. Kitazawa T, Masuo M, Somlyo AP. G protein-mediated inhibition of myosin light-chain phosphatase in vascular smooth muscle. Proc Natl Acad Sci USA 1991; 88:9307–9310.
60. Filo RS, Bohr DF, Ruegg JC. Glycerinated skeletal and smooth muscle: Calcium and magnesium dependence. Science 1965; 147:1581–1583.
61. Tanner JA, Haeberle JR, Meiss RA. Regulation of glycerinated smooth muscle contraction and relaxation by myosin phosphorylation. Am J Physiol 1988; 255:C34–C42.
62. Rembold CM, Murphy RA. Myoplasmic calcium, myosin phosphorylation, and regulation of the crossbridge cycle in swine arterial smooth muscle. Circ Res 1986; 58:803–815.
63. Rembold CM, Murphy RA. Histamine conentration and Ca^{2+} mobilization in arterial smooth muscle. Am J Physiol 1989; 257(Cell Physiol 26):C122–C128.
64. Itoh T, Ikebe M, Kargacin GJ, Hartshorne DJ, Kemp BE, Fay FS. Effects of modulators of myosin light-chain kinase activity in single smooth muscle cells. Nature 1989; 338:164–167.

65. de Lanerolle P, Strauss JD, Felsen R, Doerman GE, Paul RJ. Effects of antibodies to myosin light chain kinase on contractility and myosin phosphorylation in chemically permeabilized smooth muscle. Circ Res 1991; 68:457–465.

66. Driska SP, Aksoy MO, Murphy RA. Myosin light chain phosphorylation associated with contraction in arterial smooth muscle. Am J Physiol 1981; 240:C222–C233.

67. Dillon PF, Aksoy MO, Driska SP, Murphy RA. Myosin phosphorylation and the cross-bridge cycle in arterial smooth muscle. Science 1981; 211:495–497.

68. Aksoy MO, Mras S, Kamm KE, Murphy RA. Ca^{2+}, cAMP, and changes in myosin phosphorylation during contraction of smooth muscle. Am J Physiol 1983; 245:C255–C270.

69. Krisanda JM, Paul RJ. Phosphagen and metabolite content during contraction in porcine carotid artery. Am J Physiol 1983; 244:385–390.

70. Ratz PH, Hai C-M, Murphy RA. Dependence of stress on cross-bridge phosphorylation in vascular smooth muscle. Am J Physiol 1989; 256:C96–C100.

71. de Lanerolle P, Stull JT. Myosin phosphorylation during contraction and relaxation of tracheal smooth muscle. J Biol Chem 1980; 255:9993–10000.

72. Haeberle JR, Hott JW, Hathaway DR. Regulation of isometric force and isotonic shortening velocity by phosphorylation of the 20,000 dalton myosin light chain of rat uterine smooth muscle. Pfluegers Arch 1985; 403:215–219.

73. Bárány K, Ledvora RF, Mougios, V, Bárány M. Stretch-induced myosin light chain chosphorylation and stretch-release-induced tension development in arterial smooth muscle. J Biol Chem 1985; 260:7126–7130.

74. Silver PJ, Stull JT. Phosphorylation of myosin light chain and phosphorylase in tracheal smooth muscle in response to KCl and carbachol. Mol Pharmacol 1984; 25:267–274.

75. Silver PJ, Stull JT. Regulation of myosin light chain and phosphorylase phosphorylation in tracheal smooth muscle. J Biol Chem 1982; 257:6145–6150.

76. Hai C-M, Murphy RA. Cross-bridge dephosphorylation and relaxation of vascular smooth muscle. Am J Physiol 1989; 256:C282–C287.

77. Hoar PE, Kerrick WG. Mn^{2+} activates skinned smooth muscle cells in the absence of myosin light chain phosphorylation. Pfluegers Arch 1988; 412:225–230.

78. Gerthoffer WT. Dissociation of myosin phosphorylation and active tension during muscarinic stimulation of tracheal smooth muscle. J Pharmacol Exp Ther 1987; 240:8–15.

79. Singer HA. Protein kinase C activation and myosin light chain phosphorylation in ^{32}P-labeled arterial smooth muscle. Am J Physiol 1990; 259(Cell Physiol):C631–C639.

80. Rasmussen H, Haller H, Takuwa Y, Kelley G, Park S. Messenger Ca^{2+}, protein kinase C, and smooth muscle contraction. Prog Clin Biol Res 1990; 327:89–106.

81. Tansey MG, Hori M, Karaki H, Kamm KE, Stull JT. Okadaic acid uncouples myosin light chain phosphorylation and tension in smooth muscle. FEBS Lett 1990; 270:219–221.

82. McDaniel NL, Chen X-L, Singer HA, Murphy RA, Rembold CM. Nitrovasodilators relax arterial smooth muscle by decreasing [Ca^{2+}]$_i$, [Ca^{2+}]$_i$ sensitivity, and uncoupling stress from myosin phosphorylation. Am J Physiol 1992; 263:C461−C467.

83. O'Angello EKG, Singer HA, Rembold CM. Magnesium relaxes arterial smooth muscle by decreasing intracellular [Ca^{2+}] without changing intracellular [Mg^{2+}]. J Clin Invest 1992; 89:1988−1994.

84. Warshaw DM, Desrosiers JM, Work SS, Trybus KM. Smooth muscle myosin cross-bridge interactions modulate actin filament sliding velocity in vitro. J Cell Biol 1990; 111:453−463.

85. Singer HA, Kamm KE, Murphy RA. Estimates of activation in arterial smooth muscle. Am J Physiol 1986; 251:C465−C473.

86. Butler TM, Siegman MJ, Mooers SU, Narayan SR. Myosin-product complex in the resting state and during relaxation of smooth muscle. Am J Physiol 1990; 258(Cell Physiol):C1092−C1099.

87. Kamm KE, Stull JT. Regulation of smooth muscle contractile elements by second messengers. Annu Rev Physiol 1989; 51:299−313.

88. Ngai PK, Walsh MP. Inhibition of smooth muscle actin-activated myosin Mg^{2+}-ATPase activity by caldesmon. J Biol Chem 1984; 259:13656−13659.

89. Sobue K, Muramoto Y, Inui M, Kanda K, Kakiuchi S. Control of actin-myosin interaction of gizzard smooth muscle by calmodulin- and caldesmon-linked flip-flop mechanism. Biomed Res 1982; 2:188−196.

90. Ngai PK, Walsh MP. The effects of phosphorylation of smooth-muscle caldesmon. Biochem J 1987; 244:417−425.

91. Lash JA, Sellers JR, Hathaway DR. The effects of caldesmon on smooth muscle heavy actomeromyosin ATPase activity and binding of heavy meromyosin to actin. J Biol Chem 1986; 261:16155−16160.

92. Winder SJ, Walsh MP. Smooth muscle calponin. Inhibition of actomyosin MgATPase and regulation by phosphorylation. J Biol Chem 1990; 265:10148−10155.

93. Tsunekawa S, Takahashi K, Abe M, Hiwada K, Ozawa K, Murachi T. Calpain proteolysis of free and bound forms of calponin, a troponin T-like protein in smooth muscle. FEBS Lett 1989; 250:493−496.

94. Naka M, Kureishi Y, Muroga Y, Takahashi K, Ito M, Tanaka T. Modulation of smooth muscle calponin by protein kinase C and calmodulin. Biochem Biophys Res Commun 1990; 171:933−937.

95. Bárány M, Rokolya A, Bárány K. Absence of calponin phosphorylation in contracting or resting arterial smooth muscle. FEBS Lett 1991; 279:65−68.

96. Pohl J, Walsh MP, Gerthoffer WT. Calponin and caldesmon phosphorylation in canine tracheal smooth muscle. Biophys J 1991; 59:58a.

97. Ito M, Hartshorne DJ. Phosphorylation of myosin as a regulatory mechanism in smooth muscle. In: Sperilakis N, Wood JD, eds. Frontiers in Smooth Muscle Research. New York: Alan R Liss; 1990:57−72.

98. Colburn JC, Michnoff CH, Hsu LC, Slaughter CA, Kamm KE, Stull JT. Sites phosphorylated in myosin light chain in contracting smooth muscle. J Biol Chem 1988; 263:19166−19173.

99. Rembold CM, Murphy RA. [Ca^{2+}]-dependent myosin phosphorylation in phorbol diester stimulated smooth muscle contraction. Am J Physiol 1988; 255(Cell Physiol):C719−C723.

100. Singer HA, Oren JW, Benscoter HA. Myosin light chain phosphorylation in ^{32}P-labeled rabbit aorta stimulated by phorbol 12,13-dibutyrate and phenylephrine. J Biol Chem 1989; 264:21215–21222.
101. Karaki H. Ca^{2+} localization and sensitivity in vascular smooth muscle. TIPS 1989; 10:320–325.
102. Jiang MJ, Morgan KG. Intracellular calcium levels in phorbol ester-induced contractions of vascular muscle. Am J Physiol 1987; 253:H1365–H1371.
103. Small JV, Furst DO, De Mey J. Localization of filamin in smooth muscle. J Cell Biol 1986; 102:210–220.
104. Siegman MJ, Butler TM, Mooers SU. Energetics and regulation of crossbridge states in mammalian smooth muscle. Experientia 1985; 41:1020–1025.
105. Hai C-M, Murphy RA. Crossbridge phosphorylation and regulation of the latch state in smooth muscle. Am J Physiol 1988; 254:C99–C106.
106. Rembold CM, Murphy RA. The latchbridge model in smooth muscle: [Ca^{2+}] can quantitatively predict stress. Am J Physiol 1990; 259(Cell Physiol):C251–C257.

9
Autoimmunity Against the Nicotinic Acetylcholine Receptor and the Presynaptic Calcium Channel at the Neuromuscular Junction

ANGELO A. MANFREDI, MARIA PIA PROTTI, MATTEO BELLONE, LUCIA MOIOLA, and BIANCA M. CONTI-TRONCONI

Before noon, the stores of the spirits which influenced the muscles being almost spent, they are scarcely able to move hand or foot. . . . this person for some time speaks freely and readily enough, but after long, hasty or laborious speaking, presently she becomes mute as a fish and cannot bring forth a word.

Thomas Willis, 1672[1]

An intriguing aspect of autoimmunity is that frequently only a handful of proteins are involved in the autoimmune responses;[2] several of them are membrane proteins. The neuromuscular junction contains two frequent targets of autoimmunity, the nicotinic acetylcholine receptor (AChR) and the presynaptic ω-conotoxin-sensitive, voltage-operated calcium channel (ω-VOCC). Autoimmunity against AChR causes myasthenia gravis (MG),[3-7] and autoimmunity against ω-VOCC causes the Lambert-Eaton myasthenic syndrome (LEMS), so called because of its clinical similarity with MG.[8,9] The symptoms of both MG and LEMS are due to an impairment of neuromuscular transmission.[3]

While LEMS has a paraneoplastic nature, MG is seldom associated with cancer, although MG patients may have a thymoma. A lot is known about the molecular interactions occurring between the AChR and autoimmune T cells and antibodies, because the sequence of the AChR is known, its molecular structure is reasonably well understood, and purified AChR can easily be obtained for biophysical, biochemical, and immunologic studies. So far very little is known about the details of the molecular interaction between the ω-VOCC and specific autoimmune cells and antibodies, because the ω-VOCC structure is not known. However, the initial steps of LEMS pathogenesis are better understood than those that trigger the anti-AChR autoimmunity in MG. Combined study and comparison of the two diseases may offer useful insights on how tolerance towards a given protein is broken, and how the autoimmune process is sustained.

MG has been the subject of several excellent reviews[3-6] and monographs, such as the Proceeding of the New York Academy of Sciences

Meetings on MG, which are held every five years.[7] Excellent reviews have been recently published also on LEMS.[8,9] Readers interested in a more detailed discussion of the pathology and clinical aspects of MG and LEMS are referred to these sources.

Given the interest of our laboratory in the mechanisms of molecular recognition in autoimmune responses, and the remarkable advances in the elucidation of the molecular structure of the AChR and of its complexes with the "antigen receptors" of the autoimmune cells, the AChR-specific antibodies, and T-cell receptors the main focus of this chapter will be on the "molecular anatomy" of the interactions occurring between the AChR and the autoimmune cells in myasthenic patients. New perspectives provided by these studies toward understanding MG pathogenesis will also be discussed. In the case of LEMS, we will focus on the initial pathogenetic steps, which catalyze a generalized autoimmune response to the ω-VOCC at the neuromuscular junction, and we will discuss how a somewhat similar chain of events might occur in MG. Finally, we will compare the combined mosaic of known pathogenetic steps in MG and LEMS with what is known about the mechanism of induction and maintenance of immune tolerance, in order to assess the validity of proposed models of the origin of autoimmune phenomena.

Myasthenia Gravis and Lambert-Eaton Myasthenic Syndrome: A Brief History

Myasthenia gravis affects one in every 10,000–30,000 persons, and it is therefore a relatively uncommon disease.[5] Its importance is due to the fact that for several years MG was the only autoimmune disorder for which the molecular identity and the amino acid sequence of the autoantigen were known, making it possible to carry out sophisticated molecular studies even in human patients. MG is an excellent model for investigating how an autoimmune response against a complex protein antigen develops, and for developing experimental procedures applicable to the study of other autoimmune diseases that are less amenable to molecular investigations. For example, in 1975–1976 it was demonstrated that MG symptoms could be reproduced in experimental animals injected with immunoglobulin from MG patients.[10,11] About 10 years later, passive transfer of LEMS symptoms by treatment with antibodies from affected patients was used to demonstrate that LEMS is also mediated by autoimmune antibodies, directed against the presynaptic calcium channels that control acetylcholine release.[12] Similarly, the studies summarized below on the epitope repertoire of autoimmune anti-AChR T helper cells and antibodies should prompt similar studies toward identification of the autoimmune epitope repertoire of other autoimmune disease, for which the identity and the amino acid sequence of the main autoantigens are known, such as Graves' disease.[13]

MG was first described by Thomas Willis in 1672.[1] In 1900 Campbell and Bramwell[14] described in detail the clinical symptoms of MG, and in 1901 a relationship between MG and thymus was first noted.[15] In 1913 a case of MG associated with hyperthyroidism (possibly Graves' disease, sometimes associated with MG), which improved after thymectomy was reported.[16] In the thirties, studies on chemical transmission at the neuro-muscular junction,[17] and the observation of similarities between the symptoms of MG and curare poisoning suggested that impairment of the neuromuscular transmission was the functional defect in MG, which led Mary Walker to treat MG patients with anticholinesterase drugs.[18] In the forties, thymectomy became an accepted procedure for treatment of MG.[19] An important step in the understanding of MG pathogenesis was Simpson's insight in 1960 that MG could have an autoimmune origin.[20] By the early seventies there was general consensus that MG involved a defect in neuromuscular transmission but, based on electrophysiologic studies of MG muscles that showed a reduced size of the miniature end plate potentials, a presynaptic abnormality in the synthesis/storage/release of acetylcholine was believed to cause myasthenic symptoms.[21] Crucial for the understanding of MG pathogenesis was the discovery by Patrick and Lindstrom that rabbits immunized with purified AChR developed muscular weakness very similar to that of MG patients.[22] This fundamental study and others that followed[5] clearly demonstrated that MG symptoms can be due to an autoimmune response against muscle AChR. Later, autoimmune AChR-specific antibodies and T cells were detected in the blood of MG patients, which demonstrated that MG is indeed due to an autoimmune anti-AChR response.[3-6]

The disorder of neuromuscular transmission that occurs in LEMS in due to a decreased quantal release of acetylcholine from the motor nerve terminals,[23] which causes muscle weakness and autonomic symptoms. In 60–70% of the cases LEMS is associated with small cell lung carcinoma (SCLC). Association with other cancers, although described, is infrequent.[24,25] Although LEMS is a rare disease, it is one of the most common paraneoplastic neuromuscular disorders: about 1% of SCLC patients develop LEMS symptoms, in most cases, before the SCLC becomes clinically evident.[26,27]

Association of myasthenic symptoms with a bronchial neoplasm was first described in 1953.[27] Three years later Lambert, Eaton, and Rooke identified LEMS as an autonomous entity, distinct from MG.[28] In the following years Elmquist and Lambert demonstrated the presynaptic nature of the disorder.[23] Electrophysiologic alterations characteristics of LEMS can be passively transferred to experimental animals by injecting IgG fractions from LEMS patients,[29] and plasma exchange and immuno-suppressive drugs are beneficial for LEMS symptoms,[30] which suggests a role of autoantibodies in LEMS pathogenesis. More recently, the ω-type VOCC on the presynaptic membrane has been identified as the target of LEMS autoantibodies.[31,32]

The Autoantigen in MG: Structure of the Nicotinic Acetylcholine Receptor

MG is the best understood anti-receptor autoimmune disease, and its autoantigen, the AChR, is the best characterized neurotransmitter receptor.[33–35] This is because the electric organs of *Torpedo* and *Electrophorus* fish are an extraordinary rich source of AChR, and protein ligands of AChR are available (the α-neurotoxins from the venom of elapid snakes) that recognize the AChR with great affinity and specificity,[36] making the purification of large amounts of AChR in active form, amenable to structural and functional studies possible.[33] Because of the highly conserved AChR structure,[34,35] the approaches used and the conclusions reached from studies to *Torpedo* AChR can be extended to AChR derived from other, less rich sources. *Torpedo*, muscle, and neuronal AChRs are part of a superfamily of ligand-gated ion channel proteins, which includes the $GABA_A$ and glycine receptors of the brain.[34,35]

AChRs from peripheral tissues of vertebrates, such as those of mammalian muscle[37] and piscine electric organ,[38] and from neurons[39] are pseudo-symmetric pentameric transmembrane proteins, formed by homologous subunits that originated from a common ancestor.[38] Muscle and electroplax AChRs comsist of four different subunits. α, β, γ (or ε, see below), and δ, in a stoichiometry $\alpha_2\beta\gamma\delta$.[34,35,37,38,40,41] In neuronal tissue most or all AChRs seem to contain only two subunits, in a stoichiometry $\alpha_2\beta_3$.[39] Each subunit is glycosylated, has similar transmembrane structure, and contributes to form a channel specific for monovalent cations.[34,35] The two α subunits contain important constituent elements of the binding sites for cholinergic ligands, some of which are within sequence regions comprising amino acid residues 181–200 and 55–74.[40,41]

Mammalian muscle AChR exists in two forms, whose expression is developmentally regulated.[42,43] Embryonic or denervated muscle expresses AChR formed by α, β, γ, and δ subunits.[42,43] Upon innervation, the synthesis of the γ subunit is turned off and a gene encoding a different, homologous ε subunit is activated in its stead.[42,43] The two muscle AChRs have different functional, pharmacologic, and metabolic properties.[42,44] Corresponding subunits of peripheral AChRs from different tissues and species have a high degree of sequence identity.[34,35] The α subunits are particularly conserved: eg, the α subunits from *Torpedo* electric organ and *Homo sapiens* muscle, two species that separated approximately 400 million years ago, are 80% identical.[45] The structural similarity between mammalian and piscine AChRs has important ramifications for the study of MG, because, at least to some extent, it allowed the use of easily purified *Torpedo* AChR for studies of the anti-human AChR responses in MG patient. Indeed, *E*xperimental *A*utoimmune *M*yasthenia *G*ravis (EAMG) can be induced in mammalians by immunization against *Torpedo* or *Electrophorus* AChRs (see below).

Although the transmembrane folding of the AChR subunits is not known, they all contain four highly hydrophobic segments about 20 residues long, which may form transmembrane α helices, and which are commonly referred to as M1 to M4.[34,35] They are preceded by a long amino-terminal segment (about 200 amino acid residues), which is partially or completely involved in the formation of an extracellular domain.[34,35] The sequence region between the third and the fourth putative transmembrane segments has different length in different subunits, is less conserved, and is partially or completely involved in the formation of a cytoplasmic domain.[34,35] The fourth putative transmembrane segment is followed by a short carboxylic domain, which is rich in hydrophylic residues and must therefore be part of the AChR surface, although it is still unclear whether it has an extracellular or cytoplasmic location.[34,35]

In addition to the alternative embryonic γ and adult ε subunits, two different α subunits have been described in *Xenopus*, which coexist during muscle development.[46] Also two isoforms of the mammalian muscle β subunit exist, due to alternative splicing of the mRNA.[47] The sequence of the α, β, γ (or ε), and δ subunits of AChRs from *Torpedo* electroplax, and mouse and bovine muscle, as well as the sequence of the α, β, γ, and δ subunits of human muscle AChR are known.[34,35]

The Autoantigen in LEMS: Structure and Function of the ω-Conotoxin-Sensitive VOCC

Auto-antibodies in LEMS recognize a presynaptic voltage-operated calcium channel (VOCC),[31] a membrane protein that specifically binds ω-conotoxin (ω-CTX, a 27-residue peptide toxin from the venom of the marine snail *Conus geographus*), which is a potent blocker of acetylcholine release.[48] ω-CTX-sensitive VOCCs (ω-VOCC) have been described in cultured cell lines, in human central nervous system tissues, and in bovine chromaffin cells in the adrenal medulla,[48] as well as in endocrine cells, such as in a rat insulinoma cell line[49] and in SCLC cells.[50] The latter finding may be related to the pathogenesis of LEMS, because LEMS is frequently associated with SCLS.

At the neuromuscular junction, ω-VOCCs are present only on the presynaptic membrane, and are clustered in discrete zones that correspond to the sites of acetylcholine release.[51] Such localization is consistent with the finding that the release of acetylcholine is generally ω-CTX-sensitive.[52] ω-VOCCs may be the effectors of the localized calcium influx that triggers acetylcholine release.[48]

Little is known about the structure of ω-VOCCs. Photoaffinity studies identified several specifically labeled polypeptides in synaptic plasma membrane from rat and bovine brain.[53,54] ω-CTX binds to a site on a 230–240 kDa protein (α_1 subunit).[55,56] Purified ω-VOCC contains four other subunits with molecular weights of 40 (α_2), 110, 70 (β_1), and 60 (β_2) kDa, respectively.[9] Another VOCC, the dihydropyridine-sensitive L-type molecule, has a similar subunit composition, although it does not cross-react with LEMS antisera.[56] Monoclonal antibodies against the β subunit of L-type VOCC immunoprecipitate the ω-VOCC from rabbit brain, whereas monoclonal antibodies (mAb) against the L-type α_1, α_2, and γ subunits do not.[57] The availability of more detailed molecular informations about ω-VOCC structure will be crucial to dissect the pathogenetic mechanisms of LEMS.

The Effectors of Myasthenic Symptoms: Autoantibodies Against Membrane Proteins Involved in the Cholinergic Transmission

The symptoms of MG can be transferred to healthy animals, or reproduced in vitro with nerve/muscle preparations, by administering the IgG fraction from either the sera of MG patients or polyclonal or monoclonal anti-AChR antibodies from animals that developed EAMG.[10,11,58–61] Since no evidence of cytotoxic phenomena has been found in MG patients,[3–6] the only final effectors of MG symptoms seem to be anti-AChR antibodies.

Similarly, mice injected with IgG from LEMS patients develop abnormalities of the quantal release of acetylcholine.[12] Therefore antibodies seem to also be the final effector of the neuromuscular transmission dysfunction in LEMS.

Anti-AChR antibodies from MG patients rarely cause a direct block of the cholinergic binding site.[3–6] Instead, they cause accelerated AChR destruction either by antigenic modulation (accelerated degradation triggered by antibody cross-linking of two nearby AChR molecules), or by complement activation and complement-mediated lysis of the postsynaptic membrane.[3–6] The postsynaptic membrane eventually acquires a simplified structure without the normal postsynaptic folding and is impoverished in AChR.[62] Indeed, the first evidence that suggested a postsynaptic defect in MG was obtained in 1971 by Engel and Santa, from hystometric analysis of the ultrastructure of neuromuscular junctions of MG patients; this demonstrated the altered, simplified structure of the postsynaptic region at MG end plates.[63] Immunoglobulins and complements are present at the neuromuscular junction of MG patients.[64]

Anti-AChR antibodies in MG patients are polyclonal,[3-6] and even antibodies against an individual AChR epitope do not have preferential idiotype(s).[65,66] Therefore MG is not caused be the expansion of a single "forbidden" B cell clone. Although anti-AChR antibodies are the hallmark of MG, there is very poor or no correlation between antibody titer and severity of the disease.[3-6,67,68] This suggests that only particular subpopulations of anti-AChR antibodies have pathogenic potential, perhaps because they are better able to cause antigenic modulation and/or complement activation. In support of this possibility, a study using different mAbs against the AChR α subunit (which dominate the anti-AChR response both in MG and EAMG) demonstrated that few antibodies can cross-link AChR molecules and induce antigen modulation, perhaps because of the geometry of the epitope location on the AChR surface.[69]

MG patients without detectable anti-AChR antibodies have been described.[70,71] Since immunoglobulins from such patients, when injected into mice, caused impairment of neuromuscular transmission without detectable antibodies bound to the mouse AChR, it is possible that these patients may produce pathogenetic antibodies against non-AChR components of the neuromuscular junction.[70]

Studies on the neuromuscular transmission of nerve/muscle preparations from animals injected with LEMS antibodies found a reduced quantal release of acetylcholine.[72,73] Morphologic studies of the synaptic terminals of the neuromuscular junctions of LEMS patients and of mice treated with LEMS IgG first suggested[74] that the neuromuscular dysfunction in LEMS was due to a reduced number and altered structure of the active zone of the presynaptic membrane, where the ω-VOCC are usually clustered in an orderly fashion, and where the quantal release of acetylcholine occurs. Furthermore, IgG antibodies applied to bovine adrenal chromaffin cells uniquely impaired the voltage-dependent calcium channel currents, and intracellular administration of calcium ions restored the ability of the cells to secrete acetylcholine.[75] The demonstration that LEMS sera precipitate presynaptic ω-VOCC labeled with radioactive ω-CTX proved the existence of anti-ω-VOCC antibodies,[31,32] whose functional effects and pathogenetic mechanisms were investigated by in vitro studies of the neuromuscular transmission in diaphragms from mice treated with LEMS antibodies.[76,77] The antibodies do not affect ω-VOCC function by direct block, because when applied in vitro for a short time (3h), they failed to reproduce LEMS abnormalities.[76] When mice were chronically treated with daily injections of LEMS IgG, a long time (several weeks) was necessary both for reduced quantal content and, after antibodies administration was discontinued, for recovery to normal values.[77]

These findings, and the beautiful freeze-fracture images of Engel and coworkers,[78] which indicated a reduced number of ω-VOCC particles in both human LEMS and passively transferred experimental LEMS, suggested that, as in MG, the antibodies achieve their pathogenic effect by causing accelerated destruction of VOCC molecules. Cross-linking of VOCC particles by divalent IgG molecules was first suggested by Engel and coworkers, who observed that the earliest change in LEMS neuro-muscular junction was a reduction in the distance between nearby VOCC particles; this hypothesis was verified by the finding that divalent $F(ab')_2$ fragments were as effective as whole IgG, whereas monovalent Fab' had no effect.[78] Electrophysiologic studies on diaphragm muscles treated with LEMS IgG demonstrated that divalent $F(ab')_2$ fragments induced a decrease in the quantal content of the end-plate potential, although no effect was detected when monovalent antibody fragments were used.[79]

B and T Epitopes on the AChR Molecule

Anti-AChR Antibody Response

Main Immunogenic Region

Most anti-AChR mAbs obtained from rats which were immunized against native AChR, and most antibodies in the sera of rats which developed EAMG, are against a small region on the extracellulr surface of the α subunit; this region was therefore celled *M*ain *I*mmunogenic *R*egion (MIR).[80] The MIR is highly conserved in AChRs from different species, and anti-MIR antibodies are highly cross-reactive.[80,81] Because a single anti-MIR mAb can inhibit an average of approximately 60% of the binding of polyclonal anti-AChR antibodies in the serum of MG patients, it was concluded that this epitope—or this set of largely overlapping epitopes—also dominates the antibody response in human MG.[81] Anti-MIR antibodies have a high pathogenic potential, because they cause AChR loss when added to muscle cell cultures[69] and myasthenic symp-toms when injected into rats.[61] Anti-MIR antibodies are primarily re-sponsible for the ability of the sera from MG patients to cause AChR loss in cell cultures.[82] Although one report has challenged the existence of a true MIR epitope and has proposed that the ability of anti-MIR mAbs to inhibit a large fraction of the anti-AChR antibodies in MG sera is merely due to steric hindrance on binding of antibodies to other nearby epitopes,[83] the general consensus still supports a crucial role of the MIR in the pathogenesis of MG, and the localization of the MIR on the AChR sequence and the elucidation of its structure has been the goal of intense investigations.[84–90]

Sequence Segment α66–72

Important constituent elements of the MIR were believed to be located within residues 67–76 of both human and *Torpedo* AChR α subunits, since synthetic peptides containing this sequence segment consistently bound anti-MIR mAbs.[84,86] Another group of investigators attempted MIR localization with the use of biosynthetic α subunit fragments; they restricted MIR localization to residues 61–70 of the *Torpedo* AChR α subunit.[87] Yet another group, which also utilized synthetic peptides, restricted MIR localization to a small synthetic fragment that corresponded to residue 66–76 of the human muscle α subunit.[88] Because of the highly conserved structure of the MIR, anti-MIR antibodies frequently cross-react among AChRs of different species.[80,81] Anti-MIR mAbs recognized synthetic peptides containing the sequence α67–76 from different animal species with the same specificity shown for the native form of the corresponding AChR. This observation strongly supports the notion that the segment α67–76 indeed contains a crucial constituent loop of the MIR. The binding characteristics of different anti-MIR mAbs to synthetic peptides that contain the sequence α67–76 plus the different lengths of the flanking sequences indicate that the MIR contains different, largely overlapping epitopes.[87]

Since the MIR antibodies cross-react with synthetic peptides that contain residues α67–76, peptide analogs containing single residue substitution of the native sequence can be used to identify the amino acid residues necessary for the binding of MIR antibodies.[86] Several amino acid residues important for the binding of MIR antibodies were identified, in agreement with the accepted view that antibody epitopes include a relatively large area of protein antigens, formed by several amino acid residues, and that antibody binding occurs through a multipoint attachment.[91] All anti-MIR mAbs that were tested require residues Asn_{68} and Pro_{69} for binding. Five of them require also Asp_{71} and Tyr_{72}. Substitutions of Asp_{70} of the human sequence with an Ala residue, as in *Torpedo* AChR, affected the binding of anti-human MIR mAbs, while it was irrelevant for the binding of other anti-MIR mAbs. These other anti-MIR mAbs, although they cross-reacted with human AChR, were originally raised against, and preferentially recognized, piscine AChR. Substitution of Tyr_{67} moderately reduced the binding of some anti-MIR mAbs, while residues at positions 73, 74, 75, and 76 were of little or no importance for anti-MIR mAb binding. Because the free amino terminus of peptide α67–76 was essential for binding of most anti-MIR mAbs,[86] the side chain of the Lys residue at position 66 of the human AChR α subunit may be involved in formation of the MIR. The ε-amino group of Lys_{66} could be substituted by the amino terminus of the synthetic peptide α67–76.

All together these results restrict the residues most crucial for MIR formation to the sequence region 66–72 of the AChR α subunit. Studies on anti-MIR binding to *Torpedo* AChR mutated at residues 68 and 71 of the α subunit and expressed in *Xenopus* oocytes confirmed that the residues at positions 68 and 71 are critical to the antigenicity of the intact AChR.[89] In frog muscle AChR—the only known muscle AChR not recognized by anti-MIR antibodies[46,89]—residues Asn_{68} and Asp_{71} are non-conservatively substituted,[46,92] while they are conserved in all other known muscle AChRs. Furthermore, in a second type of α subunit expressed by frog muscle, which contributes to formation of an AChR recognized by anti-MIR antibodies, the sequence region containing the MIR and the crucial residues described above are conserved.[46]

Anti-AChR T Helper Response

Anti-AChR T Helper Cells Control Antibody Synthesis

The production of anti-AChR antibodies requires the intervention of specific T helper (Th) cells.[93,94] An important role of anti-AChR Th cells in the pathogenesis of MG, and of its experimental model, EAMG, is supported by the findings that EAMG can be prevented or ameliorated by treatment with anti-Ia or anti-CD4 antibodies,[95,96] and that the only obvious and early effect of thymectomy—a staple in MG treatment—on the anti-AChR response is an immediate and pronounced decrease in anti-AChR reactivity of circulating T cells.[97] Anti-AChR Th (CD4[+]) cells are present in the blood and thymus of MG patients, and they can be propagated in vitro from these tissues.[98–105] They stimulate production of anti-AChR antibodies in vitro[94] and recognize the AChR in association with HLA-DR molecules.[106] Unlike B cells, which recognize structural features of the soluble, native form of the protein antigens, Th cells recognize denatured antigens in association with MHC (HLA) class II molecules on the surface of antigen-presenting cells (APC).[107,108] AChR-specific Th (CD4[+]) cells from the blood of MG patients recognize equally well native or fully denatured AChR.[98] Therefore synthetic peptides that correspond to different subunits of the human AChR, whose sequence is known, can be used to propagate these autoimmune T cells in vitro, and to define the epitopes they recognize on the AChR molecule (see below).

The usefulness of synthetic peptides that correspond to AChR sequences was verified by studies which tested the response of unselected CD4[+] cells from the blood of MG patients to a pool of synthetic peptides that corresponded to the complete human muscle AChR subunits: a response to AChR peptides could only be detected in the more severely affected patients, and only after depletion of the CD8[+] cells.[109] Large amounts of activated CD8[+] T cells were present in the blood of MG patients, particularly during phases of improvement of their symptoms.[109]

Therefore CD8$^+$ cells able to exert an immunosuppressive influence on the anti-AChR response of autoreactive Th cells exist in MG patients, and their presence seems to be related to the improvement of the symptoms. The level of anti-AChR reactivity of the CD4$^+$ T cells correlated with the severity of the disease, a relationship particularly clear in longitudinal studies of MG patients.[105,109]

The results summarized above may explain the low and inconsistent T response to AChR antigens, when the total peripheral blood mononucleate cell (PBMC) population, inclusive of the CD8$^+$ T cells subset, was used for study. Furthermore, several of these studies used AChRs from fish electric organ, or non-human skeletal muscle.[93,94,98–100,110] In one study human muscles from amputated legs were used.[111] PBMC from MG patients, tested in vitro with AChR purified from these sources, responded inconsistently and weakly. In addition to the negative influences exerted by CD8$^+$ cells, the inconsistent and weak responses may have been caused by the use of antigens different from human AChR or, when human AChR was used, of an antigen that could be obtained only in proteolyzed form and in amounts insufficient for controlled, optimized studies.

Epitope Repertoire of Anti-AChR T Helper Cells

An important issue in the biology of the anti-AChR autoimmune response is the identification of the AChR sequence regions that form epitopes recognized by the autoimmune Th cells. PBMC from MG patients have been tested in vitro with synthetic peptides sequences of the human AChR, to identify sequence regins forming Th epitopes.[100,105,110,112,113] In one study, which used synthetic peptides that corresponded to a small part of the human α subunit sequence, only the synthetic sequence α252–269 induced a specific response of PBMC from 6 of the 34 MG patients tested.[100] T cell lines propagated with this peptide did not cross-react with *Torpedo* AChR.[100] In a more recent study an equimolar mixture of 18 overlapping synthetic peptides that corresponded to the putative extracellular domain of the human α subunit (residues 1–210) was used to propagate short-term T cell lines in sufficient amounts to test the profile of peptide recognition by the T cells.[113] The pattern of peptide recognition was different among the five T cell lines tested, as was expected from the known MHC restriction, since they were from patients of different MHC haplotype. However, a peptide, α146–162, was recognized by three of the five lines, and the peptide α56–71 was recognized by two lines, which suggests that at least some of the T epitopes recognized in MG may be immunodominant and be recognized in the context of different restricting DR molecules. In spite of the short propagation of the lines used in the study summarized above,[113] some bias in the selection of the T cells may have occurred, and some epitopes may have been undetected because they were poorly represented by the panel of peptides used, or because of clonal bias during the propagation of the lines. Furthermore, other epitopes may exist on the carboxyl-terminal half of the α subunit sequence, or on other AChR subunits.

To obtain a more comprehensive assessment of the epitope repertoire of anti-AChR autoimmune Th cells in MG patients, we tested CD8+-depleted, unselected and/or total PBMC CD4+ cells from the blood of a relatively large population of MG patients (22 patients) against individual overlapping synthetic peptides (20 residues long) by screening the complete sequence of the human AChR α, γ, and δ subunits[105,114] (Manfredi, et al., unpublished data). Several patients were tested twice, when they suffered from severe symptoms and during periods of remission. In agreement with our previous studies on the response to the AChR α subunit of CD8+-depleted, CD4+-enriched cells,[109] only the most severely affected patients recognized AChR epitopes, and they were mainly young women. The extent of the CD4+ cell response was facilitated by removal of the CD8+ cells. Several epitopes were identified on each AChR subunit[105,114] (Manfredi et al., unpublished data). The peptide epitopes identified within the amino-terminal half of the AChR α subunit (residues 1–210) agreed very well with those identified by Atassi and coworkers.[113] Although each patient had a unique pattern of peptide recognition, as expected from the different HLA-DR haplotype of the patients tested, four sequence regions of the α subunit (residues 48, 67, 101–137, 293–337, 308–437) were frequently recognized, even by patients of different DR haplotype. The other AChR subunits tested also contain several epitopes, some of which were recognized by a large fraction of the patients, irrespective of their DR haplotype.

CD4+ T cells from most MG patients recognize epitopes on the embryonic AChR γ subunit. This finding has interesting ramifications for the pathogenesis of MG, as will be discussed below, because the γ subunit is only expressed during the embryonic life and it is not present in the AChR of adult muscle.

Propagation of Long-Term, AChR-Specific Th Cell Lines

Long-term, AChR-specific Th cell lines can be propagated in vitro from the blood or thymus of MG patients by cycles of stimulation with AChR antigens, such as purified AChR or synthetic or biosynthetic AChR sequences. Th lines are very useful for extensive testing of anti-AChR CD4+ cells from MG patients, and for studies aimed at structural definition of the autoimmune AChR epitopes and the autoimmune antigen T cell receptors. They are highly enriched anti-AChR T cells, and in principle they could be available in unlimited amounts. They can be homogeneously specific for the AChR, or for one specific epitope. Furthermore, monoclonal lines can be obtained from polyclonal antigen-specific T cell lines, which will be necessary for structural and genetic studies of the autoimmune T cell receptors and their complexes with AChR epitopes.

Successful propagation of AChR-specific, long-term CD4$^+$ cell lines was first obtained via cycles of stimulation with *Torpedo* AChR.[93] The CD4$^+$ T cell lines thus obtained recognized AChR epitopes in a DR-restricted fashion,[106] were functional Th cells since they could stimulate the in vitro production of anti-AChR antibodies by B cells from MG patients,[94] and in most cases preferentially or exclusively recognized epitopes on the AChR α subunit.[98] This preferential recognition may occur because the α subunit is conserved more between *Torpedo* and human AChR than with the other AChR subunits,[34,35] and the use of *Torpedo* AChR may selectively propagate CD4$^+$ cells against conserved epitopes on the α subunit. In addition to the limited and biased repertoire of CD4$^+$ T cell lines propagated by stimulation with *Torpedo* AChR, the use of this antigen did not yield the large amount of T cells necessary for the large-scale testing involved in identification of their epitope repertoire, possibly due to clonal losses of the Th cells against epitopes formed by sequence regions not conserved in *Torpedo* and human AChRs. Only limited investigations could be carried out on the repertoire of two cell lines propagated with *Torpedo* AChR. One of them recognized epitope(s) formed by overlapping peptides Hα7–22 and Hα9–34, ie, close to the amino terminus of the α subunit.[115]

The shortcomings described above were overcome by the use of pools of synthetic peptides that corresponded to the complete sequence of human AChR subunits as stimulating antigens. Pools of synthetic peptides that corresponded to the complete sequence of the human α, β, γ, and δ subunits[101–104] (Moiola, Karachiunski, Howard, and Conti-Tronconi, unpublished data) were used to obtain several long-term, AChR-specific CD4$^+$ T cell lines from patients of different HLA haplotype. All lines could be grown in adequate amounts for testing with each individual peptide screening the complete sequence of the AChR subunit under study[101–104,106] (Moiola, Karachiunski, Howard, and Conti-Tronconi, unpublished data). In agreement with the results obtained using unselected CD4$^+$ cells from the patient's blood, each line had an individual profile of peptide recognition. However, immunodominant sequence regions were identified on the α and γ subunits and formed epitopes recognized by most or all lines, irrespective of their HLA-DR type[102,116] (Protti et al., unpublished data). These immunodominant regions are the same ones identified with unselected CD4$^+$ cells. Because the peptides used were 20 residues long, and T epitopes may be as short as 5–9 residues, each immunodominant synthetic sequence could accommodate several nested epitopes, as described for immunodominant regions of myelin basic protein.[117] The immunodominance of particular sequence regions might be due to their easier cleavage and release during processing of the antigen, or their folding into a conformation more agreeable for DR binding. Alternatively, a single truly immunodominant T epitope might exist, recognized in the context of different restricting DR elements, as described for other antigens.[118-120] Both alternatives—immunodominant regions or immunodominant epitopes—might occur for antigen recognition by human Th cells because of the promiscuous nature of the binding of human DR molecules to peptide epitopes; most of all human DR molecules bind a wide spectrum of peptide sequences, and about 25% of the peptide sequences tested bind to any DR molecule.[121]

Other studies have used long-term propagation of AChR-specific T cells. Harcourt and coworkers[100] propagated T cell lines with the synthetic sequence α257–269, which was specifically recognized by PBMC of MG patients. Melms and colleagues[110] propagated three Th cell lines from different MG patients using a fusion protein that corresponded to a large part of the mouse muscle AChR α subunit. The lines obtained recognized a major epitope within residues α85–142. This sequence region include one of the immunodominant sequence regions identified in one of our studies, ie, region α101–120.[109] In a third study T cell lines were raised using a large biosynthetic fragment of the human α subunit (residues α37–437), but the epitopes recognized by these lines could not be identified.[122]

Anti-AChR Th Cells in Healthy Subjects

Th cells against autoantigens have been demonstrated in normal subjects.[123-125] PBMC from normal subjects may respond to synthetic sequence segments of the AChR α subunit,[100] although they did not respond to the complete AChR molecule.[109,126] AChR-specific T cell clones can be obtained from healthy subjects using a biosynthetic fragment of mouse AChR α subunit, but much less frequently than from MG patients.[127] Propagation of specific Th cell lines from healthy controls using pools of synthetic peptides that corresponded to the complete sequence of human AChR subunits was seldom successful[101-104,116] (Protti et al., unpublished data). A CD4$^+$ δ subunit specific line was propagated for a short time from a normal subject[103] and CD4$^+$ cell lines specific for the γ subunit could be propagated for a short time from three of six healthy controls[104,116] (Prott et al., unpublished data). One of the latter responded detectably to the γ pool, but not to any individual peptide, which suggests that although CD4$^+$ T cells specific for the AChR γ subunit existed in this subject, they interacted with γ subunit epitopes with too low affinity for any one peptide epitope to be clearly recognized in vitro. The other two γ-specific lines recognized some individual peptides, but could not be propagated for longer than 4–5 weeks, a fact consistent with inefficient recognition of their epitopes.[116] Therefore, although potentially autoreactive CD4$^+$ T cells that recognize AChR sequences may exist in healthy subjects, Th cells that recognize AChR epitopes with high affinity, which can be propagated in vitro for long periods of time by stimulation with AChR antigens, are uniquely characteristic of MG patients.

The results summarized above clearly indicate that potential anti-AChR autoimmune Th cells exist in the healthy population and that, as Barkas and coworkers pointed out, "immunoregulatory imbalance rather than the presence of autoreactive T cell clones is responsible for the pathogenesis of MG."[127] Autoreactive T cells have been demonstrated in many experimental systems where the host does not normally develop autoimmune disease,[128] including EAMG.[129] BALB/c mice develop EAMG upon injection of rat AChR, which is presumably very similar, if not identical, to the autologous murine AChR,[130] and in this mouse strain $CD4^+$ T cells exist, which recognize the murine sequence region α304–322.[129] The presence of $CD4^+$ cells reactive to self antigens does not indicate a failure of tolerance, because the many autoreactive cells that are deleted from the T cell repertoire during maturation within the thymus are probably those with optimal (highest) affinity for the self antigen/MHC complex.[128] T cells with low affinity antigen receptors may escape clonal deletion,[128,131] but because of the similarity between the mechanisms of thymic clonal deletion and of peripheral T cell activation, both of which require optimal formation of the ternary complex between the T cell antigen receptor (TCR), the MHC restricting element, and the T epitope, the autoreactive T cells may never be activated at the periphery due to the low affinity of their TCRs.[128,131] The potentially autoreactive T cells can be activated under experimental conditions and cause development of autoimmune responses.

The MHC Haplotype in EAMG

EAMG can be induced in vertebrates by immunization with AChR purified from *Torpedo* electroplax (TAChR) or vertebrate muscle, with production of anti-AChR antibodies and sensitization of AChR-specific Th cells.[5] In congenic mice strains, susceptibility to EAMG correlates with their MHC class II (H-2) haplotype.[132–135] Because highly and less susceptible strains develop similar anti-AChR antibody titers,[129,132] propensity to EAMG may be related to MHC restriction of different Th epitopes on the AChR, which in turn may bias the epitope repertoire of the anti-AChR antibodies. The influence of the MHC haplotype on T helper (Th) cells epitope repertoire can be easily studied in congenic murine EAMG, while in human MG the relationship between recognition of particular Th epitopes and the severity of the disease cannot be inferred because of inter-individual variability and changes in Th response induced by immunosuppressive therapy.[105,109,134]

Overlapping synthetic peptides that corresponded to the complete sequence of the α subunits from *Torpedo* and murine AChRs were used to map Th epitopes in congenic murine strains of different H-2 haplotype.[129] Upon TAChR immunization different sequence segments of the TAChR α subunit were recognized by CD4$^+$ T helper[136] cells from strains of H-2^b (C57BL/6 and BALB/b) and H-2^d (BALB/c and CB17) haplotype. The H-2^b strains recognized epitopes formed by the sequence regions α150–169, and to a lesser extent α360–378. Th cells from C57BL/ 6 mice also responded to peptide α181–200. H-2^d strains recognized peptides α1–20 and α304–322, which are virtually identical in *Torpedo* and murine AChRs. An excellent correlation exists between CD4$^+$ epitope sequences and H-2 haplotype of the strain, and very few sequence segments of the AChR α subunit form immunodominant Th epitopes in any strain; this is in agreement with other studies on Th response to large protein antigens,[137] which demonstrated how the T response of a given individual or inbred mouse strain focuses on a limited number of sequence regions.

Another studies investigated the response of lymphocytes from several mice strains, immunized with TAChR, to synthetic peptides that corresponded to the proposed extracellular domain of TAChR α subunit (residues α1–210).[138] C57BL/6 mice recognized peptides α148–160 and α184–196,[138] in excellent agreement with the studies summarized above.[129] The other strains had individual patterns of peptide recognition, but some peptides were recognized by several strains (ie, α148–160 and α69–80).[138] On the other hand, the T repertoire of mice of H-2^b and H-2^d haplotype did not overlap.[129]

The H-2^b and H-2^d class II molecules may have extreme, opposite binding preferences for structural or sequence patterns of peptide epitopes. Other studies on the H-2 restriction of T epitopes on protein antigens consistently found that H-2^b and H-2^d strains recognized different epitopes on the same antigen, and that H-2^d class II molecule preferred peptides with an α-helical structure,[137,139,140] while the H-2^b molecule restricted epitopes which may[137] or may not[139] be in α-helical configuration. The sequence regions of the *Torpedo* AChR α subunit that formed the two epitopes recognized by the H-2^d strains had the highest predicted propensity to form amphipatic α helices,[129] which are proposed to be a frequent feature of T epitopes,[141] while the sequence regions that formed epitopes for the H-2^b strains were predicted to form a non-helical structure.[129] Studies on the secondary structure of these peptides in solution confirmed the structural predictions: The H-2^d restricted peptides had a high α-helical content, while the three H-2^b restricted peptides had a substantial β-pleated component.[142] Although the secondary structure in solution of T epitope peptides may differ from the conformation assumed upon binding to the MCH molecule, these results suggest a preference of the H-2^d haplotype for epitopes in α-helical configuration.

Another study investigated the response to selected peptide sequences of the α subunit from lymphocytes of rats immunized with TAChR:[143] different sequence segments were recognized by different strains.

Within the H-2 locus, the products of the I-A subregion are crucial for development of anti-AChR response and EAMG.[133,134,144] Anti-I-A antibodies affect the anti-AChR response in vivo and in vitro.[133,144,145] The I-A^{bm12} mutation, which involves three amino acid residues at position 67, 70, and 71 of the β subunit of the I-A^b molecule,[146-148] converts the highly susceptible C57BL/6 strain into the B6 C-H-2^{bm12} strain, which is resistant to EAMG induction, although it develops T-cell-mediated antibody responses to AChR.[134] Mutated bm12 mice lose the ability to respond to some protein antigens, whereas the response to other antigens is not affected.[149-151] F1 mice derived from breeding the H-2^{bm12} mutation into NZB mice spontaneously secrete anti-double stranded (ds) DNA auto-antibodies, whereas NZB(H-2^d) or NZB.H-2^b strains do not develop anti-dsDNA even after immunization with dsDNA.[152] The bm12 mutation therefore involves an I-A region with important immunoregulatory influence.

To further elucidate the relationship between MHC class II haplotype and Th repertoire, we investigated the bm12 mutation effect on CD4$^+$ response to AChR α subunit in C57BL/6 mice.[153] Mutant bm12 mice, immunized with *Torpedo* AChR, develop CD4$^+$ sensitization to AChR epitopes and anti-AChR antibodies, like the parental C57BL/6 strain. However, when overlapping synthetic peptides that corresponded to the complete sequence of the TAChR α subunit were used to identify their Th epitope repertoire, CD4$^+$ cells from bm12 mice did not respond to any synthetic sequence, while mice of the parental C56BL/6 strain strongly recognized three immunodominant peptide epitopes, α150-169, α181-200, and α360-378. A CD4$^+$ sensitization against peptides α181-200 and α360-378 could be induced in bm12 mice by injection with individual synthetic peptide epitopes, while CD4$^+$ could not be sensitized against peptide α150-169 under any circumstance. Infante and coworkers[154] also investigated the effect of the bm12 mutation on recognition of AChR epitopes by Th cells. They too concluded that residues within the sequence Tα148-152 form a disease-related T epitope, whose recognition is strongly influenced by the bm12 mutation. Resistance to EAMG induced by the bm12 mutation may be due to change in the epitope repertoire of AChR-specific Th cells, with a lack of recognition of otherwise immunodominant Th epitopes. For one epitope, this might be due to an absence of potentially reactive specific T clones.[153]

Pathogenetic Mechanisms in MG

Sequence Similarities Between the MIR and Other AChR T and B Epitopes

It has been proposed that immunologic cross-reactivity between viral or bacterial antigens and normal protein constituents of the host may be involved in development of autoimmune diseases, because microbial proteins and human autoantigens contain identical or similar sequence segments that may form cross-reacting epitopes.[155] Oldstone and coworkers[156] identified a sequence segment of the α subunit of human muscle AChR (residues 160–167) that shares structural similarities with a sequence region (residues 286–293) of the glycoprotein D of herpes simplex virus. Sera from approximately 15% of the MG patients tested recognized a synthetic peptide that contained the sequence region α160–167, and the sera could be purified by affinity chromatography that used a synthetic sequence α157–170 of human AChR. The affinity-purified anti-AChR antibodies thus obtained recognized the native AChR, and cross-reacted with herpes simplex virus.

Comparison of the sequence regions that contained the MIR with any other known protein sequences[157] yielded unexpected and curious similarities between the sequence segment α65–80 of human AChR and (1) the U1 small nuclear ribonucleoprotein (U1SNR), which is the marker autoantigen for two other autoimmune diseases, mixed connective tissue disease and systemic lupus erythematosus; and (2) the pol polyprotein of several human retroviruses. Eight residues (α72–80) are identical or conservatively substituted in both the MIR and U1SNR. On the amino-terminal side of this segment, other conserved residues and conservative substitutions are interspersed with non-conserved positions. The similarity of α65–80 with a sequence segment from retroviral pol polyprotein was less striking but still intriguing.

These sequence similarities do not necessarily indicate immunologic cross-reactivity, because identical six-residue sequence segments in unrelated proteins may have a different secondary structure.[158] However, it is intriguing that the epitope(s) that dominates the anti-AChR antibody response in MG bears such remarkable similarities with the marker autoantigen for two other autoimmune diseases. In this respect it should be noted that antibodies against autoantigens other than the AChR are also produced in myasthenic patients, including anti-nuclear antibodies in 20–40% of cases.[159] Furthermore MG and lupus share a high frequency of the same HLA haplotype and polymorphism in the α and β genes of the T cell receptor, and the two diseases can be found in the same patient.[160] Finally, the AChR δ subunit shares a six-residue sequence segment with DNA tropoisomerase I (the most characteristic autoantigen in diffuse systemic sclerosis) and the P30[gag] retroviral protein.[161] It is conceivable that these different autoimmune diseases start with a similar broad spectrum of autoreactivity against multiple antigens, including nuclear antigens, and in MG the response becomes focused toward the AChR and the MIR because of the structural similarity reported here. A link between the AChR and retroviral antigens is more shady. In support of a non-coincidental occurrence of these similarities is the relationship between a retroviral infection and two other neurologic diseases, spastic tropical paraparesis and multiple sclerosis.[162,163]

The availability of detailed information about the epitope repertoire of autoreactive AChR-specific Th cells from relatively large numbers of MG patients[105,114] (Manfredi et al., unpublished data) allowed the search for similarities between AChR T epitopes and unrelated protein sequences. To assess if sequence similarities with microbial proteins occur more frequently for immunogenic rather than non-immunogenic regions of the AChR, we also analyzed several "control" AChR sequences for sequence similarities with unrelated proteins, ie, sequence regions of the α, γ, and δ subunits, which were never recognized by MG patients.[103–105,114] Several T epitope sequences, as well as some non-T epitope sequences, resembled sequence regions of other proteins, including proteins of common human pathogens, such as human adenovirus, Epstein-Barr virus, influenza A virus, herpes simplex virus, etc. The similarity of two T epitopes on the embryonic AChR γ subunit with herpes simplex proteins[114] (Manfredi et al., unpublished data) is particularly intriguing, as antibodies found in 15% of MG patients recognize an epitope cross-reacting between human AChR and herpes virus (see above).[156]

The relatively high frequency of sequence similarities of AChR sequence regions with unrelated proteins is not surprising because certain short amino acid sequences (3–5 residues), which presumably correspond to specific tridimensional structural motifs, occur in the sequence of known proteins with a much higher frequency than statistically expected;[164,165] this makes it likely, and indeed frequent, that fragments of alien proteins up to 7–18 residues in length (ie, enough for T epitope formation[166–169]) will share substantial sequence similarities with self proteins fragments. A complete purging from the T repertoire of all potentially autoreactive T cells could result in a severely limited T repertoire, and in impairment of the ability to respond against pathogens. Further filters must therefore exist in the immune system to avoid widespread occurrence of autoimmune phenomena. Among them is the necessity for an alien sequence, which resembles a self protein, to be processed and presented (this may not occur if the relevant epitope is not on the protein surface, or if it is flanked by residues that are non-compatible with processing[170,171]), and the availability of potentially autoreactive T clones at the site of immune reaction. Potentially self-reactive T cells, as a result of (so far) unknown stimuli, may escape tolerance and be involved in an anti-AChR response.

The relatively large number of epitopes recognized by any patient indicates that, at least during the later phases of the anti-AChR auto-immune response, when myasthenic symptoms become obvious, the whole AChR molecule is involved in the autoimmune response; but it does not exclude the possibility that the anti-AChR sensitization may be initiated by an altered form of the AChR, as occurs in experimental systems,[131] or by mechanisms of molecular mimicry between a small region on the AChR molecule and a single epitope of viral or bacterial origin.[155] Once tolerance to a multideterminant antigen is broken for one epitope as result of molecular mimicry or altered self protein, responsiveness to multiple epitopes may occur.[131] Similarly, once tolerance is broken for a single AChR T epitope, B cells that secrete antibodies that are able to recognize the AChR with low affinity might be recruited; such B cells are not uncommon, since approximately 10% of the monoclonal immunoglobulins secreted in multiple myeloma patients bind the AChR molecules,[172] although clinical association of MG and myeloma is rare. Th cells sensitized by an epitope common to both AChR and a microbial protein could induce low-affinity B cells to start the somatic hypermutation processes, and result in secretion of antibodies that bind the AChR with intermediate or high affinity. Once an antigen is complexed with an antibody, the pattern of intracellular peptides produced during antigen processing, and therefore available for MHC binding, is affected,[173,174] and therefore the antibody repertoire may influence the extent[174] and the epitope specificity[175,176] of the T cell response in return. Such altered processing will result in "antibody-dependent" new T epitopes, which are normally not produced. T cell clones specific for such epitopes may not be deleted/inactivated during thymic ontogenesis.[131] They may therefore be activated, and result in the complex Th epitope pattern of the later phases of the anti-AChR autoimmune response, with activation of multiple AChR specific T cells, production of pathogenetic AChR antibodies, and appearance of clinical symptoms.

A similar "natural history" can be hypothesized for LEMS: the strong association with SCLC[8,9] suggests that abnormal expression by SCLC cells of ω-VOCC may trigger a break of tolerance to this molecule (see below). Anti-ω-VOCC antibody secretion could then interfere with ω-VOCC processing, and influence the pattern of resulting T-epitope sequences.

Molecule Expressed by Thymus May Be the Primary Antigen in MG

Many MG patients have Th cells that uniquely recognize epitopes on the embryonic AChR γ subunit.[104,114,116] Furthermore, MG patients produce antibodies that are able to uniquely bind the embryonic AChR, and a myasthenic patient has been described who produced antibodies only against epitopes unique to embryonic AChR.[177,178] In a study that used junctional (ie, adult) and extrajunctional (ie, embryonic) rat muscle AChR, antibodies from MG patients recognized determinants either unique to embryonic AChR or common to both forms, but never determinants unique to junctional AChR.[177] Therefore autoimmunity against embryonic AChR is a common occurrence in MG. The presence of embryonic-AChR specific Th cells[106] indicate that a true sensitization must have occurred against this form of AChR at some stages of the disease. Because embryonic AChR is normally absent from adult skeletal muscle,[42,43] this finding introduced the dilemma that either MG patients recognize a non-existing antigen, or that anti-AChR sensitization in MG originates in a tissue other than muscle.

The origin of the antigenic stimulation that triggers the anti-AChR autoimmune response is unknown. Thymus abnormalities frequently present in MG patients suggest a dysfunction of the thymus.[179] In most MG patients the thymus is hyperplastic and contains germinal centers, which are morphologic markers of active antibody production and are absent in normal thymuses. Some MG patients develop a thymoma, sometimes even before myasthenic symptoms appear.[182,183] The thymus of MG patients contains B cells that can synthesize anti-AChR antibodies in vitro, and AChR-specific $CD4^+$ cells.[99,182-185]

Wekerle et al.[186] proposed that the anti-AChR sensitization starts within the thymus. The thymus contains component(s) immunologically cross-reactive with muscle AChR[187-190] and binding sites for α-bungarotoxin (α-BTX), which specifically recognize AChR from peripheral tissue.[191-193] Cultured cells from adult human thymus can differentiate into striated muscle fibers, which bear α-BTX binding sites.[194-196]

The α-BTX binding component from thymus has the subunit structure and physico-chemical properties expected for a true AChR, and cross-reacts with antisera raised against *Torpedo* AChR.[193] It was therefore called "Thymus AChR-like protein" (T-AChR-LP). A cogent question, because of its ramification with the origin of the anti-AChR sensitization MG, is whether the T-AChR-LP is most similar to adult or to embryonic muscle AChR. The presence within the thymus of an embryonic type AChR, which is normally absent from muscle tissue, could be related to the break of tolerance that causes MG. Subunit-specific antibodies—which were raised against peptides that corresponded to unique sequence regions of the bovine muscle AChR ε and γ subunits and monoclonal antibodies and polyclonal antibodies against the α, β, and δ AChR subunit—were used to identify the subunit complement of purified bovine T-AChR-LP.[197] Subunits that were immunologically related or identical to all the subunits that formed the AChR of mammalian embryonic muscle were found, ie, α, β, γ, and δ subunits.

If the T-AChR-LP is the primary sensitizing autoantigen, it needs to be presented to the specific anti-AChR T helper cells in association with MHC class II molecules. Myoid cells do not express class II molecules, but other thymic cells do, and they could process and present AChR epitopes that were released upon damage of T-AChR-LP-bearing myoid cells. AChR-positive/class II positive cells exist in thymuses from MG patients, and they are not myoid cells.[198] They could be antigen-presenting cells in the process of active AChR presentation.

The presence in the thymus of a protein with strong structural and antigenic similarities with muscle embryonic AChR supports the notion of a primary anti-AChR sensitization within the thymus as being one of the first steps in the pathogenesis of MG. Because T-AChR-LP also contains α, β, and δ subunits, which are shared by the AChRs of normal adult muscle, it is likely that even if the anti-AChR response starts within the thymus, Th and B cells that were sensitized against epitopes formed by subunits other than the γ would be preferentially stimulated and expanded because of their cross-reactivity with corresponding subunits on the more abundant muscle AChR. Anti-γ Th cells and antibodies may, as the disease progresses, become an increasingly smaller fraction of the total anti-AChR Th cell/antibody population. Still, irrespective of their abundance and their actual role in the production of pathogenetic anti-AChR antibodies, the existence of Th cells specific for the embryonic γ subunit is an important finding, because it clearly points to a non-muscle origin of the anti-AChR autoimmune response.

Muller-Hermelink and coworkers described another AChR-related thymus protein, which may be involved in MG pathogenesis.[198] This protein is formed by a single large subunit (apparent MW 153,000; ie, about three times as big as an AChR subunit), has an isoelectric point of 5.0 (which is similar to that of the AChR), does not bind anti-MIR mAbs or α-BTX, but shares at least one antigenic determinant with muscle AChR α subunit, and is uniquely synthesized by thymomas. This protein therefore is not a true AChR, but is nevertheless a possible candidate as an AChR-like antigen involved in the pathogenesis of the cases of MG associated with thymoma. In this respect it is curious that the one MG patient from whom we could not establish and anti-γ subunit Th cell line had a thymoma.[104] MG patients with thymomas have several unique clinical and immunologic features, which may reflect a different initiation of their immune response.[3-6,199]

Role of Sensitization to Embryonic AChR in Maintenance of the Anti-AChR Response

An embryonic AChR might be involved in the maintenance of the anti-AChR response, because thymus removal has beneficial effects on MG symptoms.[200] This may be due to the removal of a source of antigenic stimulation, because the effects of thymectomy appear only after several months,[200,201] as expected from of the long lifespan of activated antigen-specific Th.[202] Circumstantial evidence that supports a persistent role of sensitization against embryonic AChR in the maintenance of MG symptoms comes from longitudinal studies on the response to γ subunit epitopes of Th cells from MG patients.[114] The CD4+ response to epitopes on the α and δ subunits, which are present in both embryonic and adult AChRs, was also determined.[105,114] The CD4+ response to all AChR epitopes, including those on the γ subunit, correlated well with the disease severity, and it was present or stronger when the symptoms were worsening. No anti-γ response could be detected at times of improvement or remission, while it was easily detected in the same patient when severe symptoms reappeared. These findings argue in favor of a continuing role of an embryonic AChR in the stimulation of Th cells: if anti-embryonic AChR sensitization only occurred at the initial stages of the anti-AChR response, the anti-γ response of unselected CD4+ cells in the blood of MG patients would not change during development of the disease, and would not be influenced by changes in the severity of symptoms.

Furthermore, a CD4+ response to γ epitopes was demonstrated in patients thymectomized more than 10 years earlier, which suggests the existence of a reservoir of embryonic AChR outside the thymus. It is possible that, although in most skeletal muscles the γ subunit is not expressed after birth and innervation, some muscle may retain even during adult life the ability to synthesize embryonic AChR, in addition to or as an alternative to adult AChR. In this respect, it is well known that muscles are differently affected by MG symptoms. The group of muscles most frequently and severely affected are the extrinsic ocular muscles. In approximately 75% of MG patients the initial symptoms—ptosis and diplopia—involve only the extrinsic muscles of the eye, and in 15% of MG patients the symptoms remain forever limited to the eye muscle, and skeletal muscle are not involved at all (ocular MG).[203] This suggests some antigenic uniqueness of AChRs in ocular muscles as compared to that expressed by all or most skeletal muscle. One possibility, particularly attractive since embryonic AChR is expressed by the thymus and is a target of the autoimmune anti-AChR response even in thymectomized patients, is that eye muscle express embryonic AChR, and embryonic γ subunit. The presence in extraocular muscles of embryonic AChRs, which have a longer mean open time, could facilitate the response to prolonged nerve stimulation, frequent in this district.[204] We investigated the AChR subunit expressed in rat and bovine extrinsic eye muscles, in polymerase chain reaction (PCR) experiments, using γ-, ε-, and α-subunit-specific primers. Preliminary results clearly indicate expression of embryonic AChR in several extrinsic eye muscles, while it could not be detected in other innervated skeletal muscles.[205]

Pathogenetic Mechanisms of LEMS

Although circumstantial, several lines of converging evidence suggest a role for neoplastic transformation in triggering the autoimmune response to ω-VOCC, at least in some LEMS patients:

1. *Strong association between SCLC and LEMS.* In 50–60% of LEMS patients the myasthenic symptoms are followed by cancer diagnosis, most frequently SCLC (21 of 24 cancer cases in a recent report[26]).

2. *Effective treatment of the cancer has a beneficial effect on LEMS symptoms.* Long-lasting (several months) improvements of the neurologic symptoms have been described after surgical or chemotherapic treatment of the cancer.[206]

3. *SCLC cells express ω-VOCC.* The histogenesis of SCLC is still obscure. However, a number of neuronal markers have been described on SCLC-derived cell lines.[207] Specific, saturable, high-affinity binding sites for ω-CTX were directly demonstrated in seven out of seven SCLC-derived cell lines,[50] confirming the results of previous electrophysiologic studies, which had reported ω-CTX-sensitive, voltage-gated calcium channels in SCLC cells.[208]

4. *LEMS autoantibodies affect ω-sensitive VOCC function in cultured SCLC cells, and the inhibition of calcium flux in SCLC correlates with disease severity.* LEMS autoantibodies precipitate ω-CTX binding sites from SCLC homogenates,[50] and a correlation has been found between the functional effect mediated by the autoantibodies on SCLC cells and severity of the LEMS symptoms.[209]

5. *The "time factor."* In most cases the SCLC becomes apparent within two years after the onset of neurologic symptoms and the serologic detection of anti-ω-VOCC antibodies;[50] this time span is compatible with the predicted time for tumoral cells to evolve to a radiologically detectable mass (1–2 inches of diameter).[210] It was therefore proposed that assay for anti-ω-VOCC antibodies may be useful for early diagnosis of cancer.[211]

Although the above findings suggest a role for SCLC in triggering autoimmunity against ω-VOCC, it is not likely that the autoantibody has any role in rejection of SCLC. The anti-ω-VOCC antibodies, although able to inhibit ion flux in cultured SCLC cells in vitro, do not seem to be toxic for tumoral cells. Furthermore, no clinical or prognostic difference between anti-ω-VOCC antibody-positive and -negative SCLC patients has been reported. Finally, while in approximately 60% of LEMS patients a SCLC is present, among SCLC patients only 1% develops LEMS.[25] It is conceivable that ω-VOCC expressed by the tumor cells of LEMS patients may be somewhat different from the normal ω-VOCCs, and as such induce an autoimmune response. Alternatively, ω-VOCCs may be identical in SCLC cells and in the normal tissues, but the expression on proliferating cancer cells, in a tissue rich in "professional" antigen-presenting cells, could result in presentation of peptides that are not produced in normal tissues, and therefore do not delete/inactivate the corresponding autoreactive T clones. Failure of tolerance to ω-VOCC may depend on different mechanisms in LEMS patients without SCLC. For the above reasons, the hypothesis that they may have rejected a clinically undetectable SCLC[12] does not appear convincing.

Possible Similarities in the Pathogenesis of LEMS and MG

MG and LEMS are seldom associated.[212] Nonetheless, strong analogies may exist in their pathogenetic mechanisms: in both diseases autoantibodies are produced that are specific for membrane proteins of the neuromuscular junction, and both of these different autoantigens have isoforms, some of which are expressed in the central nervous system, a "twilight zone" from the immunologic point of view, as neurons virtually do not express MHC molecules. The majority of symptoms in both MG and LEMS are due to failure of neuromuscular transmission, which in both diseases is due to the increasing reduction of functional autoantigen, whose increased internalization and degradation is due to the autoantibody binding. Both MG and LEMS symptoms can be passively transferred into experimental animals by injection of the patients' IgG. Production of high-affinity IgG needs antigen-specific Th cells; such cells have been demonstrated in MG patients, and they probably play a central role in LEMS pathogenesis as well. Finally, in both MG and LEMS surgical excision of an autoantigen-rich tissue (thymus in MG, SCLC in LEMS) results in an improvement of the neurologic symptoms.

It is also conceivable that the very first steps of the pathogenesis of these two autoimmune diseases, which result in the break of tolerance, may be similar. The insight we will obtain in the next few years from these two "experiments of the nature" may be crucial for our understanding of key biologic processes that occur in the induction and maintenance of tolerance, and in the induction of autoimmunity.

Acknowledgments. We thank Drs. F. Clementi and E. Sher for the enjoyable discussions. The studies from our laboratory summarized in this chapter were partially supported by the NINCDS grant NS 29919, a grant from the Muscular Dystrophy Association of America, the research grant U.S. NSF BNS-8607289, and the U.S. NIDA program project grant 5P01-DA05695 (to B.M.C-T). A.A.M. was the recipient of an MDA

Postdoctoral Fellowship, M.P.P. was the recipient of the MDA Sidney Blackmer Fellowship.

References

1. Willis T. De anima brutorum. In: Kempton JW, ed. Living with Myasthenia Gravis. Springfield, Ill.: Charles C Thomas; 1972:5.
2. Cohen IR, Young DB. Autoimmunity, Microbial immunity and the immunological homunculus. Immunol Today 1991; 12:105.
3. Engel AG. Myasthenia Gravis and myasthenic syndromes. Ann Neurol 1984; 16:519.
4. Levinson AI, Zweiman B, Lisak RP. Immunopathogenesis and treatment of Myasthenia Gravis. J Clin Immunol 1987; 7:187.
5. Lindstrom J, Shelton D, Fujii Y. Myasthenia Gravis. Adv in Immunol 1988; 42:233.
6. Schönbeck S, Chrestel S, Hohlfeld R. Myasthenia Gravis: Prototype of the antireceptor autoimmune diseases. Intl Rev Neurobiol 1990; 32:175.
7. Myasthenia Gravis and Related Disorders: Experimental and Clinical Aspects. In: Penn AS, Richman DP, Ruff RL, Lennon VA, eds. Ann NY Acad Sci (in press).
8. Engel AG, Fukuoka T, Lang B, Newsom-Davis J, Vincent A, Wray D. Lambert-Eaton myasthenic syndrome IgG: Early morphologic effects and immunolocalization at the motor endplate. Ann NY Acad Sci 1987; 505:333.
9. Sher E, Biancardi E, Passafaro M, Clementi F. Physiopathology of neuronal voltage-operated calcium channels. FASEB J 1991; 5:2677.
10. Toyka KV, Drachman DB, Pestronk A, Kao I. Mysthenia Gravis: Passive transfer from man to mouse. Science 1975; 190:397.
11. Lindstrom J, Engel MA, Seybold M, Lennon VA, Lambert EH. Pathological mechanisms in experimental autoimmune myasthenia gravis. II. Passive transfer of experimental autoimmune myasthenia gravis in rats with anti-acetylcholine receptor antibodies. J Exp Med 1976; 144:739.
12. Vincent A, Lang B, Newsom-Davis J. Autoimmunity to the voltage-gated calcium channels underlies the Lambert-Eaton myasthenic syndrome, a paraneoplastic disorder. Trends Neurol Sci 1989; 12:496.
13. Dayan CH, Londie M, Corcoran AE, Grubeck-Loebestein B, James RFL, Rapoport B, Feldmann M. Autoantigen recognition by thyroid-infiltrating T cells in Graves disease. Proc Natl Acad Sci USA 1991; 88:7415.
14. Campbell H, Bramwell E. Myasthenia Gravis. Brain 1900; 23:277.
15. Weigart G. The thymus in Myasthenia Gravis. Neurol Zentrabl 1901; 20:597.
16. Schumacher J, Roth J. Thymektomic bei einem Fall von Morbus Basedowi mit Myasthenia. Mitt Grenzebieten Med Clin 1913; 25:745.
17. Dale HH, Feldberg W. Chemical transmission at motor nerve ending in voluntary muscle? J Physiol 1934; 81:39.
18. Walker M. Treatment of Myasthenia Gravis with physostigmine. Lancet 1934; 1:1200.
19. Blalock A, Harvey AM, Ford FR, et al. Treatment of myasthenia gravis by removal of the thymus gland. JAMA 1941; 117:1529.
20. Simpson JA. Myasthenia gravis, a new hypothesis. Scott Med J 1960; 5:419.

21. Drachman DB. Myasthenia gravis: Biology and treatment. Ann NY Acad Sci 1976:274.
22. Patrick J, Lindstrom J. Autoimmune response to acetylcholine receptor. Science 1973; 180:571.
23. Elmqvist D, Lambert EH. Detailed analysis of neuromuscular transmission in a patient with the myasthenic syndrome sometimes associated with bronchogenic carcinoma. Mayo Clin Proc 1968; 43:689–713.
24. Comola M, Nemni R, Sher E, Quattrini A, Faravelli A, Comi G, Corbo M, Clementi F, Canal N. Lambert-Eaton myasthenic syndrome and polyneuropathy in a patient with epidermoid carcinoma of the lung. Eur Neurol 1993; 33:121–125.
25. Sculier JP, Feld R, Evans WK, DeBoer G, Shepherd FA, Payne DG, Pringle JF, Yeoh JL, Quirt IC, Curtis JE. Neurologic disorders in patients with small cell lung cancer. Cancer 1987; 60(9):2275.
26. O'Neill JH, Murray NMF, Newsom-Davis J. The Lambert-Eaton myasthenic syndrome: A review of 50 cases. Brain 1988; 111:577.
27. Anderson HJ, Churchill-Davidson HC, Richardson AT. Bronchial neoplasm with myasthenia: Prolonged apnea after administration of succinyl-choline. Lancet 1953; 2:1291.
28. Lambert EH, Eaton LM, Rooke ED. Defect of neuromuscular conduction associated with malignant neoplasm. Am J Physiol 1956; 187:612.
29. Fukunaga H, Engel AG, Lang B, Newsom-Davis J, Vincent A. Passive transfer of Lambert-Eaton myasthenic syndrome with IgG from man to mouse depletes the presynaptic membrane active zones. Proc Natl Acad Sci USA 1983; 80:7636.
30. Newsom-Davis J, Murray NMF. Plasma exchange and immunosuppressive drug treatment in the Lambert-Eaton myasthenic syndrome. Neurology 1984; 34:480.
31. Sher E, Gotti C, Canal N, Scopetta C, Piccolo G, Evoli A, Clementi F. Specificity of calcium channel auto-antibodies in Lambert-Eaton myasthenic syndrome. Lancet 1989; 2:640.
32. Lennon VA, Lambert EH. Autoantibodies bind solubilized calcium channel-ω-Conotoxin complexes from small cell lung carcinoma: a diagnostic aid for Lambert-Eaton myasthenic syndrome. Mayo Clinic Proc 1989; 64:1498.
33. Conti-Tronconi BM, Raftery MA. The nicotinic cholinergic receptor: Correlation of molecular structure with functional properties. Annu Rev Biochem 1982; 51:491.
34. Stroud MB, McCarthy MP, Shuster M. Nicotinic acetylcholine receptor superfamily of ligand-gated ion channels. Biochemistry 1990; 29(50):11009.
35. Claudio T. Molecular genetics of acetylcholine receptor channels. In: Glover DM, Harnes BD, eds. Frontiers in Molecular Biology: Molecular Neurobiology Volume. Oxford: IRL Press; 1989:63–142.
36. Chang CE, Lee CY. Isolation of neurotoxins from the venom of *Bungarus multicinctus* and their modes of neuromuscular blocking action. Arch Int Pharmodyn Ther 1962; 44:241.
37. Conti-Tronconi BM, Gotti CM, Hunkapiller MW, Raftery MA. Mammalian muscle acetylcholine receptor: A supramolecular structure formed by four related proteins. Science 1982; 218:1227.

38. Raftery MA, Hunkapiller HW, Strader CD, Hood LE. Acetylcholine receptor: Complex of homologous subunits. Science 1980; 208:1454.
39. Anand R, Conroy WG, Schoepfer R, Whiting P, Lindstrom J. Neuronal nicotinic acetylcholine receptors expressed in *Xenopus* oocytes have a pentameric quaternary structure. J Biol Chem 1991; 266(17):11192.
40. Conti-Tronconi BM, Tang F, Diethelm BM, Spencer SR, Reinhardt-Maelicke S, Maelicke A. Mapping of a cholinergic site by means of synthetic peptides, monoclonal antibodies and α-Bungarotoxin. Biochemistry 1990; 29(26):6221.
41. McLane KE, Wu XD, Diethelm B, Conti-Tronconi BM. Structure determinants of α-Bungarotoxin binding to the sequence segment 181–200 of the muscle nicotinic acetylcholine receptor α subunit: Effects of cysteine/cysteine modification and species-specific amino acid substitutions. Biochemistry 1991; 30(20):4925.
42. Mishina M, Takai T, Imoto K, Noda M, Takahashi H, Numa S. Molecular distinction between fetal and adult forms of muscle acetylcholine receptor. Nature 1986; 321:406.
43. Gu Y, Hall Z. Immunological evidence for a change in subunits of the acetylcholine receptor in developing and denervated rat muscle. Neuron 1988; 1:117.
44. Trautman A. Curare can open and block ionic channels associated with cholinergic receptors. Nature 1982; 298:272.
45. Noda MY, Furutani Y, Takahashi H, Toyosato M, Tanabe T, Shimizu S, Kikyotani S, Yanano T, Hirose T, Inayama S, Numa A. Cloning and sequence analysis of calf cDNA and human genomic DNA encoding α-subunit precursor of muscle acetylcholine receptor. Nature (London) 1983; 305:818.
46. Hartman DS, Claudio T. Coexpression of two distinct muscle acetylcholine receptor α subunits during development. Nature (London) 1990; 343:372.
47. Goldman D, Tanai K. Coordinate regulation of RNAs encoding two isoforms of rat muscle nicotinic acetylcholine receptor β-subunit. Nucleic Acids Res 1989; 25:3049.
48. Sher E, Clementi F. ω-conotoxin-sensitive voltage-operated calcium channels in vertebrate cells. Neuroscience 1991; 42:301.
49. Sher E, Biancardi E, Pollo A, Carbone E, Li G, Wollheim CB, Clementi F. ω-Conotoxin sensitive, voltage-operated calcium channel in insulin-secreting cells. Eur J Pharmacol 1992; 216:407.
50. Sher E, Pandiella A, Clementi F. Voltage operated calcium channels in small cell lung carcinoma cell lines: Pharmacological, functional and immunological properties. Cancer Res 1990; 50:3892.
51. Torri-Tarelli F, Passafaro M, Sher E, Clementi F. Immunolocalization of ω-Conotoxin binding sites at the frog neuromuscular junction. J Cell Biol 1990; 111:60a.
52. Miller RJ. Multiple calcium channels and neuronal functions. Science 1987; 235:46.
53. Yamaguchi T, Saisu H, Mitsui H, Abe T. Solubilization of the ω-Conotoxin receptor associated with voltage-sensitive calcium channels from bovine brain. J Biol Chem 1988; 263:9491.

54. Barhanin J, Schmid A, Lazdunski M. Properties of structure and interaction of the receptor for ω-Conotoxin, a polypeptide active on calcium channels. Biochem biophys Res Commun 1988; 150:1051.

55. Ahlijanian MK, Striessnig J, Catterall WA. Phosphorylation of an α1-like subunit of an ω-Conotoxin-sensitive brain calcium channel by cAMP-dependent protein kinase and protein kinase C. J Biol Chem 1991; 266:20192.

56. McEnery MW, Snowman AM, Sharp AH, Adams ME, Snyder SH. Purified ω-conotoxin GVIA receptor of rat brain resembles a dihydropyridine-sensitive L-type calcium channel. Proc Natl Acad Sci USA 1991; 88:11095–11099.

57. Sakamoto J, Campbell KP. A monoclonal antibody to the β subunit of the skeletal muscle dihydropyridine receptor immunoprecipitates the brain ω-conotoxin GVIA receptor. J Biol Chem 1991; 266:18914.

58. Berti F, Clementi F, Conti-Tronconi B, Folco G. A cholinoreceptor antiserum: Its pharmacological properties. Br J Pharmac 1976; 57:17–22.

59. Richman DP, Gomez MC, Berman PW, Burres SA, Fitch FW, Arnason BGW. Monoclonal anti-acetylcholine receptor antibodies can cause experimental myasthenia. Nature (London) 1980; 286:738.

60. Lennon VA, Lambert EH. Myasthenia gravis induced by monoclonal antibodies to acetylcholine receptors. Nature 1980; 285:238.

61. Tzartos S, Hochschwender S, Vasquez P, Lindstrom J. Passive transfer of experimental autoimmune myasthenia gravis by monoclonal antibodies to the main immunogenic region of the acetylcholine receptor. J Neuroimmunol 1987; 15:185.

62. Engel AG, Sahashi K, Fumagalli G. The immunopathology of acquired Myasthenia Gravis. Ann NY Acad Sci 1981; 377:258.

63. Engel AG, Santa T. Histometric analysis of the ultrastructure of the neuromuscular junction in myasthenia gravis and in the myasthenic syndrome. Ann NY Acad Sci 1971; 183:46.

64. Sahash K, Engel AG, Lambert EH, Howard FN Jr. Ultrastructural localization of the terminal and lytic ninth complement component (C9) at the motor end-plate in myasthenia gravis. Neuropathol Exp Neurol 1980; 39:160.

65. Killen JA, Hochschwender SM, Lindstrom JM. The main immunogenic region of acetylcholine receptors does not provoke the formation of antibodies of a predominant idiotype. J Neuroimmunol 1985; 9:229.

66. Dwyer DS, Bradley RJ, Oh SJ, Kearney JF. Idiotypes in myasthenia gravis. In: Dwyer DS, ed. Idiotypes in Biology and Medicine. New York: Academic Press; 1984.

67. Roses AD, Olanow CW, McAdams MW, Lane RJM. No direct correlation between serum antiacetylcholine receptor antibody levels and clinical state of individual patients with myasthenia gravis. Neurology 1981; 31:220.

68. Besinger UA, Toyka KV, Homberg M, Heininger K, Hohlfeld R, Fateh-Moghadam A. Myasthenia gravis: Long-term correlation of binding and bungarotoxin blocking antibodies against acetylcholine receptors with changes in disease severity. Neurology 1983; 33:1316.

69. Conti-Tronconi BM, Tzartos S, Lindstrom J. Monoclonal antibodies as a probe of acetylcholine receptor structure. 2. Binding to native receptor. Biochemistry 1981; 20:2181.

70. Mossman T. Myasthenia Gravis without acetylcholine receptor antibody: A distinct disease entity. Lancet 1986; 1:116.

71. Evoli A, Bartoccioni E, Barocchi AP, Scuderi F, Tonali P. Anti-AChR-negative myasthenia gravis: Clinical and immunological features. Clin Invest Med 1989; 12(2):104.

72. Lang B, Newsom-Davis J, Wray D, Vincent A, Murray N. Autoimmune etiology for myasthenic (Lambert-Eaton) syndrome. Lancet 1981; 2:224.

73. Kim YI. Passive transfer of the Lambert-Eaton myasthenic syndrome: Neuromuscular transmission in mice injected with plasma. Muscle Nerv 1985; 8:162.

74. Fukunaga H, Engel AG, Osame M, Lambert EH. Paucity and disorganization of presynaptic membrane active zones in the Lambert-Eaton myasthenic syndrome. Muscle Nerv 1982; 5:686.

75. Kim YI, Neher E. IgG from patients with Lambert-Eaton myasthenic syndrome block voltage dependent calcium channels. Science 1988; 239:405.

76. Kim YI, Sanders DB, Johns TR, Philips LH, Smith RE. Lambert-Eaton myasthenic syndrome: The lack of short-term in vitro effects of serum factors on neuromuscular transmission. J Neurol Sci 1988; 87:1.

77. Prior C, Lang B, Wray D, Newsom-Davis J. Action of Lambert-Eaton myasthenic syndrome IgG at mouse motor nerve terminals. Ann Neurol 1985; 17:587.

78. Nagel A, Engel AG, Lang B, Newsom-Davis J, Fukuoka T. Lambert-Eaton myasthenic syndrome IgG depletes presynaptic membrane active zone particles by antigenic modulation. Ann Neurol 1988; 24:552.

79. Lang B, Newsom-Davis J, Peers C, Wray DW. The action of myasthenic syndrome antibody fragments on neurotransmitter release in the mouse. J Physiol 1987; 390:173P.

80. Tzartos SJ, Lindstrom J. Monoclonal antibodies used to probe acetylcholine receptor structure: Localization of the main immunogenic region and detection of similarities between subunits. Proc Natl Acad Sci USA 1980; 77:755.

81. Tzartos SJ, Seybold M, Lindstrom J. Specificities of antibodies to acetylcholine receptors in sera from myasthenia gravis patients measured by monoclonal antibodies. Proc Natl Acad Sci USA 1982; 79:188.

82. Tzartos SJ, Sophianos D, Efthimiadis G. Role of the main immunogenic region of acetylcholine receptor in myasthenia gravis. A Fab monoclonal antibody protects against antigenic modulation by human sera. J Immunol 1985; 134:2343.

83. Lennon VA, Griesman GE. Evidence against acetylcholine receptor having a main immunogenic region as a target for autoantibodies in myasthenia gravis. Neurology 1989; 39:1069.

84. Tzartos S, Kokla A, Walgrave S, Conti-Tronconi BM. The main immunogenic region of human muscle acetylcholine receptor is localized within residues 67–76 of the α subunit. Proc Natl Acad Sci USA 1988; 85:2899.

85. Barkas T, Gabriel J-M, Mauron A, Hughes GJ, Roth B, Alliod C, Tzartos SJ. Ballivet M. Monoclonal antibodies to the main immunogenic region of the nicotinic acetylcholine receptor bind to residues 61–76 of the α subunit. J Biol Chem 1988; 263:5916.
86. Bellone M, Tang F, Milius R, Conti-Tronconi BM. The main immunogenic region of the nicotinic acetylcholine receptor: Identification of amino acid residues interacting with different antibodies. J Immunol 1989; 143:256.
87. Tzartos SJ, Loutrari HV, Tang F, Kokla A, Walgrave SL, Milius RP, Conti-Tronconi BM. Main immunogenic region of Torpedo electroplax and human muscle acetylcholine receptor: Localization and microheterogeneity revealed by the use of synthetic peptides. J Neurochem 1990; 54(1): 51.
88. Das MK, Lindstrom J. The main immunogenic region of the nicotinic acetylcholine receptor: Interaction of monoclonal antibodies with synthetic peptides. Biochem Biophys Res Commun 1989; 165:865.
89. Saedi MS, Anand R, Conroy WG, Lindstrom J. Determination of amino acid critical to the main immunogenic region in intact acetylcholine receptors by in vitro mutagenesis. FEBS Lett 1990; 267:55.
90. Papadouli I, Potamianos S, Hadjidakis I, Bairaktari K, Tsikaris V, Sakarellos C, Cung MT, Marraud M, Tzartos SJ. Antigenic role of single residues within the main immunogenic region of the nicotinic acetylcholine receptor. Biochem J 1990; 269:239.
91. Davies DR, Sheriff S, Padlan EA. Antibody-antigen complexes. J Biol Chem 1988; 262:10641.
92. Baldwin TJ, Yoshihara CM, Blackmer K, Kintner CR, Burden SJ. Regulation of acetylcholine receptor transcript expression during development in Xenopus laevis. J Cell Biol 1988; 106:469.
93. Hohlfeld R, Toyka KV, Heininger K, Grosse-Wilde H, Kalies I. Autoimmune human T-lymphocytes specific for acetylcholine receptor. Nature 1984; 310:244.
94. Hohlfeld R, Kalies I, Kohleisen B, Heininger K, Conti-Tronconi BM, Toyka KV. Myasthenia gravis: Stimulation of antireceptor autoantibodies by autoreactive T-cell lines. Neurology 1986; 36:618.
95. Waldor MK, Sriram S, McDevitt HO, Steinman L. In vivo therapy with monoclonal anti-I-A antibody suppresses immune responses to acetylcholine receptor. Proc Natl Acad Sci USA 1983; 80:2713.
96. Christados P, Dauphinee MJ. Immunotherapy for myasthenia gravis: A murine model. J Immunol 1988; 7:2437.
97. Morgutti M, Conti-Tronconi BM, Sghirlanzoni A, Clementi F. Cellular immune response to acetylcholine receptor in myasthenia gravis: II. Thymectomy and corticosteroids. Neurology 1979; 29:734.
98. Hohlfeld R, Toyka KV, Tzartos SJ, Carson W, Conti-Tronconi BM. Human T helper lymphocytes in myasthenia gravis recognize the nicotinic receptor α subunit. Proc Natl Acad Sci USA 1987; 84:5379.
99. Melms A, Schalke BCG, Kirchner T, Muller-Hermelink HK, Albert E, Wekerle H. Thymus in myasthenia gravis. Isolation of T-lymphocyte lines specific for the nicotinic acetylcholine receptor from thymuses of myasthenic patients. J Clin Invest 1988; 81:902.

100. Harcourt GC, Sommer N, Rothbard J, Wilcox HNA, Newson-Davis J. A juxta-membrane epitope on the human acetylcholine receptor recognized by T cells in Myasthenia Gravis. J Clin Invest 1988; 82:1295.
101. Protti MP, Manfredi AA, Straub C, Wu XD, Howard JF Jr, Conti-Tronconi BM. Use of synthetic peptides to establish anti-human acetylcholine receptor CD4+ cell lines from Myasthenia Gravis patients. J Immunol 1990; 144:1711.
102. Protti MP, Manfredi AA, Straub C, Howard JF Jr, Conti-Tronconi BM. Immunodominant regions for T helper sensitization on the human nicotinic receptor α subunit in myasthenia gravis. Proc Natl Acad Sci USA 1990; 87:7792.
103. Protti MP, Manfredi AA, Wu XD, Moiola L, Howad JF Jr, Conti-Tronconi BM. Myasthenia Gravis: T epitopes of the δ subunit of human muscle acetylcholine receptor. J Immunol 1991; 146:2253.
104. Protti MP, Manfredi AA, Howard JF Jr, Conti-Tronconi BM. T cells in Myasthenia Gravis specific for embryonic acetylcholine receptor. Neurology 1991; 41:1809.
105. Manfredi AA, Protti MP, Wu XD, Howard JF Jr, Conti-Tronconi BM. CD4+ T epitope repertoire on the human acetylcholine receptor α subunit in severe myasthenia gravis. A study with synthetic peptides. Neurology 1992; 42:1092–1100.
106. Hohlfeld R, Conti-Tronconi BM, Kalies I, Bertrams J, Toyka KV. Genetic restriction of autoreactive acetylcholine receptor-specific T lymphocytes in Myasthenia Gravis. J Immunol 1985; 135:2393.
107. Davis MM, Bjorkman PJ. T cell antigen receptor genes and T cell recognition. Nature 1988; 334:395.
108. Ashwell JD, Schwartz RH. T cell recognition of antigen and the Ia molecule as a ternary complex. Nature 1986; 320:176.
109. Protti MP, Manfredi AA, Straub C, Howard JF Jr, Conti-Tronconi BM. CD4+ T cell response is the human acetylcholine receptor α subunit in Myasthenia Gravis: A study with synthetic peptides. J Immunol 1990; 144:1276.
110. Melms A, Chrestel S, Schalke BCG, Wekerle H, Mauron A, Ballivet M, Barkas T. Autoimmune T lymphocytes in myasthenia gravis: Determination of target epitopes using T lines and recombinant products of the mouse nicotinic acetylcholine receptor gene. J Clin Invest 1989; 83:285.
111. McQuillen DP, Koethe SM, McQuillen MP. Cellular response to human acetylcholine receptor in patients with myasthenia gravis. J Neuroimmunol 1983; 5:59.
112. Brocke S, Brautbar C, Steinman T, Abramsky O, Rothbard J, Neumann D, Fuchs S, Moses E. In vitro proliferative responses and antibody titres specific to human acetylcholine receptor synthetic peptides in patients with Myasthenia Gravis and relation to HLA class II genes. J Clin Invest 1989; 82:1894.
113. Oshima M, Ashizawa T, Pollack MS, Atassi MZ. Autoimmune T cell recognition of human acetylcholine receptor: The sites of T cell recognition in Myasthenia Gravis on the extracellular part of the α subunit. Eur J Immunol 1990; 20:2563.

114. Manfredi AA, Protti MP, Dalton MW, Howard JF Jr, Conti-Tronconi BM. T helper cell recognition of muscle acetrylcholine receptor in myasthenia gravis. Epitopes on the γ and δ subunits. J Clin Invest 1993; in press.
115. Hohlfeld R, Toyka KY, Miner LL, Walgrave SL, Conti-Tronconi BM. Amphipathic segment of the nicotinic receptor α subunit contains epitopes recognized by T lymphocytes in Myasthenia Gravis. J Clin Invest 1988; 81:657.
116. Protti MP, Manfredi AA, Wu XD, Moiola L, Dalton MW, Howard JF Jr, Conti-Tronconi BM. Myasthenia Gravis. CD4$^+$T epitopes on the embryonic γ subunit of human muscle acetylcholine receptor. J Clin Invest 1992; 90:1558.
117. Sakai K, Sinha AA, Mitchell DJ, Zamvill SS, Rothbard JB, McDevitt HO, Steinman L. Involvement of distinct murine T-cell receptors in the autoimmune encephalitogenic response to inbred epitopes of myelin basic protein. Proc Natl Acad Sci USA 1988; 85:8608.
118. Braciale TJ, Sweetser MT, Morrison LA, Kittlesen DJ, Braciale UL. Class I major hystocompatibility complex-restricted cytotoxic T lymphocytes recognize a limited number of sites on the influenza hemagglutinin. Proc Natl Acad Sci USA 1989; 86:277.
119. Good MF, Ponibo D, Quakyi KA, Riley EM, Houghten RA, Menon A, Alling DW, Berzofsky JA, Miller LH. Human T-cell recognition of the circumsporozoite protein of plasmodium falciparum: Immunodominant T-cell domains map to the polymorphic regions of the molecule. Proc Natl Acad Sci USA 1988; 85:1199.
120. Clerici M, Stocks NI, Zajac RA, Boswell RN, Bernstein DC, Mann DL, Shearer GM, Berzofsky JA. Interleukin-2 production used to detect antigenic peptide recognition by T-helper lymphocytes from asymptomatic HIV-seropositive individuals. Nature (London) 1989; 339:383.
121. O'Sullivan P, Sidney J, Appella E, Uacker L, Phillips L, Colon SM, Miles C, Chesnut RW, Sette A. Characterization of the specificity of peptide binding to four DR haplotypes. J Immunol 1990; 145:1799.
122. Newsom-Davis J, Hartcourt G, Sommer N, Beeson D, Willcox N, Rothbard JB. T cell reactivity in Myasthenia Gravis. J Autoimmun 1989; 2:201.
123. Kartik R, Jaraquemada D, Flerlage M, et al. Fine specificity and HLA restriction of myelin basic protein-specific cytotoxic T cell lines from multiple sclerosis patients and healthy individuals. J Immunol 1990; 145: 540.
124. Pette M, Fujita K, Kitze B, et al. Myelin basic protein-specific T lymphocytes from MS patients and healthy individuals. Neurology 1990; 40:1770.
125. Sommer N, Harcourt GC, Willcox N, Beeson D, Newsom-Davis J. Acetylcholine receptor-reactive T lymphocytes from healthy subjects and myasthenia gravis patients. Neurology 1991; 41:1270.
126. Conti-Tronconi BM, Morgutti M, Sghirlanzoni A, Clementi F. Cellular immune response against acetylcholine receptor in myasthenia gravis: I. Relevance to clinical course and pathogenesis. Neurology 1979; 29:496.
127. Zhang Y, Schleup M, Frutiger S, Hughes GJ, Jaenner M, Steek A, Barkas T. Immunological heterogeneity of autoreactive T lymphocytes against the

nicotinic acetylcholine receptor in myasthenic patients. Eur J Immunol 1990; 20:2577.

128. Gammon G, Sercarz EE. Does the presence of self-reactive T cells indicate the breakdown of tolerance? Clin Immunol Immunopathol 1990; 56:287.

129. Bellone M, Ostlie N, Lei S, Conti-Tronconi BM. Experimental myasthenia gravis in congenic mice. Sequence mapping and H-2 restriction of T helper epitopes on the α subunits of *Torpedo californica* and murine acetylcholine receptors. Eur J Immunol 1991; 21:2203.

130. Granato DA, Fulpius BW, Moody JF. Experimental myasthenia in Balb/c mice immunized with rat acetylcholine receptor from rat denervated muscle. Proc Natl Acad Sci USA 1986; 73:2872.

131. Gammon G, Sercarz EE. How some T-cells escape tolerance induction. Nature (London) 1989; 342:183.

132. Berman PW, Patrick J. Linkage between the frequency of muscular weakness and loci that regulate immune responsiveness in murine experimental myasthenia gravis. J Exp Med 1980; 152:507.

133. Cristadoss P, Lennon VA, Krco CJ, Lambert EG, David CS. Genetic control of autoimmunity to acetylcholine receptors: Role of Ia molecules. Ann NY Acad Sci 1981; 377:258.

134. Cristadoss P, Lindstrom JM, Melvold RW, Tahal N. Mutation at I-A beta chain prevents experimental autoimmune myasthenia gravis. J Immunogenet 1985; 21:33.

135. Morgutti M, Conti-Tronconi BM, Sghirlanzoni A, Clementi F. Cellular immune response to acetylcholine receptor in myasthenia gravis: II. Thymectomy and corticoids. Neurology 1979; 29:734.

136. Bellone M, Ostlie N, Lei S, Manfredi AA, Conti-Tronconi BM. T helper function of CD4+ cells specific for defined epitopes on the acetylcholine receptor in congenic mice strains. J Autoimmun 1992; 5:27–46.

137. Roy S, Scherer MT, Briner TJ, Smith JA, Gefter ML. Murine MHC polymorphism and T cell specificities. Science 1989; 244:572.

138. Yokoi T, Mulac-Jericevic B, Kurisaki JL, Atassi MZ. T lymphocytes recognition of acetylcholine receptor: Localization of the full T cell receptor recognition profile on the extracellular part of the α chain of Torpedo californica acetylcholine receptor. Eur J Immol 1987; 17:1697.

139. Gao XM, Liew FY, Tite JP. Identification and characterization of T helper epitopes in the nucleoprotein of influenza A virus. Immunol 1989; 143:3007.

140. Brett SA, Cease KB, Ouyang CS, Berzofsky JA. Fine specificity of T cell recognition of the same peptide in association with different I-A molecules. J Immol 1989; 143:771.

141. Margalit H, Spouge JL, Cornette JC, Cease KB, DeLisi C, Berzofsky JA. Prediction of immunodominant helper T cell antigenic sites from the primary sequence. J Immunol 1987; 138:2213.

142. Bellone M, Bertazzon A, Milius R, Conti-Tronconi BM. Recognition of different structural motifs by H-2b and H-2d class II molecules in murine experimental myasthenia gravis. Biophysic J 1991; 59:301A.

143. Fujii Y, Lindstrom J. Specificity of the T cell immune response to acetylcholine receptor in experimental autoimmune myasthenia gravis. Response to subunits and syntheic peptides. J Immunol 1988; 140:1830.

144. Christadoss P, Lennon VA, Krco CJ, David CS. Genetic control of experimental autoimmune myasthenia gravis in mice. III. Ia molecules mediate cellular immune responsiveness to acetylcholine receptors. J Immunol 1982; 128:1141.

145. Valdor MK, Sriram S, McDevitt HO, Steinman L. In vivo theraphy with monoclonal anti-I-A antibody suppress immune responses to acetylcholine receptor. Proc Natl Acad Sci USA 1983; 80:2713.

146. Melvold RW, Kohn HI. Eight new histocompatibility mutations associated with the H-2 complex. J Immunogenet 1976; 3:185.

147. McKenzie IFC, Morgan G, Sandrin M, Michaelides MM, Melvold RW, Kohn HI. B6 C-H-2^{bm12}: A new mutation in the I region in the mouse. J Exp Med 1979; 150:1323.

148. Mengle-Gaw L, Conner S, McDevitt HO, Fathman CG. Gene conversion between class II major hystocompatibility complex loci functional and molecular evidence from the bm12 mutant. J Exp Med 1984; 160:1184.

149. Michaelides MM, Sandrin M, Morgan G, McKenzie IFC, Ashman R, Melvold RW. Ir gene function in an I-A subregion mutant B6 C-H-2.^{bm12} J Exp Med 1981; 153:464.

150. Lin CS, Rosenthal AS, Blake JT, Passmore HC, Hansen TH. Selective deletion of an antigen specific Ir gene function in the I-A mutant B6 C-H-2^{bm12} is an antigen presenting cells defect. Proc Natl Acad Sci USA 1983; 78:6404.

151. Krco CJ, Kazim DL, Atassi MZ, Melvold R, David CS. Genetic control of immune response to hemoglobin. III. VaRIPAntA (bm12) but not A,la polypeptide alters immune reactivity towards the α subunit of human hemoglobin. J Immunogenet 1981; 8:471.

152. Chang BL, Bearer E, Ansari A, Dorshkind K, Gershwin ME. The bm12 mutation and autoantibodies to dsDNA in NZB-H-2^{bm12} mice. J Immunol 1990; 145:94.

153. Bellone M, Ostlie N, Lei S, Wu XD, Conti-Tronconi BM. The I-A^{bm12} mutation, which confers resistance to experimental myasthenia gravis, drastically affects the epitope repertoire of murine CD4$^+$ cells sensitized to nicotinic acetylcholine receptor. J Immunol 1991; 147:1484.

154. Infante JA, Thompson PA, Krolick KA, Wall KA. Determinant selection in murine experimental autoimmune myasthenia gravis. Effect of the bm12 mutation on T cell recognition of acetylcholine receptor epitopes. J Immunol 1991; 146:2977.

155. Oldstone MBA. Molecular mimicry and autoimmune disease. Cell 1987; 50:819.

156. Schwimmbeck PL, Dyrberg T, Drachman DB, Oldstone MBA. Molecular mimicry and Myasthenia Gravis: An autoantigenic site of the acetylcholine receptor α-subunit that has biologic activity and reacts immunochemically with herpes simplex virus. J Clin Invest 1989; 84:1174.

157. Manfredi AA, Bellone M, Protti MP, Conti-Tronconi BM. Molecular mimicry among human autoantigens. Immunol Today 1991; 12:46.

158. Kabsch W, Sanders C. On the use of sequence homologies to predict protein structure: Identical pentapeptides can have completely different conformation. Proc Natl Acad Sci USA 1984; 81:1075.

159. Lennon VA, Howard FM. Serologic diagnosis of myasthenia gravis. In: Nakamura RM, O'Sullivan MB, eds. Clinical Laboratory and Molecular Analyses. New York: Grune and Stratton; 1985:29–44.
160. Chan MK, Britton M. Comparative features in patients with myasthenia gravis with systemic lupus erythematosus. J Rheumatol 1990; 7:838.
161. Maul GG, Jimenez SA, Riggs E, Ziemuicka-Dotula D. Determination of an epitope of the diffuse systemic sclerosis marker antigen DNA topoisomerase I: Sequence similarity with retroviral p30^{gag} protein suggests a possible cause for autoimmunity in systemic sclerosis. Proc Natl Acad Sci USA 1989; 86:8492.
162. Jacobson S, Raine CS, Mingioli ES, McFarlin DE. Amplification and molecular cloning of HTLV-I sequences from DNA of multiple sclerosis patients. Nature (London) 1988; 331:580.
163. Reddy EP, Sanberg-Wollheim M, Mettus RU, Ray PE, DeFreitas E, Kopowski H. Isolation of an HTLV-1 like retrovirus from patients with tropical spastic paraparesis. Science 1989; 243:529.
164. Ohno S. Many peptide fragments of alien antigens are homologous with host proteins, thus canalizing T-cell responses. Proc Natl Acad Sci USA 1991; 88:3065.
165. Ohno S. To be or not to be a responder in T-cell responses: Ubiquitous oligopeptides in all proteins. Immunogenet 1991; 34:215.
166. Reddehase MJ, Rothbard JB, Koszinowski UH. A pentapeptide as minimal antigenic determinant for MHC class I-restricted T lymphocytes. Nature 1989; 337:651.
167. Van Bleek GM, Nathenson SG. Isolation of an endogenously processed immunodominant viral peptide from the class I H-2Kb molecule. Nature 1990; 348:213.
168. Rudensky AY, Preston-Hurlburt P, Hong SC, Barlow A, Janeway CA Jr. Sequence analysis of peptides bound to MHC class II molecules. Nature 1991; 353:622.
169. Jardetzky TS, Lane WS, Robinson RA, Madden RA, Wiley DC. Identification of self peptides bound to purified HLA-B27. Nature 1991; 353:326.
170. Vacchio MS, Berzofsky JA, Krzych U, Smith JA, Hodes RJ, Finnegan A. Sequences outside a minimal immunodominant site exert negative effects on recognition by staphylococcal nuclease-specific T cell clones. J Immunol 1989; 143:2814.
171. Bhayani H, Carbone FR, Paterson Y. The activation of pigeon cytochrome c-specific T cell hybridomas by antigenic peptides is influenced by non-native sequences at the amino terminus of the determinant. J Immunol 1988; 141:377.
172. Eng H, Lefvert AK, Mellstedt H, Osterborg A. Human monoclonal immunoglobulins that bind the human acetylcholine receptor. Eur J Immunol 1987; 17:1867.
173. Davidson HW, Watts C. Epitope directed processing of specific antigen by B lymphocytes. J Cell Biol 1989; 109:85.
174. Manca F, Fenoglio D, Li Pira G, Kunkl A, Celada F. Effect of antigen/antibody ratio on macrophage uptake, processing and presentation to T cells of antigen complexed with polyclonal antibodies. J Exp Med 1991; 173:37.

175. Ozaki S, Berzofsky JA. Antibody conjugates mimic specific B cell presentation of antigen: relationship between T and B cell specificity. J Immunol 1987; 138:4133.
176. Manca F, Fenoglio D, Kunkl A, Cambiaggi C, Sasso M, Celada F. Differential activation of T cell clones stimulated by macrophages exposed to antigens complexed with monoclonal antibodies: A possible influence of paratope specificity on the mode of antigen processing. J Immunol 1988; 140:2893.
177. Weinberg CB, Hall JW. Antibodies from patients with myasthenia gravis recognize determinants unique to extrajunctional acetylcholine receptors. Proc Natl Acad Sci USA 1979; 76:504.
178. Schuetze SM, Vicini S, Hall ZW. Myasthenic serum selectively blocks acetylcholine receptors with long channel open times at developing rat endplates. Proc Natl Acad Sci USA 1985; 82:2533.
179. Castleman B, Norris EH. The pathology of the thymus in myasthenia gravis. Ann NY Acad Sci 1966; 135:469.
180. Kumura J, Van Allen MW. Post-thymectomy myasthenia gravis: Report a case of ocular myasthenia gravis after total removal of a thymoma and review of the literature. Neurology 1967; 17:413.
181. Cuenoud S, Feltkamp TEW, Fulpius BW, Oosterhuis JGH. Antibodies to acetylcholine receptors in patients with thymoma but without myasthenia gravis. Neurology 1980; 30:301.
182. Kamo I, Furukawa S, Tada A, Mano Y, Iwasaki Y, Furuse T, Ho N, Hayaski K, Satoyoshi E. Monoclonal antibody to acetylcholine receptor: Cell line established from thymus of patient with myasthenia gravis. Science 1982; 215:995.
183. Fujii Y, Monden Y, Nakahara K, Hashimoto J, Kawashima Y. Antibody to acetylcholine receptor in myasthenia gravis: production by lymphocytes from thymus or thymoma. Neurology 1984; 34:1182.
184. Lisak RP, Levinson AI, Zweiman B, Kornestein MJ. Antibodies to acetylcholine receptor and tetanus toxoid: in vitro synthesis by thymic lymphocytes. J Immunol 1986; 137:1221.
185. Heidenreich F, Vincent A, Willcox N, Newsom-Davis J. Anti-acetylcholine receptor antibody specificity in serum and in thymic cell culture supernatants from myasthenia gravis patients. Neurology 1988; 38:1784.
186. Wekerle H, Hohlfeld R, Ketelsen U-P, Kalden J, Kalies I. Thymic myogenesis, T lymphocytes and the pathogenesis of myasthenia gravis. Ann NY Acad Sci USA 1981; 377:455.
187. Aharonov A, Tarrab-Hazdai R, Abramsky O, Fuchs S. Immunological relationship between acetylcholine receptor and thymus: A possible significance in myasthenia gravis. Proc Natl Acad Sci USA 1975; 72:1456–1459.
188. Ueno S, Wada K, Takahashi M, Tarui S. Acetylcholine receptor in rabbit thymus: Antigenic similarity between acetylcholine receptors of muscle and thymus. Clin Exp Immunol 1980; 42:463.
189. Schleup M, Willcox N, Vincent A, Dhoot GK, Newsom-Davis J. Acetylcholine receptors in human thymic myoid cells in situ: an immunohistological study. Ann Neurol 1987; 22:212.
190. Kirchner T, Tzartos S, Hoppe F, Schalke B, Wekerle H, Muller-Hermelink HK. Pathogenesis of myasthenia gravis: Acetylcholine receptor-related

antigenic determinant in tumor-free thymuses and thymic epithelial tumors. Am J Pathol 1988; 130:268.

191. Engel WK, Trotter JL, McFarlin DE, McIntosh CL. Thymic epithelial cell contains acetylcholine receptor. Lancet 1977; 1:1310.

192. Kao I, Drachman DB. Thymus muscle cells bear acetylcholine receptors: Possible relation to myasthenia gravis. Science 1977; 195:74.

193. Kawanami S, Conti-Tronconi BM, Racs J, Raftery MA. Isolation and characterization of nicotinic acetylcholine receptor-like protein from fetal calf thymus. J Neurol Sci 1988; 87:195.

194. Feltkamp-Vroom T. Myoid cells in human thymus. Lancet 1966; I:1320.

195. Van de Velde RKL, Friedman NB. The thymic "Myoidzellen" and myasthenia gravis. JAMA 1966; 198:287.

196. Wekerle H, Paterson B, Ketelsen U-P, Feldman M. Striated muscle fibers differentiate in monolayer cultures of adult thymus reticulum. Nature (London) 1975; 256:493.

197. Nelson S, Conti-Tronconi BM. Adult thymus expresses an embryonic nicotinic acetylocholine receptor-like protein. J Neuroimmunol 1990; 29:81.

198. Marx A, O'Connor R, Tzartos S, Kalies I, Kirchner T, Muller-Hermelink H-K. Acetylcholine receptor epitope in proteins of myasthenia gravis-associated thymomas and non-thymic tissues. Thymus 1989; 14:171–178.

199. Alpert LI, Papatestas A, Kark A, Osserman RS, Osserman K. An histologic reappraisal of the thymus in myasthenia gravis: A correlative study thymic pathology and response to thymectomy. Arch Pathol 1971; 91:55.

200. Durelli L, Maggi G, Casadio C, Ferri R, Rendine S, Berganini L. Actuarial analysis of the occurrence of remissions following thymectomy for myastenia gravis in 400 patients. J Neurol Neurosurg Psychiatry 1991; 54:406.

201. Hertel G, Mertens HG, Reuther P, Ricker K. The treatment of myasthenia gravis with azathioprine. In: Dau PC, ed. Plasmapheresis and Immunobiology of Myasthenia Gravis, Vol. 27. Boston: Houghton-Mifflin; 1979:315.

202. Blackhaus BA, Nash AA. Immunological memory to herpes simplex virus type I glycoproteins B and D in mice. J Gen Virol 1990; 4:863.

203. Daroff RB. Ocular myasthenia: Diagnosis and therapy. In: Glaser J, ed. Neuroophthalmology. St Louis: CV Mosby; 1980:62.

204. Kaminski HJ, Maas E, Spiegel P, Ruff RL. Why are eye muscles frequently involved in myasthenia gravis? Neurology 1990; 40:1663.

205. Horton RM, Manfredi AA, Conti-Tronconi BM. The "embryonic" hamma subunit of the nicotinic acetylcholine receeptor is expressed in adult extraocular muscle. Neurology 1993; 43:983.

206. Chalk CH, Murray NMF, Newsom-Davis J, O'Neill JH, Spiro SG. Response of the Lambert-Eaton myasthenic syndrome to treatment of associated small-cell lung carcinoma. Neurology 1990; 40:1552.

207. Teeling ME, Carney DN. Biochemical markers of lung cancer. In: Rosen ST, Mulshine JL, Cuttitta F, Abrams PG, eds. Biology of Lung Cancer: Diagnosis and Treatment. New York: Marcel Dekker; 1988:34–58.

208. De Aizpurua HJ, Lambert E, Griesmann GE, Olivera BM, Lennon VA. Antagonism of voltage gated calcium channels in small cell carcinoma of patients with and without Lambert-Eaton myasthenic syndrome by autoantibodies, ω-Conotoxin and adenosine. Cancer Res 1988; 48:4719.

209. Lang B, Vincent A, Murray NMF, Newsom Davis J. Lambert-Eaton myasthenic syndrome: Immunoglobulin G inhibition of calcium flux in tumor cells correlates with disease severity. Ann Neurol 1989; 25:265.
210. Geddes DM. The natural history of lung cancer: a review based on rates of tumor growth. Br J Dis Chest 1979; 73:1.
211. Sher E, Comola M, Nemni R, Canal N, Clementi F. Calcium channel autoantibody and non-small-cells lung cancer in patients with Lambert-Eaton syndrome. Lancet 1990; 335:413.
212. Taphoorn MJB, Van Duijn H, Wolters ECH. A neuromuscular transmission disorder: Combined myasthenia gravis and Lambert-Eaton syndrome in one patient. J Neurol Neurosurg Psychiatry 1988; 51:880.

10
Clinical Pharmacology of Calcium Channels

TORE K. USKI, EDWARD D. HÖGESTÄTT, and
KARL-ERIK ANDERSSON

Calcium (Ca^{2+}) influx through plasmalemmal channels, driven by the electrochemical gradient for Ca^{2+}, is a crucial step in several biologic processes such as muscle contraction, neurotransmission, and hormone secretion. As regulators of transmembrane charge movements, Ca^{2+} channels may also be involved in electric events such as impulse generation and propagation. Ca^{2+} channels may be regarded as ion-selective pores composed of membrane-spanning glycoproteins, which allow Ca^{2+} to traverse the lipid bilayer in response to different stimuli.[1] The best characterized plasmalemmal Ca^{2+} channels are those activated by membrane depolarization. These channels are the principal targets of the calcium antagonists presently used in clinical therapy. Ca^{2+} channels insensitive to changes in membrane potential, such as those activated by agonist-receptor interactions, represent another category of membrane channels.

Ca^{2+} release from the sarcoplasmic reticulum is also regulated by specific channels. Patch-clamp studies of sarcoplasmic membrane vesicles have identified a high conductance, voltage-independent Ca^{2+} channel activated by Ca^{2+} itself, adenosine triphosphate (ATP), caffeine, and low concentrations of ryanodine, and inhibited by magnesium and ruthenium red.[2,3] A separate inositol-trisphosphate-regulated, G protein-coupled Ca^{2+} channel has also been proposed, although electrophysiologic evidence for its presence so far is lacking.[4-6] The existence of a voltage-gated, sarcoplasmic Ca^{2+} channel is an attractive hypothesis, but the evidence to support this is only circumstantial.[7,8] Since sarcoplasmic Ca^{2+} channels are not well defined, the clinical consequences of pharmacologic interference with these channels are unknown.

Classification of Plasmalemmal Calcium Channels

Although it has proven very useful, the original classification of Ca^{2+} channels into potential- and receptor-operated (gated) Ca^{2+} channels,

POCs and ROCs,[9,10] has turned out to be somewhat oversimplified and sometimes misleading in the light of recent experimental findings. From a molecular perspective, receptor-operated ion channels may be regarded as a superfamily of ion channels where the receptor is an integral part of the channel protein; eg, the NMDA (N-methyl-D-aspartate) receptor of the nonselective cation channel,[11] the nicotinic acetylcholine receptor of the Na^+ channel,[12] the $GABA_A$ (γ-aminobutyric acid) receptor-coupled Cl^- channel,[13] and the neuronal 5-HT$_3$ (5-hydroxytryptamine) receptor K^+ channel complex[14] are members of this family. This is an important distinction, since receptor activation can influence ion channels in several ways. In the heart, receptors coupled to adenylate cyclase can enhance the number of functional POCs and the open probability of individual channels, via protein-kinase-induced phosphorylation of specific regulatory sites on the channel protein.[15,16] Channel phosphorylation may also be effected by other kinases, such as protein kinase C in response to receptor-mediated phospholipase C activation and production of diacylglycerol.[17] Furthermore, activated G proteins (α subunits) may directly interact with POCs, leading to either an increase or a decrease of the Ca^{2+} current.[16,18–20] Regulation of Ca^{2+}-channel gating by different inositol phosphates (IP) and by cytoplasmic Ca^{2+} itself has also been suggested.[21,22] Although these different modes of channel regulation have brought new light on the complexity of Ca^{2+} channels, they have certainly complicated the task of channel classification. To avoid confusion, we suggest that the term ROC should be reserved for those channels that are directly gated by the receptor-agonist interaction, and where the receptor and ion channel reside within the same oligomeric protein complex.

Electrophysiologic evidence of ROCs is limited to that derived from patch-clamp studies of ATP-activated channels in the rabbit ear artery,[23] ADP-activated channels in platelets,[24] and the NMDA receptor-linked channel in neurones.[25,26] These channels are dihydropyridine-insensitive and less selective for Ca^{2+} than for Na^+ than are POCs. Since they fail to activate when the agonist is applied outside the patch pipette, the involvement of a diffusable second messenger has been questioned. However, the intricate coupling between the receptor and the channel protein remains to be resolved and will probably await isolation and cloning of the channels. Agonist-induced Ca^{2+} influx, resistant to Ca^{2+} antagonists, may also occur through other routes than the ROCs.[27] There is much evidence to support the existence of a pathway from the extracellular medium to the intracellular Ca^{2+} store within the sarcoplasmic reticulum.[26,28,29] Depletion of Ca^{2+} from the store activates the influx pathway, but the signal that transduces this information is unknown. Special junctions between the plasma membrane and the superficial sarcoplasmic reticulum have been observed on electron micrographs and may represent the structual counterpart of this pathway,[30] although recent studies suggest that the Ca^{2+} used for refilling the stores must go through the cytoplasm.[31] Regardless of how they actually operate, such mechanisms may easily be misinterpreted as opening of ROCs.

POCs are ubiquitously distributed among excitable and non-excitable cells.[32] The biophysical and pharmacologic properties of these channels as well as their molecular structure have been the subject of several recent reviews.[32–37] It is generally agreed that POCs can be subdivided into low-voltage-activated (LVA) and high-voltage-activated (HVA) Ca^{2+} channels with distinct biophysical properties.[36,38,39] Both channel types are rapidly activated upon depolarization. LVA channels have a lower threshold of activation than HVA channels, and consequently they have also been termed low- and high-threshold channels, respectively. The inactivation kinetics of LVA channels are fast and voltage-dependent, giving rise to a transient current and to the name T (*transient*) channels. Inactivation of HVA channels is slow and incomplete, and appears to be both voltage- and Ca^{2+}-dependent.[40] These channels were originaly denoted L channels, since they give rise to a *long-lasting* current. LVA and HVA channels can also be separated by their deactivation kinetics (following sudden repolarization), which are slower in the former than the latter channel type. Barium ions can substitute for Ca^{2+} as the charge carrier, the Ba^{2+} conductance being approximately twice that of Ca^{2+} at HVA channels but similar to the Ca^{2+} conductance at LVA channels. The unitary current measured by the patch-clamp technique in $BaCl_2$ solution is also about twice as large in HVA than in LVA channels. Studies of the pharmacology of HVA Ca^{2+} channel in central and peripheral neurons have uncovered two subsets of channels, termed N (*non* T, *non* L; *neuronal*) and P (*Purkinje*) channels,[32,36,37] which can be clearly distinguished from the L channel by selective antagonists (see below). Although P channels were first discovered in Purkinje cells in the cerebellum, they may also exist in other neurons.[41,42] T and L channels have been demonstrated in cardiac, skeletal, and smooth muscle,[32,36,43,44] whereas all four channel types (T, L, N, P) have been localized in neuronal tissue, although the distribution among neurons may differ.[33,45,46]

Whereas a high sensitivity to calcium antagonists of the dihydropyridine type may be used as a convenient marker of L channels, the usefulness of drugs blocking T, N, and P channels in terms of Ca^{2+} channel classification is less clear (see below). The lack of well-characterized, selective antagonists of, eg, T channels has hampered the investigation of their physiologic function. Clearly L channels convey the major part of the slow inward current in heart muscle and depolarization-induced Ca^{2+} influx in smooth muscle. Neuronal transmitter release, on the other hand, appears to depend more on N than on L channel-mediated Ca^{2+} influx. A differential distribution of POCs on neurons with L channels preferentially located on the cell soma and N channels at the nerve terminals is an attractive model to explain the dihydropyridine resistance of transmitter release.[33,39] The functional role of T channels is presently obscure, but these channels may be involved in pacemaker activity in the heart[47] and possibly also in some types of smooth muscle, as well as in regulation of burst firing in neurons.[48]

Isolation of the dihydropyridine receptor in skeletal muscle revealed an hetero-oligomeric protein composed of two high molecular weight subunits (α_1, α_2) and three smaller subunits (β, γ, δ), which have all been cloned and sequenced.[34,37] When incorporated into lipid bilayers, purified α_1 subunits alone can produce dihydropyridine-sensitive, functional Ca^{2+} channels,[49] the other subunits may modulate the properties of the channels.[50] The α_1 subunit is composed of a single polypeptide chain with four homologous units (I–IV), each containing six membrane-spanning, α-helical segments (S1–S6). The fourth segment (S4) in each unit is rich in positively-charged residues and may function as a voltage sensor. The C-terminal fragment and the intracellular loops connecting the homologous units contain several putative phosphorylation sites, which may be involved in channel regulation.[34,37] Homologous α_1 subunits from heart, smooth muscle, and brain have also been cloned, but there are clear differences between these proteins. For example, both the N- and C-terminal regions are larger in cardiac than in skeletal α_1 subunits. The number and localization of putative phosphorylation sites also differ between the subunits, which indicates different channel regulation.[34,37] The cytosolic loop between unit II and III is thought to be directly involved in skeletal muscle excitation-contraction coupling. This section bears little resemblance to the corresponding region of the cardiac α_1 subunit, which also cannot replace the skeletal α_1 subunit in this function.[51] Further indication of Ca^{2+} channel diversity has come from gene cloning and sequencing, which have provided evidence of at least four different cDNA families that correspond to the α_1 subunit of (1) the skeletal muscle dihydropyridine receptor, (2) the cardiac/smooth muscle/ neuronal L channels, (3) the neuronal P channel, and (4) an as-yet-unidentified neuronal channel.[34,37,52] Further divergence produced by alternative splicing within each gene family may form the basis of tissue specific isoforms of the POC.

Classification of Calcium Channel Modulators

A multitude of different agents, both organic and inorganic, share the ability of either blocking or stimulating the flux of Ca^{2+} ions through their channels in the cell membrane. These drugs, some of which can actually act in both ways (see below), can be designated calcium-channel modulators.

Trivalent and bivalent cations, such as La^{3+}, Gd^{3+}, Cd^{2+}, Mn^{2+}, Co^{2+}, Ni^{2+}, and Mg^{2+}, compete with Ca^{2+} for the binding sites in the Ca^{2+} channels and thus may inhibit the passage of Ca^{2+} through the cell membrane.[53] Ions with either an affinity for the binding sites greater than that of Ca^{2+}, or with a reasonable affinity combined with a low tendency for passage through the channels, can block all types of Ca^{2+} channels,[54] even if affinity differences between the subtypes of POCs exist.[55] In neuroblastoma cells Gd^{3+} seems to preferentially antagonize Ca^{2+} currents through N channels, leaving the other classes of POC unaffected.[56] Ni^{2+} displays some selectivity for T channels over L channels in most tissues,[32] whereas the opposite appears to be the case with Cd^{2+}.[57] Furthermore, the order of affinity for the various cations may differ between channels of the same group, eg, between L channels in cardiac myocytes and skeletal muscle.[55]

Very few organic calcium-entry modulators act specifically on ROCs. At present, only drugs blocking the NMDA-linked Ca^{2+} channels in central nervous neurons seem to have any clinical potential[58] and may eventually prove to be useful in protecting neurons during cerebral ischemia.[59] Drugs such as tetramethin, amilorid, and octanol have been used to block the T type of POC,[39,47,60] whereas ω-conotoxin GIVA (venom of the marine mollusc, Conus geographicus) preferentially inhibits N channels.[61–64] The P channel can be blocked by FTX, a low-molecular-weight toxin from the venom of the funnel web spider, Agelenopsis aperata.[41,65] Most other organic Ca^{2+} channel modulators act mainly on the L type of POC and will be the topic of the remaining sections of this chapter.

Different systems for the classification of Ca^{2+}-entry blockers acting on the L type of POC have been proposed.[66,67] These drugs may be characterized by their chemical structure, affinity to different binding sites in the Ca^{2+} channel, pharmacologic interactions at the channel, specificity of Ca^{2+} entry inhibition, tissue selectivity, or various combinations of these characteristics. By combining the classifications based on binding affinites and interactions at the Ca^{2+}-channel level as well as tissue selectivity, Godfraind[68] demonstrated that the resulting classification coincides reasonably well with the chemical structures of the various drugs. Based on their chemistry, organic Ca^{2+}-entry blockers can be subdivided into (1) dihydropyridines, (2) phenylalkylamines, (3) benzothiazepine derivatives, (4) piperazines, and (5) drugs unrelated to these main groups (Table 10.1, Figure 10.1). Compounds belonging to the piperazine group, however, do not appear to be specific for the HVA channel, since flunarizine is able to inhibit Ca^{2+} entry also through LVA channels[69] as well as the K^+ outward current in smooth muscle.[70] Furthermore, several substances belonging to the last group of organic inhibitors (Table 10.1) are not specific for Ca^{2+} channels.[66] For instance, bepridil may attenuate intracellular Ca^{2+} release[71,72] and fendiline may inhibit the contractile

TABLE 10.1. Classification of calcium L-channel modulators, based on their chemistry.

I. Calcium-entry inhibitors
  A. *Inorganic*
      Bivalent and trivalent cations (Ex Co^{2+}, Mn^{2+}, La^{3+})
  B. *Organic*
      1. Dihydropyridines (Ex nifedipine, nimodipine, felodipine, nitrendipine, nisoldipine, nicardipine, isradipine)
      2. Phenylalkylamines (Ex verapamil, D-600 = gallopamil, anipamil, ronipamil)
      3. Benzothiazepines (Ex diltiazem, fostedil, and other amidophosphonates)
      4. Piperazines (cinnarizine, flunarizine)
      5. Others (Ex bepridil, fendiline, perhexiline, SKF 525A, tiapamil)

II. Calcium-entry facilitators
  A. Dihydropyridines (Ex Bay K8644, GCP 28392, H 160/51, 202–791)
  B. Others (Ex atrotoxin, palmitoyl carnitine)

---

machinery in smooth muscle by interacting with myosin light chain kinase.[73] In addition to those belonging to the groups presented in Table 10.1, other drugs, with less well defined actions on Ca^{2+} influx, may bind to the Ca^{2+} channels. Among them are certain peptides, neuroleptics, benzodiazepines, barbiturates, and some adrenoceptor antagonists.[66]

Several years ago radioligand-binding studies led to the identification of at least three sites in the Ca^{2+} channel with varying affinities for drugs that belong to the first three groups: one with the highest affinity for dihydropyridines, one with affinity mainly to phenylalkylamines, and one that interacted with diltiazem and related substances.[74,75] Not only do drugs binding to the same site interact with each other's binding, but there also exists an allosteric interaction between the different binding sites in the channel.[68,75,76] It has been shown that while diltiazem usually increases the binding at the dihydropyridine sensitive site, verapamil has the opposite effect.[68,75–77] Other sites that bind some neuroleptic drugs such as fluspirilene,[78] as well as other structurally unrelated compounds,[76,79] have been proposed. These additional sites may also interact in an allosteric manner. As an example, the novel Ca^{2+}-entry blocker SR33557 seems to bind to a site distinct from those that are selective for either the dihydropyridines, the phenylalkylamines, or the benzothiazepines. Interaction with this site reduces binding to all three classical sites.[80] To date, at least six discrete binding sites have been characterized in the L channel.[81] These appear to have different topographic localization in the Ca^{2+} channel, with dihydropyridines binding near the orifice of the channel,[74,82] while phenylalkylamines appear to bind on the cytoplasmic side of the channel.[74,83,84]

As mentioned above, Ca^{2+} antagonists can also interact with several other structures, which may represent targets for as-yet-unexplained Ca^{2+} antagonistic effects or may be receptors for new drugs derived from

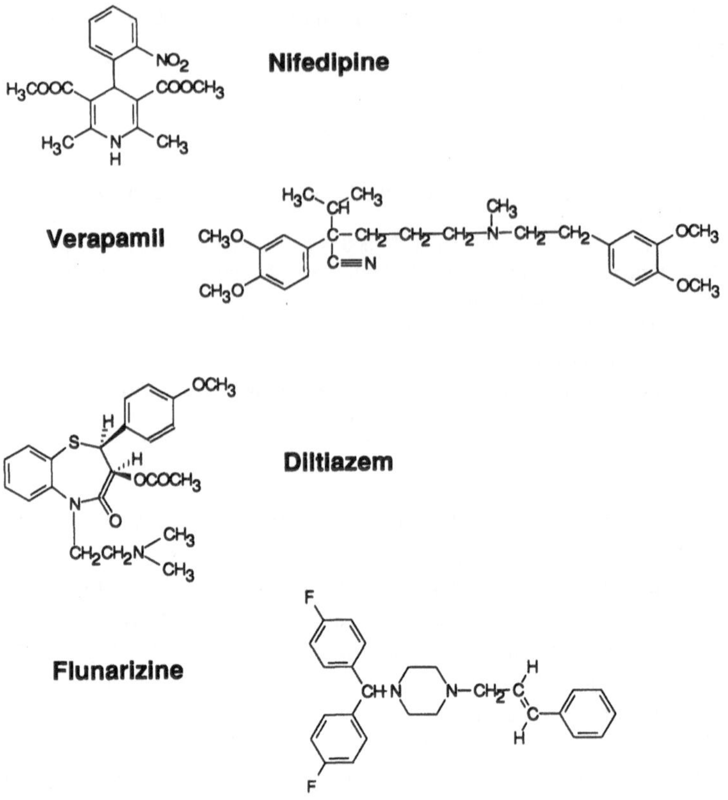

FIGURE 10.1. Formulas of representative substances belonging to the main groups of organic L-channel antagonists.

classical Ca^{2+} antagonists.[85] This may explain some of the apparent differences between Ca^{2+} antagonists, which cannot be directly linked to inhibition of L channels. Such possible additional effects are, however, only contributory to the Ca^{2+}-channel-blocking action.[85,86] Blockade of α-adrenoceptors at concentrations obtainable with the doses used clinically has been described for verapamil, but not for diltiazem or dihydropyridines. Direct inhibition of intracellular Ca^{2+} release and inhibition of calmodulin have been found. The latter mechanism can be demonstrated for dihydropyridine Ca^{2+} antagonists at 100–1000-fold higher concentrations than those required for vascular relaxation. Inhibition of cAMP phosphodiesterase (PDE), which leads to accumulation of cAMP, can be obtained with some Ca^{2+} antagonists, mainly dihydropyridines. The concentrations needed to inhibit PDE, however, are usually much higher than those required for vasodilatation. Other actions, such as direct interaction with contractile proteins, have been demonstrated for dil-tiazem, for example, but it is doubtful whether this occurs at concentrations that can be obtained with doses of the drug in clinical use.

In 1983 a dihydropyridine derivative, Bay K8644, which opposed the actions of other substances of this group in an apparent competitive manner, was described. The drug also increased cellular Ca^{2+} influx.[87,88] Later studies indicated that this "Ca^{2+}-agonistic" effect was not specific for this substance, but rather was a general feature of all dihydropyridines.[78] Subsequently a number of dihydropyridine derivatives with a preferential agonistic activity have been described (Table 10.1). However, synthetic drugs behaving principally as Ca^{2+}-channel agonists have not been found among phenylalkylamines and benzothiazepines.

The mechanisms by which specific dihydropyridines produce an agonistic or antagonistic effect have been debated. The membrane potential appears to determine whether the net effect of a drug is that of an agonist or an antagonist,[89-91] but different substances may have a preference for one or the other of these actions. An increasingly negative membrane potential promotes the agonistic effect,[92,93] whereas depolarization has the opposite effect.[93,94] In fact, even strongly "agonistic" dihydropyridines may function as Ca^{2+}-entry blockers at a sufficient degree of depolarization.[95] It has been proposed that dihydropyridines in general increase the probability of Ca^{2+}-channel inactivation and that this tendency is more pronounced for drugs which preferably act as antagonists.[93] This would result in a decrease in the probability for the Ca^{2+} channel to be open (see below). The possibility that dihydropyridines may interact with more than one binding site in the channel with different affinities for agonists and antagonists, respectively, also cannot be ruled out.[78] The apparent competitive nature of the interaction between agonists and antagonists may be the result of an allosteric interaction between these two binding sites.[95] This does not exclude an influence of the membrane potential on the net result (agonism or antagonism), as the relative affinities for the respective binding sites may be modulated by the degree of polarization.[96]

Clinical Profile and Tissue Selectivity

Organic Ca^{2+} antagonists and agonists are considered to interact with the α_1 subunit of the L-type Ca^{2+} channels, thereby selectively inhibiting or increasing Ca^{2+} inflow through the sarcolemma during membrane activation, with consequent attenuation or enhancement of muscular activity in cardiac and smooth muscle. Although Ca^{2+} antagonists presently are evaluated for treatment of disturbances in a variety of organs, they are still used mainly as cardiovascular drugs.

Cardiac Effects of Ca^{2+} Antagonists

If the mode of action of Ca^{2+} antagonists, ie, inhibition of Ca^{2+} influx through L-type Ca^{2+} channels, were the same in all cells, qualitatively similar effects of different drugs on the SA node, the AV node, the myocardium, and the coronary arteries would be expected. Experimental evidence shows that although this can be demonstrated in vitro, it is not always the case in vivo. In addition, it is known that Ca^{2+} channel number, ligand affinity, and/or function may change not only in various pathologic states, such as cardiomyopathy, myocardial ischemia, and hypertension, but also during chronic treatment with Ca^{2+} antagonists.[97] The therapeutic consequences of these changes are not known.

In isolated hearts and heart preparations, most Ca^{2+} antagonists have qualitatively similar effects on SA- and AV-nodal functions and on the myocardium.[98-101] Taira[101] compared the negative inotropic, chronotropic, and dromotropic effects of 12 different Ca^{2+} antagonists in the isolated blood-perfused papillary muscle, as well as in SA- and AV-node preparations. The coronary vasodilatory potencies were also determined. He found that dihydropyridines were more effective coronary vasodilators than modifiers of the SA- and AV-node function and papillary muscle contractility. Non-dihydropyridine Ca^{2+} antagonists, including verapamil and diltiazem, were almost equipotent in producing coronary vasodilatation and affecting SA- and AV-nodal function, and they were less potent in producing negative inotropic effects. Ning and Wit[99] have shown that at low concentrations verapamil was more effective than nifedipine in slowing the rate of impulse initiation in the isolated rabbit sinus node preparation, but at high concentrations the drugs were equieffective. Both drugs slowed conduction and prolonged refractory periods in the AV node, but verapamil was more potent than nifedipine. Diltiazem, bepridil, and perhexiline seemed to be less potent than verapamil and nifedipine in their effects on the SA and AV nodes.[98,100] In isolated perfused canine SA- and AV-node preparations local injections of isradipine produced dose-dependent decreases in spontaneous sinus rate and, at a higher dose range, a dose-dependent increase in AV-nodal conduction time was observed.[102] This is of particular interest since studies with other Ca^{2+} antagonists in the same preparation have shown that the drugs were equally effective in suppressing SA-node automaticity and AV-node conduction.[101-103] Taken together, these findings suggest that it is possible to block Ca^{2+} channels in the SA and AV nodes and in myocardial tissue with some selectivity, even if the basis for this selectivity has not been clearly determined.

All Ca^{2+} antagonists have a well-documented negative inotropic action in vitro. This has been extensively studied in various animal preparations. Generally, Ca^{2+} antagonists can cause complete excitation-contraction uncoupling, but their potencies vary, nifedipine most often being more potent than verapamil and diltiazem.[98,104,105] In the isolated human myocardium nifedipine and isradipine appear to have a more potent negative inotropic action than verapamil and diltiazem.[106] Thus, even if the qualitative effects on isolated heart tissues are the same for different Ca^{2+} antagonists, the order of potency on various cardiac structures can differ. This may be of importance for their effects in the whole individual (see below).

Vascular Effects of Ca^{2+} Antagonists

Ca^{2+} antagonists have a general vasodilating effect. However, there are differences in the sensitivity to the drugs not only between arteries and veins, but also within the vascular tree and between different vascular regions.[107] Thus, Ca^{2+} antagonists are preferentially arteriolar dilatators; they are generally more effective on smaller than on larger vessels. Furthermore, certain regions, eg, the coronary, cerebral, and renal circulations are more sensitive to their effects than other regions. Dihydropyridines also preferentially affect the skeletal muscle vasculature.[108]

Some Ca^{2+} antagonists exhibit vascular selectivity, ie, they inhibit contractile activity at lower concentrations in vascular smooth muscle than in the myocardium. Ljung[109] tested vascular vs myocardial selectivity of Ca^{2+} antagonists by comparing their effects on the isolated rat portal vein and the paced rat papillary muscle, which were studied in the same organ bath. He found that both types of tissue were equally sensitive to Ca^{2+} ion chelation with EDTA and to Ca^{2+} flux inhibition with lanthanum. Verapamil was almost equally effective on the two tissues, whereas diltiazem was 7-fold and nifedipine 14-fold more potent in causing vascular than myocardial inhibition. Among the agents tested, the dihydropyridine felodipine had the highest selectivity showing a 100-fold vascular vs myocardial preference. Several other dihydropyridines have been tested in different models and have shown a high vascular vs cardiac selectivity.[110]

Effects of Ca^{2+} Agonists

Blocking of L-type Ca^{2+} channels produces negative chronotropic, dromotropic, and inotropic effects. It could therefore be expected that drugs acting as agonists at the L channels, prolonging the duration of each open state, and decreasing the time interval between successive openings[111] would have the opposite effect. This is indeed the case.[93] Ca^{2+} antagonists of the dihydropyridine type often have a pronounced selectivity for vascular smooth muscle. The same should be true for Ca^{2+}-channel agonists. However, the strong vasoconstrictor action of all channel-activating dihydropyridines could not be separated from their positive inotropic effect.[93] The reasons for this will be discussed below.

Nevertheless, the pronounced effect on smooth muscle makes the presently available Ca^{2+} channel agonists unsuitable as therapeutic agents. Furthermore, all available channel activators produce serious central nervous system side effects in animals.[93]

Effects of Ca^{2+} Antagonists on Intact Individuals

Nakaya and coworkers[112] compared the chronotropic and inotropic effects of nifedipine, verapamil, and diltiazem in intact dogs. They found that at a similar level of systemic blood pressure decrease, heart rate was markedly increased by nifedipine, less with verapamil, and only slightly increased with diltiazem. Verapamil was found to have a negative inotropic effect, nifedipine a positive one, whereas contractility was unchanged after diltiazem. Similar findings were made by Urquhart et al.[113] Quartaroli et al.[114] who compared the hemodynamic effects of the dihydropyridines lacidipine, nitrendipine, and amlodipine with those of verapamil and diltiazem in dogs. Drugs were given intravenously as a single bolus and tested at different dose levels. All drugs produced reductions of blood pressure, although this response was short-lasting for verapamil and diltiazem. Lacidipine and nitrendipine caused a reflex increase in contractile index, whereas amlodipine, and most markedly verapamil, had a negative inotropic action. Diltiazem was practically devoid of a negative inotropic effect. A reflex increase in heart rate was observed with lacidipine, nitrendipine, and amlodipine, whereas verapamil and diltiazem caused bradycardia. Verapamil and diltiazem produced second- and third-degree AV block at the highest doses used, and surprisingly this was also found with amlodipine. It has been suggested that both the bradycardic and hypotensive effects of verapamil and diltiazem may be due, at least to some extent, to effects on the central nervous system.[115] Both drugs are lipophilic and pass the blood-brain barrier rather easily.[116] When given intracisternally to cats, verapamil and diltiazem produced relatively greater effects than after intravenous administration. The bradycardic effects after both types of administration were abolished by bilateral cervical vagotomy, and the hypotensive response was attenuated.[115]

In man, nifedipine increases and diltiazem decreases heart rate, whereas verapamil most often causes no changes.[117] Verapamil and diltiazem were shown to have more pronounced effects than nifedipine on the AV node. These effects of verapamil were observed before depression of contractility.[118,119] A negative inotropic effect of nifedipine, verapamil, and possibly diltiazem may be observed when the drugs are injected directly into the coronary arteries. However, when the drugs are given intravenously, nifedipine increases or causes no change in contractility, whereas both verapamil and diltiazem decrease contractility or cause no change.[117]

Thus in vivo there are important differences in effects on various cardiac structures, not only between dihydropyridines on one hand and verapamil and diltiazem on the other hand, but also between different dihydropyridines. In the intact individual, all Ca^{2+} antagonists reduce peripheral resistance and blood pressure. This induces a reflex activation of the sympathetic nervous system that counteracts not only the myocardial effects, but also the actions on the SA and AV nodes. As a consequence of the vascular selectivity of dihydropyridines, their direct cardiac effects, at the doses used clinically, will be less pronounced or negligible compared to those caused by verapamil and diltiazem. This was illustrated in β-adrenoceptor-blocked patients, for whom nifedipine, but not felodipine and nicardipine, had a demonstrable negative inotropic effect.[120]

Basis for Tissue Selectivity of L-Channel Antagonists

As discussed above, a decrease in membrane polarization increases the ability of dihydropyridines to inhibit calcium entry. In addition, the affinity of antagonistic dihydropyridines to their binding sites and, hence, their antagonistic action, increases with depolarization.[94,121-124] On the other hand, the binding of agonistic dihydropyridines seems essentially independent of membrane potential.[82,125] Potential dependence for binding to Ca^{2+} channels has also been demonstrated for verapamil, D-600, and diltiazem.[70,84,126] The increased affinity for Ca^{2+} channels during depolarization seems to be a general feature of Ca^{2+} antagonistic drugs with few exceptions such as pinaverium.[127]

Depending on the membrane potential, Ca^{2+} channels may exist in three hypothetical states; resting (closed and available for activation), open (activated), and inactivated (Figure 10.2).[54,74,128] The equilibrium between the three states is regulated by the membrane potential. As the potential decreases there is an increased probability for the channels to be in the activated state. However, with increasing level and duration of depolarization the probability of channels entering an inactivated state is enhanced.[54] Once inactivated, the channels are unavailable for activation until the cell has repolarized. Since almost all Ca^{2+} antagonists bind preferentially to the channels during depolarization (and, thus, to activated or inactivated channels), this could account for the voltage dependency of their antagonistic effects. Furthermore, the effects of verapamil and diltiazem have been shown to be potentiated at increasing frequency of depolarization with intervening periods of repolarization (use dependency),[123,129-132] a behavior characteristic of organs exhibiting a rhythmic activity, such as the heart.

As the frequency of depolarizing bursts increases, the probability of Ca^{2+}-channel activation is favored.[54,128] On the other hand, if the de-

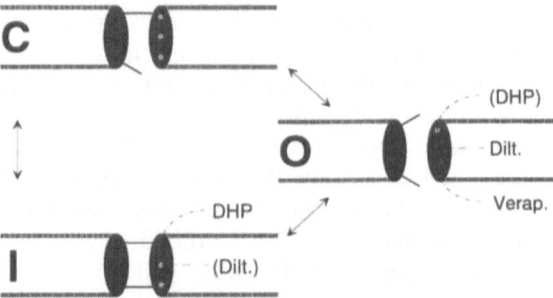

FIGURE 10.2. Schematic drawing of the three hypothetical states of L channels, closed (C) and available for activation; open (O) and activated; and inactivated (I). As the membrane potential decreases the probability of the channels to exist in the open state is favored. However, during prolonged periods of depolarization an increasing proportion of the channels will become inactivated and unavailable for activation until the cell has repolarized. The functional binding sites (which are not necessarily represented by the exact locations of the labels in this figure) for dihydropyridines (DHP), benzothiazepines (represented by diltiazem = dilt.), and phenylalkylamines (represented by verapamil = verap.) are also indicated.

polarization is maintained for a longer period of time, an increasing proportion of channels will become inactivated.[133,134] Ca^{2+} antagonists such as verapamil[130] and diltiazem[135] bind preferentially to channels in the open state. Although controversial, it has been postulated that molecules such as verapamil must enter the calcium channels in order to reach their site of action.[84] This concept is in accordance with the guarded receptor theory proposed by Hille.[136] According to his findings, local anaesthetics are able to reach their binding sites in the Na^+ channel only when the channel is in the open state. Such a mechanism may contribute to the negative chronotropic effects of verapamil and diltiazem in the heart.

Although it has been reported that dihydropyridines may bind to both activated[137] and resting[125] Ca^{2+} channels, most investigators have demonstrated a clear preference of antagonistic drugs for the inactivated channel.[78] In fact, their binding may be increased after prolonged periods of depolarization[135,137,138] and thus the effect of most dihydropyridines seem more voltage-dependent than strictly use-dependent.[123] This is especially prominent with the novel highly potent dihydropyridines such as isradipine[135] and nicardipine.[123] Such an affinity for inactivated channels may account for the minimal use dependency and, thus, minimal chronotropic effects, of this group of Ca^{2+}-entry blockers.[90] In addition, dihydropyridines may actually stimulate the voltage-dependent inactivation process of Ca^{2+} channels, thereby further augmenting this process.[90,107]

Considering that blood vessels, when activated, usually show a steady depolarization, this may contribute to the vascular selectivity of the dihydropyridines. Since not only the affinity for inactivated channels, but also the augmentation of the inactivation process, is greater for the more potent derivatives, the higher vascular preference of nisoldipine as compared to, eg, nifedipine, can also be explained.[68] In contrast, agonistic dihydropyridines are capable of binding with high affinity to both open and inactivated (and possibly also resting) channels.[93] This, as well as their low voltage dependency for binding (see above), may explain their low vascular selectivity. The interactions of piperazines and other unspecific Ca^{2+}-entry blockers with the calcium channels, as well as their dependence on membrane potential,[66] have not been adequately investigated.

In addition to an influence of the preferential binding of Ca^{2+}-entry blockers to one or the other of the three states of the Ca^{2+} channel, other factors most certainly play a role in tissue selectivity. Not only may the sensitivity of Ca^{2+} channels to dihydropyridines differ between various smooth muscle,[139] but differences in the resting membrane potential may also play a role in determining Ca^{2+}-antagonist sensitivity.[140] For example, the low negative chronotropic effect of dihydropyridines may partially be the consequence of a relatively low membrane potential in cardiac cells as compared with smooth muscle cells.[135] According to the evidence discussed above, a high degree of polarization would reduce the binding and favor the agonistic activity of the dihydropyridine derivatives. The less-negative membrane potential in smooth muscle would thus account for the vascular preference of the antagonistic action of these compounds.

The interactions between Ca^{2+} antagonists and the vascular endothelium may be of particular interest. Ca^{2+} antagonists do not seem to prevent the release of either endothelium-dependent relaxing or contracting factors. However, they facilitate the inhibition produced by the relaxing factor(s) and prevent the activation of vascular smooth muscle produced by the contracting substances, thus favoring dilatation.[141]

A factor which must be considered is the lipophility of certain Ca^{2+}-entry blockers, such as dihydropyridines and piperazines. The long duration of action of these compounds may be a consequence of this property.[82] Another example of the influence of solubility factors is the theory of "membrane catalysis" proposed by Sargent and Schwyzer.[142] According to this theory the lipid bilayer of the cell membrane may actually enhance the effect of hydrophobic substances by concentrating the drugs in the membrane and by influencing their conformation and orientation toward the receptors. Introduction of a charged side group in the otherwise hydrophobic dihydropyridine molecule has been shown to affect both the distribution and orientation of the drug in the cell membrane.[143] Thus, the effects of drugs such as dihydropyridines[66] may depend on the composition and state of the membrane. Such mechanisms, as well as the influence of polarity on the distribution if the drug, may play a role in organ selectivity of Ca^{2+} antagonists.

The availability of intracellularly-sequestered Ca^{2+} for the excitation-contraction coupling process obviously influences the susceptibility to Ca^{2+}-entry blockers. Although skeletal muscle contains a high concentration of high-affinity binding sites for dihydropyridines,[74,75,93] contraction is entirely dependent on the release of Ca^{2+} from the sarcoplasmic reticulum[144] and not on the influx via L channels.[32] It has been shown that only a fraction of the dihydropyridine-binding sites in skeletal muscle are functional Ca^{2+} channels.[145] In addition, these channels in skeletal muscle appear to have a relatively low sensitivity to Ca^{2+}-entry inhibitors.[146] This is in agreement with the recently discovered differences in the α_1 subunit of the channel protein between skeletal and cardiac muscle.[34,37] Similar mechanisms may explain the variation in Ca^{2+}-antagonist susceptibility between respiratory and vascular smooth muscle responses to some (but not all) stimulants.[67]

Differences in the effectiveness of Ca^{2+}-channel modulators as inhibitors of contraction can be demonstrated also between various types of blood vessel. In particular, cerebral arteries appear to be extremely sensitive to these drugs.[147–150] Cauvin and cowokers.[107,151] have shown that the susceptibility to Ca^{2+}-entry blockers increased with decreasing vascular size. Such differences may be explained by decreasing availability of intracellularly-sequestered Ca^{2+}.[151] Contributing to the high sensitivity of some arteries to dihydropyridines may be the presence of a dihydropyridine-sensitive subset of ROCs[107,151] or, more likely an involvement of receptor-induced intermediate steps, which ultimately lead to activation of POCs[111] (see also above). Alternative explanations may be differences in membrane potential[66,152] or in the modulatory influence of intracellular Ca^{2+} release on Ca^{2+} entry.[153]

Thus the matter of tissue selectivity of Ca^{2+}-channel modulators seems rather complex. As mentioned above the possibility of structural differences between L channels in various organs[154] must be considered. Isolation of the L channel from different tissues has indicated variations in the α_1 subunit of the L channel,[34,37,155] which contains the binding sites for calcium antagonists.[78] Such variability in the structure of Ca^{2+} channels seems to exist between skeletal muscle and cardiac and vascular muscle. However, there may be also differences between the latter two muscle types, since the sensitivity of single L channels to dihydropyridines appears to be several times greater in vascular smooth muscle than in cardiac muscle.[32,156,157] This may contribute to differences in susceptibility to dihydropyridines in various smooth muscles.[139] Finally, a number of additional factors, such as the influence of various cations (including Ca^{2+}, both extra- and intracellular), pH, protein phosphorylation,[66] and GTP and other nucleotides may contribute to the preferential action of the different Ca^{2+} entry blockers on one organ or the other, either directly or through interactions with G proteins.[18,34,158]

Clinical Use of Ca^{2+} Antagonists

Cardiac Arrhythmias

Disturbances of cardiac rhythm can be divided into two major classes: disorders of automaticity and of impulse conduction.[159] Irrespective of the mechanism, Ca^{2+} influx through L channels is likely to be involved in these events.[160] Cells in both the SA and AV nodes normally generate slow action potentials, but myocardial cells that do not ordinarily initiate action potentials can also do so when they are partially depolarized by, eg, ischemic injury. L channels are responsible for these action potentials, which thus can be blocked by Ca^{2+} antagonists. After-depolarizations are membrane potential oscillations that appear after the upstroke of the action potential. Early after-depolarizations can be induced in cardiac cells by a variety of interventions including hypoxia, excessive concentrations of catecholamines, and Ca^{2+} agonists like Bay K8644. Delayed after-polarizations can be induced by Ca^{2+} overloading induced by, eg, digitalis or high concentrations of catecholamines. Such electric activity can also be blocked by Ca^{2+} antagonists, which suggests the involvement of L channels.

Therefore all Ca^{2+} antagonists could be expected to be effective in different types of arrhythmias. However, in clinical arrhythmias, their main therapeutic role is related to their effect on the SA and AV nodes.[161,162] Only verapamil and diltiazem are widely used, even if drugs such as tiapamil, gallopamil, and anipamil have comparable anti-arrhythmic actions.[161] Verapamil and diltiazem are used for interruption and prophylaxis of supraventricular tachycardia in cases where the AV node is part of the reentry circuit. In addition, they may reduce the ventricular rate in atrial fibrillation and flutter. Ca^{2+} antagonists may reverse ischemic ventricular arrhythmias caused by coronary artery spasm, but the drugs do not provide effective therapy for controlling ventricular tachycardia in association with chronic coronary artery disease and idiopathic dilated cardiomyopathy.[161–163] Dihydropyridines seem to have no useful anti-arrhythmic effect, probably because of their vaso-selectivity and their reflex-induced stimulation of the sympathetic nervous system.[161–163]

Angina Pectoris

Ca^{2+} antagonists have been proven effective in reducing the frequency of angina both at rest and during exertion. They are considered the treatment of choice for vasospastic (variant) angina,[164] and are one of the first-line alternatives for stable angina.[165] These drugs may be useful in unstable angina and silent myocardial ischemia,[166] but there may be clinically important differences in effectiveness between different drugs. Thus Stone et al.[167] found no effect of nifedipine on any manifestation of asymptomatic ischemia, whereas diltiazem was moderately effective. Ca^{2+} antagonists are believed to exert their beneficial effects by preventing coronary vasospasm in vasospastic angina, by reducing work-load in effort angina, and, in addition by possibly inhibiting platelet aggregation in unstable angina.[165,166]

The mechanisms involved in the coronary vasospasm of variant angina have not been established,[168] but the clinical effectiveness of Ca^{2+} antagonists suggests an involvement of L channels. It has been reported that smooth muscle cells from the dog coronary artery do not normally exhibit action potentials when electrically stimulated,[169,170] and in these cells the role of POCs is less well understood. However, in enzymatically-isolated smooth muscle cells from the rabbit epicardial coronary artery, Matsuda et al.[171] were able to demonstrate the presence of L channels and Ganitkevich and Isenberg[172] found evidence for the occurrence of both T and L channels with a preponderance of L-type channels in isolated single smooth muscle cells from the circumflex guinea pig coronary artery. The same may be the case in human coronary arteries, but this remains to be conclusively demonstrated. In isolated human epicardial arteries, basal tone was found to be dependent on extracellular Ca^{2+}.[173,174] These arteries exhibit spontaneous rhythmic periods of contraction and relaxation that can be abolished by Ca^{2+} antagonists, which suggests Ca^{2+} influx through POCs. Isolated human coronary arteries, contracted by a variety of agonists, are effectively relaxed by Ca^{2+} antagonists.[168] Also in vivo Ca^{2+} antagonists have been shown to dilate not only large coronary vessels, but also resistance vessels.[175] Results from studies on collateral vessels have been contradictory.[175] The dilating action is probably the result of a direct effect on the coronary smooth muscle cells, and not due to the release of endothelial-relaxing factors. Interestingly, nifedipine and diltiazem were found to decrease ischemia-induced vasodilatation of coronary resistance vessels.[176,177]

In effort angina pectoris, which is believed to be associated with fixed obstruction, the immediate factor precipitating angina is a sudden increase in the metabolic requirements of the heart, which cannot be satisfied because of the restricted flow in the coronary arteries. The short-term effects of Ca^{2+} antagonists include a reduction in peripheral vascular resistance, an increased coronary flow and, in the case of verapamil and diltiazem, a reduction in heart rate. All these actions, which ultimately lead to a reduced metabolic demand, may well be explained by blockade of L channels. Long-term usage may in addition provide a slowing of atherogenesis and regression of cardiac hypertrophy, as well as possible protection against myocardial infarction and sudden death (see below). To what extent such effects can be attributed to inhibition of L-type Ca^{2+} channels remains to be established.

With unstable angina, which is a sudden worsening of an existing anginal state, several factors may contribute to the symptoms, including rupture of an atherosclerotic plaque, platelet aggregation and thrombus formation at the site of the ruptured plaque, coronary vasospasm, or an increase in cardiac workload. In this condition Ca^{2+} antagonists may be beneficial not only because of the actions described above, but also because they inhibit platelet aggregation.

Myocardial Protection

Pharmacologic protection against myocardial damage caused by ischemia and reperfusion of ischemic myocardium is currently a field of intensive research, partly as a consequence of the success of thrombolytic therapy in myocardial infarction. In myocardial ischemia, there is progressive damage of the myocytes, which starts with a loss of ATP and creatine phosphate associated with a loss of K^+,[67,178] and, initially, with a reduction in myocardial Ca^{2+} uptake. However, upon reperfusion Ca^{2+} uptake is markedly increased with a cytosolic accumulation of Ca^{2+}, which may have fatal effects on the cell. The ischemia-induced cytosolic Ca^{2+} overload appears to be a very complex process, which is not necessarily dependent only on Ca^{2+} entry through L channels. However, animal studies have shown that Ca^{2+} antagonists are able to increase the resistance of isolated hearts to ischemia and also to prevent the massive overloading with Ca^{2+} during reperfusion (which is believed to cause myocardial cell necrosis), provided that they are added before the myocardium is made ischemic.[67,178–180] There are several mechanisms involving L-type Ca^{2+} channels by which Ca^{2+} antagonists may exert a beneficial effect (Table 10.2), including coronary and peripheral vasodilatory effects, and slowing of the Ca^{2+} accumulation rate during reperfusion, as well as attenuation of reperfusion-induced arrhythmias.[178,180] If coronary spasm is involved in the production of ischemic necrosis, the protective action may be explained by the coronary vascular effects of the drugs. It seems, however, that at least some Ca^{2+} antagonists may exert a protective effect without producing afterload reduction or coronary vasodilatation. Therefore, nicardipine was tested on reoxygenation injury in single rat cardiac myocytes,[181] a model in which vascular effects are eliminated. It was found that at concentrations that produced similar negative inotropic effects nicardipine, but not nifedipine, had a strong protective action. As this effect was obtained in unstimulated myocytes, it was concluded that nicardipine was protective for reasons unrelated to its L-channel blocking activity.[181] Electrophysiologic studies of the ischemic myocardium have indicated that early postinfarction ventricular arrhythmias are mediated by slow-response action potentials.[182] If this is

TABLE 10.2. Mechanisms that contribute to the protective effect of calcium antagonists on the myocardium following ischemia and reperfusion.*

1. Coronary vasodilatation, which leads to an improved supply of oxygen and metabolic substrates as well as removal of products of anaerobic glycolysis.
2. Attenuation of ischemia-induced acidosis.
3. An energy-sparing effect with a slowed rate of adenosine triphosphate deletion and slowed loss of adenosine precursors.
4. Slowed release of degradative lysosomal proteases.
5. Protection of vascular endothelium.
6. Attenuation of the ischemia-reperfusion-induced mobilization of noradrenaline.
7. Protection of mitochondrial function.
8. Retardation of the early rise in cytosolic Ca^{2+}.
9. Protection of membranes against lipid-peroxidation caused by free radicals.

* Modified from Nayler.[180]

the case, Ca^{2+} antagonists may exert part of their beneficial effect by preventing the development of severe ventricular arrhythmias or fibrillation.

There is evidence that ischemia may influence the binding of Ca^{2+} antagonists to myocardial sarcolemmal fragments. There was a decrease in binding sites for both nitrendipine[179] and verapamil[183] following 60 min of ischemia by occlusion of coronary flow. Also, in cardiac membranes from guinea pig hearts made hypoxic and then reoxygenated, there was a significant reduction of [^{3}H]-nitrendipine-binding sites.[184] The functional consequences of these findings are unclear.

The attractive pharmacologic profile and the promising experimental results obtained with Ca^{2+} antagonists in different animal models of ischemia and reperfusion have motivated trials with the drugs in acute myocardial infarction (thrombolysis, coronary angioplasty), and for secondary prevention after myocardial infarction. However, Erbel et al.[185] were unable to show a beneficial effect of nifedipine (compared to placebo) given during thrombolysis on left ventricular functional recovery or clinical outcome. Similarly, Ellis et al.,[186] who retrospectively analyzed patient in-hospital survival after coronary angioplasty, found that Ca^{2+} antagonist treatment did not improve early survival. Several studies have been performed with nifedipine, diltiazem, and verapamil in the secondary prevention of myocardial infarction.[187,188] The results with nifedipine have been uniformly negative. An analysis including data from 28 randomized trials, which created a database of more than 19,000 patients, led to the conclusion that "calcium antagonists do not reduce the risk of initial or recurrent infarction or death when given routinely to patients with acute myocardial infarction."[187] However, positive results were found in a recent study with verapamil.[189] A significant reduction in major events (death or reinfarction) was demonstrated in 878 patients treated for a mean of 16 months. The positive effect was found in patients without heart failure. In a subset of patients in whom ischemic episodes were documented by 24–48 h Holter monitoring, verapamil was found to significantly reduce the number of transient ischemic episodes, to reduce the mean 24-h heart rate, and to have an anti-arrhythmic effect.[189] Such actions may have contributed to the beneficial effects of verapamil treatment.

Therefore it cannot be excluded that in selected patients with myocardial infarction, some Ca^{2+} antagonists may offer a secondary prophylactic effect, possibly comparable to those of β-adrenoceptor blockers.[188,190]

Hypertension

It is generally agreed that the main hemodynamic derangement in established essential hypertension is an increased systemic vascular resistance. In contrast to several other vasodilators, Ca^{2+} antagonists have a favorable hemodynamic profile, which makes them suitable for treatment of hypertension without causing relevant chronic sympathetic reflex activation or sodium and volume retention.[191,192]

Lederballe Pedersen et al.[193] found that aortic contractions were more dependent on extracellular Ca^{2+} in vessels of spontaneously hypertensive than of normal rats. In addition, the vessels of hypertensive animals showed increased sensitivity to the relaxing effects of nifedipine. Studies with radiolabeled Ca^{2+} in arteriolar resistance vessels of hypertensive rats revealed that the stimulatory effect of noradrenaline on Ca^{2+} influx is enhanced in these vessels, an effect that can be inhibited by Ca^{2+} antagonists.[194,195] These findings support the view that there is an increased Ca^{2+} influx in vessels of hypertensive animals that may contribute to the elevation of arteriolar tone, and that this route of Ca^{2+} entry is sensitive to Ca^{2+} antagonists. As pointed out by Nayler,[67] this may be due to an increased number of operational L channels or to an improved Ca^{2+}-ion-carrying capacity of each channel. There is some evidence that there is an increased number of L-type Ca^{2+} channels in a variety of tissues from hypertensive animals,[196,197] but whether this is a consequence of hypertension or has any importance in its pathogenesis has not been determined.

Congestive Heart Failure

Ca^{2+} antagonists, by means of a vasodilator effect that produces after-load reduction and a myocardial protective effect, may be expected to be of hemodynamic benefit to patients with chronic heart failure, particularly since 60% of them have left ventricular failure due to coronary artery disease.[198] In addition, Ca^{2+} antagonists have been shown to delay and prevent the development of heart failure in a strain of hamsters with hereditary cardiomyopathy,[199] In this model an excess of dihydropyridine-binding sites were found.[200] Although an increase in dihydropyridine-binding sites was found in human atria from hearts with hypertrophic cardiomyopathy,[201] dihydropyridine binding sites were not altered significantly in left ventricular myocardium of patients with idiopathic dilated cardiomyopathy.[202]

After reviewing the field, Packer[198] concluded that the use of Ca^{2+} antagonists should be avoided in patients with chronic heart failure, because during short-term therapy these drugs may produce deleterious hemodynamic and clinical effects, probably as a consequence of their cardiodepressant actions. During long-term therapy the drugs may increase cardiovascular morbidity and mortality, possibly as a result of their capacity to activate endogenous neurohormonal systems.[203,204] This conclusion was based on results obtained mainly with the first generation Ca^{2+} antagonists. On the other hand, Reicher-Reiss and Barasch[205] concluded that the second-generation drugs, which have more potent and selective vasodilatating properties with less negative inotropic actions, could be expected to produce benefits in patients with heart failure. However, due to the lack of large, long-term, controlled studies, no clear recommendations were given. The existing clinical trials have not proven that Ca^{2+} antagonists are superior to other vasodilators in patients with congestive heart failure.

Prevention of Atherosclerosis

In several animal models of atherosclerosis, particularly in the cholesterol-fed rabbit, experimental evidence suggests that various types of Ca^{2+} antagonists (verapamil, diltiazem, and dihydropyridines) in high doses may retard the atherogenic lesion development.[206] Other studies have suggested that in humans Ca^{2+} antagonists may delay the progression of the atherosclerotic process at the stage of early lesion.[207–210] Thus Lichtlen et al.[209] have demonstrated that treatment with nifedipine over a 3-year period suppresses the formation of new lesions in patients with moderately severe coronary artery disease. Similar results were obtained by Waters et al.,[210] who have shown that nicardipine had no effect on advanced coronary atherosclerosis, but seemed to retard the progression of minimal lesions.

The mechanisms behind the anti-atherogenic effect of these drugs are not known. Ca^{2+} antagonists may inhibit smooth muscle cell proliferation and migration, improve cellular lipoprotein metabolism in vascular cells, alter phospholipid turnover, decrease platelet adhesion in the vessel wall, reduce extracellular matrix synthesis, and protect against radical-induced cell damage.[211] Schmitz et al.[212] suggested that most of these effects are independent of Ca^{2+} flux across POCs, because smooth muscle cells were found to be the only cell type involved in atherosclerosis that expressed this type of Ca^{2+} channel. It has been proposed that owing to their lipophility, Ca^{2+} antagonists may accumulate intracellularly to a substantial degree and may possibly bind to a variety of low-affinity/high-capacity drug receptors. Many of these receptors have variable binding characteristics, and their occupation by Ca^{2+} antagonists may influence Ca^{2+} uptake and/or extrusion by non-L channel mechanisms.[85]

Cerebral Vasospasm and Ischemia

Nimodipine is commonly used for the prevention of cerebral vasospasm after aneurysmal subarachnoid hemorrhage. Despite a significant reduction of symptoms of delayed ischemic brain dysfunction and improved outcome in patients treated with nimodipine, no alleviation of vasospasm can be detected angiographically.[213] Since dihydropyridines can have a neuroprotective effect by decreasing ischemic Ca^{2+} overload,[214-217] this may theoretically account for the beneficial effect of nimodipine. However, several investigators have failed to demonstrate a neuroprotective effect of nimodipine when administered after the insult,[213,218] which is not unexpected in the light of the diversity of the Ca^{2+}-channel subtypes in central neurones (see above) and the temporal profile of the neuronal Ca^{2+} overload following ischemia.[218] Thus even if a protective effect of nimodipine on the neuronal level cannot be completely excluded, other explanations for the discrepancy between clinical effect and angiography must be considered.

It has been proposed that nimodipine may selectively dilate small cerebral arteries, which cannot be visualized angiographically,[219,220] an effect which would fit well with the proposed selectivity of Ca^{2+} antagonists for small arteries over large arteries, as observed in the peripheral circulation.[29,107,151] In fact, when nimodipine was applied locally on the cortical surface in conjunction with craniotomy for superficial temporal to middle cerebral artery bypass operations, a preferential dilatation of arteries with a diameter of less than 70 μm was observed.[221] It has been suggested that the apparent neuroprotective effect of nimodipine in experimental models of cerebral ischemia may be due to a reduced vascular Ca^{2+} overload,[217] with subsequent prevention of ischemia-induced vasoconstriction. In support of this concept, Gelmers et al.[222] observed that nimodipine induced an "inverse steal" phenomenon in patients with cerebral infarcts, which suggested an improved microcirculatory state in the penumbra zone that surrounded the irreversibly damaged ischemic core of the infarct. Despite this putative beneficial effect at the microcirculatory level and possible neuroprotective effects of L-channel blockers, the results from clinical trials in non-vasospastic cerebral ischemia have been disappointing.[218]

Conclusion

Ca^{2+} channels are present in most vertebrate cells. However, our current knowledge of their clinical pharmacology reflects almost exclusively the effects of L-type channel antagonists, notably in the cardiovascular system. Despite the fact that all Ca^{2+} antagonists bind to the α_1 subunit of the L-type channel, a certain degree of tissue selectivity of available drugs allows for effective therapy with reasonably low levels of side effects in a variety of diseases. However, not all clinically relevant actions of Ca^{2+} antagonists can be linked to their interaction with Ca^{2+} channels. The current rapid progress in the understanding of the molecular biology and electrophysiology of Ca^{2+} channels heralds the development, not only of substances with greater specificity for L channels in various organs, but also of clinically useful drugs that interact with other Ca^{2+} channel subtypes.

References

1. Hille B. Ionic channels of excitable membranes. Sunderland, Mass.: Sinauer; 1985.
2. Smith JS, Coronado R, Meissner G. Sarcoplasmic reticulum contains adenine nucleotide-activated calcium channels. Nature 1985; 16:446–449.
3. Sitsapesan R, Williams AJ. Mechanisms of caffeine activation of single calcium-release channels of sheep cardiac sarcoplasmic reticulum. J Physiol 1990; 423:425–439.
4. Gill DL, Ueda T, Chueh SH, Noel MW. Ca^{2+} release from endoplasmic reticulum is mediated by a guanine nucleotide regulatory mechanism. Nature 1986; 320:461–464.
5. Saida K, van Breemen C. GTP requirement for inositol-1,4,5-trisphosphate-induced Ca^{2+} release from sarcoplasmic reticulum in smooth muscle. Biochem Biophys Res Commun 1987; 144:1313–1316.
6. Saida K, Twort C, van Breemen C. The specific GTP requirement for inositol 1,4,5-trisphosphate-induced Ca^{2+} release from skinned vascular smooth muscle. J Cardiovasc Pharmacol 1988; 12(suppl 5):S47–S50.
7. Kobayashi S, Kanaide H, Nakamura M. K^+-depolarization induces a direct release of Ca^{2+} from intracellular storage sites in cultured vascular smooth muscle cells from rat aorta. Biochem Biophys Res Commun 1985; 129:877–884.
8. Kanaide H, Kobayashi S, Nishimura J, Hasegawa M, Shogakiuchi Y, Matsumoto T, Nakamura N. Quin2 microfluorometry and effects of verapamil and diltiazem on calcium release from rat aorta smooth muscle cells in primary culture. Circ Res 1988; 63:16–26.
9. Bolton TB. Mechanisms of action of transmitter and other substances on smooth muscle. Physiol Rev 1979; 59:606–718.
10. van Breemen C, Aaronson P, Loutzenhiser R. Na^+, Ca^{2+} interactions in mammalian smooth muscle. Pharmacol Rev 1979; 30:167–208.

11. Mayer ML, Miller RJ. Excitatory amino acid receptors, second messengers and regulation of intracellular Ca^{2+} in mammalian neurons. Trends Pharmacol Sci 1990; 11:254–260.

12. Changeux J-P, Giraudat J, Dennis M. The nicotinic acetylcholine receptor: Molecular architecture of a ligand-regulated ion channel. Trends Pharmacol Sci 1987; 8:459–465.

13. Bormann J. Electrophysiology of $GABA_A$ and $GABA_B$ receptor subtypes. Trends Neurosci 1988; 11:112–116.

14. Hartig PR. Molecular biology of 5-HT receptors. Trends Pharmacol Sci 1989; 10:64–69.

15. Brum G, Osterrieder W, Trautwein W. β-Adrenergic increase in the calcium conductance of cardiac myocytes studied with the patch clamp. Pfluegers Arch 1984; 401:111–118.

16. Trautwein W, Hescheler J. Regulation of cardiac L-type calcium current by phosphorylation and G proteins. Annu Rev Physiol 1990; 52:257–274.

17. Kostyuk PG. Calcium channels in cellular membranes. J Mol Neurosci 1990; 2:123–141.

18. Brown AM, Birnbaumer L. Direct G protein gating of ion channels. Am J Physiol 1988; 254:H401–H410.

19. Dolphin AC. G protein modulation of calcium currents in neurons. Annu Rev Physiol 1990; 52:243–255.

20. Schultz G, Rosenthal W, Hescheler J. Role of G proteins in calcium channel modulation. Annu Rev Physiol 1990; 52:275–292.

21. von Tscharner Y, Prod'hom B, Baggiolini M, Reuter H. Ion channels in human neutrophils activated by a rise in free cytosolic calcium concentration. Nature 1986; 324:369–372.

22. Kuno M, Gardner P. Ion channels activated by inositol 1,4,5-trisphosphate in plasma membrane of human T-lymphocytes. Nature 1987; 326:301–304.

23. Benham CD, Tsien RW. A novel receptor-operated Ca^{2+}-permeable channel activated by ATP in smooth muscle. Nature 1987; 328:275–278.

24. Mahaut-Smith MP, Rink TJ, Sage SO. Single channels in human platelets activated by ADP. J Physiol 1989; 415:24P.

25. Ascher P, Nowak L. Calcium permeability of the channels activated by N-methyl-D-aspartate (NMDA) in isolated mouse central neurones. J Physiol 1986; 377:35P.

26. Casteels R, Droogmans G, Missiaen L. Agonist-induced entry of Ca^{2+} in smooth muscle cells. Neurochem Int 1990; 17:297–302.

27. Hallam TJ, Rink TJ. Receptor-mediated Ca^{2+} entry: Diversity of function and mechanism. Trends Pharmacol Sci 1989; 10:8–10.

28. Nishimura J, Khalil RA, van Breemen C. Agonist-induced vascular tone. Hypertension 1989; 13:835–844.

29. van Breemen C, Saida K. Cellular mechanisms regulating $[Ca^{2+}]_i$ smooth muscle. Annu Rev Physiol 1989; 51:315–329.

30. Somlyo AV, Franzini-Armstrong C. New views of smooth muscle structure using freezing, deep-etching and rotary shadowing. Experientia 1985; 41:841–856.

31. Meldolesi J, Clementi E, Fasolato C, Zacchetti D, Pozzan T. Ca^{2+} influx following receptor activation. Trends Pharmacol Sci 1991; 12:289–292.

32. Bean BP. Classes of calcium channels in vertebrate cells. Annu Rev Physiol 1989; 51:367–384.
33. Miller RJ. Multiple calcium channels and neuronal function. Science 1987; 235:46–52.
34. Dascal N. Analysis and functional characteristics of dihydropyridine-sensitive and -insensitive calcium channel proteins. Biochem Pharmacol 1990; 40:1171–1178.
35. Nelson MT, Patlak JB, Worley JF, Standen NB. Calcium channels, potassium channels, and voltage dependence of arterial smooth muscle tone. Am J Physiol 1990; 259:C3–C18.
36. Swandulla D, Carbone E, Lux HD. Do calcium channel classifications account for neuronal calcium channel diversity? TINS 1991; 14:46–51.
37. Tsien RW, Ellinor PT, Horne WA. Molecular diversity of voltage-dependent Ca^{2+} channels. Trends Pharmacol Sci 1991; 12:349–354.
38. Dolphin AC. Regulation of calcium channel activity by GTP binding proteins and second messengers. Biochim Biophys Acta 1991; 1091:68–80.
39. Sher E, Biancardi E, Passafaro M, Clementi F. Physiopathology of neuronal voltage-operated calcium channels. FASEB J 1991; 5:2677–2683.
40. Kass RS, Sanguinetti MC. Inactivation of calcium channel current in the calf cardiac purkinje fiber. Evidence for voltage-and calcium-mediated mechanisms. J Gen Physiol 1984; 84:705–726.
41. Llinas R, Sugimori M, Lin J-W, Cherksey B. Blocking and isolation of a calcium channel from neurons in mammals and cephalopods utilizing a toxin fraction (FTX) from funnel-web spider poison. Proc Natl Acad Sci USA 1989; 86:1689–1693.
42. Regan LJ, Sah DWY, Bean BP. Ca^{2+} channels in rat central and peripheral neurons: High-threshold current resistant to dihydropyridine blockers and ω-conotoxin. Neuron 1991; 6:269–280.
43. Bean BP. Two kinds of calcium channels in canine atrial cells. Differences in kinetics, selectivity, and pharmacology. J Gen Physiol 1985; 86:1–30.
44. Benham CD, Hess P, Tsien RW. Two types of calcium channels in single smooth muscle cells from rabbit ear artery studied with whole-cell and single-channel recordings. Circ Res 1987; 61:I10–I16.
45. Nowycky MC, Fox AP, Tsien RW. Three types of neuronal calcium channel with different calcium agonist sensitivity. Nature 1985; 316:440–443.
46. Tsien RW, Lipscombe D, Madison DV, Bley KR, Fox AP. Multiple types of neuronal calcium channels and their selective modulation. Trends Neurosci 1988; 11:431–438.
47. Hagiwara N, Irisawa H, Kameyama M. Contribution of two types of calcium currents to the pacemaker potentials of rabbit sino-atrial node cells. J Physiol 1988; 395:233–253.
48. Jahnsen H, Llinas R. Electrophysiological properties of guinea-pig thalamic neurones: An in vitro study. J Physiol 1984; 349:205–226.
49. Pelzer D, Grant AO, Cavalie A, Pelzer S, Sieber M, Hofmann F, Trautwein W. Calcium channels reconstituted from the skeletal muscle DHP receptor protein complex and its α_1 peptide subunit in lipid bilayers. Ann NY Acad Sci 1989; 560:138–154.
50. Singer D, Biel M, Lotan I, Flockerzi V, Hofmann F, Dascal N. The roles of the subunits in the function of the calcium channel. Science 1991; 253:1553–1557.

51. Tanabe T, Beam KG, Adams BA, Niidome T, Numa S. Regions of the skeletal muscle dihydropyridine receptor critical for excitation-contraction coupling. Nature 1990; 346:567–569.
52. Koch WJ, Ellinor PT, Schwartz A. cDNA cloning of a dihydropyridine-sensitive calcium channel from rat aorta. Evidence for the existence of alternatively spliced forms. J Biol Chem 1990; 265:17786–17791.
53. Kostyuk PG. Calcium channels in the neuronal membrane. Biochim Biophys Acta 1981; 650:128–150.
54. Hurwitz L. Pharmacology of calcium channels and smooth muscle. Ann Rev Pharmacol Toxicol 1986; 26:225–258.
55. Tsien RW, Hess P, McCleskey EW, Rosenberg RL. Calcium channels: Mechanisms of selectivity, permeation, and block. Ann Rev Biophys Biophys Chem 1987; 16:265–290.
56. Docherty RJ. Gadolinium selectively blocks a component of calcium current in rodent neuroblastoma × glioma hybrid (NG108-15) cells. J Physiol 1988; 398:33–47.
57. Fox AP, Nowycky MC, Tsien RW. Kinetic and pharmacological properties distinguishing three types of calcium currents in chick sensory neurones. J Physiol 1987; 394:149–172.
58. Siesjö BK, Bengtsson F. Calcium fluxes, calcium antagonists, and calcium-related pathology in brain ischemia, hypoglycemia, and spreading depression: A unifying hypothesis. J Cereb Blood Flow Metab 1989; 9:127–140.
59. Buchan AM. Do NMDA antagonists protect against cerebral ischemia: Are clinical trials warranted? Cerebrovasc Brain Metab Rev 1990; 2:1–26.
60. Llinas R, Yarom Y. Specific blockage of the slow threshold calcium channel by high molecular weight alcohols. Soc Neurosci Abst 1986; 12:174.
61. Reynolds IJ, Wagner JA, Snyder SH, Thayer SA, Olivera BM, Miller RJ. Brain voltage-sensitive calcium channel subtypes differentiated by ω-conotoxin fraction GVIA. Proc Natl Acad Sci USA 1986; 83:8804–8807.
62. McCleskey EW, Fox AP, Feldman DH, Cruz LJ, Olivera BM, Tsien RW, Yoshikami D. Omega-conotoxin: direct and persistent blockade of specific types of calcium channels in neurons but not muscle. Proc Natl Acad Sci USA 1987; 84:4327–4331.
63. Sher E, Clementi F. ω-conotoxin-sensitive voltage-operated calcium channels in vertebrate cells. Neuroscience 1991; 42:301–307.
64. Suzuki N, Yoshioka T. Differential blocking action of synthetic ω-conotoxin on components of Ca^{2+} channel current in clonal GH3 cells. Neurosci Lett 1987; 75:235–239.
65. Lin J-W, Rudy B, Llinas R. Funnel-web spider venom and a toxin fraction block calcium current expressed from rat brain mRNA in Xenopus oocytes. Proc Natl Acad Sci USA 1990; 87:4538–4542.
66. Janis RA, Silver PJ, Triggle DJ. Drug action and cellular calcium regulation. Adv Drug Res 1987; 16:309–591.
67. Nayler WG. Calcium antagonists. London: Academic Press; 1988.
68. Godfraind T. Classification of calcium antagonists. Am J Cardiol 1987; 59:11B–23B.
69. Tytgat J, Vereecke J, Carmeliet E. Differential effects of verapamil and flunarizine on cardiac L-type and T-type Ca channels. Naunyn-Schmiedeberg's Arch Pharmacol 1988; 337:690–692.

70. Terada K, Ohya Y, Kitamura K, Kuriyama H. Actions of flunarizine, a Ca^{++} antagonist, on ionic currents in fragmented smooth muscle cells of the rabbit small intestine. J Pharmacol Exp Ther 1987; 240:978–983.

71. Leboeuf J, Leoty C, Lamar J-C, Massingham R. Comparative effects of bepridil, its quaternary derivative CERM 11888 and verapamil on caffeine-induced contracture in ferret hearts. Br J Pharmacol 1989; 98:119–126.

72. John GW, Fabregues E, Kamal M, Massingham R. Caffeine-induced contractions in rabbit isolated renal artery are differentially inhibited by calcium antagonists. Eur J Pharmacol 1991; 196:307–312.

73. Schächtele C, Wagner B, Rudolph C. Effect of Ca^{2+} entry blockers on myosin light-chain kinase and protein kinase C. Eur J Pharmacol 1989; 163:151–155.

74. Glossmann H, Ferry DR, Lübbecke F, Mewes R, Hofmann F. Calcium channels: Direct identification with radioligand binding studies. Trends Pharmacol Sci 1982; 3:431–437.

75. Glossmann H, Ferry DR, Goll A, Striessnig J, Schober M. Calcium channels: Basic properties as revealed by radioligand binding studies. J Cardiovasc Pharmacol 1985; 7(suppl 6):S20–S30.

76. Triggle DJ. Calcium, calcium channels, and calcium channel antagonists. Can J Physiol Pharmacol 1990; 68:1474–1481.

77. DePover A, Grupp IL, Grupp G, Schwartz A. Diltiazem potentiates the negative inotropic action of nimodipine in heart. Biochem Biophys Res Commun 1983; 114:922–929.

78. Hosey MM, Lazdunski M. Calcium channels: Molecular pharmacology, structure and regulation. J Membr Biol 1988; 104:81–105.

79. Triggle DJ. Calcium antagonists. History and perspective. Stroke 1990; 21:IV-49–IV-58.

80. Nokin P, Clinet M, Beaufort P, Meysmans L, Laruel R, Chatelain P. SR33557, a novel calcium entry blocker. II. Interactions with 1,4-dihydropyridine, phenylalkylamine and benzothiazepine binding sites in rat heart sarcolemmal membranes. J Pharmacol Exp Ther 1990; 255:600–607.

81. Rampe D, Triggle DJ. New ligands for L-type Ca^{2+} channels. Trends Pharmacol Sci 1990; 11:112–115.

82. Godfraind T, Govoni S. Increasing complexity revealed in regulation of Ca^{2+} antagonist receptor. Trends Pharmacol Sci 1989; 10:297–301.

83. Hescheler J, Pelzer D, Trube G, Trautwein W. Does the organic calcium channel blocker D600 act from inside or outside on the cardiac cell membrane? Pfluegers Arch 1982; 393:287–291.

84. Hering S, Bolton TB, Beech DJ, Lim SP. Mechanism of calcium channel block by D600 in single smooth muscle cells from rabbit ear artery. Circ Res 1989; 64:928–936.

85. Zernig G. Widening potential for Ca^{2+} antagonists: non-L-type Ca^{2+} channel interaction. Trends Pharmacol Sci 1990; 11:38–44.

86. Andersson K-E. Pharmacodynamic profiles of different calcium channel blockers. Acta Pharmacol Toxicol 1986; 58(suppl II):31–42.

87. Schramm M, Thomas G, Towart R, Franckowiak G. Novel dihydropyridines with positive inotropic action through activation of Ca^{2+} channels. Nature 1983; 303:535–537.

88. Schramm M, Thomas G, Towart R, Franckowiak G. Activation of calcium channels by novel 1,4-dihydropyridines. A new mechanism for positive inotropics or smooth muscle stimulants. Arzneim-Forsch/Drug Res 1983; 33:1268–1272.

89. Sanguinetti MC, Kass RS. Voltage-dependent block of calcium channel current in the calf cardiac purkinje fiber by dihydropyridine calcium channel antagonists. Circ Res 1984; 55:336–348.

90. Terada K, Nakao K, Okabe K, Kitamura K, Kuriyama H. Action of the 1,4-dihydropyridine derivative, KW-3049, on the smooth muscle membrane of the rabbit mesenteric artery. Br J Pharmacol 1987; 92:615–625.

91. Sanguinetti MC, Krafte DS, Kass RS. Voltage-dependent modulation of Ca channel current in heart cells by Bay K8644. J Gen Physiol 1986; 88:369–392.

92. Hamilton SL, Yatani A, Brush K, Schwartz A, Brown AM. A comparison between the binding and electrophysiological effects of dihydropyridines on cardiac membranes. Mol Pharmacol 1987; 31:221–231.

93. Bechem M, Hebisch S, Schramm M. Ca^{2+} agonists: New, sensitive probes for Ca^{2+} channels. Trends Pharmacol Sci 1988; 9:257–261.

94. Kunze DL, Hamilton SL, Hawkes MJ, Brown AM. Dihydropyridine binding and calcium channel function in clonal rat adrenal medullary tumor cells. Mol Pharmacol 1987; 31:401–409.

95. Triggle DJ, Rampe D. 1,4-dihydropyridine activators and antagonists: Structural and functional distinctions. Trends Pharmacol Sci 1989; 10:507–511.

96. Kokubun S, Prod'hom B, Becker C, Porzig H, Reuter H. Studies on Ca channels in intact cardiac cells: Voltage-dependent effects and cooperative interactions of dihydropyridine enantiomers. Mol Pharmacol 1986; 30:571–584.

97. Ferrante J, Triggle DJ. Drug- and disease-induced regulation of voltage-depenent calcium channels. Pharmacol Rev 1990; 42:29–44.

98. Ono H, Hashimoto K. In vitro tissue effects of calcium flux inhibition. In: Stone PH, Antman EM, eds. Calcium Channel Blocking Agents in the Treatment of Cardiovascular Disorders. Mount Kisco, NY: Futura Publishing; 1983:155–175.

99. Ning W, Wit AL. Comparison of the direct effects of nifedipine and verapamil on the electrical activity of the sinoatrial and atrioventricular nodes of the rabbit heart. Am Heart J 1983; 106:345–355.

100. Goto J, Sperelakis N. Depression of automaticity of the rabbit SA-node by bepredil and nifedipine. Eur J Pharmacol 1984; 99:227–231.

101. Taira N. Differences in cardiovascular profile among calcium antagonists. Am J Cardiol 1987; 59:24B–29B.

102. Wada Y, Satoh K, Taira N. Separation of the coronary vasodilator from cardiac effects of PN 200-110, a new dihydropyridine calcium antagonist, in the dog heart. J Cardiovasc Pharmacol 1985; 7:190–196.

103. Satoh K, Kawada M, Wada Y, Taira N. Cardiovascular actions of the dihydropyridine calcium antagonist nimodipine in the dog. Arzneim-Forsch/Drug Res 1984; 34:563–568.

104. Lathrop DA, Valle-Aguilera JR, Millard RW, Gaum WE, Hannon DW, Francis PD, Nakaya H, Schwartz A. Comparative electrophysiologic and

coronary hemodynamic effects of diltiazem, nisoldipine and verapamil on myocardial tissue. Am J Cardiol 1982; 49:613–620.

105. Henry PD. Comparative cardiac pharmacology of calcium blockers. In: Flaim SF, Zelis R, eds. Calcium Blockers: Mechanisms of action and clinical application. Munich: Urban & Schwarzenberg; 1982:135–153.

106. Schwinger RHG, Bohm M, Erdmann E. Negative inotropic properties of isradipine, nifedipine, diltiazem, and verapamil in diseased human myocardial tissue. J Cardiovasc Pharmacol 1990; 15:892–899.

107. Cauvin C, Loutzenhiser R, van Breemen C. Mechanisms of calcium antagonist-induced vasodilation. Ann Rev Pharmacol Toxicol 1983; 23:373–396.

108. Hof RP, Salzmann R, Siegl H. Selective effects of PN 200-110 (Isradipine) on the peripheral circulation and the heart. Am J Cardiol 1987; 59:30B–36B.

109. Ljung B. Vascular selectivity of felodipine. Drugs 1985; 29 (suppl 2):46–58.

110. Struyker-Boudier HAJ, Smits JFM, De Mey JGR. The pharmacology of calcium antagonists: A review. J Cardiovasc Pharmacol 1990; 15(suppl 4):S1–S10.

111. Hess P, Lansman JB, Tsien RW. Different modes of Ca channel gating behaviour favoured by dihydropyridine Ca agonists and antagonists. Nature 1984; 311:538–544.

112. Nakaya HA, Schwartz A, Millard RW. Reflex chronotropic and inotropic effects of calcium channel-blocking agents in conscious dogs. Diltiazem, verapamil and nifedipine compared. Circ Res 1983; 52:302–311.

113. Urquhart J, Patterson RE, Bacharach SL, Green MV, Speir EH, Aamodt R, Epstein SE. Comparative effects of verapamil, diltiazem, and nifedipine on hemodynamics and left ventricular function during acute myocardial ischemia in dogs. Circulation 1984; 69:382–390.

114. Quartaroli M, Gambini F, Tarter G, Micheli D, Trist DG, Gaviraghi G. The hemodynamic effects of lacidipine in anesthetized dogs: Comparison with nitrendipine, amlodipine, verapamil, and diltiazem. J Cardiovasc Pharmacol 1991; 18:326–336.

115. Roychowdhary AK, Gurtu S, Dhawan KN, Sinha JN, Gupta GP. Evidence for a central component in the cardiovascular effects of calcium channel blockers. J Cardiovasc Pharmacol 1991; 17:1015–1018.

116. Kulkarni SK, Shukla KV. Extracardiac actions of calcium channel blockers. Drugs 1988; 24:303–325.

117. Soward AL, Vanhaleweyk GLJ, Serruys PW. The haemodynamic effects of nifedipine, verapamil and diltiazem in patients with coronary artery disease. A review. Drugs 1986; 32:66–101.

118. Rowland E, Evans T, Krikler D. Effect of nifedipine on atrioventricular conduction as compared with verapamil. Intracardiac electrophysiologic study. Br Heart J 1979; 42:124–127.

119. Mitchell LB, Schroeder JS, Mason JW. Comparative clinical electrophysiologic effects of diltiazem, verapamil and nifedipine. A review. Am J Cardiol 1982; 49:629–635.

120. Sheridan DJ, Thomas P. Vascular versus myocardial selectivity of calcium antagonists. J Cardiovasc Pharmacol 1987; 10(suppl 1):S165–S168.

121. Bean BP. Nitrendipine block of cardiac calcium channels: High-affinity binding to the inactivated state. Proc Natl Acad Sci USA 1984; 81:6388–6392.
122. Cognard C, Romey G, Galizzi J-P, Fosset M, Lazdunski M. Dihydropyridine-sensitive Ca^{2+} channels in mammalian skeletal muscle cells in culture: Electrophysiological properties and interactions with Ca^{2+} channel activator (Bay K8644) and inhibitor (PN 200-110). Proc Natl Acad Sci USA 1986; 83:1518–1522.
123. Terada K, Kitamura K, Kuriyama H. Blocking actions of Ca^{2+} antagonists on the Ca^{2+} channels in the smooth muscle cell membrane of rabbit small intestine. Pfluegers Arch 1987; 408:552–557.
124. Dacquet C, Pacaud P, Loirand G, Mironneau C, Mironneau J. Comparison of binding affinities and calcium current inhibitory effects of a 1,4-dihydropyridine derivative (PN 200-110) in vascular smooth muscle. Biochem Biophys Res Commun 1988; 152:1165–1172.
125. Hering S, Beech DJ, Bolton TB, Lim SP. Action of nifedipine or BAY K8644 is dependent on calcium channel state in single smooth muscle cells from rabbit ear artery. Pfluegers Arch 1988; 411:590–592.
126. Ganitkevich VY, Shuba MF, Smirnov SV. Potential-dependent calcium inward current in a single isolated smooth muscle cell of the guinea-pig *taenia caeci*. J Physiol 1986; 380:1–16.
127. Beech DJ, MacKenzie I, Bolton TB, Christen MO. Effects of pinaverium on voltage-activated calcium channel currents of single smooth muscle cells isolated from the longitudinal muscle of the rabbit jejunum. Br J Pharmacol 1990; 99:374–378.
128. Reuter H. Calcium channel modulation by neurotransmitters, enzymes and drugs. Nature 1983; 301:569–574.
129. Lee KS, Tsien RW. Mechanism of calcium channel blockade by verapamil, D600, diltiazem and nitrendipine in single dialysed heart cells. Nature 1983; 302:790–794.
130. Hondeghem LM, Katzung BG. Antiarrhythmic agents: The modulated receptor mechanism of action of sodium and calcium channel-blocking drugs. Ann Rev Pharmacol Toxicol 1984; 24:387–423.
131. Ehara T, Daufmann R. The voltage- and time-dependent effects of (−)-verapamil on the slow inward current in isolated cat ventricular myocardium. J Pharmacol Exp Ther 1978; 207:49–55.
132. Pelzer D, Trautwein W, McDonald TF. Calcium channel block and recovery from block in mammalian ventricular muscle treated with organic channel inhibitors. Pfluegers Arch 1982; 394:97–105.
133. Fox AP. Voltage-dependent inactivation of a calcium channel. Proc Natl Acad Sci USA 1981; 78:953–956.
134. Fukushima Y, Hagiwara S. Voltage-gated Ca^{2+} channel in mouse myeloma cells. Proc Natl Acad Sci USA 1983; 80:2240–2242.
135. Wibo M. Mode of action of calcium antagonists: voltage-dependence and kinetics of drug-receptor interaction. Pharmacol Toxicol 1989; 65:1–8.
136. Hille B. Local anesthetics: Hydrophilic and hydrophobic pathways for the drug-receptor reaction. J Gen Physiol 1977; 69:497–515.
137. Cohen CJ, McCarthy RT. Nimodipine block of calcium channels in rat anterior pituitary cells. J Physiol 1987; 387:195–225.

220   T.K. Uski, E.D. Högestätt, and K.-E. Andersson

138. Godfraind T, Morel N, Wibo M. Modulation of the action of calcium antagonists in arteries. Blood Vessels 1990; 27:184–196.
139. Yousif FB, Triggle DJ. Inhibitory actions of a series of Ca^{2+} channel antagonists against agonist and K^+ depolarization induced responses in smooth muscle: An assessment of selectivity of action. Can J Physiol Pharmacol 1986; 64:273–283.
140. Jones AW. Vascular smooth muscle and alterations during hypertension. In: Bülbring E, Brading AF, Jones AW, Tomita T, eds. Smooth muscle: An assessment of current knowledge. London: E Arnold; 1981:379–429.
141. Vanhoutte PM. Vascular endothelium and Ca^{2+} antagonists. J Cardiovasc Pharmacol 1988; 12(suppl 6):S21–S28.
142. Sargent DF, Schwyzer R. Membrane lipid phase as catalyst for peptide-receptor interactions. Proc Natl Acad Sci USA 1986; 83:5774–5778.
143. Bäuerle H-D, Seelig J. Interaction of charged and uncharged calcium channel antagonists with phospholipid membranes: Binding equilibrium, binding enthalpy, and membrane location. Biochemistry 1991; 30:7203–7211.
144. Nayler WG. Influx and efflux of calcium in the physiology of muscle contraction. Clin Orthoped Related Res 1966; 46:157–182.
145. Schwartz LM, McCleskey EW, Almers W. Dihydropyridine receptors in muscle are voltage-dependent but most are not functional calcium channels. Nature 1985; 314:747–751.
146. Almers W, McCleskey EW. Non-selective conductance in calcium channels of frog muscle: Calcium selectivity in a single-file pore. J Physiol 1984; 353:585–608.
147. Shimizu K, Ohta T, Toda N. Evidence for greater susceptibility of isolated dog cerebral arteries to Ca antagonists than peripheral arteries. Stroke 1980; 11:261–266.
148. Towart R. The selective inhibition of serotonin-induced contractions of rabbit cerebral vascular smooth muscle by calcium-antagonistic dihydropyridines: An investigation of the mechanism of action of nimodipine. Circ Res 1981; 48:650–657.
149. Uski TK, Andersson K-E. Effects of prostanoids on isolated feline cerebral arteries. II. Roles of extra- and intracellular calcium for the prostaglandin $F_{2\alpha}$-induced contraction. Acta Physiol Scand 1984; 120:197–205.
150. Uski TK, Andersson K-E, Brandt L, Ljunggren B. Characterization of the prostanoid receptors and of the contractile effects of prostaglandin $F_{2\alpha}$ in human pial arteries. Acta Physiol Scand 1984; 121:369–378.
151. Cauvin C, Saida K, van Breemen C. Extracellular Ca^{2+} dependence and diltiazem inhibition of contraction in rabbit conduit arteries and mesenteric resistance vessels. Blood Vessels 1984; 21:23–31.
152. Godfraind T, Morel N, Wibo M. Modulation of the action of calcium antagonists in arteries. Blood Vessels 1990; 27:184–196.
153. Meisheri KD, Sage II GP, Cipkus-Dubray LA. Factors affecting rabbit mesenteric artery smooth muscle sensitivity to calcium antagonists. J Pharmacol Exp Ther 1990; 252:1167–1174.
154. Glossmann H, Striessnig J. Structure and pharmacology of voltage-dependent calcium channels. In: ISI ed. Atlas of Science: Pharmacology. Philadelphia: ISI; 1988; 2:202–210.

155. McKenna E, Koch WJ, Slish DF, Schwartz A. Toward an understanding of the dihydropyridine-sensitive calcium channel. Biochem Pharmacol 1990; 39:1145–1150.
156. Bean BP, Sturek M, Puga A, Hermsmeyer K. Calcium channels in muscle cells isolated from rat mesenteric arteries: modulation by dihydropyridine drugs. Circ Res 1986; 59:229–235.
157. Yatani A, Seidel CL, Allen J, Brown AM. Whole-cell and single-channel calcium currents of isolated smooth muscle cells from saphenous vein. Circ Res 1987; 60:523–533.
158. Sperelakis N. Properties of calcium channels in cardiac muscle and vascular smooth muscle. Molec Cell Biochem 1990; 99:97–109.
159. Hoffman BF, Dangman KH. Mechanisms for cardiac arrhythmias. Experientia 1987; 43:1049–1056.
160. Levy MN. Role of calcium in arrhythmogenesis. Circulation 1989; 80(suppl IV):23–30.
161. Singh BN, Nademanee K. Use of calcium antagonists for cardiac arrhythmias. Am J Cardiol 1987; 59:153B–162B.
162. Akhtar M, Tchou P, Jazayeri M. Use of calcium channel entry blockers in the treatment of cardiac arrhythmias. Circulation 1989; 80(suppl IV):IV-31–IV-39.
163. Singh BN, Nademanee K, Baky S. Calcium antagonists: Uses in the treatment of cardiac arrhythmias. Drugs 1983; 25:125–164.
164. Ardissino D, Savonitti S, Mussini A, Zanini P, Rolla A, Barberis P, Sardina M, Specchia G. Felodipine (once daily) versus nifedipine (four times daily) for Prinzmetal's angina pectoris. Am J Cardiol 1991; 68:1587–1592.
165. Krikler DM. Calcium antagonists for chronic stable angina pectoris. Am J Cardiol 1987; 59:95B–100B.
166. Stone PH. Calcium antagonists for Prinzmetal's variant angina, unstable angina, and silent myocardial ischemia: Therapeutic tool and probe for identification of pathophysiologic mechanisms. Am J Cardiol 1987; 59:101B–115B.
167. Stone PH, Gibson RS, Glasser SP, DeWood MA, Parker JD, Kawanishi DT, Crawford MH, Messineo FC, Shook TL, Raby K, Curtis DG, Hoop RS, Young PM, Braunwald E, the ASIS Study Group. Comparison of propranolol, diltiazem, and nifedipine in the treatment of ambulatory ischemia in patients with stable angina: Differential effects on ambulatory ischemia, exercise performance, and anginal symptoms. Circulation 1990; 82:1962–1972.
168. Andersson K-E, Brandt L, Ljungren B. Spasm of cerebral and coronary vessels: effects of calcium antagonists. In: Refsum H, Sulg IA, Rasmussen K, eds. Heart & Brain, Brain & Heart. Berlin: Springer-Verlag; 1989: 3–19.
169. Harder DR, Belardinelli L, Sperelakis N, Rubio R, Berne RM. Differential effects of adenosine and nitroglycerin on the action potentials of large and small coronary arteries. Circ Res 1979; 44:176–182.
170. Mekata F. Electrophysiological studies of the smooth muscle cell membrane of the dog coronary artery. J Physiol 1980; 298:205–212.
171. Matsuda JJ, Volk KA, Shibata EF. Calcium currents in isolated rabbit coronary arterial smooth muscle myocytes. J Physiol 1990; 427:657–680.

172. Ganitkevich VY, Isenberg G. Contribution of two types of calcium channels to membrane conductance of single myocytes from guinea-pig coronary artery. J Physiol 1990; 426:19–42.
173. Ginsburg R.The isolated human epicardial coronary artery. Am J Cardiol 1983; 52:61A–66A.
174. Vedernikov YP. Mechanisms of coronary spasm of isolated human epicardial coronary segments excised 3 to 5 hours after sudden death. J Am Coll Cardiol 1986; 8:42A–49A.
175. Schwartz JS, Bache RJ. Pharmacologic vasodilators in the coronary circulation. Circulation 1987; 75(suppl 1):162–167.
176. Bache RJ, Dymek DJ. Local and regional regulation of coronary vascular tone. Progr Cardiovasc Dis 1982; 24:191–212.
177. Bache RJ, Tockman BA. Effect of nitroglycerin and nifedipine on subendocardial perfusion in the presence of a flow-limiting coronary stenosis in the awake dog. Circ Res. 1982; 50:678–687.
178. Nayler WG. Calcium antagonists and the ischaemic myocardium. Int J Cardiol 1987; 15:267–285.
179. Nayler WG, Elz JS. Reperfusion injury: Laboratory artifact or clinical dilemma. Circulation 1986; 74:215–221.
180. Nayler WG. Basic mechanisms involved in the protection of the ischaemic myocardium. Role of calcium antagonists. Drugs 1991; 42(suppl 2):21–27.
181. Hano O, Silvermann HS, Blank PS, Mellits ED, Baumgardner R, Lakatta EG, Stern MD. Nicardipine prevents calcium loading and "oxygen paradox" in anoxic single rat myocytes by a mechanism independent of calcium channel blockade. Circ Res 1991; 69:1500–1505.
182. Clusin WT, Buchbinder M, Ellis AK, Kernoff KI, Giacomini JC, Harrison DC. Reduction of ischemic depolarization by the calcium channel blocker diltiazem. Correlation with improvement of ventricular conduction and early arrhythmias in the dog. Circ Res 1984; 54:10–20.
183. Dillon JS, Nayler WG. [^{3}H]Verapamil binding to rat cardiac sarcolemmal membrane fragments: An effect of ischaemia. Br J Pharmacol 1987; 90:99–109.
184. Matucci R, Benardini F, Sciamarella ML, Baccaro C, Stenardi I, Franconi F, Giotti A. [^{3}H]Nitrendipine binding in membranes obtained from hypoxic and reoxygenated heart. Biochem Pharmacol 1987; 36:1059–1062.
185. Erbel R, Pop T, Meinertz T, Olshausen KV, Treese N, Henrichs KJ, Schuster CJ, Rupprecht HJ, Schlurmann W, Meyer J. Combination of calcium channel antagonist and thrombolytic therapy in acute myocardial infarction. Am Heart J 1988; 15:529–538.
186. Ellis SG, Muller DW, Topol EJ. Possible survival benefit from concomitant beta- but not calcium-antagonist therapy during reperfusion for acute myocardial infarction. Am J Cardiol 1990; 66:125–128.
187. Held PH, Yusuf S, Furberg CD. Calcium channel blockers in acute myocardial infarction and unstable angina: An overview. Br Med J 1989; 299:1187–1192.
188. Persson S. Calcium antagonists in secondary prevention after myocardial infarction. Drugs 1991; 42(suppl 2):54–60.
189. Vaage-Nielsen M, Rasmussen V, Fisher Hansen J, Hagerup L, Borring Sørensen M, Pedersen Bjergard O, Mellemgaard K, Holländer NH, Nielsen

I, Sigurd B, the Danish Verapamil Infarction Trial II Study Group. Effect of verapamil on ischaemia and ventricular arrhythmias after an acute myocardial infarction: Prognostic implications. J Cardiovasc Pharmacol 1991; 18(suppl 6):S26–S29.

190. Fischer Hansen J. Calcium antagonists and myocardial infarction. Cardiovasc Drug Ther 1991; 5:665–670.

191. Frishman WH, Stroh JA, Greenberg S, Suarez T, Karp A, Peled H. Calcium channel blockers in systemic hypertension. Med Clin N Am 1988; 72:449–499.

192. Kiowski W, Bolli P, Erne P, Müller FB, Hulthén U, Bühler FR. Mechanism of action and clinical use of calcium antagonists in hypertension. Circulation 1989; 80(suppl IV):IV-136–IV-144.

193. Lederballe Pedersen O, Mikkelsen E, Andersson K-E. Effects of extracellular calcium on potassium and noradrenaline induced contractions in the aorta of spontaneously hypertensive rats—increased sensitivity to nifedipine. Acta Pharmacol Toxicol 1978; 43:137–144.

194. van Breemen C, Cauvin C, Johns A, Leijten P, Yamamoto H. Ca^{2+} regulation of vascular smooth muscle. Fed Proc 1986; 45:2746–2751.

195. Cauvin C, Hwang O, Yamamoto H, van Breemen C. Effect of dihydropyridines on tension and calcium-45 influx in isolated mesenteric resistance vessels from spontaneously hypertensive and normotensive rats. Am J Cardiol 1987; 59:116B–122B.

196. Chatelain P, Demol D, Roba J. Comparison of [³H] nitrendipine binding to heart membranes of normotensive and spontaneously hypertensive rats. J Cardiovasc Pharmacol 1984; 6:220–223.

197. Ishii K, Kano T, Ando J, Yoshida H. Binding of [³H] nitrendipine to cardiac and cerebral membranes from normotensive and spontaneously hypertensive rats. Eur J Pharmacol 1986; 123:271–278.

198. Packer M. Pathophysiological mechanisms underlying the adverse effects of calcium channel-blocking drugs in patients with chronic heart failure. Circulation 1989; 80(suppl IV):IV-59–IV-67.

199. Rouleau J-L, Chuck LHS, Hollosi G, Kidd P, Sievers RE, Wikman-Coffelt J, Parmley WW. Verapamil preserves myocardial contractility in the hereditary cardiomyopathy of the Syrian hamster. Circ Res 1979; 50:405–412.

200. Wagner JA, Reynolds IJ, Weisman HF, Dudeck P, Weisfeldt ML, Snyder SH. Calcium antagonist receptors in cardiomyopathic hamster: Selective increases in heart, muscle, brain. Science 1986; 232:515–517.

201. Wagner JA, Sax FL, Weisman HF, Porterfield J, McIntosh C, Weisfeldt ML, Snyder SH, Epstein SE. Calcium-antagonist receptors in the atrial tissue of patients with hypertrophic cardiomyopathy. N Engl J Med 1989; 320:755–761.

202. Rasmussen RP, Minobe W, Bristow MR. Calcium antagonist binding sites in failing and nonfailing human ventricular myocardium. Biochem Pharmacol 1990; 39:691–696.

203. Elkayam U, Amin J, Mehra A, Vasquez J, Weber L, Rahimtoola SH. A prospective, randomized, double-blind, crossover study to compare the efficacy and safety of chronic nifedipine therapy with that of isosorbide dinitrate and their combination in the treatment of chronic congestive heart failure. Circulation 1990; 82:1954–1961.

204. Packer M. Calcium channel blockers in chronic heart failure. The risks of "physiologically rational" therapy. Circulation 1990; 82:2254–2257.
205. Reicher-Reiss H, Barasch E. Calcium antagonists in patients with heart failure. A review. Drugs 1991; 42:343–364.
206. Henry PD. Anti-atherogenic effects of calcium channel blockers: Possible mechanisms of action. Cardiovasc Drug Ther 1990; 4(suppl 5):1015–1020.
207. Gottlieb SO, Brinker JA, Mellits D. Effect of nifedipine on the development of coronary bypass graft stenoses in high-risk patients: A randomized, double-blind, placebo-controlled trial. Circulation 1989; 80:II-228.
208. Loaldi A, Polese A, Montorsi P. Comparison of nifedipine, propranolol and isosorbide dinitrate on angiographic progression and regression of coronary arterial narrowing in angina pectoris. Am J Cardiol 1989; 64:433–439.
209. Lichtlen PR, Hugenholtz PG, Rafflenbeul W, Hecker H, Jost S, Deckers JW. Retardation of angiographic progression of coronary artery disease by nifedipine: Results of International Nifedipine Trial on Anti-atherosclerotic Therapy (INTACT). Lancet 1990; 335:1109–1113.
210. Waters D, Lesperance J, Francetich M, Causey D, Theroux P, Chiang YK, Hudon G, Lemarbre L, Reitman M, Joyal M, Gosselin G, Dyrda I, Macer J, Havel RJ. A controlled clinical trial to assess the effect of a calcium channel blocker upon the progression of coronary atherosclerosis. Circulation 1990; 82:1940–1953.
211. Schmitz G, Hankowitz J, Kovacs EM. Cellular processes in atherogenesis: Potential targets of Ca^{2+} channel blockers. Atherosclerosis 1991; 88:109–132.
212. Schmitz G, Hankowitz J, Brennhausen B, Schmutte C. Der Einfluss von Calcium-Antagonisten auf den zellulären Lipidstoffwechsel. Arzneim-Forsch/Drug Res 1990; 40:366–372.
213. Langley MS, Sorkin EM. Nimodipine: A review of its pharmacodynamic and pharmacokinetic properties, and therapeutic potential in cerebrovascular disease. Drugs 1989; 37:669–699.
214. Germano IM, Bartkowski HM, Cassel ME, Pitts LH. The therapeutic value of nimodipine in experimental focal cerebral ischemia: Neurological outcome and histopathological findings. J Neurosurg 1987; 67:81–87.
215. Fujisawa A, Matsumoto M, Matsuyama T, Ueda H, Wanaka A, Yoneda S, Kimura K, Kamada T. The effect of the calcium antagonist nimodipine on the gerbil model of experimental cerebral ischemia. Stroke 1986; 17:748–752.
216. Steen PA, Gisvold SE, Milde JH, Newberg LA, Scheithauer BW, Lanier WL, Michenfelder JD. Nimodipine improves outcome when given after complete cerebral ischemia in primates. Anesthesiology 1985; 62:406–414.
217. Lazarewicz JW, Pluta R, Salinska E, Puka M. Beneficial effect of nimodipine on metabolic and functional disturbances in rabbit hippocampus following complete cerebral ischemia. Stroke 1989; 20:70–77.
218. Grotta, JC. Clinical aspects of the use of calcium antagonists in cerebrovascular disease. Clin Neuropharmacol 1991; 14:373–390.
219. Ljunggren B, Brandt L, Säveland H, Nilsson P-E, Cronqvist S, Andersson K-E, Vinge E. Outcome in 60 consecutive patients treated with early

aneurysm operation and intravenous nimodipine. J Neurosurg 1984; 61:864–873.

220. Petruk KC, West M, Mohr G, Weir BKA, Benoit BG, Gentili F, Disney LB, Khan MI, Grace M, Holness RO, Karwon MS, Ford RM, Cameron GS, Tucker WS, Purves GB, Miller JDR, Hunter KM, Richard MT, Durity FA, Chan R, Clein LJ, Maroun FB, Godon A. Nimodipine treatment in poor-grade aneurysm patients. Results of a multicenter double-blind placebo-controlled trial. J Neurosurg 1988; 68:505–517.

221. Auer LM, Oberbauer RW, Schalk HV. Human pial vascular reactions to intravenous nimodipine-infusion during EC-IC bypass surgery. Stroke 1983; 14:210–213.

222. Gelmers HJ. Effect of nimodipine (Bay e 9736) on postischaemic cerebrovascular reactivity, as revealed by measuring regional cerebral blood flow (rCBF). Acta Neurochir 1982; 63:283–290.

11
Hormonal Modulation of Sodium Pump Activity: Identification of Second Messengers

DOUGLAS R. YINGST

The sodium pump actively transports ions of sodium and potassium across the plasma membrane of most animal cells using energy acquired from the hydrolysis of ATP. Electrochemical gradients of sodium and potassium are formed that initiate the subsequent diffusion of sodium and potassium, which helps establish the resting membrane potential. Under normal physiologic conditions three Na^+ ions are expelled in exchange for two K^+ ions, which produces a net charge movement that also contributes to the resting membrane potential. The sodium pump regulates cell volume and the sodium gradient itself subserves the secondary transport of other amino acids, sugars, calcium ions, and protons. In more general terms the sodium pump is the major mechanism responsible for the ability of the body to regulate levels of sodium and potassium. The energy spent in this capacity is a major source of heat production.

The sodium pump has been studied intensely. It is now well established that the same protein that hydrolyzes ATP in a sodium- and potassium-dependent manner (the Na,K-ATPase) transports sodium and potassium.[1] The pump belongs to the E_1, E_2 class of transport proteins, which when phosphorylated (normally from ATP) change conformation and thus participate in the mechanism of ion transport across the cell membrane. The pump is composed of at least two subunits, α and β.[2] The α subunit is responsible for all of the transport and enzymatic functions of the pump.[3] The β subunit may facilitate the accumulation of the α subunit in the plasma membrane.[4] Three isoforms of α ($\alpha1$, $\alpha2$, and $\alpha3$)[5,6] and four of β have been identified[7] and show a cell- and tissue-specific pattern of distribution. Functional differences between various isoforms are being investigated.[7]

The magnitude of the ion transport via the sodium pump in a typical cell is determined by three factors: the number of pumps, the type of pump isoforms, and the rate of transport per pump. In some cells hormones such as insulin[8] can change the number of pumps via endocytosis and exocytosis within minutes,[9] while steroid (aldosterone,

226

corticosterone, and cortisol) and thyroid hormones can alter the number of pumps and the type of pump isoform over a period that ranges from hours to days via synthesis of new protein.[6,9-11]

The rate of transport per pump responds rapidly to changes in the concentrations of pump substrates, especially ATP and intracellular sodium. These effects are well understood. For instance, the concentration of intracellular sodium in most cells is in the range of 10–20 mM, which is close to the $K_{1/2}$ for sodium activation of the Na,K-ATPase. Therefore the rate of pumping easily responds to changes in the intracellular sodium concentration. The rate of transport per pump can also be modified in a matter of seconds or minutes by specific hormones (insulin, glucagon, vasopressin, or catecholamines)[9] and neurotransmitters (catecholamines and dopamine), which bind to extracellular receptors.[12,13] Some of these effects are the result of hormone-induced changes in the ionic environment of the pump. For instance, insulin stimulates the pump in fibroblasts by increasing the concentration of intracellular sodium.[14] In other cases hormones appear to affect the sodium pump without apparently altering the ionic environment.[15] Such presumptive direct effects have been difficult to prove, because they are most readily observed in whole cells and are often lost when the cell is disrupted for further experimental manipulation, and because even when the most careful measurement is used; it is difficult to rule out any effect of hormone-induced changes in the ion gradients. Nonetheless, there has been a persistent expectation that the short-term activity of the sodium pump is regulated by hormones and neurotransmitters. In addition, it has been proposed that rapid changes in the activity of the pump may play a role in the control of secretion[16] in the regulation of plasma potassium during exercise,[17,18] in the renal reabsorption of sodium,[19] in the modulation of neuronal activity,[20] and in the contraction of skeletal,[17] cardiac,[21] and smooth muscle.[22] The rate of transport per pump may also be inhibited by endogenous ouabain-like compounds that bind to specific receptors on the pump itself.[23]

Direct effects of hormones and neurotransmitters on the sodium pump may be mediated by classical second messengers. For instance, there is considerable evidence that the stimulation of the sodium pump by catecholamines is receptor mediated, a mechanism that should not be confused with the nonspecific stimulation of Na,K-ATPase activity in broken membrane preparations by high catecholamine concentrations that chelate Ca^{2+} or other inhibitory ions. Both α and β receptors have been implicated in this stimulation, but the intracellular mechanisms have not yet been elucidated.[12,13,24]. Although β-adrenergic agonists stimulate adenylate cyclase, it is tempting to speculate that cAMP mediates some of the effects of catecholamines on the pump. The primary sequence of the α subunit of all three pump isoforms includes a potential phosphorylation site for a cAMP-dependent kinase,[25] cAMP has been observed to inhibit

Na,K-ATPase activity in rat brain[26] and rabbit iris ciliary body.[27] In shark rectal gland[28] and Swiss 3T3 cells[29] the observed effects suggest that cAMP may also stimulate the pump. In Friend erythroleukemia cells the α subunit of the Na,K-ATPase is phosphorylated by an unidentified kinase, although there is no suggestion that cAMP is implicated or that the activity of the Na,K-ATPase is thereby altered.[30]

It has been known for some time that calcium, a well-established second messenger, can affect both the ability of the pump to hydrolyze ATP[31] and to transport sodium.[32] Concentrations of free calcium above the physiologic range (0.05–5 mM) interfere with the normal effects of sodium, potassium, magnessium, and ATP on the Na,K-ATPase and may inhibit its activity through competitive[31,33,34] and noncompetitive interactions with magnesium.[35] Calcium can substitute for magnesium in the formation of the phosphoenzyme that is part of the normal reaction mechanism of the pump,[35] even though the phosphoenzyme formed in this manner has altered properties.[36] Calcium can also inhibit the pump by forming Ca-ATP, which is not as good a substrate for the Na,K-ATPase as Mg-ATP.[33,37] Similarly, calcium inhibits by interfering competitively with the ability of sodium to activate the Na,K-ATPase.[33] Finally, calcium can occupy the potassium-binding sites[38] on the inside surface of the sodium pump and be occluded in the membrane as if it were being transported, but it cannot be released at extracellular sites.[39]

In addition, the Na,K-ATPase is responsive to calcium in the range over which it is functions as a second messenger (0.05 to 10 μM).[37,40–45] These effects appear to involve intracellular proteins that make the Na,K-ATPase more sensitive to inhibition by calcium.[41,44,46] In rat myometrial cells high-affinity calcium inhibition of the Na,K-ATPase may be mediated by a cAMP-dependent kinase.[47] Protein kinase C (PKC), calmodulin, and perhaps calnaktin may also be involved.[44,48,49] PKC can phosphorylate the α subunit of the purified Na,K-ATPase.[50] However, this does not result in a change of enzymatic activity, perhaps because relatively few (maximum 10%) Na,K-ATPase molecules are phosphorylated. On the other hand, in most cells, agonist binding to hormone-specific extracellular receptors causes an increase in the intracellular concentration of calcium and diacylglycerol (DAG) and a consequent increase in the activities of calcium-dependent kinases and of PKC.[51,52] DAG can be further broken down to fatty acids, some of which could affect the activity of the Na,K-ATPase.[53–55] Indeed, in diabetes an altered metabolism of myoinositol inhibits the activity of the Na,K-ATPase, perhaps by decreasing the activity of PKC.[56,57] In the liver, catecholamines stimulate the sodium pump via α-adrenergic receptors. Release from inhibition by calcium[58] and activation of PKC following an increased production of DAG are the suggested intracellular mechanisms.[59]

Calmodulin binds calcium with $K_{1/2}$ of $5-10\,\mu M$ and modifies the activity of several enzymes when the concentration of free calcium is in the range of 0.1 to $1\,\mu M$,[60] among them the Na,K-ATPase.[44,61,62] For example, the Na,K-ATPase from red cells can be inhibited up to 80% in a dose-dependent manner by 15 to 100 nM calmodulin.[63] However, since calmodulin itself has no effect on the purified Na,K-ATPase,[64] it is unlikely that calmodulin directly affects the pump. We have suggested that calmodulin potentiates the effects of calnaktin,[44] a proposed but yet unidentified protein that mediates the effects of physiologic free calcium on the Na,K-ATPase. Calnaktin could be a calmodulin-dependent phosphatase or kinase that could modify the proposed regulatory phosphorylation site on the Na,K-ATPase (see above) or alter the phosphorylation of a cytoskeleton protein, such as ankyrin, that interacts with the Na,K-ATPase.[65] Recent studies in our laboratory (data not shown) have shown that the effects of calmodulin on the Na,K-ATPase can be blocked by KN-62, a selective calcium calmodulin-dependent protein kinase II inhibitor.[66] This result suggests that the effects of low concentrations of free calcium on the Na,K-ATPase may be mediated by a calmodulin-dependent protein kinase. However, calmodulin could also interact with other cell signaling mechanisms that affect the pump. For instance, calmodulin may after the level of cAMP by inhibiting or stimulating either adenylate cyclase or phosphodiesterase, depending on the cell in question. Likewise, calmodulin may inhibit PKC.[67,68]

In the following pages I will review a number of papers published since 1988,[44] which provide still-debatable evidence that hormones regulate the short-term activity of the pump directly. The emphasis is on possible second messengers that might mediate such regulation, and possible tissue differences.

Adrenal Glomerulosa Cells

Angiotensin II inhibits ouabain-sensitive ^{86}Rb uptake up to 50% in rat adrenal glomerulosa cells ($K_i = 0.3\,nM$).[16] Inhibition is observed only in the presence of extracellular calcium and is blocked by DuP 753 ($IC_{50} = 200\,nM$), a selective antagonist of AT_1 angiotensin II receptors. The AT_2-receptor antagonist PD 123177 is ineffective. According to the authors, occupancy of the AT_1 receptors in glomerulosa cells is coupled to phospholipase C activation, increased hydrolysis of phosphoinositol and consequent increase in intracellular calcium and PKC. However, this explanation may be in doubt because phorbol-12-myristate acetate (PMA) has no effect on the response of the pump, which indicates that PKC may not be involved, while the addition of ionomycin, a calcium ionophore, fails to mimic the effects of angiotensin II, which also suggests that calcium alone is not responsible for the inhibition. The matter cannot be settled because intracellular free calcium was not measured in these experiments, nor was the intracellular concentration of sodium. Thus one cannot rule out the possibility that angiotensin II affected the rate of ^{86}Rb uptake by altering the concentration of intracellular sodium. Likewise it is difficult to judge if the effect of angiotensin II was on the V_{max} of the pump or on the affinity of the pump for sodium or potassium. The authors suggest that angiotensin II regulates the secretion of aldosterone by inhibiting the sodium pump and depolarizing the membrane, thereby triggering an increase in intracellular calcium and stimulating hormone secretion.

Heart

Guinea Pig Ventricular Myocytes

The ouabain-sensitive sodium pump current of the guinea pig ventricular myocytes is inhibited by isoproterenol when the intracellular free calcium is less than 150 nM, but is stimulated at higher calcium concentrations.[69] Both effects are mediated by β receptors because they are blocked by propranolol. Isoproterenol appears to affect the transport V_{max} rather than the pump affinity for sodium and potassium. At 1.4 µM free calcium concentration, isoproterenol stimulates the pump 25% with a $K_{1/2}$ of 11.6 µM. This effect is eliminated by H-7, a nonspecific protein kinase inhibitor. Isoproterenol appears to act primarily on the α_1 isoform of ATPase, although the myocytes contain also the α_2 isoform.

Canine Purkinje Myocytes

Shah et al.[21] have observed that phenylephrine (0.1 µM) stimulates sodium pump currents in canine myocytes. The effect is blocked by pertussis toxin and appears to be mediated by α_1 receptors. The authors suggest that this stimulation, which would tend to hyperpolarize the membrane, is in part responsible for an observed decrease in the spontaneous firing rate. Earlier studies have demonstrated that norepinephrine produces a decrease in intracellular sodium that is consistent with stimulation of the sodium pump.[70,71] The effects of norepinephrine may be mediated by β receptors[72] and an increase in cAMP.[71] Thus in these fibers the effects of catecholamines on the sodium pump may be mediated by both α and β receptors.

Sheep Purkinje Cells

Norepinephrine stimulates the sodium pump in electrically stimulated Purkinje fibers, which decreases intracellular sodium, prolongs the action potential, and hyperpolarizes the diastolic membrane potential.[73]

Interestingly, norepinephrine has no observed effect on the pump in quiescent fibers, a difference that appears to depend on the value of the membrane potential and perhaps is due to the voltage dependency of the pump. Neither the type of receptors (α or β) nor the intracellular mechanism for these effects have been established.

Skeletal Muscle

Insulin stimulates the sodium pump in rat soleus muscle, an action that is independent of changes in the concentration of intracellular sodium.[17,74–76] In the experiments of Weil et al.[76] the stimulation of ouabain-sensitive ^{86}Rb uptake was accompanied by an increase in ouabain binding, which indicated that the effect was due either to an increase in the rate of ouabain binding or to an increase in the number of pumps. An increase in the rate of binding could be consistent with an increase in the rate of transport.[77] However, the relationship between the rate of transport and the rate of ouabain binding is complex, with the rate of ouabain binding (under some conditions of internal and external sodium and potassium) also being inversely related to the transport rate.[78] In frog skeletal muscle insulin can also rapidly increase the number of pumps in the plasma membrane by recruitment from intracellular vesicles.[8] A recent critical review by Zierler[79] concludes that insulin may not have any direct effect on sodium pump activity in this tissue, because all of the observed effects could be explained by changes in the pump environment.

Liver

When liver slices are exposed to phenylephrine (PE) (10 µM) the rate of ^{86}Rb transport is either stimulated or inhibited, depending on the length of exposure and the concentration of extracellular calcium.[80] Exposure of the cells to PMA during the transport measurements altered the ^{86}Rb transport, which suggests that PKC is involved in the effect of PE, although the complexity of the results do not allow any conclusion on the mechanism of this involvement, especially since measurements of intracellular sodium or calcium were not made. To test if phenylephrine directly affected the activity of the pump, cells were exposed to phenylephrine in situ and Na,K-ATPase activity later measured after homogenization of the cells and harvesting of the membranes. Prior exposure to phenylephrine and calcium did appear to influence the subsequent activity of the Na,K-ATPase, which suggests a direct effect on the sodium pump. However, the results were again complex and cannot easily be compared to the transport measurements. The observation that phenylephrine could either stimulate or inhibit the pump depending on the concentration of extracellular calcium could indicate that the effects depend on the relative concentration of intracellular free calcium.[69]

HeLa Cells Transfected with the 5-HT$_{1A}$ Receptor

5-HT (EC$_{50}$ ~3 µM) moderately stimulates ouabain-sensitive ^{86}Rb uptake in HeLa cells transfected with a cloned 5-HT$_{1A}$ receptor,[81] but had no effect in nontransfected HeLa cells. The onset of the response is rapid (5 to 10 min) and short-lived (<30 min), and is greatest a low concentrations of extracellular potassium (38% stimulation of 2 mM K with 10 µM 5-HT). 5-HT has no effect on the maximum rate of pumping and appears to act by increasing the affinity of the pump for extracellular potassium. Intracellular sodium was not measured in these experiments, so a stimulation of the pump by an increase in intracellular sodium cannot be ruled out. It is believed that occupancy of the 5-HT receptor activates a G protein, which inhibits adenylate cyclase and activates PIP$_2$ hydrolysis, leading to an increase in intracellular calcium. Indeed, 10 µM 5-HT increased intracellular free calcium by approximately 50 nM. When this increase is prevented, the effect of 5-HT on the sodium pump is abolished. Although the effect of 5-HT on the pump is blocked by pertussis toxin, stimulation of the sodium pump does not appear to involve inhibition of adenylate cyclase, because the effect of 5-HT can still be demonstrated in the presence of dibutryl cAMP and isobutylmethyl-xanthine. If intracellular calcium is increased by either A23187 or thapsigargin, which presumably bypass the 5-HT receptor, the pump is also stimulated. These agents have no effect if the change in free calcium is prevented. In addition, neither PMA nor phorbol-12,13 butyrats (PbBU) have an effect on the response of the pump to 5-HT, which suggests that PKC is not involved. These observations suggest that the effects of 5-HT on the sodium pump may involve an unidentified protein kinase that is activated by an increase in free calcium.

Platelets

Turaihi et al.[82] reported that isoproterenol and epinephrine stimulate, and norepinephrine inhibits, ouabain-sensitive ^{86}Rb uptake in human platelets; they suggested that the stimulation may be due to a β-adrenergic effect (and associated increase in cAMP) and inhibition to an α-adrenergic receptor-mediated mechanism. Thrombin increases intracellular calcium and alters intracellular sodium, two effects which appear to be related, at least in part, to thrombin-induced inhibition of the sodium pump.[83] The mechanism of this inhibition has not been established but can be mimicked by the addition of phorbol ester 12-O-tetradecanoylphorbol-13-acetate, a stimulant of PKC.

Neurons

Dopamine inhibits Na,K-ATPase activity of neostriatal neurons[20] through an apparently synergistic effect of the D_1 and D_2 receptors. Both D_1 and D_2 agonists are needed for inhibition in dopamine-depleted neurons and either a D_1 or a D_2 antagonist can block dopamine-induced inhibition. Dopamine inhibition is concentration-dependent from $0.1\,nM$ to $1\,\mu M$, with a half-maximal response at $1\,nM$. The intracellular pathways that result in pump inhibition have not been established. However, recent observations with cloned D_2 receptors in Chinese hamster ovary cells suggest that the synergistic effects of D_1 and D_2 receptors in neurons may be due to enhanced arachidonic acid release that results from an elevation of intracellular Ca.[84]

Kidney

Dopamine inhibits Na,K-ATPase activity in rat proximal tubules.[85] According to Bertorello and Aperia,[86] activation of both DA_1 and DA_2 receptors are needed for inhibition. These authors conclude that the effect is due to a synergistic action of the two receptors and suggest that although DA_1 may act to increase cAMP levels, DA_2 may be linked to an unknown phosphatase inhibitor capable of sustaining the response initiated by DA_1. Support for this hypothesis derives from the recent observation that a phosphopeptide that corresponds to residues 8–38 of the protein phosphatase inhibitor DARPP-32 inhibits the Na,K-ATPase in cells of the ascending limb of the loop of Henle.[19] On the other hand, a linkage between DA_2 receptors and different types of renal cells has not yet been established and may involve different second-messenger systems in different cells. The possible role of calcium in the synergistic effects of DA_1 and DA_2 receptors on the Na,K-ATPase in these cells has not yet been investigated.

Charlton and Baylis[87] have reported that arginine vasopressin (AVP) stimulates Na,K-ATPase activity in the thick ascending limb of the loop of Henle of the rat renal medulla and, using a cytochemical assay in tissue slices, have shown that the response to AVP is blocked by V_2/V_1 antagonists, but not by a V_1 antagonist. The authors conclude that AVP stimulates the pump via a V_2 receptor-mediated stimulation of adenylate cyclase.

Rat Brain Synaptosomes

High concentrations of insulin ($K_i = 100–200\,nM$) stimulate ouabain-sensitive ^{86}Rb uptake in rat brain synaptosomes when the intracellular concentration of sodium is low ($<20\,mM$), but not when it is high

(>100 mM).[88] The synaptosomes contain α_1 and α_2 isoforms of the Na,K-ATPase, which differ in their apparent affinity for intracellular sodium in the absence of insulin. The authors conclude that insulin preferentially stimulates α_2 by increasing its affinity for intracellular sodium. Much lower concentrations of insulin stimulate the pump in adipocytes by a similar mechanism,[89,90] although these conclusions have recently been modified[15] (see below). In synaptosomes disruption of the membranes to measure Na,K-ATPase activity increases the affinity of α_2 for sodium and abolishes the effect of insulin.[90] Thus it is difficult to disprove that insulin affects the pump by altering the ion gradients. The transduction mechanism by which insulin alters the affinity of α_2 for sodium is unknown.

Adipocytes

In rat adipocytes, insulin stimulates the pump by a mechanism other than increasing intracellular sodium or increasing the number of pumps.[89,91,92] Adipocytes contain both α_1 and α_2 isozymes, and it had been suggested that the increase in pump activity was due to a selective increase in the affinity of α_2 for intracellular Na.[89,90] However, more recent investigations have demonstrated that the procedures used in the earlier studies altered the contribution of intracellular sodium[93] and that when the concentration of the latter is kept constant, insulin stimulates transport by increasing the affinity of both α_1 and α_2 for sodium and by increasing the V_{max} of α_2.[15] Insulin has no effect on ouabain-sensitive transport in adipocyte ghosts or on isolated adipocyte membranes.[93] The biochemical mechanisms by which insulin alters the kinetic properties of α_1 and α_2 are unknown.

Summary

Recent studies provide additional support for the hypothesis that the short-term activity of individual pumps is regulated by hormones. Calcium and cAMP appear to function as second messengers in a number of cell types. In some cells the effect of calcium may be associated with the activation of PKC. In other cells the effects of calcium may be mediated by a calcium-dependent kinase not yet characterized. In red cells the effects of physiologic concentrations of free calcium on the Na,K-ATPase may be due to the activation of a calmodulin-dependent protein kinase. It is not yet known if hormones regulate short-term activity of the sodium pump under physiologic conditions. To answer this question it will be necessary to demonstrate that hormones have direct receptor-mediated effects on the pump and that these effects result in meaningful physiologic responses, such as secretion. Only then will it be possible to identify the second messengers and the biochemical mechanisms that mediate these physiologic responses.

Acknowledgments. This work was supported by NSF grant DCB-8817269.

References

1. Skou JC. The identification of the sodium-pump as the membrane-bound Na$^+$/K$^+$-ATPase. Biochim Biophys Acta 1989; 1000:435–438.
2. Jorgensen PL. Mechanism of the Na,K-pump. Biochim Biophys Acta 1982; 694:26–68.
3. Pedemonte CH, Kaplan JH. Chemical modification as an approach to elucidation of sodium pump structure-function relations. Am J Physiol 1990; 258:C1–C23.
4. Kawamura M, Naguchi S. Possible role of the β-subunit in the expression of the sodium pump. In: Kaplan JH, DeWeer P, eds. The Sodium Pump: Structure, Mechanism, and Regulation. New York: Rockefeller University Press; 1991:45–61.
5. Sweadner KJ. Isozymes of the Na$^+$/K$^+$-ATPase. Biochim Biophys Acta 1989; 988:185–220.
6. Lingrel JB, Orlowski J, Price EM, Pathak BG. Regulation of the α-subunit genes of the Na,K-ATPase and determinants of cardiac glycoside sensitivity. In: Kaplan JH, DeWeer P, eds. The Sodium Pump: Structure, Mechanism, and Regulation. New York: Rockefeller University Press; 1991:1–16.
7. Sweadner KJ. Overview: Subunit diversity in the Na,K-ATPase. In: Kaplan JH, DeWeer P, eds. The Sodium Pump: Structure, Mechanism, and Regulation. New York: Rockefeller University Press; 1991:63–76.
8. Omatsu-Kanbe M, Kitasato H. Insulin stimulates the translocation of Na$^+$/K$^+$-dependent ATPase molecules from intracellular stores in the plasma membrane in frog skeletal muscle. Biochem J 1990; 272:727–733.
9. Rossier BC, Geering K, Kraehenbuhl JP. Regulation of the sodium pump: How and why? TIBS 1987; 12:483–487.
10. Fambrough DM, Wolitzky BA, Taormino JP, Tamkun MM, Takeyasu K, Somerville D, Renaud KJ, Lemas MV, Lebovitz RM, Kone BC, Hamrick M, Rome J, Inman EM, Barnstein A. A cell biologists perspective on sites of Na,K-ATPase regulation. In: Kaplan JH, DeWeer P, eds. The Sodium Pump: Structure Mechanism, and Regulation. New York: Rockefeller University Press; 1991:17–30.
11. Gick GG, Ismail-Beigi F, Edelman IS. Hormonal regulation of Na,K-ATPase. Prog Clin Biol Res 1988; 268B:277–295.
12. Hernandez-R J. Na$^+$/K$^+$-ATPase regulation by neurotransmitters. Neurochem Int 1992; 20:1–10.
13. Vizi ES, Oberfrank F. Na$^+$/K$^+$-ATPase, its endogenous ligands and neurotransmitter release. Neurochem Int 1992; 20:11–17.
14. Brodsky JL. Characterization of the (Na$^+$ + K$^+$)-ATPase from 3T3-F442A fibroblasts and adipocytes: Isozymes and insulin sensitivity. J Biol Chem 1990; 265:10458–10465.
15. McGill DL, Guidotti G. Insulin stimulates both the alpha 1 and the alpha 2 isoforms of the rat adipocyte (Na$^+$,K$^+$) ATPase. Two mechanisms of stimulation. J Biol Chem 1991; 266:15824–15831

16. Hajnoczky G, Csordas G, Hunyady L, Kalapos MP, Balla T, Enyedi P, Spat A. Angiotensin-II inhibits Na^+/K^+ pump in rat adrenal glomerulosa cells: Possible contribution to stimulation of aldosterone production. Endocrinology 1992; 130:1637–1644.

17. Clausen T, Everts ME. Regulation of the Na,K-pump in skeletal muscle. Kid Inter 1989; 35:1–13.

18. Clausen T. Significance of Na^+-K^+ pump regulation in skeletal muscle. News Physiol Sci 1990; 5:148–151.

19. Aperia A, Fryckstedt J, Svensson L, Hemmings HC Jr, Nairn AC, Greengard P. Phosphorylated Mr 32,000 dopamine- and cAMP-regulated phosphoprotein inhibits Na^+,K^+-ATPase activity in renal tubule cells. Proc Natl Acad Sci USA 1991; 88:2798–2801.

20. Bertorello AM, Hopfield JF, Aperia A, Greengard P. Inhibition by dopamine of (Na^+,K^+)-ATPase activity in neostriatal neurons through D1 and D2 dopamine receptor synergism. Nature 1990; 347:386–388.

21. Shah A, Cohen IS, Rosen MR. Stimulation of cardiac alpha receptors increases Na/K pump current and decreases gK via a pertussis toxin-sensitive pathway. Biophysic J 1988; 54:219–225.

22. Sasaguri T, Watson SP. Phorbol esters inhibit smooth muscle contractions through activation of Na^+-K^+-ATPase. Br J Pharmacol 1990; 99:237–242.

23. Schoner W. Endogenous digitalis-like factors. TIPS 1991; 12:209–211.

24. Phillis JW. Na^+/K^+-ATPase as an effector of synpatic transmission. Neurochem Int 1992; 20:19–22.

25. Shull GE, Greeb J, Lingrel JB. Molecular cloning of three distinct forms of the Na^+,K^+-ATPase α-subunit from rat brain. Biochemistry 1986; 25:8125–8132.

26. Lingham RB, Sen AK. Regulation of rat brain $(Na^+ + K^+)$-ATPase activity by cyclic AMP. Biochim Biophys Acta 1982; 688:475–485.

27. Delamere NA, Socci RR, King KL. Alteration of sodium, potassium-adenosine triphosphatase activity in rabbit ciliary processes by cyclic adenosine monophosphate-dependent protein kinase. Invest Ophth Vis Sci 1990; 31:2164–2170.

28. Marver DS, Lear S, Marver LT, Silva P, Epstein EH. Cyclic AMP-dependent stimulation of Na,K-ATPase in shark rectal gland. J Membr Biol 1986; 94:205–215.

29. Paris S, Rozengurt E. Cyclic AMP stimulation of Na,K pump activity in quiescent Swiss 3T3 cells. J Cell Physiol 1982; 112:273–280.

30. Ling L, Cantley L. The (Na,K)-ATPase of Friend erythroleukemia cells is phosphorylated near the ATP hydrolysis by an endogenous membrane-bound kinase. J Biol Chem 1984; 259:4089–4095.

31. Skou JC. The influence of some cations on an adenosine triphosphatase from peripheral nerves. Biochim Biophys Acta 1957; 23:394–401.

32. Hoffman JF. Cation transport and structure of the red cell plasma membrane. Circulation 1962; 26:1201–1213.

33. Robinson JD. Nucleotide and divalent cation interactions with the $(Na^+ + K^+)$-dependent ATPase. Biochim Biophys Acta 1974; 341:232–247.

34. Apell HJ, Marcus MM. $(Na^+ + K^+)$-ATPase in artificial lipid vesicles: Influence of the concentration of mono- and divalent cations on the pumping rate. Biochim Biophys Acta 1986; 862:254–264.

35. Tobin T, Akera T, Baskin I, Brody TM. Calcium ion and sodium- and potassium-dependent adenosine triphosphatase: Its mechanism of inhibition and identification of the E_1-P intermediate. Mol Pharmacol 1973; 9:336–349.
36. Fukushima Y, Post RL. Binding of divalent cation to phosphoenzyme of sodium- and potassium-transport adenosine triphosphatase. J Biol Chem 1978; 253:6853–6862.
37. Haag M, Gevers W, Bohmer RG. The interaction between calcium and the activation of the Na^+,K^+-ATPase by noradrenaline. Mol Cell Biochem 1985; 66:111–116.
38. Vasallo PM, Post RL. Calcium ion as a probe of the monovalent cation center of sodium, potassium ATPase. J Biol Chem 1986; 261:16957–16962.
39. Forbush B. Rapid release of ^{45}Ca from an occluded state of the Na,K-pump. J Biol Chem 1988; 263:7970–7978.
40. Winkler BS, Riley MV. Influence of calcium on retinal ATPases. Invest Ophth Vis Sci 1980; 19:562–564.
41. Yingst DR, Marcovitz MJ. Effect of hemolysate on calcium inhibition of the $(Na^+ + K^+)$-ATPase of human red blood cells. Biochem Biophys Res Commun 1983; 111(3):970–979.
42. Yingst DR, Hoffman JF. Ca-induced K transport in resealed human red cells containing arsenazo III: Transmembrane effects of Na and K and the relationship to the functioning Na-K pump. J Gen Physiol 1984; 84:19–45.
43. Turi A, Somogyi J, Muller N. The effect of micromolar Ca^{2+} on the activities of the different Na^+/K^+-ATPase isozymes in the rat myometrium. Biochem Biophys Res Commun 1991; 174:969–974.
44. Yingst DR. Modulation of the Na,K-ATPase by Ca and intracellular proteins. Annu Rev Physiol 1988; 50:291–303.
45. McGeoch JEM. The α_2 isomer of the sodium pump is inhibited by calcium at physiological levels. Biochem Biophys Res Commun 1990; 173:99–105.
46. Aksentsev SL, Novitskaia NA, Okun' IM, Lyskova TI, Konev SV. Membrane regulation of brain Na,K-ATPase inhibition by calcium ions. Biokhimiia 1980; 45:679–682.
47. Turi A, Somogyi J. Possible regulation of the myometrial Na^+/K^+-ATPase activity by Ca^{2+} and cAMP-dependent protein kinase. Biochim Biophys Acta 1988; 940:77–84.
48. Komabayashi T, Izawa T, Suda K, Shinoda S, Tsuboi M. Effects of Ca^{2+} and calmodulin antagonists on the Na^+ pump activity induced by pilocarpine in dog submandibular gland. Res Commun Chem Path Pharm 1987; 57:273–276.
49. Vasilets LA, Schmalzing G, Madefessel K, Haase W, Schwarz W. Activation of protein kinase C by phorbol ester induces downregulation of the Na^+/K^+-ATPase in oocytes of Xenopus laevis. J Membr Biol 1990; 118:131–142.
50. Lowndes JM, Hokin-Neaverson M, Bertics PJ. Kinetics of phosphorylation of Na^+/K^+-ATPase by protein kinase C. Biochim Biophys Acta 1990;1052:143–151.
51. Nishizuka Y. Studies and perspective of protein kinase C. Science 1986; 233:305–312.
52. Berridge MJ. Inositol trisphosphate and diacylglycerol: Two interacting second messengers. Ann Rev Biochem 1987; 56:159–193.

53. Oishi K, Zheng B, Kuo JF. Inhibition of Na,K-ATPase and sodium pump by protein kinase C regulators sphingosine, lysophosphatidylcholine, and oleic acid. J Biol Chem 1990; 265:70–75.

54. Nishikawa T, Tomori Y, Yamashita S, Shimizu S. Inhibition of Na^+,K^+-ATPase activity by phospholipase A_2 and several lysophospholipids: Possible role of phospholipase A_2 in noradrenaline release from cerebral cortical synaptosomes. J Pharm Pharmacol 1989; 41:450–458.

55. Askari A, Xie ZJ, Wang YH, Periyasamy S, Huang WH. A second messenger role for monoacylglycerols is suggested by their activating effects on the sodium pump. Biochim Biophys Acta 1991; 1069:127–130.

56. Kim J, Rushovich EH, Thomas TP, Ueda T, Agranoff BW, Greene DA. Diminished specific activity of cytosolic protein kinase C in sciatic nerve of streptozocin-induced diabetic rats and its correction by dietary myoinositol. Diabetes 1991; 40:1545–1554.

57. Kowluru R, Bitensky MW, Kowluru A, Dembo M, Keaton PA, Buican T. Reversible sodium pump defect and swelling in the diabetic rat erythrocyte: Effects on filterability and implications for microangiopathy. Proc Natl Acad Sci USA 1989; 86:3327–3331.

58. Berthon B, Capiod T, Claret M. Effect of noradrenaline, vasopressin and angiotensin on the Na-K pump in rat isolated liver cells. Br J Pharmacol 1985; 86:151–161.

59. Lynch CJ, Wilson RB, Blackmore PF, Exton JH. The hormone-sensitive hepatic Na^+-pump: Evidence for regulation by diacylglycerol and tumor promoters. J Biol Chem 1985; 261:14551–14556.

60. Klee CB. Interaction of calmodulin with Ca^{2+} and target proteins. In: Cohen P, Klee CB, eds. Calmodulin. New York: Elsevier; 1988:35–56.

61. Cirillo M, David-Bufilho M, Devynck MA. Calmodulin reduces ouabain-sensitive ATPase of cardiac sarcolemmal membranes: High reduction in spontaneously hypertensive rats. Clin Sci 1984; 67:535–540.

62. David-Dufilho M, Cirillo M, Beugras JP, Meyer P, Devynck MA. Active calcium and sodium transport by cardiac plasma membranes in the genetically hypertensive rat [Fre]. Archives des Maladies du Coeur et des Vaisseaux 1984; 77:1261–1265.

63. Yingst DR, Ye-Hu J, Chen H, Barrett V. Calmodulin increases Ca-dependent inhibition of the Na,K-ATPase in human red cells. Arch Biochem Biophys 1992; 295:49–54.

64. Huang WH, Askari A. Ca^{2+}-dependent activities of $(Na^+ + K^+)$-ATPase. Arch Biochem Biophys 1982; 216:741–750.

65. Nelson WJ, Veshnock P. Ankyrin binding to $(Na^+ + K^+)$ATPase and implications for the organization of membrane domains in polarized cells. Nature 1987; 328:533–536.

66. Tokumitsu H, Chijiwa T, Hagiwara M, Mizutani A, Terasawa M, Hidaka H. KN-62, 1-[N,O-bis(5-isoquinolinesulfonyl)-N-methyl-L-tyrosyl]-4-phenyl-piperazine, a specific inhibitor of Ca^{2+}/calmodulin-dependent protein kinase II. J Biol Chem 1990; 265:4315–4320.

67. Albert KA, Wu C-S, Nairn AC, Greengard P. Inhibition by calmodulin of calcium/phospholipid-dependent protein phosphorylation. Proc Natl Acad Sci USA 1984; 81:3622–3625.

68. Kruger H, Schroder W, Buchner K, Hucho F. Protein kinase C inhibition by calmodulin and its fragments. J Protein Chemistry 1990; 9:467–473.
69. Gao J, Mathias RT, Cohen IS, Baldo GJ. $[Ca^{2+}]_i$ determines the effects of isoproterenol on the Na/K pump in ventricular myocytes. Biophysic J 1992; 61:A445.
70. Lee CO, Vassalle M. Modulation of intracellular Na^+ activity and cardiac force by norepinephrine and Ca^{2+}. Am J Physiol 1983; 244:C110–C114.
71. Pecker MS, Im WB, Sonn JK, Lee CO. Effects of norepinephrine and cyclic AMP on intracellular sodium ion activity and contractile force in canine cardiac Purkinje fibers. Circ Res 1986; 59:390–397.
72. Wasserstrom JA, Schwartz DJ, Fozzard HA. Catecholamine effects on intracellular sodium activity and tension in dog heart. Am J Physiol 1982; 243:H670–H675.
73. Chae SW, Wang DY, Gong QY, Lee CO. Effect of norepinephrine on Na^+-K^+ pump and Na^+ influx in sheep cardiac Purkinje fibers. Am J Physiol 1990; 258:C713–722.
74. Clausen T. Regulation of active Na^+-K^+ transport in skeletal muscle. Physiol Rev 1986; 66:542–480.
75. Clausen T, Van Hardeveld C, Everts ME. Significance of cation transport in control of energy metabolism and thermogenesis. Physiol Rev 1991; 71:733–774.
76. Weil E, Sasson S, Gutman Y. Mechanism of insulin-induced activation of Na^+-K^+-ATPase in isolated rat soleus muscle. Am J Physiol 1991; 261:C224–230.
77. Joiner CH, Lauf PK. Modulation of ouabain binding and potassium fluxes by cellular sodium and potassium in human and sheep erythrocytes. J Physiol 1978; 283:177–196.
78. Bodeman HH, Hoffman JF. Side-dependent effects of internal versus external Na and K on ouabain binding to reconstituted human red blood cell ghosts. J Gen Physiol 1976; 67:497–525.
79. Zierler K. Insulin, membrane polarization and ionic currents. In: Cuatrecasas P, Jacobs S, eds. Insulin: Handbook of Experimental Pharmacology. Berlin: Springer-Verlag; 1990:421–450.
80. Smart JL, Deth RC. Influence of alpha 1-adrenergic receptor stimulation and phorbol esters on hepatic Na^+/K^+-ATPase activity. Pharm 1988; 37:94–104.
81. Middleton JP, Raymond JR, Whorton AR, Dennis W. Short-term regulation of Na^+/K^+ adenosine triphosphatase by recombinant human serotonin 5-HT1A receptor expressed in HeLa cells. J Clin Invest 1990; 86:1799–1805.
82. Turaihi K, Khokher MA, Barradas MA, Mikhailidis DP, Dandona P. $^{86}Rb(K)$ influx and $[^3H]$ouabain binding by human platelets: Evidence for beta-adrenergic stimulation of Na-K ATPase activity. Metabolism: Clinical & Experimental 1989; 38:773–776.
83. Borin M, Siffert W. Further characterization of the mechanisms mediating the rise in cytosolic free Na^+ in thrombin-stimulated platelets. Evidence for inhibition of the Na^+, K^+-ATPase and for Na^+ entry via a Ca^{2+} influx pathway. J Biol Chem 1991; 266:13153–13160.

84. Piomelli D, Pilon C, Giros B, Sokoloff P, Martres MP, Schwartz JC. Dopamine activation of the arachidonic acid cascade as a basis for D_1/D_2 receptor synergism. Nature 1991; 353:164–167.
85. Aperia A, Bertorello A, Seri I. Dopamine causes inhibition of Na^+-K^+-ATPase activity in rat proximal convoluted tubule segments. Am J Physiol 1987; 252:F39–F45.
86. Bertorello A, Aperia A. Inhibition of proximal tubule Na^+-K^+-ATPase activity requires simultaneous activation of DA1 and DA2 receptors. Am J Physiol 1990; 259:F924–F928.
87. Charlton JA, Baylis PH. Stimulation of rat renal medullary Na^+/K^+-ATPase by arginine vasopressin is mediated by the V2 receptor. J Endocrinol 1990; 127:213–216.
88. Brodsky JL. Insulin activation of brain Na^+-K^+-ATPase is mediated by alpha 2-form of enzyme. Am J Physiol 1990; 258:C812–C817.
89. Resh MD, Nemenoff RA, Guidotti G. Insulin stimulation of (Na^+,K^+)-adenosine triphosphatase-dependent $^{86}Rb^+$ uptake in rat adipocytes. J Biol Chem 1980; 255:10938–10945.
90. Lytton J, Lin JC, Guidotti G. Identification of two molecular forms of (Na^+,K^+)-ATPase in rat adipocytes: Relation to insulin stimulation of the enzyme. J Biol Chem 1985; 260:1177–1184.
91. Lytton J. Insulin affects the sodium affinity of the rat adipocyte $(Na^+ + K^+)$-ATPase. J Biol Chem 1985; 260:10075–10080.
92. Clausen T, Hansen O. Active Na-K transport and the rate of ouabain binding: The effect of insulin and other stimuli on skeletal muscle and adipocytes. J Physiol 1977; 270:415–430.
93. McGill DL. Characterization of the adipocyte ghost (Na^+,K^+) pump. Insights into the insulin regulation of the adipocyte (Na^+,K^+) pump. J Biol Chem 1991; 266:15817–15823.

Note added in proof. Recent studies strengthen the evidence that the Na,K-ATPase is regulated via protein kinases.[94–98]

94. Bertorello AM, Aperia A, Walaas SI, Nairn AC, Greengard P. Phosphorylation of the catalytic subunit of Na^+,K^+-ATPase inhibits the activity of the enzyme. Proc Natl Acad Sci 1991; 88:11359–11362.
95. Vasilets LA, Schwarz W. Regulation of endogenous and expressed Na^+/K^+ pumps in Xenopus cocytes by membrane potential and stimulation of protein kinases. J Memb Biol 1992; 125:119–243.
96. Chibalin AV, Vasilets LA, Hennekes H, Pralong D, Geering K. Phosphorylation of Na,K-ATPase alpha-subunits in microsomes and in homogenates of Xenopus oocytes resulting from the stimulation of protein kinase A and protein kinase C. J Biol Chem 1992; 267:22378–22384.
97. Chibalin AV, Lopina OD, Petukhov SP, Vasilets LA. Phosphorylation of the Na,K-ATPase by Ca, phospholipid-dependent and cAMP-dependent protein kinases. Mapping of the region phosphorylated by Ca, phospholipid-dependent protein kinase. J Bioenerg Biomemb 1993; 25:61–66.
98. Ibarra F, Aperia A, Svensson LB, Eklof AC, Greengard P. Bidirectional regulation of Na^+,K^+-ATPase activity by dopamine and an alpha-adrenergic agonist. Proc Natl Acad Sci USA 1993; 90:21–24.

12
Endogenous Regulation of Sodium Pump Activity

Peter A. Doris

Following volume expansion, the plasma acquires the capacity to cause natriuresis, inhibit membrane sodium transport, and increase vascular reactivity. The discovery of the atrial natriuretic factor (ANF) in 1981 provided a partial explanation for the first of these phenomena, but it was rapidly recognized that ANF was not an inhibitor of sodium transport,[1] nor was it able to increase vascular reactivity.[2] Thus it became apparent that a 20-year effort to identify a plasma-borne substance that was natriuretic by virtue of its ability to inhibit cellular sodium transport and which might increase vascular reactivity as a result of a similar action on vascular smooth muscle had not yet succeeded. Ten years later, in 1991, this effort appeared to have achieved its first major breakthrough.

In the period following the second World War, the prevailing view of renal sodium handling was that changes in extracellular fluid volume produced alterations in sodium excretion either by a change in glomerular filtration rate or a change in the plasma levels of mineralocorticoid. This view proved inadequate, in part as a result of human studies that demonstrated that artificial elevations of plasma mineralocorticoid levels in subjects with a constant sodium intake did not prevent relatively rapid alterations in sodium excretion in response to an increase in sodium balance, and that the prolonged administration of mineralocorticoids was eventually followed by a return of urinary sodium excretion to normal levels.[3,4] This observation has come to be known as the "escape" phenomenon.

The fact that increased sodium excretion occurred without elevation in glomerular filtration rate and must therefore have resulted from a change in tubular reabsorption of sodium was convincingly demonstrated by De Wardener and colleagues.[5] Glomerular filtration rate was lowered by the inflation of a balloon in the thoracic aorta in dogs treated with mineralocorticoids and vasopressin. Although filtered sodium was markedly reduced, an infusion of saline into these animals resulted in a rise of urinary sodium excretion. For the first time convincing data had been obtained, showing that a "third" factor could influence the excretion of sodium by the kidney. Much work followed to investigate the mechanism of this change in sodium excretion. One possibility was that, under these circumstances, increased sodium excretion might result from the diluting effects of saline on the circulating blood. However, De Wardener's cross-circulation experiments in vasopressin- and mineralocorticoid-treated dogs revealed that natriuresis following saline infusion occurred even when there were no significant changes in arterial blood pressure, hematocrit, plasma protein, or plasma sodium concentration.

This chapter will examine the pursuit of a hypertensinogenic third factor and the current status of research in this area.

Volume Expansion and Sodium Pump Inhibitor

Ten years after these and other convincing experiments suggested the existence of an endogenous natriuretic factor, two papers provided evidence that acute volume expansion was accompanied by an increase in the ability of plasma to inhibit sodium, potassium-ATPase (NKA). Buckalew and colleagues showed that dialysates and ultrafiltrates of plasma from salt-loaded dogs inhibited sodium transport in the toad bladder preparation,[6] while De Wardener's laboratory made similar observations using renal tubule fragments as the assay model.[7] Since NKA is a major force in returning filtered sodium from the tubular lumen to the interstitial fluid, and since NKA is known to be inhibited by cardiac glycosides, increased plasma levels of an NKA inhibitor would result in natriuresis through inhibition of renal NKA. It was assumed from these findings that the natriuretic capacity of the plasma of volume-expanded subjects and its ability to inhibit NKA were due to the same substance, generally termed "natriuretic hormone."

In the early 1980s, De Bold's work on the natriuretic properties of atrial extracts led to the characterization of atrial natriuretic peptide (ANP).[8] Studies quickly followed to determine the mechanism of its action on renal sodium handling, and it soon became clear that the atrial natriuretic factor did not inhibit NKA.[1] The resistance of science to accept new breakthroughs has perhaps been illustrated nowhere more clearly than in the case of ANP. Many workers in the field were reluctant to accept the relevance of this substance to the physiologic regulation of sodium balance. Consequently the concept of a natriuretic hormone that acted by inhibiting NKA continued to be pursued, even though well-known inhibitors of NKA, such as ouabain and digoxin, were scarcely natriuretic in intact animals, especially in comparison with ANP. The chief consequence of this contradiction appears to have been largely semantic: the term natriuretic hormone (as applied to endogenous inhibitors of NKA) fell out of fashion in favor of more specific terms ("endogenous digitalis-like factor," "ouabain-like substance").

By this time the endogenous opiates were well understood, which provided an interesting precedent. Although they are endogenous substances, they are analogous to a group of plant-derived pharmaceuticals whose therapeutic activity is based on their property to cross-react with the endorphin receptors. Similarly, the plant cardiac glycosides, such as digitalis, ouabain, strophanthidin, and others have been used therapeutically for hundreds of years and their biochemistry and pharmacology are very well understood. They consist of a steroid nucleus which typically is modified to contain an unsaturated cyclic group attached to the C17 atom and one or several sugars attached in the C3 position. They bind to a receptor, which is an integral portion of the α subunit of the NKA protein, and as a result they inhibit the membrane ion-translocating action of NKA. It became attractive to speculate that this receptor had evolved to mediate the activity of an endogenous, not a botanic, agent. Thus the concept of an endogenous digitalis-like substance (EDLF) or ouabain-like substance became firmly established in the literature. There followed a rapid development of radioimunoassays for plasma glycoside levels in patients using these drugs, which led to the fortuitous finding in patients (who had not received cardiac glycosides) of demonstrable plasma glycoside immunoreactivity, providing possible evidence of EDLF.

Sodium Pump Inhibitor and Hypertension

Interest in inhibitors of NKA grew during the late 1970s, in no small part because of observations suggesting that hypertension was associated with alteration in ion handling by the sodium pump. Haddy's group reported that renal hypertension in dogs was accompanied by suppression of NKA activity.[9] Similarly, myocardial NKA activity was found to be suppressed in a rodent model of low renin hypertension.[10] A wide range of studies now strengthen the association between several forms of experimental and hereditary hypertension and elevation of plasma levels of a material able to inhibit NKA. These studies have antecedents in the literature, which are surprisingly ancient. In 1940, Solandt and colleagues reported that the cross-circulation of blood between a renal hypertensive dog and a bilaterally nephrectomized dog led to the onset of hypertension in the recipient dog after about one hour.[11] Gordon et al., reported similar findings in cross-circulation experiments between a renal hypertensive and a salt-loaded, normotensive rabbit.[12] Perhaps the most notable of these early observations were the parabiosis studies of Dahl et al.,[13] in which hereditary salt-sensitive animals were connected to salt-resistant hypertensive animals. The experiments revealed that the salt-resistant animal became hypertensive when both animals were fed salt. The investigators suggested that a diffusible saliuretic substance was responsible for the hypertension in the otherwise salt-resistant animal.

Cellular Mechanisms Linking Hypertension, Fluid Balance, and the Sodium Pump

Haddy and Overbeck extended the hypothesis that a blood-borne factor might be responsible for linking volume-expanded states to hypertension by suggesting that blood volume expansion is accompanied by a rise in plasma levels of a sodium transport inhibitor.[14] A year later Blaustein outlined a hypothetical mechanism by which such an inhibitor of NKA could result in hypertension.[15,16] This hypothesis, which was founded on the supposition that an agent which produced natriuresis by inhibition of NKA in the kidney would also inhibit the activity of NKA and the transport of ions across the cell membrane in other tissues, had several ramifications. The mammalian cell membrane is "leaky" to sodium and sodium passes down its sharp concentration gradient from extracellular to the intracellular fluid. Normally this sodium is extruded by the sodium pump. Inhibition of the sodium pump in volume-expanded states facilitates the excretion of sodium by preventing pumping of sodium out of the renal tubular cell into the interstitial fluid. Although such an action is inherently hypotensive by virtue of reducing the extracellular fluid volume and thereby the circulating volume, it may result in hypertension, at least under certain conditions.

Blaustein explains this apparent anomaly by assuming that NKA inhibition extends beyond the kidney to other tissues, including vascular smooth muscle (VSM). Inhibition of NKA in VSM results in a net intracellular accumulation of sodium ions. In addition, the VSM cells possess the ability to transfer sodium ions across the cell membrane in exchange for calcium ions. The sodium-calcium exchanger can mediate net calcium entry or exit, depending on the electrochemical gradient produced by sodium. Increased intracellular sodium ion concentration results in net calcium entry and as a result ionized calcium levels in the vascular smooth muscle cells increase.[17] This calcium is sequestered by the sarcoplasmic reticulum and serves to increase the available pool of calcium for smooth muscle contraction, thus increasing vascular reactivity. Resting tension may also increase if the increase in intracellular ionized calcium levels is incompletely buffered by the sarcoplasmic reticulum.[16]

There is a great deal of epidemiologic evidence for an association between salt intake and hypertension in human populations, but a pathophysiologic explanation has been lacking. The volume-expanding property of increased sodium intake and the hypertensive effects of an endogenous NKA inhibitor provide an attractive possibility. However, a little caution is in order. Although increased NKA release in response to salt intake may be a necessary component of such a mechanism, it is unlikely to account entirely for the appearance of hypertension, since not everyone responds to salt loading with a significant elevation in blood pressure. At present there are no data to indicate whether an elevation of NKA inhibitor level in plasma in response to salt intake is predictive of a hypertensive response. Furthermore, it is possible that the response of plasma NKA inhibitor to equivalent salt loads may vary from individual to individual, perhaps because of hereditary or acquired differences in renal function. Although NKA inhibition may be a common pathway that leads to hypertension in individuals with increased salt intake, the cause may be a renal defect in excreting sodium, which in turn places greater reliance on NKA inhibition to achieve sodium balance. Until these issues are resolved, NKA inhibition remains an attractive, but hypothetical, explanation for the link between salt intake and blood pressure.

The Search for an Endogenous Digitalis-Like Factor (EDLF)

The relationship between cellular ion handling, NKA inhibition, and blood pressure has been widely investigated in both clinical and laboratory studies. The first account of an abnormality in sodium transport in patients with essential hypertension appears to be that of Losse and colleagues, who reported a reduction in sodium transport in the red cells of these patients.[18] Subsequent work suggested a disturbance of membrane sodium transport in various blood cells incubated with sera from hypertensive subjects,[19,20] but the results were inconsistent. Poston and colleagues reported that incubation with sera from hypertensive patients lowered sodium transport in leukocytes of normotensive subjects to approximately the level observed in leukocytes of hypertensive patients.[21] Edmonson and MacGregor produced evidence that an abnormality in sodium transport was most prominent in hypertensive patients whose blood pressure was most likely to be influenced by their sodium balance,[22] and that the impairment of ion handling by red cells from hypertensive subjects was inversely proportional to the increase in plasma renin that occurred when these subjects were deprived of sodium, which established a correlation between low renin hypertension and an abnormality in the handling of sodium.

Following the report that the activity of the sodium pump was altered in dogs and rodents with hypertension,[9,10] attempts were made to measure materials with properties expected of NKA inhibitors and to associate their plasma levels with blood pressure and fluid and electrolyte status. Thus Gruber and colleagues[23] reported an association between hypertension and increased plasma levels in primates of a material they termed "endigen." Endigen was measured as a material that cross-reacted with digoxin antibody. The same investigators had previously demonstrated that volume expansion in dogs was accompanied by an increase in plasma levels of digoxin cross-reactive material.[24] Using the analogy of the endorphins and arguing that immunoreactivity was evidence of shared receptors, the endogenous ligand for the glycoside receptor of NKA was named "endigen." Increased plasma levels of digoxin-like material were subsequently found also in spontaneously hypertensive rhesus monkeys, in Goldblatt-hypertensive vervet monkeys, but not in similarly prepared animals which did not become hypertensive. These two sets of experiments in dogs and monkeys suggested that the immunologic detection of digoxin-like material in plasma might be useful in studies of the role of the substance in volume regulation as well as in blood pressure control.

Hamlyn and his colleagues in Baltimore used a different approach toward the detection and quantification of NKA inhibitor in hypertensive patients,[25] and they measured the effect of plasma extracts on the activity of NKA prepared from kidney. This assay couples the hydrolysis of ATP to the oxidation of NADH, and the resulting reduction in NADH is measured by spectroscopy. These studies revealed a positive correlation between a wide range of mean arterial pressure values and the degree of inhibition of NKA produced by plasma extracts. The level of NKA inhibition correlated well with red cell sodium content, but no correlation could be shown with plasma digoxin immunoreactivity.

Further experimental confirmation of a change in NKA activity in hypertension came from studies by Haddy's group. Using a model of low renin, volume-dependent hypertension produced in rats by a 75% reduction in renal mass and replacement of drinking water with 1% saline, these investigators showed an increase in the ability of boiled plasma extracts to inhibit ouabain-sensitive, but not ouabain-insensitive, NKA activity in tail arteries,[26,27] as measured by the inhibition of ^{86}Rb uptake. Canrenone, a competitive inhibitor of ouabain binding to NKA, was found to reduce both the inhibition of vascular NKA and blood pressure in animals with reduced renal mass hypertension.[28] Unlike the findings of Hamlyn and colleagues,[25] these investigators failed to find a correlation between the magnitude of hypertension and the ability of plasma extracts to inhibit NKA. Haddy's group also found that lesions of the anteroventral region of the third ventricle, a hypothalamic region involved in fluid and electrolyte balance, lowered the hypertensive response to reduced renal mass, and that plasma from these lesioned animals was less effective in inhibiting NKA activity in vascular tissue than were plasma extracts from hypertensive animals without the hypothalamic lesion.[27] Lesions of this region have been shown to modify other forms of low renin experimental hypertension.[29] Central sympathectomy was found to abate both hypertension and the increase in plasma NKA inhibition activity,[27] which indicates that central adrenergic pathways may be involved in controlling the release of the plasma-borne NKA inhibitor.

Plasma-borne inhibitor of NKA activity has been studied also in the spontaneously hypertensive rat (SHR). For example, De Wardener measured this activity in acetone extract of hypothalamus by means of two assays.[30] The first used glucose-6-phosphate dehydrogenase stimulation as an index of NKA inhibition. The second, which was more direct, determined NKA inhibition in fresh guinea pig proximal tubules. Either way, much greater NKA inhibition was obtained with plasma and hypothalamic extracts from spontaneously hypertensive animals than from normotensive controls. Similarly an increase in the plasma and hypothalamic levels of NKA inhibitor was demonstrated recently in reduced renal-mass, saline-drinking hypertensive rats.[31] Wauquier and colleagues[32] have also reported that the plasma level of NKA-inhibitory, digoxin-like immunoreactive material was twice as high in SHR rats as in normotensive controls.

Hamlyn used an erythrocyte ^{86}Rb flux assay to determine NKA inhibitory activity before deoxycorticosterone acetate (DOCA) treatment and during both the pre-"escape" and post-"escape" periods of mineralocorticoid treatment in pigs.[33] NKA inhibitory activity increased promptly after DOCA treatment, peaked immediately prior to the commencement of renal sodium escape, fell during the initial period of escape, and then rose again to high levels. Acute expansion of plasma volume in both DOCA and non-DOCA treated animals was accompanied by increased levels of plasma NKA inhibitory activity.

Possible Sources of Endogenous Digitalis-Like Factor

These convincing demonstrations linking the level of a plasma-borne substance to sodium balance, vascular NKA activity, ion handling, and blood pressure have stimulated much interest in its source, identity, and biosynthesis. An incomplete list of tissues that have been investigated as a potential source of EDLF includes: heart, skeletal muscle, whole brain, hypothalamus, hypophysis, adrenals, kidney, and body fluids. Some of these investigations will be briefly reviewed for the purpose of demonstrating not only the accomplishments, but also some of the obstacles encountered in this pursuit.

Skeletal Muscle

Josephson and Cantley were the first to isolate and define a potent inhibitor of NKA from an endogenous mammalian source, performing a detailed and rigorous examination of an inhibitor of NKA activity present in commercial preparations of equine muscle ATP.[34] The inhibitor was highly potent and had similar inhibitory effects on NKA from various sources, but was ineffective against other ATPases. Subsequently these investigators identified the inhibitor as the vanadate ion.[35] The vanadate ion shares physicochemical properties with the phosphate ion which, of course, is produced by the hydrolysis of ATP (which NKA catalyzes). This observation was presented with the implicit assumption that vanadate was not a physiologic regulator of NKA. The recent discovery that simple inorganic compounds can be important regulators of vascular tone may indicate that this assumption should be reevaluated.

Brain

Interest in the brain as a source of EDLF was stimulated by two papers published simultaneously in 1979. In the first, Haupert and Sancho reported that acetone and ether extracts of bovine hypothalamus, subsequently fractionated by gel chromatography, contained a material that inhibited sodium transport in isolated toad bladder preparations[36] and competed with ouabain for binding to frog urinary bladder sites. The material also inhibited the production of inorganic phosphate by NKA purified from rabbit kidney. In the second paper, Fishman reported the presence of EDLF in guinea pig brain extracts.[37] After fractionation by gel chromatography and desalting, a material was obtained that displaced tritiated ouabain from brain microsome NKA and inhibited ^{86}Rb uptake by erythrocytes. Digitalis-like immunoreactivity was also observed in extracts of brain tissue from guinea pigs.[38] Subsequent work by Haupert generated a substantial body of data that indicated that the hypothalamus may be a source of EDLF,[39-41] which appears to be a low molecular weight, nonpeptidic substance. The activity is resistant to acid, but not alkaline hydrolysis. Ionic requirements for NKA inhibition by this hypothalamic material were not identical to those of ouabain. The presence of magnesium ions alone increased inhibition; however, the addition of sodium ions and ATP led to an increase in activity in the presence of the inhibitor. Furthermore, ouabain increases phosphorylation of NKA, whereas the hypothalamic factor does not. While this evidence does not rule out the possibility that the hypothalamic factor may be a physiologic regulator of NKA, it suggests that, if it is, it has a different mechanism of action. Although the hypothalamic factor described by Haupert, as well as a brain NKA inhibitor reported by Lichtstein's group,[42,43] appear to inhibit the binding of ouabain to NKA, no Scatchard binding analysis of this inhibition has been reported. Other studies have revealed that the brain-derived NKA inhibitors have inotropic activity on heart muscle[39] and increase vascular reactivity in aortic strip preparations.[45]

Mir and colleagues have added an interesting and important element to this picture,[44,45] demonstrating that the incubation medium of cultured hypothalamic cells is enriched with EDLF activity, which indicates that the EDLF found in tissue extracts of hypothalamus may actually arise there. Unfortunately, EDLF is not secreted into the blood stream, although it does appear in the cerebrospinal fluid.[46,47] One paper has reported that EDLF can be demonstrated immunohistochemically in the supraoptic and paraventricular neurons of the hypothalamus, that sodium loading increases urinary excretion of EDLF and blood pressure, that lesioning of the adjacent hypothalamus inhibited these responses, and that plasma levels of EDLF could be influenced by sodium infusion into the cerebral ventricles.[48] Unfortunately, others have produced evidence that the hypothalamic EDLF is not a peptide[39-41] and so the immunochemical evidence of its presence in peptidergic brain nuclei is somewhat anomalous.

The studies described above reveal a major difficulty in defining a physiologically relevant EDLF, the fact that various potential EDLFs do not show the same range of activity in different assays. This poses the problem of which assays to use in the process of purification and characterization. Another biologically relevant problem is the shortage of evidence that brain-derived EDLF enters the circulation. Therefore it is not possible to correlate changes in the physiologic state of the organism with changes in the circulating levels of these materials. Although this fact calls into question the biologic relevance of brain-derived EDLF, it should be pointed out that the brain itself is a major site of NKA activity and may be endowed with its own intrinsic system of NKA inhibition.

Adrenal Tissue

Thin layer chromatography of adrenal extracts by Schreiber and co-workers produced the first evidence that an immunoassayable EDLF was present in adrenal tissue.[49] This material was shown to be different from corticosterone or aldosterone. Castenada-Hernandez and Godfraind also found digoxin-like immunoreactivity in the adrenal gland and demonstrated an increase in plasma levels during sodium loading.[50] Extensive purification of adrenal extracts lead Ng and colleagues[51] to conclude that ascorbic acid was an adrenal inhibitor of NKA, but that any physiologic role it might have was independent of the cardiac glycoside-binding site. Other laboratories have also reported finding adrenal EDLF.[52-58] This material cross-reacts with a commercial digoxin antibody (New England Nuclear). Hypophysectomy reduced plasma levels, while in adrenalectomized animals the material was almost absent from plasma.[56] HPLC purification and preliminary mass spectroscopic analysis of an adrenal EDLF has been reported by Inagami and collaborators.[58,59] The substance, purified to homogeneity, inhibited NKA hydrolysis, displaced tritiated ouabain from purified NKA, and inhibited ^{86}Rb uptake by erythrocytes. This material had a molecular mass of 336 Da and was not a peptide in nature. Complete chemical identification has not been published.

Evidence that the adrenal secretes EDLF has been obtained in my laboratory as well. Using a digoxin antiserum, I demonstrated the presence of EDLF-like immunoreactivity in rat plasma. This was elevated by sodium intake.[60] Adrenal tissue contained the highest levels of EDLF[52] and explanted adrenal tissue was shown to secrete EDLF.[61] Adrenocortical tumor cells have also been shown to release this material into the culture medium.[53] A blockade of cholesterol side-chain cleavage alone, or in conjunction with blockers of pregnenolone metabolism, failed to consistently inhibit EDLF release and appeared to stimulate release, which suggests that cholesterol side-chain cleavage is not directly involved in the biosynthesis of EDLF.[53] Material from adrenal explants and adrenocortical tumor cultures has been assayed in the erythrocyte ouabain-binding assay, and Scatchard analysis indicates the presence of a single specific competitor of tritiated ouabain binding to the erythrocyte glycoside receptor. Erythrocyte ^{86}Rb uptake is also inhibited by this material and the EDLF activities in each of the three assays appear to partially co-elute in HPLC (P.A. Doris, unpublished data, 1992).

Most recently, Valdes and colleagues reported the purification and identification of digoxin in mammalian adrenal extracts.[57] Digoxin-like immunoreactivity was found in adrenal extracts from human and rat adrenals, and its concentration in dogs was substantially higher in the blood of the adrenal vein than in that of the infra-renal inferior vena cava. Immunoreactive EDLF was found also in bovine adrenal medulla, although the cortex contained nearly 10 times as much. HPLC purification of EDLF from bovine adrenal extracts co-eluted with digoxin, and purified material had the mass spectral properties of digoxin. These results are at variance with those obtained in my laboratory, which indicated that the primary peaks of digoxin immunoreactivity in adrenal conditioned medium did not coincide with those of pure digoxin, but did coincide with the peak elution of material that showed EDLF receptor-binding properties and inhibition of NKA ion transport. One reason for the discrepancy may be that the material in extracts from adrenal tissue from rats and humans, as well as in effluent blood from canine adrenal glands, may be accounted for by cross-reacting steroids. Another problem is that the source of digoxin purified from bovine adrenals has not been shown to be endogenous. Plant sources of cardiac glycosides are widespread and herbivorous animals may consume some of them. There have also been accounts of livestock being injected with cardiac glycosides prior to transportation in the belief that these drugs reduce loss of weight during trucking. Clearly, evidence showing that bovine adrenal cells in tissue culture can synthesize digoxin would be compelling.

Body Fluids

EDLF has been found in the urine of various species. Using very large quantities of human urine, Goto and colleagues[62-65] extracted and HPLC purified material with ouabain-like properties. The material had similar polarity to ouabain, was highly soluble in water and methanol, showed dose-dependent inhibition of purified NKA activity, competed with ouabain for binding to brain synaptosomes, inhibited ^{86}Rb uptake by human erythrocytes, and increased calcium influx and decreased calcium efflux in cultured vascular smooth muscle cells. Preliminary mass spectral analysis indicated a molecular weight of 343 Da, which suggests that the material was not ouabain. Furthermore, the material had a distinctly different UV absorbance spectrum from ouabain. A later report indicated the presence of two distinct inhibitors in urine extracts, one more polar than the other.[66] The more polar compound increased cytosolic free calcium in cultured rat aortic cells while the less polar compound resembled digoxin in its UV absorbance spectrum and immunoreactivity to digoxin antibody. Recently the same authors isolated a material from human urine extracts that appears to be digoxin.[65] Its mobility in a variety of thin-layer and high-performance liquid chromatography systems is identical to digoxin; it has digoxin-like immunoreactivity, as well as mass spectroscopy and nuclear magnetic resonance properties identical to those of digoxin. Unfortunately, these studies have not yet excluded the possibility that the digoxin material may be derived from dietary or medicinal sources.

The strongest line of evidence for the existence of endogenous mammalian cardiac glycosides originating in the adrenals was provided by Hamlyn and colleagues,[55,67-70] who have identified ouabain as an endogenous mammalian steroid. In view of the botanic origins of ouabain, this finding was initially surprising, perhaps no less to the investigators who reported it than to others working in the same area. However, it has been substantiated by a series of rigorous studies. The essential findings are that volume-expanded human plasma contains a material which can be purified by dialysis, affinity chromotography, and HPLC. This material has a mass identical to that of ouabain and produces fragments with a mass identical to those of ouabain. An immunoassay for ouabain indicated the presence of ouabain in the plasma of rats, dogs, and humans (Figure 12.1). Analysis of tissue extracts showed ouabain to be present in the adrenal gland of the rat (Figure 12.2), as well as in adrenal tissue of several other mammalian species. Adrenalectomy lowered the plasma level, while uninephrectomy or DOCA treatment increased it dramatically. Ouabain immunoreactivity was found to be released from cultured bovine adrenocortical cells. Since ouabain is not orally active, the possibility that the material measured originated in the diet or from medication is negligible. These findings

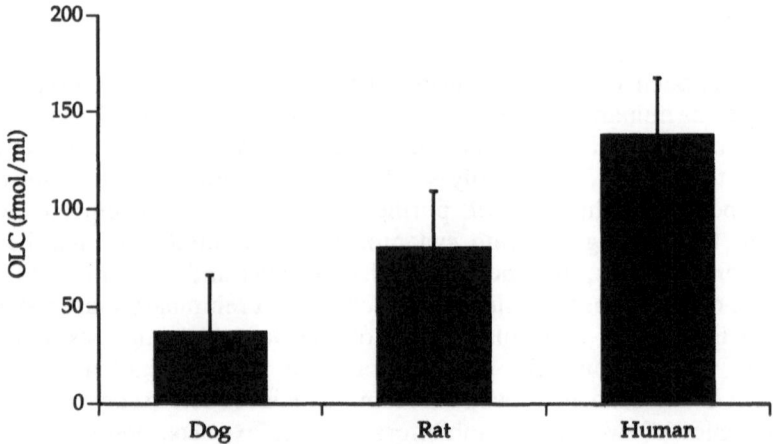

FIGURE 12.1. Ouabain immunoreactivity (OLC) in plasma. (Adapted with permission from Hamlyn and colleagues.[55])

require confirmation. Only then it will be possible to fully investigate the role of endogenous digoxin and ouabain in the physiopathology of fluid and electrolyte balance, of renal function and in the regulation of arterial blood pressure.

Sodium Pump Inhibition in the Clinical Setting

The foregoing discussion has dealt primarily with the role of endogenous NKA inhibitors in salt and water balance and in blood pressure control.

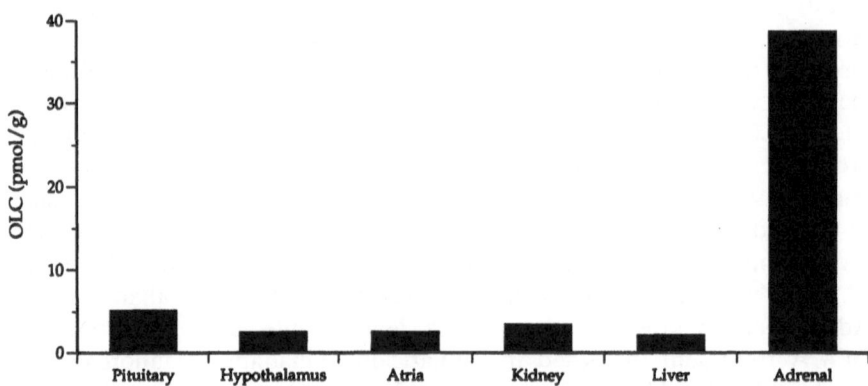

FIGURE 12.2. Distribution of ouabain-like immunoreactivity in extracts of fresh rat tissues. (Adapted with permission from Hamlyn and colleagues.[55])

However, there are many reports suggesting that other clinical situations may be correlated with changes in the plasma or urinary levels of EDLF. These are briefly reviewed below.

Endocrine Disease

Acromegaly is often associated with volume expansion and sodium retention, and consequently arterial hypertension. Using ouabain-binding and NKA activity assays Deray and colleagues found that acromegalic patients had significantly higher levels of plasma EDLF, and the EDLF level was correlated with the degree of plasma volume expansion. Ng and colleagues have reported that ouabain-sensitive sodium efflux from leukocytes of patients with newly diagnosed acromegaly was increased.[71,72,73] These findings suggest overactivity of the sodium pump, followed by a compensatory increase in NKA inhibition by EDLF. Soszynski and co-workers[74,75] have confirmed and extended these findings using erythrocyte ^{86}Rb uptake and digoxin-like immunoreactivity as indices of EDLF. The former index was significantly elevated in acromegalic patients, more so in patients with hypertension. No significant changes were found in the levels of immunoassayable digoxin.

Insulin-induced hypoglycemia appears to be also associated with elevated plasma immunoreactive EDLF[76] and an increased overnight urinary excretion of immunoreactive EDLF has been found in insulin-dependent diabetic patients.[77] The urinary EDLF level correlated with body mass and urinary excretion rates of sodium, potassium, and creatinine in both diabetic patients and healthy controls. The effect of hyperinsulinemia from exogenous insulin on sodium reabsorption by the kidney may account for the increased release of EDLF.

Heart Disease

Little work has been done to evaluate the effects of acute (infarction) or chronic (failure) heart disease on plasma NKA inhibitory activity. One report suggests that plasma concentration and urinary excretion of EDLF measured by digoxin immunoassay are elevated in patients with congestive heart failure patients who had not been treated with cardiac glycosides.[78] Similarly, patients with acute transmural myocardial infarction have been shown to have lower erythrocyte NKA activity, especially if the patients also had ventricular fibrillation.[79] Boiled plasma supernatants from these patients inhibited erythrocyte NKA activity, and this inhibition was antagonized by digoxin antibodies. Plasma immunoreactive EDLF was also higher than normal in patients with myocardial infarction. The chemical identification of NKA inhibitors will allow a better assessment of the effects of myocardial disease on EDLF.

Pregnancy

An increase of plasma EDLF concentration has been observed in pregnancy, pregnancy-induced hypertension, and neonatal hyponatremia.

During pregnancy there is a progressive rise in maternal serum levels of immunoreactive EDLF,[80] and newborns have elevated plasma EDLF, as measured in a variety of assays.[81-84] Levels are higher in pre-eclampsia than in normotensive pregnancy.[85-87] This phenomenom does not appear to be accounted for by increased levels of adrenocortical steroids[81] and the source of this material is not entirely clear, though both the fetal adrenal cortex[88] and the placenta[81] have been implicated. It is possible that this EDLF-like immunoreactive material may not be identical to similar material that showed EDLF properties in other types of detection systems. Ringel and colleagues found increased plasma digoxin immuno-reactivity in normal and hypertensive pregnancy, but the release of digitalis from erythrocytes was not associated with a greater increase in NKA activity in patients who had elevated plasma digoxin immuno-reactivity.[87] Delivery is associated with a lowering of maternal plasma digoxin levels.[81,89]

Neonates also appear to have elevated levels of EDLF in plasma.[81-84] Indeed, Valdes and colleagues used several digoxin radioimmunoassays to show that digoxin was readily detectable in plasma of neonates as well as in amniotic fluid,[84] but could not be detected in adult plasma. The amniotic fluid immunoreactivity appears to remain constant from the 16th to the 33rd week of gestation, and then it rises until term. Hyponatremia is relatively common in newborns and a weak correlation between fractional sodium excretion and plasma EDLF has been shown in infants.[90] In addition, leukocyte sodium transport appears to be inhibited in neonates and in mothers during the late gestational period.[81,91]

Conclusion

Important progress has been made toward the determination of the chemical nature of endogenous regulators of NKA activity with potentially relevant physiologic functions. A major issue to be addressed is whether ouabain is the only endogenous regulator produced in mammals or whether different cardiac glycosides or other materials can serve as physiologic regulators of NKA activity. With respect to ouabain, two major issues remain: what are the mechanisms of its synthesis and release, and what is its role in normal and deranged control of sodium balance and vascular tone? Interesting evidence that suggests a role of the hypothalamus in regulating plasma levels of NKA inhibitor is worth pursuing, as is the possibility that the hypothalamus may represent the apex of an endocrine control system analogous to that controlling the secretion of glucocorticoid. The roles of cholesterol as a precursor of steroid modifying enzymes and of drugs that inhibit adrenocortical steroidogenesis in the biosynthesis of ouabain also merit attention.

Among the many interesting clinical problems that can and should be investigated are whether circulating levels of EDLF can influence the toxicity of and the response to therapeutic doses of cardiac glycosides; whether normal changes in dietary sodium intake have a role in regulating the release of ouabain; whether normal fluctuations in plasma ouabain levels have a physiologic role in renal sodium handling and how it interacts with other known mechanisms; and what, if any, role does ouabain have in the pathogenesis of low renin hypertension, in volume-expanded states, impaired renal function, and other biologic processes influenced by membrane ion transport.

Acknowledgments. The author is indebted to the South Plains Foundation for financial support of his research. The helpful critique of this manuscript by Dr. Lorenz O. Lutherer is gratefully acknowledged.

References

1. Thibault G, Garcia R, Cantin M, Genest J. Atrial natriuretic factor, characterization and partial purification. Hypertension 1983; 5:I-75–I-80.
2. Pamnani MB, Clough DL, Chen JS, Link WT, Haddy FJ. Effects of atrial extract on sodium transport and blood pressure in the rat. Proc Soc Exptl Biol Med 1984; 176:123–131.
3. August JT, Nelson DH, Thorn GW. Response of normal subjects to large amounts of aldosterone. J Clin Invest 1958; 37:1549–1555.
4. Relman AS, Schwartz WB. The effect of DOCA on electrolyte balance in normal man and its relation to sodium chloride intake. Yale J Biol Med 1952; 24:540–558.
5. De Wardener HE, Mills IH, Clapham WF, Hayter CJ. Studies on the efferent mechanism of the sodium diuresis which follows the administration of intravenous saline in the dog. Clin Sci 1961; 21:249–258.
6. Buckalew VA, Nelson DB. Natriuretic and sodium transport inhibiting activity in plasma of volume expanded dogs. Kid Inter 1974; 5:12–22.
7. Clarkson EM, Talner LB, De Wardener HE. The effect of plasma from blood volume expanded dogs on sodium, potassium and PAH transport of renal tubule fragments. Clin Sci 1970; 38:617–627.
8. De Bold AJ. Atrial natriuretic factor of the rat heart. Studies on isolation and properties. Proc Soc Exptl Biol Med 1982; 170:133–138.
9. Overbeck HW, Pamnani MB, Akera T, Brody TM, Haddy FJ. Depressed function of a ouabain sensitive sodium-potassium pump in blood vessels from renal hypertensive dogs. Circ Res 1976; 38:II-48–II-52.
10. Clough DL, Huot SJ, Pamnani MB, Haddy FJ. Decreased myocardial Na,K-ATPase activity in rats with reduced renal mass-saline hypertension. J Hypertension 1985; 3:583–589.

11. Solandt DY, Nassim R, Cowan CR. Hypertensive effect of blood from hypertensive dogs. Lancet 1940; 1:873–874.
12. Gordon DB, Drury DR, Schapiro S. The salt-fed animal as a test object for pressor substances in the blood of hypertensive animals. Am J Physiol 1953; 175:123–128.
13. Dahl LK, Knudsen KD, Iwai J. Humoral transmission of hypertension: Evidence from parabiosis. Circ Res 1969; 24/25:I-21–I-33.
14. Haddy FJ, Overbeck HW. The sodium, potassium pump in volume expanded hypertension. Life Sci 1976; 19:935–948.
15. Blaustein MP. Sodium ions, calcium ions, blood pressure regulation, and hypertension: a reassessment and a hypothesis. Am J Physiol 1977; 232:C167–C173.
16. Blaustein MP, Ashida T, Hamlyn JM. Sodium metabolism and hypertension: How are they linked? Klin Wochenschr 1987; 65:21–32.
17. Rasgado-Flores H, Blaustein MP. Calcium influx and sodium efflux mediated by the Na/Ca exchanger in giant barnacle muscle cells are promoted by intracellular Ca. Biophysic J 1986; 49:546a.
18. Losse H, Wehmeyer H, Wessels F. The water and electrolyte content of erythrocytes in arterial hypertension. Klin Wochenschr 1960; 38:393–395.
19. MacGregor G, De Wardener H. Is a circulating sodium transport inhibitor involved in the pathogenesis of essential hypertension? In: Fregly M, Kare M, eds. The role of salt in cardiovascular hypertension. New York: Academic Press; 1982:331–343.
20. Poston L. Endogenous sodium pump inhibitors: A role in essential hypertension? Clin Sci 1987; 72:647–655.
21. Poston L, Sewell RB, Wilkinson SP, Richardson PJ, Williams R, Clarkson EM, MacGregor GA, De Wardener HE. Evidence for a circulating sodium transport inhibitor in essential hypertension. Br Med J 1981; 282:847–849.
22. Edmondson RPS, MacGregor GA. Leucocyte cation transport in essential hypertension: its relation to the renin-angiotensin system. Br Med J 1981; 282:1267–1269.
23. Gruber K, Rudel LL, Bullock BC. Increased circulating levels of an endogenous digoxin-like factor in hypertensive monkeys. Hypertension 1982; 4:348–354.
24. Gruber KA, Whitaker JM, Buckalew VM. Endogenous digitalis-like substance in plasma of volume-expanded dogs. Nature 1980; 287:743–745.
25. Hamlyn JM, Ringel R, Schaeffer J, Levinson PD, Hamilton BP, Kowarski AA, Blaustein M. A criculating inhibitor of (Na,K)-ATPase associated with essential hypertension. Nature 1982; 300:650–652.
26. Huot S, Pamnani MB, Clough DL, Haddy FJ. The role of sodium intake, the Na-K-pump and a ouabain-like humoral agent in the genesis of reduced renal mass hypertension. Am J Nephrol 1983; 3:92–99.
27. Huot SJ, Pamnani MB, Clough DL, Buggy J, Bryant HJ, Harder DR, Haddy FJ. Sodium-potassium pump activity in reduced renal mass hypertension. Hypertension 1983; 5:I-94–I-100.
28. Pamnani MB, Whitehorn WV, Clough DL, Haddy FJ. Effects of canrenone on blood pressure in rats with reduced renal mass. Am J Hypertens 1990; 3:188–195.

29. Brody MJ, Johnson AK. Role of the anteroventral third ventricle region in fluid and electrolyte balance, arterial pressure regulation, and hypertension. In: Martini L, Ganong WF, eds. Frontiers in Neuroendocrinology. New York: Raven Press; 1980:249–292.

30. Millett JA, Holland SA, Alaghband-Zadeh J, De Wardener HE. Na,K-ATPase-inhibiting and glucose-6-phosphate dehydrogenase-stimulating activity of plasma and hypothalamus of the Okamoto spontaneously hypertensive rat. J Endocrinol 1986; 108:69–73.

31. Holland S, Millett J, Alaghband-Zadeh J, De Wardener H, Pamnani M, Haddy F. Cytochemically detectable glucose-6-phosphate dehydrogenase-stimulating/Na-K-ATPase-inhibiting activity of plasma and hypothalamus in reduced renal mass hypertension. Am J Hypertens 1991; 4:315–320.

32. Wauquier I, Pernollet M-G, Grichois M-L, Lacour B, Meyer P, Devynck M-A. Endogenous digitalislike circulating substances in spontaneously hypertensive rats. Hypertension 1988; 12:108–116.

33. Hamlyn JM. Increased levels of a humoral digitalis-like factor in deoxycorticosterone acetate-induced hypertension in the pig. J Endocrinol 1989; 122:409–420.

34. Josephson L, Cantley LC. Isolation of a potent (Na,K)-ATPase inhibitor from striated muscle. Biochemistry 1977; 16:4572–4578.

35. Cantley LC, Josephson L, Warner R, Yanagisawa M, Lechene C, Guidotti G. Vanadate is potent (Na,K)-ATPase inhibitor found in ATP derived from muscle. J Biol Chem 1977; 252:7421–7423.

36. Haupert GT, Sancho JM. Sodium transport inhibitor from bovine hypothalamus. Proc Natl Acad Sci USA 1979; 76:4658–4660.

37. Fishman MC. Endogenous digitalis-like activity in mammalian brain. Proc Natl Acad Sci USA 1979; 76:4661–4663.

38. Godfraind T, Castenada-Hernandez G. Properties of a digitalis-like factor extracted from guinea pig brain. Arch Int Pharmacodyn 1981; 250:316–317.

39. Hallaq HA, Haupert GT. Positive inotropic effects of the endogenous Na/K-transporting ATPase inhibitor from the hypothalamus. Proc Natl Acad Sci USA 1989; 86:10080–10084.

40. Haber E, Haupert GT. The search for a hypothalamic Na,K-ATPase inhibitor. Hypertension 1987; 9:315.

41. Carilli CT, Berne M, Cantley LC, Haupert GT. Hypothalamic factor inhibits the (Na,K)-ATPase from the extracellular surface. J Biol Chem 1985; 260:1027–1031.

42. Shimoni Y, Gotsman M, Deutsch J, Kachalsky S, Lichtstein D. Endogenous ouabain-like compound increases heart muscle contractility. Nature 1984; 307:369–371.

43. Lichtstein D, Samuelov S. Endogenous "ouabain like" activity in rat brain. Biochem Biophys Res Commun 1980; 96:1518–1523.

44. Morgan K, Lewis MD, Spurlock G, Collins PA, Foord SM, Southgate K, Scanlon MF, Mir MA. Characterization and partial purification of the sodium-potassium-ATPase inhibitor released from cultured rat hypothalamic cells. J Biol Chem 1985; 260:13595–13600.

45. Mir MA, Chappell SP, Morgan K, Lewis MD, Scanlon MF, Lewis MJ. Hypothalamic sodium transport inhibitor and vascular reactivity. J Hypertension 1986; 4:S233–S235.

46. Halperin J. Schaffer R, Galvez L, Malare S. Ouabain-like activity in human cerebrospinal fluid. Proc Natl Acad Sci USA 1983; 80:6101–6104.

47. Lichtstein D, Minc D, Bourrit A, Deutsch J, Karlish SJD, Belmaker H, Rimon R, Palo J. Evidence for the presence of a ouabain-like compound in human cerebrospinal fluid. Brain Res 1985; 325:13–19.

48. Takahashi H, Matsusawa M, Suga K, Ikegaki I, Nishimura M, Yoshimura M, Ihara N, Yamada H, Sano Y. Hypothalamic digitalis-like substance is released with sodium loading in rats. Am J Hypertens 1988; 1:147–151.

49. Schreiber V, Stepan J, Gregorova I, Krejcikova J. Crossed digoxin immunoreactivity in chromatography fractions of rat adrenal extract. Biochem Pharmacol 1981; 30:805.

50. Castenada-Hernandez G, Godfraind T. Effect of high sodium intake on tissue distribution of endogenous digitalis-like material in the rat. Clin Sci USA 1984; 66:225.

51. Ng Y-C, Akera T, Han C-S, Braselton WE, Kennedy RH, Temma K, Brody T, Sato P. Ascorbic acid: An endogenous inhibitor of isolated Na,K-ATPase. Biochem Pharmacol 1985; 34:2525–2530.

52. Doris PA. Immunological evidence that the adrenal gland is the source of an endogenous digitalis-like factor in the rat. Endocrinology 1988; 123:2440–2444.

53. Doris PA, Kilgore MW, Durham D, Alberts D, Stocco DM. An endogenous digitalis-like factor derived from the adrenal gland: Studies of adrenocortical tumor cells. Endocrinology 1989; 125:2580–2586.

54. Gault MH, Vasdev S, Longerich L, Johnson E, Farid N, Legal Y, Prabhakaran V, Fine A. Evidence for an adrenal contribution to plasma digitalis-like factors. Clin Physiol Biochem 1988; 6:253–261.

55. Hamlyn JM, Blaustein MP, Bova S, DuCharme DW, Harris DW, Mandel F, Mathews WR, Ludens JH. Identification and characterization of a ouabain-like compound from human plasma. Proc Natl Acad Sci USA 1991; 88:6259–6263.

56. Pernollet MG, Ali RM, Meyer P, Devynck MA. Are the circulating digitalis-like compounds of adrenal origin? J Hypertension 1986; 4(suppl 6):382–384.

57. Shaikh IM, Lau BWC, Siegfried BA, Valdes R. Isolation of digoxin-like immunoreactive factors from mammalian adrenal cortex. J Biol Chem 1991; 266:13672–13678.

58. Tamura M, Lam T-T, Inagami T. Isolation and characterization of a specific endogenous Na,K-ATPase inhibitor from bovine adrenal. Biochemistry 1988; 27:4244–4253.

59. Tamura M, Lam T-T, Inagami T. Specific endogenous Na,K-ATPase inhibitor purified from bovine adrenal. Biochem Biophys Res Commun 1987; 149:468.

60. Doris PA. Digoxin-like immunoreactivity in rat plasma: Effect of sodium and calcium intake. Life Sci 1988; 42:783–790.

61. Doris PA, Stocco DM. An endogenous digitalis-like factor derived from the adrenal gland: Studies of adrenal tissue from various sources. Endocrinology 1989; 125:2573–2579.

62. Goto A, Yamada K, Ishii M, Yoshioka M, Ishiguro T, Eguchi C, Sagimoto T. Purification and characterization of human urine-derived digitalis-like factor. Biochem Biophys Res Commun 1988; 154:847–853.

63. Goto A, Yamada K, Ishii M, Yoshioka MIT, Sugimoto T. The effects of urinary digitalis-like factor on cultured vascular smooth muscle cells. Hypertension 1988; II:645–650.
64. Goto A, Yamada K, Ishii M, Yoshioka M, Ishiguro T, Eguchi C, Sugimoto T. Urinary sodium pump inhibitor raises cytosolic free calcium concentration in rat aorta. Hypertension 1989; 13:916–921.
65. Goto A, Ishiguro T, Yamada K, Ishii M, Yoshioka M, Eguchi C, Shimora M, Sugimoto T. Isolation of a urinary digitalis-like factor indistinguishable from digoxin. 1990; 173:1093–1101.
66. Goto A, Tamada K, Ishii M, Sugimoto T. Does digoxin-like immunoreactivity really represent natriuretic hormone? Nephron 1990; 54:99–100.
67. Mathews WR, DuCharme DW, Hamlyn JM, Harris DW, Mandel F, Clark MA, Ludens JH. Mass spectral characterization of an endogenous digitalislike factor from human plasma. Hypertension 1991; 17:930–935.
68. Ludens JH, Clark MA, DuCharme DW, Lutzke BS, Mandel F, Mathews WR, Sutter DM, Hamlyn JM. Purification of an endogenous digitalislike factor from human plasma for structural analysis. Hypertension 1991; 17:923–929.
69. Harris DW, Clark MA, Fisher JF, Hamlyn JM, Kolbasa KP, Ludens JH, DuCharme DW. Development of an immunoassay for endogenous digitalislike factor. Hypertension 1991; 17:936–943.
70. Bova S, Blaustein MP, Ludens JH, Harris DW, DuCharme DW, Hamlyn J. Effects of an endogenous ouabainlike compound on heart and aorta. Hypertension 1991; 17:944–950.
71. Deray G, Rieu M, Devynck MA, Pernollet M-G, Chanson P, Luton JP, Meyer P. Evidence of an endogenous digitalis-like factor in the plasma of patients with acromegaly. N Engl J Med 1987; 316:575–580.
72. Ng LL, Evans DJ. Leucocyte sodium transport in acromegaly. Clin Endocrinol 1987; 26:471–480.
73. Ng LL, Hockaday TDR. Endogenous digitalis-like factors in acromegaly. N Engl J Med 1987; 317:572–573.
74. Soszynski P, Slowinska-Szrednicka J, Kasperlik-Zaluska A, Zgliczynski S. Endogenous natriueretic factors: Atrial natriuretic hormone and digitalis-like substance in Cushing's syndrome. J Endocrinol 1990; 129:453–458.
75. Soszynski P, Slowinska-Srzednicka J, Zgliczynski S. Increased activity of digoxin-like substance in low-renin hypertension on acromegaly. Clin Exp Hypertens 1990; A12:533–549.
76. Graves SW, Adler G, Stuenkel C, Sharma K, Brena A, Majoub J. Increases in plasma digitalis-like factor activity during insulin-induced hypoglycemia. Neuroendocrinology 1989; 49:586–591.
77. Giampetro O, Cleirco A, Gregori G, Bertoli S, Del Chicca MG, Miccoli R, Luchetti A, Cruschelli L, Navalesi R. Increased urinary excretion of digoxin-like immunoreactive substance by insulin-dependent diabetic patients: A linkage with hypertension? Clin Chem 1988; 34:2418–2422.
78. Shilo L, Adawi A, Soloman G, Shenkman L. Endogenous digoxin-like immunoreactivity in congestive heart failure. Br Med J 1987; 295:415–416.
79. Bagrov AY, Fedorova OV, Maslova MN, Roukoyatkina NI, Ukhanova MV, Zhabko EP. Endogenous plasma Na,K-ATPase inhibitory activity and

digoxin like immunoreactivity after acute myocardial infarction. Cardiovasc Res 1991; 25:371–377.

80. Kerkez S, Poston L, Wolfe CD, Quartero HW, Carabelli P, Petruckevitch A, Hiltom PJ. A longitudinal study of maternal digoxin-like immunoreactive substances in normotensive pregnancy and pregnancy-induced hypertension. Am J Obstet Gynecol 1990; 162:783–787.

81. Morris JF, McEachern MD, Poston L, Smith SE, Mulvany MJ, Hilton PJ. Evidence for an inhibitor of leucocyte sodium transport in the serum of neonates. Clin Sci 1987; 73:291–297.

82. Ebarra H, Suzuki S, Nagashima K, Shimano S, Kuruome T. Digoxin-like immunoreactive substances in urine and serum from preterm and term infants: Relationship to renal excretion of sodium. J Pediatrics 1986; 108:760–762.

83. Seccombe DW, Pudek MR, Whitfield MF, Jacobson BE, Wittmann BK, King JF. Perinatal changes in a digoxin-like immunoreactive substance. Pediatric Research 1984; 18:1097–1099.

84. Valdes R, Graves SW, Brown BA, Landt M. Endogenous substance in newborn infants causing false positive digoxin measurements. J Pediatrics 1983; 102:947–950.

85. Delva P, Capra C, Degan M, Minuz P, Covi G, Milan L, Steele A, Lechi A. High plasma levels of a ouabain-like factor in normal pregnancy and in pre-eclampsia. Eur J Clin Invest 1989; 19:95–100.

86. Fievet P, Gregoire I, Fourmier A, Roth D, Siegenthaler G, El Esper N, Favre H, De Bold A. Ouabain-like natriuretic factor and atrial natriuretic factor in pregnancy. Kid Inter 1988; 34:S89–S92.

87. Ringel R, Pinkas G, Hamlyn J, Mullins L, Hamilton B. Endogenous inhibition of red blood cell Na,K-ATPase in essential and pregnancy-induced hypertension. Clin Exp Hypertens 1989; A11:587–601.

88. Goodlin R, Makowski EL. Fetal endotoxins and complications of pregnancy. Western J Med 1988; 148:590–592.

89. Witherspoon L, Shuler S, Alyea K, Fegueroa J, Neely H. Digoxin-like substance in term pregnancy, newborns, and renal failure. J Nucl Med 1986; 27:1418–1422.

90. Kovacs L, Lichardus B, Bircak J, Ponecova M, Masurova A, Stefanilova J, Bruchac D, Sulyok E. Endogenous digoxin-like immunoreactivity in urine of preterm infants with late hyponatremia. Contrib Nephrol 1988; 67:145–148.

91. Morris JF, Poston L, Wolfe CD, Hilton PJ. A comparison of endogenous digoxin-like immunoreactivity and sodium transport inhibitory activity in umbilical arterial and venous serum. Clin Sci 1988; 75:577–579.

13
Structure, Gating, and Clinical Implications of the Potassium Channel

GIANFRANCO PRESTIPINO, MARIO NOBILE, and EGIDIO MAESTRONE

Potassium ions are of paramount importance in living organisms. They play an essential role in cell metabolism, particularly in the synthesis of protein and glycogen, and in many enzymatic reactions, such as: (1) the hydrolysis of ATP, which supplies energy necessary for the active transport of ions and other substances; (2) the synthesis of acetyl-CoA, which is involved in the tricarboxylic acid cycle; and (3) the activity of pyruvic phosphokinase, which is necessary for the transfer of phosphate from phosphopyruvate to ADP. In addition, potassium ions maintain resting potentials across cell membranes, and thus control nerve and muscle tissue excitability. Although several factors may influence the redistribution of potassium between the intra- and the extracellular compartment, an exhaustive discussion is beyond the scope of this chapter.

Clinical manifestations of potassium alteration, apart from metabolic derangements, usually reflect changes in membrane resting potential and therefore mainly affect the cardiovascular system (rhythm disturbances), the CNS and the neuromuscular system (muscle weakness, cramps, etc.). Occasionally more complex phenomena of drug "interaction" have been reported. For example, hypokalemia in the early postoperative period may favor "recurarization;" that is, the reappearance of paralysis by nondepolarizing muscle relaxants.[1] Indeed, experiments in vitro and in vivo have demonstrated that a reduction in the concentration of potassium increases the potency of nondepolarizing muscle relaxant drugs and enhances the dose of neostigmine – is required as an antagonist.[2-4] Conversely, an increased concentration of potassium decreases the sensitivity of the neuromuscular junction to the same blockers.[5] When different electrolyte disturbances occur concomitantly, the clinical consequences may be more complex. For example, hypokalemia cannot be appropriately corrected if it is associated with hypomagnesemia, unless the latter is corrected first.[6] The mechanisms of these phenomena are part of the "traditional" knowledge regarding electrolyte disorders. Recommended therapeutic guidelines have been based historically on clinical experience since thorough characterization of potassium (and other ionic) channels are a relatively recent achievement. Cook,[7] in a proposed taxonomy, suggested the existence of 11 different types of potassium channels, each serving a separate function, a classification that proved to be useful:

- to gain more knowledge of physio-pathologic mechanisms;
- to reexamine the mechanism of action of established drugs;
- to develop new agents that are able to act selectively on the appropriate channels without interfering with the function of other channel subtypes, and thus provide more powerful pharmacologic tools with fewer side effects.

An Approach to the Study of Potassium Channels

Potassium channels are the most widespread ion channels, found in all cell types. Spanning the lipid bilayer, they form a pore that is selectively permeable to potassium ions across the cell membrane and capable of transporting discreet amounts of K ions through the membrane, thus playing a crucial role in stabililizing and restoring electrophysiologic properties of cells. The biophysical features of potassium transport, the kinetics of activation and inactivation, and the sensitivity to voltage, conductance, ion selectivity, and modulation by calcium, neurotransmitters, and second messengers demonstrate the striking diversity of potassium channels.

An open potassium channel drives the membrane potential toward the potassium equilibrium potential and helps restore the resting state after depolarization. Thus potassium channels decrease excitability by stabilizing the hyperpolarized state, which is characterized by low opening probability of depolarization-dependent calcium and sodium channels. These voltage-dependent potassium channels have been known for a long time.

More recently, the use of patch-clamp recording techniques[8] has led to the discovery of many new potassium channels that are said to display novel functions, which have not yet been completely elucidated. Indeed, these different kinds of potassium channels may be present in a single cell, each of them affecting the whole cell current. For example, four species of potassium channels have been characterized at a single vertebrate smooth muscle cell,[9] while in the mammalian heart eight different potassium channels have been described.[10] The density and ratio of different types of potassium channels on the cell membrane probably account for the macroscopic currents. On the other hand, the simultaneous presence of different types of potassium channels, in general, does not explain the correlation between macroscopic currents and single-channel currents. Therefore it seems appropriate to ascertain the contribution and the individual biophysical features of each type of channel in different tissues. Three experimental methods have been developed in recent years to study channel proteins at the molecular level:

1. Biochemical techniques have made it possible to isolate integral membrane channel-forming proteins, which are solubilized from their native membranes and purified from other membrane constituents. These proteins, recombined with specific phospholipids and reassembled into a well-defined planar bilayer membrane, exhibit functional activity.[11] This experimental approach is based on the high-affinity binding of toxins and drugs to particular channel proteins. Thus the classical delayed rectifier of the squid axon membrane[12] and a 76–80 kDa polypeptide of rat brain[13] were successfully reconstituted following the discovery of specific inhibitors, noxiustoxin from scorpion venom[14] and dendrotoxin from snake venom.[15] Several other toxins from bee (apamin), scorpion (charybdotoxin), and snake venoms, with affinity with different types of potassium channels, have been characterized. Castle et al.[16] and Dreyer[17] reviewed the last relevant peptide toxins known to act on the K^+ channels family. These compounds are useful, not only to characterize channels proteins, but also to help identify a given potassium channel by inhibiting its contribution to the total current.

2. Voltage-clamp recording techniques are a powerful multipurpose tool to study potassium channels. The patch clamp, developed by Neher and Sakmann[18] in 1976, made it possible to monitor currents through a single channel molecule in a patch of native membrane, and together with the reconstitution technique, to investigate the elementary quantal events that underlie macroscopic currents; thus single channel properties such as conductance, subconductance states, ion selectivity, open probability, and kinetics analysis of dwell times can be described.

3. Techniques of molecular genetics led to the isolation of complementary DNA (cDNA) clones encoding the primary structure of the potassium-channel subunits, while information on the amino acid sequence has allowed predictions about the secondary and tertiary structure of the membrane channel proteins, leading to a better understanding of the structure-function relationships of the channel. For this purpose *Xenopus* oocytes injected with cDNA-derived mRNA have been used to express channel proteins, which are functionally integrated in the oocyte membrane. This preparation is particularly attractive for electrophysiologic studies because it is relatively large (1.2 mm in diameter) and easy to penetrate by microelectrodes for either the standard two-electrode voltage-clamp or patch-clamp measurements. Poor results from probes aimed at recognizing potassium channels led to greater use of the genetic approach, and to the identification and

sequence analysis of potassium channel genes in *Drosophila*,[19] rat brain,[20] and rat heart;[21] the usage of site-specific mutations has made it possible to localize regions of the transmembrane proteins involved in specific channel functions.

In the next section we will deal primarily with the voltage-dependent potassium channels, because their biophysical properties are the best known among all potassium channels. These properties will be discussed in relation to the structure of the channel subunits, with the purpose of elucidating the molecular basis of electrical excitability.

Biophysical Properties

The Pore

The K ion crosses the cell membrane through the channel pore, a transmembrane structure that is formed by an integral amphiphilic membrane protein with hydrophobic side chains that face the lipid core and polar groups that are aligned in a hydrophilic pathway and probably form the wall of the channel itself. The membrane lipid bilayer is an electric insulator and represents a large energy barrier to the flow of ions. The arrangement of charged or hydrophilic side groups that span the membrane provides a low energy pathway for the transport of K ions across the membrane.[22] From this point of view the channel protein can be considered to be an enzyme that catalyzes transmembrane ion movement. Indeed, as pointed out by Latorre and Miller,[23] channels are characterized by high rates of catalysis with 10^6–10^9 ions being transported per second. Furthermore, the channel protein displays typical Michaelis-Menten enzyme kinetics: (1) saturation at high substrate (ion) concentration; (2) competitive inhibition by substrate analogs (blockers); and (3) high specificity for a particular substrate (ionic selectivity). Moreover, channels represent dynamic structures that are subjected to rapid (allosteric) conformational changes between conducting (open) and nonconducting (closed) states. The pore faces both sides of the membrane, displaying a polar well that extends into external and internal solutions. The inner and outer entrances of potassium channels are wide, but the channels become progressively narrower in the region inside the pore. Blockade with tetraethylammonium (TEA$^+$) and its derivatives[24] have been used to measure the dimensions of specific regions of the pore. Thus the inner mouth in the potassium channel of the sarcoplasmic reticulum and in the large-conductance calcium-dependent potassium channels have been estimated to be 0.8 nm, which, by the way, is the diameter of the TEA$^+$ ion or the hydrated K$^+$ ion. The deeper narrow region is a tunnel about 0.6 nm long, with a width approximately the diameter of a nonhydrated K^+ ion. In other words, the shape of the potassium channel resembles that of an hourglass.

Ionic Selectivity

The primary function of a potassium channel is to conduct K^+ ions across the membrane, while preventing Na and other ions with different charge, size, and degree of hydration from using the same pathway. This selective passage may occur through two different mechanisms: (1) the specific binding of the ion to sites inside the channel or (2) the rejection of undesired ions, with the pore acting in the manner of a sieve. In some cases selectivity depends on a combination of both mechanisms. In the ideal pore, as described above, the short narrow tunnel is the region where K^+ ions, which interact intimately with the channel, are filtered. Armstrong[25] proposed that the narrow part of the potassium-channel pore of squid axon membrane may have a diameter of 0.26 nm to 0.3 nm and may contain also oxygen atoms. This would indicate that the pore is permeable to ions with a crystal radius of 0.133 nm (K^+) to 0.148 nm (Rb^+, NH_4), but is impermeable to smaller (Na^+, 0.095; Li^+, 0.068) and larger ions (Cs^+, 0.165 nm). Transit of an ion across the filter region requires loss of the water shell. Therefore the channel must compensate for the lost energy through electrostatic and chemical interactions. The polar groups of the filter wall provide an aqueous environment where K^+ ions have an energy a little higher than they would in water, while ions smaller than K^+ are rejected because their potential energy is much higher in the pore than in water, and thus they cannot interact favorably with charged walls of the pore. On the other hand, large ions with a crystal diameter greater than 0.3 nm are excluded sterically. The selectivity sequence of monovalent ions through the potassium channel in myelinated nerve fiber of *Rana pipiens* has been found to be:

$$Tl^+ > K^+ > Rb^+ > NH_4^+ > Cs^+, Li^+, Na^+.$$

The short length of the narrow region of the selective filter in the channel serves the function of reducing the electrical resistance of the pore.

The Gate

The mechanism by which a potassium channel can control its ionic flux is called *gating*. It is an intrinsic property of the channel capable of sensing appropriate stimuli from the surrounding environment. Hille[26] has extensively reviewed the gating of different ionic channels and has described the many ways in which ionic currents can be modulated. The best-known gate is that of the voltage-dependent channels, where the potassium currents are activated by changes in membrane potential. Among these channels are the classic outward-delayed rectifying channel

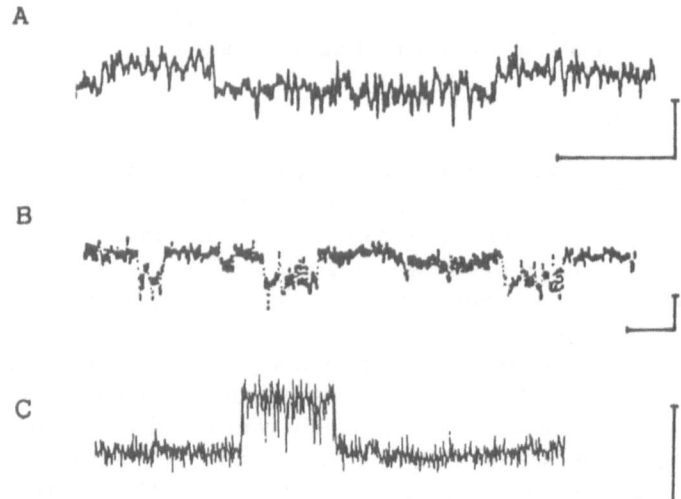

FIGURE 13.1. Gating behavior of delayed rectifying potassium channels during prolonged depolarization. (A) Current recorded from a potassium channel purified from squid axon membrane and reconstituted into planar bilayer. Membrane potential is 100 mV. (B) Recording of patch current at −25 mV membrane potential from the cytoplasmic side of squid axon membrane. (C) Single-channel current recorded at 0 mV from membrane patch of oocytes injected with potassium channel-forming protein (RCK 3) RNA. Amplitude calibration and time scale are 2 pA and 50 ms, respectively. In (A) and (C) upward deflections correspond to channel opening. In (B) downward deflections correspond to channel opening. (Reprinted from (A) Prestipino, G., unpublished data. (B) Nature 285:142, © 1980 Macmillan Magazines Ltd.: and (C) EMBO J 8:3241, 1989. IRL Press Ltd., with permission.

described by Hodgkin and Huxley,[27] the fast transient potassium channel,[28] and the inward or anomalous rectifying potassium channel.[29] Single-channel current recordings indicate that channels can exist in two discrete states, closed or open. Activation of the gating mechanism opens the channels during membrane depolarization, while its inactivation spontaneously closes the channels during depolarization. This process arises from the movement of charges or dipole reorientations that accompany the conformational transitions between the open and the closed state and that represent the gating current, as recently measured in potassium channels expressed in *Xenopus* oocytes.[30] Figure 13.1 shows discrete current events obtained in the voltage-dependent delayed rectifying potassium channel using combined biochemical, biophysical, and molecular genetics methods. Another type of gate is expressed in the calcium-activated potassium channels, which are opened by changes in intracellular Ca^{2+} concentrations. Their properties will be briefly described in the next section. A third class of potassium channels is gated by neurotransmitters and second messengers. These potassium channels are receptor-controlled and are activated by specific agonists, such as acetylcholine (ACh)[31] or kainate (KA) and α-amino-3-hydroxy-5-methylisoxazole-4-propionate (AMPA).[32] In addition, these three mechanisms of gating can be modulated by phosphorylation and by guanyl nucleotide-binding (G) proteins.

Channel Blockers

The interaction of a blocker with a channel requires the presence of a receptor and a binding site. Indeed, blocking agents have long been the main tool for studying potassium channel structure, more so since blockers purified from arthropod or snake venoms have demonstrated the existence of different potassium channels and have provided probes for their biochemical purification.[12,13] The action of TEA^+ and of its derivatives, the classical specific blockers of potassium channels, have been well characterized on the delayed rectifying potassium channel of the squid axon. When perfused internally, these substances block the channel in a voltage-dependent manner by occlusion of the pore.[33] The mechanism of the block shows that the gate is located at the inner end of the channel. Actually, the drug does not interact with the normal opening of the channel during depolarization; rather, it enters the open pore, blocks the current, and inactivates the channel. The rate of inactivation, at different blocker concentrations, indicates a one-to-one drug/channel interaction. The affinity of TEA^+ for the channel increases as the membrane becomes more positive on the cytoplasmic side. The membrane field drives the positively charged TEA^+ into the pore, and thereby increases the block. The effect, however, is not very strong. In fact, large organic blockers have a low-voltage dependence, which indicates that they only move a short distance into the pore. According to this model small ions penetrate deeper into the conduction pathway and block the channel with a higher voltage dependence. Sometimes these ions compete with the K^+ ions, which suggests that they may actually enter the short narrow passage. The voltage dependence allows one to estimate the electrical distance of binding sites within the pore and to show that the blocking site of divalent ions, such as Ba^{2+}, is near the channel's outer end.[34]

Molecular Structure

With the cloning of transient A-type potassium channels from the *Shaker* mutant of the fruitfly *Drosophila melanogaster*,[12] the molecular biology approach is commonly used to study the structure of potassium-channel proteins. The complete amino acid sequences of the protein-coding region of this and of other potassium-channel clones are now available, so that models can be built and direct comparisons with the sequences at sodium and calcium voltage-activated channels can be made. The deduced amino acid sequence of these clones suggests an integral membrane protein of 70^{12} and 56^{13} kDa, with sizes comparable to those of noxiustoxin-and dendrotoxin-binding proteins purified from squid axon $(60 kDa)^5$ and mammalian brain $(76-80 kDa),^6$ respectively. The molecular structure of the potassium channel assessed by genetic studies is that of a protein with a molecular weight one-fourth as large as that of the sodium and calcium voltage-activated channels. In fact, a simple copy of the peptide (one repeat) is found instead of the four repeats found in the sodium and calcium voltage-activated channels. The A-type and the delayed rectifying potassium channels appear to be oligomers arranged in a tetrameric assembly. Thus, if potassium channels are hetero-oligomers, assembly of subunits in various combinations could result in a great number of channel subtypes.[35] Each repeat (subunit) has six hydrophobic transmembrane segments (S1–S6), which would span the membrane in the form of an α helix with amino- and carboxy-terminal sequences most likely facing the cytoplasmic side of the membrane. The topology and the amino acid sequence of potassium channel repeats show high homology to the repeats of sodium channel. The S5 and S6 segments are hydrophobic and without charged residues. In this model the S4 segment contains a characteristic sequence of positive charges and it is probable that this is the voltage sensor of the voltage-dependent channels. It has been suggested that the region linking S5 and S6 forms part of the pore. Indeed, mutations in this region affect TEA^+ blockade and ion permeation in voltage-activated potassium channels.[36] Detailed discussion of this model has been presented by Guy and Conti.[37]

Different Types of Potassium Channel Conductance

This section briefly describes the best-characterized potassium-selective channels. An exhaustive discussion of their properties, pharmacology, and tissue distribution may be found in recent reviews.[38–40]

Delayed Rectifier K^+ Current

Delayed rectifying outward potassium currents (I_K) have been identified in excitable and nonexcitable cells. They are activated by depolarization and remain activated as long as the depolarization is maintained. They have been described in detail by Hodgkin and Huxley[27] in the squid axon, in which this is the major current responsible for the repolarization phase of an action potential. The term "delayed rectifiers" has been appropriately used to characterize voltage-dependent potassium currents that display different inactivation, kinetic, and pharmacologic properties that are distinct from typical "A currents" and are not activated by a rise in intracellular calcium concentration. Many cells and most neurons contain at least one type of I_K, and some cells may contain more than one type. For example, in a pheochromocytoma cells line (PC12), macroscopic current kinetics and pharmacologic studies recently have shown at least two different types of potassium currents.[41,42] Both are considered "delayed rectifiers" because they contribute to the total delayed rectifying current, and, by single-channel analysis, seem to represent distinct channel types. The delayed rectifier potassium current can be blocked by the classical potassium channel blockers TEA$^+$, 4-aminopyridine (4-AP), Cs$^+$ ions, and quinine. Apart from noxiustoxin, mentioned earlier, no selective pharmacologic agonists are known.

Transient Potassium or A Current

The transient potassium current, often called A current, differs from the delayed rectifying potassium current in its activation and inactivation kinetics. The A current activates and inactivates faster during the depolarizing phase. It is activated by depolarizing pulses from holding potentials that are more negative than the resting potential. Thus this current operates in the subthreshold region for action potential generation and can regulate the frequency of repetitive firing when a neuron is spontaneously active, or it fires repetitively in response to tonic depolarization. The current is more sensitive to 4-AP[43] than I_K and the channel activity can be modulated by cAMP. Currents with similar properties have been described in neurons of several species, as well as in other cells. However, some differences have been observed in the voltage dependence and in the rate of inactivation. Using the latter parameter alone, Rudy[38] distinguished three groups of A currents.

Calcium-Activated Potassium Channels

The section above described "delayed rectifiers" and "A currents" as being due to the activity of voltage-dependent channels. Other types of potassium channels (K_{Ca}) are calcium-dependent: an increase in intracellular Ca^{2+} concentration opens the channels, although in the case of the so-called maxi calcium-activated potassium or BK channels, a voltage dependence may also be observed, perhaps because the binding of Ca^{2+} to the channel is in itself voltage-dependent. In addition, the voltage-dependence of the calcium-activated potassium channel may be due to the entry of Ca^{2+} through calcium channels, which are themselves voltage-dependent. Indeed, acting as a second messenger, Ca^{2+} can regulate the potassium channels by stimulating protein kinase C and calmodulin-dependent kinases.

Since their original discovery in molluscan neurons, calcium-activated potassium currents have been found to be widely distributed, particularly in secretory and excitable cells, where they are involved in regulating firing frequency, and in certain cells they contribute to the resting potential. In secretory cells their activation causes the closing of voltage-dependent calcium channels, thus providing a negative feedback that regulates calcium entry. In this context they play an important role in the metabolism of the cell. Calcium-activated potassium channels are common targets for modulation by neurotransmitters and second messengers.

A second type of potassium channel (SK) has been characterized in excitable cells. It shows little or no voltage-dependence and has small unit conductance (10–14 pS). Potassium channels can be differentiated further according to their sensitivity to blocking agents, such as charybdotoxin, TEA^+, and apamin. The SK channel is also called the AHP channel because it is responsible for the slow after-hyperpolarization (AHP) that follows an action potential. So far AHP channels have been described in excitable, but not in nonexcitable cell membranes.

A third type of calcium-activated potassium channel, with a conductance that ranges from 20 to 120 pS, has been described in both excitable and nonexcitable cells. The Ca^{2+} sensitivity, voltage-dependence, and unitary conductance of these channels lie between those of the two channel types mentioned above. The activity of these channels is modulated by cAMP-dependent metabolic pathways. Furthermore, a calcium-activated transient potassium current, or calcium-activated A current, has been described in some preparations, but requires further characterization.

Inward or Anomalous Rectifying Potassium Current

Ohmori et al.[44] described an inward (anomalous) rectifying potassium current using the patch-clamp technique in both excitable and nonexcitable cells. Like the M current (described below) and the transient potassium current, this current contributes significantly to the restoration of the resting membrane potential in vertebrate and invertebrate neurons. In some cases intracellular Mg^{2+} may act as the blocking agent of the inward potassium current. Several types of inward rectifiers have been described. In particular, two types are present in cardiac ventricular cells: a muscarinic ACh-activated and an ACh-independent one. It has been proposed that the first potassium channel type may be coupled to a G protein. Other inward rectifiers are modulated by transmitters, such as serotonin (5-HT), adenosine, and peptides, such as substance P and somatostatin.

M Current

The M current is a small voltage-dependent outward potassium current. Its threshold of activation lies at potentials negative to the resting potential. Since this current does not inactivate, it is believed to contribute to the normal resting membrane current. It may also limit repetitive activity and the pattern of spike discharge. The M current, first characterized in frog lumbar sympathetic ganglia,[45] was later found in a variety of ganglia, neurons, muscle cells, and neuroblastoma cells. It can be inhibited by several neurotransmitters, cholinergic and peptidergic agonists, and muscarinic ACh-receptor agonists, including muscarine itself.

S Current

A class of serotonin (5-HT)-sensitive potassium channels was first investigated by Kandel and Schwartz[46] in sensory neurons of *Aplysia californica*. In the presence of 5-HT they manifest a decrease of the outward current, which seems to be weakly voltage-dependent and independent of internal calcium. These channels are open at resting membrane potential and therefore, like the M current, contribute to the resting membrane conductance, excitability, and repolarization. They are inhibited by activation of the cyclic adenosine monophosphate (cAMP)-dependent protein kinase and occur following activation of the serotonin receptor. In contrast, an increase of the channel open probability is observed after external application of the neuropeptide Phe-Met-Arg-Phe-NH$_2$ (FMRF-amide), which activates the arachidonic acid cascade. Recently, Volterra and Siegelbaum[47] reported that the coupling of the 5-HT receptor to adenylate cyclase requires a pertussis toxin-insensitive G protein, whereas a toxin-sensitive G protein couples the FMRF-amide receptor to phospholipase A$_2$, causing its activation and the release of arachidonic acid.

ATP-Sensitive Potassium Channels

Potassium channels sensitive to intracellular ATP (K$_{ATP}$) are located in the heart, pancreatic islet β cells, and skeletal muscle. At physiologic resting levels of intracellular ATP these channels are closed and open only when the cytoplasmic concentration of ATP falls below a critical level (0.2 mM in heart tissue). An increase of ATP$_i$ decreases the channel open probability and affects both the channel's open and closed times. No effect of ATP on single-channel amplitude has been reported. Activation of these channels is believed to counteract myocardial overexcitability and to shorten the time during which the heart is susceptible to potentially dangerous arrhythmias. Insulin release from pancreatic islet β cells is regulated by these channels because glucose increases cytoplasmic ATP, which results in their closure.

Clinical Implications

Several drugs, currently in use or undergoing clinical trial, appear to act by opening or closing potassium channels.

Potassium Channel Openers

These agents constitute a chemically heterogeneous group and, according to some preliminary evidence, they appear to induce vascular relaxation and endocrine changes by causing cell membrane hyperpolarization (which in turn prevents Ca^{2+} entry through voltage-dependent calcium channels). These effects are presently being investigated because of their therapeutic potential. Due to the chemical diversity of these compounds,[48] it seems appropriate to ascertain whether they act by opening a unique potassium-channel type and whether they all act on the same site of the same channel. For example, even though cromakalim and diazoxide can activate K_{Ca} channels, they have a greater selectivity for the K_{ATP} channels, an effect that is inhibited by glibenclamide, a powerful blocker of these channels. Although the use of potassium-channel modifiers may help establish a link between the metabolic state of a cell and its response,[48] the mechanisms involved are far from clear for several reasons.[49] Among them are the facts that, to date, at least 13 major types of potassium channels and several subgroups have been identified; that ligand-binding and patch-clamp studies have been performed mostly in smooth muscle and are far from exhaustive; and that some of the drugs investigated display significant tissue selectivity (for example, diazoxide is 100 times less potent than cromakalim in activating the K_{ATP} channel in pancreatic islets,[50] while other drugs are not active at all[51]). Furthermore, these drugs may be classified into seven distinct groups and often have mechanisms of action other than those that depend on the K_{ATP} channel.

The first group includes:[49] benzopyrene compounds (cromakalim, an inhibitor of the Ca^{2+} channels of β cells in pancreatic islets,[52] and lemakalim), SDZ PCO 400, RO 316930, WAY 120491,[53] SO121 (with a selective relaxant effect on the smooth muscle of the ureter), and SR46142A, a putative antidepressant.

Another group consists of the guanidine derivatives, such as LY22675, LY211808, EP-A0354553, and pinacidil. The latter has interesting properties. One of them, shared with lemakalim and nicorandil, is to inhibit GABA release from the substantia nigra (an effect antagonized by glibenclamide); another property, shared with cromakalim, is to relax the smooth muscles of the trachea, intestine, uterus,[7] and blood vessels.[54] This vasodilatation persists in the presence of various potassium-channel blockers, which suggests the possibility of an additional effect on Ca^{2+} extrusion.

The third group consists of the pyridine derivatives (nicorandil, SG-209, SG-103, and KRN 2391). Nicorandil is considered a hybrid molecule and its relaxant effect on the blood vessel seems to be due, at least partially, to guanylate cyclase activation by the nitroxy side chain. It has been reported that nicorandil may also open the K_{Ca} channel in rat portal vein.[55]

Pyrimidine derivatives (minoxidil and SKF 11197), make up the fourth group, but definitive evidence concerning their role as potassium-channel openers is still lacking. The last three groups include the benzothiazines (diazoxide is the prototype substance), the thioformamides (RP 52891), and the 1–4 dihydropyridines, such as niguldipine. These agents prolong the action of calcium-entry blockers.

As mentioned above, the potassium-channel openers exhibit a degree of tissue selectivity. Thus their effects on regional blood flows may vary.[56] Vasodilatation has been observed in liver, stomach, peritoneum, and skeletal muscles, but no consistent effects on renal blood flow have been noted. Possible beneficial actions of some of these compounds on coronary flow have been investigated in healthy, conscious dogs and in the isolated ischemic rat heart.[57] Cromakalim and pinacidil dilate large and small coronary arteries, and mostly the large epicardial vessels.[58] In addition, these compounds appear to have a direct cardioprotective action, which is blocked by glyburide. This may happen because hyperpolarization reduces the expenditure of energy during ischemia and thereby preserves the myocardial tissue. However, pinacidil, in the presence of severe coronary artery stenosis, may cause "endocardial steal" in the ischemic areas.[59] Further studies are clearly necessary in this area.

Although cromakalim can induce relaxation of the dog middle cerebral artery, the response is rather weak; one possible explanation could be the low density of K_{ATP} channels in this artery.[60] This finding further stresses that variability of the response may be tissue- and/or drug-dependent. Potassium-channel openers cause membrane hyperpolarization and block calcium channels, and so a certain decrease of contractility may be expected. Even though this is still a matter of debate, it seems that negative inotropic action is achieved only with doses far greater than those required to produce vasodilatation.[56]

In addition, there is some concern about the possibility that these agents, by decreasing the action potential and the duration of the refractory period in Purkinje cells and ventricular myocites, may not have the claimed antidisrhythmic effect and, in fact, facilitate disrhythmias.[61–63] Pinacidil appears to cause disrhythmia in dogs and, when used as a antihypertensive drug in humans, it may cause changes in the "T" wave. Moreover, glibenclamide and tolbutamide, which are known to antagonize K_{ATP} currents, have been shown to have an antidisrhythmic effect on rat heart, an effect which can be counteracted by pinacidil and cromakalim.[64]

In spite of this controversy, a clinical trial[56] has demonstrated that cromakalim (0.5–2 mg a day, for a week) can lower arterial pressure significantly, although problems with both efficacy and tolerance and the possibility of major side effects (such as lesions of cardiac papillary muscle[49]) have caused the interruption of these human studies. Clinical trials are now in progress with lemakalim, an active enantiomer of cromakalim. Five clinical studies have demonstrated that pinacidil effectively lowers diastolic pressure, an action reinforced by the simultaneous administration of established antihypertensive agents such as propranolol, hydrochlorthiazide, hydralazine, and α-methyldopa. However, pinacidil, at dosages higher than 2.5 mg a day, may also have serious side effects, such as tachycardia, flushing, headache, edema with weight gain greater than 1 kg, and activation of the renin-angiotensin system, which may require separate treatment. Minoxidil has similar side effects, while diazoxide, sometimes used in an emergency to treat a hypertensive crisis, may cause hyperglycemia, headache, vomiting, and antidiuresis.

In conclusion, although some of these compounds show promise, none of them fit all the criteria of clinically useful drugs. Three lines of investigation appear promising:[48] (1) the opening of K_{ATP} channels in skeletal muscle for the treatment of myopathies associated with membrane depolarization; (2) the blockage of these channels, whose openings may cause disrhythmia, for the protection of the ischemic heart; and (3) the role of the K_{ATP} channels in the release of excitatory amino acids in the brain and in the mechanism of hyperpolarization and CNS depression induced by certain general anesthetics.

Potassium Channel Blockers

Drugs that Affect ATP-Sensitive Potassium Channels

Of the several drugs believed to act as potassium-channel blockers, only the sulphonylureas (tolbutamide and glibenclamide) may reasonably be considered to inhibit potassium currents.[7] These agents, applied to resting pancreatic β cells, cause depolarization of the cell membrane, which in turn triggers the rise of intracellular Ca^{2+} and the secretion of insulin. However, the action of sulphonylureas is not confined to the endocrine pancreas; in fact, there are high-affinity binding sites for (^{3}H)-glibenclamide in rat brain (substantia nigra and globus pallidus) and heart membranes,[65] and it has been shown that GABA release from substantia nigra is stimulated by gliquidone, a sulphonylurea,[66] and inhibited by pinacidil, lemakalim, and nicorandil.[49] Anecdotal evidence relates the anti-arrhythmic action of class III agents (according to Vaughan-Williams classification[67]) to K_{ATP}-channel blockade. Thus sotalol prolongs the action potential in guinea pig papillary muscle, and this effect is counteracted by the K_{ATP} opener cromakalim.[68] Tedisamil, a novel anti-arrhythmic derivative of sparteine, appears to block the slowly-developing, time-dependent delayed rectifier and transient potassium current in both rat and guinea pig ventricular myocytes and mouse astrocytes.[69] Rather intriguing is the observation that this drug does not exhibit any voltage or use dependence.

Drugs that Affecting Other Potassium Channels

Quinidine, a Vaughan-Williams class I anti-arrhythmic compound, has been shown to cause a time- and voltage-dependent block of delayed potassium current in rabbit sinoatrial and atrioventricular nodes. This is considered to be relevant to the mechanism of action of quinidine, which inhibits the automaticity of nodal tissues.[70] Forskolin,[71] first described as a cardioactive drug that acts by stimulating adenyl cyclase and therefore enhancing intracellular cAMP level, seems to have also cAMP-independent actions. Thus forskolin inhibits delayed rectifier potassium currents in mouse pancreatic β cells, PC12 cells, and in human T lymphocytes, effects that deserve exploration for their therapeutic potential. Several neuromuscular-blocking drugs (tubocurarine, atracurium, pancuronium, and vecuronium), in concentrations higher than those achieved clinically, have been shown to mimic the effect of apamin, an octapeptide in bee venom that blocks low-conductance K_{Ca} channels.[72] More information about the ability of these drugs to block K_{Ca} channels would be of interest, especially for the care of patients after receiving large doses.

The selectivity of these compounds is also of interest. Thus charybdo-toxin, a potent inhibitor of the K_{Ca} channel, abolishes guinea pig tracheal relaxation caused by isoproterenol or salbutamol,[73] but not the tracheal relaxation caused by K_{ATP}-channel activators (pinacidil and cromakalim). The outward K^+ currents of rat ventricular myocytes[74] are inhibited by noradrenaline (and possibly by α-adrenergic agents), an action abolished by phentolamine, an α-adrenoceptor blocker.[74] The potassium channel involved has not been characterized.

4-aminopyridine (4-AP) is a potent blocker of transient outward A current, which has been studied extensively in humans and in animal models. Marketed under the trade name of "pymadine" in Bulgaria,[75] this drug has been used clinically to antagonize the non-depolarizing neuromuscular blockade at the end of anesthesia. Other clinical studies have shown[76] that 4-AP and neostigmine (an anticholinesterase agent) act synergistically in antagonizing the neuromuscular block. It has also been shown that 4-AP reverses diazepam anesthesia in human volunteers[77] (in other words it has an analeptic action) and has a beneficial effect on the hemodynamic depression induced by verapamil (a calcium-entry blocker) intoxication in cats.[78] Unfortunately 4-AP itself has an excessive CNS stimulatory action and therefore a rather low therapeutic index.

Thus the search for better compounds continues. Among established drugs, the general anesthetics appear to block potassium conductance in a variety of excitable tissues and in a range of concentrations comparable to those found in patients undergoing surgery. Halothane, isoflurane, and enflurane reversibly depress the amplitude of delayed rectifiers in canine cardiac Purkinje cells.[79] In the squid giant axon, general anesthetics produce[80,81] a depolarization of about 4–5 mV. This is considered to be due to a block of a component of the potassium conductance and to contribute significantly to the resting potential. Anesthetics exert qualitatively similar effects in the rabbit cornea,[82] but excitatory phenomena caused by clinical concentrations of anesthetics are more reminiscent of those produced by 4-AP. In other words, it is possible that in mammals these compounds may use different potassium channels. Phencyclidine (PCP) is a psychomimetic that inhibits voltage-dependent potassium currents in murine thymocytes, which in turn produce lymphokines.[83] Thus it may seem appropriate to investigate the role of drugs that affect potassium conductance in the modulation of the immunologic system.

Conclusions

The therapeutic usefulness of many drugs (eg, general anesthetics), has been discovered as a result of careful, albeit empirical, observations; in spite of their widespread use, their mechanisms of action are still not fully understood. A different approach has been followed in the development of drugs that affect potassium channels, whose clinical potential has been investigated only after exhaustive biochemical, biophysical, and pharmacodynamic investigations. This approach has provided other benefits; indeed, the actions of some oral antidiabetic agents and certain side effects of general anesthetic agents may now be explained by an interaction of these drugs with certain potassium channels. Many promising compounds that are capable of modifying the activity of these channels are in the pipeline awaiting the results of pharmacokinetic and biopharmaceutic studies.

References

1. Feldman SA. Effect of changes in electrolytes, hydration and pH upon the reactions to muscle relaxants. Br J Anaesth 1963; 35:546–551.

2. Miller RD, Roderick LL. Pancuronium induced neuromuscular blockade. Anesthesiology 1977; 46:333–335.
3. Miller RD, Roderick LL. Diuretic induced hypokaelemia, pancuronium neuromuscular blockade and its antagonism by neostigmine. Br J Anaesth 1978; 50:541–544.
4. Waud BE, Mookerjee A, Waud DR. Chronic potassium depletion and sensitivity to tubocurarine. Anesthesiology 1982; 57:111–115.
5. Waud BE, Waud DR. Interaction of calcium and potassium with neuromuscular blocking agents. Br J Anaesth 1980; 52:863–866.
6. Horn B. Magnesium deficiency causing persistent hypokalemia. Anesthesiology 1977; 46:310.
7. Cook NS. The pharmacology of potassium channels and their therapeutic potential. Trends Pharmacol Sci 1988; 9:21–28.
8. Hamill OP, Marty A, Neher E, Sackmann B, Sigworth FJ. Improved patch-clamp techniques for high-resolution current recording from cells and cell-free membrane patches. Pfluegers Arch 1981; 391:85–100.
9. Kirber MT, Ordway RW, Clapp LH, Sims SM, Walsh JV Jr, Singer JJ. Voltage, ligand and mechanically gated channels in freshly dissociated single smooth muscle cells. In: Colatsky TJ, ed. Potassium Channels. New York: Wiley-Liss; 1990:123–143.
10. Osterrieder W, Waterfall JF. Therapeutic potential of K^+ channel modulation in heart. In: Cook NS, ed. Potassium Channels. Chicester, England: Ellis Horwood; 1990:337–347.
11. Montal M. Reconstitution of channel proteins from excitable cells in planar lipid bilayer membranes. J Membr Biol 1987; 98:101–115.
12. Prestipino G, Valdivia HH, Lievano A, Darszon A, Ramirez AN, Possani LD. Purification and reconstitution of potassium channel from squid axon membranes. FEBS Lett 1989; 250:570–574.
13. Rehm H, Pelzer S, Cochet C, Chembaz E, Tempel BL, Trautwain W, Pelzer D, Lazdunski M. Dendrotoxin-binding brain membrane protein displays a K^+ channel activity that is stimulated by both cAMP-dependent and endogenous phosphorilations. Biochemistry 1989; 28:6455–6460.
14. Carbone E, Prestipino G, Spadavecchia L, Franciolini F, Possani LD. Blocking of the squid giant axon K^+ channel by noxiustoxin: A toxin from the venom of the scorpion Centruroides noxius. Pfluegers Arch 1987; 408: 423–431.
15. Halliwell JV, Othman IB, Pelchen-Matthews A, Dolly J. Central action of dendrotoxin: Selective reduction of a transient K^+ conductance in hippocampus and binding to localized acceptors. Proc Natl Acad Sci USA 1986; 83:493–497.
16. Castle NA, Haylett DG, Jenkinson DH. Toxins in the characterization of potassium channels. Trends Neurosci 1989; 12:59–65.
17. Dreyer F. Peptide toxins and potassium channel. Rev Physiol Biochem Pharmacol 1990; 115:93–136.
18. Neher E, Sakmann B. Single channel currents recorded from membrane at denervated frog muscle membrane. Nature 1976;260:799–802.
19. Tempel BL, Papazian DM, Schwarz TL, Jan YN, Jan LY. Sequence of a probable potassium channel component encoded at Shaker locus of Drosophila. Science 1987; 237:770–775.

20. Frech GC, VanDongen AMJ, Schuster G, Brown AM, Joho RH. A novel potassium channel with delayed rectifier properties isolated from rat brain by expression cloning. Nature 1989; 340:642–645.
21. Roberds S, Tomkins MM. Cloning and tissue-specific expression of five voltage-gated potassium channel cDNAs expressed in rat heart. Proc Natl Acad Sci USA 1991; 88:1798–1802.
22. Eisenman G, Dani JA. An introduction to molecular architecture and permeability of ion channels. Ann Rev Biophys Chem 1987; 16:205–226.
23. Latorre R, Miller C. Conduction and selectivity in potassium channels. J Membr Biol 1983; 71:11–30.
24. Miller C. Bis-Quaternary ammonium blockers as structual probes of the sarcoplasmic reticulum K^+ channel. J Gen Physiol 1982; 79:869–891.
25. Armstrong CM. Ionic pores, gates, and gating currents. Quart Rev Biophysics 1975; 7:179–210.
26. Hille B. Gating mechanisms. In: Hille B. ed. Ionic Channels of Excitable Membranes. Sunderland, Mass.: Sinauer; 1984:329–353.
27. Hodgkin AL, Huxley AF. Currents carried by sodium and potassium ions through the membrane of the giant axon of Loligo. J Physiol (Lond) 1952a; 116:449–472.
28. Neher E. Two fast transient current components during voltage clamp on snail neurons. J Gen Physiol 1971; 58:36–53.
29. Hagiwara S, Takahashi K. The anomalous rectification and cation selectivity of the membrane of a starfish egg cell. J Membr Biol 1974; 18:61–80.
30. Stühmer W, Conti F, Stocker M, Pongs O, Heinemann SH. Gating currents of inactivating and non-inactivating potassium channels expressed in *Xenopus* oocytes. Pfluegers Arch 1991; 418:423–429.
31. Kurachi Y, Nakajima T, Sugimoto T. On the mechanisms of activation of muscarinic K^+ channel by adenosine in isolated atrial cells: Involvement of GTP-binding proteins. Pfluegers Arch 1986b; 407:264–274.
32. Ambrosini A, Barnard EA, Prestipino G. AMPA and kainate-operated channels reconstituted in artificial bilayers. FEBS Lett 1991; 281:27–29.
33. Woodhull AM. Ionic blockage of sodium channel in nerve. J Gen Physiol 1973; 61:687–708.
34. Miller C, Latorre R, Reisin I. Coupling of voltage-dependent gating and Ba^{++} block in the high-conductance, Ca^{++}-activated K^+ channel. J Gen Physiol 1987; 90:427–499.
35. Schawrz TL, Tempel BL, Papazian DM, Jan YN, Jan LY. Multiple potassium-channel components are produced by alternative splicing at the *Shaker* locus in *Drosophila*. Nature 1988; 331:137–142.
36. MacKinnon R, Yellen G. Mutations affecting TEA^+ blockade and ion permeation in voltage-activated K^+ channels. Science 1990; 250:276–279.
37. Guy HR, Conti F. Pursuing the structure and function of voltage-gated channels. TINS 1990; 13:201–206.
38. Rudy B. Diversity and ubiquity of K^+ channels. Neuroscience 1988; 25:729–749.
39. Kolb HA. Potassium channels in excitable and non-excitable cells. Rev Physiol Biochem Pharmacol 1990; 115:51–91.
40. Cook NS, Quast U. Potassium channel pharmacology. In: Cook NS, ed. Potassium Channels. Chicester, England: Ellis Horwood; 1990:181–231.

41. Hoshi T, Aldrich RW. Voltage-dependent K^+ currents and underlying single K^+ channels in phochromocytoma cells. J Gen Physiol 1988; 91:73–106.

42. Magnelli V, Nobile M, Maestrone E. K^+ channels in PC12 are affected by propofol. Pfluegers Arch 1992; 420:393–398.

43. Taylor PS. Selectivity and patch measurements of A-current in *Helix aspersa* neurons. J Physiol (Lond) 1987; 388:437–447.

44. Ohmori H, Yoshida S, Hagiwara S. Single K^+ channel currents of anomalous rectification in cultured rat myotubes. Proc Natl Acad Sci USA 1981; 78:4960–4964.

45. Brown DA, Adams PR. Muscarinic suppression of a novel voltage-sensitive K^+ current in a vertebrate neuron. Nature 1980; 283:673–676.

46. Kandel ER, Schwartz JH. Molecular biology of learning: Modulation of transmitter release. Science 1982; 218:433–443.

47. Volterra A, Siegelbaum SA. Role of two different guanine nucleotide-binding proteins in the antagonistic modulation of the S-type K^+ channels by cAMP and arachidonic acid metabolites in *Aplysia* sensory neuron. Proc Natl Acad Sci USA 1988; 85:7810–7814.

48. Quast U, Cook NS. Moving together: K^+ channel openers and ATP-sensitive K^+ channels. Trends Pharmacol Sci 1989; 10:431–435.

49. Edwards G, Weston AH. Structure activity relationship of K^+ channel openers. Trends Pharmacol Sci 1990; 11:417–422.

50. Quast U, Cook NS. "In vitro" and "In vivo" comparison of two K^+ channel openers, diazoxide and cromakalim and their inhibition by glibenclamide. J Pharmacol Exp Ther 1989; 250:261–271.

51. Lebrun P, Devreux U, Herman M, Herculez A. Similarities between the effects of pinacidil and diazoxide on ionic and secretory events in rat pancreatic islets. J Pharmacol Exp Ther 1989; 250:1011–1018.

52. Lebrun P, Antoine MH, Devreux U, Hermann M, Herculez A. Paradoxical inhibitory effect of cromakalim on $^{86}Rb^+$ outflow from pancreatic islet cells. J Pharmacol Exp Ther 1990; 255:948–954.

53. Lodge NJ, Cohen RB, Havens CN, Colatsky TJ. The effects of the putative potassium channel activator WAY 120,491 on $^{86}Rb^+$ efflux from rabbit aorta. J Pharmacol Exp Ther 1991; 256:639–644.

54. Meisheri KD, Swirtz MA, Purhoit SS, Cipkus-Dubray LA, Khan SA, Oleynek JJ. Characterization of K^+ channel-dependent as well as independent components of pinacidil-induced vasodilatation. J Pharmacol Exp Ther 1991; 256:492–499.

55. Kajioka S, Oike M, Kitamura K. Nicorandil opens a calcium-dependent potassium channel in smooth muscle cells of the rat portal vein. J Pharmacol Exp Ther 1990; 254:905–913.

56. Giudicelli JF, Richer C, Berdeux A. Les activateurs des canaux potassiques. Perspectives dans le traitment de l'hypertension arterielle. La Presse Medicale 1991; 20:75–90.

57. Giudicelli JF, Drieu La Rochelle C, Berdeux A. Effects of cromakalim and pinacidil on large epicardial and small coronary arteries in conscious dogs. J Pharmacol Exp Ther 1990; 255:836–842.

58. Grover GJ, McCullough JR, Henry DE, Conder ML, Sleph PG. Anti ischemic effects of potassium channel activators pinacidil and cromakalim and

the reversal of these effects with the potassium channel blocker glyburide. J Pharmacol Exp Ther 1989; 251:98–104.

59. Sakamoto S, Liang CL, Stone CK, Wood WB Jr. Effects of pinacidil on myocardial blood flow and infarct size after acute left anterior descending coronary artery occlusion and reperfusion in awake dogs with and without coexisting left circumflex coronary artery stenosis. J Cardiovasc Pharmacol 1989; 14:747–756.

60. Masuzawa K, Asano M, Matsuda T, Imaizumi Y, Watanabe M. Possible involvement of ATP sensitive K^+ channels in the relaxant response of dog middle cerebral artery to cromakalim. J Pharmacol Exp Ther 1990; 255:818–825.

61. Bril A, Man R. Effects of the potassium channel activator BRL 34915 on the action potential characteristics of canine cardiac Purkinje fibers. J Pharmacol Exp Ther 1990; 253:1090–1096.

62. Ripoll C, Lederer WJ, Nichols CG. Modulation of ATP sensitive K^+ channel activity and contractile behaviour in mammalian ventricle by the potassium channel openers cromakalim and RP49356. J Pharmacol Exp Ther 1990; 255:429–435.

63. Hiraoka M, Fan Z. Activation of ATP sensitive outward K^+ current by nicorandil (2-nicotinamidoethyl nitrate) in isolated ventricular myocytes. J Pharmacol Exp Ther 1989; 250:278–285.

64. Kantor PF, Coetzee WA, Carmeliet EE. Reduction of ischemic K^+ loss and arrhytmias in rat hearts: Effects of glibenclamide, a solphonylurea. Circ Res 1990; 66:478–483.

65. Miller JA, Velayo NL, Dage RC, Rampe D, Lazdunski M. High affinity (^{3}H)glibenclamide binding sites in rat neuronal and cardiac tissue: Localization and developmental characteristics. J Pharmacol Exp Ther 1991; 256:358–364.

66. Amoroso S, Schmid-Antonmarchi H, Fosset M. Glucose solphonylureas and neurotransmitters release: Role of ATP sensitive K^+ channels. Science 1990; 247:852–859.

67. Vaughan-Williams EM. Classification of antidysrhythmic drugs. Pharmacol Ther 1975; 1:115–138.

68. Ijzerman AP, Soudijn W. The antiarrhythmic properties of adrenoceptor antagonists. Trends Pharmacol Sci 1989; 10:31–35.

69. Dukes ID, Cleeman L, Morand J. Tedisamil blocks the transient and delayed rectifier K^+ currents in mammalian cardiac and glial cells. J Pharmacol Exp Ther 1990; 254:560–569.

70. Furukawa T, Tsujimura Y, Kitamura K, Tanaka H, Habuchi Y. Time and voltage dependent block of the delayed K^+ current by quinidine in rabbit sinoatrial and atrioventricular nodes. J Pharmacol Exp Ther 1989; 251:756–763.

71. Laurenza A, McHugh-Sutkowski E, Seamon KB. Forskolin: A specific stimulator of adenylyl cyclase or a diterpene with multiple sites of action? Trends Pharmacol Sci 1989; 10:442–443.

72. Bowman WC. Pharmacology of Neuromuscular Function. London: Wright; 1990:29–32.

73. Jones TR, Charette L, Garcia ML, Kaczorowski J. Selective inhibition of relaxation of guinea pig trachea by charybdotoxin, a potent Ca^{++} activated K^+ channel inhibitor. J Pharmacol Exp Ther 1990; 255:697–706.

74. Ravens U, Wang XL, Wettwer E. Adrenoceptor stimulation reduces outward currents in rat ventricular myocytes. J Pharmacol Exp Ther 1989; 250:364–370.
75. Stojanov E, Vulchev P, Shtrubova M, Marinova M. Clinical electro-myomechanographic and electromyographic studies in decurarization with pymadine. Anaesth Resus Inten Ther 1976; 4:139–143.
76. Miller RD, Booij LHDJ, Agoston S, Crul JF. 4-aminopyridine potentiates neostigmine and pyridostigmine in man. Anesthesiology 1979; 50:416–420.
77. Agoston S, Salt PJ, Erdmann W, Hilkemeijer T, Bencini A, Langrehr D. Antagonism of ketamine-diazepam anaesthesia by 4-aminopyridine in human volunteers. Br J Anaesth 1980; 52:367–370.
78. Agoston S, Maestrone E, van Ezik EJ, Ket JM, Houwertjes MC, Uges DRA. Effective treatment of verapamil intoxication with 4-aminopyridine in the cat. J Clin Invest 1984; 73:1291–1295.
79. Supan F, Buljubasic N, Eskinder H, Kampine JP, Bosnjak ZJ. Effects of halothane, isoflurane and enflurane on K^+ current in canine cardiac Purkinje cells. Anesth Analg 1991; 72:S286.
80. Haydon DA, Requena J, Simon AJB. The potassium conductance of the resting squid axon and its blockage by clinical concentrations of general anaesthetics. J Physiol 1988; 402:362–374.
81. Haydon DA, Simon AJB. Excitation of the squid giant axon by general anaesthetics. J Physiol 1988; 402:375–379.
82. MacIver MB, Tanellian DL. Volatile anesthetics excite mammalian nociceptor afferents recorded "in vitro." Anesthesiology 1990; 72:1022–1030.
83. Fiorica-Howells E, Gambale F, Horn R, Osses L, Spector S. Phencyclidine blocks voltage-dependent potassium currents in murine thymocytes. J Pharmacol Exp Ther 1990; 252:610–615.

14
Potassium Channels in Skeletal Muscle

RALF WEIK

Several different potassium-selective channels have been described[1,2] since 1902, when Julius Bernstein postulated the existence of a selective potassium permeability in excitable cell membranes.[3] This diversity results from the expression of different or related genes, from the alternative splicing of a primary transcript, or from posttranslational modifications. In addition, the assembling of different channel proteins into heteromultimeric species and tissue-specific expression of different channel types have been proposed.[4,5] Potassium channels can be found in almost all eukaryotic cells. Various types of potassium channels may be present in the same cell, while different cells may contain similar types. The functional role of all types of potassium channels is to lower the excitability of the cell[1] and, although different channels play different roles in stabilizing the cell membrane, the individual contribution of a specific channel type to the total ionic current is often difficult to determine.

Many methods have been developed to study potassium currents in living cells. In skeletal muscle several devices have been constructed to control the voltage across the membrane and to measure the current. In 1970 Adrian et al. introduced the three microelectrode technique.[6] They inserted microelectrodes near the end of a fiber in the frog sartorius muscle, one for delivering current and the other two for measuring membrane potential and membrane current. Two years later, Ildefonse and Rougier[7] used the double-sucrose gap method to study early membrane currents in skeletal muscle fibers. Hille and Campbell[8] perfected the vaseline gap voltage-clamp method introduced by Frankenhaeuser and colleagues[9] to describe currents in frog muscle. Almers et al.[10] developed an improved loose-patch technique to study ion currents in human skeletal muscle. They used a concentric arrangement of two micropipettes to electrically isolate small-diameter (10–15 μm) membrane patches on the surface of the muscle.

All these methods allow the measurement of macroscopic currents, ie, currents from large areas of cell membranes, which represent the sum of ion fluxes through many channels and makes it difficult to estimate the contribution of different channel types to the total current. Two more recent developments have made it possible to study properties of single channels, which are macromolecular pores in cell membranes. The "patch-clamp" technique introduced by Neher and Sakmann[11-13] allows the characterization of single-channel proteins within the cell membrane. Briefly, a patch of membrane is isolated electrically from the external solution and the currents flowing through single channels in the patch are recorded. The second approach to the analysis of single channels is the reconstitution of purified channel proteins into planar bilayer membranes.[14,15] Two aqueous chambers are separated by a phospholipid bilayer, in which channel proteins are inserted.[15] The advantage of this method is that channels in otherwise inaccessible membranes, such as sarcoplasmic reticulum, can be studied on the single-channel level. The two methods have been successfully combined.[16] The same type of potassium channel, studied with the patch-clamp technique[17] or the reconstitution technique,[18] showed similar characteristics.

The significant advantage of single-channel studies is that different types of potassium channels can be classified according to the electrophysiologic properties of one pore-forming protein. Some of the features frequently used to distinguish the various types of channels are listed below.

One channel characteristic is the electrical conductance (expressed in siemens, or S), which measures the ease of current flow and is equal to the ratio of current and voltage.[1] Different potassium channel types exhibit different conductances (Table 14.1).

In some potassium-channel types the membrane conductance changes with voltage, a property called rectification. In symmetrical potassium-containing solutions, at $-40\,mV$ membrane potential single-channel currents of the so called "inward rectifier" can be seen, whereas at $+40\,mV$ currents are no longer visible (Figure 14.1a). The current-voltage curve is not a straight line (Figure 14.1b,c).

The probability of a channel being open can depend on the voltage across the membrane. The delayed rectifier, for instance, is activated by membrane depolarization[19] (Figure 14.2), whereas the inward rectifier shows openings at a resting membrane potential[20] (Figure 14.1).

Cytoplasmic factors such as calcium or adenosine-5'-triphosphate (ATP) may affect certain types of potassium channels. Calcium-activated potassium channels exhibit an increased open probability if the calcium concentration on the cytoplasmic side of the channel is increased (Figure 14.3). Intracellular ATP blocks the ATP-dependent potassium channel (Figure 14.4).

Moreover, a variety of pharmacologic agents reduce or increase the open probability of different potassium-channel types (Table 14.1). For instance, N-ethylmaleimide and chloramine T irreversibly close ATP-

TABLE 14.1. Types of single potassium channels in skeletal muscle.

Type	Function	Single channel conductance	Physiologic Blockers	Physiologic Activators	Pharmacologic Blockers	Pharmacologic Activators	Refs
ATP-dependent K channel	Might contribute to K efflux in metabolically exhausted muscle	60, 74 pS	ATP, Mg, Na		Tolbutamide		20, 57
		42–48 pS[a]	ATP, ADP	Low pH	AMP-PNP, TEA		30, 46, 163, 164
			ATP		Cs, Ba, 4-AP		165, 166
		67 pS[b]	ATP			GTP-γ-S, AlF_4	47
		21 pS[c]	ATP				167
			ATP				21
					NEM		
					Chloramine-T		
			ATP			Cromakalim, RP 49356, pinacidil	67
			ATP			EMD 52692	35, 68
			ATP			$V_{10}O_{28}^{6-}$	43
Ca-activated K channels							
BK	Might stabilize and repolarize membrane after action potential	187–218 pS		Ca			17, 20, 79
		250 pS		Ca	Polymyxin B		168
				Ca		Cd, Sr, Mn, Fe, Co	169
		250–290 pS		Ca	NBA		170
				Ca	TEA, QA		25, 93, 94
				Ca	TMO		171, 172
				Ca	Ba		173–176
				Ca	Charybdotoxin		80, 177
				Ca	Iberiotoxin		89, 90
				Ca	DTX-I, BPTI		178, 179
				Ca			180
SK	Responsible for long-lasting after-hyperpolarization	10–14 pS	Mg	Ca	Apamin		23
			Apamin-like factor				103, 104

Channel	Function	Conductance	Activation	Blockers	References
Delayed rectifier	Takes part in action potential repolarization and duration	28–30 pS	Depolarization	TEA	19, 22, 113
Inward rectifier	Provides part of resting conductance	9–10; 20–26 pS	Na, Mg	Ba, Rb, Cs	20, 120, 123, 181
SR K channel	Facilitates Ca release in sarcoplasmic reticulum	130 pS^d 150^d, 50 pS^{d,e}		Cs bisG10, bisQ6, bisQ10 Ag, Hg, Cu, Cd, Pb, Zn MOC Cs	78, 182, 183 125, 184, 185 186 187 188

The single channel conductance was estimated in solutions containing 120–160 mM potassium on both sides of the membrane unless otherwise stated.

[a] Extracellular KCl was 60 mM.

[b] 250 mM internal, 50 mM external KCl.

[c] 103 mM KCl internal, no KCl external.

[d] 100 mM K on both sides.

[e] Second open state in frog muscle.

Abbreviations: ADP = adenosine-5'-diphosphate; AMP-PNP = adenylylimidodiphosphate; 4-AP = 4-aminopyridine; ATP = adenosine-5'-triphosphate; bisG10 = 1,10-*bis*-guanidino-*n*-decane; bisQ6 = hexamethonium; bisQ10 = decamethonium; BPTI = bovine pancreatic trypsin inhibitor; DTX-I = dendrotoxin-I; GTP-γ-S = guanosine 5'-O-(3-thiotriphosphate); MOC = monovalent organic cations; NBA = *N*-bromoacetamide; NEM = *N*-ethylmaleimide; QA = quaternary ammonium; TEA = tetraethylammonium; TMO = trimethyloxonium; $V_{10}O_{28}^{6-}$ = decavanadate.

dependent potassium channels, whereas calcium-activated potassium channels are not modified.[21]

Single-channel studies allow the exact classification of different potassium channel types. Their physiologic roles can be revealed by making summed records of many single-channel events. The time course of these summed currents sometimes closely resembles that of macroscopic currents measured in large membrane areas.[22] A second approach is to use specific blockers for certain types of K channels to modify macroscopic currents. The neurotoxin apamin from bee venom, which blocks a small-conductance, calcium-activated potassium channel,[23] abolishes the long-lasting after-hyperpolarization that follows an action potential in cultured rat muscle.[24,25]

The aim of this chapter is to describe the various potassium-channel types found in skeletal muscle, which have been characterized on the single-channel level, and to discuss their putative physiologic and pathophysiologic roles.

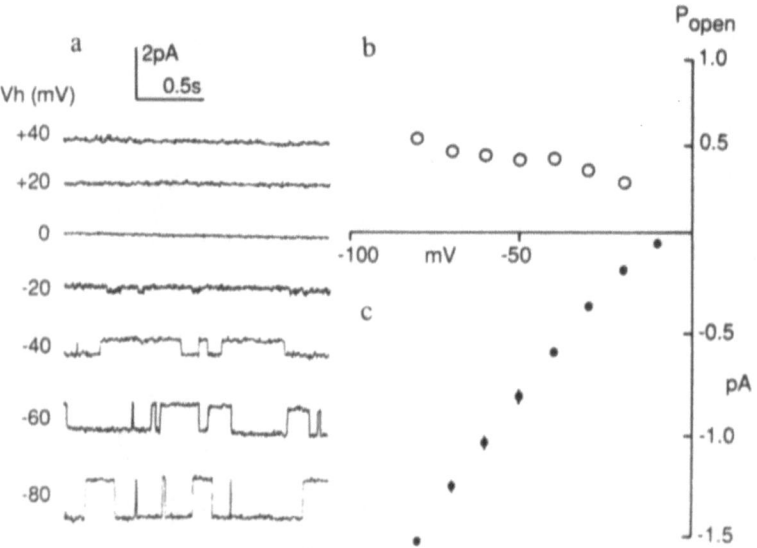

FIGURE 14.1. Single-channel currents of inward rectifiers in a membrane patch of rat sarcolemma. (a) Channel openings at different holding potentials, as indicated. Inward currents are downward. Records were made in symmetrical potassium-rich (140 mM) solutions. The cytoplasmic side of the membrane was exposed to 2 mM MgCl₂. (b) Open probability-voltage relationship obtained from measurements of successive channel open and closed intervals. The probability that a channel is open is about 0.5 and shows only a minor voltage dependence. (c) Current-voltage relationship obtained in symmetrical 140 mM potassium-containing solutions. (Reproduced with permission of John Wiley & Sons Inc., from reference 20.)

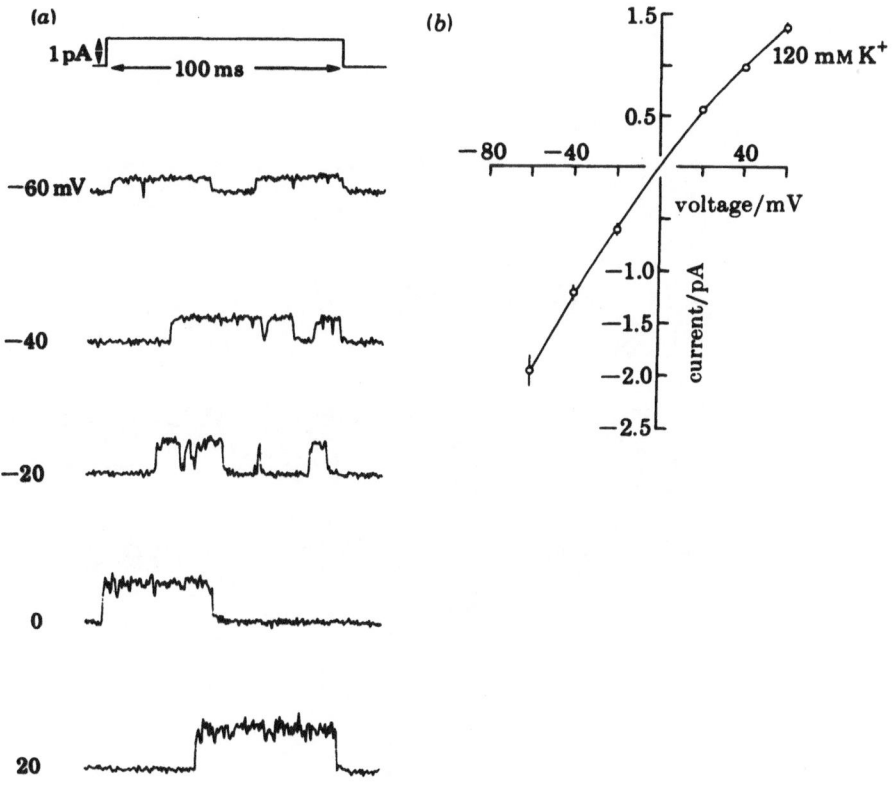

FIGURE 14.2. Voltage dependence of unitary potassium currents from delayed rectifiers. (a) The membrane potential was increased stepwise from the holding potential ($-100\,\text{mV}$) to the value given at the left of each record. The schematic voltage pulse, shown above the current records, also gives current and time calibrations. The extracellular potassium concentration was 2.5 mM. (b) Mean single-channel current-voltage relationships; symmetrical potassium concentration was 120 mM. (Reproduced with permission from reference 22.)

Ionic Channels in Skeletal Muscle

Excitation and contraction in skeletal muscle involve the movement of ions through ionic channels. The cations Na^+, K^+, Ca^{2+}, and the anion Cl^- seem to be responsible for these actions and move through different classes of channels with specific and selective ionic permeabilities.

Sodium channels are needed for the generation of propagated action potentials. After a suprathreshold membrane depolarization, sodium channels open and an inward current depolarizes the membrane, the equilibrium potential of sodium being the electromotive force. These currents are sufficient to excite neighboring membrane areas, thereby

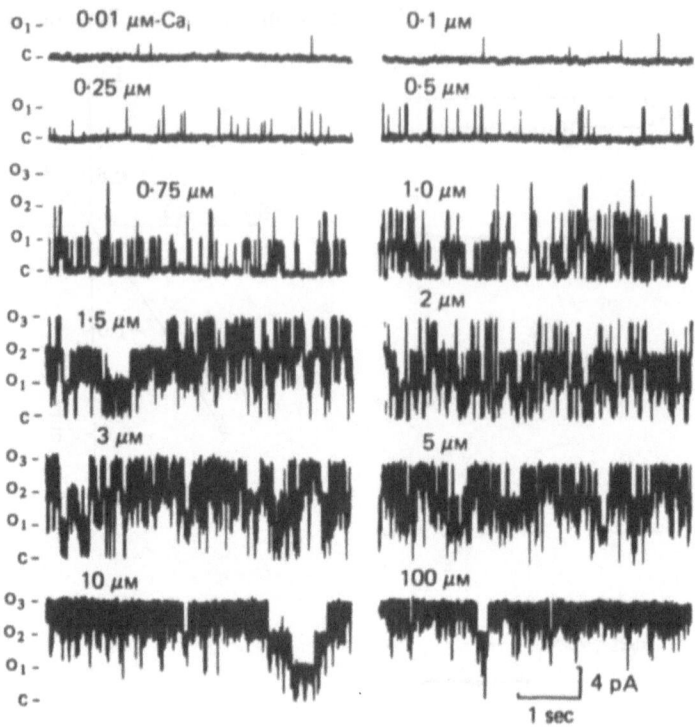

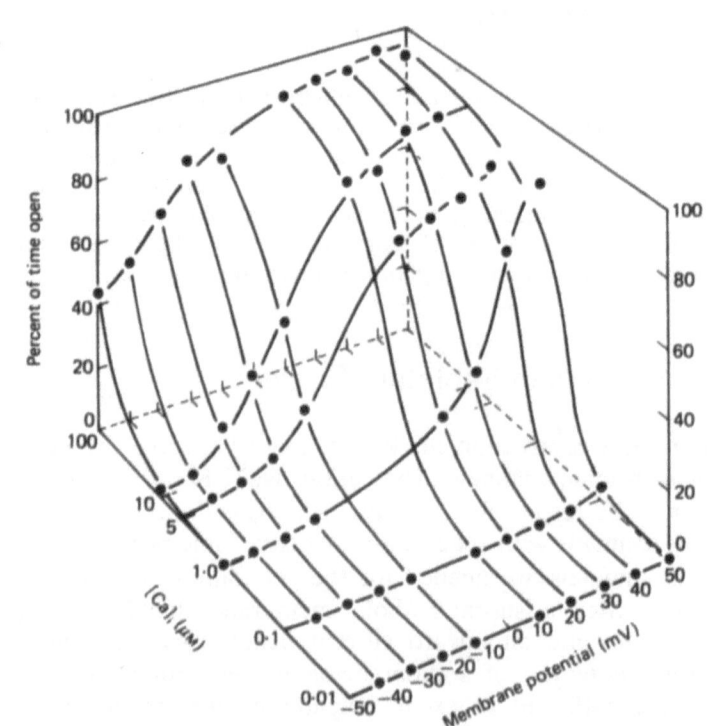

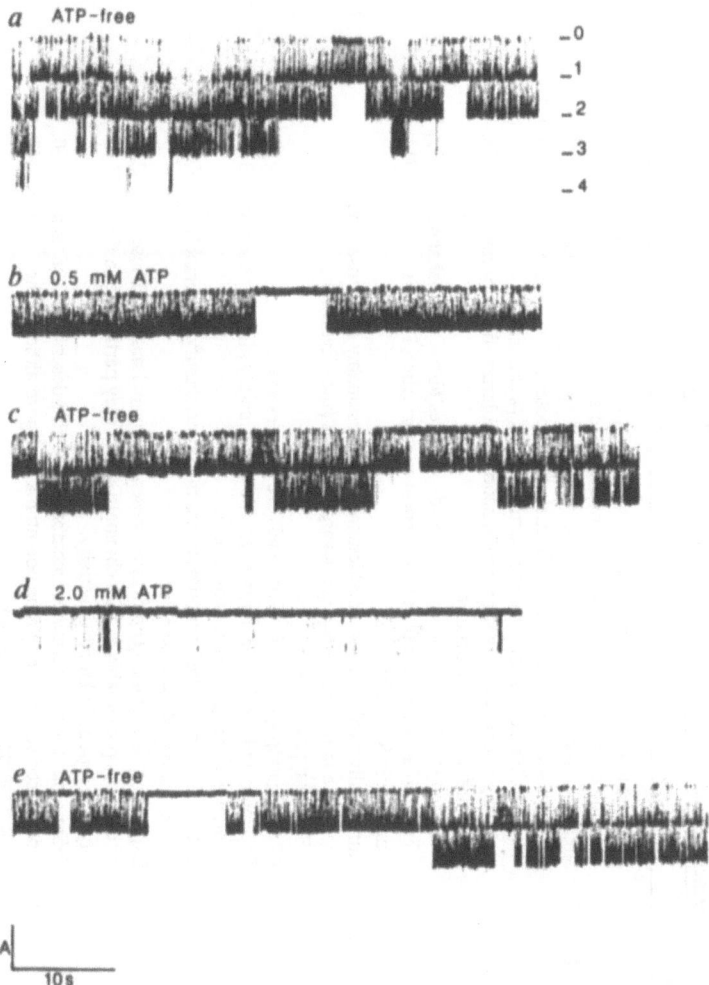

FIGURE 14.4. Effect of ATP on single potassium-channel currents from frog muscle. Current levels that correspond to zero, one, two, three, and four open channels are indicated at the right of record (a). The cytoplasmic surface of the membrane was exposed to different ATP concentrations as indicated. The membrane potential was -60 mV thoughout. (Reproduced with permission from reference 163, copyright 1985 Macmillan Magazines, Ltd.)

FIGURE 14.3. Effect of internal calcium on single-channel activity of calcium-activated potassium channels. (a) Records of membrane currents at the indicated calcium concentration are shown. Three channels were present in this membrane patch. One, two, and three channels open simultaneously are indicated by o_1, o_2, and o_3, respectively; c indicates all channels were closed. Membrane potential was $+20$ mV. (b) Plot of percentage of time channels were open (percent of time open) vs membrane potential for the indicated internal calcium concentration. (Reproduced with permission from reference 79.)

TABLE 14.2. Pathophysiologic and pharmacologic involvement of potassium channels in selected muscle diseases.

Disease	Channel type involved	Pathophysiology and pharmacology	Refs
McArdle's syndrome	Presumably K_{ATP}	Reduced muscle phosphorylation potential facilitates channel openings during exercise, thereby producing hyperkalemia.	38
Hyperkalemic periodic paralysis (adynamia episodica hereditaria) with myotonia	α Subunit of sodium channel	Slight increase in extracellular potassium induces openings of sodium channels with no inactivation. Cromakalim repolarizes otherwise irreversible depolarized fibers by activating K_{ATP} channels.	132, 133 129, 131 69, 70
Hypokalemic periodic paralysis	K Conductance	Lowering of extracellular K or insulin action depolarizes fibers because of a reduced K conductance.	136, 137
	Na Conductance	Increased steady-state Na conductance depolarizes fibers. Cromakalim hyperpolarizes muscle fibers and improves contraction force by activating K_{ATP} channels.	135, 137 71
Myotonia congenita	Cl Conductance	Abnormally small chloride conductance and, presumably compensatory, increase in K current through inward rectifiers (recessively inherited Becker type).	
	Na Conductance	Altered gating kinetics of Na channels are also described. Tocainide and mexiletine, antiarrhythmic drugs mainly affecting the fast sodium currents, are drugs of choice. EMD 52962 and cromakalim suppresses myotonic activity presumably by activating K_{ATP} channels.	189 130, 190 70
Myotonic dystrophy (Steinert disease)	Presumably K_{Ca}	Expression of apamin receptors not present in normal muscle. These receptors presumably are SK channels, which may participate in the typical repetitive bursts of activity. Other alterations of the muscle membrane are also described, such as a decreased resting membrane potential and abnormalities in the inactivation of Na channels.	100 130, 139 142

propagating the action potential. The α subunit of a sodium channel present in rat skeletal muscle has been isolated and characterized.[26] More and more evidence indicates that alterations of sodium channels are responsible for certain muscle diseases (Table 14.2).

Action potentials mediated by sodium channels in the plasma membrane depolarize the transverse tubule membrane, thereby activating voltage-gated calcium channels. Proposed protein–protein interactions between these voltage-gated calcium channels in the transverse tubule membrane and the calcium-release channels situated in the sarcoplasmic reticulum membrane may be responsible for the release of calcium in the sarcoplasmic reticulum. This release results in a contraction of the muscle fiber.[27] The primary structure of both channel types has been described.[27]

Chloride channels in skeletal muscle maintain a sufficiently high background conductance, thereby stabilizing the membrane resting potential and reducing excitability.[1,28,29] A reduced chloride conductance results in hyperexcitability (Table 14.2).

Potassium channels are most varied. So far, at least six different types have been described in skeletal muscle.

ATP-Dependent Potassium Channels

One type of potassium channel that is found in high densities in the sarcolemma is the so-called ATP-dependent potassium channel, K_{ATP}.[30] Intracellular ATP inhibits this channel (Figure 14.4). As a consequence, opening of the channel at physiologic intracellular ATP concentrations in skeletal muscle occur only rarely and briefly.[30]

The physiologic role of this channel type in skeletal muscle is not well understood.[31] Measurements of macroscopic currents in metabolically poisoned and mechanically exhausted frog skeletal muscle fibers revealed an increase in potassium conductance.[32,33] In normal, untreated fibers potassium contributes two-fifths and chloride three-fifths of the total conductance of the resting membrane, whereas in exhausted fibers potassium dominates with a ratio of $5:1$.[32] Castle and Haylett[34] used specific channel blockers to demonstrate that the increased potassium conductance during exhaustion was predominantly due to activation of ATP-dependent channels. Therefore these channels might hold the membrane potential near the equilibrium potential (E_K) during exhaustion to reduce the possibility of electric excitation of the fiber.[30] Sauviat and co-workers[35] tested the effects of a pharmacologic opener of K_{ATP} channels and demonstrated a reduction of the spike after potential together with a reduced muscle twitch tension in frog skeletal muscle. These authors speculated that the reduction of muscle contractile force underlying fatigue could be associated with the activation of these channels. However, at least in humans, central mechanisms are involved in regulating and limiting the motorneuron firing rates for maximum force generation, thereby optimizing motor control during muscular fatigue.[36]

Opening of the K_{ATP} channels results in an enhanced efflux of potassium. This may contribute significantly to vasodilation of skeletal muscle arterioles,[30,31] as well as to excitation of arterial chemoreceptors and to an increase of ventilation.[37,38]

To summarize, K_{ATP} channels seem to play an important role during prolonged contractions of skeletal muscle, but evidence showing that changes in cellular ATP are sufficient to modulate channel activity is still lacking. Total muscle cell ATP levels are not significantly lowered during exercise in normal subjects and patients with McArdle's disease.[39,40] The cytoplasmic ATP concentration of 3.4 mmol/kg in frog skeletal muscle can be reduced only to 2.6 mmol/kg by repetitive stimulation.[41,42] In the presence of metabolic poisons such as 1-fluoro-2,4-dinitrobenzene (FDNB), the ATP concentration does not fall below 0.6 mM,[33] whereas the concentration needed for half-maximal channel blockage in cytosol-free membrane patches is 0.135 mM for frog[30] and 0.008 mM[43] for mouse skeletal muscle.

The recent finding that cytosolic adenosine-5′-diphosphate (ADP) reduces the affinity of ATP for its receptor, first described in pancreatic β cells,[44,45] helps explain this discrepancy. Other possibilities may be an ATP-concentration gradient within the cell with much lower levels that is present near the plasma membrane, or unknown intracellular factors, which are lost after patch excision. Experiments by Davies[46] showed that intracellular pH is important in regulating the activity of K_{ATP} channels since lowering the pH from 7.2 to 6.3 resulted in a ninefold increase in the open probability of these channels.

The planar bilayer recording technique provided evidence that reconstituted K_{ATP} channels of transverse tubular membrane may be regulated by a nucleotide-binding G protein.[47] This doesn't seem to be the case for K_{ATP} channels of the surface membrane (R. Weik and B. Neumcke, unpublished data). The same type of channel situated in different parts of the membrane may therefore exhibit different properties.

Since K_{ATP} channels with similar electrophysiologic characteristics can be found in a variety of different tissues such as pancreatic β cells,[48] cardiac muscle,[49] and smooth muscle,[50] it is of pharmacologic interest to compare the putative functions of these channels in skeletal muscle and other tissues.

In ventricular cells, otherwise silent K_{ATP} channels seem to be activated during ischemia. This may increase extracellular potassium and therefore lead to vasodilation and increased blood supply. In addition, activation of K_{ATP} channels seems to shorten the duration of action potentials, resulting in a decreased contractile force. As a consequence, cellular ATP levels are preserved.[51,52] On the other hand, shortening of the action potential and of the refractory period also leads to arrhythmia and fibrillation, a rather deleterious effect.[31,53]

In pancreatic β cells K_{ATP} channels play a key role in the insulin secretory response.[51-54] In unstimulated β cells, unlike cardiac and skeletal muscle cells, K_{ATP} channels are principally responsible for the maintenance of the resting potential. An increase in extracellular glucose concentration evokes closure of the K_{ATP} channels and, as a consequence, membrane depolarization. If the depolarization is sufficient to activate voltage-dependent calcium channels, a calcium influx into the cell occurs. The increased calcium concentration in the cell is then an important stimulus for insulin exocytosis. This mechanism underlies also the action of the sulphonylureas, a class of hypoglycemic drugs, which blocks K_{ATP} channels and is used in the treatment of non-insulin-dependent diabetes mellitus.[55,56] This block ultimately increases insulin release. Fortunately, this pharmacologic effect of the sulphonylureas used clinically is restricted mainly to β cells, whereas K_{ATP} channels present in other tissues do not seem to be affected. This can be explained by the different role of K_{ATP} channels in different tissues. In pancreatic β cells K_{ATP} channels must stay open to maintain the resting potential. Therefore, they can be blocked by sulphonylureas. In muscle K_{ATP} channels do not seem to be open under physiologic conditions and therefore blocking agents have no effect. Furthermore, the affinity of the sulphonylureas for their receptors varies in different tissues. For example, the concentrations of tolbutamide required for half-maximal inhibition in skeletal and cardiac muscle are about 60 and 380 μM, respectively,[57,58] whereas in the β cells 3–18 μM are sufficient.[51] Thus there are highly specific drugs directed against one specific type of channel in a given tissue.

Smooth muscle also possess K_{ATP} channels.[50] These channels do not seem to be open under resting conditions,[59] but are activated by acetylcholine and by the vasoactive intestinal polypeptide (VIP),[50] as well as by the calcitonin gene-related peptide, thereby inducing vasodilation.[60]

The search for drugs that act as potassium-channel openers in smooth muscle is of special interest, since these drugs are hypotensive and bronchodilator agents.[61-65] So far seven distinct groups of potassium-channel openers with different chemical and pharmacologic characteristics have been identified.[66] As stated above, the K_{ATP} channels of different tissues have different affinities for potassium-channel openers. For instance, cromakalim has a very low threshold concentration of 0.01–0.1 μM in smooth muscle, whereas the sensitivity is much lower in skeletal and cardiac muscle and in pancreatic β cells.[63]

Despite the fact that many potassium-channel openers show only a low affinity for K_{ATP} channels in skeletal muscle,[67,68] they may be useful in the treatment of myopathies associated with membrane depolarization.[69-71]

Calcium-Activated Potassium Channels

Calcium-activated potassium channels couple the membrane potential to the intracellular calcium concentration. An increase in intracellular calcium results in an efflux of potassium and as a consequence in a hyperpolarization of the membrane.[72] Calcium-activated potassium currents can be found in many different cell types.[2,73,74] A modulation of potassium permeability by calcium was first described by Gardos[75] in red blood cells. So far, two different types of calcium-activated potassium channels have been described for skeletal muscle (Table 14.1).

Big Unitary Conductance Channels (BK)

The first calcium-activated potassium channels (K_{Ca}) observed at the single-channel level were found in chromaffin cells[76] and cultured rat skeletal muscle.[17] These channels could be easily detected since they had a single-channel conductance of about 200–300 pS. Therefore they were called big unitary conductance "BK"[77] or "maxi-K" channels.[78] Intracellular calcium activates this channel type (Figure 14.3A), whereas extracellular calcium has no effect.[79] In addition, BK channels are modulated by the membrane potential.[17,79] Depolarization activates the channels at a constant intracellular calcium concentration (Figure 14.3B). Miller and colleagues[80] found that a minor component of the venom of the Israeli scorpion, *Leiurus quinquestriatus*, could reversibly block this type of K_{Ca} channel with a very high affinity. With reference to the location of its binding site in a channel structure similar to a "giant whirlpool" the active peptide was called "charybdotoxin". This observation was interesting, because it was thought to provide an example of a toxin binding specifically and with high affinity to a given type of ion channels. Unfortunately, later work revealed that charybdotoxin also blocked numerous other potassium channels, such as type *n* and *n'* voltage-gated potassium channels in human and murine T lymphocytes,[81] intermediate conductance calcium-activated potassium channels in *Aplysia* neurons[82] and human red blood cells,[83] dendrotoxin-sensitive voltage-activated potassium channels of rat dorsal root ganglion cells,[84] calcium-activated and calcium-independent potassium channels in rat brain synaptosomes,[85] and *Shaker* potassium channels of *Zrosophila*.[86,87] The three-dimensional structure of charybdotoxin has recently been determined.[88]

A newly described toxin from the scorpion *Butus tamulus*, called iberiotoxin, seems to have a high selectivity for large-conductance K_{Ca} channels. Other potassium channels modified by charybdotoxin are unaffected by iberiotoxin.[89,90]

The physiologic function of the large-conductance calcium-activated potassium channels in skeletal muscle has not been clarified yet. Since these channels are activated at large positive potentials (Figure 14.3B), as would occur during the upstroke of an action potential, it is tempting to speculate that they stabilize and repolarize muscle membranes following an action potential.[91] As a consequence they would prevent the cell from a prolonged depolarization accompanied by an increase of the internal calcium concentration.[25] Furthermore, these channels might be activated by opened voltage-activated calcium channels located in the near vicinity, even at low mean cytoplasmic calcium concentrations.[91]

Recently, a *Drosophila slowpoke* locus, a gene that encodes a structural component of calcium-activated potassium channels present in larval and adult muscle,[92] has been cloned.

Small-Conductance Channels (SK)

Under the conditions described below a second type of calcium-activated potassium channel can be detected in skeletal muscle. The features of these channels are quite different from those of the large-conductance calcium-activated potassium channels (BK)[2] (Table 14.1). Their single-channel conductance is much lower and, therefore they are named small-conductance (SK) calcium-activated potassium channels.[23] Moreover, these channels are more calcium-sensitive than the BK channels at negative membrane potentials[23] (Figure 14.5), and they show low or no voltage-dependence in contrast to BK channels.[2] Furthermore, their pharmacology also differs from that of BK channels. Externally applied tetraethylammonium (TEA) shows half-maximal blockage of BK channels at concentrations lower than 0.5 mM,[93,94] whereas 5 mM and even higher concentrations of TEA do not block SK channels.[23,25] Apamin, an octadecapeptide found in honey bee venom (*Apis mellifera*),[95,96] blocks SK channels at very low concentrations,[23–25,97] whereas the BK channels are not modified.[25] Therefore in skeletal muscle charybdotoxin and apamin specifically block either one or the other type of calcium-activated potassium channels. The high affinity and specificity of apamin was exploited for the purification of the channel proteins.[98]

Apamin, used to study the physiologic role of SK channels, led to the observation that it blocked the long-lasting after-hyperpolarization (AHP) that followed the action potential.[24,25] According to Barrett and colleagues,[99] it is likely that the slow potassium conductance responsible for this AHP acts as a pacemaker in noninnervated muscle, regulating the rate of spontanous discharge. At less negative resting membrane potentials, repetitive-action potentials can be generated, because after each of them this potassium conductance permits a transient hyper-polarization, thereby providing sufficient time for sodium channels to recover from inactivation.[91,100,101] This spontaneous contractile activity may help to maintain the muscle in a "healthy state".[99] In the innervated muscle, this spontaneous activity is unnecessary and the SK channels

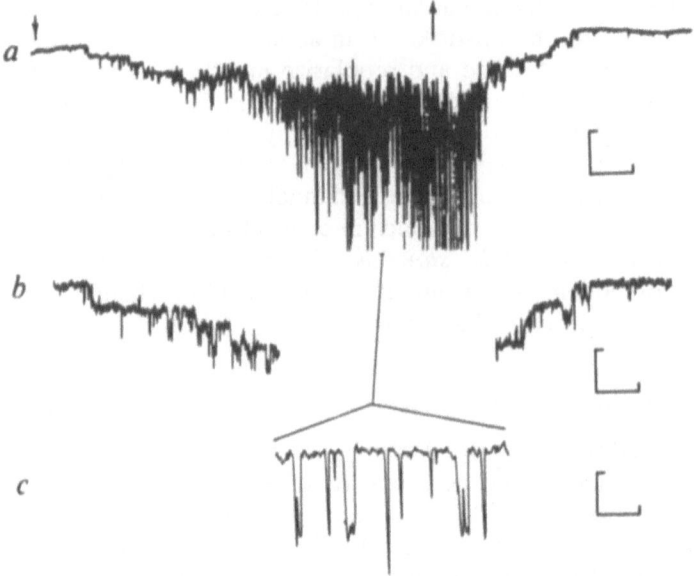

FIGURE 14.5. Calcium activation of SK and BK channels in the same membrane patch. (a) The cytoplasmic calcium concentration rose from 0 to 1 µM at the downward arrow and returned to zero at the upward arrow. Small and large downward current steps arose from SK and BK channels, respectively. It can be seen that with increasing calcium concentration SK channels were activated before BK channels. (b) Currents through SK channels recorded at higher gain and (c) currents through BK channels recorded at lower gain. Membrane potential was 40 mV. Vertical bar (pA): (a) 2; (b) 1; (c) 5. Horizontal bar (s): (a) 5; (b) 2.5; (c) 0.05. (Reproduced with permission from reference 23, copyright 1986 Macmillan Magazines, Ltd.)

responsible for the AHP are no longer present[102] (Figure 14.6). Indeed, neurons that form stable synapses on muscle cells elicit several changes within the muscle cell membrane, presumably with the help of small endogenous peptides. Endogenous apamin-like factors have been isolated from pig brain[103] and chick spinal cord.[104] These peptides antagonize apamin binding to its receptor and can act as after-hyperpolarization blockers in cultured skeletal muscle.

A change in the expression rate of SK channels is also found during muscle ontogeny.[102] Apamin receptors disappear in rat skeletal muscle as in utero and postnatal development proceed, and become undetectable from the fifth day of life to the adult age. At the same time, the number of voltage-activated sodium and calcium channels increases, and reaches a maximum between 2 and 3 weeks after birth.[105,106]

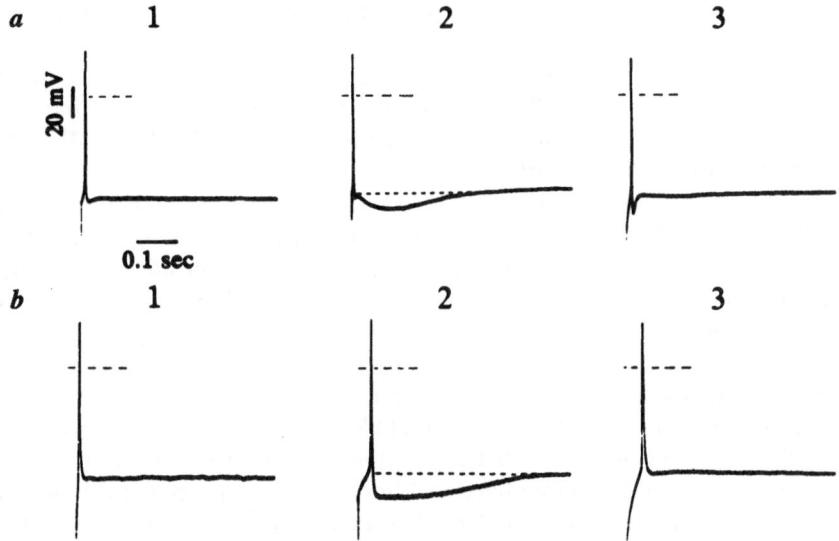

FIGURE 14.6. Action potentials and AHP evoked by anodal break stimulation from rat extensor digitorum longus (EDL) muscle (a) and from rat myotubes in culture (b). (a1) Innervated EDL muscle with no AHP detectable; (a2) EDL muscle 10 days after denervation shows an AHP; (a3) same EDL muscle as in (a2) 10 min after application of 10 nM apamin. AHP is no longer visible. (b1) Rat myotubes co-cultured with nerve cells from spinal cord; (b2) nerve-free rat myotube in culture and (b3) same myotube 10 min after application of 10 nM apamin. (Reproduced with permission from reference 102.)

Delayed Rectifiers

Using the three-electrode technique, Adrian et al. found a delayed rectifier current that was activated with a delay upon membrane depolarization and rose more slowly than the sodium current.[6] This outward current inactivated fairly rapidly and completely during a prolonged depolarization. A second component, which was activated by depolarization but at a rate that was one or two orders of magnitude slower, was also identified.[107] Some years later Duval and Leoty used the double sucrose-gap voltage-clamp method to describe the delayed rectifier in slow and fast fibers from rat muscle.[108–110] They found that in slow fibers the delayed rectifier current inactivated with a time constant about 10 times larger than that seen in fast fibers. In both types there was a fast component in the outward current, whereas a large slower component could only be found in slow fibers.[109] Denervated fast fibers developed a slow component similar to that of slow innervated muscles.[111] Furthermore, denervation modified the sensitivity for the channel blockers 4-aminopyridine (4-AP) and TEA.[111] Different human muscle cells grown in culture expressed non-inactivating and inactivating delayed rectifiers.[112] Almers and colleagues, using the loose-patch voltage-clamp technique,[10] found in human and rat muscle that the potassium outward currents were variable and remarkably small compared to those of frog muscle.

Single-channel analysises of delayed rectifiers have been performed by Standen and colleagues[19,22,113] (Figure 14.2). These channels had a single-channel conductance around 15 pS, measured in a physiologic extracellular potassium concentration. The time course of summed records of many single-channel events closely resembled that of macroscopic currents, activating with a delay and inactivating subsequently under maintained depolarization.

Koren and colleagues[114] cloned a cDNA for a delayed rectifier from rat muscle, which was identical to a potassium channel from rat brain. These channels, expressed in a myoblast cell line, exhibited properties similar to the delayed rectifiers described by Standen et al.[19] Both channel types had unitary conductances of about 15 pS. In addition, single-channel current-voltage relations were similar[22,114] and both channel types were sensitive for externally applied TEA.[19,115] One main difference is the slower inactivation of the cloned channel type, with time constants of several seconds.[114]

In a more recent paper Matsubara and colleagues[115] cloned a second cDNA coding for a delayed rectifier potassium channel from rat skeletal muscle. The expression of these channels is developmentally regulated. The mRNA for this channel type is detected mainly in 4- and 12-week-old rats, but not in 6-month-old ones. The single-channel conductance of the cloned delayed rectifier is about 8 pS, much lower than that of delayed rectifiers found in adult frog skeletal muscle.[19] In addition, externally applied TEA blocks delayed rectifiers in adult frog muscle, whereas there is almost no block in cloned rectifiers. Therefore this cloned-delayed rectifier seems not to be identical with that expressed in adult frog muscle.

The potential dependence and kinetics of the delayed rectifiers reveal their function: they are responsible for the rapid repolarization during the falling phase of the action potential.[2,107]

Inward Rectifiers

As early as 1949 Katz[116] found that the potassium current in membranes of frog striated muscle showed rectification. With a muscle preparation placed in an isotonic potassium sulphate solution, he could demonstrate that the only ion capable of carrying a substantial current across the membrane was potassium. Despite the fact that the potassium concentration on both sides of the membrane was the same, he could measure a conductance about 20 times greater for inward as for outward current. This rectifying current was blocked by replacing potassium by rubidium.[117] Later work revealed the existence of an inward rectifier that inactivated with time.[107,109] With small hyperpolarizations, this decline of the current with time is due to depletion of potassium from the lumen of the transverse tubular system. With larger hyperpolarizations, a voltage-dependent decrease of the membrane permeability predominates.[107,118,119] The fall in potassium permeability was explained by the binding of sodium ions to sites within the inward rectifier, thereby blocking inward movement of potassium ions at negative potentials.[120,121]

The developmental changes of inward-rectifier currents in fibers of mouse skeletal muscle were studied by Gonoi and Hasegawa.[122] Two different components of inward-rectifying currents developed within 3 weeks after birth. Denervation of muscle fibers reduced the currents, which suggested that innervation plays a key role in their induction.

Ohmori and colleagues[123] described a single channel with a conductance of 10 pS as a candidate for the inward-rectifying current. Later work revealed the existence of a channel type of about 25 pS with characteristics similar to that of macroscopic currents.[20,120] Openings of these channels can only be detected at negative membrane potentials, using a magnesium-containing bath solution (Figure 14.1). Without magnesium, noisy outward current events can be detected. A channel with a similar conductance of 21 pS in a symmetrical potassium-containing solution without magnesium, which showed on rectification, has been described.[68] Therefore intracellular magnesium seems to be responsible for the rectification. Rubidium and cesium reduced the open probability,[20] whereas ATP, the potassium-channel openers cromakalim, RP 49356, EMD 52692, and the sulphonylurea glibenclamide had no effect.[68] The function of inward rectifiers in skeletal muscle is not very clear.[1] They clamp the membrane potential near the equilibrium potential of potassium,[124] they may serve for potassium reentry from potassium-loaded transverse tubules after an action potential, and they may prevent the loss of dangerous amounts of potassium into the blood.

SR Potassium Channels

So far I have described potassium channels, which can be found in the surface membrane and the tubular system of skeletal muscle, and which play an important role in controlling the action potentials and resting potential of the cell (Table 14.1).

Upon a depolarization of the transverse tubule membrane, the change in electrical potential causes the release of calcium stored in the sarcoplasmic reticulum (SR). This massive release of calcium must be compensated by other ions in order to avoid a charge imbalance. It has been suggested that the SR potassium channel may be a good candidate for this task.[15,125] Because of the almost complete inaccessibility of the sarco-plasmic reticulum membrane, it was difficult to obtain precise information on the conductance pathways in these membranes. Miller and Racker, however, developed a method for the fusion of isolated SR vesicles with an artificial planar bilayer membrane,[126] which allowed them to assay the incorporation of permeability pathways from the SR membrane by measuring the conductance of the artificial membrane. In further ex-periments Miller demonstrated that a voltage-dependent potassium con-ductance was present in SR membrane vesicles.[14] Under conditions that allowed only a small number of ionophores to enter the artificial mem-brane, he was able to measure single-channel currents of potassium channels. The precise structure of these channels was extensively inves-tigated by Miller and colleagues.[15] The unusual combination of a high conductance (Table 14.1) and high selectivity was explained with a highly localized active site within the channel. The short distance, where ion selectivity takes place, allows the conductance to remain high.

Recent work by Abramcheck and Best that used an optical technique with the calcium-sensitive dye antipyrylazo III supports the hypothesis that potassium ions flowing through SR potassium channels are important counter-ions for calcium during its release from the sarcoplasmic reticulum.[125]

Involvement of Potassium Channels in Selected Muscle Diseases

The following section describes selected muscle diseases to illustrate the pathophysiologic involvement of potassium channels in certain syndromes and gives future perspectives for a possible pharmacologic treatment of muscle diseases that activates potassium channels.

McArdle's Syndrome

Subjects suffering from McArdle's syndrome are unable to catabolize muscle glycogen.[127] During exercise these patients show a dramatic ele-vation of inorganic phosphate in the muscle that is associated with a decline in phosphocreatine.[40] The ATP concentration is only slightly altered, but the muscle phosphorylation potential is reduced[128] and may be responsible for opening the ATP-dependent potassium channels at the start of the exercise. As a consequence the potassium efflux increases, which leads to an abnormally high hyperkalemic response to a given workload.

As mentioned before, a reduction in intracellular pH, occurring during sustained muscular activity, results in enhanced activity of ATP-dependent potassium channels.[46] In contradiction to this observation, McArdle's subjects showed a progressive alkalosis in the blood during exercise, and presumably also in the muscle, which is associated with the greater hyperkalemic response.[38] Therefore intracellular pH seems not be the main coupling factor between cellular metabolism and K_{ATP}-channel activity in McArdle's subjects.

Hyperkalemic Periodic Paralysis

Hyperkalemic periodic paralysis is a autosomal-dominant disease, which typically begins in the first decade of life with attacks of muscle weakness induced by cold, resting after exercise, fasting, or ingestion of potassium.[129] Three clinical variants can be found, one with myotonia, one without myotonia, and a third one with paramyotonia.[130] Lehmann-Horn and colleagues described the pathophysiology of the variant with myotonia[131] and found that the potassium and chloride currents were normal in such fibers, but that a sodium current could be triggered by depolarizing the muscle fiber in a potential range only slightly below the resting value. This persistent sodium current seemed to be large enough to cause inexcitability but not large enough to reach the threshold for contracture. Indeed, single-channel studies revealed the existence of a small fraction of sodium channels that persistently reopened and did not inactivate in the presence of moderately elevated extracellular potassium.[132,133] The genetic defect responsible for this behavior seems to be located in the α subunit of the sodium channel.[129] Furthermore, this gene seems to be the site of a defect in another autosomal-dominant muscle disease, paramyotonia congenita.[134]

The potassium-channel opener cromakalim was able to repolarize otherwise irreversible depolarized fibers.[69] Figure 14.7 demonstrates the effect of cromakalim in fibers from a patient with hyperkalemic periodic paralysis and suggests that this drug could improve muscle contractile force. Indeed, cromakalim was shown to preferentially increase the twitch force of fiber bundles from a patient with hypokalemic periodic paralysis, but not that of control fibers.[71] Single-channel experiments demonstrated that cromakalim and some other potassium-channel openers were able to reopen K_{ATP} channels that were blocked by ATP applied to the cytoplasmic side of the membrane (Figure 14.8). In addition, in human skeletal muscle these openers may modify another ATP-independent channel with lower conductance.[68]

Hypokalemic Periodic Paralysis

Hypokalemic periodic paralysis is the oldest known and most frequent form of periodic paralysis.[135] Patients develop episodic muscle weakness associated with a reduced potassium serum concentration.[136] It has been suggested that a depolarized membrane potential is responsible for the inexcitability.

Kao and Gordon[136] used potassium-depleted rats as a model for this disease. They speculated that the inward rectifier is involved in hypokalemic periodic paralysis (Table 14.2). They found that insulin depolarized the skeletal muscle due to a decrease in the potassium conductance. The depolarization occurred only in low extracellular potassium solutions and not at the normal potassium concentration. This finding was explained by the inward-rectifying property of the resting conductance of potassium. At a low external potassium concentration the driving force for potassium is outward and resting potassium conductance becomes small. A further reduction of the potassium conductance by insulin depolarizes the membrane. However, using the same animal model, Otsuka and Ohtsuki[137] revealed that an increase of the sodium permeability is responsible for the depolarization. They concluded that insulin activated the $(Na^+ + K^+)$-ATPase (NKA), which is indeed the

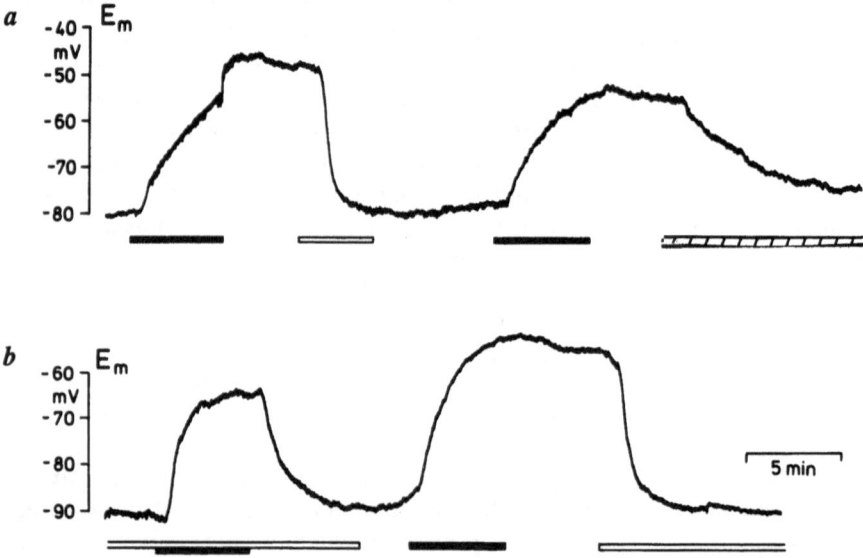

FIGURE 14.7. Membrane potential of fibers from a patient with hyperkalemic periodic paralysis. (a) Addition of 7 mM potassium to the normal bath solution resulted in a depolarization of the membrane (■■). Lowering the potassium concentration did not lead to a repolarization. Cromakalim (100 μM) restored the normal resting level (▭). Tetrodotoxin (TTX, 0.6 μM), a paralytic poison of some fishes of the order Tetraodontiformes[1] that blocks sodium channels with high specificity, also repolarized the membrane (▨). (b) When 7 mM potassium were added in the presence of cromakalim, no excessive membrane depolarization was detected. (Reproduced with permission of Springer-Verlag Inc. from reference 69.)

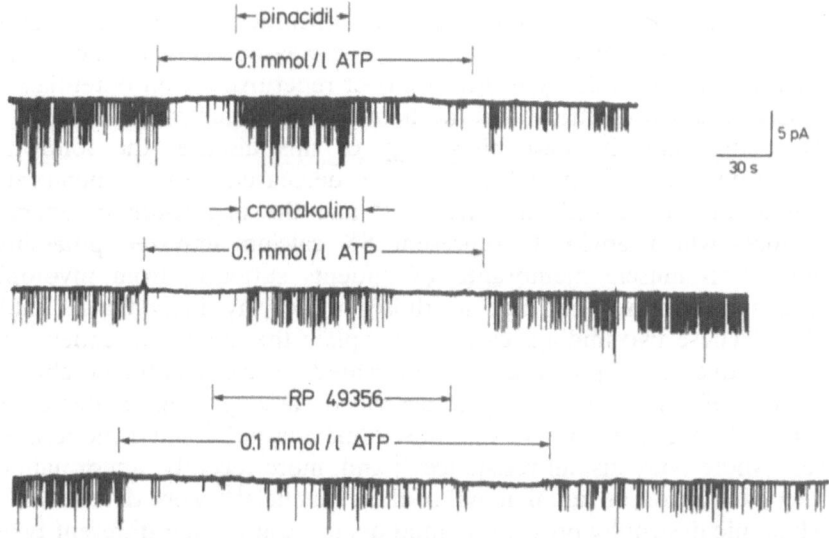

FIGURE 14.8. Effects of some potassium-channel openers on the open probability of single-channel currents from ATP-dependent potassium channels. Pinacidil (0.4 mM), cromakalim (0.2 mM), and RP 49356 (0.4 mM) were applied to the cytoplasmic side of the patch membrane in the presence of 0.1 mM ATP. It is clearly visible that the channels closed by ATP reopen in the presence of the different openers. Channel openings are plotted downwards. Membrane potential −50 mV. (Reproduced, with permission of Springer-Verlag Inc. from reference 67.)

case,[138] thereby further decreasing the lowered extracellular potassium concentration. This results in a reduced permeability for potassium. In parallel the permeability for sodium is increased. Both effects may be responsible for the depolarization of the membrane.

Rüdel and colleagues[135] also found an increased steady-state conductance of sodium in human skeletal muscle from patients with hypokalemic periodic paralysis. In these patients insulin was not a necessary factor for paralysis when the extracellular potassium concentration was low. In addition, these investigators could not detect an abnormally reduced potassium conductance under conditions of low extracellular potassium concentrations. These results are inconclusive at the present time, but it seems to be obvious that a single defect cannot account for all the described symptoms.

Myotonic Dystrophy

This dominantly inherited disease is characterized by a late onset and a steady progression of all symptoms, muscular dystrophy being the most severe one.[130] Myotonic stiffness and difficulties in relaxation after contractions are obviously due to an increased muscle excitability associated with a tendency to fire trains of repetitive action potentials in response to direct electrical and mechanical stimulation.[100]

To understand the pathophysiology of this disease, the following observations are of special interest: a decreased resting membrane potential of muscle cells in vitro[130,139] and the expression of apamin receptors (which obviously represent SK calcium-activated potassium channels) in muscle membranes of patients suffering from myotonic muscular dystrophy, but not in those of healthy individuals (Table 14.2).[100] These two findings can easily explain the increased tendency to fire repetitive action potentials by denervated muscle, mentioned above.

Other defects have also been described, among them a decreased number of sodium pumps in the muscle membrane[140] and a moderately severe whole body insulin resistance[141] and, more recently, abnormalities in the inactivation of sarcolemmal sodium channels[142] were detected.

Myotonic dystrophy provides a good demonstration that different types of potassium channels have different tasks in muscle membranes. The pathophysiology of this disease seems to be based at least in part on the fact that the normally absent SK channels are expressed in the muscle membrane. These channels, together with a reduced resting membrane potential, may be responsible for the myotonic symptoms. Another type of potassium channel, the ATP-dependent potassium channel, can be used as pharmacologic targets to stop contractions. Indeed, the potassium-channel openers cromakalim and EMD 52962 can open otherwise silent K_{ATP} channels and thereby suppress spontaneous contractions[70] that are due to a repolarization of the membrane.

To summarize, the increasing number of potassium-channel openers promise to become useful therapeutic agents for the reduction of symptoms associated with muscle diseases. Diseases associated with abnormalities of the sodium channel, such as hyperkalemic periodic paralysis and paramyotonia congenita, and disorders such as hypokalemic periodic paralysis, which seems to be associated with a reduced potassium conductance, can be treated with channel openers to repolarize the membrane and to improve muscle contraction force (Table 14.2). In addition, patients with myotonia congenita, which is due in part to a reduced chloride conductance that leads to the typical myotonic excitability, might be treated with potassium-channel openers to suppress myotonic activity.[70] The channel openers tested so far do not seem to be suitable for clinical application, because their affinity for K_{ATP} channels localized in skeletal muscle is much lower than for the K_{ATP} channels in other tissues.[69] Early clinical data have indeed revealed severe side effects.[143] The development of a new class of potassium-channel openers that preferentially act on K_{ATP} channels of skeletal muscle would be highly desirable.

Conclusions

In the last few years new concepts about the mechanisms responsible for the excitability of different tissues have emerged. Methods for measuring currents in the range of picoamperes made it possible to record openings and closings of single ion channels and to describe different classes of channels with typical physiologic and pharmacologic characteristics. This biophysical approach, combined with a variety of new techniques in molecular genetics,[144,145] has greatly increased our understanding of how ion channels are built and how they work.[146] Several different potassium-channel types of mammalian brain have been cloned and successfully expressed in oocytes.[147–150] The potassium currents measured in the expressed systems resemble both the transient and the delayed rectifier current.[149–151] Using a coding region probe specific for one type of potassium channel normally expressed in rat brain, Beckh and Pongs[4] found an mRNA species in skeletal muscle that might represent a species from another member of the already cloned RCK rat brain gene family. Philipson and co-workers[152] identified the sequence of a human fetal skeletal muscle potassium channel which, when expressed in oocytes, showed properties similar to transient potassium-type currents.[149,152] This channel type, as well as other cloned delayed rectifiers from rat skeletal muscle[114,115] are identical with certain types of brain potassium channels.

Information is now available about the role of specific amino acid residues in the binding of TEA[153] and in the activation[154] and inactivation[155] of potassium channels. A certain part of the channel protein, the segment between S5 and S6, was identified as the pore-forming sequence that is responsible for the ion-selectivity.[156–158] According to increasing evidence, the diversity of potassium channels can be partly explained by the assembly of different polypeptides that results in heteromultimeric molecules.[159,160] It can be expected that in the next few years we will have a greater understanding of the molecular structure of the different types of ion channels present in skeletal muscle. This knowledge should enable us to identify the sites of primary defects responsible for some muscle diseases. For instance, the calcium-release channel gene is thought to be a candidate for the predisposition to malignant hyperthermia.[161] An ultimate result of this effort might be the production of new classes of drugs, such as potassium-channel openers, which are directed against a specific receptor substructure and thereby provide a tool to specifically alter the excitability of muscle cells. Another approach might be the insertion of genes that are defective in certain diseases into skeletal muscle, which would lead to the expression of otherwise defective or missing protein products. This direct gene transfection has been successfully performed with dystrophin-deficient mice.[162]

References

1. Hille B. Ionic Channels of Excitable Membranes. Sunderland, Mass.: Sinauer; 1984.
2. Rudy B. Diversity and ubiquity of K channels. Neurosci 1988; 25:729–749.
3. Bernstein J. Untersuchungen zur Thermodynamik der bioelektrischen Ströme: Erster Theil. Pfluegers Arch 1902; 92:521–562.
4. Beckh S, Pongs O. Members of the RCK potassium channel family are differentially expressed in the rat nervous system. EMBO J 1990; 9:777–782.
5. Luneau CJ, Williams JB, Marshall J, Levitan ES, Oliva C, Smith JS, Antanavage J, Folander K, Stein RB, Swanson R, Kaczmarek LK, Buhrow SA. Alternative splicing contributes to K$^+$ channel diversity in the mammalian central nervous system. Proc Natl Acad Sci USA 1991; 88:3932–3936.
6. Adrian RH, Chandler WK, Hodgkin AL. Voltage clamp experiments in striated muscle fibres. J Physiol 1970; 208:607–644.
7. Ildefonse M, Rougier O. Voltage-clamp analysis of the early current in frog skeletal muscle fibre using the double sucrose-gap method. J Physiol 1972; 222:373–395.
8. Hille B, Campbell DT. An improved vaseline gap voltage clamp for skeletal muscle fibers. J Gen Physiol 1976; 67:265–293.
9. Frankenhaeuser B, Lindley BD, Smith RS. Potentiometric measurement of membrane action potentials in frog muscle fibres. J Physiol 1966; 183:152–166.
10. Almers W, Roberts WM, Ruff RL. Voltage clamp of rat and human skeletal muscle: Measurements with an improved loose-patch technique. J Physiol 1984; 347:751–768.
11. Neher E, Sakmann B. Single-channel currents recorded from membrane of denervated frog muscle fibres. Nature 1976; 260:799–802.
12. Hamill OP, Marty A, Neher E, Sakmann B, Sigworth FJ. Improved patch-clamp techniques for high-resolution current recording from cells and cell-free membrane patches. Pfluegers Arch 1981; 392:85–100.
13. Sakmann B, Neher E. Single-Channel Recording. New York: Plenum; 1983.
14. Miller C. Voltage-gated cation conductance channel from fragmented sarcoplasmic reticulum: Steady-state electrical properties. J Membr Biol 1978; 40:1–23.
15. Miller C, Bell JE, Garcia AM. The potassium channel of sarcoplasmic reticulum. Curr Top Membr Transp 1984; 21:99–132.
16. Tank DW, Miller C. Patch-clamped liposomes. In: Sakmann B, Neher E, eds. Single-Channel Recording. New York: Plenum; 1983:91–105.
17. Pallotta BS, Magleby KL, Barrett JN. Single-channel recordings of Ca^{2+}-activated K$^+$ currents in rat muscle cell culture. Nature 1981; 293:471–474.
18. Latorre R, Vergara C, Hidalgo C. Reconstitution in planar lipid bilayers of a Ca^{2+}-dependent K$^+$ channel from transverse tubule membranes isolated from rabbit skeletal muscle. Proc Natl Acad Sci USA 1982; 79:805–809.
19. Standen NB, Stanfield PR, Ward TA. Properties of single potassium channels in vesicles formed from the sarcolemma of frog skeletal muscle. J Physiol 1985; 364:339–358.

20. Burton F, Dörstelmann U, Hutter OF. Single-channel activity in sarcolemmal vesicles from human and other mammalian muscles. Muscle Nerve 1988; 11:1029–1038.

21. Weik R, Neumcke B. ATP-sensitive potassium channels in adult mouse skeletal muscle: Characterization of the ATP-binding site. J Membr Biol 1989; 110:217–226.

22. Standen NB, Stanfield PR, Ward TA, Wilson SW. A new preparation for recording single-channel currents from skeletal muscle. Proc R Soc Lond 1984; B221:455–464.

23. Blatz AL, Magleby KL. Single apamin-blocked Ca-activated K^+ channels of small conductance in cultured rat skeletal muscle. Nature 1986; 323:718–720.

24. Hugues M, Schmid H, Romey G, Duval D, Frelin C, Lazdunski M. The Ca^{2+}-dependent slow K^+ conductance in cultured rat muscle cells: Characterization with apamin. EMBO J 1982; 1:1039–1042.

25. Romey G, Lazdunski M. The coexistence in rat muscle cells of two distinct classes of Ca^{2+}-dependent K^+ channels with different pharmacological and different physiological functions. Biochem Biophys Res Commun 1984; 118:669–674.

26. Trimmer JS, Cooperman SS, Tomiko SA, Zhou J, Crean SM, Boyle MB, Kallen RG, Sheng Z, Barchi RL, Sigworth FJ, Goodman RH, Agnew WS, Mandel G. Primary structure and functional expression of a mammalian skeletal muscle sodium channel. Neuron 1989; 3:33–49.

27. Catterall WA. Excitation-contraction coupling in vertebrate skeletal muscle: A tale of two calcium channels. Cell 1991; 64:871–874.

28. Franciolini F, Petris A. Chloride channels of biological membranes. Biochim Biophys Acta 1990; 1031:247–259.

29. Bretag AH. Muscle chloride channels. Physiol Rev 1987; 67:618–724.

30. Spruce AE, Standen NB, Stanfield PR. Studies of the unitary properties of adenosine-5'-triphosphate-regulated potassium channels of frog skeletal muscle. J Physiol 1987; 382:213–236.

31. Davies NW, Standen NB, Stanfield PR. ATP-dependent potassium channels of muscle cells: Their properties, regulation, and possible functions. J Bioenergetics and Biomembranes 1991; 23:509–535.

32. Fink R, Lüttgau HC. An evaluation of the membrane constants and the potassium conductance in metabolically exhausted muscle fibres. J Physiol 1976; 263:215–238.

33. Fink R, Hase S, Lüttgau HCh, Wettwer E. The effect of cellular energy reserves and internal calcium ions on the potassium conductance in skeletal muscle of the frog. J Physiol 1983; 336:211–228.

34. Castle NA, Haylett DG. Effect of channel blockers on potassium efflux from metabolically exhausted frog skeletal muscle. J Physiol 1987; 383:31–43.

35. Sauviat M-P, Ecault E, Faivre J-F, Findlay I. Activation of ATP-sensitive K channels by a K channel opener (SR 44866) and the effect upon electrical and mechanical activity of frog skeletal muscle. Pfluegers Arch 1991; 418:261–265.

36. Bigland-Ritchie B, Woods JJ. Changes in muscle contractile properties and neural control during human muscular fatigue. Muscle Nerve 1984; 7:691–699.

37. Paterson DJ, Robbins PA, Conway J. Changes in arterial plasma potassium and ventilation during exercise in man. Resp Physiol 1989; 79:323–330.

38. Paterson DJ, Friedland JS, Bascom DA, Clament ID, Cunningham DA, Painter R, Robbins PA. Changes in arterial K^+ and ventilation during exercise in normal subjects and subjects with McArdle's syndrome. J Physiol 1990; 429:339–348.

39. Ball-Burnett M, Green HJ, Houston ME. Energy metabolism in human slow and fast twitch fibres during prolonged cycle exercise. J Physiol 1991; 437:257–267.

40. Lewis SF, Haller RG, Cook JD, Nunnally RL. Muscle fatigue in McArdle's disease studied by ^{31}P-NMR: Effect of glucose infusion. J Applied Physiol 1985; 59:1991–1994.

41. Dawson MJ, Gadian DG, Wilkie DR. Muscular fatigue investigated by phosphorus nuclear magnetic resonance. Nature 1978; 274:861–866.

42. Dawson MJ, Gadian DG, Wilkie DR. Mechanical relaxation rate and metabolism studied in fatiguing muscle by phosphorus nuclear magnetic resonance. J Physiol 1980; 299:465–484.

43. Neumcke B, Weik R. Vanadate as an activator of ATP-sensitive potassium channels in mouse skeletal muscle. Eur Biophys J 1991; 19:119–123.

44. Kakei M, Kelly RP, Ashcroft SJH, Ashcroft FM. The ATP-sensitivity of K^+ channels in rat pancreatic B-cells is modulated by ADP. FEBS Lett 1986; 208:63–66.

45. Misler S, Falke LC, Gillis K, McDaniel ML. A metabolite-regulated potassium channel in rat pancreatic B cells. Proc Natl Acad Sci USA 1986; 83:7119–7123.

46. Davies NW. Modulation of ATP-sensitive K^+ channels in skeletal muscle by intracellular protons. Nature 1990; 343:375–377.

47. Parent L, Coronado R. Reconstitution of the ATP-sensitive potassium channel of skeletal muscle. J Gen Physiol 1989; 94:445–463.

48. Cook DL, Hales CN. Intracellular ATP directly blocks K^+ channels in pancreatic B-cells. Nature 1984; 311:271–273.

49. Noma A. ATP-regulated K^+ channels in cardiac muscle. Nature 1983; 305:147–148.

50. Standen NB, Quayle JM, Davies NW, Brayden JE, Huang Y, Nelson MT. Hyperpolarizing vasodilators activate ATP-sensitive K^+ channels in arterial smooth muscle. Science 1989; 245:177–180.

51. Ashcroft SJH, Ashcroft FM. Properties and functions of ATP-sensitive K-channels. Cellular Signalling 1990; 2:197–214.

52. Ashcroft FM. Adenosine 5'-triphosphate-sensitive potassium channels. Ann Rev Neurosci 1988; 11:97–118.

53. de Weille JR, Lazdunski M. Regulation of the ATP-sensitive potassium channel. In: Narahashi T, ed. Ion Channels, Volume 2. New York: Plenum; 1990:205–222.

54. Petersen OH. Control of potassium channels in insulin-secreting cells. ISI Atlas of Science: Biochemistry 1988; 1:144–149.

55. Trube G, Rorsman P, Ohno-Shosaku T. Opposite effects of tolbutamide and diazoxide on the ATP-dependent K^+ channel in mouse pancreatic β-cells. Pfluegers Arch 1986; 407:493–499.

56. Sturgess NC, Ashford MLJ, Cook DL, Hales CN. The sulphonylurea receptor may be an ATP-sensitive potassium channel. Lancet 1985; 8453: 474–475.
57. Woll KH, Lönnendonker U, Neumcke B. ATP-sensitive potassium channels in adult mouse skeletal muscle: Different modes of blockage by internal cations, ATP and tolbutamide. Pfluegers Arch 1989; 414:622–638.
58. Belles B, Hescheler J, Trube G. Changes of membrane current in cardiac cells induced by long whole-cell recordings and tolbutamide. Pfluegers Arch 1987; 409:582–588.
59. Hamilton TC, Weir SW, Weston AH. Comparison of the effects of BRL 34915 and verapamil on electrical and mechanical activity in rat portal vein. Br J Pharmacol 1986; 88:103–111.
60. Nelson MT, Huang Y, Brayden JE, Hescheler J, Standen NB. Arterial dilations in response to calcitonin gene-related peptide involve activation of K^+ channels. Nature 1990; 344:770–773.
61. Cook NS. The pharmacology of potassium channels and their therapeutical potential. Trends Pharmacol Sci 1988; 9:21–28.
62. Hamilton TC, Weston AH. Cromakalim, nicorandil and pinacidil: Novel drugs which open potassium channels in smooth muscle. Gen Pharmac 1989; 20:1–9.
63. Quast U, Cook NS. Moving together: K^+ channel openers and ATP-sensitive K^+ channels. Trends Pharmacol Sci 1989; 10:431–435.
64. Weston AH. Smooth muscle K^+ channel openers: Their pharmacology and clinical potential. Pfluegers Arch 1989; 414(suppl 1):S99–S105.
65. Duty S, Weston AH. Potassium channel openers: Pharmacological effects and future uses. Drugs 1990; 40:785–791.
66. Edwards G, Weston AH. Structure-activity relationships of K^+ channel openers. Trends Pharmacol Sci 1990; 11:417–422.
67. Weik R, Neumcke B. Effects of potassium channel openers on single potassium channels in mouse skeletal muscle. Naunyn-Schmiedeberg's Arch Pharmacol 1990; 342:258–263.
68. Quasthoff S, Franke C, Hatt H, Richter-Turtur M. Two different types of potassium channels in human skeletal muscle activated by potassium channel openers. Neurosci Lett 1990; 119:191–194.
69. Spuler A, Lehmann-Horn F, Grafe P. Cromakalim (BRL 34915) restores in vitro the membrane potential of depolarized human skeletal fibres. Naunyn-Schmiedeberg's Arch Pharmacol 1989; 339:327–331.
70. Quasthoff S, Spuler A, Spittelmeister W, Lehmann-Horn F, Grafe P. K^+ channel openers suppress myotonic activity of human skeletal muscle in vitro. Eur J Pharmacol 1990; 186:125–128.
71. Grafe P, Quasthoff S, Strupp M, Lehmann-Horn F. Enhancement of K^+ conductance improves in vitro the contraction force of skeletal muscle in hypokalemic periodic paralysis. Muscle Nerve 1990; 13:451–457.
72. McManus OB. Calcium-activated potassium channels: Regulation by calcium. J Bioenergetics and Biomembranes 1991; 23:537–560.
73. Behrens MI, Vergara C, Latorre R. Calcium-activated potassium channels of large unitary conductance. Brazilian J Med Biol Res 1988; 21:1101–1117.
74. Latorre R, Oberhauser A, Labarca P, Alvarez O. Varieties of calcium-activated potassium channels. Annu Rev Physiol 1989; 51:385–399.

75. Gardos G. The function of calcium in the potassium permeability of human erythrocytes. Biochim Biophys Acta 1958; 30:653–654.
76. Marty A. Ca-dependent K channels with large unitary conductance in chromaffin cell membranes. Nature 1981; 291:497–500.
77. Marty A. Ca^{2+}-dependent K^+ channels with large unitary conductance. Trends Neurosci 1983; 6:262–265.
78. Latorre R, Miller C. Conductance and selectivity in potassium channels. J Membr Biol 1983; 71:11–30.
79. Barrett JN, Magleby KL, Pallotta BS. Properties of single calcium-activated potassium channels in cultured rat muscle. J Physiol 1982; 331:211–230.
80. Miller C, Moczydlowski E, Latorre R, Phillips M. Charybdotoxin, a protein inhibitor of single Ca^{2+}-ativated K^+ channels from mammalian skeletal muscle. Nature 1985; 313:316–318.
81. Sands SB, Lewis RS, Cahalan MD. Charybdotoxin blocks voltage-gated K^+ channels in human and murine T lymphocytes. J Gen Physiol 1989; 93:1061–1074.
82. Hermann A, Erxleben C. Charybdotoxin selectively blocks small Ca-activated K channels in *Aplsyia* neurons. J Gen Physiol 1987; 90:27–47.
83. Castle NA, Strong PN. Identification of two toxins from scorpion (*Leiurus quinquestriatus*) venom which block distinct classes of calcium-activated potassium channel. FEBS Lett 1986; 209:117–121.
84. Schweitz H, Stansfeld CE, Bidard J-N, Fagni L, Maes P, Lazdunski M. Charybdotoxin blocks dendrotoxin-sensitive voltage-activated K^+ channels. FEBS Lett 1989; 250:519–522.
85. Schneider MJ, Rogowski RS, Krueger BK, Blaustein MP. Charybdotoxin blocks both Ca-activated K channels and Ca-independent voltage-gated K channels in rat brain synaptosomes. FEBS Lett 1989; 250:433–436.
86. MacKinnon R, Reinhart PH, White MM. Charybdotoxin block of *Shaker* K^+ channels suggests that different types of K^+ channels share common structural features. Neuron 1988; 1:997–1001.
87. MacKinnon R, Miller C. Mutant potassium channels with altered binding of charybdotoxin, a pore-blocking peptide inhibitor. Science 1989; 245:1382–1385.
88. Massefski W Jr, Redfield AG, Hare DR, Miller C. Molecular structure of charybdotoxin, a pore-directed inhibitor of potassium channels. Science 1990; 249:521–524.
89. Galvez A, Gimenez-Gallego G, Reuben JP, Roy-Contancin L, Feigenbaum P, Kaczorowski GJ, Garcia ML. Purification and characterization of a unique, potent, peptidyl probe for the high conductance calcium-activated potassium channel from venom of the scorpion *Buthus tamulus*. J Biol Chem 1990; 265:11083–11090.
90. Garcia ML, Galvez A, Garcia-Calvo M, King VF, Vazquez J, Kaczorowski GJ. Use of toxins to study potassium channels. J Bioenergetics and Biomembranes 1991; 23:615–646.
91. Blatz AL, Magleby KL. Calcium-activated potassium channels. Trends Neurosci 1987; 10:463–467.
92. Atkinson NS, Robertson GA, Ganetzky B. A component of calcium-activated potassium channels encoded by the *Drosophila slo* locus. Science 1991; 253:551–555.

93. Villarroel A, Alvarez O, Oberhauser A, Latorre R. Probing a Ca^{2+}-activated K^+ channel with quaternary ammonium ions. Pfluegers Arch 1988; 413:118–126.

94. Blatz AL, Magleby KL. Ion conductance and selectivity of single calcium-activated potassium channels in cultured rat muscle. J Gen Physiol 1984; 84:1–23.

95. Moczydlowski E, Lucchesi K, Ravindran A. An emerging pharmacology of peptide toxins targeted against potassium channels J Membr Biol 1988; 105:95–111.

96. Castle NA, Haylkett DG, Jenkinson DH. Toxins in the characterization of potassium channels. Trends Neurosci 1989; 12:59–65.

97. Hugues M, Romey G, Duval D, Vincent JP, Lazdunski M. Apamin as a selective blocker of the calcium-dependent potassium channel in neuroblastoma cells: Voltage-clamp and biochemical characterization of the toxin receptor. Proc Natl Acad Sci USA 1982; 79:1308–1312.

98. Leveque C, Marqueze B, Couraud F, Seagar M. Polypeptide components of the apamin receptor associated with a calcium activated potassium channel. FEBS Lett 1990; 275:185–189.

99. Barrett JN, Barrett EF, Dribin LB. Calcium-dependent slow potassium conductance in rat skeletal myotubes. Develop Biol 1981; 82:258–266.

100. Renaud J-F, Desnuelle C, Schmid-Antomarchi H, Hugues M, Serratrice G, Lazdunski M. Expression of apamin receptor in muscles of patients with myotonic muscular dystrophy. Nature 1986; 319:678–680.

101. Thesleff S, Ward MR. Studies on the mechanism of fibrillation potentials in denerved muscle. J Physiol 1975; 244:313–323.

102. Schmid-Antomarchi H, Renaud J-F, Romey G, Hugues M, Schmid A, Lazdunski M. The all-or-none role of innervation in expression of apamin receptor and of apamin-sensitive Ca^{2+}-activated K^+ channel in mammalian skeletal muscle. Proc Natl Acad Sci USA 1985; 82:2188–2191.

103. Fosset M, Schmid-Antomarchi H, Hugues M, Romey G, Lazdunski M. The presence in pig brain of an endogenous equivalent of apamin, the bee venom peptide that specifically blocks Ca^{2+}-dependent K^+ channels. Proc Natl Acad Sci USA 1984; 81:7228–7232.

104. Suarez-Isla BA, Cosgrove JW, Thompson JM, Rapoport SI. A soluble factor (<4000 Da) from chick spinal cord blocks slow hyperpolarizing afterpotentials in cultured rat muscle cells. Develop Brain Res 1986; 30:274–277.

105. Lombet A, Kazazoglou T, Delpont E, Renaud J-F, Lazdunski M. Ontogenic appearance of Na^+ channels characterized as high affinity binding sites for tetrodotoxin during development of the rat nervous and skeletal muscle system. Biochem Biophys Res Commun 1983; 110:894–901.

106. Kazazoglou T, Schmid A, Renaud J-F, Lazdunski M. Ontogenic appearance of Ca^{2+} channels characterized as binding sites for nitrendipine during development of nervous, skeletal and cardiac muscle systems in the rat. FEBS Lett 1983; 164:75–79.

107. Adrian RH, Chandler WK, Hodgkin AL. Slow changes in potassium permeability in skeletal muscle. J Physiol 1970; 208:645–668.

108. Duval A, Leoty C. Ionic currents in mammalian fast skeletal muscle. J Physiol 1978; 278:403–423.

109. Duval A, Leoty C. Ionic currents in slow twitch skeletal muscle. J Physiol 1980; 307:23–41.
110. Duval A, Leoty C. Comparison between the delayed outward current in slow and fast twitch skeletal muscle in the rat. J Physiol 1980; 307:43–57.
111. Duval A, Leoty C. Changes in the ionic currents sensitivity to inhibitors in twitch rat skeletal muscles following denervation. Pfluegers Arch 1985; 403:407–414.
112. Trautmann A, Delaporte C, Marty A. Voltage-dependent channels of human muscle cultures. Pfluegers Arch 1986; 406:163–172.
113. Spruce AE, Standen NB, Stanfield PR. Rubidium ions and the gating of delayed rectifier potassium channels of frog skeletal muscle. J Physiol 1989; 411:597–610.
114. Koren G, Liman ER, Logothetis DE, Nadal-Ginard B, Hess P. Gating mechanism of a cloned potassium channel expressed in frog oocytes and mammalian cells. Neuron 1990; 2:39–51.
115. Matsubara H, Liman ER, Hess P, Koren G. Pretranslational mechanisms determine the type of potassium channels expressed in the rat skeletal and cardiac muscles. J Biol Chem 1991; 266:13324–13328.
116. Katz B. Les constantes electriques de la membrane du muscle. Arch Sci Physiol 1949; 3:285–300.
117. Adrian RH. Rectification in muscle membrane. Prog Biophys Mol Biol 1969; 19:340–369.
118. Almers W. Potassium conductance changes in skeletal muscle and the potassium concentration in the transverse tubules. J Physiol 1972; 225:33–56.
119. Almers W. The decline of potassium permeability during extreme hyperpolarization in frog skeletal muscle. J Physiol 1972; 225:57–83.
120. Matsuda H, Stanfield PR. Single inwardly rectifying potassium channels in cultured muscle cells from rat and mouse. J Physiol 1989; 414:111–124.
121. Standen NB, Stanfield PR. Potassium depletion and sodium block of potassium currents under hyperpolarization in frog sartorius muscle. J Physiol 1979; 294:497–520.
122. Gonoi T, Hasegawa S. Postnatal induction and neural regulation of inward rectifiers in mouse skeletal muscle. Pfluegers Arch 1991; 418:601–607.
123. Ohmori H, Yoshida S, Hagiwara S. Single K^+ channel currents of anomalous rectification in cultured rat myotubes. Proc Natl Acad Sci USA 1981; 78:4960–4964.
124. Hodgkin AL, Horowicz P. The influence of potassium and chloride ions on the membrane potential of single muscle fibres. J Physiol 1959; 148:127–160.
125. Abramcheck CW, Best PM. Physiological role and selectivity of the in situ potassium channel of the sarcoplamic reticulum in skinned frog skeletal muscle fibers. J Gen Physiol 1989; 93:1–21.
126. Miller C, Racker E. Ca^{++}-induced fusion of fragmented sarcoplasmic reticulum with artificial planar bilayers. J Membr Biol 1976; 30:283–300.
127. McArdle B. Myopathy due to a defect in muscle glycogen breakdown. Clin Sci 1951; 10:13–33.
128. Lewis SF, Haller RG. The pathophysiology of McArdle's disease: Clues to regulation in exercise and fatigue. J Applied Physiol 1986; 61:391–401.

129. Fontaine B, Khurana TS, Hoffman EP, Bruns GA, Haines JL, Trofatter JA, Hanson MP, Rich J, McFarlane H, Yasek DMcK, Romano D, Gusella JF, Brown RH Jr. Hyperkalemic periodic paralysis and the adult muscle sodium channel α-subunit gene. Science 1990; 250:1000–1002.
130. Rüdel R, Lehmann-Horn F. Membrane changes in cells from myotonia patients. Physiol Rev 1985; 65:310–356.
131. Lehmann-Horn F, Küther G, Ricker K, Grafe P, Ballanyi K, Rüdel R. Adynamia episodica hereditaria with myotonia: A non-inactivating sodium current and the effect of extracellular pH. Muscle Nerve 1987; 10:363–374.
132. Cannon SC, Brown RH Jr, Corey DP. A sodium channel defect in hyperkalemic periodic paralysis: Potassium-induced failure of inactivation. Neuron 1991; 6:619–626.
133. Lehamann-Horn F, Iaizzo PA, Hatt H, Franke Ch. Altered gating and conductance of Na+ channels in hyperkalemic periodic paralysis. Pfluegers Arch 1991; 418:297–299.
134. Ptacek LJ, Trimmer JS, Agnew WS, Roberts JR, Petajan JH, Leppert M. Paramyotonia congenita and hyperkalemic periodic paralysis map to the same sodium channel locus. Neurology 1991; 41:1163.
135. Rüdel R, Lehmann-Horn F, Ricker K, Küther G. Hypokalemic periodic paralysis: In vitro investigation of muscle fiber membrane parameters. Muscle Nerve 1984; 7:110–120.
136. Kao I, Gordon AM. Mechanism of insulin-induced paralysis of muscles from potassium-depleted rats. Science 1975; 188:740–741.
137. Otsuka M, Ohtsuki I. Mechanism of muscular paralysis by insulin with special reference to periodic paralysis. Am J Physiol 1970; 219:1178–1182.
138. Flatman A, Clausen T. Combined effects of adrenaline and insulin on active electrogenic Na+-K+ transport in rat soleus muscle. Nature 1979; 281:580–581.
139. Merickel M, Gray R, Chauvin P, Appel S. Cultured muscle from myotonic muscular dystrophy patients: Altered membrane electrical properties. Proc Natl Acad Sci USA 1981; 78:648–652.
140. Desnuelle C, Lombert A, Serratrice G, Lazdunski M. Sodium channel and sodium pump in normal and pathological muscles from patients with myotonic muscular dystrophy and lower motor neuron impairment. J Clin Invest 1982; 69:358–367.
141. Moxley RT, Corbett AJ, Minaker KL, Rowe JW. Whole body insulin resistance in myotonic dystrophy. Ann Neurol 1984; 15:157–162.
142. Franke Ch, Hatt H, Iaizzo PA, Lehmann-Horn F. Characteristics of Na+ channels and Cl− conductance in resealed muscle fibre segments from patients with myotonic dystrophy. J Physiol 1990; 425:391–405.
143. Spittelmeister W, Quasthoff S, Grafe P, Lehmann-Horn F. Hypokalemic periodic paralysis: In vitro and clinical data about the K+ channel openers pinacidil. J Neurol Sci 1990; 98(suppl):370.
144. Miller C. Genetic manipulation of ion channels: A new approach to structure and mechanism. Neuron 1989; 2:1195–1205.
145. Catterall WA. Structure and function of voltage-sensitive ion channels. Science 1988; 242:50–61.
146. Miller C. 1990: Annus mirabilis of potassium channels. Science 1991; 252:1092–1096.

147. Baumann A, Grupe A, Ackermann A, Pongs O. Structure of the voltage-dependent potassium channel is highly conserved from *Drosophila* to vertebrate central nervous system. EMBO J 1988; 7:2457–2463.
148. Stühmer W, Stocker M, Sakmann B, Seeburg P, Baumann A, Grup A, Pongs O. Potassium channels expressed from rat brain cDNA have delayed rectifier properties. FEBS Lett 1988; 242:199–206.
149. Stühmer W, Ruppersberg JP, Schröter KH, Sakmann B, Stocker M, Giese KP, Perschke A, Baumann A, Pongs O. Molecular basis of functional diversity of voltage-gated potassium channels in mammalian brain. EMBO J 1989; 8:3235–3244.
150. Christie MJ, Adelman JP, Douglass J, North RA. Expression of a cloned rat brain potassium channel in *Xenopus* oocytes. Science 1989; 244:221–224.
151. Schröter K-H, Ruppersberg JP, Wunder F, Rettig J, Stocker M, Pongs O. Cloning and functional expression of a TEA-sensitive A-type potassium channel from rat brain. FEBS Lett 1991; 278:211–216.
152. Philipson LH, Schaefer K, LaMendola J, Bell IG, Steiner DF. Sequence of a human fetal skeletal muscle potassium channel cDNA related to RCK4. Nucleic Acids Res 1990; 18:7160.
153. MacKinnon R, Yellen G. Mutations affecting TEA blockade and ion permeation in voltage-activated K^+ channels. Science 1990; 250:276–279.
154. Papazian DM, Timpe LC, Jan YN, Jan LY. Alteration of voltage-dependence of *Shaker* potassium channel by mutations in the S4 sequence. Nature 1991; 349:305–310.
155. Hoshi T, Zagotta WN, Aldrich RW. Biophysical and molecular mechanisms of *Shaker* potassium channel inactivation. Science 1990; 250:533–538.
156. Yool AJ, Schwarz TL. Alteration of ionic selectivity of a K^+ channel by mutation of the H5 region. Nature 1991; 349:700–704.
157. Hartmann HA, Kirsch GE, Drewe JA, Taglialatela M, Joho RH, Brown AM. Exchange of conduction pathways between two related K^+ channels. Science 1991; 251:942–944.
158. Yellen G, Jurman ME, Abramson T, MacKinnon R. Mutations affecting internal TEA blockade identify the probable pore-forming region of a K^+ channel. Science 1991; 251:939–942.
159. Isacoff EY, Jan YN, Jan LY. Evidence for the formation of heteromultimeric potassium channels in *Xenopus* oocytes. Nature 1990; 345:530–534.
160. Ruppersberg JP, Schröter KH, Sakmann B, Stocker M, Sewing S, Pongs O. Heteromultimeric channels formed by rat brain potassium-channel proteins. Nature 1990; 345:535–537.
161. MacLennan DH. Molecular tools to elucidate problems in excitation-contraction coupling. Biophysic J 1990; 58:1355–1365.
162. Acsadi G, Dickson G, Love DR, Jani A, Walsh FS, Gurusinghe A, Wolff JA, Davies KE. Human dystrophin expression in mdx mice after intramuscular injection of DNA constructs. Nature 1991; 352:815–818.
163. Spruce AE, Standen NB, Stanfield PR. Voltage-dependent ATP-sensitive potassium channels of skeletal muscle membrane. Nature 1985; 316:736–738.
164. Davies NW, Spruce AE, Standen NB, Stanfield PR. Multiple blocking mechanisms of ATP-sensitive potassium channels of frog skeletal muscle by tetraethylammonium ions. J Physiol 1989; 413:31–48.

165. Quayle JM, Standen NB, Stanfield PR. The voltage-dependent block of ATP-sensitive potassium channels of frog skeletal muscle by caesium and barium ions. J Physiol 1988; 405:677–697.
166. Davies NW, Pettit AI, Agarwal R, Standen NB. The flickery block of ATP-dependent potassium channels of skeletal muscle by internal 4-aminopyridine. Pfluegers Arch 1991; 419:25–31.
167. Stein P, Palade P. Patch clamp of sarcolemmal spheres from stretched skeletal muscle fibers. Am J Physiol 1989; 256:C434–C440.
168. Weik R, Lönnendonker U. Polymyxin B as a highly effective gating modifier of high-conductance Ca^{2+}-activated K^+ channels in mouse skeletal muscle. Pfluegers Arch 1990; 415:671–677.
169. Oberhauser A, Alvarez O, Latorre R. Activation by divalent cations of a Ca^{2+}-activated K^+ channel from skeletal muscle membrane. J Gen Physiol 1988; 92:67–86.
170. Pallotta BS. N-Bromoacetamide removes a calcium-dependent component of channel opening from calcium-activated potassium channels in rat skeletal muscle. J Gen Physiol 1985; 86:601–611.
171. MacKinnon R, Latorre R, Miller C. Functional modification of a Ca^{2+}-activated K^+ channel by trimethyloxonium. Biochemistry 1989; 28:8087–8092.
172. MacKinnon R, Latorre R, Miller C. Role of surface electrostatics in the operation of a high-conductance Ca^{2+}-activated K^+ channel. Biochemistry 1989; 28:8092–8099.
173. Vergara C, Latorre R. Kinetics of Ca^{2+}-activated K^+ channels from rabbit muscle incorporated into planar bilayers. J Gen Physiol 1983; 82:543–568.
174. Miller C, Latorre R, Reisin I. Coupling of voltage-dependent gating and Ba^{++} block in the high conductance, Ca^{++}-activated K^+ channel. J Gen Physiol 1987; 90:427–449.
175. Neyton J, Miller C. Potassium blocks barium permeation through a calcium-activated potassium channel. J Gen Physiol 1988; 92:549–567.
176. Neyton J, Miller C. Discrete Ba^{2+} block as a probe of ion occupancy and pore structure in the high-conductance Ca^{2+}-activated K^+ channel. J Gen Physiol 1988; 92:569–586.
177. Miller C. Competition for block of a Ca^{2+}-activated K^+ channel by charybdotoxin and tetraethylammonium. Neuron 1988; 1:1003–1006.
178. Lucchesi KJ, Moczydlowski E. On the interaction of bovine pancreatic trypsin inhibitor with maxi Ca^{2+}-activated K^+ channels. J Gen Physiol 1991; 97:1295–1319.
179. Lucchesi KJ, Moczydlowski E. Subconductance behavior in a maxi Ca^{2+}-activated K^+ channel induced by dendrotoxin-I. Neuron 1990; 2:141–148.
180. Ferguson WB. Competitive Mg^+ block of a large-conductance, Ca^{2+}-activated K^+ channel in rat skeletal muscle. J Gen Physiol 1991; 98:163–181.
181. Burton FL, Hutter OF. Sensitivity to flow of intrinsic gating in inwardly rectifying potassium channel from mammalian skeletal muscle. J Physiol 1990; 424:253–261.
182. Coronado R, Rosenberg RL, Miller C. Ionic selectivity, saturation, and block in a K^+-selective channel from sarcoplasmic reticulum. J Gen Physiol 1980; 76:425–446.

183. Coronado R, Miller C. Voltage-dependent caesium blockade of a cation channel from fragmented sarcoplasmic reticulum. Nature 1979; 280:807–810.
184. Garcia AM, Miller C. Channel-mediated monovalent cation fluxes in isolated sarcoplasmic reticulum vesicles. J Gen Physiol 1984; 83:819–839.
185. Coronado R, Miller C. Decamethonium and hexamethonium block K^+ channels of sarcoplasmic reticulum. Nature 1980; 288:495–497.
186. Miller C, Rosenberg RL. A voltage-gated cation conductance channel from fragmented sarcoplasmic reticulum. Effects of transition metal ions. Biochemistry 1979; 18:1138–1145.
187. Coronado R, Miller C. Conduction and block by organic cations in a K^+-selective channel from sarcoplasmic reticulum incorporated into planar phospholipid bilayers. J Gen Physiol 1982; 79:529–547.
188. Labarca PP, Miller C. A K^+-selective, three-state channel from fragmented sarcoplasmic reticulum of frog leg muscle. J Membr Biol 1981; 61:31–38.
189. Franke CH, Iaizzo PA, Hatt H, Spittelmeister W, Ricker K, Lehmann-Horn F. Altered Na^+ channel activity and reduced Cl^- conductance cause hyperexcitability in recessive generalized myotonia (Becker). Muscle Nerve 1991; 14:762–770.
190. Rüdel R, Ricker K, Lehmann-Horn F. Transient weakness and altered membrane characteristic in recessive generalized myotonia (Becker). Muscle Nerve 1988; 11:202–211.

15
The Role of Potassium Ions in the Control of Heart Function

Fritz Markwardt

The rhythmic contractions of the heart are controlled by periodic changes of the membrane potential (E_m) of cardiac myocytes, called action potentials. The cardiac action potential consists of five phases. Phase 0, which lasts only a few milliseconds, is the time of rapid depolarization. A short and small repolarization (phase 1) is followed by a long (about 100 to 400 ms) plateau at a depolarized level (phase 2). The repolarization of the plateau potential is called phase 3. The final phase 4 continues to the next rapid depolarization. The action potential is the result of a concerted action of inward (depolarizing) and outward (hyperpolarizing) ionic currents. The outward component is carried by K^+ ions through potassium-permeable transmembrane proteins, the potassium channels.

The energy for the potassium current is provided by the Na^+,K^+-ATPase (NKA) of the plasma membrane. Fueled by cytosolic ATP, it maintains the extracellular and intracellular K^+ concentrations at about 5 mM and 135 mM, respectively.[1,2] Given the chemical activities of K^+ inside and outside the cell, the potassium equilibrium potential E_K (the voltage across an exclusively potassium-permeable membrane) is in the range of -85 mV, as calculated by the Nernst equation. This provides the outward driving force for K^+ ions, since at physiologic as well as supraphysiologic extracellular K^+ concentrations (up to 30 mM), the membrane potential of the cardiomyocytes is more positive than E_K.[3]

Control of the main mechanical properties of the myocardium, such as chronotropy (rate of contractions), dromotropy (interval between excitation of atria and ventricles that determines the pause between contraction of the these chambers), and inotropy (force of contraction) by K^+ ions is exerted through their action on three different tissues. In the sinoatrial node (SA node), potassium current affects heart rhythm through its involvement in diastolic depolarization, the phase between two action potentials.[4,5] In the atrioventricular node (AV node), potassium permeability determines the time required for depolarization

317

to reach the threshold necessary to elicit an action potential, which is then conducted to the ventricle.[6,7]

In the ventricular myocardium, contractility depends on the length of the action potential, which is also controlled by potassium channels. During the plateau phase of the action potential, calcium-permeable channels are open. The Ca^{2+} ions flowing through these channels into the cytosol directly activate the contractile machinery of the cardiomyocyte, enter the sarcoplasmic reticulum (SR) via its calcium pump (refilling effect), or trigger the release of additional Ca^{2+} ions from this intracellular Ca^{2+} store. Thus, in general, any event that leads to a shortening of the plateau phase as a result of increased K^+ permeability of the sarcolemma causes a decrease in contractility. Potassium currents also control the height of the action potential plateau. Up to a potential of at least $+20\,mV$, a positive shift of the plateau voltage due to decreased potassium conductance augments force of contraction. This is partly due to an increased calcium influx through L-type calcium channels. Another component determining the Ca^{2+} balance is the sodium-calcium exchanger, which exchanges one Ca^{2+} ion for three Na^+ ions across the sarcolemma. The exchanger can work in two directions, ie, it can move Ca^{2+} inwards or outwards depending on the concentration of the transported ions on the two sides of the membrane and the membrane potential. Since the transport is electrogenic, a depolarization promotes Ca^{2+} influx or hampers Ca^{2+} efflux, and thus increases the net inward transport of Ca^{2+} ions.[8-12]

The shape of cardiac action potentials, ie, $E_m(t)$, varies considerably depending on species, age, extracellular ion and hormone concentrations, and metabolic state, as well as on the localization of the cardiomyocyte within heart muscle. This is brought about mainly by the diversity of potassium-channel types, whose opening and closing (called *gating*) is influenced specifically by the above-mentioned parameters and by the membrane potential.

Detailed studies of distinct potassium-channel types have been carried out on enzymatically isolated cardiomyocytes incubated in controlled extracellular media, using the patch-clamp technique for voltage control of isolated potassium channels and the methods of molecular biology to investigate structure-function relationships of channel proteins.[13-15]

This chapter will review the properties of cardiac potassium-channel types as revealed by these new methods and will show that distinct potassium channels are involved in physiologic as well as pathophysiologic processes. Whenever possible, the clinical relevance of these studies will be discussed.

Permeability

Although extremely different in their gating behavior, the potassium channels considered here are very similar with respect to the properties of their ion-conducting pores. All these channels are highly selective for K^+ ions, in contrast to Na^+, Ca^{2+}, Mg^{2+}, and all anions. In heart muscle cells, this selectivity is illustrated by the strong and specific influence of the concentration of K^+ in the extracellular fluid on the reversal potential of currents through potassium channels. Indeed, the ion permeability sequence for nearly all potassium-selective channels in heart and other tissues was found to be as follows:[16-17]

$$Tl^+ > K^+ > Rb^+ > NH_4^+ \gg Na^+, Li^+, Cs^+$$

Because of their similar permeabilities, the different potassium-channel types are classified according to their gating behavior.

Rectification

The voltage dependence of the current flowing through an ensemble of potassium channels of a particular type is usually not linear. If the current changes direction from inward to outward, the conductance may be smaller (inward rectification) or greater (outward rectification) than one would expect from an ohmic behavior. The concentration gradient of K^+ ions across the membrane causes an outward rectification of the open channel according to the Goldman-Hodgkin-Katz equation.[3] On the other hand, the outward current of most potassium channels is blocked by intracellular cations, although the extent of this phenomenon varies between potassium-channel types. Thus the inward rectification is strongly dependent on the dissociation constant of the binding site for these cations located near the internal opening of the pore. Furthermore, voltage-dependent gating of the channel protein itself can cause inward or outward rectification.[18]

The Delayed Rectifier (IK_{dr})

Outward potassium currents that increase with time (activation) after a stepwise depolarization of the cell membrane were first described in the multicellular Purkinje fiber preparations.[19] In general these currents were named I_K. Because of doubts about the selectivity for K^+ ions in multicellular preparations, they are also referred to as I_x.[20-21]

Gating

After membrane depolarization, currents through delayed rectifier (IK_{dr}) channels are activated with time constants between 100 ms and several seconds.[21-28] Inactivation, ie, closing of IK_{dr} channels during maintained depolarization, is very slow (in the range of seconds).[29-31] After return of the membrane potential to negative values, the channels also close (deactivation) with time constants in the seconds range.[4,28,32-33] Individual channels show considerable variations in the mean open times (between 0.4 and several hundred ms).[26,30,34] The distribution of channel closed times can be described satisfactorily only by more than one exponential (time constants between about 0.5 and 40 ms),[23,26,35] which indicates the existence of at least two closed-channel states.

Localization and Function

Currents through IK_{dr} channels have been measured in SA node, AV node, and atrium, as well as ventricle. Channel densities of 0.025 and $0.7\,\mu m^{-2}$ have been reported.[14,27,35-38] The activating current leads to the final repolarization of the action potential. In pacemaker tissue, the slow deactivation of IK_{dr} contributes to the diastolic depolarization.[4-5,22,29,37,39]

Conductance

With extracellular K^+ concentrations between 1.3 and 4.5 mM, single channel conductances γ for outward currents of between 5.4 and 60 pS (picosiemens, $1 S = 1/\Omega$) have been reported. Subconductance states of 15, 30, and 60 pS of an outward current with delayed rectifier properties have been reported in chick embryo heart.[38] At a concentration of 150 mM $[K^+]_o$ inward single-channel currents with deactivation kinetics that are similar to IK_{dr} and conductances of 10 and 3 pS have been demonstrated.[26,35] Other authors found patch currents with activation and deactivation behavior similar to that of macroscopic IK_{dr}, but were unable to resolve single-channel events. This indicates that a channel with $\gamma > 1\,pS$ contributes to the delayed rectifier properties of their preparation.[40]

Due to the voltage dependence of IK_{dr} gating, the whole-cell current shows an outward rectification in the activation range, which gives the delayed rectifier the second part of its name. For the fully activated (ie, gating-independent) current, linear or slightly inward-rectifying current-voltage relationships have been reported.[21,23,25,30,32,35,41] In contrast to other potassium channels, IK_{dr} does not seem to be blocked in a voltage-dependent manner by intracellular Mg^{2+}.[18]

Modulation

The probability that IK_{dr} channels are open increases after activation of protein kinases A and C (PKA, PKC). Although the role of PKC is not yet clearly understood, the increase of IK_{dr} by catecholamines occurs via phosphorylation of the channel protein by PKA and subsequent increase of the open probability. This may prevent an excessive lengthening of the action potential as the delayed rectifier could thus counteract the simultaneous stimulation of L-type calcium-channel current by these hormones.[25,28,41–46] Acetylcholine decreases IK_{dr}, especially after previous stimulation by β-adrenergic agonists.[47–48]

Structure

In recent years the primary structure of delayed rectifier channel proteins has been resolved mainly by analyzing clones of genes encoding IK_{dr} channels from rat and mouse brain.[49–52] Recently the cDNA of a delayed rectifier channel from human ventricle has also been cloned.[14] A sequence homology to a rat brain delayed rectifier channel of 86% implies that heart and brain IK_{dr} channels have a similar structure. Hydropathy profiles and site-directed mutagenesis studies allow the following conclusions about the conformation and function of the channel protein: the channel protein is a tetramer of four similar subunits or domains around the ion-conducting pore. Each domain has a molecular weight of about 70 kDa, consists of about 650 amino acid residues that form six transmembrane α helices (S1–S6) with a highly conserved amino acid sequence. The N-terminal as well as the C-terminal end of the polypeptide are located intracellularly and vary considerably between different subunit types. The fourth α-helix (S4-α-helix) contains several positively charged amino acids and is believed to be involved in the activation process. The link between S5 and S6 (the fifth hydrophobic or H5 region), formerly assumed to lie completely outside the cell, dips into the membrane in the form of a β hairpin. This structure is believed to be a part of the channel pore. The extracellular beginning and end of this β loop bind the potassium-channel blocker charybdotoxin and extracellular tetraethylammonium ions (TEA). The intracellular tip of H5 binds intracellularly applied TEA.

Block and Pharmacology

Delayed-rectifier channels are blocked by the classic potassium-channel blockers Cs^+, Ba^{2+}, TEA, and 4-aminopyridine (4-AP) in millimolar concentrations. In general, IK_{dr} is more sensitive to TEA and less to 4-AP than the transient outward current IK_{to}.[16–17,29,30,53] Intracellular Mg^{2+} ions block IK_{dr}, but the mechanism is different from the blocking effect of Mg^{2+} on other potassium channels and is not completely understood.[33,41]

Class III anti-arrhythmic drugs such as solatol, E 4031, clofilium, and amiodarone, which prolong the cardiac action potential as well as the refractory period,[54] block IK_{dr} in the micromolar range.[22,39,55-62] Substances, which according to their blocking action on cardiac sodium channels were originally categorized as class I anti-arrhythmic drugs, may also block the delayed rectifier. Examples of this rather unspecific drug action are quinidine and pirmenol.[24,59,63-64]

The Inward Rectifier ($I\dot{K}_{ir}$)

The resting conductance of heart as well as skeletal muscle displays an inward rectification, ie, current can flow more easily in the inward than in the outward direction. This phenomenon was also called anomalous rectification because the concentration gradient of K^+ ions across the membrane should promote outward rather than inward potassium currents, and was defined the dominant time-independent background current IK_1 in various cardiac preparations.[65]

Gating

The striking difference between outward and inward current through inward rectifier (IK_{ir}) channels is caused by the blocking action of intracellular divalent cations. Under physiologic conditions, the main blocking ion is Mg^{2+}, because the dissociation constant for blocking (K_D = $2-20\,\mu M$) is well below its normal intracellular concentration (0.5– $3.5\,mM$).[66-69] Intracellular Ca^{2+} ions are also able to block the open channel, but in diastole the magnesium block prevails because contrary to $[Mg^{2+}]_i$, $[Ca^{2+}]_i$ is well in the range of the dissociation constant ($K_D(Ca)$ $\approx 0.1\,\mu M$). However, during the plateau of the action potential, the blocking effect of Ca^{2+} ions may be more pronounced since the calcium influx increases $[Ca^{2+}]$ on the inner side of the membrane.[70-71]

Inward currents through IK_{ir} channels show a marked decay, which is caused by a voltage-dependent block by external Na^+, Ca^{2+}, and Mg^{2+} ions in the millimolar range.[72] For single channels, long mean open times ($50-400\,ms$) have been observed for inward currents, while closed time measurements indicate two populations of mean values ($1-10\,ms$ and $10-1000\,ms$). Therefore gating schemes with one open and two closed states have been proposed.[65,73-79]

Localization and Function

The inward rectifier channels are most abundant in ventricular myocytes, where, due to their gating behavior, they are responsible for the maintenance of the resting membrane potential as well as for the high input resistance during the action potential. During ontogeny, channel density increases from about $0.1 \mu M^{-1}$ to $0.5-2 \mu M^{-1}$ in adult ventricular myocytes.[70,80-83] Ventricular myocytes from hypertrophied cat heart have higher than normal densities of IK_{ir}.[84] In atrial cells, current through the less strongly rectifying IK_{Ach} channels (see below) also affect the diastolic and plateau potentials. This is why the action potentials in the atrium are shorter than in the ventricle.[70,74-75,85]

In pacemaker tissue, inward rectifier channels are virtually absent. Therefore, cells of SA and AV node have no stable resting potential.[5,81]

Another function of IK_{ir} channels may be a "spatial buffering" of locally increased extracellular K^+ concentration. A local increase of $[K^+]_o$ in intercellular clefts or transversal tubules would create a K^+ equilibrium potential more positive than the neighboring membrane potential. This would allow the extra K^+ ions to enter the myocyte through IK_{ir} channels. This buffering of localized extracellular hyperkalemia may prevent cardiac arrhythmias.[82]

Conductance

Single-channel conductances for inward currents of between about 25 and 30 pS at room temperature and with $[K^+]_o$ of 150 mM have been reported.[71,73-76,78-79,81,86-87] At more physiologic potassium concentrations of 1.3 and 4.5 mM $[K^+]_o$, the single-channel conductance is reduced to 2–3 or 4 pS, respectively.[38,88]

Modulation

Hormonal regulation of IK_{ir} channels has not been observed. The atrial myocyte channels are insensitive to isoproterenol.[25] The phospholipid platelet-activating factor (PAF) at subnanomolar concentrations (10^{-12}– 10^{-11} M) depresses IK_{ir} activity in ventricular cells.[79]

Structure

To date, nothing is known about the detailed structure of the inward rectifier channel protein. On the other hand, single-channel currents often show equally spaced subconductance states, which can be blocked independently by intracellular Mg^{2+}. This has led to the hypothesis that the inward rectifier channel is composed of a cluster of channels with individual conductance equal to the subconductance levels.[67,71,76,81-82]

Block and Pharmacology

The IK_{ir} channels are also blocked by Na^+, Mg^{2+}, and Ca^{2+} ions (see above), and, in a voltage-dependent manner by extracellular Cs^+

and Ba^{2+} ions, with half-blocking concentrations in the micromolar range.[89–96] It has been reported that quinidine blocks IK_{ir} in micromolar concentrations by keeping the channel in a closed state.[88] A block of single IK_{ir} currents by $2–5\,\mu M$ amiodarone appears to be the result of an action of the drug on the inside of the membrane.[97]

The Transient Outward Current (IK_{to})

Outward currents, transiently opened by a depolarizing voltage-clamp step, have been described in several cardiac preparations. A variety of terms, such as transient outward (I_t, I_{to}), brief outward (I_{bo}), early outward (I_{eo}), long-lasting outward (I_{lo}), or I_{qr} (according to the name of the gating variables) have been used to describe these currents. Some transient outward currents seem to be dependent on intracellular Ca^{2+}, other not. In this section, only the calcium-independent type will be reviewed. Because of its similarity to transient outward currents in neurones, it has also been called I_A.[29]

Gating

Like the delayed rectifier, the transient outward current turns on after a depolarizing voltage step. But the activation of IK_{to} is faster, ie, the activation time constants are usually smaller ($2–10\,ms$) than those of IK_{dr}.[29,98–105] The more striking difference, however, to the delayed rectifier lies in the marked inactivation behavior of the transient outward current with inactivation time constants between 20 and $350\,ms$. Whereas the activation process is significantly accelerated with stronger depolarizations, the inactivation time constants are not voltage-dependent or only slightly reduced at more positive test potentials. The description of the time course of inactivation often needs two exponentials, ie, a fast and a severalfold slower inactivating component.[29,99–109] During repolarization to negative holding potentials the IK_{to} channels recover from inactivation with time constants of 20 to $1500\,ms$.[29,99–104,109]

Few kinetic analyses of single channel data are available. In mouse ventricular cells, single-channel currents with three different conductances have been described. For all three channels the open times could be fitted monoexponentially with voltage-independent mean open times of 4, 24, and $1\,ms$, for a 27, 12, and $5\,pS$ channel, respectively. The closed time distributions needed two exponentials with voltage-independent fast components τ_{c1} (about 2 to $8\,ms$) and slow components of about 50 to $100\,ms$.[110–111]

Localization and Function

Usually the transient outward current is found in cardiac tissue where the action potential displays a "notch" before the plateau phase, which suggests that IK_{to} is the main component responsible for the early repolarization (phase 1) of the action potential. Transient outward currents have been described in heart tissue of sheep, rats, dogs, guinea pigs, chickens, rabbits, and humans, but seem to be absent in frogs.[34,38,104,109,112–114] Within the ventricle, higher densities of IK_{to} were found in epicardial than in endocardial myocytes.[115–117]

Conductance

In extracellular solutions with low K^+ concentration of between 1.3 and 5.4 mM single-channel conductances range from 5 to 27 pS.[34,38,102,110–111,113,115]

Modulation

The prolongation of the action potential in rabbit atrium and rat ventricle after α-adrenergic stimulation was shown to be caused by a blockade of IK_{to}. In rat ventricle, the inhibition of IK_{to} was attributed to stimulation of protein kinase C, since the effect can be mimicked by the stimulators of PKC, phorbol 12-myristate-13-acetate (PMA) and 1-oleoyl-2-acetylglycerol (OAG).[118–119] The blocking action of catecholamines on IK_{to} in canine Purkinje cells, on the other hand, is mediated by β-adrenergic stimulation of PKA via adenylate cyclase and cAMP.[108]

Structure

Molecular cloning and functional expression of two potassium channel cDNAs from rat and human heart, which display gating characteristics of transient outward current, have recently been reported.[14,105] The sequences of the cDNAs of these cardiac channels are over 90% homologous to a cDNA derived from rat brain, which indicates a structural similarity.

The main structural differences between the IK_{to} channel and the nearly non-inactivating IK_{dr} channel are in the N- and C-terminal sequences of the cDNAs, which suggests that these structures encode the amino acid sequences responsible for the inactivation process. Indeed, mutations in the cDNA region that encode the N-terminal endings of the IK_{to} channel polypeptides strikingly alter the inactivation kinetics of the channel. The recent model for the inactivation process of IK_{to} proposes that the NH_2 terminus of the channel domains is a charged "ball" connected to the transmembraneous part of the domain by a "chain" of hydrophobic amino acids. When the channel is opened by a depolarizing voltage pulse, the ball moves into the inner mouth of the channel, thereby plugging the pore and inactivating the channel.[50]

Block and Pharmacology

Whereas Ba^{2+} and Cd^{2+} ions are able to block IK_{to} in the millimolar range, Hg^{2+} is effective in micromolar concentrations.[98,101,107] In contrast to the delayed rectifier current, the transient outward current is much more sensitive to the blocking action of 4-AP ($K_D < 3\,mM$) than to TEA ($K_D > 50\,mM$).[29,99,101,103–104,106–107,109,120] It has been reported that the class I anti-arrhythmic drug bupivacaine blocks IK_{to} with a K_D of about $20\,\mu M$ without any effect on IK_{ir}, up to $1\,mM$.[121]

The Calcium-Dependent Potassium Current (IK_{Ca})

It has long been assumed that the transient outward current in the heart includes a component dependent on intracellular Ca^{2+}, variously named IK_{Ca}, I_{to2}, or I_{bo}.[104,122]

Gating

Using a variety of cardiac preparations, it has been shown that one part of the transient outward current is suppressed by blockers of L-type calcium channels, such as Co^{2+}, Mn^{2+}, D600,[100,104,120,122–123] by inhibiting the release of Ca^{2+} from intracellular stores with ryanodine or caffeine or by replacing extracellular Ca^{2+} with Sr^{2+}.[100,104,106–107,120,123–125] The kinetics of IK_{Ca} is similar to that of the calcium-independent transient outward current. Activation time constants of about 5 to 25 ms and inactivation time constants of about 6 to 100 ms have been reported.[100,106,126] The time course of IK_{Ca} probably reflects time-dependent changes of the intracellular Ca^{2+} concentration.[123] Single-channel recordings revealed mean currents through channels, which open transiently after a depolarizing voltage pulse at sufficiently high ($1\,\mu M$) constant intracellular Ca^{2+} concentration, but stay closed at $[Ca^{2+}]_i$ of about $1\,nM$.[112,126]

An alternative explanation for the calcium-dependence of the gating of the transient outward current has recently been proposed. It was found that the IK_{to} is independent of intracellular Ca^{2+}. However, an increase in extracellular Ca^{2+} or other divalent cations shifts the voltage dependence of activation as well as of inactivation to more positive potentials. Thus at membrane potentials positive to $-60\,mV$, this shift may increase the transient outward current by reducing the steady-state inactivation. Caffeine and ryanodine, apart from their interaction with the calcium release from SR, have direct blocking effects on the IK_{to} channel protein.[127–128]

Localization and Function

Calcium-activated potassium currents have been described in Purkinje fibers as well as in atrial and ventricular myocytes. Like IK_{to}, they may serve for the initial rapid repolarization of the cardiac action potential (see references in the section "Gating").

Conductance

In cow, sheep, and rabbit Purkinje cells at 70 mM $[K^+]_o$, single-channel conductances of 140, 17, and 26 pS have been found.[112,126]

Block and Pharmacology

The calcium-dependent potassium currents do not seem to be blocked by the classical IK_{to} blocker 4-AP (4 mM) or by TEA (50 mM).[100,104,106–107,120,123] Indirect blocking effects follow the inhibition of a rise in $[Ca^{2+}]_i$ (see the section "Gating").

The $[ATP]_i$-Dependent Potassium Current (IK_{ATP})

It has long been known that blocking the ATP production in myocardial cells evokes a time-independent outward current for membrane potentials positive to E_K, which leads to a shortening of the action potential.[129,130] Later it was shown that a potassium channel, normally blocked by intracellular ATP, is the carrier of this current.[131–134]

Gating

The potassium conductance in cardiomyocytes induced by metabolic depression due to hypoxia, treatment with dinitrophenol or CN^-, or intracellular ATP-deficiency does not change its amplitude after stepping up the voltage of the membrane.[135–139] Therefore it can be concluded that the gating of the channels carrying this current is not voltage-dependent. Reversal of the metabolic blockade or restoration of the intracellular ATP depresses the current.[135,137] Since in the working myocardium, intracellular ATP is probably functionally compartmentalized, the ATP produced by glycolytic enzymes beneath the $[ATP]_i$-dependent potassium channels is more efficient in preventing channel opening than the ATP supplied by mitochondrial oxidative phosphorylation.[138,140–142]

The action of intracellular ATP on single channels has been studied in detail by numerous investigators. The depressing effect of $[ATP]_i$ on the open probability of single-channel currents, described by Hill equations, has a Hill coefficient of 1 to 4 and a dissociation constant of about 10 to 600 µM.[132,133,141,143-147] Phosphorylation is not involved in the blocking action of ATP, because magnesium is not required[143,145] and the free energy of ATP hydrolysis does not affect IK_{ATP}.[148] Other nucleotides, if applied intracellularly, also block IK_{ATP} with an order of effectiveness of

$$ATP > AMP\text{-}PNP > ADP > CTP \approx GDP \approx AMP$$
$$\approx ITP \approx UTP \approx GTP.^{132,145-146}$$

It has been observed that $Mg \cdot ADP$ in submillimolar concentrations increased K_D ([ATP]$_i$), which led to the conclusion that the ATP/ADP ratio, and not the intracellular ATP concentration, controls the open probability of IK_{ATP} channels. The effect of $Mg \cdot ADP$ may be due to inhibition of the cooperativity between two ATP-binding sites, due to binding of $Mg \cdot ADP$ to one of these sites.[146,149]

In the absence of intracellular ATP the channels show a "bursting" behavior. Fast transitions between an open and a closed channel state (bursts) are interrupted by longer periods of closure. Within the burst, mean open times of 1 to 5 ms and mean closed times of 0.1 to 4 ms have been measured, which are hardly voltage-dependent but rather are influenced by the magnitude of the current flowing through the channels. The long mean closed time between the bursts is in the range of 100 ms. An increase in [ATP]$_i$ reduces the mean open time and increases the mean shut time, which leads to decreased open probability.[74,145,147,150-153]

When ATP concentration jumps were applied to the inside of multichannel patches, a monoexponential time course of the ensemble average current with an ATP-dependent closing and an ATP-independent opening rate constant was found. Both rate constants were voltage-independent. Thus it may be concluded that one binding site for intracellular ATP outside the electrical field of the membrane dominates the gating behavior of ATP-dependent potassium channels.[147]

Localization and Function

ATP-dependent potassium channels have been found in ventricular, atrial, and AV node cells of humans, dogs, cats, rats, mice, guinea pigs, rabbits, and frogs.[74,137,142,145-146,148,151,154-155]

The number of open channels in a patch at high open probability,[138,145-146,156-159] divided by the approximate area of the patch (calculated according to reference 160) yields channel densities well above $1 \, \mu m^{-2}$.[141]

The [ATP]$_i$-sensitive potassium channels are responsible for the shortening of the cardiac action potential under hypoxic or ischemic conditions. These changes in the configuration of the action potential reduce calcium influx, [Ca^{2+}]$_i$, the amplitude of contraction, and the energy consumption of the cardiomyocytes. This may prevent metabolic exhaustion and prolong survival of heart tissue.[137,151,161-162]

However, under conditions of sustained metabolic stress, the increased potassium outward current may lead to an increased extracellular K^+ concentration that causes membrane depolarizations. This may be the reason for cardiac arrhythmias observed under prolonged hypoxia. On the other hand, elevated $[K^+]_o$ may be beneficial because of its vaso-dilatatory action and the generation of pain, which may cause cessation of physical activity.[131-132,163]

Conductance

Single-channel conductance at 150 mM $[K^+]_o$ is about 80 pS.[131] Values of 52 to 83 pS at temperatures of 20 to 37°C have been reported.[74,133-134,150,152-154,161,164] Under symmetrical K^+ concentrations ($[K^+]_i = [K^+]_o$), the open channel current displays a slight inward rectification. The diminution of outward currents is caused by a voltage-dependent block by intracellular cations. Hill plots, at 0 mV membrane potential, revealed K_D values of 2 and 50 mM for one Mg^{2+} and two Na^+, respectively.[74,150,154,165]

Modulation

In the absence of ATP the activity of $[ATP]_i$-dependent potassium channels declines within minutes. This "rundown" can be prevented by the administration of $Mg \cdot ATP$, which suggests that opening of the channels requires phosphorylation. Since the rundown is accelerated by intracellular Ca^{2+}, an involvement of intracellular calcium-activated phosphatases may be assumed. However, the time course of rundown can not be decelerated by inhibitors of phosphatases or proteases.[78,149,164,166-167]

Stimulation of IK_{ATP} by adenosine via the GTP-binding protein G_i has been observed in rat ventricular myocytes, in harmony with the observation that stimulators of G-proteins, such as GTP-γ-S or fluoride ions, can stimulate IK_{ATP}.[78,153,168]

A decrease in intracellular pH increases the open probability of $[ATP]_i$-dependent potassium channels. The mechanism underlying this effect, probably relevant in ischemic muscle, is not yet clear.[146,169]

Block and Pharmacology

The $[ATP]_i$-dependent channels are blocked by cesium, TEA, 4-AP, and quinidine in the millimolar range, and by barium at micromolar concentrations.[145,151,154,170,171]

The potassium-channel openers cromakalim, pinacidil, nicorandil, and RP 49356 activate IK_{ATP} by shifting the K_D for ATP to higher values. A simple competition of these drugs with ATP at the intra-cellular binding sites has been proposed and, more recently, questioned.[155–156,158–159,172–179] Although the cardioprotective effect of potassium-channel openers may be due in part to their action on the IK_{ATP} of heart muscle cells, it may more directly related to vasodilatatory action secondary to stimulation of potassium efflux in vascular smooth muscle.[180]

Sulfonylureas, such as glibenclamide and tolbutamide, block cardiac $[ATP]_i$-sensitive potassium channels, but at concentrations about two orders of magnitude higher than those known to block IK_{ATP} in the pancreatic β cells.[139,163,181–183]

The ACh-Sensitive Potassium Current (IK_{ACh})

It has long been known that acetylcholine (ACh), released from stimulated vagal nerve endings, slows down the heart rate. This, as well as the negative dromotropic and negative inotropic effects of ACh, are brought about by inhibition of the cardiac calcium conductance and by the activation of a special potassium current: the ACh-sensitive or muscarinic K^+ current, IK_{ACh}. Recent investigations have elucidated the detailed gating and conduction properties of the channels carrying this current.[184–186]

Gating

Activation of IK_{ACh} requires binding of ACh to the muscarinic ACh receptor, probably subtype M_2.[187] This receptor belongs to a broad family of transmembranous proteins, which couple to membrane-bound GTP-binding proteins. After binding of ACh, the receptor accelerates the separation of GDP from the α subunit of the G protein, which thereby facilitates the exchange of GTP for GDP. Binding of GTP causes dissociation of the β and γ subunit from the α subunit. In this way the α subunit becomes active, ie, it becomes able to open a potassium-selective channel directly. It has been suggested that the activating effect of $α_i$ is achieved by removing an intracellular gating particle, which occludes the channel pore of the unstimulated channel. Deactivation is due to the hydrolysis of GTP to GDP by the GTPase activity of the α subunit. The GDP-α-subunit complex reassociates with βγ, thereby becoming unable to open ACh-dependent potassium channels. It has been shown that the subtypes $α_{i1}$, $α_{i2}$, and $α_{i3}$ of the α subunit, named after their inhibitory action on adenylate cyclase, are involved in the muscarinic activation of IK_{ACh}. The complete G protein, including the $α_i$ and the rather unspecific β and γ, is called G_i.[186,188–198] G_i can be irreversibly stimulated by binding non-hydrolyzable analogs of GTP, such as GTP-γ-S, instead of GTP.[199-200] Another mechanism of α_i activation is a transphosphorylation of α_i-bound GDP by an intracellular localized nucleoside diphosphate kinase (NDPK), which uses intracellular ATP.[201-202]

The possibility that the β-γ-subunits may also be able to activate IK_{ACh} has given rise to controversy.[190-191,203-207] One possible mechanism of βγ-stimulation of IK_{ACh} is an activation of phospholipase A_2 (PLA_2), which releases arachidonic acid (AA). Lipooxygenase products of AA are known to stimulate IK_{ACh}.[208-209]

In the heart G_i is activated not only by the M_2-receptor, but also by the somatostatin receptor, the adenosine receptor P_1, and, at least in frog and bovine atrial cells, by the adenosine receptor P_2. Receptors for endothelin or calcitonin gene-related peptide are also able to induce IK_{ACh}.[6,210-220] The muscarinic stimulation of IK_{ACh} displays a homologous desensitization, ie, the current declines during prolonged exposure to ACh. A heterologous desensitization by adenosine has also been observed. These observations led to the conclusion that desensitization occurs at the level of the G_i protein or is due to calcium-dependent dephosphorylation of the channel protein.[215,221-224]

Single Ach-activated channels have monoexponentially distributed open times with mean values of about 1 ms (0.9–2.5 ms). The closed times can best be fit biexponentially with time constants τ_{c1} of about 1 ms (0.5–1.4 ms) and τ_{c2} between 4 and 80 ms.[74,190,196,205,207,215,223,225-228] Increasing the extracellular ACh concentration decreases the shut times. Desensitization is accompanied by an increase of the shut times and a diminution of the mean open times.[223,229]

Localization and Function

ACh-sensitive channels are mainly found in pacemaker and atrial tissue, although ACh-induced potassium currents in sheep and rabbit Purkinje fibers and in ferret ventricular muscle have also been reported.[47,89,221] It is assumed that in the atrium IK_{ACh} channels contribute to the membrane conductance at rest as well as at plateau potentials. Since the IK_{ACh} inward rectification is weaker than the IK_{ir} rectification, which mainly contributes to the resting and plateau conductance in the ventricle, the membrane conductance in the atrium is higher than in ventricular muscle at plateau potentials. This may be one reason for the shorter action potential in atrium in comparison to ventricle.[75,214] Under vagal stimulation the released ACh increases IK_{ACh}, shortening the action potential and decreasing contractility (see beginning of this chapter and references 221, 230). In the SA node, activation of IK_{ACh} leads to a hyperpolarizing shift of the diastolic potential and a slowing of the diastolic depolarization.[5,231-232] In AV-node cells the hyperpolarization and increased membrane conductance, due to stimulation of IK_{ACh}, decelerates the depolarization, which results in a negative dromotropic effect.[7]

The potassium conductance of the membrane can be enhanced under hypoxic conditions Via IK_{ATP} as well as by the increased extracellular adenosine concentration, that stimulate IK_{ACh}.[6,218,220]

Conductance

The single-channel conductance at 150 mM $[K^+]_o$ and at room temperature is about 40 pS. Values of between 34 and 46 pS for the inward current have been reported.[74,215–217,227,229,233] Single-channel currents display an inward rectification, which is caused by a voltage-dependent block of the channel pore by the general cytosolic potassium channel blocker Mg^{2+}. The inward rectification is weaker than that of IK_{ir}, but stronger than that observed for IK_{ATP}.[165]

Modulation

Stimulation of PKA increases IK_{ACh} if the current has been activated by ACh beforehand.[226] Derivatives of arachidonic acid exert different influences on IK_{ACh}. Cyclooxygenase products inhibit, while lipooxygenase products stimulate, ACh-dependent potassium channels.[208–209,234]

Block and Pharmacology

The activation of IK_{ACh} can be inhibited at different stages of the cascade. The classic blockers of muscarinic receptors, atropine and scopolamine in submicromolar concentrations, prevent stimulation of IK_{ACh} by ACh at the receptor level.[89,227] IK_{ACh} may be inhibited by pertussis toxin, heparin, ras p21, N-ethylmaleimide, and GTPase-activating protein by action on the G_i-protein.[228,235–237]

The channel itself is blocked by Cs^+ in millimolar, and by Ba^{2+} in micromolar, concentrations.[89]

The Na-Dependent Potassium Current (IK_{Na})

In 1976 it was demonstrated that a large increase in intracellular Na^+ enhances the potassium conductance of neurones. Later a similar phenomenon was described in ventricular myocytes.[238]

Gating

The sodium-dependent potassium channel is activated by intracellular Na^+ ($K_D \approx 50$ mM) and is independent of $[Ca^{2+}]_i$. The binding of at least two Na^{2+} ions is probably necessary for opening the channel. Single-channel currents appear in bursts, which means that as for other potassium channels, the distribution of open times is described by one exponential, whereas the fit of shut time distributions requires two. Within the bursts, mean open times of 2 to 12 ms and mean closed times of 0.3 to 7 ms have been reported.[152,239–241] The channel gating is not obviously voltage-dependent.

Localization and Function

So far, cardiac sodium-dependent potassium channels have been described only in guinea pig ventricular muscle. They may become activated during the upstroke of the cardiac action potential due to Na^+ influx or by intracellular Na^+ overload caused by glycosides. Although the dissociation constant for sodium is high, this channel might contribute to the resting potassium conductance since the maximal IK_{Na} is about 10 times higher than the maximal calcium current.[152,240–241]

Conductance

The single-channel conductance is relatively high: values of 150 to 210 pS have been reported for inward currents. Outward currents are somewhat smaller due to a voltage-dependent block by intracellular Mg^{2+} ($K_D \approx$ 10 mM). Single-channel currents display multiple subconductance levels.[152,239–240,242]

Block and Pharmacology

It has been demonstrated that the sulfonylurea R 56865 blocks IK_{Na}. The Hill plots reveal a K_D of about 1 µM and a Hill coefficient of 1.[240–241]

The Arachidonic Acid-Activated Potassium Current (IK_{AA})

In neonatal rat atrial cells arachidonic acid, as well as linoleic acid, at a concentration of 10 µM activate a potassium-selective channel. This channel is not activated by derivatives of fatty acids or by other fatty acids such as oleic, palmitic, stearic, or myristic acid. The open probability is also independent of intracellular Ca^{2+}, Mg^{2+}, or ATP. The mean open time of single-channel currents is about 1 ms. In symmetrical 150 mM [K^+] the single-channel conductance for inward currents is 70 pS, for outward currents 160 pS. The current is blocked by extracellular Cs^+ ions, but enhanced by intracellular protons. Because ischemia leads to a decrease of intracellular pH and to a release of arachidonic acid, the channel may be involved in the increased potassium conductance under this pathophysiologic condition.[243]

The Plateau-Potential Potassium Current (IK_P)

In guinea pig ventricular cells a potassium-conducting channel, which is activated at depolarized membrane potentials, has been described. Depolarizing voltage steps quickly activate a potassium current with time constants <10 ms. No inactivation was observed for times of up to 600 ms. Mean open times are between 2 and 5 ms, and increase with depolarization. The single-channel conductance amounts to 14 pS at 1 mM $[K^+]_o$ and decreases to about 7 pS if the extracellular K^+ concentration is increased to 95 mM, which distinguishes it from IK_{dr}. Single-channel currents can be observed at extracellular concentrations of TEA of up to 135 mM. The channel might carry substantial outward current during the plateau of the cardiac action potential.[244]

Summary

Many different potassium channels that are involved in the regulation of the heartbeat under physiologic and pathologic conditions have been described. Others may remain to be discovered. Specific pharmacologic treatment of pathophysiologic states requires comprehensive knowledge of cardiac ion channels and drug action. The situation is complicated because potassium channels presumably exist in all mammalian cells and many drugs with affinity for the potassium channels may influence cardiac contraction indirectly through an interaction with cardiac neurones or blood vessels, or even with noncardiac tissues.

References

1. Eisner DA. The Na-K pump in cardiac muscle. In: Fozzard HM, ed. The Heart and Cardiovascular System. New York: Raven Press; 1986:489–507.
2. Walker JL. Intracellular inorganic ions in cardiac tissue. In: Fozzard HM, ed. The Heart and Cardiovascular System. New York: Raven Press; 1986:561–572.
3. Fozzard HA, Arnsdorf MF. Cardiac electrophysiology. In: Fozzard HM, ed. The Heart and Cardiovascular System. New York: Raven Press; 1986:1–30.
4. Giles W, Shibata EF. Voltage clamp of bullfrog cardiac pacemaker cells: A quantitative analysis of potassium currents. J Physiol (Lond) 1985; 368:265–292.
5. Irisawa H, Hagiwara N. Pacemaker mechanisms of mammalian sinoatrial node cells. Prog Clin Biol Res 1988; 257:33–52.

6. Belardinelli L. Modulation of atrioventricular transmission by adenosine. Prog Clin Biol Res 1987; 230:109–118.
7. Nishimura M, Habuchi Y, Hiromasa S, Watanabe Y. Ionic basis of depressed automaticity and conduction by acetylcholine in rabbit AV node. Am J Physiol 1988; 255:H7–H14.
8. Beuckelmann DJ, Wier WG. Mechanism of release of calcium from sarcoplasmic reticulum of guinea-pig cardiac cells. J Physiol (Lond) 1988; 405:233–255.
9. Egan TM, Noble D, Noble SJ, Powell T, Spindler AJ, Twist VW. Sodium-calcium exchange during the action potential in guinea-pig ventricular cells. J Physiol (Lond) 1989; 411:639–661.
10. Gibbons WR. Cellular control of cardiac contraction. In: Fozzard HM, ed. The Heart and Cardiovascular System. New York: Raven Press; 1986:747–778.
11. Näbauer M, Callewaert G, Cleemann L, Morad M. Regulation of calcium release is gated by calcium current, not gating charge, in cardiac myocytes. Science 1989; 244:800–803.
12. Winegrad S. Membrane control of force generation. In: Fozzard HM, ed. The Heart and Cardiovascular System. New York: Raven Press; 1986:703–730.
13. Hamill OP, Marty A, Neher E, Sakmann B, Sigworth FJ. Improved patch-clamp techniques for high-resolution current recording from cells and cell-free membrane patches. Pfluegers Arch 1981; 319:85–100.
14. Tamkun MM, Knoth KM, Walbridge JA, Kroemer H. Molecular cloning and characterization of two voltage-gated K^+ channel cDNAs from human ventricle. FASEB J 1991; 5:331–337.
15. Trube G. Enzymatic dispersion of heart and other tissues. In: Sakmann B, Neher E, eds. Single Channel Recording. New York: Plenum; 1983:69–76.
16. Adams DJ, Nonner W. Voltage-dependent potassium channels: Gating, ion permeation and block. In: Cook NS, ed. Potassium Channels: Structure, Classification, Function and Therapeutic Potential. Chichester: Ellis Horwood; 1990:40–69.
17. Kolb HA. Potassium channels in excitable and non-excitable cells. Rev Physiol Biochem Pharmacol 1990; 115:51–91.
18. Carmeliet E. K^+ channels in cardiac cells: Mechanism of activation, inactivation, rectification and K^+_e sensitivity. Pfluegers Arch 1989; 414:88–92.
19. Noble D, Tsien RW. Outward membrane currents activated in the plateau range of potentials in cardiac Purkinje fibres. J Physiol (Lond) 1969; 200:205–231.
20. Cohen IS, Datyner NB, Gintant GA, Kline RP. Time-dependent outward currents in the heart. In: Fozzard HM, ed. The heart and Cardiovascular System. New York: Raven Press; 1986:637–669.
21. Matsuura H, Ehara T, Imoto Y. An analysis of the delayed outward current in single ventricular cells of the guinea-pig. Pfluegers Arch 1987; 410:596–603.
22. Arena JP, Walsh KB, Kass RS. Measurement, block, and modulation of potassium channel currents in the heart. In: Cook NS, ed. Potassium Channels: Structure, Classification, Function and Therapeutic Potential. Chichester: Ellis Horwood; 1990:43–63.

23. Duchatelle-Gourdon I, Hartzell HC. Single delayed rectifier channels in frog atrial cells: Effect of β-adrenergic stimulation. Biophysic J 1990; 57:903–909.

24. Furukawa T, Tsujimura Y, Kitamura K, Tanaka H, Habuchi Y. Time- and voltage-dependent block of the delayed K^+ current by quinidine in rabbit sinoatrial and atrioventricular nodes. J Pharmacol Exp Ther 1989; 251:756–763.

25. Giles W, Nakajima T, Ono K, Shibata EF. Modulation of the delayed rectifier K^+ current by isoprenaline in bull-frog atrial myocytes. J Physiol (Lond) 1989; 415:233–249.

26. Horie M, Hayashi S, Kawai Ch. Two types of delayed rectifying K^+ channels in atrial cells of guinea pig heart. Jap J Physiol 1990; 40:479–490.

27. Nakayama T, Kurachi Y, Noma A, Irisawa H. Action potential and membrane currents of single pacemaker cells of the rabbit heart. Pfluegers Arch 1984; 402:248–257.

28. Walsh KB, Begenisich TB, Kass RS. β-Adrenergic modulation of cardiac ion channels: Differential temperature sensitivity of potassium and calcium currents. J Gen Physiol 1989; 93:841–854.

29. Apkon M, Nerbonne JM. Characterization of two distinct depolarization-activated K^+ currents in isolated adult rat ventricular myocytes. J Gen Physiol 1991; 97:973–1011.

30. Clapham DE, Logothesis DE. Delayed rectifier K^+ current in embryonic chick heart ventricle. Am J Physiol 1988; 254:H192–H197.

31. Folander K, Smith JS, Antanavage J, Bennett C, Stein RB, Swanson R. Cloning and expression of the delayed-rectifier I_{sK} channel from neonatal rat heart and diethylstilbestrol-primed rat uterus. Proc Natl Acad Sci USA 1990; 87:2975–2979.

32. Clay JR, Hill CE, Roitman D, Shrier A. Repolarization current in embryonic chick atrial heart cells. J Physiol (Lond) 1988; 403:525–537.

33. Tarr M, Trank JW, Goertz KK. Intracellular magnesium affects I_K in single frog atrial cells. Am J Physiol 1989; 257:H1663–H1669.

34. Balser JR, Bennett PB, Roden DM. Time-dependent outward current in guinea pig ventricular myocytes: Gating kinetics of the delayed rectifier. J Gen Physiol 1990; 96:835–863.

35. Shibasaki T. Conductance and kinetics of delayed rectifier potassium channels in nodal cells of the rabbit heart. J Physiol (Lond) 1987; 387:227–250.

36. Boyle WA, Nerbonne JM. A novel type of depolarization-activated K^+ current in isolated adult rat atrial myocytes. Am J Physiol 1991; 260:H1236–H1247.

37. Hume J, Uehara A, Hadley RW, Harvey AL. Comparison of K^+ channels in mammalian atrial and ventricular myocytes. In: Cook NS, ed. Potassium Channels: Structure, Classification, Function and Therapeutic Potential. Chichester: Ellis Horwood; 1990:17–41.

38. Mazzanti M, DeFelice LJ. K channel kinetics during the spontaneous heart beat in embryonic chick ventricle cells. Biophys J 1988; 54:1139–1148.

39. Sanguinetti MC, Jurkiewicz NK. Two components of cardiac delayed rectifier K^+ current: Differential sensitivity to block by Class III antiarrhythmic agents. J Gen Physiol 1990; 96:195–215.

40. Walsh KB, Arena JP, Kwok WM, Freeman L, Kass RS. Delayed-rectifier potassium channel activity in isolated membrane patches of guinea pig ventricular myocytes. Am J Physiol 1991; 260:H1390–H1393.
41. Duchatelle-Gourdon I, Hartzell HC, Lagrutta AA. Modulation of the delayed rectifier potassium current in frog cardiomyocytes by beta-adrenergic agonists and magnesium. J Physiol (Lond) 1989; 415:251–274.
42. Bennett PB, Begenisich TB. Catecholamines modulate the delayed rectifying potassium current (I_K) in guinea pig ventricular myocytes. Pfluegers Arch 1987; 410:217–219.
43. Bennett PB, Kass RS, Begenisich TB. Nonstationary fluctuation analysis of the delayed rectifier K channel in cardiac Purkinje fibers: Actions of norepinephrine on single-channel current. Biophysic J 1989; 55:731–738.
44. Tohse N, Kameyama M, Irisawa H. Intracellular Ca^{2+} and protein kinase C modulate K^+ current in guinea pig heart cells. Am J Physiol 1987; 253:H1321–H1324.
45. Tohse N, Kameyama M, Sekiguchi K, Shearman MS, Kanno M. Protein kinase C activation enhances the delayed rectifier potassium current in guinea-pig heart cells. J Mol Cell Cardiol 1990; 22:725–734.
46. Walsh KB, Kass RS. Regulation of a heart potassium channel by protein kinase A and C. Science 1988; 242:67–69.
47. Carmeliet E, Mubagwa K. Changes by acetylcholine of membrane currents in rabbit cardiac Purkinje fibres. J Physiol (Lond) 1986; 371:201–217.
48. Yazawa K, Kameyama M. Mechanism of receptor-mediated modulation of the delayed outward potassium current in guinea-pig ventricular myocytes. J Physiol (Lond) 1990; 421:135–150.
49. Chandy KG, Williams CB, Spencer RH, Aguilar BA, Ghanshani S, Tempel BL, Gutman GA. A family of three mouse potassium channel genes with intronless coding regions. Science 1990; 247:973–975.
50. Miller C. 1990: Annus mirabilis of potassium channels. Science 1991; 252:1092–1096.
51. Pongs O. Molecular basis of potassium channel diversity. Pfluegers Arch 1989; 414:71–75.
52. Stühmer W, Ruppersberg JP, Schröter KH, Sakmann B, Stocker M, Giese KP, Perschke A, Baumann A, Pongs O. Molecular basis of functional diversity of voltage-gated potassium channels in mammalian brain. EMBO J 1989; 8:3235–3244.
53. Cook NS. The pharmacology of potassium channels and their therapeutic potential. Trends Pharmacol Sci 1988; 9:21–28.
54. Woosley RL. Antiarrhythmic drugs. Ann Rev Pharmacol Toxicol 1991; 31:427–455.
55. Arena JP, Kass RS. Block of heart potassium channels by clofilium and its tertiary analogs: Relationship between drug structure and type of channel blocked. Mol Pharmacol 1988; 34:60–66.
56. Balser JR, Hondeghem LM, Roden DM. Amiodarone reduces time dependent I_K activation. Circulation 1987; 76:IV-151. Abstract.
57. Carmeliet E. Electrophysiologic and voltage clamp analysis of the effects of solatol on isolated cardiac muscle and Purkinje fibers. J Pharmacol Exp Ther 1985; 232:817–825.

58. Colatsky TJ, Follmer CH, Starmer CF. Channel specificity in antiarrhythmic drug action: Mechanism of potassium channel block and its role in suppressing and aggravating cardiac arrhythmias. Circulation 1990; 82:2235–2242.

59. Rials SJ, Friehling TD, Marinchak RA, Kowey PR. Potassium channels in cardiac arrhythmias: Focus on antiarrhythmic drug action. Prog Clin Biol Res 1990; 334:111–121.

60. Sanguinetti MC, Jurkiewicz NK. Lanthanum blocks a specific component of I_K and screens membrane surface charge in cardiac cells. Am J Physiol 1990; 259:H1881–H1889.

61. Sanguinetti MC, Jurkiewicz NK. Delayed rectifier outward K^+ current is composed of two currents in guinea pig atrial cells. Am J Physiol 1991; 260:H393–H399.

62. Sanguinetti MC, Jurkiewicz NK, Scott AL, Siegl PKS. Isoproterenol antagonizes prolongation of refractory period by the Class III antiarrhythmic agent E-4031 in guinea pig myocytes: Mechanism of action. Circ Res 1991; 68:77–84.

63. Reichardt B, Konzen G, Hauswirth O. Pirmenol, a new antiarrhythmic drug with potassium- and sodium-channel blocking activity: A voltage-clamp study in rabbit Purkinje fibres. Naunyn-Schmiedeberg's Arch Pharmacol 1990; 341:462–471.

64. Roden DM, Bennett PB, Snyders DJ, Balser JR, Hondeghem LM. Quinidine delays I_K activation in guinea pig ventricular myocytes. Circ Res 1988; 62:1055–1058.

65. Kurachi Y. Voltage-dependent activation of the inward-rectifier potassium channel in the ventricular cell membrane of guinea-pig heart. J Physiol (Lond) 1985; 366:365–385.

66. Ishihara K, Mitsuiye T, Noma A, Takano M. The Mg^{2+} block and intrinsic gating underlying inward rectification of the K^+ current in guinea-pig cardiac myocytes. J Physiol (Lond) 1989; 419:297–320.

67. Matsuda H. Open-state substructure of inwardly rectifying potassium channels revealed by magnesium block in guinea-pig heart cells. J Physiol (Lond) 1988; 397:237–258.

68. Matsuda H, Saigusa A, Irisawa H. Ohmic conductance through the inwardly rectifying K channel and blocking by internal Mg^{2+}. Nature 1987; 325:156–159.

69. Vandenberg CA. Inward rectification of a potassium channel in cardiac ventricular cells depends on internal magnesium ions. Proc Natl Acad Sci USA 1987; 84:2560–2564.

70. Mazzanti M, DeFelice LJ. Ca modulates outward current through I_{K1} channels. J Membr Biol 1990; 116:41–45.

71. Mazzanti M, DiFrancesco D. Intracellular Ca modulates K-inward rectification in cardiac myocytes. Pfluegers Arch 1989; 413:322–324.

72. Biermans G, Vereecke J, Carmeliet E. The mechanism of the inactivation of the inward-rectifying K current during hyperpolarizing steps in guinea-pig ventricular myocytes. Pfluegers Arch 1987; 410:604–613.

73. Clark RB, Nakajima T, Giles W, Kanai K, Momose Y, Szabo G. Two distinct types of inwardly rectifying K^+ channels in bull-frog atrial myocytes. J Physiol (Lond) 1990; 424:229–251.

74. Heidbüchel H, Vereecke J, Callewaert G, Carmeliet E. Three different potassium channels in human atrium. Circ Res 1990; 66:1277–1286.
75. Hume JR, Uehara A. Ionic basis of the different action potential configurations of single guinea-pig atrial and ventricular myocytes. J Physiol (Lond) 1985; 368:525–544.
76. Matsuda H, Matsuura H, Noma A. Triple-barrel structure of inwardly rectifying K^+ channels revealed by Cs^+ and Rb^+ block in guinea-pig heart cells. J Physiol (Lond) 1989; 413:139–157.
77. Sakmann B, Trube G. Voltage-dependent inactivation of inward-rectifying single-channel currents in the guinea-pig heart cell membrane. J Physiol (Lond) 1984; 347:659–683.
78. Trube G, Hescheler J. Inward-rectifying channels in isolated patches of the heart cell membrane: ATP-dependence and comparison with cell-attached patches. Pfluegers Arch 1984; 401:178–184.
79. Wahler GM, Coyle DE, Sperelakis N. Effects of platelet-activating factor on single potassium channel currents in guinea pig ventricular myocytes. Molec Cell Biochem 1990; 93:69–76.
80. DeHaan RL, Satin J. The role of local cues in physiological differentiation of the embryonic heart: Ion channel development. In: Robert R, Schneider M, eds. Molecular Biology of the Cardiovascular System. New York: Alan R. Liss; 1990:73–98.
81. Josephson IR, Sperelakis N. Developmental increases in the inward-rectifying K^+ current of embryonic chick ventricular myocytes. Biochim Biophys Acta 1990; 1052:123–127.
82. Kell MJ, DeFelice LJ. Surface charge near the cardiac inward-rectifier channel measured from single-channel conductance. J Membr Biol 1988; 102:1–10.
83. Stanfield PR. Intracellular Mg^{2+} may act as a co-factor in ion channel function. Trends Neurosci 1988; 11:475–477.
84. Kleiman RB, Houser SR. Outward currents in hypertrophied feline ventricular myocytes. Prog Clin Biol Res 1990; 334:65–83.
85. Giles WR, Imaizumi Y. Comparison of potassium currents in rabbit atrial and ventricular cells. J Cardiovasc Pharmacol 1988; 405:123–145.
86. Burnashev NA, Edwards FA, Verkhratsky AN. Patch-clamp recording on rat cardiac muscle slices. Pfluegers Arch 1990; 417:123–125.
87. Sakmann B, Trube G. Conductance properties of single inwardly rectifying potassium channels in ventricular cells from guinea-pig heart. J Physiol (Lond) 1984; 347:641–657.
88. Balser JR, Roden DM, Bennett PB. Single inward rectifier potassium channels in guinea pig ventricular myocytes: Effects of quinidine. Biophysic J 1991; 59:150–161.
89. Carmeliet E, Mubagwa K. Characterization of the acetylcholine-induced potassium current in rabbit cardiac Purkinje fibres. J Physiol (Lond) 1986; 371:219–237.
90. Delmar M, Jalife J. Low Ba-induced pacemaker current in well-polarized cat papillary muscle. Am J Physiol 1987; 252:H258–H268.
91. Harvey RD, Ten Eick RE. Voltage-dependent block of cardiac inward-rectifying potassium current by monovalent cations. J Gen Physiol 1989; 94:349–361.

92. Hirano Y, Hiraoka M. Changes in K^+ currents induced by Ba^{2+} in guinea pig ventricular muscle. Am J Physiol 1986; 251:H24–H33.

93. Imoto Y, Ehara T, Matsuura H. Voltage- and time-dependent block of i_{K1} underlying Ba^{2+}-induced ventricular automaticity. Am J Physiol 1987; 252:H325–H333.

94. Isenberg G. Cardiac Purkinje fibers: Cesium as a tool to block inward rectifying potassium currents. Pfluegers Arch 1976; 365:99–106.

95. Kilborn MJ, Fedida D. A study of the developmental changes in outward currents of rat ventricular myocytes. J Physiol (Lond) 1990; 430:37–60.

96. Tourneur Y, Mitra RL, Morad M, Rougier O. Activation properties of the inward-rectifying potassium channel on mammalian heart cells. J Membr Biol 1987; 97:127–135.

97. Sato R, Hisatome I, Singer DH. Amiodarone blocks the inward-rectifier K^+ channel in guinea-pig ventricular myocytes. Circulation 1987; 76:IV-150. Abstract.

98. Benndorf K, Nilius B. Different blocking effects of Cd^{++} and Hg^{++} on the early outward current in myocardial mouse cells. Gen Physiol Biophys 1988; 7:345–452.

99. Giles WR, van Ginneken ACG. A transient outward current in isolated cells from the crista terminalis of rabbit heart. J Physiol (Lond) 1985; 368:243–264.

100. Hiraoka M, Kawano S. Calcium-sensitive and insensitive transient outward current in rabbit ventricular myocytes. J Physiol (Lond) 1989; 410:187–212.

101. Josephson IR, Sanchez-Chapula J, Brown AM. Early outward current in rat single ventricular cells. Circ Res 1984; 54:157–162.

102. Nakayama T, Irisawa H. Transient outward current carried by potassium and sodium in quiescent atrioventricular node cells of rabbits. Circ Res 1985; 57:65–73.

103. Simurda J, Simurdova M, Cupera P. 4-aminopyridine sensitive transient outward current in dog ventricular fibres. Pfluegers Arch 1988; 411:442–449.

104. Tseng GN, Hoffman BF. Two components of myocytes. Circ Res 1989; 64:633–647.

105. Tseng-Crank JL, Tseng GN, Schwartz A, Tanouye MA. Molecular cloning and functional expression of a potassium channel cDNA isolated from a rat cardiac library. FEBS Lett 1990; 268:63–68.

106. Coraboeuf E, Carmeliet E. Existence of two transient outward currents in sheep cardiac Purkinje fibers. Pfluegers Arch 1982; 392:352–359.

107. Escande D, Coulombe A, Faivre JF, Deroubaix E, Coraboeuf E. Two types of transient outward currents in adult human atrial cells. Am J Physiol 1987; 252:H142–H148.

108. Nakayama T, Fozzard HA. Adrenergic modulation of the transient outward current in isolated canine Purkinje cells. Circ Res 1988; 62:162–172.

109. Shibata EF, Drury T, Refsum H, Aldrete V, Giles W. Contributions of a transient outward current to repolarization in human atrium. Am J Physiol 1989; 257:H1773–H1781.

110. Benndorf K. Three types of single K channels contribute to the transient outward current in myocardial mouse cells. Biomed Biochim Acta 1988; 47:401–416.

111. Benndorf K, Markwardt F, Nilius B. Two types of transient outward currents in cardiac ventricular cells of mice. Pfluegers Arch 1987; 409:641–643.
112. Carmeliet E, Biermans G, Callewaert G, Vereecke J. Potassium current in cardiac cells. Experientia 1987; 43:1175–1184.
113. Josephson IR, Sperelakis N. Two types of outward K^+ channel currents in early embryonic chick ventricular myocytes. J Dev Physiol 1989; 12:201–207.
114. Kenyon JL, Gibbons WR. Influence of chloride, potassium, and tetraethylammonium on the early outward current of sheep cardiac Purkinje fibers. J Gen Physiol 1979; 73:117–138.
115. Fedida D, Giles WR. Regional variations in action potentials and transient outward current in myocytes isolated from rabbit left ventricle. J Physiol (Lond) 1991; 442:191–209.
116. Furukawa T, Myerburg RJ, Furukawa N, Bassett AL, Kimura S. Differences in transient outward currents of feline endocardial and epicardial myocytes. Circ Res 1990; 67:1287–1291.
117. Litovsky SH, Antzelvitch C. Transient outward current prominent in canine ventricular epicardium but not endocardium. Circ Res 1988; 62:116–126.
118. Apkon M, Nerbonne JM. α_1-Adrenergic agonists selectively suppress voltage-dependent K^+ currents in rat ventricular myocytes. Proc Natl Acad Sci USA 1988; 85:8756–8760.
119. Fedida D, Shimoni Y, Giles W. α-Adrenergic modulation of the transient outward current in rabbit atrial myocytes. J Physiol (Lond) 1990; 423:257–277.
120. Tseng GN, Robinson RB, Hoffman BF. Passive properties and membrane currents of canine ventricular myocytes. J Gen Physiol 1987; 90:671–701.
121. Castle NA. Bupivacaine inhibits the transient outward K^+ current but not the inward rectifier in rat ventricular myocytes. J Pharmacol Exp Ther 1990; 255:1038–1046.
122. Siegelbaum SA, Tsien RW, Kass EN. Role of intracellular calcium in the transient outward current of calf Purkinje fibres. Nature 1977; 269:611–613.
123. Baro I, Escande D. A long lasting Ca^{2+}-activated outward current in guinea-pig atrial myocytes. Pfluegers Arch 1989; 415:63–71.
124. Albitz R, Gainullin R, Kukushkin N, Nilius B, Saxon M. Contribution of a Ca-dependent component to the transient outward current in rabbit ventricular fibres. Biomed Biochim Acta 1988; 47:1077–1080.
125. Kenyon JL, Sutko JL. Calcium- and voltage-activated plateau currents of cardiac Purkinje fibers. J Gen Physiol 1987; 89:921–958.
126. Callewaert G, Vereecke J, Carmeliet E. Existence of a calcium-dependent potassium channel in the membrane of cow cardiac Purkinje cells. Pfluegers Arch 1986; 406:424–426.
127. Agus ZA, Dukes ID, Morad M. Divalent cations modulate the transient outward current in rat ventricular myocytes. Am J Physiol 1991; 261:C310–C318.
128. Dukes ID, Morad M. The transient K^+ current in rat ventricular myocytes: Evaluation of its Ca^+ and Na^+ dependence. J Physiol (Lond) 1991; 435:395–420.

129. Trautwein W, Gottstein U, Dudel J. Der Aktionsstrom der Myokardfaser im Sauerstoffmangel. Pfluegers Arch 1954; 260:40–60.
130. Vleugels A, Vereecke J, Carmeliet E. Ionic currents during hypoxia in voltage-clamped cat ventricular muscle. Circ Res 1980; 47:501–508.
131. Ashcroft SJH, Ashcroft FM. Properties and functions of ATP-sensitive K-channels. Cell Signal 1990; 2:197–214.
132. Davies NW, Standen NB, Stanfield PR. ATP-dependent potassium channels of muscle cells—Their properties, regulation, and possible functions. J Bioenergetics and Biomembranes 1991; 23:509–535.
133. Noma A. ATP-regulated K^+ channels in cardiac muscle. Nature 1983; 305:147–148.
134. Trube G, Hescheler J. Potassium channels in isolated patches of cardiac cell membrane. Naunyn-Schmiedeberg's Arch Pharmacol 1983; 322:R64. Abstract.
135. Friedrich M, Benndorf K, Schwalb M, Hirche H. Effects of anoxia on K and Ca currents in isolated guinea pig cardiocytes. Pfluegers Arch 1990; 416:207–209.
136. Isenberg G, Vereecke J, Heyden G, Carmeliet E. The shortening of the action potential by DNP in guinea-pig ventricular myocytes is mediated by an increase of a time-independent K conductance. Pfluegers Arch 1983; 397:251–259.
137. Noma A, Shibasaki T. Membrane current through adenosine-triphosphate-regulated potassium channels in guinea-pig ventricular cells. J Physiol (Lond) 1985; 363:463–480.
138. Weiss JN, Lamp ST. Cardiac ATP-sensitive K^+ channels: Evidence for preferential regulation by glycolysis. J Gen Physiol 1989; 94:911–935.
139. Wilde AAM, Escande D, Schumacher CA, Thuringer D, Mestre M, Fiolet JWT, Janse MJ. Potassium accumulation in the globally ischemic mammalian heart: A role for the ATP-sensitive potassium channel. Circ Res 1990; 67:835–843.
140. Nakamura S, Kiyosue T, Arita M. Glucose reverses 2,4-dinitrophenol induced changes in action potentials and membrane currents of guinea pig ventricular cells via enhanced glycolysis. Cardiovasc Res 1989; 23:286–294.
141. Nichols CG, Lederer WJ. The regulation of ATP-sensitive K^+ channel activity in intact and permeabilized rat ventricular myocytes. J Physiol (Lond) 1990; 423:91–110.
142. Pilsudski R, Rougier O, Tourneur Y. Reversible activation of the ATP-dependent potassium current with dialysis of frog atrial cells by micromolar concentrations of GDP. J Membr Biol 1990; 117:223–231.
143. Findlay I. ATP^{4-} and ATP·Mg inhibit the ATP-sensitive K^+ channel of rat ventricular myocytes. Pfluegers Arch 1988; 412:37–41.
144. Furukawa T, Kimura S, Furukawa N, Bassett AL, Myerburg RJ. Role of cardiac ATP-regulated potassium channels in differential responses of endocardial and epicardial cells to ischemia. Circ Res 1991; 68:1693–1702.
145. Kakei M, Noma A, Shibasaki T. Properties of adenosine-triphosphate-regulated potassium channels in guinea-pig ventricular cells. J Physiol (Lond) 1985; 363:441–462.
146. Lederer WJ, Nichols CG. Nucleotide modulation of the activity of rat heart ATP-sensitive channels in isolated membrane patches. J Physiol (Lond) 1989; 419:193–211.

147. Qin D, Takano M, Noma A. Kinetics of ATP-sensitive K^+ channel revealed with oil-gate concentration jump method. Am J Physiol 1989; 257:H1–H10.
148. Albitz R, Kammermeier H, Nilius B. Free energy of ATP-hydrolysis fails to affect ATP-dependent potassium channels in isolated mouse ventricular cells. J Mol Cell Cardiol 1990; 22:183–190.
149. Findlay I. Effects of ADP upon the ATP-sensitive K^+ channel in rat ventricular myocytes. J Membr Biol 1988; 101:83–92.
150. Fan Z, Nakayama K, Hiraoka M. Pinacidil activates the ATP-sensitive K^+ channel in inside-out and cell-attached patch membranes of guinea-pig ventricular myocytes. Pfluegers Arch 1990; 415:387–394.
151. Kakei M, Noma A. Adenosine-5′-triphosphate-sensitive single potassium channel in the atrioventricular node cell of the rabbit heart. J Physiol (Lond) 1984; 352:265–284.
152. Sanguinetti MC. Na^+_i-activated and ATP-sensitive K^+ channels in the heart. Prog Clin Biol Res 1990; 334:85–109.
153. Zilberter Y, Burnashev NA, Papin A, Portnov V, Khodorov BI. Gating kinetics of ATP-sensitive single potassium channels in myocardial cells depends on electromotive force. Pfluegers Arch 1988; 411:584–589.
154. Cameron JS, Kimura S, Jackson-Burns DA, Smith BD, Bassett AL. ATP-sensitive K^+ channels are altered in hypertrophied ventricular myocytes. Am J Physiol 1988; 255:H1254–H1258.
155. Tseng GN, Hoffman BF. Actions of pinacidil on membrane currents in canine ventricular myocytes and their modulation by intracellular ATP and cAMP. Pfluegers Arch 1990; 415:414–424.
156. Escande D, Thuringer D, Le Guern S, Courteix J, Laville M, Cavero I. Potassium channel openers act through an activation of ATP-sensitive K^+ channels in guinea-pig cardiac myocytes. Pfluegers Arch 1989; 414:669–675.
157. Kim D, Duff RA. Regulation of K^+ channels in cardiac myocytes by free fatty acids. Circ Res 1990; 67:1040–1046.
158. Nakayama K, Fan Z, Marumo F, Hiraoka M. Interrelation between pinacidil and intracellular ATP concentrations on activation of the ATP-sensitive K^+ current in guinea pig ventricular myocytes. Circ Res 1990; 67:1124–1133.
159. Thuringer D, Escande D. Apparent competition between ATP and the potassium channel opener RP 49356 on ATP-sensitive K^+ channels of cardiac myocytes. Mol Pharmacol 1989; 36:897–902.
160. Sakmann B, Neher E. Geometric parameters of pipettes and membrane patches. In: Sakmann B, Neher E, eds. Single Channel Recording. New York: Plenum; 1983:37–51.
161. Benndorf K, Friedrich M, Hirche H. Anoxia opens ATP regulated K-channels in isolated heart cells of the guinea pig. Pfluegers Arch 1991; 419:108–110.
162. Faivre JF, Findlay I. Action potential duration and activation of ATP-sensitive potassium current in isolated guinea-pig ventricular myocytes. Biochim Biophys Acta 1990; 1029:167–172.
163. Escande D. The pharmacology of ATP-sensitive K^+ channels in the heart. Pfluegers Arch 1989; 414:93–98.
164. Findlay I. ATP-sensitive K^+ channels in rat ventricular myocytes are blocked and inactivated by internal divalent cations. Pfluegers Arch 1987; 410:313–320.

165. Horie M, Irisawa H, Noma A. Voltage-dependent magnesium block of adenosine-triphosphate-sensitive potassium channel in guinea-pig ventricular cells. J Physiol (Lond) 1987; 387:251–272.

166. DeWeille JR, Lazdunski M. Regulation of the ATP-sensitive potassium channel. In: Narahashi T, ed. Ion Channels, Vol. 2. New York: Plenum; 1990:205–222.

167. Findlay I. Calcium-dependent inactivation of the ATP-sensitive K^+ channel of rat ventricular myocytes. Biochim Biophys Acta 1988; 943:297–304.

168. Kirsch GE, Codina J, Birnbaumer L, Brown AM. Coupling of ATP-sensitive K^+ channels to A_1 receptors by G proteins in rat ventricular myocytes. Am J Physiol 1990; 259:H820–H826.

169. Cuevas J, Bassett AL, Cameron JS, Furukawa T, Myerburg RJ, Kimura S. Effect of H^+ on ATP-regulated K^+ channels in feline ventricular myocytes. Am J Physiol 1991; 261:H755–H761.

170. Haworth RA, Goknur AB, Berkoff HA. Inhibition of ATP-sensitive potassium channels of adult rat heart cells by antiarrhythmic drugs. Circ Res 1989; 65:1157–1160.

171. Randle JCR, Oliet SHR, Renaud JF. Alkali cation permeability and caesium blockade of cromakalim-activated current in guinea-pig ventricular myocytes. Br J Pharmacol 1991; 103:1795–1801.

172. Arena JP, Kass RS. Activation of ATP-sensitive K channels in heart cells by pinacidil: Dependence on ATP. Am J Physiol 1989; 257:H2092–H2096.

173. Fan Z, Nakayama K, Hiraoka M. Multiple actions of pinacidil on adenosine triphosphate-sensitive potassium channels in guinea-pig ventricular myocytes. J Physiol (Lond) 1990; 430:273–295.

174. Hiraoka M, Fan Z. Activation of ATP-sensitive outward K^+ current by nicorandil (2-nicotinamidoethyl nitrate) in isolated ventricular myocytes. J Pharmacol Exp Ther 1989; 250:278–285.

175. Martin CL, Chinn K. Pinacidil opens ATP-dependent K^+ channels in cardiac myocytes in an ATP- and temperature-dependent manner. J Cardiovasc Pharmacol 1990; 15:510–514.

176. Pilsudski R, Rougier O, Tourneur Y. Action of cromakalim on potassium membrane conductance in isolated heart myocytes of frog. Br J Pharmacol 1990; 100:581–587.

177. Ripoll C, Lederer WJ, Nichols CG. Modulation of ATP-sensitive K^+ channel activity and contractile behaviour in mammalian ventricle by the potassium channel openers cromakalim and RP49356. J Pharmacol Exp Ther 1990; 255:429–435.

178. Robertson DW, Steinberg MI. Potassium channel modulators: Scientific applications and therapeutic promise. J Med Chem 1990; 33:1530–1541.

179. Tung RT, Kurachi Y. On the mechanism of nucleotide diphosphate activation of the ATP-sensitive K^+ channel in ventricular cell of guinea-pig. J Physiol (Lond) 1991; 437:239–256.

180. Richer C, Pratz J, Mulder P, Mondot S, Guidicelli JF, Cavero I. Cardiovascular and biological effects of K^+ channel openers, a class of drugs with vasorelaxant and cardioprotective properties. Life Sci 1990; 47:1693–1705.

181. DeWeille JR, Fosset M, Mourre C, Schmid-Antomarchi H, Bernardi H, Lazdunski M. Pharmacology and regulation of ATP-sensitive K^+ channels. Pfluegers Arch 1989; 414:80–87.

182. Faivre JF, Findlay I. Effects of tolbutamide, glibenclamide and diazoxide upon action potentials recorded from rat ventricular muscle. Biochim Biophys Acta 1989; 984:1–5.

183. Wilde AAM, Escande D, Schumacher CA, Thuringer D, Mestre M, Fiolet JWT. Glibenclamide inhibition of ATP-sensitive K^+ channels and ischemia-induced K^+ accumulation in the mammalian heart. Pfluegers Arch 1989; 414:S176. Abstract.

184. Löffelholz K, Pappano AJ. The parasympathetic neuro-effector junction of the heart. Pharmacol Rev 1985; 37:1–24.

185. Schimerlik MI. Structure and regulation of muscarinic receptors. Annu Rev Physiol 1989; 51:217–227.

186. Szabo G, Otero AS. G protein mediated regulation of K^+ channels in heart. Annu Rev Physiol 1990; 52:293–305.

187. Brown AM. Regulation of heartbeat by G protein-coupled ion channels. Am J Physiol 1990; 259:H1621–H1627.

188. Brown AM, Birnbaumer L. Ionic channels and their regulation by G protein subunits. Annu Rev Physiol 1990; 52:197–213.

189. Brown AM, Yatani A, Codina J, Birnbaumer L. Gating of atrial muscarinic K^+ channels by G proteins. Prog Clin Biol Res 1990; 334:303–312.

190. Cerbai E, Klöckner U, Isenberg G. Ca-antagonistic effects of adenosine in guinea pig atrial cells. Am J Physiol 1988; 255:H872–H878.

191. Codina J, Yatani A, Grenet D, Brown AM, Birnbaumer L. The α subunit of the GTP binding protein G_k opens atrial potassium channels. Science 1987; 236:442–445.

192. Inomata N, Ishihara T, Akaike N. Activation kinetics of the acetylcholine-gated potassium current in isolated atrial cells. Am J Physiol 1989; 257:H646–H650.

193. Kirsch GE, Brown AM. Trypsin activation of atrial muscarinic K^+ channels. Am J Physiol 1989; 257:H334–H338.

194. Mattera R, Yatani A, Kirsch GE, Graf R, Okabe K, Olate J, Codina J, Brown AM, Birnbaumer L. Recombinant α_i-3 subunit of G protein activates G_k-gated K^+ channels. J Biol Chem 1989; 264:465–471.

195. Pfaffinger PJ, Martin JM, Hunter DD, Nathanson NM, Hille B. GTP-binding proteins couple cardiac muscarinic receptors to a K channel. Nature 1985; 317:536–538.

196. Sakmann B, Noma A, Trautwein W. Acetylcholine activation of single muscarinic K^+ channels in isolated pacemaker cells of the mammalian heart. Nature 1983; 303:250–253.

197. Yatani A, Codina J, Brown AM, Birnbaumer L. Direct activation of mammalian atrial muscarinic potassium channels by GTP regulatory protein G_k. Science 1987; 235:207–211.

198. Yatani A, Mattera R, Codina J, Graf R, Okabe K, Padrell E, Iyengar R, Brown AM, Birnbaumer L. The G protein-gated atrial K^+ channel is stimulated by three distinct G_i alpha-subunits. Nature 1988; 336:680–682.

199. Breitwieser GE, Szabo G. Uncoupling of cardiac muscarinic and β-adrenergic receptors from ion channels by a guanine nucleotide analogue. Nature 1985; 317:538–540.

200. Sorota S, Hoffman BF. Role of G-proteins in the acetylcholine-induced potassium current of canine atrial cells. Am J Physiol 1989; 257:H1516–H1522.

201. Heidbüchel H, Callewaert G, Vereecke J, Carmeliet E. ATP-dependent activation of atrial muscarinic K$^+$ channels in the absence of agonist and G-nucleotides. Pfluegers Arch 1990; 416:213–215.
202. Otero AS, Breitwieser GE, Szabo G. Activation of muscarinic potassium currents by ATP S in atrial cells. Science 1988; 242:443–445.
203. Kurachi Y, Ito H, Sugimoto T, Katada T, Michio U. Activation of atrial muscarinic K$^+$ channels by low concentrations of βγ subunits of rat brain G protein. Pfluegers Arch 1989; 413:325–327.
204. Logothesis DE, Kim D, Northup JK, Neer EJ, Clapham DE. Specificity of action of guanine nucleotide-binding regulatory protein subunits on the cardiac muscarinic K$^+$ channel. Proc Natl Acad Sci USA 1988; 85:5814–5818.
205. Logothesis DE, Kurachi Y, Galper J, Neer EJ, Clapham DE. The βγ subunits of GTP-binding proteins activate the muscarinic K$^+$ channel in heart. Nature 1987; 325:321–326.
206. Okabe K, Yatani A, Evans T, Ho YK, Codina J, Birnbaumer L, Brown AM. βγ Dimers of G proteins inhibit atrial muscarinic K$^+$ channels. J Biol Chem 1990; 265:12854–12858.
207. Yatani A, Okabe K, Birnbaumer L, Brown AM. Detergents, dimeric Gβγ, and eicosanoid pathways to muscarinic atrial K$^+$ channels. Am J Physiol 1990; 258:H1507–H1514.
208. Kim D, Lewis DL, Graziadei L, Neer EJ, Bar-Sagi D, Clapham DE. G-protein βγ-subunits activate the cardiac muscarine K$^+$-channel via phospholipase A$_2$. Nature 1989; 337:557–560.
209. Scherer RW, Breitwieser GE. Arachidonic acid metabolites alter G protein-mediated signal transduction in heart: Effects on muscarinic K$^+$ channels. J Gen Physiol 1990; 96:735–755.
210. Bailey JC, Rardon DP. Electrophysiological effects of adenosine and dipyridamole on cardiac Purkinje fibers and ventricular myocardium. Prog Clin Biol Res 1987; 230:119–133.
211. Belardinelli L, Giles W, West GA. Ionic mechanisms of adenosine actions in pacemaker cell from rabbit heart. J Physiol (Lond) 1988; 405:615–633.
212. Friel DD, Bean BP. Two ATP-activated conductances in bullfrog atrial cells. J Gen Physiol 1988; 91:1–27.
213. Friel DD, Bean BP. Dual control by ATP and acetylcholine of inwardly rectifying K$^+$ channels in bovine atrial cells. Pfluegers Arch 1990; 415:651–657.
214. Isenberg G, Cerbai E, Klöckner U. Ionic channels and adenosine in isolated heart cells. In: Gerlach E, ed. Topics and Perspectives in Adenosine Research. Berlin: Springer-Verlag; 1987:323–335.
215. Kim D. Calcitonin-gene-related peptide activates the muscarinic-gated K$^+$ current in atrial cells. Pfluegers Arch 1991; 418:338–345.
216. Kim D. Endothelin activation of an inwardly rectifying K$^+$-current in atrial cells. Circ Res 1991; 69:250–255.
217. Kurachi Y, Nakajima T, Sugimoto T. On the mechanism of activation of muscarinic K$^+$ channels by adenosine in isolated atrial cells: Involvement of GTP-binding proteins. Pfluegers Arch 1986; 407:264–274.
218. Lerman BB, Belardinelli L. Cardiac electrophysiology of adenosine—Basic and clinical concepts. Circulation 1991; 83:1499–1509.

219. Pelleg A, Mitamura H, Mitsuoka T, Mazgalev T, Michelson EL, Dreifus LS. Interactive negative chronotropic actions of adenosine and verapamil on the canine sinus node in vivo. Prog Clin Biol Res 1987; 230:235–252.
220. West GA. Actions of adenosine on the sinus node. Prog Clin Biol Res 1987; 230:97–108.
221. Boyett MR, Kirby MS, Orchard CH, Roberts AB. The negative inotropic effect of acetylcholine on ferret ventricular myocardium. J Physiol (Lond) 1988; 404:613–635.
222. Carmeliet E, Mubagwa K. Desensitization of the acetylcholine-induced increase of potassium conductance in rabbit cardiac Purkinje fibres. J Physiol (Lond) 1986; 371:239–255.
223. Kim D. Modulation of acetylcholine-activated K^+ channel function in rat atrial cells by phosphorylation. J Physiol (Lond) 1991; 437:133–155.
224. Kurachi Y, Nakajima T, Sugimoto T. Short-term desensitization of muscarinic K^+ channel current in isolated atrial myocytes and possible role of GTP-binding proteins. Pfluegers Arch 1987; 410:227–233.
225. Kaibara M, Nakajima T, Irisawa H, Giles W. Regulation of spontaneous opening of muscarinic K^+ channels in rabbit atrium. J Physiol (Lond) 1991; 433:589–613.
226. Kim D. β-Adrenergic regulation of the muscarinic-gated K^+ channel via cyclic AMP-dependent protein kinase in atrial cells. Circ Res 1990; 67:1292–1298.
227. Soejima M, Noma A. Mode of regulation of the ACh-sensitive K-channel by the muscarinic receptor in rabbit atrial cells. Pfluegers Arch 1984; 400:424–431.
228. Yatani A, Polakis P, Halenbeck R, McCormick F, Brown AM. Ras p21 and GAP inhibit coupling of muscarinic receptors to atrial K^+ channels. Cell 1990; 61:769–776.
229. Sato R, Hisatome I, Wasserstrom JA, Arentzen CE, Singer DH. Acetylcholine-sensitive potassium channels in human atrial myocyte. Am J Physiol 1990; 259:H1730–H1735.
230. Shumaker JM, Clark JW, Giles W, Szabo G. A model of the muscarinic receptor-induced changes in K^+-current and action potentials in the bullfrog atrial cell. Biophysic J 1990; 57:567–576.
231. Irisawa H. Membrane currents in cardiac pacemaker tissue. Experientia 1987; 43:1131–1240.
232. Noma A, Peper K, Trautwein W. Acetylcholine-induced potassium current fluctuations in the rabbit sino-atrial node. Pfluegers Arch 1979; 381:255–262.
233. Lewis DL, Clapham DE. Somatostatin activates an inward rectifying K^+ channel in neonatal rat atrial cells. Pfluegers Arch 1989; 414:492–494.
234. Kurachi Y, Ito H, Sugimoto T, Shimizu T, Miki I, Ui M. Arachidonic acid metabolites as intracellular modulators of the G protein-gated cardiac K^+ channel. Nature 1989; 337:555–557.
235. Ito H, Takikawa R, Iguchi M, Hamada E, Sugimoto T, Kurachi Y. Heparin uncouples the muscarinic receptors from G_K protein in the atrial cell membrane of the guinea-pig heart. Pfluegers Arch 1990; 417:126–128.
236. Nakajima T, Irisawa H, Giles W. N-ethyl-maleimide uncouples muscarinic receptors from acetylcholine-sensitive potassium channels in bullfrog atrium. J Gen Physiol 1990; 96:887–903.

237. Nakajima T, Kaibara M, Irisawa H, Giles W. Inhibition of the muscarinic receptor-activated K^+ current by N-ethylmaleimide in rabbit heart. Naunyn-Schmiedeberg's Arch Pharmacol 1991; 343:14–19.
238. Martin AR, Dryer SE. Potassium channels activated by sodium. Quarterly Jour Exp Physiol 1989; 74:1033–1041.
239. Kameyama M, Kakei M, Sato R, Shibasaki T, Matsuda H, Irisawa H. Intracellular Na^+ activates a K^+ channel in mammalian cardiac cells. Nature 1984; 309:354–356.
240. Luk HN, Carmeliet E. Na^+-activated K^+ current in cardiac cells: Rectification, open probability, block and role in digitalis toxicity. Pfluegers Arch 1990; 416:766–768.
241. Rodrigo GC, Chapman RA. A sodium-activated potassium current in intact ventricular myocytes isolated from the guinea-pig heart. Exp Physiol 1990; 75:839–842.
242. Wang Z, Kimitsuki T, Noma A. Conductance properties of the Na^+-activated K^+ channel in guinea-pig ventricular cells. J Physiol (Lond) 1991; 433:241–257.
243. Kim D, Clapham DE. Potassium channels in cardiac cells activated by arachidonic acid and phospholipids. Science 1989; 244:1174–1176.
244. Yue DT, Marban E. A novel cardiac potassium channel that is active and conductive at depolarized potentials. Pfluegers Arch 1988; 413:127–133.

16
Nonrenal Potassium Homeostasis: Hypokalemia and Potaasium Depletion—Role of Skeletal Muscle Potassium-Pump (Na$^+$,K$^+$-ATPase)

Keld Kjeldsen

Venous plasma-potassium concentration is one of the most frequently determined values in clinical practice. However, although its long-term regulation by the kidney is well known, its potential for modulation within seconds to minutes due to exchange of potassium across skeletal muscle membranes has not hitherto been well recognized. Moreover, venous plasma potassium is not a very accurate indicator of body potassium homeostasis. Accordingly, in certain clinical situations venous plasma-potassium values may be difficult to interpret and even misleading. This may be the reason why the meaning of mild hypokalemia and its dangers are often questioned and the need for potassium supplementation debated. However, recently research on the regulation of skeletal muscle potassium pumps Na$^+$,K$^+$-ATPase (NKA) have shed light on the rapid regulation of plasma potassium. The results of this work may have important clinical implications. Hence it is the purpose of the present chapter to discuss the physiology of potassium homeostasis in relation to clinical experience and the possible basis for a rational approach to disturbances of potassium homeostasis in clinical practice.

Body Potassium Homeostasis

The average potassium intake in normal adult human subjects is approximately 100 mmol/day, derived primarily from meat, vegetables, and fruit. Potassium is absorbed mainly in the upper gastrointestinal tract. Total daily secretion of potassium into the gut is around 60 mmol/day. Most of it is reabsorbed and only around 5 mmol/day are usually excreted in the stool. The remaining 95 mmol/day are excreted in urine and represent the balance between the total amount filtered (around 700 mmol/day), the amount reabsorbed in the proximal tubules, and the amount secreted by the distal tubules and collecting ducts. This secretion occurs in competition with H$^+$ ions and is stimulated by aldosterone. The minimum amount of potassium excreted in the urine has not been well defined, but is probably around 25 mmol/day.[1] Thus, by adjusting potassium excretion, the kidneys are the main regulator of body potassium content.

The total quantity of potassium in a normal adult human subject is around 3500 mmol. About 2% of it or 65 mmol are located in the extracellular fluid at concentrations of 3.5–5.0 mmol/l; the rest is located within the cells, mainly in skeletal muscles, liver, bone, and erythrocytes, at a concentration of 140 to 150 mmol/l of cell water. Since the skeletal muscle constitutes around 40% of body weight in normal adult human subjects, this tissue constitutes the largest body potassium compartment, containing around 75% of body potassium. Thus minor changes in potassium content in the skeletal muscle pool may result in major changes in plasma potassium concentration. It can be calculated that an increase or a decrease in muscle potassium concentration of 1% could result in an increase or a decrease of 50% in the concentration of plasma potassium. In practice, however, such changes in plasma potassium are dampened by translocation of potassium across the cell membrane. Indeed, it has been found that 70–80% of an acute potassium load is translocated rapidly into cells to maintain relatively constant plasma-potassium values.[2] Thus the potassium pool in the skeletal muscle is the main basis for regulation of potassium distribution among compartments.

Skeletal Muscles and Potassium Homeostasis

A major challenge to potassium homeostasis occurs during exercise since large amounts of potassium leak out of muscle cells during contraction (Figure 16.1). The potassium concentration in the interstitial space of human skeletal muscles may during exercise reach values of up to 8–15 mmol/l.[3,4] Nevertheless the concentration of potassium in venous as well as arterial plasma does not increase more than 1–2 mmol/l during moderate muscle activity.[5–8] However, after prolonged heavy exercise venous plasma-potassium concentrations up to 10 mmol/l have been reported[9] while during the first minutes the arterial blood plasma-potassium concentration may increase by 3–4 mmol/l.[10,11] Furthermore, within the first couple of minutes of rest after physical activity the plasma-potassium concentration may drop 0.5 mmol/l below the steady-state level.[7,12,13] These changes reflect the importance of extracellular potassium levels in the regulation of function. Thus the extracellular concentration of potassium in skeletal muscle must be kept low enough during contraction to prevent muscle fatigue,[14–16] but high enough to increase blood flow to the working muscle by causing vasodilation[3,14,16] and to stimulate ventilation and heart rate trough afferent nervous fibers.[16,17] Further-more, since the heart is exposed to the shifts in plasma potassium concentration, potassium shifts must be kept within limits in order to prevent

Exercising Muscles

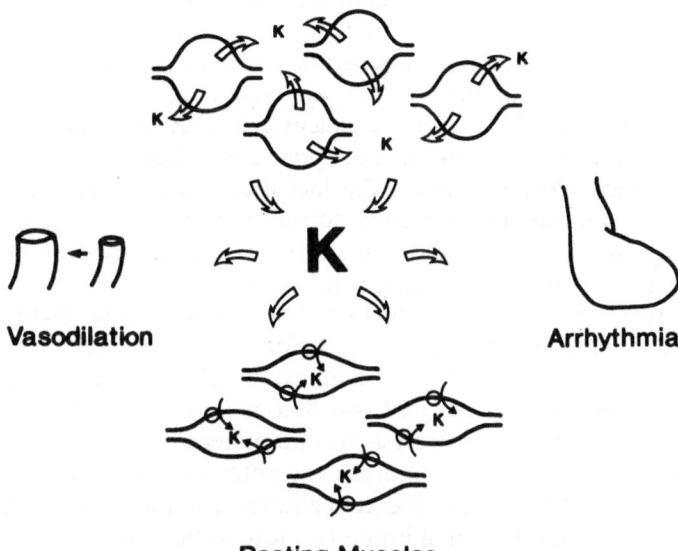

Vasodilation   Arrhythmia

Resting Muscles

FIGURE 16.1. Schematic diagram of the potassium homeostasis during muscle activity. The nervous impulse depolarizes skeletal muscle plasma membrane, which causes potassium to leak out of contracting cells. Potassium not taken up again into resting cells equilibrate with plasma. The vasodilatatory effect of increased interstitial potassium concentration, as well as the arrhythmic effect of subsequent increase in plasma-potassium concentration on the heart, are indicated.

cardiac arrhythmias. Thus a failure of skeletal muscle potassium homeostasis that may be tolerated at rest may become dangerous during exercise.[18] This may be very important in patients with heart disease whose myocardium may be especially vulnerable to shifts in electrolyte concentration.[19]

The amount of potassium that leaves the skeletal muscle depends on the balance between potassium influx and efflux. Potassium efflux is of the order of 10 nmol/g wet wt./contraction.[20] Since normal muscle activity involves about 1500 contractions/min,[14] the leak of potassium from skeletal muscles may amount to 15 mmol/kg of muscle/min. Accordingly, in a normal adult human subject the maximum potassium leakage obtained when all muscles are working is in the range 420 mmol/min, an amount potentially capable of doubling the plasma potassium in about 10 sec. Potassium influx in resting muscles is around 0.5 μmol/g wet wt./min of which approximately 35% is due to active transport.[21] During maximum stimulation the active transport system has the capacity to increase its potassium uptake to around $5 \mu mol/g$ wet wt./min.[22] If this could be obtained in a normal adult human subject, the skeletal muscle pool could clear 140 mmol potassium/min, which corresponds to a capacity to remove the entire amount of potassium from the extracellular space in about 30 sec. When all the muscles are active and potassium influx is maximally stimulated, it can be calculated that the skeletal muscle potassium pool may show a net potassium loss of 280 mmol/min (420 mmol/min − 120 mmol/min). The fact that muscle work can continue without a deleterious rise in plasma potassium is probably due to the fact that during work some muscle fibers are working and release potassium while other are resting and taking it in[15] (Figure 16.1).

The active transport of potassium across the plasma membrane in skeletal muscles and other tissues is performed by the membrane-bound Na^+,K^+-ATPase or sodium, potassium pump (Na,K pump).[23] The pump transports three Na ions out and two K ions into the cell per cycle, and derives energy from the hydrolysis of one molecule of ATP. Aided by a higher permeability of the membrane for potassium than for sodium, the pump creates and maintains the concentration gradients of sodium and potassium across the cell membrane (ie, low sodium and high potassium concentrations inside, and vice versa outside, the cell), as well as the membrane potential, which is essential for excitability and contractility. The sodium gradient provides energy for the coupled sodium, calcium and sodium, hydrogen exchange (countertransport) and for the sodium, amino acid exchange (cotransport) across the cell membrane, ie, for the cellular calcium and hydrogen clearance and the amino acid uptake (Figure 16.2). Thus if the sodium gradient across the muscle cell membrane decreases, the capacity for calcium and hydrogen clearance and for amino acid uptake are also reduced.

In human subjects the Na,K pump concentration in skeletal muscles has been found to be in the range 200−400 pmol/g wet wt.[24] Assuming a mean concentration of 300 pmol/g wet wt. and a molecular activity of the Na,K pump of 8000 cycles per minute at 37°C, which corresponds to a transport of 16,000 potassium ions/min, it can be calculated that the human skeletal muscles have a maximum capacity for active potassium uptake of $4.8 \mu mol/g$ wet wt./min.[25] Thus the total skeletal muscle pool in a healthy adult human subject has a total capacity for active potassium clearance of 134 mmol/min. This value is in good agreement with that calculated on the basis of flux studies (see above), which indicates good agreement between measurements of transport and Na,K-pumps. The high capacity for active potassium uptake by skeletal muscles indicates a need for a precise regulation of the Na,K pumps. This is achieved by a rapid regulation of the activity of existing Na,K pumps and a long-term regulation of the number of Na,K pumps. Whereas changes in pump activity can take place within seconds or minutes, changing the number of pumps requires days or weeks since it involves changing the rates of

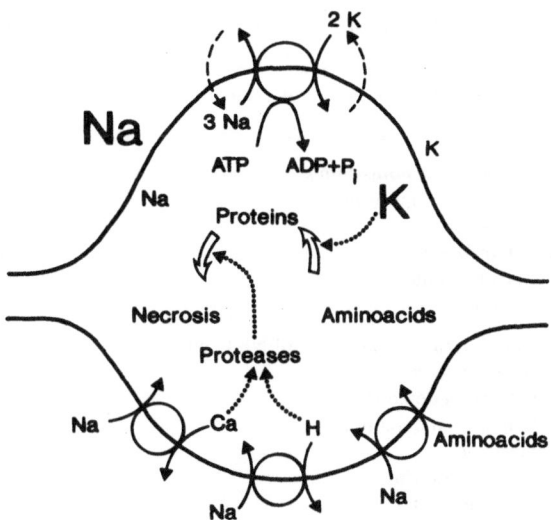

FIGURE 16.2. Schematic diagram of the Na,K pump in skeletal muscles and the effect of potassium depletion on protein synthesis and degradation. (See text for details.)

pump synthesis and degradation. The main factors that regulate skeletal muscle Na,K pumps are listed in Table 16.1. (For greater details the reader may consult references 25, 26, and 27.)

Rapid Regulation of the Na,K Pump in Skeletal Muscle

Local Factors

The local concentrations of sodium, potassium, and ATP are of major importance for the activity of the Na,K pumps.[23] In particular, the pumps are stimulated by high intracellular concentrations of sodium and high extracellular concentrations of potassium. Due to such stimulation a reduced number of pumps may at rest be compensated for by increased activity of the remaining ones. This is possible since skeletal muscle fibers at steady state use only a small portion of their total active Na,K-transport capacity.[25] Hence a reduced Na,K-pump concentration may be disclosed most easily when the need for recruitment is high. Furthermore, excitation of muscle fibers by electrical stimulation has been shown to activate skeletal muscle Na,K pumps. This is caused partly by the increasing intracellular sodium concentration and partly by the electrical stimulation per se.[28,29] A major decrease in intracellular ATP concentrations reduces Na,K-pump function.[23] This may be of

TABLE 16.1. Major factors that regulate skeletal muscle Na,K pumps.

Regulation of activity of Na,K pumps
Local factors
  Intracellular sodium
  Extracellular potassium
  Adenosine triphosphate
  Excitation
Systemic factors
  Catecholamines
  Insulin

Regulation of concentration of Na,K pumps
Downregulation
  Potassium depletion
  Magnesium depletion
  Starvation
  Hypothyroidism
  Physical inactivity
Upregulation
  Potassium loading
  Hyperthyroidism
  Physical conditioning

importance for the release of potassium from muscle tissue during severe ischemia. However, other factors such as oxygen free radicals may influence the Na,K pumps during ischemia.[30] The putative existence of an endogenous digitalis-like regulator of the Na,K pump is often discussed. Although it may exist, there is at present no firm evidence for its physiologic significance in the regulation of potassium homeostasis.[31]

Catecholamines

Both adrenaline and noradrenaline at physiologic concentrations stimulate potassium uptake by the Na,K pump in skeletal muscles via β_2-adrenoceptor-mediated synthesis of cyclic adenosine 3′,5′-phosphate (cAMP).[32] This phenomenon has several beneficial and non beneficial clinical implications. Thus β_2-adrenoceptor agonists that cause a reduction of plasma-potassium concentration of around 1 mmol/l may be used to suppress attacks of hyperkalemic periodic paralysis[33] and as acute provisional treatment of hyperkalemia in renal insufficiency.[34,35] Conversely, the muscle fatigue experienced by patients who receive β-adrenoceptor antagonists (especially non-cardioselective) may be related to the increased rise in interstitial potassium concentration in skeletal muscles, as mirrored in the additional rise in venous plasma-potassium concentration of around 0.5 mmol/l during exercise and in the delayed return to baseline during the rest period, which are caused by suppression of the catecholamine-induced activation of the pumps.[36-38]

Both β_2-adrenoceptor agonists and theophylline are widely used bronchodilators, or as in combination. Treatment with these drugs is associated with cAMP-mediated lowering of plasma potassium.[39,40] It should be noted that severe hypokalemia has been suggested as a possible cause of sudden, unexpected death in asthmatic patients.[41] Thus meticulous monitoring of venous plasma potassium seems mandatory during intensive bronchodilator therapy of asthma, especially in patients already at risk for hypokalemia.

The concentration of plasma catecholamines is known to increase during exercise and training has been found to increase this response.[42,43] This may explain the reduced rise in venous plasma-potassium concentration during and following exercise after physical conditioning.[6] On the other hand, some cases of sudden death that occurred shortly after the sudden interruption of severe exercise in otherwise healthy subjects have been attributed to catecholamine-induced hypokalemia.[13] Thus the observation that ingestion of a potassium-glucose solution during the recovery period after exercise raises plasma-potassium concentration is of particular interest.[16]

In acute myocardial infarction hypokalemia seems to increase the incidence of serious ventricular arrhythmias.[44] In this case hypokalemia may be caused by stress-induced catecholamine release. The beneficial prophylactic effect of β-adrenoceptor antagonist on morbidity and mortality in acute myocardial infarction, in addition to the well-known protective effect against rupture of the myocardium, may be associated with reduced catecholamine-induced hypokalemia.[45] It should be noted that catecholamine-induced hypokalemia may be further aggravated by a preexisting reduction in the plasma or skeletal muscle potassium concentration.

Insulin

Insulin, independently of its action on glucose metabolism, stimulates the active uptake of potassium in skeletal muscles by increasing Na,K-pump activity.[46] Conversely, hyperkalemia stimulates insulin secretion,[47] which creates a feedback system where hyperkalemia stimulates the secretion of insulin, which in turn promotes potassium uptake by the skeletal muscles. The effect of insulin on the Na,K pump in skeletal muscles and the resulting translocation of potassium from the extracellular to the intracellular compartment provide the rationale for acute provisional treatment of hyperkalemia with infusions of insulin and glucose, a procedure which may temporarily reduce plasma potassium concentration by approximately 1 mmol/l until permanent removal of potassium from the body can be achieved. In agreement with the concept that catecholamines and insulin activate skeletal muscle Na,K pumps by different mechanisms, the hypokalemic effect of β_2-adrenoceptor agonists and of insulin appear to be additive.[35] This may be of benefit in clinical treatment of acute hyperkalemia, but it may also have undesirable effects since insulin-induced hypoglycemia may stimulate catecholamine secretion and thus cause severe hypokalemia.[27]

Long-Term Regulation of the Na,K Pump in Skeletal Muscle

Potassium

The earliest and most pronounced change seen in experimental animals during potassium depletion is a decrease in the concentration of plasma potassium that develops within a few days. During the following weeks this is accompanied by a marked reduction in the potassium content and a concomitant increase in the sodium content of the skeletal muscles. The heart shows similar, although more modest changes, whereas the content of potassium and sodium in the brain, liver, and erythrocytes remains almost normal even during severe and prolonged potassium depletion.[48–50]

The loss of skeletal muscle potassium is associated with a progressive reduction in the concentration of Na,K pumps in skeletal muscles to around 30% of normal value[49,50] and a minor decrease in the myocardial Na,K-pump concentration,[51,52] whereas the Na,K-pump concentration of peripheral nerves[53] and brain does not change.[54,55] An upregulation of the NKA concentration has been found in animal[56] and human[57] erythrocytes and in cultured tumor cells[58] and myocytes[59] exposed to low potassium concentrations. The oral administration of potassium to severely potassium-depleted rats normalizes plasma as well as muscle potassium concentration within 24 h, whereas normalization of skeletal muscle Na,K-pump concentration requires 6 days of potassium repletion.[49,50] The studies on experimental potassium depletion and Na,K pumps have been reviewed in detail elsewhere.[25,26,32]

In patients loss of body potassium is also followed by a substantial decrease in the skeletal muscle potassium concentration. Thus in 25 patients who received diuretics, a 19% reduction in muscle potassium concentration and a 18% reduction in Na,K-pump concentration in skeletal muscle biopsies was observed.[60] It should be noted that only one of these patients had hypokalemia (plasma potassium 3.4 mmol/l). It is also of major interest that these changes were observed in spite of the fact that 52% of the patients received potassium supplementation.

The decrease in Na,K-pump concentration in skeletal muscles during potassium depletion, which leads to a selective major loss of potassium from skeletal muscles, may act as an important protective potassium-sparing mechanism for more vital organs, at least for some time. Given the large ion exchange during contraction, skeletal muscle may be the first tissue to feel the brunt of hypokalemia. Potassium loss, in turn, may directly or indirectly decrease synthesis of Na,K pumps. In resting muscles the increase in sodium content that accompanies potassium depletion may stimulate the activity of Na,K pumps, which compensates for the decrease in pump concentration and keeps the pump-mediated potassium uptake at normal levels,[21] while during maximal pump stimulation the Na,K-pump deficit may become decompensated and the pump-mediated potassium uptake may decrease.[22] Thus during potassium depletion the capacity for potassium reuptake during muscle activity may be insufficient, which leads to further potassium loss and to the establishment of a vicious cycle. In rats potassium depletion has been found to be associated with a reduced capacity for clearance of a potassium load,[61] which leads to the suggestion that the rapid replenishment of a major long-lasting potassium deficit in human subjects may be dangerous due to decreased clearance from plasma. Since normalization of muscle potassium may occur almost immediately, whereas the synthesis of new pumps requires several days, potassium depleted subjects may remain sensitive to major challanges to their potassium homeostasis, such as exercise or anesthesia, for some days after normalization of plasma- and muscle-potassium concentrations.

Starvation is also associated with a decrease in skeletal muscle NKA concentration.[62] Although this effect is minute compared to that of potassium depletion, it may perhaps indicate that starvation can predispose to potassium depletion. On the other hand, it has been found that in rats hyperkalemia is associated with an upregulation of skeletal muscle Na,K-pump concentration[63,64] and with a concomitant improved capacity to clear potassium loads.[63]

Magnesium

In animals dietary magnesium depletion is associated with a loss of potassium and secondary decrease in skeletal muscle Na,K-pump concentration.[65]

Diuretic therapy in patients lead to a 20% decrease in the concentration of magnesium in biopsy samples of skeletal muscle unaccompanied by hypomagnesemia and is associated with a decrease in the concentration of potassium and Na,K pumps.[60] Thus during potassium supplementation a concomitant magnesium depletion may be expected to interfere with potassium repletion, while magnesium substitution in human subjects who receive diuretics and potassium supplementation may lead to an improvement of the disturbances in muscle magnesium, potassium, and Na,K-pump concentrations.[66,67] A major problem with oral magnesium supplementation is, however, that it must be used in doses high enough to induce magnesium repletion and, at the same time, in doses low enough to avoid the gastrointestinal loss of potassium and magnesium that is due to a laxative effect.

Thyroid Status

Thyroid hormones markedly influence the concentration of Na,K pumps in skeletal muscles. Thus the skeletal muscle Na,K-pump concentration in hyperthyroid animals may be up to 10 times greater than that in hypothyroid animals.[68]

In patients a 50% decrease and 68% increase in Na,K-pump concentration have been found in hypothyroidism and hyperthyroidism, respectively.[69] The increase in pump concentration associated with hyperthyroidism may explain the development of thyreotoxic hypokalemic periodic paralysis. Plasma potassium has been found to be in the range 1.2–3.2 mmol/l during such attacks, as a result of potassium translocation into skeletal muscles without a change in total body content.[70,71] The attacks can be treated with potassium and may be prevented by potassium supplementation or by β-adrenoceptor antagonists until the thyroid status can be normalized.[72]

Physical Condition

A reversible increase in skeletal muscle Na,K-pump concentration was observed in vigorously swimming trained animals, while a reduction was observed following rigid inactivation by denervation, tenotomy, or plaster cast immobilization.[73,74]

Although moderate training in conscripts did not induce any changes in the Na,K-pump concentration in skeletal muscle biopsies,[75] it significantly reduced exercise-induced hyperkalemia.[6] On the other hand, the Na,K-pump concentration in skeletal muscles has been found to be higher in trained than in untrained subjects.[76] Thus it seems that physical conditioning can improve extrarenal potassium homeostasis without changing skeletal muscle Na,K-pump concentration, probably (as mentioned above) as a result of increased cathecolamine action. What is not fully clear is whether moderate changes in pump concentration in human subjects are associated with changes in the capacity for muscle performance, although there is evidence that major changes in performance may be associated with moderate regulation of skeletal muscle Na,K pumps. In patients with heart failure and a reduced capacity for exercise the skeletal muscle Na,K-pump concentration has been found to be reduced,[77] which suggests that this may either be a part of the heart failure syndrome or a common feature of any debilitating illness.

In this context it is of interest that an excessive rise in plasma-potassium concentration after exercise has been observed in patients with muscular dystrophy.[78] Thus it could be summarized that skeletal muscle dysfunction due to muscle disease, inactivity, or general disease may expose the heart to dangerous shifts in potassium concentration during heavy exercise, and that physical training may have protective effect.

Etiology of Hypokalemia and Potassium Depletion

Hypokalemia is usually defined as a condition where plasma-potassium concentration is below 3.5 mmol/l, whereas potassium depletion is a condition where the total amount of body potassium is reduced and the muscle potassium concentration is reduced below 75 μmol/g wet wt.[79] It should be noted that hypokalemia may occur with or without concomitant potassium depletion and that potassium depletion may occur with or without concomitant hypokalemia. In general, hypokalemia may result from potassium uptake to the cells, insufficient dietary intake, excessive loss of potassium, or combinations thereof. Except in the case of cellular uptake of potassium, hypokalemia represents potassium deficiency. Causes of hypokalemia related to the Na,K pump are listed in Table 16.2.

Factors that cause redistribution of potassium between the extra- and intracellular fluid due to activation or inhibition of skeletal muscle Na,K pumps have already been discussed in detail. A common cause of changes in renal as well as extrarenal potassium homeostasis is a disturbance in acid–base balance. Thus acidosis increases and alkalosis decreases plasma-potassium concentration.[2] In general a change in acid–base balance of 0.1 pH unit may be associated with a change in plasma-potassium concentration of 0.2 mmol/l,[80] although in clinical practice this correlation may vary considerably. These events may be the result of a combined action on Na,H exchange and Na,K pumps. Thus, alkalinization of plasma may increase the extrusion of protons from the cells, which buffers the plasma and causes a rise in intracellular sodium

TABLE 16.2. Causes of hypokalemia related to the Na,K pump.

Redistribution of potassium
Insulin
Catecholamines
Hyperthyroidism
Reduced total body potassium
Deficient diet
Gastrointestinal loss
Diuretic therapy

concentration. This, in turn, may activate the Na,K pumps, which results in potassium uptake and reduces the plasma-potassium level. Acidification of plasma would cause the opposite results. Potassium depletion in humans has been found to be associated with a rise in plasma bicarbonate concentration[81] that is probably caused by a decreased capacity for extrusion of H^+ ions in skeletal muscles and the ensuing intracellular acidosis (Figure 16.2). In agreement with this hypothesis, acid loading in potassium-depleted dogs resulted in a smaller decrease in plasma pH and rise in plasma-potassium concentration.[82] This acidosis-induced rise in plasma-potassium concentration, which occurs in spite of potassium depletion, may contribute to the potassium loss seen when acidosis is sustained, eg, in diabetic ketoacidosis. Conversely, potassium depletion may be expected to enhance the increase in plasma pH and decrease in plasma-potassium concentration associated with alkalosis.

Nutritional surveys indicate that certain groups of impoverished people, especially the elderly, consume a diet very poor in potassium.[83] For example, the potassium content of white bread, cheese, coffee, and tea is only around 0.01 mmol/g, and people living on such a diet would have to consume large quantities of food to avoid potassium depletion. In addition, as mentioned above, dietary-induced hypokalemia may be further accentuated by concomitant magnesium depletion.[60,65] This is especially true in malnourished infants and children of developing countries.[84] It is evident that people living on diets containing close to the minimum requirements of potassium and magnesium are especially vulnerable to increased loss of these elements.

Gastrointestinal loss of potassium is due mainly to vomiting, diarrhea, or drainage through fistulae. The potassium concentration in gastric secretion is usually only 10 mmol/l; thus vomiting contributes little to hypokalemia directly. However, the renal loss of potassium due to the associated metabolic alkalosis may be significant. The concentration of potassium in stool water is around 75 mmol/l;[85] diarrhea from any cause such as infection, inflammation, malabsorption, or cancer frequently leads to hypokalemia, which may be especially fatal in children with infectious diarrhea, concomitant malnutrition, and relatively small potassium stores. It must be remembered that increased loss of potassium through stool is often associated with increased loss of magnesium.

Increased urinary potassium loss is a well-known effect of thiazide and loop diuretics. This facilitated loss of potassium through the kidney is caused by a number of factors, such as increased distal tubular flow, augmented potassium secretion in the distal nephron (in turn due to increased aldosterone secretion caused by sodium depletion), and decreased potassium reabsorption in the thick ascending limb.[86] The degree of hypokalemia varies with the patient, the dietary sodium content, and the specific diuretic drug; the renal loss of potassium may be limited by using angiotensin-converting enzymes (ACE) inhibitors, which decrease the synthesis of aldosterone and thereby limit the renal loss of potassium. Concomitant treatment with ACE inhibitors may thus reduce the degree of hypokalemia in patients on diuretics, but may also predispose to hyperkalemia should potassium intake increase or renal function decline.[87-89] As in the case of gastrointestinal loss, the increased loss of potassium in urine is often associated with an increased loss of magnesium.

Methods for the Assessment of Hypokalemia and Potassium-Depletion

Potassium homeostasis is usually assessed by measurements of venous plasma-potassium concentration. Sampling, sample handling, and potassium determination are, however, potential sources of errors.[90,91] A major one is the leak of potassium from skeletal muscle cells into the blood just prior to sampling and from the blood cells during sampling or storage. Clenching of the fist or pumping the blood during blood sampling by exercising the arm muscles distal to the venous puncture may increase the potassium concentration by as much as 25 and 80%, respectively. Similarly, release of potassium from platelets explains why the serum potassium is generally about 0.5 mmol/l higher than heparinized plasma potassium. Hemolysis, which is suggested by pink discoloration of serum, increases the potassium concentration and may mask an existing hypokalemia. In general, low plasma-potassium values are more reliable than high values. Furthermore, since plasma potassium may be normal in spite of reduced potassium content in the skeletal muscles,[60,92] even accurate determinations of plasma potassium may not disclose potassium depletion. Similarly, plasma magnesium may be normal in spite of a reduced skeletal muscle magnesium content,[60] and plasma magnesium determinations may not disclose magnesium depletion.

The use of urinary potassium to evaluate potassium homeostasis is also fraught with complications. In general, a potassium excretion of less than 25 mmol/day suggests a diagnosis of potassium depletion. It should be noted, however, that in patients with hyperaldosteronism a high urinary potassium excretion may coexist with severe hypokalemia and potassium depletion. On the other hand, a urinary excretion of potassium greater then 25 mmol/day in patients with hypokalemia suggests ongoing potassium depletion. A major source of error, the incomplete collection of urine, may be corrected by simultaneous measurements of creatinine excretion. Thus the excretion of 25 mmol potassium/day in urine normally corresponds to an excretion of about 1.5 mmol potassium/mmol creatinine.[92]

In view of these methodologic errors, it may be suggested that the state of potassium homeostasis and especially whether or not potassium depletion prevails may be evaluated more accurately by muscle biopsies, which are easily obtained by using a Bergström biopsy needle.[93] Following local anesthesia of skin, subcutaneous tissues, and fascia, biopsies of around 50 mg wet wt. may be taken from the vastus lateralis muscle approximately 10 cm above the knee, without signicant complications or major discomfort. The potassium content is then determined using flame photometry or atomic absorption. If muscle potassium content is normal, it may be reasonably assumed that potassium deficiency does not exist. In addition to measurements of potassium and other ions, eg, magnesium, muscle biopsies allow quantification of Na,K pumps by vanadate-facilitated ^{3}H-ouabain binding to intact tissue samples.[24,25,94]

Most studies that characterize NKA in the skeletal muscle are based upon measurements of NKA activity in membrane fractions that contain but a few percent of the total enzyme activity of the starting material. It is not known whether these fractions represent a true random sample of the plasma membrane of the tissue or whether they arise from subspecialized regions such as the transverse tubules.[25,94-96] Although such studies may be of interest of qualitative studies of NKA in skeletal muscles, they are not applicable to quantitative studies. During the last few years it has, however, become possible to quantify the Na,K pump concentration in skeletal muscles with high accuracy and precision by vanadate-facilitated ^{3}H-ouabain binding to intact samples of a few milligrams wet weight, which avoids purification and the ensuing loss of enzyme activity.[24,25,96]

It should be emphasized that body potassium homeostasis cannot be evaluated by studying cells that do not represent the major potassium reservoir, namely the skeletal muscles. Thus, even though blood cells are easy to obtain, their potassium homeostasis, although of interest for understanding their functional state, may not reflect overall body potassium homeostasis. Hence, in contrast to human skeletal muscles, human erythrocytes maintain normal potassium content[56] and upregulate Na,K-pump concentration[57] during potassium depletion. Similarly, it has been shown that the regulatory response of cultured cells to potassium depletion is different from that of human skeletal muscle tissues.[58,59]

Measurement of potassium transport is as yet not possible in biopsies of human skeletal muscle since only intact cells and an intact metabolism would yield values relevant to the in vivo situation. Thus meaningful studies of this parameter can be conducted only by using intact skeletal muscle fibers from small animals.[22,50] Although techniques are available for noninvasive studies of aspects of potassium homeostasis using isotopes and radioactivity counting, such measurements have hitherto not gained general use. It may be, however, that positron-emissions tomography may become a tool of importance in future noninvasive studies of human potassium homeostasis. Measurements of shifts in plasma-potassium concentration in conditions that challange potassium homeostasis, eg, during exercise or the administration of β-adrenoceptor agonists or antagonists, may prove clinically useful.

Consequences of Hypokalemia and Potassium Depletion

The major effects of hypokalemia and potassium depletion are disturbances of cardiovascular and skeletal muscle function. Other consequences have been reviewed elsewhere.[86,90,97]

Hypokalemia induces marked changes in resting membrane potential and action potential of the myocardium, altering excitability and conduction speed[98] and causing electrocardiographic changes of the ST segment and of the T wave and arrhythmias. Potassium inhibits the binding of digitalis glycosides to the Na,K pump.[23] Accordingly, hypokalemia enhances digitalis toxicity and may induce arrhythmias, whereas the effects of digitalis intoxication are diminished by potassium infusion.[99] Furthermore, ventricular arrhythmias may be seen more often in hypokalemic than in normokalemic patients with myocardial infarction,[19,44] perhaps because of the additional reduction in plasma-potassium concentration caused by stress-induced catecholamine response.[45] It may be that shifts in plasma-potassium concentration, rather than sustained deviations, are dangerous to the heart. This may explain cases of sudden unexpected death in association with physical activity[13] or with the treatment of asthma.[41]

Hypokalemia-induced skeletal muscle myopathy is characterized clinically by weakness and histologically by rhabdomyolysis. The weakness first affects the proximal muscles of the lower extremities and is often accompanied by general feeling of fatigue, loss of energy, and malaise.[100] Deep tendon reflexes are decreased and may be absent. Patients may complain of restless legs, cramps, and paresthesia. If plasma potassium falls to about 1 mmol/l, quadriplegia, respiratory failure, muscle fasciculation, and tetany may develop. Smooth muscle function may also be impaired, causing paralytic ileus and postural hypotension. Several factors probably contribute to the rhabdomyolysis associated with severe hypokalemia.[101,102] Decreased release of potassium may decrease exercise-induced vasodilation, leading to skeletal muscle ischemia. The decreased sodium gradient across the cell membrane may cause intracellular accumulation of Ca^{2+} and H^+ ions. This, in turn, may activate proteolytic enzymes that cause protein degradation and subsequent cell necrosis (Figure 16.2). Furthermore, protein synthesis may be reduced because of decreased cellular uptake of amino acids and because potassium is a cofactor for protein synthesis (Figure 16.2). In agreement with this hypothesis, protein synthesis in the intact organism has been found to be very sensitive to potassium deficiency.[103] Indeed, plasma-potassium concentrations below 2.0 mmol/l may be associated with muscle ischemia, increased plasma creatinine phosphokinase, rhabdomyolysis, and myoglobinuria.[100] It should be noted that as a result of rhabdomyolysis, potassium may leak out of muscle cells and mask hypokalemia, or even induce hyperkalemia.

Therapy of Hypokalemia and Potassium Depletion

Treatment of patients with plasma potassium concentration below 3.0 mmol/l and with symptomatic hypokalemia is mandatory. Unless the concentration is lower than 2.5 mmol/l, the symptoms are serious or life-threatening, or the patient is unable to take oral medication, it is generally better to administer potassium supplementation orally to avoid the risk of serious arrhythmias should erroneously large amounts of potassium be quickly infused.[104] Oral administration is safer given the adaptive capacity of the normal kidney to excrete potassium following chronically increased intake amounts to approximately 700 mmol/day.[105] However, if parenteral potassium administration becomes necessary the patient must be carefully monitored for signs of hyperkalemia, changes in the electrocardiogram, and inappropriate rise in plasma-potassium level.

On the other hand, the need for potassium supplementation in asymptomatic patients with plasma-potassium levels between 3.0 and 3.5 mmol/l has been questioned.[106–109] After the administration of moderate doses of furosemide or thiazide the plasma potassium shows falls of around 0.3 to 0.6 mmol/l[110] and it has been reported that 20–80% of patients who receive diuretics have hypokalemia.[111] Thus mild hypokalemia represents a quantitatively large clinical problem, especially since it may become dangerous if aggravated by further redistribution of potassium from the plasma into the skeletal muscles. Hence conditions that favor the redistribution of potassium may be especially hazardous when the muscle potassium concentration is already reduced by potassium depletion. For these reasons it is recommended that hypokalemia and potassium depletion be treated, whether considered severe or mild.

A practical clinical guide to the treatment and prevention of mild hypokalemia and potassium depletion is given in Table 16.3. Treatment may start with a diet designed to increase daily potassium intake followed, if necessary, by oral potassium supplementation. Magnesium depletion often associated with potassium depletion must be corrected by oral magnesium supplementation. If the results are unsatisfactory or the oral potassium supplementation has reached unacceptable amounts, the use of potassium-sparing diuretics must be considered. Furthermore, therapy that causes hypokalemia as well as potassium depletion must be constantly reviewed to minimize potassium loss. It should be remembered that potassium, as well as magnesium depletion, may be masked by normal plasma values. Thus in situations where potassium and magnesium depletion might be a possibility in spite of normal plasma leves, treatment it should be considered for subjects who complain of otherwise unexplained muscle fatigue, as outlined above. In selected cases, that is in patients in whom occult but clinically significant

TABLE 16.3. Practical clinical approach to treatment and prophylaxis of mild hypokalemia/potassium depletion.

Dietary guidelines
Potassium supplementation
Magnesium supplementation
Use of potassium-sparing diuretics
Revision of therapy

potassium depletion is a threat, treatment may have to be guided by skeletal muscle biopsies.

Whenever potassium supplementation or other therapies that may cause potassium retention are used, it is mandatory to institute regular venous plasma potassium determinations to avoid hyperkalemia. Such measurements must especially be carried out with only small intervals at the beginning of treatment. This is of special importance in patients with renal dysfunction, elderly patients, and patients treated with potassium-sparing diuretics, ACE inhibitors, and nonsteroidal anti-inflammatory drugs.[112]

As mentioned above, during diuretic therapy the concentration of potassium in skeletal muscles may be reduced considerably in spite of almost normal plasma-potassium level.[60,93] Furthermore, erroneously high plasma-potassium values may also have led to an erroneous allocation of patients in studies that evaluated the clinical significance of hypokalemia and potassium depletion, as well as its treatment. The further establishment of firm therapeutic guidelines requires additional information on potassium content and Na,K-pump concentration in skeletal muscle biopsies, on the shifts in plasma-potassium concentration during a challenge to potassium homeostasis, and on the effects of potassium supplementation on these parameters.

Conclusion

The papers reviewed in this chapter, in agreement with others reviewed elsewhere,[2,16,32,79,97,113] establish the fact that the concentration of potassium in the plasma reflects a dynamic equilibrium between plasma and skeletal muscles. Thus measurements of steady state venous plasma-potassium concentrations are not sufficient to evaluate in depth extrarenal body potassium homeostasis. This may explain the conflicting opinions about whether or not mild hypokalemia should be treated and prevented. Recognition of this dynamic equilibrium indicates that hypokalemia and potassium depletion should be treated and prevented, whether mild or severe. The measurement of potassium content and Na,K pump concentration in muscle biopsies and the determination of plasma potassium shifts during challenges to the extrarenal potassium homeostasis may provide a rational basis for future research on hypokalemia and potassium depletion, and may become clinically important in patients in whom occult but clinically significant potassium depletion is a threat.

Acknowledgment. We wish to thank the Danish Heart Foundation and Novo Foundation for supporting in part the work which provided the basis for this chapter. Thanks are due also to Stig Haunsø for valuable discussions.

References

1. Wright FS, Giebish G. Regulation of potassium excretion. In: Seldin DW, Giebish G, eds. The Kidney: Physiology and Pathophysiology. New York: Raven Press; 1985:1223–1249.
2. DeFronzo RA, Bia M. Extrarenal potassium homeostasis. In: Seldin DW, Giebish G, eds. The Kidney: Physiology and Pathophysiology. New York: Raven Press; 1985:1179–1207.
3. Kjellmer I. The role of potassium ions in exercise hyperaemia. Medical Experience 1961; 5:56–60.
4. Vyskocil F, Hnik P, Rehfeldt H, Vejsada R, Ujec E. The measurement of K^+_e concentration changes in human muscle during volitional concentrations. Pfluegers Arch 1983; 399:235–237.
5. Saltin B, Blomqvist G, Mitchel JH, Johnson RL, Wildenthal K, Chapman CB. Response to exercise after bed rest and after training. Circulation 1968; 7(suppl):1–78.
6. Kjeldsen K, Nørgaard A, Hau C. Exercise induced hyperkalemia is in human subjects reduced by moderate training without change in skeletal muscle Na,K-ATPase concentration. Eur J Clin Invest 1990; 20:642–647.
7. Juel C, Bangsbo J, Graham T, Saltin B. Lactate and potassium fluxes from human skeletal muscle during and after intense, dynamic, knee extensor exercise. Acta Physiol Scand 1990; 140:147–159.
8. Lindinger MI, Heigenhauser GJF, McKelvie RS, Jones NL. Blood ion regulation during repeated maximal exercise and recovery in humans. Am J Physiol 1992; 262:R126–R136.
9. McKechnie JK, Leary WP, Joubert SM. Some electrocardiographic and biochemical changes recorded in marathon runners. South African Medical Journal 1967; 41:722–725.
10. Band DM, Lim M, Linton RAF, Wolf CB. Changes in arterial plasma potassium concentration during exercise. J Physiol 1982; 328:74P–75P.
11. Medbø JI, Sejersted OM. Plasma potassium changes with high intensity exercise. J Physiol 1990; 421:105–122.
12. Beaumont van W, Underkofler S, Baumont van S. Erythrocyte volume, plasma volume, and acid-base changes in exercise and heat dehydration. J Applied Physiol 1981; 50:1255–1262.

13. Brady HR, Kinirons M, Lynch T, Ohman EM, Tormay W, O'Malley KM, Horgan JH. Heart rate and metabolic response to competitive squash in veteran players: Identification of risk factors for sudden cardiac death. Eur Heart J 1989; 10:1029–1035.

14. Sjøgaard G, Adams RP, Saltin B. Water and ion shifts in skeletal muscle of humans with intense dynamic knee extension. Am J Physiol 1985; 248:190R–196R.

15. Sjøgaard G. Water and electrolyte fluxes during exercise and their relation to muscle fatigue. Acta Physiol Scand 1986; 556(suppl):129–136.

16. Lindinger MI, Sjøgaard G. Potassium regulation during exercise and recovery. Sports Med 1991; 11:382–401.

17. Patterson DJ. Potassium and ventilation in exercise. J Applied Physiol 1992; 72:811–820.

18. Kjeldsen K. Dysfunction of skeletal muscle Na,K-pumps may expose the heart to arrhythmic potassium concentrations during exercise. Can J Sports Med 1990; 11:304–309.

19. Gettes LS. Electrolyte abnormalities underlying lethal and ventricular arrhythmias. Circulation 1992; 85:170–176.

20. Clausen T, Everts ME. Is the Na,K-pump capacity in skeletal muscle inadequate during sustained work. In: Skou JC, Nørby JN, Maunsbach AB, Esmann M, eds. Progress in Clinical and Biological Research, Volume 268B. The Na^+,K^+-Pump. Part B: Cellular Aspects. New York: Allan R. Liss; 1988:239–244.

21. Kjeldsen K, Nørgaard A, Clausen T. Effects of ouabain, age and K-depletion on K-uptake in rat soleus muscle. Pfluegers Arch 1985; 404:365–373.

22. Clausen T, Everts ME, Kjeldsen K. Quantification of maximum capacity for active sodium-potassium transport in rat skeletal muscle. J Physiol 1987; 388:163–181.

23. Skou JC. Enzymatic basis for active transport of Na^+ and K^+ across cell membrane. Physiol Rev 1965; 45:596–617.

24. Nørgaard A, Kjeldsen K, Clausen T. A method for the determination of the total number of ^3H-ouabain binding sites in biopsies of human skeletal muscle. Scand J Clin Lab Invest 1984; 44:509–518.

25. Kjeldsen K. Regulation of the concentration of ^3H-ouabain binding sites in mammalian skeletal muscle: Effects of age, K-depletion, thyroid status and hypertension. Dan Med Bull 1987; 34:15–46.

26. Clausen T, Kjeldsen K. Effects of K-deficiency on Na,K-homeostasis and Na,K-ATPase in muscle. In: Giebish G, ed. Current Topics in Membranes and Transport. Potassium Transport: Physiology and Pathophysiology. Orlando: Academic Press 1987; 28:403–419.

27. Clausen T, Everts ME. Regulation of the Na,K-pump in skeletal muscle. Kid Inter 1989; 35:1–13.

28. Hazeyama Y, Sparks HV. A model of potassium ion efflux during exercise of skeletal muscle. Am J Physiol 1979; 236:83R–90R.

29. Everts ME, Rettersbøl K, Clausen T. Effects of adrenalin on excitation-induced stimulation of the sodium-potassium pump in rat skeletal muscle. Acta Physiol Scand 1988; 134:189–198.

30. Kim MS, Akera T. O_2 free radicals: Cause of ischemia-reperfusion injury to cardiac Na^+-K^+-ATPase. Am J Physiol 1987; 252:252H–257H.
31. Kelley RA, Smith TW. The search for the endogenous digitalis: An alternative hypothesis. Am J Physiol 1989; 256:937C–950C.
32. Clausen T. Regulation of active Na^+-K^+ transport in skeletal muscle. Physiol Rev 1986; 66:542–580.
33. Wang P, Clausen T. Treatment of attacks in hyperkalemic familial periodic paralysis by inhalation of salbutamol. Lancet 1976; i:221–227.
34. Montoliu J, Lens XM, Revert L. Potassium-lowering effect of albuterol for hyperkalemia in renal failure. Arch Int Med 1987; 147:713–717.
35. Allon M, Copkney C. Albuterol and insulin for the treatment of hyperkalemia in hemodialysis patients. Kid Inter 1990; 38:869–872.
36. Carlsson E, Fellenius E, Lundborg P, Svensson L. Beta-adrenoceptor blockers, plasma-potassium, and exercise. Lancet 1978; ii:424–425.
37. Lundborg P. The effect of adrenergic blockade on potassium concentration in different conditions. Acta Med Scand 1983; 672(suppl):121–125.
38. Fletcher GF, Sweeney ME, Fletcher BJ. Blood magnesium and potassium alterations with maximal treadmill exercise testing: Effects of beta-adrenergic blockade. Am Heart J 1990; 121:105–108.
39. Whyte KF, Reid C, Addis GJ, Whitesmith R, Reid JL. Salbutamol induced hypokalemia: The effect of theophylline alone and in combination with adrenaline. Br J Clin Pharmacol 1988; 25:571–578.
40. Lipworth BJ, McDevitt DG, Struthers AD. Prior treatment with diuretic augments the hypokalemic and electrocardiographic effects of inhaled albuterol. Am J Med 1989; 86:653–657.
41. Epelbaum S, Benhamou PH, Pautard JC, Devoldere C, Kremp O, Piussan C. Arrêt respiratoire chez un enfant asthmatique traitée par bêta-2-mimétiques et théophylline. Rôle possible de l'hypokaliémie dans les décès subits des asthmatiques. Annales de Pédiatrie (Paris) 1989; 36:473–475.
42. Christensen NJ, Galbo H. Sympathetic nervous activity during exercise. Annu Rev Physiol 1983; 45:139–153.
43. Kjær M, Farrel PA, Christensen NJ, Galbo H. Increased epinephrin response and inaccurate glucoregulation in exercising athletes. J Applied Physiol 1986; 61:1693–1700.
44. Nordrehaug JE, von der Lippe G. Hypokalemia and ventricular fibrillation in acute myocardial infarction. Br Heart J 1983; 50:525–529.
45. Simpson E, Rodger JC, Raj SM, Wong C, Wilkie L, Robertson C. Pre-treatment with beta-blockers and the frequency of hypokalemia in patients with acute chest pain. Br Heart J 1987; 58:499–504.
46. Moore RD. Effects of insulin upon ion transport. Biochem Biophys Acta 1983; 737:1–49.
47. Hiatt N, Davidson MB, Bonorris G. The effect of potassium chloride infusion on insulin secretion in vivo. Horm Metab Res 1972; 4:64–68.
48. Heppel LA. The electrolytes of muscle and liver in potassium-depleted rats. Am J Physiol 1939; 127:385–392.
49. Nørgaard A, Kjeldsen K, Clausen T. Potassium depletion decreases the number of ^{3}H-ouabain binding sites and the Na-K transport in skeletal muscle. Nature 1981; 293:739–741.

50. Kjeldsen K, Nørgaard A, Clausen T. Effect of potassium-depletion on ^{3}H-ouabain binding and sodium-potassium-contents in mammalian skeletal muscle. Acta Physiol Scand 1984; 122:103–117.

51. Nørgaard A, Kjeldsen K, Hansen O. K^+-dependent 3-O-methylfluorescein phosphatase activity in crude muscle homogenates of rodent heart ventricle. Eur J Pharmacol 1985; 113:373–382.

52. Brown L, Wagner G, Hug E, Erdmann E. Ouabain binding and inotropy in acute potassium depletion in guinea pig. Cardiovasc Res 1986; 20:286–293.

53. Kjeldsen K, Nørgaard A. Quantification of rat sciatic nerve Na,K-ATPase by measurements of ^{3}H-ouabain binding in intact nerve samples. J Neurol Sci 1987; 79:205–219.

54. Larsen JS, Schmidt TA, Kjeldsen K. Quantification of rat cerebral cortex Na,K-ATPase with high recovery. Evaluation of age and K-depletion. In: Kaplan JH, De Weer P, eds. The Sodium Pump: Recent Developments. New York: Rockefeller University Press; Society of General Physiologists, Series 46 (II); 1990:585–589.

55. Schmidt TA, Larsen JS, Kjeldsen K. Quantification of rat cerebral cortex Na,K-ATPase. Effect of age and potassium depletion. J Neurochem 1992; 59:2094–2104.

56. Chan PC, Sanslone WR. The influence of a low potassium diet on rat erythrocyte membrane adenosine triphosphatase. Arch Biochem Biophys 1969; 134:48–52.

57. Erdmann E, Krawietz W. Increased number of ouabain binding sites in human erythrocyte membranes in chronic hypokalemia. Acta Biol Med German 1977; 36:879–883.

58. Lamb JF, McCall D. Effect of prolonged ouabain treatment on Na, K, Cl and Ca concentration and fluxes in cultured human cells. J Physiol 1972; 225:599–617.

59. Werdan K, Wagenknecht B, Zwissler O, Brown L, Krawietz W, Erdmann E. Cardiac glycoside receptors in cultured heart cells: Characterization of one single class of high affinity receptors in heart muscle from chicken embryos. Biochem Pharmacol 1984; 33:55–70.

60. Dørup I, Skajaa K, Clausen T, Kjeldsen K. Reduced concentrations of potassium, magnesium, and sodium-potassium pumps in human skeletal muscle during treatment with diuretics. Br Med J 1988; 67:455–458.

61. Sadre M, Sheng HP, Fiorotto M, Nocholos BL. Electrolyte composition changes of chronically K-depleted rats after K loading. J Applied Physiol 1987; 63:765–769.

62. Kjeldsen K, Everts ME, Clausen T. Effect of semi-starvation and potassium deficiency on the concentration of ^{3}H-ouabain binding sites and sodium and potassium contents in rat skeletal muscle. Br J Nutrition 1986; 56:519–532.

63. Blachley JD, Crider BP, Johnson JH. Extrarenal potassium adaptation: Role of skeletal muscle. Am J Physiol 1986; 251:313F–318F.

64. Kjeldsen K, Everts ME, Nørgaard A. Na,K-ATPase concentration in skeletal muscle: Quantification, regulation and significance. In: Skou JC, Nørby JN, Maunsbach AB, Esmann M, eds. Progress in Clinical and Biological Research, Volume 268B. The Na^+,K^+-Pump. Part B: Cellular Aspects. New York: Allan R. Liss; 1988:251–256.

65. Kjeldsen K, Nørgaard A. The effect of magnesium depletion on ^{3}H-ouabain binding site concentration in rat skeletal muscle. Magnesium 1987; 6:55–60.
66. Dyckner T, Wester PO, Widman L. Effect of peroral magnesium on plasma and skeletal muscle electrolytes in patients on long-term diuretic therapy. Int J Cardiol 1988; 19:81–87.
67. Dørup I, Skajaa K. Magnesium and potassium depletion during long term diuretic treatment. Res Clin Forums 1989; 11:19–25.
68. Kjeldsen K, Everts ME, Clausen T. The effects of thyroid hormones on ^{3}H-ouabain binding site concentration, Na-K-contents and ^{86}Rb-efflux in rat skeletal muscle. Pfluegers Arch 1986; 406:529–535.
69. Kjeldsen K, Nørgaard A, Gøtzsche CO, Thomassen A, Clausen T. Effect of thyroid function on number of Na-K-pumps in human skeletal muscle. Lancet 1984; ii:8–10.
70. Feldman DL, Goldberg WM. Hyperthyroidism and periodic paralysis. Can Med Assoc J 101:667–671.
71. Miller D, DelCastillo J, Tsang TK. Severe hypokalemia in thyrotoxic periodic paralysis. Am J Emer Med 1989; 7:584–587.
72. McFadzean AJS, Yeung R. Periodic paralysis complicating thyreotoxicosis in Chinese. Br Med J 1967; 1:451–455.
73. Kjeldsen K, Richter EA, Galbo H, Lortie G, Clausen T. Training increases the concentration of ^{3}H-ouabain binding sites in rat skeletal muscle. Biochim Biophys Acta 1986; 860:708–712.
74. Kjeldsen K, Bjerregaard P, Richter EA, Thomsen PEB, Nørgaard A. Na$^+$,K$^+$-ATPase concentration in rodent and human heart and skeletal muscle: Apparent relation to muscle performance. Cardiovasc Res 1988; 22:95–100.
75. Kjeldsen K, Nørgaard A, Hau C. Human skeletal muscle Na,K-ATPase concentration quantified by ^{3}H-ouabain binding to intact biopsies before and after moderate physical conditioning. Int J Sports Med 1990; 11:304–309.
76. Klitgaard H, Clausen T. Increased total concentration of Na-K pumps in vastus lateralis muscle of old trained human subjects. J Applied Physiol 1989; 67:2491–2494.
77. Nørgaard A, Bjerregaard P, Baandrup U, Kjeldsen K, Reske-Nielsen E, Thomsen PEB. The concentration of the Na,K-pump in skeletal and heart muscle in congestive heart failure. Int J Cardiol 1990; 26:185–190.
78. Wevers RA, Joosten MG, Biezenbos JBM, Theewes GM, Veerkamp JH. Excessive plasma K$^+$ increase after ischemic exercise in myotonic muscular dystrophy. Muscle Nerv 1990; 13:27–32.
79. Nørgaard A, Kjeldsen K. Interrelation of hypokalaemia and potassium depletion and its implications: A re-evaluation based on studies of the skeletal muscle sodium, potassium-pump. Clin Sci 1991; 81:449–455.
80. Sterns RH, Cox M, Feig PU, Singer I. Internal potassium balance and the control of the plasma potassium concentration. Medicine 1981; 60:339–354.
81. Jones JW, Sebastian A, Hulter HN, Schambelan M, Sutton JM, Biglieri EG. Systemic and renal acid-base effect of chronic dietary potassium depletion in humans. Kid Inter 1982; 21:402–410.
82. Vaamonde CA, Ostler JR, Alpert HC, Rodriguez GR. Effect of potassium depletion on acidosis-induced changes in plasma potassium concentration. Min Elect Metab 1985; 11:381–388.

83. Knochel JP. Hypokalemia. Adv Int Med 1984; 30:317–335.
84. Michaelsen KF, Clausen T. Inadequate supplies of potassium and magnesium in relief food—Implications and countermeasures. Lancet 1987; i:1421–1423.
85. Wrong O, Metcalfe-Gibson A, Morrison RBI, Ng ST, Howard AV. In vivo dialysis of faeces as a method of stool analysis. Clin Sci 1975; 28:357–375.
86. Richardson RMA, Kunau RT Jr. Potassium deficiency and intoxication. In: Seldin DW, Giebish G, eds. The Kidney: Physiology and Pathophysiology. New York: Raven Press; 1985:1251–1267.
87. Webster MWI, Fitzpatrick MA, Nicholls MG, Ikram H, Wells JE. Effect of enalapril on ventricular arrhythmias in congestive heart failure. Am J Cardiol 1985; 56:566–569.
88. Packer M, Lee WH. Provocation of hyper- and hypokalemic sudden death during treatment with and withdrawal of converting-enzyme inhibition in severe chronic congestive heart failure. Am J Cardiol 1986; 57:347–348.
89. Johnston RT, de Bono DP, Nyman CR. Preventable sudden death in patients receiving angiotensin converting enzyme inhibitors and loop/ potassium sparing diuretics. Int J Cardiol 1992; 34:213–215.
90. Zull DN. Disorders of potassium metabolism. Endocrine Metab Emer 1989; 7:771–794.
91. Moore D, Walker P, Ismail A. The alteration of serum potassium level during sample transit. Practitioner 1989; 233:395–397.
92. Kamel KS, Ethier JH, Richardson RMA, Bear RA, Halperin ML. Urine electrolytes and osmolality: When and how to use them. Am J Nephrol 1990; 10:89–102.
93. Bergström J. Muscle electrolytes in man. Scand J Clin Lab Invest 1962; 14(suppl 68):1–110.
94. Hansen O, Clausen T. Quantitative determination of Na^+-K^+-ATPase and other sarcolemmal component in muscle cells. Am J Physiol 1988; 245:1C–7C.
95. Jones LR, Besh HR. Isolation of canine cardiac sarcolemmal vesicles. Meth Pharmacol 1984; 5:1–12.
96. Kjeldsen K. Complete quantification of the total concentration of rat skeletal muscle Na,K-ATPase by measurements of ^{3}H-ouabain binding. Biochem J 1986; 240:725–730.
97. Krishna GG. Hypokalemic states: Current clinical issues. Seminars Nephrol 1990; 10:515–524.
98. Fisch C. Relation of electrolyte disturbances to cardiac arrhythmias. Circulation 1973; 47:408–419.
99. Smith TW. Digitalis: Mechanisms of action and clinical use. N Engl J Med 1988; 318:358–365.
100. Knochel JP. Neuromuscular manifestations of electrolyte disorders. Am J Med 1982; 72:521–535.
101. Knochel JP, Schlein EM. On the mechanism of rhabdomyolysis in potassium depletion. J Clin Invest 1972; 51:1750–1758.
102. Knochel JP. Rhabdomyolysis and effects of potassium deficiency on muscle structure and function. Cardiovasc Med 1978; 3:247–261.
103. Dørup I, Clausen T. Effects of potassium deficiency on growth and protein synthesis in skeletal muscle and the heart of rats. Br J Nutrition 1989; 62:269–284.

104. Lawson DH. Adverse reactions to potassium chloride. Quarterly J Med 1974; 171:433–440.
105. Williams ME. Hyperkalemia. Crit Care Clinics 1991; 7:155–174.
106. Kaplan NM. Our appropriate concern about hypokalemia. Am J Med 1984; 77:1–4.
107. Kaplan NM, Carnegie A, Raskin P, Heller JA, Simmons M. Potassium supplementation in hypertensive patients with diuretic-induced hypokalemia. N Engl J Med 1985; 312:746–749.
108. Harrington JT, Isner JM, Kassirer JP. Our national obsession with potassium. Am J Med 1982; 73:155–159.
109. Kassirer JP, Harrington JT. Fending off the potassium pushers. N Engl J Med 1985; 312:785–787.
110. Morgan DB, Davidson C. Hypokalemia and diuretics: An analysis of publications. Br Med J 1980; 59:905–908.
111. Knochel JP. Diuretic-induced hypokalemia. Am J Med 1984; 77:18–27.
112. Swales JD. Salt substitutes and potassium intake: Too much potassium may be disastrous for some. Br Med J 1991; 303:1084–1085.
113. Sterns RH, Spital A. Disorders of internal potassium balance. Seminars Nephrol 1987; 7:399–415.

17
The Clinical Pharmacology of Potassium Channels

P.N. STRONG

Role of Potassium Channels in Cell Function

Potassium channels form a remarkably diverse group of ion channel structures.[1] More subtypes of potassium channels, with different biophysical and/or pharmacologic properties, are known than of any other type of ion channel. One of the most important roles of these channels is the regulation of membrane potential and excitability in nerve and muscle cells. Many potassium channels in turn are regulated by second messengers and other molecules of intermediary metabolism, such as G proteins, cAMP, ATP, inositol trisphosphate, and calcium.[2] In this manner potassium channels provide an important link between membrane potential and cell metabolism.

Potassium channels do not act in isolation. Indeed, their opening causes cell hyperpolarization and a reduction in ion currents flowing through voltage-sensitive calcium, sodium, and chloride channels. Conversely, the blockade of an (activated) potassium channel generally results in depolarization, which subsequently leads to the opening of other voltage-activated channels (most notably calcium). However, it should be remembered that potassium-channel blockade will not necessarily have physiologic consequences in the resting state, since most potassium channels are closed at resting membrane potentials. Thus potassium channels can be considered as both direct and indirect targets for therapeutic intervention. In recent years an important new class of potassium channel-activating molecules has been recognized. These compounds may have significant clinical benefit in cardiovascular disorders and asthma. In addition, the discovery that anti-diabetic sulphonylureas and class III anti-arrhythmic drugs block different potassium channel subtypes has increased awareness of the clinical pharmacology of potassium channels and their therapeutic potential.

Potassium-Channel Subtypes

There is no single satisfactory classification of potassium channels. The most convenient ones are based on their biophysical or their pharmacologic properties. The two are not mutually exclusive and each has advantages and drawbacks. When the structure of more potassium channels becomes known, a molecular classification might become possible. However, for any classification to be useful, it is extremely important that the parameter of choice be linked to ion-channel function.

Biophysical Classification

To be precise, any biophysical classification must be based on ion currents flowing through a channel,[3] rather than on the channel structure itself. The first potassium channel to be described was the classical voltage-activated (delayed rectifier) channel found in the squid giant axon.[4] This channel is activated by depolarization and is responsible for the repolarizing phase of an action potential. A second type of potassium channel (inward or anomalous rectifier[5]) is activated by hyperpolarization at voltages around the resting potential and may be involved in the maintenance of an action potential plateau. It is often modulated by intracellular metabolites and second messengers. A third class of potassium channels, first described in molluscan neurones,[6] produces transient outward currents (A currents). These channels are outward rectifiers and are inactivated at normal resting potentials. They are thought to play a role in neurons that are spontaneously active or fire repetitively in response to tonic depolarization. A final category of potassium channels, characterized by the ionic currents that flow through them, are those activated by internal calcium ions. The maxi-channel (also known as the BK channel[7]) is a calcium-activated potassium channel with a single conductance often in excess of 200 pS. A second type of calcium-activated potassium channel (SK channel[8]) has a small single-channel conductance (approximately 10 pS), and a third type (IK channel[9]) has an intermediate conductance (20–50 pS). Calcium-activated potassium currents occur in both excitable and nonexcitable cells. In neurons they contribute to spike repolarization and regulate repetitive firing frequency.[10] In particular, SK currents are responsible for the slow after-hyperpolarization (AHP) in nerve and muscle cells that follows an action potential. In many nonexcitable cells, SK and BK channels play a vital role in controlling secretion.[11]

Pharmacologic Classification

It is also possible to classify potassium channels on a pharmacologic or regulatory basis.[12] Although this method is not as precise as that based on biophysical criteria, it offers the frequent advantage of bringing an element of function into consideration. Reference has already been made to a subgroup of channels that are activated by internal calcium. M currents were first identified as slowly activating potassium currents in muscarinic neurones[13] and other subtypes of potassium channels can be activated by sodium and by changes in cell volume.[14] Potassium currents activated by internal ATP are extremely important in regulating insulin secretion in pancreatic β cells.[15] ATP-dependent potassium channels are also found in all types of muscle[16] (skeletal, cardiac, smooth), as well as in the brain. They are very important in hypoxic and ischaemic situations because they serve as metabolic sensors for energy-deprived cells. More recently, these channels have been identified as the probable target of a class of clinical agents known as potassium-channel openers (see below).

Another large category of potassium channels includes those activated by neurotransmitters. Although these currents can be activated by direct interaction with neurotransmitter receptor molecules, second-messenger systems are often involved. For example, the activation of potassium currents by serotonin and noradrenaline is probably mediated by cAMP.[17] Other neurotransmitter-activated potassium currents, such as those activated by acetylcholine and adenosine and which mediate negative chronotropic responses in the atria, are coupled to pertussis-sensitive G proteins.[18]

Hypoglycemic Sulphonylureas

Many sulphonylurea compounds are hypoglycemic agents and are commonly used to treat type II (noninsulin-dependent) diabetes.[19] Their hypoglycemic activity, discovered over 40 years ago while investigating the antibacterial properties of sulphonamides, led to the synthesis of effective drugs such as tolbutamide, tolazamide, and chlorpropamide (Figure 17.1) and later glibenclamide[20] and glipizide. Depolarization of β cells and the consequent decrease in potassium permeability lead to activation of voltage-sensitive calcium channels, calcium influx, and the initiation of stimulus-secretion coupling that results in the release of insulin.[21] Patch-clamp experiments have shown that sulphonylureas stimulate insulin release by specifically blocking ATP-dependent potassium channels.[22] High-affinity binding sites for glibenclamide (glyburide) are also very abundant in the brain. Indeed, putative ATP-dependent potassium-channel proteins have been purified not only from insulinoma cells, but also from brain tissue by glibenclamide binding,[23,24] and one may expect that the identification of other, less abundant binding sites will be helped by the recent synthesis of a radioiodinated, biologically active analogue of this compound.[25,26] Sulphonylureas can inhibit the channel from the outside and from the inside, which suggests

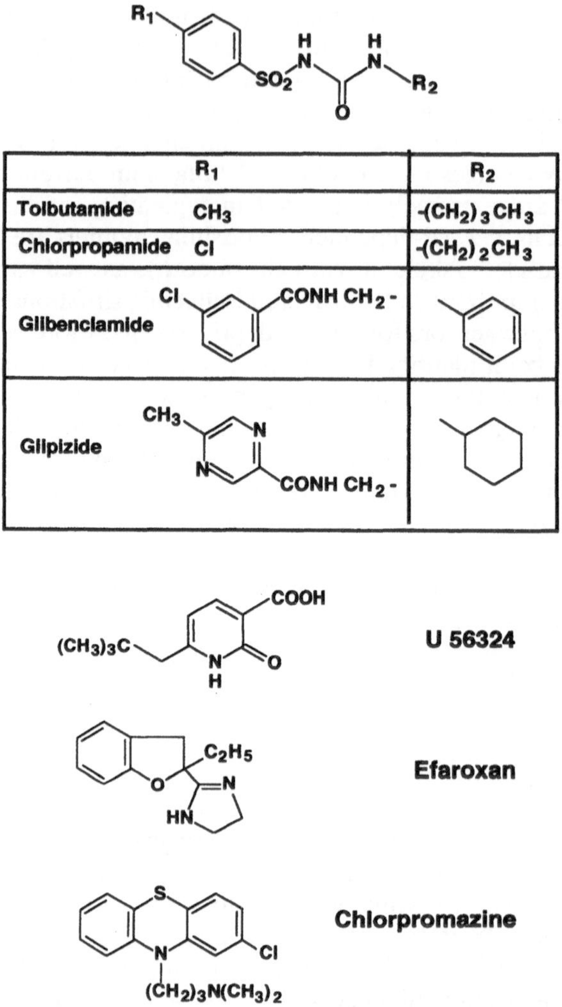

FIGURE 17.1. ATP-dependent potassium-channel blockers.

that the binding site may be reached through the lipid phase of the
membrane. This is consistent with the hydrophobic nature of these
molecules and the remarkably long time required for reaching binding
equilibrium.[27] Glibenclamide-sensitive channels in brain and in all types
of muscle (cardiac, smooth, and skeletal) are of great importance in
hypoxic and ischaemic situations because they serve as metabolic sensors.
In cardiac cells, glibenclamide reduces the shortening of the action
potential or refractory period that is produced by hypoxia.[28,29] Although
it is not known at present whether glibenclamide is able to prevent the
induction of ventricular fibrillation in vivo, the intravenous administration

of glibenclamide to anesthetised dogs reduces the buildup of potassium within the myocardium, following coronary occlusion. Sulphonylurea-binding sites in the brain appear to be most abundant in regions associated with motor function (eg, motor cortex and substantia nigra). Microinfusion of tolbutamide into the rat substantia nigra (specifically the cells of the pars compacta) associated with a systemically administered challenge dose of amphetamine, results in a circular motor behavior, which suggests that ATP-dependent potassium channels in the pars compacta may play a selective role in modifying the activity of the nigrostriatal pathway and therefore be involved in the control of movement. Glucose antagonizes morphine analgesia, possibly because of an effect on intracellular ATP levels.[30] More recently, it has been shown that glibenclamide antagonizes the same analgesic effects, which again suggests that ATP-dependent potassium channels may play an important role in modulating morphine analgesia.[31]

Other ATP-Dependent Potassium-Channel Blockers

Other hypoglycemic agents structurally related to nicotinic acid, such as Upjohn's U-56324, also block ATP-dependent potassium channels in pancreatic β cells,[32] while certain imidazoline-based α antagonists (eg, efaroxan) mimic the effects of glucose and promote insulin secretion in the absence of α-adrenoceptor stimulation by blocking ATP-dependent channels.[33] Finally, chlorpromazine and related phenothiazines directly inhibit the same channel in insulinoma cells,[34] although the biologic significance of this intriguing observation is not clear.

Class III Anti-Arrhythmics

Drugs that prolong the action potential in cardiac cells are known as class III anti-arrhythmic drugs. One mechanism whereby these drugs exert their action is to block cardiac-potassium channels, thus prolonging the time required for ventricular repolarization. Among these drugs are sotalol, which blocks both inward and delayed rectifier currents in cardiac cells;[35] clofilium, a quaternary derivative of tetraethylammonium (TEA), which blocks the cardiac delayed rectifier potassium channel; and acetylprocainamide (see Figure 17.2). Bretylium tosylate and bethanidine sulphate, two drugs that suppress ventricular fibrillation, are also believed to block cardiac-potassium channels.[36] However, most of the first-generation drugs described above have secondary effects; for example, bretylium has adrenergic-neuronal blocking activity,[37] while racemic sotalol also acts as a blocker of the β-adrenergic receptors. The selectivity of second-generation class III anti-arrhythmic drugs, such as UK 68,798,[38] semantilide,[39] Wy48,986,[40] and E-4031[41] (Figure 17.2), is much better.

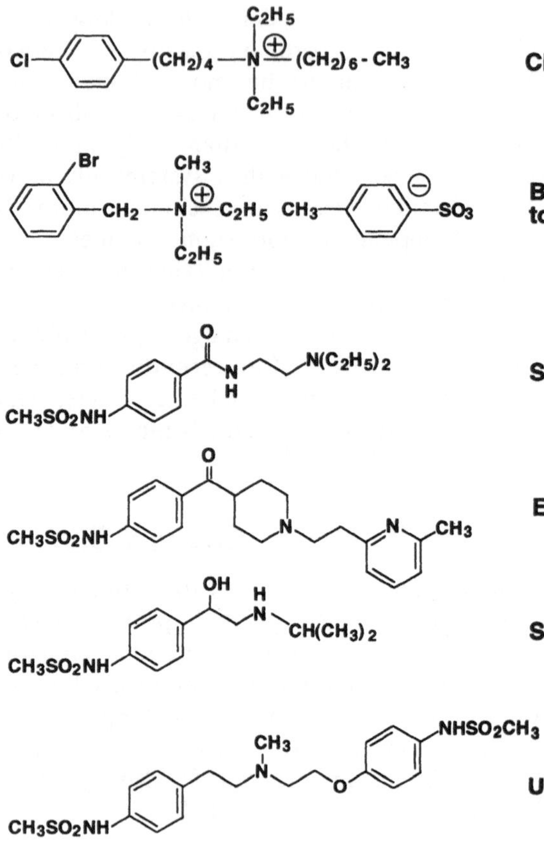

Clofilium

Bretylium tosylate

Sematilide

E - 4031

Sotalol

UK 68798

FIGURE 17.2. Class III anti-arrhythmic agents.

Blockade of delayed rectifier potassium channels during the chronic phase of myocardial infarction with these compounds results in an effective prolongation of refractoriness in the infarcted myocardium. In addition, these drugs suppress re-entrant atrial arrhythmias and prevent ventricular fibrillation caused by either coronary artery ligation or ischaemia.

A few complications are associated with the use of class III anti-arrhythmic drugs. A major one is that the blocking of potassium channels can cause after-depolarization which, in turn, induces a characteristic form of ventricular tachychardia[42] known as *torsade de pointes*. Another complication is the phenomenon of reverse-use dependence.[43] Because the magnitude of the action potential prolongation becomes smaller with increased frequency of stimulation, class III compounds could increase the tendency to develop arrhythmias at low heart rates.

Drug Design Based on the Structure of Potassium-Channel Toxins

Classic pharmacologic drugs and natural venoms that block potassium channels are ideal starting molecules for the design of new potentially useful agents. Among the classic blockers, 4-aminopyridine (4-AP) and TEA ion (Figure 17.3) are the least specific, lowest-affinity blockers of many voltage-activated potassium channels. 4-AP primarily affects fast transient currents, while TEA blocks the delayed rectifier and the M currents as well as the BK channels. 4-AP has been used to treat patients with myasthenia gravis and with myasthenic (Lambert-Eaton) syndrome.[44,45] Since 4-AP prolongs action potentials in experimentally demyelinated nerve fibres (thereby improving nerve conduction),[46] it has been suggested that potassium-channel blockers may have a role in the treatment of demyelinating disorders. However, both blockers have had limited clinical success because they are too toxic, although a few compounds modelled on the TEA molecule have had some success. Among them are bretylium tosylate and clofilium (see above).

Snake, scorpion, and bee venoms contain many different peptide and protein toxins that block specific potassium-channel subtypes.[47–50] Mast cell-degranulating (MCD) peptide, which is isolated from the venom of the European honey bee, and dendrotoxin, which is isolated from the venom of African mambas (Figure 17.4A,B,C), block voltage-activated potassium channels. Injections of MCD peptide into the brain produce long-lasting hippocampal rhythms that are associated with an increased level of wakefulness, followed by the induction of epileptic discharges, especially at higher doses.[51] The epileptic discharges can, in turn, be inhibited by potassium-channel openers such as cromakalim[52] (see

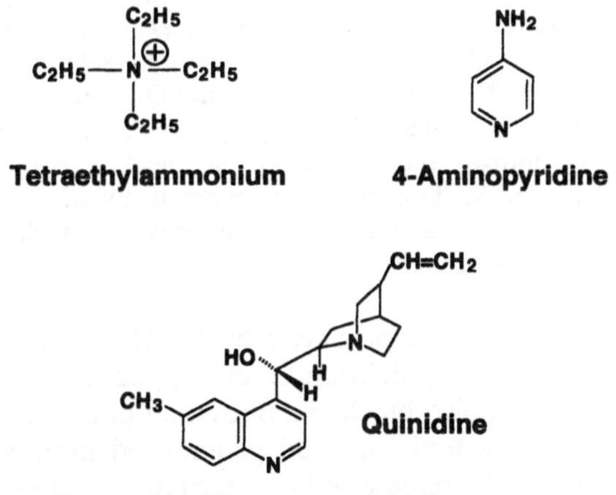

FIGURE 17.3. Miscellaneous potassium-channel blockers.

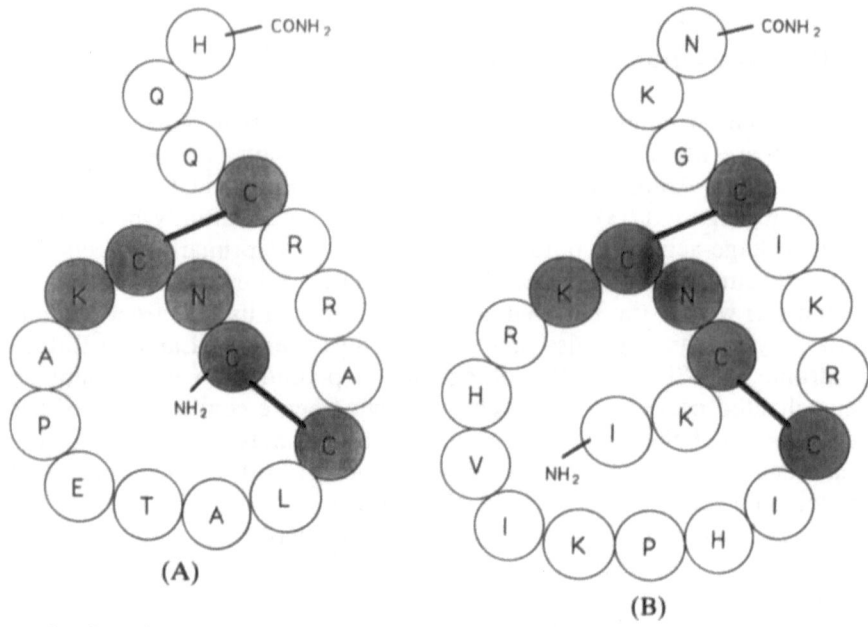

(A)

(B)

Dendrotoxin

ZPRRKLCILHRNPGRCYDKIPAFYYNQKKKQCERFDWSGCGGNSNRFKTIEECRRTCIG

Iberiotoxin
Charybdotoxin

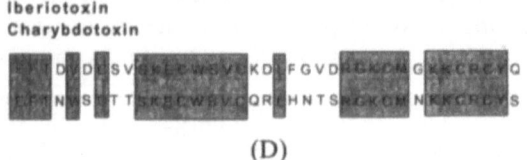

(D)

FIGURE 17.4. Potassium-channel toxins: (A) apamin, (B) MCD peptide, (C) dendrotoxin, (D) iberiotoxin and charybdotoxin. Standard single letter amino acid abbreviations used: A, alanine; C, cysteine; D, aspartic acid; E, glutamic acid; F, phenylalanine; G, glycine; H, histidine; I, isoleucine; K, lysine; L, leucine; M, methionine; N, asparagine; P, proline; Q, glutamine; R, arginine; S, serine; T, threonine; V, valine; W, tryptophan; Y, tyrosine; N-termini of dendrotoxin, charybdotoxin and iberiotoxin are blocked as pyroglutamic acid. C-termini of apamin and MCD peptide are amidated. Invariant residues between apamin and MCD peptide and between iberiotoxin and charybdotoxin, are shaded and boxed in, respectively.

below). The state of arousal produced by MCD peptide appears to be associated with a lasting potentiation of hippocampal synaptic transmission.[53] Dendrotoxin also blocks hippocampal potassium currents; this suggests that there is a toxin-induced generalized increase in neuronal activity, which affects the release of both excitatory and inhibitory neurotransmitters.[54] The convulsive effects of MCD peptide (and of dendrotoxin) can probably be ascribed to the inhibition of a hyperpolarizing current, which normally supresses excitability. It has been suggested that potassium-channel inactivation may be a significant factor in the enhanced secretion during high frequency firing in neurosecretory neurons. Support for this hypothesis has come from the observation that dendrotoxin enhances the electrically evoked secretion of vasopressin and oxytocin from isolated intermediate lobes.[55] Dendrotoxin has also been used as a specific, high-affinity ligand to isolate and characterize a voltage-activated potassium channel protein.[56,57]

It has recently been demonstrated that MCD peptide is not a true antiinflammatory agent as previously believed, but that it exerts its activity through the degranulation of mast cells (ie, acts as a counter-irritant).[58] The ability of MCD peptide to trigger histamine release is rather nonspecific and is probably due to its polycationic structure—a feature shared by other degranulating agents—and is not related to its potassium-channel-blocking activity.

All calcium-activated potassium currents are blocked by quinine, a low affinity ligand. In addition, several toxins from bee and scorpion venoms are selective for low and high conductance calcium-activated channels. Apamin (Figure 17.4A), a neurotoxin isolated from the same bee venom as MCD peptide and a specific high-affinity blocker of SK currents,[59,60] is a centrally acting neurotoxin that produces motor hyperactivity and convulsions. In spite of its size, apamin crosses the blood-brain barrier and it is three orders of magnitude more toxic when injected intraventricularly than when it is injected intravenously.[61] Chemical modification of guanidino moieties has suggested that two arginine residues are part of the active site of apamin.[62] A series of neuromuscular-blocking agents, which contain two positively charged nitrogen atoms separated by 11 Å, competitively block toxin binding sites and toxin-sensitive potassium fluxes.[63] These observations have led to direct attempts to identify the apamin pharmacophore.[64] Although none of the bis-guanidino molecules synthesized to date have shown significant apamin-like activity, this represents a promising general approach to the identification of new, synthetic potassium-channel blockers.

Scorpion venoms contains several toxins (eg, charybdotoxin and iberiotoxin) (Figure 17.4D), which have high affinity for BK channels. Both toxins increase the contractility of tracheal smooth muscle after relaxation by either cAMP-dependent or cAMP-independent bronchodilators.[65,66] This suggests that drugs that modulate BK channels might be valuable for the therapy of conditions that affect smooth muscle contractility. Charybdotoxin also inhibits the irreversible dehydration of sickle cells in vitro by blocking the erythrocyte IK channels,[67] which suggests that the structure of this toxin might serve as a starting point for the design of drugs effective in sickle cell anaemia. Other similarly designed drugs may be effective in brain ischaemia, since it has been shown that charybdotoxin reduces the formation of brain oedema that results from an experimentally-induced ischaemic insult.[68] Charybdotoxin also blocks certain voltage-activated potassium channels, notably those in human T lymphocytes, with high affinity. Whether charybdotoxin inhibits interleukin-2 production, T-cell activation, and mitogenesis[69,70] is still controversial. Charybdotoxin-sensitive, voltage-activated channels in the brain are also blocked by dendrotoxin; binding of the two toxins is mutually competitive.[71]

Putative Endogenous Potassium-Channel Blockers

Several putative endogenous blockers of different potassium-channel subtypes have been identified in the brain, using either radioimmuno-assay or competitive binding techniques. Identification of such molecules may be extremely important, both for drug design and in order to understand the modulation of potassium-channel function. Thus, an apamin-like molecule has been identified in pig brain and has been purified to homogeneity.[72] The purified molecule inhibits also the after-hyperpolarization (AHP) in cultured rat embryonic muscle cells. Brain contains very small amounts of this peptide (less than 0.75 pmol per pig brain) and its role is intriguing. One would assume that regions of high apamin-binding activity would also contain the putative endogenous apamin molecule. One of these regions is the hypothalamic supraoptic nucleus. Here the secretion of oxytocin and vasopressin is controlled by the electrical activity of magnocellular neurosecretory cells, whose AHPs produced after bursts of action potentials are blocked by apamin.[73] Thus one possible function of an endogenous apamin-like molecule could be the control of oxytocin and vasopressin secretion in the neural lobe.

Using similar radioimmunoassay techniques, minute quantities of an endogenous MCD peptide-like molecule has also been purified from pig brain. It also inhibits MCD peptide binding to synaptosomal membranes.[74] In the light of the behavioral effects of MCD peptide (see above), it is fascinating to speculate that this endogenous molecule may have a role in memory function. Preliminary evidence for the existence of endogeneous peptide that inhibits [^{3}H]-glipizide binding in the brain (and presumably binds to a putative ATP-dependent potassium channel) has also been obtained.[75]

Potassium-Channel Openers

The molecules now considered as typical channel openers were originally used as vasodilatory drugs with an ill-defined mechanism of action. An ability to hyperpolarize smooth muscle membranes by increasing potassium conductance was demonstrated for nicorandil.[76] Subsequent

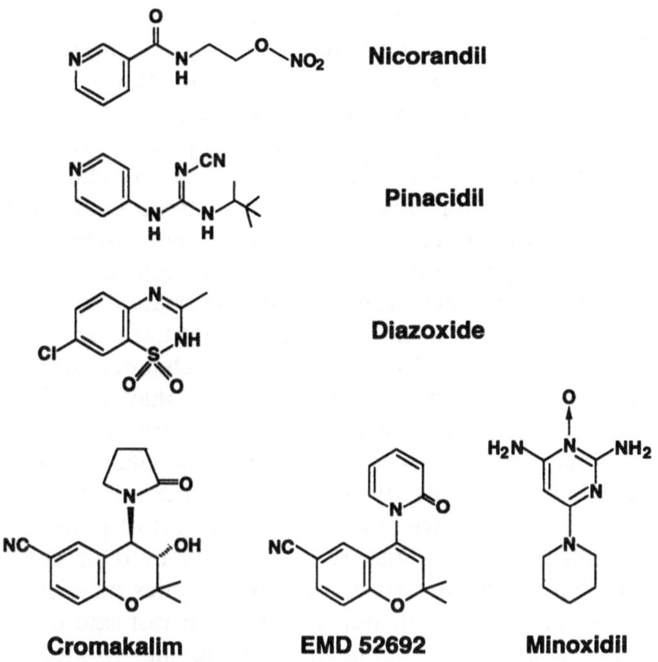

FIGURE 17.5. Potassium-channel openers.

electrophysiologic and ^{86}Rb efflux studies with two other vasodilators, cromakalim[77] (formerly known as BRL 34915) and pinacidil,[78] firmly established this class of drugs as potassium-channel openers (sometimes called channel activators or agonists). More recent studies have shown that nicorandil (unlike cromakalim and pinacidil) also acts through a cGMP-guanylate cyclase pathway.[79] Nicorandil, cromakalim, and pinacidil have distinct chemical structures (Figure 17.5). The pharmacologic activity of cromakalim (a benzopyran derivative), is due almost exclusively to the optically active (−)(3S,4R) enantiomer (lemakalim). EMD 52692 (Merck), a cromakalim derivative, is probably the most potent hypertensive agent characterized to date. Pinacidil (a thiourea derivative) and nicorandil (a pyridine derivative) have also been used as the starting molecules for extensive drug development studies.

Other classes of molecules, subsequently shown to be potassium-channel openers, are certain pyrimidines (minoxidil sulphate) and benzothiadiazines (diazoxide). Both minoxidil sulphate[80] and diazoxide[81] are clinically established anti-hypertensive agents. The reader is referred to Edwards and Weston[82] for an overview of these structure-activity studies. The peripheral vasodilatory activity of potassium-channel openers has prompted studies of their potential use in cardiovascular disorders such as angina pectoris, congestive heart failure, and hypertension. (For example, pinacidil produces regression in ventricular hypertrophy and lowers serum cholesterol;[83,84] nicorandil suppresses ischaemic S-T wave depression during anginal attacks and selectively dilates coronary vessels without affecting other hemodynamic parameters.[85,86]) Indeed, potassium-channel openers may prove to be extremely useful drugs, especially if it could be shown that their prolonged use does not result in other undesirable side effects characteristic of direct-acting vasodilators (eg, edema and reflex tachycardia). The powerful smooth muscle relaxant properties of potassium-channel openers can be demonstrated using bronchial preparations in vitro.[87] This suggests that these drugs may be used for the treatment of asthma, especially since potassium-channel openers may also inhibit the release of sensory neuropeptides that contribute to the inflammatory reaction in the airways of asthmatic patients.[88] The ability of potassium-channel openers to relax gastrointestinal smooth muscle[89] has prompted investigations of these compounds in the treatment of irritable bowel syndrome, while the relaxing effect of potassium-channel openers on uterine smooth muscle[90] may lead to their clinical use in premature labour. Both in vivo and in vitro studies have confirmed the ability of potassium channel openers to relax detrusor smooth muscle,[91,92] which suggests a promising application in the management of urinary incontinence that results from bladder hypertrophy.

Although the effects of potassium-channel openers on the central nervous system is less well known, there is reason to expect that molecules of this type may be of therapeutic benefit. Thus, cromakalim alters the excitability of hippocampal neurons[93] while potassium-channel openers activate $^{86}Rb^+$ efflux and decrease GABA release in the substantia nigra in a sulphonylurea-dependent manner, which suggests a link between activation of ATP-dependent potassium channels and inhibition of transmitter release.[94] Since activation of ATP channels can counteract the effects of cyanide-induced anoxia in the substantia nigra,[95] it is possible that potassium-channel openers may be useful in the treatment of neurodegenerative disorders, such as Parkinson's and Alzheimer's disease.[96] In addition, it has been shown that diazoxide reduces anoxic glutamate release from the presynaptic nerve terminals of hippocampal mossy fibres; potassium-channel openers may therefore prove useful in preventing the excitotoxic effects of excess glutamate during anoxia and ischaemia.[97] Intracerebroventricular injection of cromakalim attenuates seizures in genetically epileptic rats[98] and inhibits epileptiform seizures induced by MCD peptide.[99] This suggests a potential use of potassium-channel openers in the prophylactic treatment of epilepsy; indeed, the clinically established anti-epileptic drug carbamazepine increases potassium currents in rat cortical neurons.[100]

Cromakalim also restores the membrane potential to experimentally depolarized isolated human muscle fibres,[101] while cromakalim and EMD

52962 suppress in vitro the myotonic activity of muscle from patients with neuromuscular disorders, such as myotonic dystrophy and myotonia congenita.[102] Cromakalim also improves the in vitro contractile force of skeletal muscle samples from patients with hypokalemic periodic paralysis.[103] Unfortunately the specific effects of cromakalim on skeletal muscle in vitro occur at concentrations which in vivo would cause hyperglycemia and hypotension. Other future potential uses for potassium-channel openers include the treatment of impotence, ischaemia, cerebral vasospasms, glaucoma, and hair growth.[104,105] However, it is unlikely that the three original molecules discussed above have significant therapeutic potential. What is needed is the development of new, tissue-selective drugs based on these prototype structures.

Site of Action of Potassium-Channel Openers

The mechanism of action of potassium-channel openers is a matter of controversy. Evidence suggests that some interact with ATP-dependent potassium channels. The sulphonylurea drugs (including glibenclamide) act as competitive inhibitors of potassium channel opener-induced smooth muscle relaxation and $^{86}Rb^+$ efflux.[106,107] Diazoxide is an activator of ATP-dependent potassium channels in pancreatic β cells (inhibiting the release of insulin), and acts as a potassium-channel opener in vascular smooth muscle.[108] Direct evidence is provided by the observation that cromakalim increases the open probability of ATP-dependent channels in arterial smooth muscle.[109] On the other hand, the channel opened by cromakalim in cultured aortic smooth muscle does not appear to be ATP-dependent; rather, it is indistinguishable from the charybdotoxin-sensitive, large-conductance, calcium-activated potassium channel.[110] This is in contrast with the observation that charybdotoxin has no effect on cromakalim-induced $^{86}Rb^+$ fluxes in blood vessels.[111] A third possibility has been proposed as a result of detailed and thorough studies of potassium currents in whole smooth muscle cells from portal vein.[112] In these cells the cromakalim-sensitive current mostly resembles delayed outward rectifier current. Finally, the matter is complicated further by a recent report that describes an ATP-sensitive, low conductance, calcium-activated potassium channel that is opened by nicorandil.[113] Binding studies have not helped resolve these problems. Potassium-channel openers appear to inhibit a low affinity sulphonylurea-binding site on smooth muscle, but do not compete for high affinity [^{3}H]-glibenclamide-binding sites.[27] To date no binding studies that used potassium-channel openers have been published, possibly because of problems created by the extreme hydrophobicity of these molecules. The one water-soluble potassium-channel opener, minoxidil sulphate, is only active as a pro-drug. A related fundamental question that binding studies might help to answer is whether the potassium-channel openers act directly on the channel and, if so, whether the site of action is extracellular or cytoplasmic. It is not clear how these controversies are going to be resolved. There is no *a priori* reason why potassium-channel openers should affect only one type of channel. Indeed, glibenclamide-sensitive potassium channels that are insensitive to ATP have recently been described.[114,115] The one unifying factor to date is that all channels activated by potassium-channel openers appear to be glibenclamide-sensitive. One must also consider the possibility that the sensitivity of a potassium channel to channel openers may be altered by subtle changes in the ion channel structure, such as those caused by the enzymatic dispersion of cells, which are often required for electrophysiologic experiments.

Naturally Occurring Potassium-Channel Openers

There is currently much interest in the possible role of endogenous vasodilators that modulate blood pressure.[116] The presence of an endothelium-derived hyperpolarizing factor (EDHF) in blood vessels (which is distinct from nitric oxide, an endothelium-derived relaxing factor [EDRF]) has been unambiguously demonstrated.[117] It has been suggested that one stimulus for the release of EDHF may be hemodynamic shear stress.[118] Patch-clamp experiments have directly demonstrated that calcitonin gene-related peptide (CGRP) activates single potassium channels in mesenteric arteries, and that CGRP-induced membrane hyperpolarization and its associated relaxation are antagonized by glibenclamide.[119] Vasoactive-intestinal peptide (VIP) and adenosine can act in a similar fashion, while somatostatin and galanin can activate ATP-dependent potassium channels in insulin-secreting cells[120,121] through a GTP-dependent mechanism. G proteins have also been shown to activate ATP-dependent potassium channels in cardiac muscle,[122] and it is possible that they may also mediate the effects of CGRP, VIP, and adenosine.[123] Finally, it has been demonstrated that the intracellular application of lactate to patch-clamped cardiac myocytes results in the activation of ATP-dependent potassium currents. This suggests that lactate produced by glycolysis during myocardial ischaemia may contribute to the opening of ATP-dependent channels and may shorten the action potential, even when the levels of ATP have been lowered by a moderate to severe ischaemia.[124]

Conclusion

The concept that potassium channels are useful targets for therapeutic drugs is a recent one. Roles for drugs that either block or activate potassium channels can be envisioned. The diversity of potassium-channel function in excitable and nonexcitable cells means that an extremely wide range of disorders might be amenable to new drugs that are specifically designed to interact with particular channel subtypes in different tissues.

Acknowledgments. I thank Barry Brewster and Jonathan Wadsworth for their constructive comments during the preparation of this manuscript. I acknowledge the Muscular Dystrophy Group of Great Britain and the Wellcome Trust for financial support.

References

1. Rudy B. Diversity and ubiquity of K channels. Neuroscience 1988; 25: 729–749.
2. Levitan IB, Kaczamarek LK. The Neuron. New York: Oxford University Press; 1991:193–272.
3. Yellen G. Permeation in potassium channels: Implications for channel structure. Ann Rev Biophys Biophys Chem 1987; 16:227–246.
4. Hodgkin AL, Huxley AF. A quantitative description of membrane current and its application to conduction and excitation in nerve. J Physiol 1952; 117:500–544.
5. Adrian RH. Rectification in muscle membrane. Prog Biophys Mol Biol 1969; 19:340–369.
6. Hagiwara S, Kusano K, Saito N. Membrane changes of *Onchidium* nerve cell in potassium-rich media. J Physiol 1961; 275:357–376.
7. Marty A. Ca^{2+}-dependent K^+ channels with large unitary conductance. Trends Neurosci 1983; 6:262–265.
8. Blatz AL, Magleby KL. Single apamin-blocked Ca^{2+}-activated K^+ channels of small conductance in cultured rat skeletal muscle. Nature 1986; 323:718–720.
9. Hermann A, Erxleben C. Charybdotoxin selectivity blocks small Ca^{2+}-activated K^+ channels in *Aplysia* neurons. J Gen Physiol 1987; 90:27–47.
10. Meech RW. Calcium-dependent potassium activation in nervous tissues. Ann Rev Biophys Bioeng 1978; 7:1–18.
11. Petersen OH, Maruyama Y. Calcium-activated potassium channels and their role in secretion. Nature 1984; 307:693–696.
12. Cook NS. The pharmacology of potassium channels and their therapeutic potential. Trends Pharmacol Sci 1988; 9:21–28.
13. Adams PR, Brown DA, Constanti A. Pharmacological inhibition of the M-current. J Physiol 1982; 332:223–262.
14. Richards NW, Dawson DC. Single potassium channels blocked by lidocaine and quinine in isolated turtle colon epithelial cells. Am J Physiol 1986; 251:C85–C89.
15. Dunne MJ, Petersen OH. Potassium selective ion channels in insulin-secreting cells: Physiology, pharmacology and their role in stimulus-secretion coupling. Biochim Biophys Acta 1991; 1071:67–82.
16. Davies NW, Standen NB, Stanfield PR. ATP-dependent potassium channels of muscle cells: Their properties, regulation and possible functions. J Bioenerg Biomembr 1991; 32:509–535.

17. Belardetti F, Siegelbaum SA. Up and down modulation of single K^+ channels function by distinct second messengers. Trends Neurosci 1988; 11:232–238.

18. Dunlap K, Holz GG, Rane SG. G proteins as regulators of ión channel function. Trends Neurosci 1987; 10:241–244.

19. Loubatières A. Effects of sulfonylureas on the pancreas. In: Volk BW, Wellman KE, eds. The Diabetic Pancreas. New York: Plenum Press; 1977: 489–515.

20. Schmid-Antomarchi H, de Weille JR, Fosset M, Lazdunski M. The antidiabetic sulphonylurea glibenclamide is a potent blocker of the ATP-modulated K^+ channel in insulin secreting cells. Biochem Biophys Res Commun 1987; 146:21–25.

21. Gylfe E, Hellman B, Sehlin J, Taljedal LB. Interaction of sulphonylurea with the pancreatic B cell. Experientia 1984; 40:1126–1134.

22. Sturgess NC, Kozlowoski RZ, Carrington CA, Hales CN, Ashford MLJ. Effects of sulphonylureas and diazoxide on insulin secretion and nucleotide-sensitive channels in an insulin-secreting cell line. Br J Pharmacol 1988; 95:83–94.

23. Bernardi H, Fosset M, Lazdunski M. Characterization, purification and affinity labeling of the brain [³H]glibenclamide-binding protein, a putative neuronal ATP-regulated K^+ channel. Proc Natl Acad Sci USA 1988; 85:9816–9820.

24. Aguilar-Bryan L, Nelson DA, Vu QA, Humphrey MB, Boyd AE III. Photoaffinity labeling and partial purification of the B cell sulphonylurea receptor using a novel biologically active glyburide analog. J Biol Chem 1990; 14:8218–8224.

25. Robertson DW, Schober DA, Krushinski JH, Mais DE, Thompson DC, Gehlert DR. Expedient synthesis and biochemical properties of an ¹²⁵I-labelled analogue of glyburide, a radioligand for ATP-inhibited potassium channels. J Med Chem 1990; 33:3124–3126.

26. Gehlert DR, Gackenheimer SL, Mais DE, Robertson DW. Quantitative autoradiography of the binding sites for [¹²⁵I]iodoglyburide, a novel, high affinity ligand for ATP-sensitive potassium channels in rat brain. J Pharmacol Exp Ther 1991; 257:901–907.

27. Gopalakrishnan M, Johnson DE, Janis RA, Triggle DJ. Characterization of binding of the ATP-sensitive potassium channel ligand [³H]glyburide, to neuronal and muscle preparations. J Pharmacol Exp Ther 1991; 257:1162–1171.

28. Beckheit SS, Restivo M, Boutjdir M, Henhin R, Gooyandeh K, Assadi M, Khatib S, Gough W, El-Sherif N. Effects of glyburide on ischaemia-induced changes in extracellular potassium and local myocardial activation: A potential new approach to the management of ischaemia-induced malignant ventricular arrythmias. Am Heart J 1990; 119:1025–1033.

29. Kantor PFW, Coetzee WA, Carmeliet EE, Dennis SC, Opie LH. Reduction of ischaemic K^+ loss and arrythmias in rat hearts. Circ Res 1990; 66:478–485.

30. Singh IS, Chaterjee TK, Ghosh JJ. Modification of morphine antinociceptive response by blood glucose status: Possible involvement of cellular energetics. Eur J Pharmacol 1983; 90:437–439.

31. Ocana M, Del Pozo E, Barrios M, Robles LI, Baeyens JM. An ATP-dependent potassium channel blocker antagonizes morphine analgesia. Eur J Pharmacol 1990; 186:377–378.
32. Hopkins WF, Fatherazi S, Cook DL. The oral hypoglycemic agent, U-56324, inhibits the activity of ATP-sensitive potassium channels in cell-free membrane patches from cultured mouse pancreatic B-cells. FEBS Lett 1990; 277:101–104.
33. Chan SLF, Dunne MJ, Stillings MR. The α_2-adrenoceptor antagonist efaroxan modulates ATP-dependent potassium channels in insulin-secreting cells. Eur J Pharmacol 1991; 204:41–48.
34. Muller M, De Weille JR, Lazdunski M. Chlorpromazine and related phenothiazines inhibit the ATP-sensitive K^+ channel. Eur J Pharmacol 1991; 198:101–104.
35. Carmeliet E. Electrophysiologic and voltage clamp analysis of the effects of sotalol on isolated cardiac muscle and purkinje fibres. J Pharmacol Exp Ther 1985; 232:817–825.
36. Bacaner MB, Clay JR, Shrier A, Brochu RM. Potassium channel blockade: A mechanism for suppressing ventricular fibrillation. Proc Natl Acad Sci USA 1986; 83:2223–2227.
37. Heisenbuttel RH, Bigger JT. Bretylium tosylate—A newly available drug for ventricular arrythmias. Ann Int Med 1979; 91:229–238.
38. Gwilt M, Arrowsmith JE, Blackburn KJ, Burges RA, Cross PE, Dalrymple HW, Higgins AJ. UK-68, 798: A novel potent and highly selective class III antiarrhythmic agent which blocks potassium channels in cardiac cells. J Pharmacol Exp Ther 1991; 256:318–324.
39. Chi L, Mu D-X, Drisoll EM, Lucchesi BR. Antiarrhythmic and electrophysiologic actions of CK-3579 and sematilide in conscious canine model of sudden coronary death. J Cardiovasc Pharmacol 1990; 16:312–324.
40. Colatsky TJ, Jurkiewicz NK, Follmer CH, Bird LB. Antiarrhythmic efficacy of Wy48,986, a novel class III antiarrhythmic agent, on ventricular arrhythmias induced by coronary ligation in dogs and pigs: Effects on acute, subacute and chronic phase post-ligation arrhythmias. J Mol Cell Cardiol 1989; 21(suppl 12):S10.
41. Katoh H, Ogawa S, Furuno I, Yoh S, Saeki K, Nakamura Y. Electrophysiologic effects of E-4031, a class III antiarrhythmic agent, on reentrant ventricular arrhythmias in a canine 7 day old infarction model. J Pharmacol Exp Ther 1990; 253:1077–1082.
42. Carlsson L, Almgren O, Duker G. QTU prolongation and *torsade de pointes* induced by putative class III antiarrhythmic agents in the rabbit: Etiology and interventions. J Cardiovasc Pharmacol 1990; 16:276–285.
43. Hondeghem LM, Snyders DJ. Class III antiarrhythmics have a lot of potential but a long way to go. Reduced effectiveness and dangers of reverse use dependence. Circulation 1990; 81:686–690.
44. Murray NMF, Newsom-Davis J. Treatment with oral 4-aminopyridine in disorders of neuromuscular transmission. Neurology 1981; 31:265–271.
45. Lundh H, Nilsson O, Rosen I. Effects of 4-aminopyridine on normal and demyelinated mammalian nerve fibres. Nature 1980; 283:570–572.
46. Sherratt RM, Bostock H, Sears TA. Effects of 4-aminopyridine on normal and demyelinated mammalian nerve fibres. Nature 1980; 283:570–572.

47. Strong PN. Potassium channel toxins. Pharmacol Ther 1990; 46:137–162.
48. Castle NS, Haylett DG, Jenkinson DH. Toxins in the characterization of potassium channels. Trends Neurosci 1989; 12:59–65.
49. Moczydlowski E, Lucchesi K, Ravindran A. An emerging pharmacology of peptide toxins targeted against potassium channels. J Membr Biol 1988; 105:95–111.
50. Brewster BS, Strong PN. Naturally occurring potassium channel blockers. In: Weston AH, Hamilton TC, eds. Potassium channel modulators: Pharmacological, molecular and clinical aspects. Oxford: Blackwell; 1992:592.
51. Bidard JN, Gandolfo G, Mourre C, Gottesmann C, Lazdunski M. The brain response to the bee venom peptide MCD: Activation and densensitisation of a hippocampal target. Brain Res 1987; 418:235–244.
52. Gandolfo G, Gottesmann C, Bidard JN, Lazdunski M. K^+ channel openers prevent epilepsy induced by the bee venom peptide MCD. Eur J Pharmacol 1989; 159:329–330.
53. Cherubini E, Ben-Ari Y, Goh M, Bidard JN, Lazdunski M. Long term potentiation of synaptic transmission in the hippocampus induced by a bee venom peptide. Nature 1987; 328:70–73.
54. Halliwell JV, Othman IB, Pelchen-Matthews A, Dolly JO. Central action of dendrotoxin: Selective reduction of a transient K conductance in hippocampus and binding to localized acceptors. Proc Natl Acad Sci USA 1986; 83:493–497.
55. Bondy CA, Gainer H, Russell JT. Effects of stimulus frequency and potassium channel blockade on the secretion of vasopressin and oxytocin from the neurohypophysis. Neuroendocrinology 1987; 46:258–267.
56. Rehm H, Lazdunski M. Purification and subunit structure of a putative K^+-channel protein: Identification by its binding properties for dendrotoxin I. Proc Natl Acad Sci USA 1988; 85:4919–4923.
57. Parcej DN, Dolly JO. Dendrotoxin acceptor from bovine synaptic plasma membranes. Binding properties, purification and subunit composition of a putative constituent of certain voltage-activated K^+ channels. Biochem J 1989; 257:899–903.
58. Banks BEC, Dempsey CE, Vernon CA, Warner JA, Yamey J. Anti-inflammatory activity of bee venom peptide 401 (mast cell degranulating peptide) and compound 48/80 results from mast cell degranulation in vivo. Br J Pharmacol 1990; 99:350–354.
59. Blatz AL, Magleby KL. Single apamin-blocked Ca-activated K^+ channels of small conductance in cultured rat skeletal muscle. Nature 1986; 323:718–720.
60. Capiod T, Ogden DC. The properties of calcium-activated potassium channels in guinea-pig isolated hepatocytes. J Physiol 1989; 409:285–295.
61. Habermann E. Neurotoxicity of apamin and MCD peptide upon central application. Naunyn-Schmiedeberg's Arch Pharmacol 1977; 300:189–191.
62. Vincent JP, Schweitz H, Lazdunski M. Structure-function relationships and the site of action of apamin, a neurotoxic polypeptide of bee venom with an action on the central nervous system. Biochemistry 1975; 14:2521–2525.
63. Cook NS, Haylett DG. Effects of apamin, quinine and neuromuscular blockers on calcium-activated potassium channels in guinea-pig hepatocytes. J Physiol 1985; 358:373–394.

64. Demonchaux P, Ganellin CR, Dunn PM, Haylett DG, Jenkinson DH. Search for the pharmacophore of the potassium channel blocker, apamin. Eur J Med Chem 1991; 26:915–920.

65. Suarez-Kurtz G, Garcia ML, Kaczorowski GJ. Effects of charybdotoxin and iberitoxin on the spontaneous motility and tonus of different guinea pig smooth muscle tissues. J Pharmacol Exp Ther 1991; 259:439–443.

66. Jones TR, Charette L, Garcia ML, Kaczorowski GJ. Selective inhibition of guinea pig trachea by charybdotoxin, a potent calcium-activated potassium channel inhibitor. J Pharmacol Exp Ther 1990; 256:697–706.

67. Ohnishi ST, Katagi H, Katagi C. Inhibition of the in vitro formation of dense cells and of irreversibly sickled cells by charybdotoxin, a specific inhibitor of calcium-activated potassium efflux. Biochim Biophys Acta 1989; 1010:199–203.

68. Tominaga T, Katagi H, Ohnishi ST. Is calcium-activated potassium efflux involved in the formation of ischaemic brain oedema? Brain Res 1988; 460:376–378.

69. Price M, Lee SC, Deutsch C. Charybdotoxin inhibits proliferation and interleukin 2 production in human peripheral blood lymphocytes. Proc Natl Acad Sci USA 1989; 86:10171–10175.

70. Gelfand EW, Or R. Charybdotoxin-sensitive, calcium-dependent membrane potential changes are not involved in human T or B cell activation and proliferation. J Immunol 1991; 147:3452–3458.

71. Schweitz H, Bidard JN, Maes P, Lazdunski M. Charybdotoxin is a new member of the K^+ channel toxin family that includes dendrotoxin I and mast cell degranulating peptide. Biochemistry 1989; 28:9708–9714.

72. Fosset M, Schmid-Antomarchi H, Hugues M, Romey G, Lazdunski M. The presence in pig brain of an endogenous equivalent of apamin, the bee venom peptide that specifically blocks Ca^{2+}-channels. Proc Natl Acad Sci USA 1984; 81:7228–7232.

73. Bourque CW, Brown DA. Apamin and d-tubocurarine block the after-hyperpolarization of rat supraoptic neurosecretory neurons. Neurosci Lett 1987; 82:185–190.

74. Cherubini E, Ben-Ari Y, Goh M, Bidard JN, Lazdunski M. Long term potentiation of synaptic transmission in the hippocampus induced by a bee venom peptide. Nature 1987; 328:70–73.

75. Virsolvy-Vergine A, Bruck M, Dufour M, Cauvin A, Lupo B, Bataille D. An endogenous ligand for the central sulphonylurea receptor. FEBS Lett 1988; 242:65–69.

76. Furukawa K, Itoh T, Kajiwara M, Kitamura K, Suzuki H, Ito Y, Kuriyama H. Vasodilating actions of 2-nicotinamidoethyl nitrate on porcine and guinea-pig coronary arteries. J Pharmacol Exp Ther 1981; 218:248–259.

77. Hamilton TC, Weir SW, Weston AH. Comparison of the effects of BRL 34915 and verapamil on electrical and mechanical activity in rat portal vein. Br J Pharmacol 1986; 88:103–111.

78. Bray KM, Newgreen DT, Small RC, Southerton JS, Taylor SG, Weir SW, Weston AH. Evidence that the mechanism of the inhibitory action of pinacidil in rat and guinea pig smooth muscle differs from that of glyceryl trinitrate. Br J Pharmacol 1987; 91:421–429.

79. Holzman S. Cyclic GMP as a possible mediator of coronary arterial relaxation by nicorandil. J Cardiovasc Pharmacol 1983; 5:364–370.

80. Meisheri K, Cipkus LA, Taylor CJ. Mechanism of action of minoxidil sulphate-induced vasodilation: A role for increased K^+ permeability. J Pharmacol Exp Ther 1988; 245:751–760.
81. Gross F. Drugs acting on arteriolar smooth muscle—Diazoxide. In: Gross F, ed. Antihypertensive agents. Berlin: Springer-Verlag; 1977:430–443.
82. Edwards G, Weston AH. Structure-activity relationships of K^+ channel openers. Trends Pharmacol Sci 1990; 11:417–422.
83. Steensgard-Hansen F, Carlsen JE. Effects of long term treatment with pinacidil and nifedipine on left ventricular anatomy and function in patients with mild to moderate systemic hypertension. Drugs 1988; 36(suppl 7):70.
84. Goldberg MR, Rockhold RW. Beneficial effects of pinacidil on plasma lipids. J Hypertension 1986; 4(suppl 5):575.
85. Sakai K, Shiraka Y, Nabuta H. Cardiovascular effect of a new coronary vasodilator 2-nicotinamide ethyl nitrate (SG-75): Comparison with nitroglycerin and diltiazem. J Cardiovasc Pharmacol 1981; 3:139–150.
86. Kishida H, Murao S. Effect of a new coronary vasodilator, nicorandil, on variant angina pectoris. Clin Pharmacol Ther 1987; 42:166–174.
87. Allen SL, Boyle JP, Cortijo J, Foster RW, Morgan GP, Small RC. Electrical and mechanical effects of BRL34915 in guinea-pig isolated trachealis. Br J Pharmacol 1986; 89:395–405.
88. Black JL, Barnes PJ. Potassium channels and airway function: New therapeutic prospects. Thorax 1990; 45:213–218.
89. den Hertog A, van den Akker J, Nelemans A. Effect of cromakalim on smooth muscle cells of guinea pig taeni caeci. Eur J Pharmacol 1989; 174:287–291.
90. Piper I, Minshall E, Downing SJ, Hollingsworth M, Sadraei H. Effect of several potassium channel openers and glibenclamide on the uterus of the rat. Br J Pharmacol 1990; 101:901–907.
91. Malmgren A, Anderson KE, Sjogren C, Andersson PO. Effects of pinacidil and cromakalim (BRL34915) on bladder function in rats with detrusor instability. J Urol 1989; 142:1134–1138.
92. Malgrem A, Andersson K-E, Andersson PO, Fovaeus M, Sjogren C. Effects of cromakalim (BRL34915) and pinacidil on normal and hypertrophied rat detrusor in vitro. J Urol 1990; 143:828–834.
93. Alzheimer C, ten Bruggencate G. Actions of BRL 34915 (cromakalim) upon convulsive discharges in guinea-pig hippocampal slices. Naunyn-Schmiedeberg's Arch Pharmacol 1988; 337:429–434.
94. Schmid-Antomarchi H, Amoroso S, Fosset M, Lazdunski M. K^+ channel openers activate brain sulphonylurea-sensitive K^+ channels and block neurosecretion. Proc Natl Acad Sci USA 1990; 87:3489–3492.
95. Murphy KPSJ, Greenfield SA. ATP-sensitive potassium channels counteract anoxia in neurones of the substantia nigra. Exp Brain Res 1991; 84: 355–358.
96. Amoroso S, Schmid-Antomarchi H, Fosset M, Lazdunski M. Glucose, sulphonylureas and neurotransmitter release: Role of ATP-sensitive K^+ channels. Science 1990; 247:852–854.
97. Krnjevic K. Adenosine triphosphate sensitive potassium channels in anoxia. Stroke 1991; 21:190–193.

98. Gandolfo G, Romettino S, Gottesmann C, van Luijtelaar G, Counen A, Bidard JN, Lazdunski M. K$^+$ channel openers decrease seizures in genetically epileptic rats. Eur J Pharmacol 1989; 167:181–183.

99. Gandolfo G, Gottesmann C, Bidard JN, Lazdunski M. K$^+$ channel openers prevent epilepsy induced by the bee venom peptide MCD. Eur J Pharmacol 1989; 159:329–330.

100. Zona C, Tancredi V, Palma E, Pirroni GC, Avoli M. Potassium currents in rat cortical neurons in cultures are enhanced by the antiepileptic drug carbamazepine. Can J Physiol Pharmacol 1990; 68:545–547.

101. Spuler A, Lehmann-Horn F, Grafe P. Cromakalim (BRL 34915) restores in vitro the membrane potential of depolarized human skeletal muscle fibres. Naunyn-Schmiedeberg's Arch Pharmacol 1989; 339:327–331.

102. Quasthoff S, Spuler A, Spittelmeister W, Lehmann-Horn F, Grafe P. K$^+$ channel openers suppress myotonic activity of human skeletal muscle in vitro. Eur J Pharmacol 1990; 186:125–128.

103. Grafe P, Quasthoff S, Strupp M, Lehmann-Horn F. Enhancement of K$^+$ conductance improves in vitro the contraction force of skeletal muscle in hypokalemic periodic paralysis. Muscle Nerv 1990; 13:451–457.

104. Duty S, Weston AH. Potassium channel openers: Pharmacological effects and future uses. Drugs 1990; 6:758–791.

105. Evans JM, Longman SD. Potassium channel activators. Ann Rep Med Chem 1991; 26:73–82.

106. Buckingham RE, Hamilton RC, Howleytt DR, Mooto S, Wilson C. Inhibition by glibenclamide of the vasorelaxant action of cromakalim in the rat. Br J Pharmacol 1989; 97:57–64.

107. Sanguinetti MC, Scott AL, Zingare CJ, Siegl PKS. BRL 34915 (cromakalim) activates ATP-sensitive K$^+$ current in cardiac muscle. Proc Natl Acad Sci USA 1988; 85:8360–8364.

108. Dunne MJ, Yule DI, Gallacher DV, Petersen OH. Comparative study of the effects of cromakalim and diazoxide on membrane potential, [Ca]$_i$, and ATP-sensitive potassium currents in insulin-secreting cells. J Membr Biol 1990; 114:53–60.

109. Standen NB, Quayle JM, Davies NW, Brayden JE, Huang Y, Nelson MT. Hyperpolarizing vasodilators activate ATP-sensitive K$^+$ channels in arterial smooth muscle. Science 1989; 245:177–180.

110. Gelband CH, McCoullough JR, van Breemen C. Modulation of vascular Ca^{2+}-activated K$^+$ channels by cromakalim, pinacidil and glyburide. Biophysic J 1990; 57:509a.

111. Strong PN, Weir SS, Beech DJ, Hiestand P, Kocher HP. Potassium channel toxins from Leiurus quinquestriatus hebraeus venom: Purification of charybdotoxin and a second toxin which inhibits cromakalim-stimulated ^{86}Rb$^+$ efflux from aortic smooth muscle. Br J Pharmacol 1989; 98:817–826.

112. Beech DJ, Bolton TB. Properties of the cromakalim-induced potassium conductance in smooth muscle cells isolated from the rabbit portal vein. Br J Pharmacol 1989; 98:851–864.

113. Kajioka S, Oike M, Kitamura K. Nicorandil opens a calcium-dependent potassium channel in smooth muscle cells of the rat portal vein. J Pharmacol Exp Ther 1990; 245:905–913.

114. Nakao K, Bolton TB. Cromakalim-induced potassium currents in single dispersed smooth muscle cells of rabbit artery and vein. Br J Pharmacol 1991; 102:155P.
115. Quasthoff S, Franke C, Hatt H, Richter-Turter M. Two different types of potassium channels in human skeletal muscle activated by potassium channel openers. Neurosci Lett 1990; 119:191–194.
116. Weston AH, Edwards G. Recent progress in potassium channel opener pharmacology. Biochem Pharmacol 1992; 43:47–54.
117. Chen G, Suzuki H, Weston AH. Acetylcholine releases endothelium-derived hyperpolarising factors and EDRF from rat blood vessels. Br J Pharmacol 1988; 95:1165–1174.
118. Olesen SP, Clapham DE, Davies PF. Haemodynamic shear stress activates a K^+ current in vascular endothelial cells. Nature 1988; 331:168–170.
119. Nelson MT, Huang Y, Brayden JE, Hescheler J, Standen NB. Activation of K^+ channels is involved in arterialdilations to calcitonin gene-related peptide. Nature 1990; 344:770–773.
120. De Weille J, Schmid-Antomarchi H, Fosset M, Lazdunski M. ATP-sensitive K^+ channels that are blocked by hypoglycemia-inducing sulfonylureas in insulin-secreting cells are activated by galanin, a hyperglycemia-inducing hormone. Proc Natl Acad Sci USA 1988; 85:1312–1316.
121. De Weille JR, Schmid-Antomarchi H, Fosset M, Lazdunski M. Regulation of ATP-sensitive K^+ channels in insulinoma cells: Activation by somatostatin and protein kinase C and the role of cAMP. Proc Natl Acad Sci USA 1989; 86:2971–2975.
122. Kirsch GE, Codina J, Birnbaumer L, Brown AM. Coupling of ATP-sensitive K^+ channels to A_1 receptors by G proteins in rat ventricular myocytes. Am J Physiol 1990; 259:H820–H826.
123. Nelson MT, Huang Y, Brayden JE, Heschler J, Standen NB. Atrial dilations in response to calcitonin gene-related peptide involve activation of K^+ channels. Nature 1990; 344:770–773.
124. Keung EC, Li Q. Lactate activates ATP-sensitive potassium channels in guinea-pig ventricular myocytes. J Clin Invest 1991; 88:1772–1777.

18
Ion Transport in Vascular Smooth Muscle and the Pathogenesis of Hypertension

Christian Aalkjær and Keld Kjeldsen

Over the last 10–15 years a number of reports have documented that hypertension is associated with changes in a variety of ion transport systems, although it has been difficult to define the significance of these changes. One of the reasons for this difficulty is that in human essential hypertension evidence for changes in the ion transport over cell membranes stems from observations in blood cells. If these abnormalities are to play a pathogenetic role they should, however, be present in vascular smooth muscle cells (VSMC), neurons, or the kidney from patients with essential hypertension as well. Confirmation of this by direct measurements has been difficult for technical and ethical reasons. Information concerning ion transport abnormalities in VSMC in hypertension has, however, been obtained from experiments with different types of experimental hypertension, although the extent to which this information is relevant in essential hypertension is open for discussion. With respect to the animals (mainly rats) with genetic forms of hypertension the choice of normotensive controls is a problem (see below), while in the case of secondary forms of experimental hypertension a major problem is the lack of genetic predisposition, which is an important aspect of essential hypertension. Another difficulty is the poor understanding of the physiologic importance of the various ion transport systems in VSMC. Hence the pathogenetic significance of any abnormality is difficult to predict.

So far the demonstration of ion transport abnormalities in blood cells from patients with essential hypertension has contributed little to our understanding of the pathogenesis of essential hypertension. On the other hand, the presence of these abnormalities has stimulated a lot of research directed toward an understanding of the ion transport pathways in VSMC. One purpose of this chapter is to review the recently acquired information about these matters.

It is evident that an understanding of the possible role of an abnormality in VSMC ion transport in essential hypertension requires clarification of the pathophysiologic and pathoanatomic changes in the resistance vasculature.[1] Therefore a second purpose of this chapter will be to briefly review the information concerning the VSMC changes in essential hypertension. A final section will discuss whether it is possible to find a connection between these changes and changes in ion transport.

Ion Transport in Vascular Smooth Muscle and How this Relates to Contraction

Definition of Ion Transport Pathways

A characteristic of cell membranes is that they are permeable to some substances and impermeable to others. To subserve this function, various types of transport systems are available. A practical way to describe them is to divide them into (a) active transport, (b) facilitated transport, and (c) passive diffusion.

In *active transport* the translocation of ions is coupled directly to the hydrolysis of adenosine triphosphate (ATP). The energy derived from this hydrolysis enables active transport pathways to move ions against a chemical and electrical gradient, and thus maintain an electrochemical gradient for the transported ion. This transport is often said to be mediated through ion pumps.

In *facilitated transport* ions can also be transported against an electrochemical gradient. In contrast to active transport, the required energy is derived from the movement of another ion, which moves down its electrochemical gradient. The transport can either be a cotransport, where the ions move in the same direction, or a countertransport (exchange), where the ions move in opposite directions. In facilitated transport, the influx of sodium frequently provides the energy for the transport of a number of other substances.

The term *passive diffusion* describes the movement of an ion moving down its electrochemical gradient without directly coupling to the transport of another ion. In most cases the structural equivalent of passive diffusion is the ion channel. Using this classification we have summarized the major characteristics of the ion transport pathways in VSMC in Table 18.1.

Monovalent Ion Transport

A number of monovalent ion transport pathways (in addition to ion channels), well characterized in other cell types, have also been demonstrated in VSMC (Figure 18.1). Most of these systems involve the transport of sodium.

Na,K Pump

VSMC, like all other cell types, possess a Na,K pump,[2] which by pumping sodium out of the cells and potassium into the cells has an important role in setting up the gradients for these ions. These ion gradients, in turn, provide the energy for transport of other ions and larger molecules, and help establish the membrane potential. The presence of a Na,K pump in vascular tissue has been documented by the demonstration of ^{3}H-ouabain-binding and of ouabain-sensitive ^{22}Na and ^{42}K or ^{86}Rb fluxes,[2] and of ouabain-sensitive ATP hydrolysis in vesicles of plasmalemma;[3] this has been confirmed by electrophysiologic evidence.[4] There is no evidence that the characteristics of the Na,K pump in the VSMC are different from those of the Na,K pump in other tissues. In particular the concentration of ^{3}H-ouabain-binding sites appears to be of the same order of magnitude as in nonvascular smooth muscles.[5] On the other hand, flux studies indicate that a substantial part of the sodium efflux (which is a transport against the electrochemical gradient and therefore energy-consuming) is ouabain-insensitive and also occurs in the absence of external potassium,[6] which suggests that a substantial part of it is not mediated by the Na,K pump. Clearly an understanding of the mechanism responsible for this uphill transport, which could potentially influence the sodium gradient, is of substantial interest.

It has been suggested that in essential hypertension the presence of a ouabain-like substance contributes to the increase in peripheral resistance through inhibition of the VSMC Na,K pump.[7-9] This substance has recently been purified and structurally identified.[10,11] Twelve µg of the compound were isolated from 85 l of plasma from patients who underwent plasmapheresis. The compound was indistinguishable from the cardenolide ouabain by mass spectrometry, and by its capacity to inhibit ^{86}Rb uptake and displace ^{3}H-ouabain from its red-cell receptor. Although the observation that plasma concentrations of the compound are identical in patients on parenteral nutrition and in normal subjects may be taken as evidence against its dietary source, contamination with cardenolide cannot be completely excluded, considering the large number of patients and large quantities of plasma necessary to isolate minute amounts. Furthermore, the physiologic and pathophysiologic importance of such a minute concentration may be questioned.

In this regard it should be noted that neither prolonged washing of skeletal muscle samples before vanadate-facilitated ^{3}H-ouabain binding in vitro, nor in vivo binding of the isotope to rat skeletal muscle have provided evidence for the existance of significant quantities of a ouabain-like factor that is capable of significantly decreasing the binding of the isotope to skeletal muscle.[12,13] Prolonged washing of human myocardial samples in digoxin-antibody prior to vanadate-facilitated ^{3}H-ouabain binding showed only a tendency to an increase in isotope binding of 9%

TABLE 18.1. Ion transport pathways in vascular smooth muscle cells.

Name	Function
Active transport	
Na,K pump	Through hydrolysis of ATP, the Na,K pump extrudes sodium from the cell and pumps potassium into the cell against their electrochemical gradients. Through this mechanism the Na,K pump sets up gradients for sodium and potassium, which are used as energy sources in different facilitated transport pathways. In addition, the ion gradients set up by the Na,K pump contribute to the establishment of the membrane potential.
Ca pump	Through hydrolysis of ATP, the Ca pump pumps calcium out of the cell against the electrochemical gradient, keeping the intracellular calcium activity around 100 nM under resting conditions or below the activity needed for activation of many calcium dependent processes.
Facilitated transport	
Na,H exchange	Uses the sodium gradient to extrude protons against their electrochemical gradient, thus taking part in the control of intracellular pH.
Na,HCO$_3^-$ cotransport and Na,HCO$_3^-$,Cl exchange	Uses the sodium gradient to effectively extrude protons against their electrochemical gradient, thus taking part in the control of intracellular pH. The Na,HCO$_3^-$,Cl exchange may also take part in the control of the intracellular chloride concentration.
Cl,HCO$_3^-$ exchange	In some vascular smooth muscle cells the Cl,HCO$_3^-$ exchange takes part in the recovery from an intracellular alkaline load. It may also take part in the control of the intracellular chloride concentration.
Na,Ca exchange	Uses the sodium gradient to extrude calcium against the electrochemical gradient. In doing so it assists the Ca pump in maintaining the intracellular calcium activity in the nanomolar range. The importance of this transport in the regulation of $[Ca^{2+}]_i$ is debated.
Na,K,Cl cotransport	Has not been well characterized in VSMC. It probably controls the osmotic activity and therefore the VSMC volume.
Na,Na exchange and Na,Li exchange	Can be demonstrated experimentally but may be a manifestation of Na,Ca exchange or Na,H exchange. The physiologic role, if any, is unknown.
Passive diffusion	
Ca channel	Allows calcium to move into the cells along the electrochemical gradient. Channels that are voltage-sensitive and inhibited by dihydropyridines are well characterized. The role of voltage-insensitive calcium channels, which are regulated by receptor-mediated mechanisms, are being investigated.
K channel	A variety of K channels, which are important in the determination of the membrane potential, have been characterized. These channels are regulated by the intracellular activity of calcium and ATP, the membrane potential, the intracellular pH, and possibly by receptor-mediated mechanisms.

TABLE 18.1 *Continued*

Name	Function
Cl channel	The importance of chloride channels for the membrane potential during agonist-induced activation has recently been demonstrated. A more precise definition of the role played by these channels is desirable.
Cation-unspecific channel	These channels carry sodium and calcium and are controlled by receptor-mediated mechanisms. Whether, under physiologic conditions, they play a role as sodium channels or calcium channels is uncertain.

$(P > 0.2)$.[14] Thus these studies indicate that a digoxin-like factor, at most, occupies a small fraction of myocardial ouabain receptors. This should be compared with digoxin receptor occupancy in the human heart during digitalization of 30–40%.[14,15] Furthermore, a relatively minor blockage of Na,K pumps by a ouabain-like compound should be compared with the major downregulation observed during various physiologic and pathophysiologic conditions. For example, during

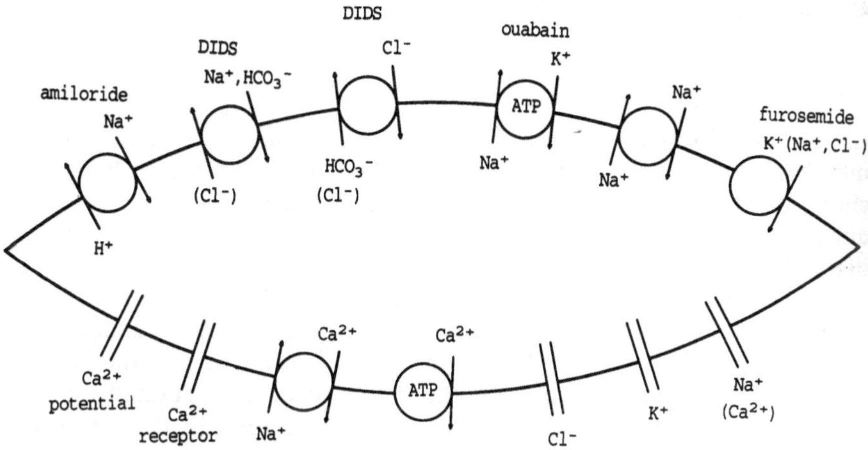

FIGURE 18.1. A schematic drawing of ion transport pathways and of their inhibitors in vascular smooth muscle cells. *Upper row*: Na,H exchange (inhibited by amiloride); Na,HCO$_3^-$ cotransport and Na,HCO$_3^-$, Cl exchange (inhibited by DIDS); Cl,HCO$_3^-$ exchange and Cl,Cl exchange (inhibited by DIDS); Na,K pump (inhibited by ouabain); Na,Na exchange; and Na,K,Cl cotransport (inhibited by furosemide). *Lower row*: Voltage-gated calcium channel (inhibited by calcium channel antagonists); receptor-gated calcium channel; Na,Ca exchange; Ca pump; Cl channel; K channel; and nonspecific cation channel.

potassium depletion the ^{3}H-ouabain receptor concentration in skeletal muscle, myocardium, and resistance vessels has been reduced by as much as 78, 13, and 37%, respectively.[12,16,17]

The mechanism through which ouabain exerts a vasoactive effect in resistance arteries is of interest, especially because (despite an increase in the intracellular sodium concentration) ouabain does not cause a sustained increase in the tone of resting resistance arteries over several hours.[18,19] This is in contrast to the rapid development of tone of conduit arteries after the addition of ouabain.[20] In precontracted isolated resistance arteries, ouabain causes a transient enhancement of tone while the vascular resistance of the forearm increases during intra-arterial infusion of ouabain. It has been proposed that this potentiating effect of ouabain is due to a rise of intracellular sodium with a consequent rise in intracellular calcium through a Na,Ca exchange.[21] Another possibility is that the depolarization caused by ouabain induces an influx of calcium through voltage-operated calcium channels. Evidence in favor of the latter mechanism comes from the observation that the potentiation of tone with ouabain is inhibited by calcium antagonists,[22,23] which exert their action through inhibition of the voltage-operated calcium channels rather than through an effect on the Na,Ca exchange.

Na,Ca Exchange

In several tissues exchange of sodium for calcium across the cell membrane plays an important role in the control of intracellular calcium concentration ($[Ca^{2+}]_i$).[24] It is generally accepted that this mechanism is important for the inotropic effect of digitalis.[25] Thus during digitalization a fraction of the Na,K pumps are inhibited, which causes a rise in intracellular Na^+ concentration and a subsequent reduction in extrusion of Ca^{2+} from the myocytes by the Na,Ca exchange. This in turn causes a rise in $[Ca^{2+}]_i$ and an activation of contractile proteins. There is also evidence for the presence of a functionally important Na,Ca exchange in the control of vascular tone in the aorta and other large arteries.[20,26] Furthermore, it has been shown that in isolated VSMC from conduit arteries changes in the sodium gradient affect $[Ca^{2+}]_i$ in a way that is consistent with the presence of Na,Ca exchange.[27]

In smaller arteries evidence for the presence of Na,Ca exchange has been provided by the demonstration of a reduction of the efflux of ^{45}Ca from rat mesenteric resistance vessels[28] following the substitution of extracellular sodium with sucrose, and by the observation that substitution of extracellular calcium with manganese causes a reduction in the efflux of ^{22}Na from sodium-loaded vessels.[29] The functional role of the exchange system in these vessels, however, is controversial.[24,30] On the one hand, it has not been possible to find a correlation between changes in the sodium electrochemical gradient and noradrenaline- or potassium-induced vessel tone in rat mesenteric resistance arteries.[22] On the other hand, if extracellular sodium is reduced to 25 mM or less and intracellular sodium is increased at the same time, this nonphysiologic shift of the sodium gradient induces tone that is not inhibited by calcium antagonists[22] and is associated with an increase in $[Ca^{2+}]_i$.[31] Similar observations with respect to the mechanical effects of changes in the sodium gradient have recently been reported in human small arteries.[19] Furthermore, a reduction of extracellular sodium reduces the rate of relaxation after washout of an agonist, or may abolish it altogether.[28,32] These mechanical effects are probably mediated by the Na,Ca exchange. However, the large changes in the sodium gradient necessary to have a substantial effect on the tone make it unlikely that the Na,Ca exchange plays an important role in the control of $[Ca^{2+}]_i$ and tone in resistance arteries in vivo.

Na,K,Cl Cotransport

A Na,K,Cl cotransport, which in some cells is involved in the regulation of cell volume,[33] has been demonstrated in the rat and rabbit aorta[34,35] and in the rabbit ear artery,[36] using measurements of ^{36}Cl turnover and ^{86}Rb uptake. In rat mesenteric resistance arteries, the ouabain-insensitive ^{86}Rb uptake is inhibited by furosemide (C. Aalkjær and M.J. Mulvany, unpublished data, 1990), which suggests that a Na,K,Cl cotransport may be present also in these arteries. There is, however, no information on the role played by the Na,K,Cl cotransport in the VSMC. Interestingly, in vitro experiments have shown that furosemide has vasodilator effect in conduit arteries from the rabbit,[35,37] although the relationship of this effect to inhibition of the Na,K,Cl cotransport is uncertain.

Na,H Exchange

It is well established that a maintained reduction of pH causes vasodilation and that the control of intracellular pH (pH_i) therefore is an important homeostatic task for a metabolically active cell type, such as the VSMC. In most cells one mechanism for the control of pH_i is the amiloride-sensitive Na,H exchange. In rat resistance arteries an amiloride-sensitive and sodium-dependent recovery of intracellular pH_i after an acid load can be demonstrated under HCO_3^--free conditions.[38,39] Furthermore, an intracellular acid load causes an increase in sodium influx that can be inhibited with amiloride and 5-(N-ethyl-N-isopropyl)-amiloride,[38] a more specific inhibitor of the Na,H exchange. These observations provide clear evidence for the presence of Na,H exchange in resistance arteries. Similar observations have been made in cultured VSMC.[40,41] The importance of this mechanism in the regulation of pH_i is far from understood. In rat mesenteric resistance arteries it has been difficult to demonstrate an important role of Na,H exchange for the control of pH_i in the presence of bicarbonate, although in young rats the Na,H exchange may play a more significant role even in the presence of bicarbonate,[42] which suggests the interesting possibility that the importance of Na,H exchange depends upon the stage of development. In the absence of HCO_3^- there is little doubt that Na,H exchange plays the dominant role in the control of pH_i, although this is obviously a nonphysiologic situation to which the vessels are never exposed in vivo. It is possible that under experimental conditions the Na,H exchange may operate as Na,Na exchange or as a Na,Li exchange.

Na,HCO_3^- Cotransport

The importance of bicarbonate transport to the control of pH_i has not received the same attention as the Na,H-exchange. The importance of bicarbonate transport in smooth muscle has been pointed out by Aickin,[43] while a diisothio-cyanato-stilbene-disulphonic (DIDS) acid-sensitive, sodium-coupled uptake of bicarbonate has been shown to be of primary importance for the recovery from an intracellular acid load in rat mesenteric resistance arteries.[38] This sodium-coupled bicarbonate uptake is not associated with an efflux of chloride, which suggests that the mechanism is a Na,HCO_3^- cotransport.[44] Of particular interest is the observation that in the small arteries[44] (but also in the smooth muscle of the guinea pig ureter[43] and in a smooth muscle cell line [BC3H-1][45]) the Na,HCO_3^- cotransport appears to be electroneutral, which is in contrast to what has been reported for nonsmooth muscle cells.[46,47] Although a Na,HCO_3^- cotransport has been demonstrated in several smooth muscle types, it is not ubiquitous since in two cell lines derived from smooth muscles (A10 and A7r5) no sodium-dependent bicarbonate transport was found.[48,49] The Na,HCO_3^-, Cl exchange may also take part in the control of the intracellular chloride concentration.

Seven to 8 years ago studies on the hormonal activation of transmembranal proton transport led to the suggestion that an increase in pH_i, as a consequence of hormonal activation of Na,H exchange, serves as a second messenger that mediates the response to growth factors.[50] Recently it has been proposed that the activation not only of Na,H exchange, but also of bicarbonate transport, occurs in anticipation of a metabolic load,[51,52] although it is unknown how the cells anticipate a stimulus. We found that in resistance arteries from the rat, activation with vasoconstrictor hormones causes no change in pH_i, although both Na,HCO_3^- cotransport and Na,H exchange were activated by the vasoconstrictors.[53] This suggests that the acid production associated with contraction is pumped out by activation of acid extrusion mechanisms in a very precise feedback control of pH_i. In contrast, during depolarization with high potassium pH_i falls, which indicates that the homeostatic control only operates during stimulation with "physiologic" activators.[53] As indicated above, changes in pH affects the tone of resistance arteries (and conduit arteries) and the peripheral resistance. Thus it appears that the smooth muscles of resistance arteries (which, in contrast to skeletal muscles, are in a tonic state that changes in response to exogenous stimuli) can efficiently and appropriately extrude the metabolically produced acid, which prevents the acid load per se from interfering with the control of the tone.

Cl,HCO_3^- Exchange

There is also evidence for the presence of Cl,HCO_3^- exchange in VSMC. Thus in rat mesenteric arteries, omission of chloride from the external medium causes a substantial DIDS-sensitive increase in pH_i of rat mesenteric resistance arteries in the presence of bicarbonate, but not in its absence.[44] Furthermore, omission of chloride reduces the ^{36}Cl efflux by about 75%, but only when bicarbonate is absent. These findings are consistent with the presence in these vessels of a Cl,HCO_3^- exchange that can operate in a Cl,Cl exchange mode.[44] The findings further suggest that the major part of the ^{36}Cl efflux under resting conditions is mediated by the exchange mechanism, and only a smaller fraction is mediated by a conductive chloride transport (a chloride channel). This conclusion is supported also by previous observations on the vas deferens,[54] which indicates that in resting smooth muscle cells chloride permeability and thus the contribution of chloride distribution to the membrane potential is smaller than previously believed. Under resting conditions, as well as during an alkaline load, the transmembrane gradients of chloride and bicarbonate result in net efflux of bicarbonate by Cl,HCO_3^- exchange. In many tissues the Cl,HCO_3^- exchange probably extrude an alkaline load. However, in mesenteric resistance vessels from WKY rats, the recovery of pH_i after an alkaline load is not significantly affected by DIDS or by the omission of bicarbonate.[38,44] This indicates that the Cl,HCO_3^- exchange may not be quantitatively important for the recovery of pH_i from an alkakine load, which is probably mediated by a bicarbonate-independent mechanism.[44] This conclusion is supported by the rapid recovery of pH_i seen after an alkaline load induced by omission of CO_2 and bicarbonate.[38] On the other hand, in similar vessels from Wistar rats the recovery from an alkaline load is reduced in bicarbonate-free medium and by DIDS (C. Aalkjær, unpublished), which suggests that in these animals the Cl,HCO_3^- exchange has an important role in protecting the VSMC against an alkaline load. These contrasting observations point to the difficulty of making general statements about the role of various transport systems, even in the same small arteries from different strains of rats.

Calcium Translocation

The tone of VSMC is to a large extent dependent on $[Ca^{2+}]_i$. This does not exclude the possible role of calcium-independent mechanisms in the control of tone. Indeed, recent evidence obtained using calcium-sensitive fluorescent dyes to assess $[Ca^{2+}]_i$ in VSMC suggests that there is a very important calcium-independent component in the control of tone.[55-59]

Ca Influx

A large inwardly directed electrochemical gradient drives calcium into the cell through calcium channels. Several years ago it was suggested[60,61] that two separate calcium channels may exist in VSMC. One channel would open when the membrane depolarized (voltage-operated channel) while another channel, not affected by the membrane potential, would open in response to an agonist (receptor-operated channel or nonspecific cation channels [see below]). This hypothesis is supported by the observations that (1) agonists cause an influx of ^{45}Ca and a development of force that is additive to that seen after a full depolarization of the VSMC; (2) calcium antagonists more or less selectively affect tone induced by depolarization rather than the agonist-induced tone;[62,63] (3) tone develops at a more negative membrane potential during activation with agonists than during depolarization with potassium;[64,65] and (4) more recently, cation-nonspecific ion channels that are activated by extracellular ATP[66] or noradrenaline[67] but unaffected by membrane potential have been demonstrated in patch-clamp experiments. It is important to note, however, that in most instances agonists also cause a depolarization of the membrane, which indicates they may also open the voltage-operated calcium channels with a consequent influx of calcium. Furthermore, it has been argued[68] that the evidence put forward in the favor of two different channels can be explained by also assuming the existence of voltage-operated calcium channels that can be modulated by agonists. Thus it has been demonstrated that noradrenaline increases the open probability of voltage-operated calcium channels,[23] a phenomenon that could explain the effect of agonists on ^{45}Ca influx and tone in depolarized tissues in terms of a single channel. The recent observations that agonists enhance the effect of intracellular calcium on the contractile filaments (see above) may also explain why agonists can potentiate the tone of a fully depolarized VSMC and why agonists may still produce a substantial amount of force at relatively negative membrane potentials that may allow the entry of only small amounts of calcium. Thus the mechanisms through which agonists mediate an influx of calcium into VSMC are still controversial. More information will be obtained using the patch-clamp technique combined with measurements of $[Ca^{2+}]_i$ with fluorescent dyes.[69]

Ca Efflux

The efflux of calcium from VSMC has not been well characterized. As discussed above, one possibility is that calcium is extruded in exchange for sodium (Na,Ca exchange), although the physiologic role of this mechanism in resistance arteries remains controversial. A more likely possibility is that calcium is extruded by a Ca-ATPase. Indeed, a calcium-activated ATPase activity has been demonstrated in VSMC,[70] although the lack of specific inhibitors for this transport and for the Na,Ca exchange makes it difficult to evaluate their relative importance for calcium extrusion and for $[Ca^{2+}]_i$ under different experimental conditions.

Monovalent Ion Channels

As discussed above, the membrane potential plays an important role in the control of $[Ca^{2+}]_i$ and consequently of VSMC tone. The membrane potential is to a large extent determined by the permeability to potassium, chloride, and sodium through channels and by the transmembrane gradients of these ions in accordance with the Goldman equation. Furthermore, changes in the membrane potential are to a large extent mediated by changes in the permeability to these monovalent ions. In addition there is an important contribution by the electrogenic effect of the Na,K pump, which, by pumping three Na ions out in exchange for two K ions, may bring about a hyperpolarization of about 5 to 15 mV.

K Channels

The effects of change in the transmembranal potassium gradient on membrane potential, $[Ca^{2+}]_i$, and force in VSMC suggest that the potassium permeability or potassium channels play an important role in determining the tone of VSMC. Given that the equilibrium potential for potassium is about −80 to −90 mV in VSMC, a resting membrane potential of about −60 mV[71] further underlines the importance of the potassium permeability for the membrane potential under resting conditions.

On the basis of electrophysiologic and other evidence, at least five groups of potassium channels have been described in VSMC:[71,72] (1) an inward rectifier potassium channel, (2) a delayed rectifier potassium channel, (3) a large conductance calcium-sensitive potassium channel, (4) a small conductance calcium-sensitive potassium channel, and (5) an ATP-sensitive potassium channel. The inward rectifier channel and the delayed rectifier channel are voltage-sensitive potassium channels. The inward rectifier channel opens at hyperpolarized potentials, while the delayed rectifier channel opens with a delay when the cells depolarize.

The relative importance of these channels for the control of the membrane potential, and the membrane potentials at which their open probability changes, may vary from preparation to preparation.[71] In contrast to the rectifier channels, the calcium-sensitive potassium channels are relatively insensitive to the membrane potential. The probability of these channels being open is, however, increased by an increase in $[Ca^{2+}]_i$, and it has been suggested that these channels may prevent the development of action potentials in VSMC, where indeed action potentials are rarely seen. Recently a lot of interest has been focused on ATP-sensitive potassium channels. This interest started with the observation that a number of chemically unrelated drugs caused a vasodilation that was inhibited by glibenclamide and other sulfonylureas, which are know to block ATP-sensitive potassium channels in the β cells of the pancreas.[73,74,75,76] The open probability of ATP-sensitive potassium channels is increased when the ATP concentration is reduced and the results are hyperpolarization and vasodilation. It has also been suggested that endogenous vasodilators, such as the calcitonin gene-related peptide (CGRP), may work through an opening of these channels.[77]

Chloride Channels and Nonspecific Cation Channels

Although chloride channels and sodium channels may play an important role in the generation of membrane potential in some tissues, knowledge about these channels in VSMC is too limited to allow definite statements on the role of chloride and sodium permeability in excitation-contraction coupling in VSMC.

Recently, however, electrophysiologic evidence has suggested that chloride channels may have an important role in the depolarization induced by noradrenaline in VSMC from the rabbit ear artery and the portal vein,[78] where under voltage-clamp conditions noradrenaline induces a current with the characteristics of a chloride current. Since the chloride equilibrium potential in VSMC is about $-30\,mV$, the opening of these channels could contribute to a depolarization of the cells.[78,79] Very recently endothelin, an endothelium-derived vasoconstrictor, has also been shown to activate a chloride current, which could contribute to the depolarization induced by endothelin.[80,81] Thus there is increasing evidence to suggest that chloride channels are important for the depolarization induced by vasoactive substances.

Most VSMC do not appear to be sensitive to the action of tetrodotoxin, which suggests that they do not possess sodium channels with characteristics similar to those observed in neural tissue and in the heart. On the other hand, evidence has been provided for the presence of cation-nonspecific ion channels that are activated by ATP or noradrenaline.[67,69,82,83] Under physiologic conditions these channels may behave as sodium channels, and since they have a reversal potential of about 0 mV (with physiologic ion gradients), they may therefore contribute to the depolarization during activation with agonists.[69]

As pointed out earlier, there is increasing evidence that suggests a role for chloride channels and possibly sodium channels, in addition to that played by potassium channels, in determining the depolarization induced by agonists and consequently the influx of calcium through voltage-operated calcium channels. It appears, however, that the importance of the various channels (or permeabilities) varies from preparation to preparation, and we are still far from understanding the complex interaction between these different channels and how they play a role in the electrical activities of the VSMC.

Pathophysiologic Changes in the Vasculature in Essential Hypertension

Structural Aspects

The importance of structural changes, such as an increased thickness of the vascular wall in relation to the diameter of the lumen as a cause of increased peripheral resistance in essential hypertension, has been stressed.[1] Indeed, many studies have demonstrated that during maximal vasodilation the vascular resistance is higher in patients with hypertension, which suggests either a structurally determined reduction of the average lumen diameter of the resistance arteries[1] or a decrease in the number of vessels (rarefaction). Not only is the resistance seen at maximal dilation higher, but its increase during infusion of vasoconstrictors is steeper, even though the sensitivity to these agonists is unchanged.[84] According to an analysis proposed by Folkow,[85,86] this is consistent with an important role of an increased wall thickness-to-lumen diameter ratio in essential hypertension. In addition, there is evidence that in the most distal part of the arterial bed, vessel closure may lead to a functional and later structural reduction of arteries,[87-92] which further contributes to the increased peripheral resistance.

Direct evidence for a structurally based reduction in the lumen diameter and for an increased wall thickness-to-lumen diameter ratio in essential hypertension has been provided by in vitro experiments.[93-96] This change in the vascular geometry is often ascribed to growth of the vascular wall. However, an increased wall thickness-to-lumen diameter ratio need not be a consequence of an increase in the volume of the vascular wall, as it could also be a consequence of reorganization of a constant amount of wall material, as pointed out several years ago by Short[94] and more recently by Baumbach and Heistad.[97,98] An evaluation of the relative importance of these two mechanisms from patients with essential hypertension is difficult, it seems though that although growth may play a role in the larger arteries,[87,99] growth plays a smaller role for the altered vascular structure in the small arteries.[94,96] Furthermore, in the smallest arteries no evidence for an increase in the media thickness-to-lumen diameter ratio has been reported,[88,91] although as pointed out above rarefaction may occur. The structural changes are thus complex and not uniform along the vascular tree.

Function of the Vascular Smooth Muscles

In the spontaneously hypertensive rat (SHR), which is the most widely used model of human essential hypertension, an increased sensitivity of the resistance arteries to noradrenaline[100–102] and to calcium agonists[103] has been reported. Some—although not all[12]—reports have also suggested that various ion transport systems, including those for potassium,[104–107] sodium,[108–111] chloride,[105,106] protons,[39,112] and calcium[70,113,114] are altered in the VSMC of spontaneously hypertensive rats. This multitude of transport abnormalities has led to the idea that a defect of the lipid membrane of the cells may be the fundamental disorder in hypertension.[103,115,116] However attractive, this hypothesis has some weak points. Thus in some cases the differences between spontaneously hypertensive and normotensive control rats disappear if normotensive controls of a different strain are used. This underlines the problem that some characteristics bred into the spontaneously hypertensive rats but are unrelated to the development of high blood pressure.

The increased pressor response seen when agonists are infused intra-arterially in patients with essential hypertension[117,118] can most easily be explained by the altered vascular geometry discussed above. The few in vitro experiments done with arteries from such patients have also failed to demonstrate any increase in the sensitivity to agonists or in maximal responsiveness, except for what could be accounted for by the altered vascular geometry. On the other hand the relaxation of the forearm vasculature during intra-arterial infusion of verapamil was increased and the relaxation to nitroprusside was decreased in patients with essential hypertension,[119] which suggests that there may be some abnormality in the function of the VSMC in essential hypertension. A similar conclusion was reached also following in vitro experiments, which showed that the sensitivity to extracellular calcium was lower in isolated subcutaneous small arteries from patients with essential hypertension than in vessels from normotensive controls.[96] More recently experiments designed to assess the control of pH_i in subcutaneous arteries from patients with essential hypertension have shown that neither resting pH_i nor the activity of the Na,H exchange were abnormal,[112] in contrast with the observation that in leukocytes from patients with essential hypertension the Na,H exchange activity is increased[120] and the steady state pH_i is reduced. These disparate observations may underline the danger of extrapolating results obtained with blood cells in an attempt to assess the importance of a change in ion transport in VSMC for the pathogenesis of essential hypertension.

Conclusion

Although much new information about several types of ion transport systems in VSMC has been presented in the last decade, the function of these transport pathways in relation to VSMC contractility, growth, and structural organization is still limited, and attempts to set VSMC ion transport into a hemodynamic context is very difficult. One problem that has become apparent is that the importance of the different ion transport pathways and ion channels varies in different parts of the vasculature; therefore it is difficult to extrapolate results obtained in one vascular district to others. Other basic questions concerning ion transport in VSMC also remain unanswered: Is it meaningful to use a model that assumes the existence of two classes of calcium channels, one that is potential-sensitive and one that is receptor-operated? Is the efflux of calcium mediated mainly through Na,Ca exchange or by a Ca-ATPase? To what extent do calcium-independent mechanisms control tone in VSMC? What is the role of the various monovalent ion channels in determining the membrane potential, particularly during activation with hormones? Do potassium and possibly chloride permeabilities play a dominant role for this (as seems to be the case), and does sodium permeability play a role?

This lack of knowledge constitutes a major obstacle in the formulation of a hypothesis that links an abnormality of ion transport in the VSMC and essential hypertension, even if any such abnormality could be documented. This question can also, however, be addressed by using the functional and morphologic changes of the VSMC in essential hypertension in an attempt to predict any abnormality in ion transport. There is no evidence that essential hypertension is associated with an increased excitability of the resistance arteries. Thus the question whether an abnormality of VSMC ion transport may lead to an increased excitability of these cells is moot. There is still evidence, though, for functional changes in the VSMC, and changes in the control of calcium may be present. The best-documented changes in essential hypertension are those involving the geometry of the resistance arteries and the growth of the VMSC in the large arteries. Understanding of the mechanisms responsible for these structural changes and the possible role of ion transport in these changes requires more knowledge about the basic mechanisms that control vascular geometry.

Acknowledgments. We thank Michael J. Mulvany and Stig Haunsø for valuable discussions. The research in the authors' laboratories is supported in part by The Danish Heart Foundation, The Danish Medical Research Council, and The Novo Foundation.

References

1. Folkow B. Physiological aspects of primary hypertension. Physiol Rev 1982; 62:347–504.
2. Jones AW. Content and fluxes of electrolytes. In: Bohr DF, Somlyo AP, Sparks HV, eds. Handbook of Physiology. The Cardiovascular System: Vascular Smooth Muscle. Washington, DC: Am Physiol Soc; 1980:253–299.
3. Wei J-W, Janis RA, Daniel EE. Isolation and characterization of plasma membrane from rat mesenteric arteries. Blood Vessels 1976; 13:279–292.
4. Hermsmeyer K. Electrogenic ion pumps and other determinants of membrane potential in vascular muscle. Physiologist 1982; 25(6):454–465.
5. Aalkjær C, Mulvany MJ. Effect of ouabain on tone, membrane potential and sodium efflux compared with [^{3}H]ouabain binding in rat resistance vessels. J Physiol (Lond) 1985; 362:215–231.
6. Aalkjær C, Mulvany MJ. Ouabain insensitive extrusion in resistance vessels. In: Bevan JA, Majewski R, Maxwell RA, Story DF, eds. Vascular Neuroeffector Mechanisms. Washington, DC: IRL; 1988:201–208 (ICSU Symp Ser).
7. Overbeck HW, Derifield RS, Ramnani MB, Sözen T. Attenuated vasodilator responses in K$^+$ in essential hypertensive men. J Clin Invest 1974; 53:678–686.
8. DeWardener HE, MacGregor GA. Dahl's hypothesis that a saluretic substance may be responsible for a sustained rise in arterial pressure: Its possible role in essential hypertension. Kid Inter 1980; 18:1–9.
9. Blaustein MP, Hamlyn JM. Role of a natriuretic factor in essential hypertension: A hypothesis. Ann Int Med 1983; 98(2):785–792.
10. Hamlyn JM, Blaustein MP, Bova S, DuCharme DW, Harris DW, Mandel F, Mathews WR, Ludens JH. Identification and characterization of a ouabain-like compound from human plasma. Proc Natl Acad Sci USA 1991; 88:6259–6263.
11. Welcome to ouabain—A new steroid hormone. Lancet 1991; 338:543–544. Editorial.
12. Aalkjær C, Kjeldsen K, Nørgaard A, Clausen T, Mulvany MJ. Ouabain binding and Na$^+$ content in resistance vessels and skeletal muscles of spontaneously hypertensive rats and K$^+$-depleted rats. Hypertension 1985; 7:277–286.
13. Kjeldsen K, Nørgaard A. The ^{3}H-ouabain binding capacity of skeletal muscle from spontaneously hypertensive rats show marginal changes in comparison to the major effects of differentiation, K-depletion and thyroid status. Scand J Clin Lab Invest 1986; 46(suppl 180):72–81.
14. Schmidt TA, Holm-Nielsen P, Kjeldsen K. No upregulation of digitalis glycoside receptor (Na,K-ATPase) concentration in human heart left

ventricle samples obtained at necropsy after long term digitalisation. Cardiovasc Res 1991; 25:684–691.

15. Rasmussen HH, Okita GT, Hartz RS, Eik RET. Inhibition of electrogenic Na-pumping in isolated atrial tissue from patients treated with digoxin. J Pharmacol Exp Ther 1990; 252:60–64.

16. Nørgaard A, Kjeldsen K, Clausen T. Potassium depletion decreases the number of ^{3}H-ouabain binding sites and the active Na-K-transport in skeletal muscles. Nature 1981; 293:739–741.

17. Nørgaard A, Kjeldsen K, Hansen O. K$^+$-dependent 3-O-methylfluorescein phosphatase activity in crude homogenate of rodent heart ventricle: Effect of K$^+$ depletion and change in thyroid status. Eur J Pharmacol 1985; 113:373–382.

18. Mulvany MJ, Aalkjær C, Nilsson H, Korsgaard N, Petersen TT. Raised intracellular sodium consequent to sodium-potassium-dependent ATPase inhibition does not cause myogenic contractions of 150-μm arteries from rat and guinea pig. Clin Sci 1982; 63:45–48.

19. Woolfson RG, Hillerton PJ, Poston L. Effects of ouabain and low sodium on contractility in human resistance arteries. Hypertension 1990; 15:583–590.

20. Karaki H, Ozaki H, Urakawa N. Effects of ouabain and potassium-free solution on the contraction of isolated blood vessels. Eur J Pharmacol 1978; 48:439–443.

21. Blaustein MP. Sodium ions, calcium ions, blood pressure regulation, and hypertension: A reassessment and a hypothesis. Am J Physiol 1977; 232:C165–C173.

22. Mulvany MJ, Aalkjær C, Petersen TT. Intracellular sodium, membrane potential, and contractility of rat mesenteric small arteries. Circ Res 1984; 54:740–749.

23. Nelson MT, Standen NB, Brayden JE, Worley JF III. Noradrenaline contracts arteries by activating voltage-dependent calcium channels. Nature 1988; 336:382–385.

24. Blaustein MP. Sodium/calcium exchange and the control of contractility in cardiac muscle and vascular smooth muscle. J Cardiovasc Pharmacol 1988; 12(5):56–68.

25. Akera T. Effects of cardiac glycosides on Na$^+$, K$^+$-ATPase. In: Greeff K, ed. Handbook of Experimental Pharmacology. Berlin: Springer-Verlag; 1981:287–336.

26. Ozaki H, Urakawa N. Involvement of a Na-Ca exchange mechanism in contraction induced by low-Na solution in isolated guinea-pig aorta. Pfluegers Arch 1981; 390:107–112.

27. Goldman WF, Blaustein MP. Stimulation-induced regional alteration of Ca^{2+} levels in single arterial smooth muscle cells. J Cardiovasc Pharmacol 1988; 12(5):13–19.

28. Cauvin C. The effects of Na$^+$ replacement on ^{45}Ca fluxes in isolated rat mesenteric resistance vessels. In: Halpern W, Pegram B, Brayden J, Mackey K, McLaughlin M, Osol G, eds. Resistance Arteries. Ithaca, NY: Perinatology Press; 1988:195–211.

29. Aalkjær C. Some aspects of sodium metabolism in arterial resistance vessels: A Review. Aarthus, Denmark: Aarhus University; 1988. Thesis.

30. Mulvany MJ. Changes in sodium pump activity and vascular contraction. J Hypertension 1985; 3:429–436.
31. Mulvany MJ, Aalkjær C, Jensen PE. Sodium, calcium-exchange in vascular smooth muscle. Ann NY Acad Sci 1991; 639:498–504.
32. Petersen TT, Mulvany MJ. Effect of sodium gradient on the rate of relaxation of rat mesenteric small arteries from potassium contractures. Blood Vessels 1984; 21:279–289.
33. Hoffmann EK, Simonsen LO. Membrane mechanisms in volume and pH regulation in vertebrate cells. Physiol Rev 1989; 69:315–382.
34. Kreye VAW, Bauer PK, Villhauer I. Evidence for furosemide-sensitive active chloride transport in vascular smooth muscle. Eur J Pharmacol 1981; 73(1):91–95.
35. Deth RC, Payne RA, Peecher DM. Influence of furosemide on rubidium-86 uptake and alpha-adrenergic responsiveness of arterial smooth muscle. Blood Vessels 1987; 24:321–333.
36. Tian R, Aalkjær C, Andreasen F. Mechanisms behind the relaxing effect of furosemide on the isolated rabbit ear artery. Pharmacol Toxicol 1991; 68:406–410.
37. Andreasen F, Christensen JH. The effect of furosemide on vascular smooth muscle is influenced by plasma protein. Pharmacol Toxicol 1988; 63:324–326.
38. Aalkjær C, Cragoe EJ Jr. Intracellular pH regulation in resting and contracting segments of rat mesenteric resistance vessels. J Physiol (Lond) 1988; 402:391–410.
39. Izzard AS, Heagerty AM. The measurement of internal pH in resistance arterioles: Evidence that intracellular pH is more alkaline in SHR than WKY animals. J Hypertension 1989; 7:173–180.
40. Little PJ, Cragoe EJ, Bobik A. Na-H exchange is a major pathway for Na influx in rat vascular smooth muscle. Am J Physiol 1986; 251:C707–C712.
41. Weissberg PL, Little PJ, Cragoe EJ, Bonik A. Na-H antiport in cultured rat aortic smooth muscle: Its role in cytoplasmic pH regulation. Am J Physiol 1987; 253:C193–C198.
42. Izzard AS, Heagerty AM. Resting intracellular Ph in mesenteric resistance arteries from spontaneously hypertensive and Wistar-Kyoto rats: Effects of amiloride and $4,4^1$-diisothiocyanatostilbene-$2,2^1$-disulphonic acid. J Hypertension 1989; 7(6):S128–S129.
43. Aickin CC. Mechanisms involved in control of intracellular pH in smooth muscle. V. Symposium: Intracellular pH. Verh Dtsch Zool Ges 1989; 82:121–129.
44. Aalkjær C, Hughes A. Chloride and bicarbonate transport in rat resistance arteries. J Physiol 1991; 436:57–73.
45. Putnam RW. pH regulatory transport systems in a smooth muscle-like cell line. Am J Physiol 1990; 258:C470–C479.
46. Jentsch TJ, Korbmacher C, Janicke I, Fischer DG, Stahl F, Helbig H, Hollwede H, Cragoe EJ Jr, Keller SK, Wiederholt M. Regulation of cytoplasmic pH of cultured bovine corneal endothelial cells in the absence and presence of bicarbonate. J Membr Biol 1988; 103:29–40.
47. Boron WF, Boulpaep EL. Intracellular pH regulation in the renal proximal tubule of the salamander. J Gen Physiol 1983; 81:29–52.

48. Korbmacher C, Helbig H, Stahl F, Wiederholt M. Evidence for Na/H exchange and Cl/HCO$_3$ exchange in A10 vascular smooth muscle cells. Pfluegers Arch 1988; 412:29–36.
49. Vigne P, Breittmayer J-P, Frelin C, Lazdunski M. Dual control of the intracellular pH in aortic smooth muscle cells by a cAMP-sensitive HCO$_3$/Cl$^-$ antiporter and a protein kinase C-sensitive Na$^+$/H$^+$ antiporter. J Biol Chem 1988; 263:18023–18029.
50. Moolenaar WH, Yarden Y, De Laat SW, Schlessinger J. Epidermal growth factor induces electrically silent Na$^+$ influx in human fibroblasts. J Biol Chem 1982; 257:8502–8506.
51. Ganz MB, Boyarsky G, Sterzel RB, Boron WF. Arginine vasopressin enhances pH$_i$ regulation in the presence of HCO$_3^-$ by stimulating three acid-base transport systems. Nature 1989; 337:648–651.
52. Thomas RC. Cell growth factors, bicarbonate and pH$_i$ response. Nature 1989; 337:601.
53. Aalkjær C, Mulvany MJ. Steady-state effects of arginine vasopressin on force and pH$_i$ of isolated mesenteric resistance arteries from rats. Am J Physiol 1991; 261:C1010–C1017.
54. Aickin CC, Brading AF. Measurement of intracellular chloride in guinea-pig vas deferens by ion analysis, ^{36}chloride efflux and micro-electrodes. J Physiol 1982; 326:139–154.
55. Morgan JP, Morgan KG. Stimulus-specific patterns of intracellular calcium levels in smooth muscle of ferret portal vein. J Physiol 1984; 351:155–167.
56. Himpens B, Matthijs G, Somlyo AV, Butler TM, Somlyo AP. Cytoplasmic free calcium, myosin light chain phosphorylation, and force in phasic and tonic smooth muscle. J Gen Physiol 1988; 92:713–729.
57. Ruzycky AL, Morgan KG. Involvement of the protein kinase C system in calcium-force relationship in ferret aorta. Br J Pharmacol 1989; 97:391–400.
58. Nishimura J, Khalil RA, Drenth JP, van Breemen C. Evidence for increased myofilament Ca^{2+} sensitivity in norepinephrine-activated vascular smooth muscle. Am J Physiol 1990; 259:H2–H8.
59. Jensen PE, Mulvany MJ, Aalkjær C. Free cytosolic calcium measured with fura-2, and the tension-[Ca^{2+}]$_i$ relationship in rat mesenteric small arteries. Pfluegers Arch (in press).
60. Bolton TB. Mechanisms of action of transmitters and other substances on smooth muscle. Physiol Rev 1979; 59:606–718.
61. Aronson P, Vanbreemen C. Effects of sodium gradient manipulation upon cellular calcium, ^{45}Ca fluxes and cellular sodium in the guinea-pig taenia coli. J Physiol (Lond) 1981; 443–461.
62. Nyborg NCB, Mulvany MJ. Effect of felodipine, a new dihydropyridine vasodilator on contractile responses to potassium, noradrenaline, and calcium in mesenteric resistance vessels of the rat. J Cardiovasc Pharmacol 1984; 6:499–505.
63. Rüegg UT, Wallnöfer A, Weir S, Cauvin C. Receptor-operated calcium permeable channels in vascular smooth muscle. J Cardiovasc Pharmacol 1989; 14(6):49–58.
64. Mulvany MJ, Nilsson H, Flatman JA. Role of membrane potential in the response of rat mesenteric arteries to exogenous noradrenaline stimulation. J Physiol (Lond) 1982; 332:363–373.

65. Casteels R, Kitamura K, Kuriyama H, Suzuki H. The membrane properties of the smooth muscle cells of the rabbit main pulmonary artery. J Physiol (Lond) 1977; 271:41–61.
66. Benham CD, Tsien RW. A novel receptor-operated Ca^{2+}-permeable channel activated by ATP in smooth muscle. Nature 1987; 328:275–278.
67. Amédée T, Benham CD, Bolton TB, Byrne NG, Large WA. Potassium, chloride and nonselective cation conductances opened by noradrenaline in rabbit ear artery cells. J Physiol 1990; 423:551–568.
68. Nelson MT, Patlak JB, Worley JF, Standen NB. Calcium channels, potassium channels, and voltage dependence of arterial smooth muscle tone. Am J Physiol 1990; 259:C3–C18.
69. Benham CD. ATP-activated channels gate calcium entry in single smooth muscle cells dissociated from rabbit ear artery. J Physiol 1989; 419:689–701.
70. Kwan C-Y, Daniel EE. Calcium handling by membranes isolated from vascular smooth muscle in hypertension. In: Aoki K, Frohlich ED, eds. Calcium in Essential Hypertension. Japan: Academic Press; 1989:201–230.
71. Hirst GDS, Edwards FR. Sympathetic neuroeffector transmission in arteries and arterioles. Physiol Rev 1989; 69:546–604.
72. Bolton TB, Beech DJ, Komori S, Prestwich SA. Voltage- and receptor gated channels. Prog Clin Biol Res 1990; 327:229–243.
73. Hamilton TC, Weir SW, Weston AH. Comparison of the effects of BRL 34915 and verapamil on electrical and mechanical activity in rat portal vein. Br J Pharmacol 1986; 88:103–111.
74. Videbæk LM, Aalkjær C, Mulvany MJ. Pinacidil opens K^+-selective channels causing hyperpolarization and relaxation of noradrenaline contractions in rat mesenteric resistance vessels. Br J Pharmacol 1988; 95:103–108.
75. Quast U, Cook NS. Moving together: K^+ channel openers and ATP-sensitive K^+ channels. TIPS 1989; 10:431–435.
76. Daut J, Maier-Rudolph W, Backerath N, Mehrke G, Günther K, Goedel-Meinen L. Hypoxic dilatation of coronary arteries is mediated by ATP-sensitive potassium channels. Science 1990; 247:1341–1344.
77. Nelson MT, Huang Y, Brayden JE, Hescheler JK, Standen NB. Arterial dilations in response to calcitonin gene-related peptide involve activation of K^+ channels. Nature 1990; 344:770–773.
78. Amédée T, Large WA, Wang Q. Characteristics of chloride currents activated by noradrenaline in rabbit ear artery cells. J Physiol 1990; 428:501–516.
79. Amédée T, Large WA. Microelectrode study on the ionic mechanisms which contribute to the noradrenaline-induced depolarization in isolated cells of the rabbit portal vein. Br J Pharmacol 1989; 97:1331–1337.
80. Iijima K, Lin L, Nasjletti A, Goligorsky S. Intracellular ramification of endothelin signal. Am J Physiol 1991; 260:C982–C992.
81. Klockner U, Isenberg G. Endothelin depolarizes myocytes from porcine coronary and human mesenteric arteries through a Ca-activated chloride current. Pfluegers Arch 1991; 418(1–2):168–175.
82. Benham CD, Tsien RW. A novel receptor-operated Ca^{2+}-permeable channel activated by ATP in smooth muscle. Nature 1987; 328:275–278.

83. Benham CD, Bolton TB, Byrne NG, Large WA. Action of externally applied adenosine triphosphate on single smooth muscle cells dispersed from rabbit ear artery. J Physiol (Lond) 1987; 387:473–488.
84. Sivertsson R. The hemodynamic importance of structural vascular changes in essential hypertension. Acta Physiol Scand 1970; 343(suppl):1–56.
85. Folkow B, Grimby G, Thulesius O. Adaptive structural changes of the vascular walls in hypertension and their relation to the control of the peripheral resistance. Acta Physiol Scand 1958; 44:255–272.
86. Folkow B, Gurevich M, Hallbäck M, Lundgren Y, Weiss L. The hemodynamic consequences of regional hypertension in spontaneously hypertensive and normotensive rats. Acta Physiol Scand 1971; 100: 270–272.
87. Short DS, Thomson AD. The arteries of the small intestine in systemic hypertension. J Pathol 1959; 78:311–334.
88. Hartper RN, Moore MA, Marr MC, Watts LE, Hutchins PM. Arteriolar rarefaction in conjunctiva of human essential hypertensives. Microvasc Res 1978; 16:369–372.
89. Hutchins PM, Darnell AE. Observation of a decreased number of small arterioles in spontaneously hypertensive rats. Circ Res 1974; 34–35:1–161.
90. Heinrich H, Hertel R, Assmann R. Structural differences in the mesentery microcirculation between normotensive and spontaneously hypertensive rats. Pfluegers Arch 1978; 375:153–159.
91. Sullivan JM, Prewitt RL, Josephs JA. Attenuation of the microcirculation in young patients with high-output borderline hypertension. Hypertension 1983; 5:844–851.
92. Greene AS, Tonellato PJ, Lui J, Lombard JH, Cowley AW Jr. Microvascular rarefaction and tissue vascular resistance in hypertension. Am J Physiol 1989; 256:H126–H131.
93. Foa PP, Foa NL, Pett MM. Arteriolar lesions in hypertension: A study of 350 consecutive cases treated surgically. An estimation of the prognostic value of muscle biopsy. J Clim Invest 1943; 22:727–742.
94. Short D. Morphology of the intestinal arterioles in chronic human hypertension. Br Heart J 1966; 28:184–192.
95. Furuyama M. Histometrical investigations of arteries in reference to arterial hypertention. Tohoku J Exp Med 1962; 76:388–414.
96. Aalkjær C, Heagerty AM, Petersen KK, Swales JD, Mulvany MJ. Evidence for increased media thickness, increased neuronal amine uptake, and depressed excitation-contraction coupling in isolated resistance vessels from essential hypertensives. Circ Res 1987; 66:181–186.
97. Baumbach GL, Heistad DD. Remodeling of cerebral arterioles in chronic hypertension. Hypertension 1989; 13:968–972.
98. Baumbach GL, Heistad DD. Adaptive changes in cerebral blood vessels during chronic hypertension. J Hypertension 1991; 9:987–991.
99. Barrett AM. Arterial measurements in the interpretation of cardiomegaly at necropsy: Cardiac hypertrophy and myocardial infarction. J Pathol Bact 1963; 86:9–20.
100. Whall CW, Myers MM, Halpern W. Norepinephrine sensitivity, tension development and neuronal uptake in resistance arteries from spontaneously hypertensive and normotensive rats. Blood Vessels 1980; 17:1–15.

101. Mulvany MJ, Hansen PK, Aalkjær C. Direct evidence that the greater contractility of resistance vessels in spntaneously hypertensive rats is associated with a narrowed lumen, a thickened media, and an increased number of smooth muscle cell layers. Circ Res 1978; 43:854–864.

102. Mulvany MJ, Nyborg NCB. An increased calcium sensitivity of mesenteric resistance vessels in young and adult spontaneously hypertensive rats. Br J Pharmacol 1980; 71:585–596.

103. Dominiczak AF, Bohr DF. Cell membrane abnormalities and the regulation of intracellular calcium concentration in hypertension. Clin Sci 1990; 79:415–423.

104. Jones AW. Altered ion transport in vascular smooth muscle from spontaneously hypertensive rats, influences of aldosterone, norepinephrine, and angiotensin. Circ Res 1973; 33:563–572.

105. Jones AW. Reactivity of ion fluxes in rat aorta during hypertension and circulatory control. Fed Proc 1974; 33:133–137.

106. Jones AW. Altered ion transport in large and small arteries from spontaneously hypertensive rats and the influence of calcium. Circ Res 1974; 34(1):117–122.

107. Jones AW, Miller LA. Ion transport in tonic and phasic vascular smooth muscle and changes during deoxycorticosterone hypertension. Blood Vessels 1978; 15:83–92.

108. Jones AW. Kinetics of active sodium transport on aortas from control and deoxycorticosterone hypertensive rats. Hypertension 1981; 3:631–640.

109. Friedman SM, Friedman CL. Cell permeability, sodium transport, and the hypertensive process in the rat. Circ Res 1976; 39:433–441.

110. Friedman SM. Evidence for enhanced sodium transport in the tail artery of the spontaneously hypertensive rat. Hypertension 1979; 1:572–582.

111. Friedman SM. Cellular ionic perturbations in hypertension. J Hypertension 1983; 1:109–114.

112. Izzard AS, Cragoe EJ Jr, Heagerty AM. Intracellular pH in human resistance arteries in essential hypertension. Hypertension 1991; 17:780–786.

113. Cauvin C, Johns A, Yamamoto M-K, Hwang O, Gelband C, van Breemen C. Ca^{2+} movements in vascular smooth muscle and their alterations in hypertension. In: Kwan C-Y, ed. Membrane Abnormalities in Hypertension. Florida: CRC Press; 1989:146–179.

114. Rusch NJ, Hermsmeyer K. Calcium currents are altered in the vascular muscle cell membrane of spontaneously hypertensive rats. Circ Res 1988; 63:997–1002.

115. Postnov YV, Orlov SN. Ion transport across plasma membrane in primary hypertension. Physiol Rev 1985; 65(4):904–945.

116. Heagerty AM, Riozzi A, Brand SC, Bing RF, Thurston H, Swales JD. Membrane transport of ions in hypertension: A review. Scand J Clin Lab Invest 1986; 46(180):54–64.

117. Doyle AE, Black H. Reactivity to pressor agents in hypertension. Circ Res 1955; 12:974–980.

118. Egan B, Schork N, Panis R, Hinderliter A. Vascular structure enhances regional resistance responses in mild essential hypertension. J Hypertension 1988; 6:41–48.

119. Robinson BF, Dobbs RJ, Bayley S. Response of forearm resistance vessels to verapamil and sodium nitropruside in normotensive and hypertensive men: Evidence for a functional abnormality of vascular smooth muscle in primary hypertension. Clin Sci 1982; 63:33–42.
120. Ng LL, Harker M, Abel ED. Mechanisms of leucocyte sodium influx in essential hypertension. Clin Sci 1988; 75:521–526.

19
Calcium Ion Homeostasis in the Aging Brain: Regulation of Voltage-Dependent Calcium Channels

FIORENZO BATTAINI, STEFANO GOVONI, MARIA SANDRA MAGNONI, and MARCO TRABUCCHI

Calcium Ion Homeostasis at the Neuronal Level

The tuning of neuronal networks is controlled by a number of processes that are in turn strictly controlled by cytosolic free Ca^{2+} ions. Short-term functions such as neurotransmitter release and enzymatic activity, and long-term mechanisms such as cytoskeletal integrity, neuronal plasticity, and even gene expression, depend on a rise in free intraneuronal calcium levels.[1] Although a "transient" rise in calcium concentration is of physiologic significance, a prolonged calcium increase in the cytoplasm, if not buffered properly, can lead to cell membrane derangement (overstimulation of calcium-dependent kinases, lipases, and proteases) that may end in cell death.[2] Intracellular calcium "oscillations" represent an additional mechanism for neuronal and endocrine signaling,[3] which has been investigated in particular in nonexcitable cells.[4]

The increase in free intracellular calcium above basal levels from submicromolar concentrations (10^{-7} M) is effected by fluxes from two different compartments in which the ion is present in millimolar concentrations, ie, the extracellular space and intracellular storage sites.[5] Calcium enters the cell through differentially regulated channels.[6] Inside the cell, calcium is released from intracellular vesicular stores through channels that are activated by calcium itself or by inositol 1,4,5-trisphosphate (IP_3) that is produced by the receptor-dependent hydrolysis of membrane phosphatidyl inositol 4,5-bisphosphate.[7,8]

Once calcium has exerted its physiologic function, it must be either transported to the extracellular space by exchange (sodium-dependent) and transport (ATP-dependent) systems, sequestered by an ATP-dependent mechanism into the vesicular system, or buffered in the intracellular compartment in a membrane- or protein-bound form.[5] Calcium enters the neuron by two main types of Ca^{2+} channel: one is controlled and activated by a change in membrane voltage (the voltage-dependent calcium channels, VDCC),[9] and the other is activated by specific receptors (receptor-operated calcium channels, ROC).[10] VDCC

418

Ca++

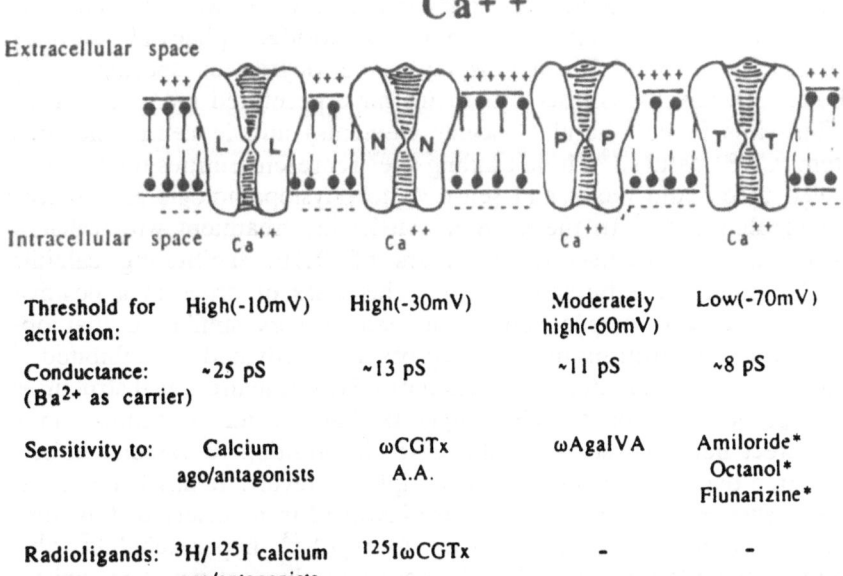

Threshold for activation:	High(-10mV)	High(-30mV)	Moderately high(-60mV)	Low(-70mV)
Conductance: (Ba²⁺ as carrier)	~25 pS	~13 pS	~11 pS	~8 pS
Sensitivity to:	Calcium ago/antagonists	ωCGTx A.A.	ωAgaIVA	Amiloride* Octanol* Flunarizine*
Radioligands:	³H/¹²⁵I calcium ago/antagonists	¹²⁵IωCGTx	-	-

FIGURE 19.1. Electrophysiologic and pharmacologic characteristics of the various subtypes of voltage-dependent calcium channels. Abbreviations: ω-CGTx, ω-conotoxin fraction GVIA (from the mollusk *Conus geographus*); A.A., aminoglycoside antibiotics; ω AgaIVA, ω-agatoxin fraction type IV (from the spider *Agelenopsis aperta*); * nonselective compounds. Further details in reference 15.

are classified according to activation threshold (low or high)[11] or by electrophysiologic and pharmacologic properties, as L, N, and T channels. According to the original classification proposed by Nowicky, Fox, and Tsien[12] in 1985 (Figure 19.1), "L" channel stands for "*Long*" lasting, "T" stands for "*Transient*," and "N" is for "*Neither L nor T*," having intermediate electrophysiologic characteristics. A newly discovered channel, termed "P" according to its localization (first described in rat cerebellar "*Purkinje cells*"), has had its electrophysiology and pharmacology characterized recently.[13] (See Figure 19.1.)

As is the case for other channels (Na⁺ and K⁺), natural substances may be used to interfere with specific Ca²⁺ channels ω-conotoxin, ωCGTx, from the marine snail *Conus geographus* with N[14] and ω-agatoxin, AGa IV A, from the funnel web spider *Agelenopsis aperta* with P-type channels[15]). Drugs of organic or inorganic origin interfere with L and T channels. Calcium antagonist drugs such as dihydropyridines (DHPs) (nitrendipine, isradipine, and nimodipine) phenylalkylamines (PHAs) (verapamil, gallopamil, and desmethoxy-verapamil) and benzothiazepines (diltiazem) specifically interact with

L-type channels at specific sites.[16] T channels are less well characterized because no specific inhibitors are known; amiloride, diphenylalkylamines (flunarizine), and Ni^{2+} ions display some degree of specificity.[17,18] Labeled calcium antagonists and toxins have permitted identification of binding sites in peripheral (vascular elements) and in central nervous system (CNS) tissues.[19] These binding sites represent entities functionally coupled to calcium fluxes and sensitive to physiopathologic and toxicologic regulation.[20,21] In the CNS in particular, treatment with calcium antagonists and agonists (enantiomers of DHP, facilitating calcium fluxes) regulates the binding sites for this class of drugs in a positive and negative way, respectively, in a manner very similar to that observed when neurotransmitter receptors are activated or inhibited.[19] Moreover, the preferential localization of DHPs in neuron/glial structures as compared to blood vessels[22] suggests that calcium antagonists may have direct neuronal action in the CNS, in addition to vascular effects (the latter being prominent at the peripheral level). It has been tentatively suggested that brain VDCC are involved in a variety of functions such as neuronal repetitive firing (T-type),[23] hyperexcitability,[24,25] peptide transmitter release,[26-28] (L-type) and classical neurotransmitter release[29,30] (N-type). P-type channels control cerebellar function and invertebrate neurotransmitter release (squid giant neurons).[13]

Calcium and the Aging Nervous System

Age-related changes in brain calcium metabolism have been reported since the beginning of this century,[31] but it is only in the last decade that a number of calcium-dependent functions such as neurotransmitter release, enzymatic activities, and transport systems have been found to be impaired in old age.[32] Indeed a "calcium theory of brain aging" has been proposed to explain a variety of age-related declines in brain function.[33,34] This has also stimulated the search for age-dependent modifications in functions indirectly linked to the control of calcium, such as ion fluxes[35] and neuronal excitability.[36,37]

The functions involved in controlling neuronal calcium homeostasis are accessible to neurochemical analysis. Calcium fluxes are measurable both with single-channel or whole-cell recording techniques[38] and the free concentration of the ion is measurable in single cells.[39] These technologic advances have been useful for the study of cells in culture, but are not yet applicable to studies of aging. The extension of these studies to tissue slices[40] will allow a better cellular description of calcium homeostasis during aging.

Pharmacologic studies have demonstrated that antagonizing calcium channels and fluxes may ameliorate some behavioral and electrophysiologic changes observed during aging. For example, DHP calcium antagonists have been reported to improve associative learning[41] and to ameliorate hippocampal-impaired excitability in aged rodents.[42] In addition, DHPs are reported to improve a number of behavioral functions related to aging (water maze, exploratory behavior, and sensory motor function[43]). Moreover, calcium antagonists are able to improve neuronal survival after axotomy in vivo[44] and can counteract the in vitro toxic effects of the gp 120 protein from the HIV-I virus.[45]

In this chapter the status of VDCC during aging will be analyzed and when possible, a correlation with functional studies will be attempted. A final section will be devoted to the use of this basic knowledge as a rationale for the design of therapeutic pharmacologic interventions for the aging brain.

VDCC and Brain Aging

L-Type Channels

Initial studies on brain VDCC utilized tritiated DHPs and PHAs as specific probes for the L-type channel expression. In Sprague Dawley rat cortical membranes, nitrendipine-labeled channels (B_{max} values) increase gradually from postnatal day 3 up to 3 months of age and then plateau, being similar in 10- and 24-month-old rats. Binding affinity values (K_d) remain constant up to 10 months of age and almost double in 24-month-old rats, a decrease in binding affinity of more than 80%.[46] Because Ca^{2+} ions are needed in vitro to maximally express nitrendipine association with its specific site in the channel structure,[47] calcium sensitivity of nitrendipine binding to Ca^{2+} channels prepared from cortical membranes of animals of different ages was investigated. The calcium reconstitution curve of nitrendipine binding is bell-shaped at all ages, peaking at $300 \mu M$, $400 \mu M$, and $500 \mu M$ when using membranes prepared from rats of 3, 10, and 24 months of age, respectively. Indeed, the data on calcium sensitivity and those showing an age-related decrease in binding affinity agree with reduced DHP-binding efficiency with aging and suggest a modification of Ca^{2+} channels during aging.[46]

It is known from biochemical[48] and photoaffinity labeling studies[49] that DHPs and PHAs specifically interact with L-type channels at distinct but allosterically coupled drug receptor sites, and that Ca^{2+} ions modulate such complex interactions. Investigation of the competition between various calcium antagonists and the DHP calcium agonist Bay K8644 on [^{3}H]-nitrendipine binding to cortical membranes of adult (3 month) and aged (24 month) rats has indicated that verapamil displaces bound nitrendipine by a maximum of roughly 30% in adult rats with an IC_{50} of 60 nM and by 55% with a significantly lower IC_{50} (28 nM) in the aged rat

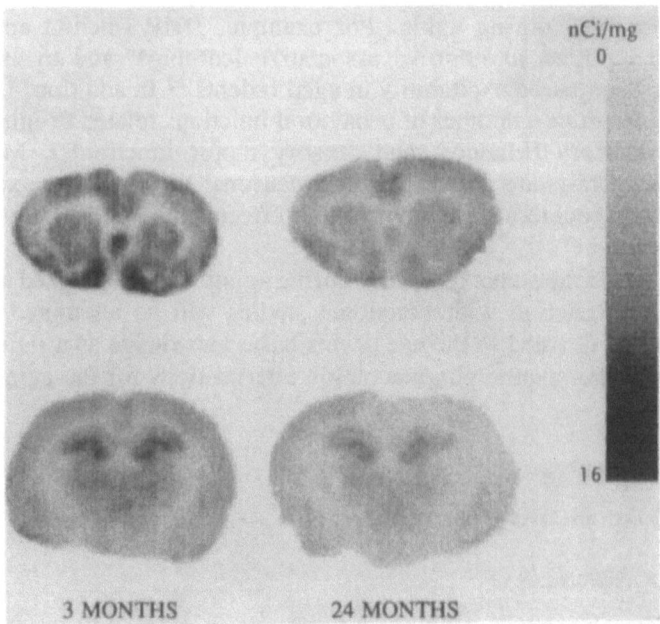

FIGURE 19.2. Autoradiography of [3H]-isradipine binding to Wistar rat brain. 12 μm brain sections on gelatin-coated slides were treated with 1.5 nM [3H]-isradipine as previously reported in detail in reference 56. The slides were exposed to [3H]-Hyperfilm for 7 days and analyzed with the Image 1.37 program of Dr. Wayne Rasband, NIMH, Bethesda, Md.

membranes.[50] These data suggest the possibility of an age-related increase in affinity or number of binding sites to L channels for [3H]-verapamil. Kinetic analysis of PHA binding to the channel indicated that the maximal binding capacity (B_{max}) increased gradually with age, being higher by 30% and 70% in 10- and 24-month-old rats than in the 3-month-old age group.[50] No changes in affinity are evident with increasing age. The data on DHP and PHA binding of L channels during aging indicate that the concomitant decrease in DHP-binding affinity and increase in PHA-binding B_{max} is in accord with biochemical data that indicates verapamil in vitro decreases DHP's binding affinity.[48,51] However, the possibility cannot be excluded that the increased PHA-binding B_{max} in aged rat brain cortical membranes may be correlated with changes occurring in glial cells that display significant levels of DHP binding[52] and of L-type calcium currents.[53] Other groups have analyzed DHP binding during aging in the strain of rats used by us[54] and in other strains such as the Fisher 344.[55] These authors find age-related, area-dependent decreases in [3H]-isradipine-binding B_{max} in various brain

TABLE 19.1. Microdensitometric analysis of [³H]-isradipine binding to rat brain sections.

Area	Isradipine binding (nCi/mg tissue)*		% of control (3 months) in 24-month-old rats
	3 months	24 months	
Frontal cortex	1.79 ± 0.1	0.99 ± 0.3	55
Parietal cortex	1.20 ± 0.3	0.86 ± 0.2	72
Striatum	1.83 ± 0.3	0.93 ± 0.3	51
Nucleus accumbens	1.90 ± 0.3	0.94 ± 0.2	49
Septum	1.84 ± 0.2	1.10 ± 0.4	59
Olfactory tubercle	2.10 ± 0.2	1.33 ± 0.3	63
Hippocampus	1.64 ± 0.4	1.10 ± 0.2	69

* The values in nCi/mg tissue were calculated using Amersham [³H]-microscales as standards. The values are means ± SD. Four rats, with 2–4 measurements for each area, were used.

areas. We have also analyzed autoradiographically the binding of [³H]-isradipine, a higher affinity ligand for DHP-binding sites, in the Wistar rat brain (Figure 19.2).

The data indicate that in 24-month-old rats, isradipine binding is reduced in areas such as the cortex, striatum, nucleus accumbens, septum, olfactory tubercle, and hippocampus (Table 19.1). [³H]-nitrendipine-labeled channels have also been analyzed in a different rodent model of aging. The senescent accelerated mouse (SAM) shows early onset of senescence;[57] in particular, at the age of 8 months, a subset of these animals (the so-called prone mouse, P8) show signs of pathologic aging (brain amiloidosis) and memory deterioration when compared to age-matched control animals (the resistant mouse, R1). Analysis of [³H]-nitrendipine binding indicates that in the cortex of the P8 mice the B_{max} values increase (+24%), while at the hippocampal level the B_{max} values decrease by almost 20% compared to the R1 animals.[57] These data indicate that an imbalance in DHP-sensitive Ca^{2+} channels exists in the brain in conditions characterized by memory deterioration, and may provide some insight into Ca^{2+} channel modifications related to pathologic aging. This aspect has also been analyzed from a clinical point of view. Colvin and co-workers[58] compared the binding of [³H]-nitrendipine in frontal cortex tissues from Alzheimer's patients, normal age-matched controls, and patients with non-Alzheimer dementia. No changes in K_d and B_{max} parameters were observed. Similar data have been reported using autoradiographic distribution of binding sites for nimodipine and D888 and analyzing cortical and subcortical structures.[59] No data were presented on age-dependent values in normal aging in either study, so this point remains to be investigated in greater detail.

N-Type Channels

Ω-conotoxin binds and inhibits N-type channels with high affinity and specificity.[14] Calcium antagonists interacting with L- or T-type channels (DHPs, PHAs, and diphenylalkylamines such as flunarizine) do not compete in vitro with binding of the iodinated toxin [^{125}I]-ωCGTx to brain membranes. In turn, ωCGTx does not interfere with DHP- and PHA-binding sites,[60] confirming, as observed with electrophysiologic measurements, that these ligands interact with different channels.[61] [^{125}I]-ωCGTx binding was investigated using single-point measurements in rats of three different ages (2 weeks, 2 and 18 month, Sprague Dawley rats). Binding is decreased in the aged rats in the striatum and spinal cord (in this case compared to the 2-week-old rats), while in other structures such as the hippocampus, hypothalamus, and cerebellum, binding does not change with age.[62] Analysis of iodinated ωCGTx binding saturation curves for membranes from different areas of adult and aged rats indicate that B_{max} values decrease in aged rats in the striatum and cortex, while no modification is observed in the hippocampus. K_d values are not modified as a function of age in all of the areas investigated.[63] These data are in accord with observations made in the Fisher 344 strain,[55] which also present an age-dependent decrease in binding in all the brain areas studied, including the hippocampus. The latter discrepancy could be due to the different strain of rats used. Recent studies have analyzed ωCGTx binding to human cortex in pathologic aging (Alzheimer's disease), a condition in which brain calcium homeostasis[64,65] and calcium-dependent enzymes are impaired.[66] No changes in N-type channels were observed in Alzheimer cortical and hippocampal structures in comparison to age-matched controls.[67] Data on human physiologic aging are not yet available.

T- and P-Type Channels

Since there are no specific probes or toxins for the analysis of T channels from a biochemical point of view, the only possible characterization is electrophysiologic,[68] and will be discussed in the section on the functional expression of VDCC (see below).

As far as P channels are concerned, labeled funnel web spider toxin is not yet available. Studies with specific antibodies have been carried out only in young animals.[69]

Functional Expression of VDCC During Aging

Calcium Fluxes

The effect of aging on depolarization-dependent calcium fluxes in the brain has been investigated by different groups. All studies revealed an age-dependent decrease in $^{45}Ca^{2+}$ uptake.[69-72] Both the "fast phase" of uptake (controlled by VDCC opening and limited to the first seconds of depolarization) and the "slow phase" (sustained by Na^+-Ca^{2+} exchange and activated with longer than 3 s depolarizations) were found to decrease.[73] These data may explain the decrease in K^+ or electrically stimulated neurotransmitter release observed in synaptosomes[74,75] and brain slices.[76] Calcium buffering capacity during aging has been investigated in detail by Giovannelli and Pepeu.[72] Cortical synaptosomes from 24-month-old rats buffer intracellular calcium at a slower rate than synaptosomes from adults rats, which gives rise to sustained higher ion levels lasting several minutes in vitro. This sustained increase in calcium is totally abolished by pretreating the synaptosomes with verapamil, which suggests that increased calcium passing through VDCC contributes to the decreased buffering capacity with aging. This observation is in accord with data that indicates an age-related increase in cortical $[^3H]$-PHA binding.[50]

Calcium Currents

The time course of calcium fluxes, measured with either $^{45}Ca^{2+}$ or fluorimetric probes (such as fura-2), is in the range of seconds to minutes and no distinction among the various calcium currents is possible. Such distinction can be made with voltage-clamp measurements. Initial difficulties encountered in culturing cells from donors of different ages have been partially overcome by refinements in nerve cell dissociations and by the use of tissue slice techniques.[77,78] T-type currents have been investigated in CA1 hippocampal neurons from 2-, 8-, and 44-week-old Wistar rats.[79] No age-related changes were observed in their activation/inactivation properties and the sensitivity to inhibition by Ni^{2+} ions was also unmodified, which suggests this type of current is preserved with aging.

Calcium currents of the L-type have been studied in dentate gyrus hippocampal neurons by Reynolds and Carlen.[80] The peak amplitude of this current was decreased in 24-month-old rats in comparison with adult (6 month) animals. This difference was abolished when the recording electrode contained the calcium chelator EGTA, raising the possibility that an increase in intracellular free calcium was responsible for the age-dependent L-current inactivation.

In the same experiments, a decrease in amplitude of the DHP-insensitive N-type current was also noted during aging. These results do not conflict with the previously observed lack of changes in hippocampal N channels, because it is possible that the number of N channels available for opening and/or the simple channel conductance may be decreased, although there is no change in their total number, as suggested by the $[^{125}I]$-ωCGTx in vitro binding data.[63] The electrophysiologic data can be reconciled with biochemical data that shows an increase in free intracellular calcium with aging;[81] in fact, the decreased currents could be the result of a calcium-dependent inactivation due to an effect of calcium from inside the cell.

Conclusions

Analysis of VDCC during brain aging indicates that opposite modifications may take place at pre- and postsynaptic sites. At the presynaptic sites (nerve terminals) calcium fluxes coupled to neurotransmitter release are impaired as a consequence of aging. The N-type channels controlling such processes are generally decreased in number. Evidence for a deficient calcium homeostasis at the presynaptic level is strengthened by observations that drugs such as 3,4-diaminopyridine, which increase calcium influx through inhibition of potassium efflux, are able to counterbalance the decreased neurotransmitter release and to improve certain memory deficits associated with aging in the rat.[82] It must, however, be pointed out that defects in mechanisms beyond the channels could contribute to impaired neurotransmitter release, as indicated by the reduced ability of the calcium ionophore A23187 to promote acetylcholine release from cortical synaptosomes from aged rats.[83] Moreover, other mechanisms involved in the control of calcium homeostasis are impaired during aging. Ion extrusion and compartmentalization into intracellular stores (endoplasmic reticulum and mitochondria) are depressed as a consequence of aging.[71,84] Also, the mechanisms responsible for calcium release from intracellular stores induced by IP_3 and the receptor for IP_3 itself are impaired in aged rats.[85,86] At the postsynaptic level (dendrites and neuronal cell bodies) electrophysiologic and spectrofluorimetric measurements indicate that Ca^{2+} ion concentration may be increased in aging.[35,81] In hippocampus, calcium-dependent processes such as afterhyperpolarization (AHP)[35] and protein kinase C activity[87] are increased. The decreased neuronal excitability consequent to prolonged AHP may be counteracted with a number of interventions aimed at decreasing calcium fluxes, such as feeding the animals a magnesium-rich diet[88] or treating the brain slices with DHP calcium antagonists.[42] The latter observation is reminiscent of the ability of verapamil to reestablish the calcium buffering capability of aged cortical synaptosomes.[72] As in the case of verapamil binding in the cortex,[50] an increase in binding of DHP in the hippocampus of aged rats was expected. Quantitation of the autoradiographic data indicates, however, that in the hippocampus of aged rats [³H]-DHP binding is decreased (Table 19.1). These observations are in accordance with data obtained with in vitro [³H]-DHP binding to tissue homogenates.[42,54,55] The explanation for this discrepancy with the electrophysiologic data could be that in vitro DHPs bind to the channel in an "inactivated" form[89] so that a decreased binding with aging could be interpreted as an increased proportion of the channel in the "activated" form, which thus allows a higher calcium influx and increased calcium levels.

The differential modification of L- and N-type channels with aging could also be related to the fact that L channels are preferentially concentrated on neuronal cell bodies while N channels are located preferentially on nerve terminals, an observation strengthened by studies of lesioned and pathologic material. Indeed, lesions of the striatal neurons produced by kainic acid in the rat are associated with a decrease in striatal [³H]-DHP binding (which is consistent with a postsynaptic localization[90,91] and is analogous to Huntington's disease), where the density of striatal DHP-binding sites is decreased compared to age-matched controls, which suggests a striatal intrinsic neuronal degeneration. In contrast, no changes in DHP-binding sites are observed in Parkinson's disease characterized by a striatal presynaptic degeneration.[59,92,93]

The diversity of VDCC is further underlined by studies performed in the last few years on the molecular composition of the L-channel subunit. The original studies, performed with skeletal muscle, demonstrated the presence of two domains of high molecular weight ($>200\,kDa$), termed α_1 and α_2/δ, and two lower molecular weight domains termed β and γ.[94] Such structural composition has been confirmed in heart and brain tissues.[95,96] Calcium antagonists interfere specifically with the α_1 subunit. This has the basic properties necessary to function as a Ca^{2+} channel, the other subunits control α_1 and its interactions with drugs and gating properties.[97] Four different α_1 subunits have been described in rat brain and at least eight different Ca^{2+} channel transcripts may be present at this level.[96] VDCC resembling the P-type have been also characterized at the molecular level;[98] the other types of channels (T and N) have not been described in molecular terms so far. No studies are as yet available on the molecular expression of these channels during aging using nucleotide hybridization techniques, nor is their expression in *Xenopus* oocytes after injection with mRNA purified from aged animals known. Understanding this diversity in young or aged animals will be important in the development of new pharmacologic tools for clinical use.

Of particular relevance to studies on aging is the observation that Ca^{2+} channels contribute significantly to the cascade of phenomena leading to cell death.[2] Various studies have demonstrated the involvement of VDCC in conditions such as in vivo[99] and in in vitro models of ischemia.[100]

For instance, in rats binding of [³H]-DHP to the homolateral hippocampus is increased 1 h following right carotid occlusion.[101] This might be related to a previously observed increase in calcium fluxes and intracellular free ion concentrations under ischemic conditions.[102] The relevance of this observation is underscored by data showing that L-type channels represent only a very small fraction of the VDCC present in the hippocampus.[103] Accordingly, it has been reported that DHP reduce neuronal death after ischemia, either in brain as a whole or in hippocampal cell cultures.[104,106] The action of DHP calcium antagonist drugs at the neuronal level is the object of intense investigation for their possible antagonizing effect on age and/or lesion-related learning and memory impairment[43] and on the changes in brain excitability and metabolism. Analysis of compounds acting on non-DHP-sensitive channels may reveal new strategies to affect presynaptic functions, such as neurotransmitter release (N-type channels)[27] and neuronal plasticity (P-type channels could be involved in such phenomena in cerebellum,[98] while T-type could modulate excitability and bursting activity of central neurons[107]).

From a pharmacologic perspective, it is possible that certain characteristics of diphenylbutylpiperidine drugs such as pimozide, fluspirilene, and penfluridol (namely the effect of these drugs on the negative symptoms of schizophrenia) may be related to an intrinsic calcium antagonist property[108,109] in addition to their "classic" antidopaminergic activity.

Moreover, recent data have demonstrated that gene expression may be modulated by drugs acting on VDCC. In primary cortical cultures, the expression of various immediate early genes is modulated by L-type DHP drugs. Calcium antagonists (isradipine) inhibit, while calcium agonists (Bay K8644) increase the basal expression of transcription factor genes zif268, c-*fos*, jun-B, and fos-B. DHPs also modulate kainate induction of neuronal excitability and of c-*fos* expression in a similar way.[110] Interestingly, c-*fos* expression is not modified by N-type channel blockade with ω-conotoxin.[110] The possibility of age-dependent changes in gene expression,[111] related in particular to calcium influx through VDCC of the L type in the brain, has not yet been investigated. These studies may lead to a better interpretation of neuronal plasticity during aging in particular because zif268 appears to be involved in the phenomena of long-term synaptic potentiation in the young animal.[112,113]

Acknowledgments. We wish to thank Mrs. L. Fusar Imperatore for expert secretarial assistance.

References

1. Pietrobon D, Di Virgilio F, Pozzan T. Structural and functional aspects of calcium homeostasis in eukaryotic cells. Eur J Biochem 1990; 193:599–622.
2. Choi DW. Calcium-mediated neurotoxicity: Relationship to specific channel types and role in ischemic damage. TINS 1988; 11:465–469.
3. Miller RJ. Calcium signalling in neurons. TINS 1988; 11:415–419.

4. Berridge MJ. Inositol trisphosphate, calcium, lithium, and cell signaling. JAMA 1989; 262(13):1834–1841.
5. Carafoli E. Intracellular calcium homeostasis. Annu Rev Biochem 1987; 56:395–433.
6. Meldolesi J, Pozzan T. Pathways of Ca^{2+} influx at the plasma membrane: Voltage-, receptor-, and second messenger-operated channels. Exp Cell Res 1987; 171:271–283.
7. Neering IR, McBurney RN. Role for microsomal Ca storage in mammalian neurones? Nature 1984; 309:158–160.
8. Berridge MJ, Irvine RF. Inositol phosphates and cell signalling. Nature 1989; 341:197–205.
9. Hess P. Calcium channels in vertebrate cells. Nature 1990; 13:1337–1356.
10. Ascher P, Nowak L. Electrophysiological studies of NMDA receptors. TINS 1987; 10:284–298.
11. Tsien RW, Elliot PT, Horne WA. Molecular diversity of voltage-dependent calcium channels. TIPS 1991; 12:349–354.
12. Nowicky MC, Fox AP, Tsien RW. Three types of neuronal calcium channels with different calcium agonist sensitivity. Nature 1985; 316:443–446.
13. Llinas R, Sugimori M, Lin JW, Cherskey B. Blocking and isolation of a calcium channel from neurons in mammals and cephalopods utilizing a toxin fraction (FTX) from funnel-web spider poison. Proc Natl Acad Sci USA 1989; 86:1689–1693.
14. Olivera BM, Gray WR, Zeikus R, McIntosh JM, Varga J, Rivier J, Santos V, Cruz LJ. Peptide neurotoxin from fish hunting cone snails. Science 1985; 230:1338–1343.
15. Miller RJ. Voltage sensitive calcium channels. J Biol Chem 1992; 267:1403–1406.
16. Glossman H, Striessnig J. Molecular properties of calcium channels. Rev Physiol Biochem 1990; 114:1–105.
17. Tytgat J, Vereccke J, Cornelliet E. Differential effect of verapamil and flunarizine on cardiac L-type and T-type channels. Arch Pharmacol 1990; 337:690–692.
18. Bean BP. Classes of calcium channels in vertebrate cells. Annu Rev Physiol 1989; 51:367–384.
19. Triggle DJ, Janis RA. Calcium channels ligands. Ann Rev Pharmacol Toxicol 1987; 27:347–369.
20. Govoni S, Battaini F, Magnoni MS, Lucchi L, Rius RA, Trabucchi M. Plasticity of neuronal L-type calcium channels. Ann NY Acad Sci 1988; 522:187–202.
21. Triggle DJ, Langs DA, Janis RA. Calcium channels ligands: Structure-function relationships of the 1,4-dihydropyridines. Med Res Rev 1989; 9:123–180.
22. Gould RJ, Murphy KMM, Snyder SH. In vitro autoradiography of (^{3}H)nitrendipine localizes calcium channels to synaptic rich zone. Brain Res 1985; 330:217–223.
23. Hess P, Tsien RW. Mechanism of ion permeation through calcium channels. Nature 1984; 309:453–456.
24. Ramkumar V, El-Fakahany EE. Morphine treatment increases nimodipine binding sites in rat brain. Ann NY Acad Sci 1988; 522:207–209.

25. Littleton JM, Little HJ. Dihydropyridine-sensitive Ca channels in brain are involved in the CNS hyperexcitability associated with alcohol withdrawal states. Ann NY Acad Sci 1988; 522:199–202.

26. Govoni S, Goss I, Di Giovine S, Battaini F, Trabucchi M. Calcium antagonists inhibit met-enkephalin immunoreactive material release: in vitro and ex vivo experiments. J Neural Transm 1990; 80:1–8

27. Hirning LD, Fox AP, McCleskey B, Olivera BM, Thayer SA, Miller RJ, Tsien RW. Dominant role of N type calcium channels in evoked release of norepinephrine from sympathetic neurons. Science 1988; 239:57–60.

28. Rane SG, Holz GG, Dunlap K. Dihydropiridine inhibition of neuronal calcium currents and substance P release. Pfluegers Arch 1987; 409:361–366.

29. Dayanithi G, Martin-Moutot N, Barlier S, Colin DA, Kretz-Zoepfel M, Couraud F, Nordmann JJ. The calcium channel antagonist ω conotoxin inhibits secretion from peptidergic nerve terminals. Biochem Biophys Res Commun 1988; 156:255–262.

30. Dooley DJ, Lupp A, Hertting G. Inhibition of central neurotransmitter release by omega-conotoxin GVIA, a peptide modulator of the N-type voltage-sensitive calcium channel. Arch Pharmacol 1987; 336:467–470.

31. Novi I. Calcium et le magnesium du cerveau des differents ages. Arch Ital Biol 1912; 58:333–336.

32. Gibson GE, Peterson C. Calcium and the aging nervous system. Neurobiol Aging 1987; 8:329–344.

33. Khatchaturian Z. Towards theories of brain aging. In: Kay DW, Burrows GD, eds. Handbook of Studies in Psychiatry and Old Age. New York: Elsevier; 1984:7–30.

34. Khachaturian Z. The role of calcium regulation in brain aging: Reexamination of a hypothesis. Aging 1989; 1:17–34.

35. Landfield PW, Pitler TA. Prolonged calcium-dependent afterhyperpolarization in hippocampal neurons of aged rats. Science 1984; 226:1089–1091.

36. Barnes CA. Memory deficits associated with senescence: A neurophysiological and behavioral study in the rat. J Comp Physiol Psychol 1979; 93:74–104.

37. Landfield PW, McGaugh JL, Lynch G. Impaired synaptic potentiation process in the hippocampus of aged, memory-deficient rats. Brain Res 1978; 150:85–101.

38. Marty A, Neher E. Tight-seal whole cell recording. In: Sakmann B, Neher E, eds. Single Channel Recording. New York: Plenum; 1983:107–132.

39. Tsien RY. Fluorescent indicators of ion concentrations. Meth Cell Biol 1989; 30:127–156.

40. Landfield PW, Lynch G. Impaired monosynaptic potentiation in in vitro hippocampal slices from aged memory deficient rats. J Gerontol 1977; 32:523–533.

41. Deyo RA, Straube KT, Disterhoft J. Nimodipine facilitates trace conditioning of the eye-blink response in aging rabbits. Science 1989; 243:809–811.

42. Landfield PW. Nimodipine modulation of aging-related increases in hippocampal calcium currents. In: Traber J, Gispen WH, eds. Nimodipine and CNS Function: New Vistas. Stuttgart: Shattaner; 1989:227–238.

43. Scriabine A, Schuurman T, Traber J. Pharmacological basis for the use of nimodipine in central nervous system disorders. FASEB J 1989; 3:1799–1806.
44. Rich KM, Hollowell JP. Flunarizine protects neurons from death after axotomy or NGF deprivation. Science 1990; 248:1419–1421.
45. Dreyer EB, Kaiser PK, Offermann JT, Lipton SA. HIV-1 coat protein neurotoxicity prevented by calcium channel antagonists. Science 1990; 248:364–367.
46. Govoni S, Rius RA, Battaini F, Bianchi A, Trabucchi L. Age-related reduced affinity in (^{3}H)-Nitrendipine labeling of brain voltage-dependent calcium channels. Brain Res 1985; 33:374–377.
47. Gould RJ, Murphy KMM, Snyder SH. (^{3}H)-nitrendipine labeled calcium channels discriminate inorganic calcium agonists and antagonists. Proc Natl Acad Sci USA 1982; 79:3656–3660.
48. Boles RG, Yamamura HI, Schoemaker H, Roeske WR. Temperature dependent modulation of (^{3}H)nitrendipine binding by the calcium channel antagonists verapamil and diltiazem in rat brain synaptosomes. J Pharmacol Exp Ther 1984; 229:333–339.
49. Striessnig J, Glossmann H, Catterall WA. Identification of a phenylalkyl-amine binding region within the alfal subunit of skeletal muscle Ca^{2+} channels. Proc Natl Acad Sci USA 1990; 87:9108–9112.
50. Battaini F, Govoni S, Rius RA, Trabucchi M. Age-dependent increase in (^{3}H)-Verapamil binding to rat cortical membranes. Neurosci Lett 1985; 61:67–71.
51. Battaini F, Govoni S, Del Vesco R, DiGiovine S, Trabucchi M. Concomitant regulation of hippocampal calcium antagonist receptors and calcium uptake by substance P. Biochem Biophys Res Commun 1987; 114:1135–1142.
52. Hertz L. Functional interactions between neurons and glial cells. In: Battaini F, Govoni S, Magnoni MS, Trabucchi M, eds. Regulatory Mechanisms of Neurons to Vessel Communication in the Brain. NATO ASI Series, Vol. 33. Heidelberg: Springer-Verlag; 1989:271–306.
53. Barres BA, Chun LLY, Corey DP. Calcium current in cortical astrocytes: Induction by cAMP and neurotransmitters and permissive effect of serum factors. J Neurosci 1989; 9:3169–3175.
54. Huguet F, Huchet AM, Gerad P, Narcisse G. Characterization of dihydropiridine binding sites in the rat brain: Hypertension and age-dependent modulation of (^{3}H)(+)-PN 200-110 binding. Brain Res 1987; 412:125–130.
55. Bangalore R, Ferrante J, Hawthorn M, et al. The regulation of neuronal calcium channels. In: Paoletti R, Vanhoutte PM, Govoni S, eds. Calcium Antagonists: Pharmacology and Clinical Research. Medical Science Symposia, Vol 3, Dordrecht, Kluwer 1993: 221–230.
56. Ferry DR, Goll A, Gadow C, Glossman H. ^{3}H Desmethoxyverapamil labeling of putative calcium channels in brain: Autoradiographic distribution and allosteric coupling to 1,4 dihydropyridine and diltiazem binding sites. Arch Pharmacol 1984; 321:80–83.
57. Kitamura Y, Zhao XZ, Ohnuki T, Nomura Y. Ligand-binding characteristics of (^{3}H)QNB, (^{3}H)prazosin, (^{3}H)rauwolscine, (^{3}H)TCP and

(^{3}H)nitrendipine to cerebral cortical and hippocampal membranes of senescence accelerated mouse. Neurosci Lett 1989; 106:334–338.

58. Colvin RA, Williams RG, Eagle DT, Allen RA, Oibo SA, Ibok I. Neuronal binding of (^{3}H)-nitrendipine in dementia. In: Miner GD, Richter RW, Blass JP, Valentine JL, Winters-Miner LA, eds. Familial Alzheimer's Disease: Molecular Genetics and Clinical Perspectives. New York: Marcel Dekker; 1989:325–330.

59. Quirion R, Nair NPV. Dihydropyridine and phenylakylamines binding sites in Alzheimer disease and other neurological disorders. In: Traber J, Gispen WH, eds. Nimodipine and CNS Function: New Vistas. Stuttgart: Shattaner; 1989:257–265.

60. Wagner JA, Snowman AM, Biswas A, Olivera BM, Snyder SH. Omega-conotoxin GVIA binding to a high-affinity receptor in brain: Characterization, calcium sensitivity, and solubilization. J Neurosci 1988; 8:3345–3359.

61. Regan LJ, Sah DWY, Bean BP. Ca^{2+} channels in rat central and pheripheral neurons: High-threshold current resistant to dihydropyridine blockers and omega-conotoxin. Neuron 1991; 6:269–280.

62. Dooley DJ, Lickert M, Lupp A, Osswald H. Distribution of (^{125}I)omega-conotoxin GVIA and (^{3}H)isradipine binding sites in the central nervous system of rats of different ages. Neurosci Lett 1988; 93:318–323.

63. Moresco RM, Govoni S, Battaini F, Trivulzio S, Trabucchi M. Omega-conotoxin binding decreases in aged rat brain. Neurobiol Aging 1990; 11:433–436.

64. Deary IJ, Hendrickson AE. Calcium and Alzheimer's disease. Lancet 1986; 8491:1219.

65. Peterson C, Ratan RR, Shelanski ML, Goldman JE. Cytosolic free calcium and cell spreading decreases in fibroblasts from aged and Alzheimer donors. Proc Natl Acad Sci USA 1986; 83:7999–8001.

66. Masliah E, Cole GM, Hansen LA, Mallory M, Albright T, Terry RD, Saitoh T. Protein kinase C alteration is an early biochemical marker in Alzheimer's disease. J Neurosci 1991; 1:2759–2767.

67. Colvin RA, Allen RA, Williams RG, Eagle DT, Oibo JA, Miner GD. (^{125}I)-Omega conotoxin binding to human frontal cortex from normal, Alzheimer's and non-Alzheimer's dementia patients. Neurobiol Aging 1990; 11:151–153.

68. Carbone E, Lux HD. A low voltage activated, fully inactivating calcium channel in vertebrate sensory neurones. Nature 1984; 310:501–511.

69. Hillman D, Chen S, Aung TT, Cherskey B, Sugimori M, Llinas RR. Localization of P-type calcium channels in the central nervous system. Proc Natl Acad Sci USA 1991; 88:7076–7080.

70. Leslie SW, Chandler LJ, Barr EM, Farrar RP. Reduced calcium uptake by rat brain mitochondria and synaptosomes in response to aging. Brain Res 1985; 329:177–183.

71. Vitorica J, Satrustegui J. Involvement of mitochondria in the age-dependent decrease in calcium uptake of rat brain synaptosomes. Brain Res 1986; 378:36–48.

72. Giovannelli L, Pepeu G. Effect of age on K$^+$-induced cytosolic Ca^{2+} changes in rat cortical synaptosomes. J Neurochem 1989; 53:392–398.

73. Martínez A, Vitórica J, Bogónez E, Satrustégui J. Differential effect of age on the pathways of calcium influx into nerve terminals. Brain Res 1987; 435:249–257.

74. Gibson GE, Peterson C. Aging decreases oxidative metabolism and the synthesis and release of acetylcholine. J Neurochem 1981; 37:978–984.

75. Meyer EM, Onge St E, Crews FT. Effects of aging on rat cortical presynaptic cholinergic processes. Neurobiol Aging 1984; 5:315–317.

76. Pedata F, Giovannelli L, Spignoli G, Giovannini MG, Pepeu G. Phosphatidylserine increases acetylcholine release from cortical slices in aged rats. Neurobiol Aging 1985; 6:337–339.

77. Scott BS. Adult neurons in cell culture: Electrophysiological characterization and use in neurobiological research. Prog Neurobiol 1982; 19:187–211.

78. Mody I, Salter MW, MacDonald JF. Whole-cell voltage-clamp recording in granule cells acutely isolated from hippocampal slices of adult or aged rats. Neurosci Lett 1989; 96:70–75.

79. Takahashi K, Tateishi N, Kaneda M, Akaike N. Comparison of low-threshold calcium currents in the hippocampal CA1 neurons among the newborn, adult and aged rats. Neurosci Lett 1989; 103:29–33.

80. Reynolds JN, Carlen PL. Diminished currents in aged hippocampal dentate gyrus granule neurones. Brain Res 1989; 384–390

81. Martinez A, Vitorica J, Satrustegui J. Cytosolic free calcium levels increase with age in rat brain synaptosomes. Neurosci Lett 1988; 88:336–342.

82. Peterson C, Gibson GE. Amelioration of age-related neurochemical and behavioral deficits by 3,4-diaminopyridine. Neurobiol Aging 1983; 4:25–30.

83. Meyers EM, Crews FT, Otero DH, Larsen K. Aging decreases the sensitivity of rat cortical synaptosomes to calcium ionophore-induced acetylcholine release. J Neurochem 1986; 47:1244–1246.

84. Michaelis ML, Johe K, Kitos TE. Age-dependent alteration in synaptic membrane systems for calcium regulation. Mech Age Dev 1984; 25:215–225.

85. Burnett DM, Daniell LC, Zahniser NR. Decreased efficacy of inositol 1,4,5-trisphosphate to elicit calcium mobilization from cerebrocortical microsomes of aged rats. Mol Pharmacol 1990; 37:566–571.

86. Li PP, Vecil GG, Green MA, Warsh JJ. Inositol 1,4,5-trisphosphate receptor in developing and senescent rat cerebellum. Neurobiol Aging 1992; 13:89–92.

87. Battaini F, Del Vesco R, Govoni S, Trabucchi M. Regulation of phorbol ester binding and protein kinase C activity in aged rat brain. Neurobiol Aging 1990; 11:563–566.

88. Landfield PW, Pitler TA, Applegate MD. The effects of high Mg to Ca ratios on frequency potentiation in hippocampal slices of young and aged rats. J Neurophys 1986; 56:797–811.

89. Bean BP. Nitrendipine block of cardiac calcium channels: High affinity binding to the inactivated states. Proc Natl Acad Sci USA 1984; 81:6388–6392.

90. Skattebol A, Hruska RE, Hawthorn M, Triggle DJ. Kainic acid lesions decrease striatal dopamine receptors and 1,4-dihydropyridine sites. Neurosci Lett 1988; 89:85–89.

91. Bolger GT, Basile AS, Janowsky AJ, Paul SM, Skolnick P. Regulation of dihydropyridine calcium antagonist binding sites in the rat hippocampus following neurochemical lesions. J Neurosci Res 1987; 17:285–290.

92. Watson DL, Carpenter CL, Marks SS, Greenberg DA. Striatal calcium channel antagonist receptors in Huntington's disease and Parkinson's disease. Ann Neurol 1988; 23:303–305.

93. Piggott MA, Candy JM, Perry RH. (^{3}H)Nitrendipine binding in temporal cortex in Alzheimer's and Huntington's disease. Brain Res 1991; 565:42–47.

94. Tanabe T, Takeshima H, Mikami A, Flokerzi V, Takahashi H, Kangawa K, Kojima M, Matsuo H, Hirose T, Numa S. Primary structure of the receptor for calcium channel blockers from scheletal muscle. Nature 1987; 328:313–318.

95. Mikami A, Imoto K, Tanabe T, Niidome T, Mori Y, Takeshima H, Narumiya S, Numa S. Primary structure and functional expression of the cardiac dihydropyridine-sensitive calcium channel. Nature 1989; 340:230–233.

96. Snutch TP, Leonard JP, Gilbert MM, Lester HA, Davidson N. Rat brain expresses a heterogeneous family of calcium channels. Proc Natl Acad Sci USA 1990; 87:3391–3395.

97. Tsien RW, Elliot PT, Horne WA. Molecular diversity of voltage-dependent Ca^{++} channels. TIPS 1991; 12:349–354.

98. Mori Y, Friedrich T, Kim MS, Mikami A, Nakai J, Ruth P, Bosse E, Hofmann I, Flockerzi V, Furuichi T, Mikoshiba K, Imoto K, Tanabe T, Numa S. Primary structure and functional expression from complementary DNA of a brain calcium channel. Nature 1991; 350:398–402.

99. Siesjo BK. Calcium ischemia and death of brain cells. Ann NY Acad Sci 1988; 522:638–661.

100. Magnoni MS, Trabucchi M, Battaini F, Govoni S. The role of calcium in cerebral ischemic damage. In: Anghileri LJ, ed. The Role of Calcium in Biological Systems. Vol. V. Boca Raton, Fla.: CRC Press; 1989:197–215.

101. Magnoni MS, Govoni S, Battaini F, Trabucchi M. L-type calcium channels are modified in rat hippocampus by short term experimental ischemia. J Cereb Blood Flow Metab 1988; 8:96–99.

102. Simon RP, Griffiths T, Evans MC, Swan JH, Meldrum BS. Calcium overload in selectivity vulnerable neurons of the hippocampus during and after ischemia: An electron microscopy study in the rat. J Cereb Blood Flow Metab 1984; 4:351–360.

103. Docherty RJ, Brown DA. Interaction of 1,4 dihydropyridines with somatic Ca^{++} currents in hippocampal CA1 neurones of the guinea pig in vitro. Neurosci Lett 1986; 70:110–115.

104. Uematsu D, Greenberg JH, Hickey WF, Reinich M. Nimodipine attenuates both increase in cytosolic free calcium and histologic damage following focal cerebral ischemia and reperfusion in cats. Stroke 1989; 20:1531–1537.

105. Weiss JH, Hartley DM, Koh J, Choi DW. The calcium channel blocker nifedipine attenuates slow excitatory amino acid neurotoxicity. Science 1990; 247:1474–1477.

106. Abele AE, Scholz KP, Scholz WK, Miller RJ. Excitoxicity induced by enhanced excitatory neurotransmission in cultured hippocampal pyramidal neurons. Neuron 1990; 4:413–419.

107. Schroeder JE, Fischbach PS, McCleskey EW. T-type calcium channels: Heterogenesous expression in rat sensory neurons and selective modulation by phorbol esters. J Neurosci 1990; 10(3):947–951.
108. Gould RJ, Murphy KMM, Reynolds IJ, Snyder SH. Antischizophrenic drugs of the diphenylbutylpiperidine type act as calcium antagonists. Proc Natl Acad Sci USA 1983; 80:5122–5125.
109. Galizzi JP, Fosset G, Romey P, Laduron P, Lazdunski M. Neuroleptics of the diphenylbutylpiperidine series are potent calcium channel inhibitors. Proc Natl Acad Sci USA 1986; 83:7513–7517.
110. Murphy TH, Worley PF, Baraban JM. L-type voltage-sensitive calcium channels mediate synaptic activation of immediate early genes. Neuron 1991; 7:625–635.
111. Finch CE, Morgan DG. RNA and protein metabolism in the aging brain. Annu Rev Neurosci 1990; 13:75–88.
112. Cole AJ, Saffen DW, Baraban JM, Worley PF. Rapid increase of an immediate early gene meassenger RNA in hippocampal neurons by synaptic NMDA receptor activation. Nature 1989; 340:474–476.
113. Wisden W, Errington ML, Williams S, Dunnett C, Hitchcock D, Evan G, Bliss TVP, Hunt SP. Differential expression of immediate early genes in the hippocampus and spinal cord. Neuron 1990; 4:603–614.

20
Interactions of Ethanol with Ion Channels: Possible Implications for Mechanisms of Intoxication and Dependence

Wojciech Kostowski

There is much evidence that ethyl alcohol (ethanol, ET-OH), like other anesthetics, affects the central nervous system (CNS) by a direct nonspecific physical action on the excitable membranes. Thus ET-OH molecules may insert into the lipid layer, thereby affecting the mobility of the membrane lipids and the hydrophobic portion of the membrane proteins. It is possible that ET-OH may also act directly upon nerve membrane proteins, ie, ion channels and ion pumps.[1-3] Thus several mechanisms of ET-OH action have been suggested: changes in membrane fluidity, direct action upon the receptor-ionophore complex, and action upon the proteins that regulate the function of this complex. Although no specific ET-OH effects on ion channels has been described to date, several interesting effects of ET-OH on ion currents and ion channels have been reported. The γ-aminobutyric acid- (GABA) activated Cl^- channel, glutamate receptor-activated ion currents, voltage-dependent Ca^{2+} channels, and sodium-, potassium-dependent, magnesium-activated adenosine triphosphatase (Na^+,K^+-ATPase or NKA) are among those affected. It has been suggested by many investigators that the action of ET-OH on neuronal membranes may contribute substantially to ET-OH intoxication, tolerance, and dependence. These problems will be discussed below.

Ethanol and Cellular Calcium Mechanisms

There is evidence that ET-OH interferes with the metabolism of Ca^{2+}, a fundamental cation in the central nervous system.[4] This interaction seems to be of particular importance in the central actions of ET-OH because of the involvement of calcium in mechanisms of neurotransmission. Many putative steps in transmitter release are believed to be calcium-dependent.[5-7] Thus the ET-OH effects on neurotransmission may result, at least in part, from changes in intracellular Ca^{2+} $[Ca^{2+}]_i$. In most studies single high doses of ET-OH inhibit release of transmitters while chronic ET-OH either stimulates or reduces these processes in various brain structures.[8-11] Wu, Fan, and Naranjo[10] showed that synaptic transmission was more active in Wistar rats with high and moderate preference for ET-OH that in nonpreferring animals. This result may indicate that intrinsic differences exist in calcium-dependent neurotransmission among laboratory animals showing different degree of preference for ET-OH.

The action of ET-OH upon calcium-dependent cellular mechanisms is complex, and multiple targets for this action may exist. Interference by ET-OH with the function of any of the calcium-regulating systems (eg, Ca^{2+} channels, intracellular uptake of calcium, Na^+-Ca^{2+} exchange system, etc.) in neurons could markedly change calcium homeostasis and, in turn, influence the calcium-dependent cellular processes.

Ethanol, Calcium Uptake, and Cellular Concentration

A number of investigators have reported that ET-OH reduces calcium uptake by synaptosomes in vitro in a dose- (concentration) dependent manner.[12-14] Others, however, have shown that ET-OH, particularly when administered chronically, increases rather than decreasing calcium uptake.[15-16] More recent studies suggest that ET-OH may influence neuronal calcium homeostasis by increasing free $[Ca^{2+}]_i$. High ET-OH concentrations in vitro and single high doses in vivo attenuate calcium influx, but increase its cytosolic concentration.[15,17] It appears that this ET-OH induced increase occurs at least in part as the result of a decrease in calcium binding to cytoplasma buffers and an inhibition of calcium uptake by mitochondria.[15] Interestingly, chronic ET-OH treatment results in an adaptation to the inhibitory ET-OH effect on voltage-dependent calcium uptake within a time frame similar to that of the development of behavioral tolerance to ET-OH.[17] Therefore it appears that the intoxicating effects of single ET-OH doses are mediated, at least in part, by an inhibition of neuronal calcium entry. Tolerance development may result from a reduction in the ability of ET-OH to attenuate calcium influx. Furthermore, chronic ET-OH treatment increases the number of membrane Ca^{2+} channels (see below) and further increases $[Ca^{2+}]_i$ concentrations.[18]

As mentioned above, ET-OH affects not only membrane calcium transport but also intracellular calcium mechanisms such as uptake into intracellular stores. Garret and Ross[12] found that acute administration of a high dose of ET-OH (4 g/kg IP) inhibits the sequestration of calcium in the endoplasmatic reticulum. Furthermore, they found an inhibitory action of ET-OH upon both calcium-stimulated ATPase hydrolysis and ATP-dependent calcium uptake in the vesicles. These effects occured at the time of behavioral intoxication (loss of righting reflex). Interestingly, upon recovery of righting reflex the Ca^{2+}-ATPase returned to normal levels, while the ATP-dependent uptake was still reduced. Furthermore, Lynch and Littleton[19] reported that Ca^{2+},Mg^{2+}-ATPase activity is higher in synaptosomes isolated from ET-OH-tolerant rats, which suggests adaptation to long-term ET-OH exposure.

Ethanol and Na^+-Ca^{2+} Exchange Activity

The synaptic membrane Na^+-Ca^{2+} exchange carrier is thought to be particularly active in neuronal membranes. This carrier system, unlike the ATP-dependent Ca^{2+} transport system, is highly sensitive to the effects of ET-OH in vitro.[20-22] Recent studies have shown that repeated administration of ET-OH produces alteration in Na^+-Ca^{2+} antiporter in Sprague-Dawley rats, which is possibly indicative of an adaptive response to ET-OH. Results of these studies revealed that the administration of ET-OH for 3 weeks brought about a nearly twofold increase in the sodium-dependent calcium transport activity of the fraction enriched in synaptic junctional complexes.[22] The fact that ET-OH-induced alterations in carrier activity were observed, particularly in the junction-enriched membrane fraction, suggests that exposure to ET-OH was more disruptive to the function of nerve terminals than to the function of other parts of the neuron. It is worthy to note that nerve endings contain a high density of voltage-sensitive Ca^{2+} channels; thus these regions are quite susceptible to the disruptive effects of an inhibitor of the calcium extrusion mechanism.[23]

Ethanol, Ca^{2+} Channels, and Their Inhibitors

It is generally believed that ET-OH affects calcium fluxes in neurons. As mentioned previously, when administered acutely ET-OH decreases calcium influx across membrane Ca^{2+} channels[24] while increasing Ca_i concentration, probably by reducing the intracellular sequestration of the ion and/or facilitating the mobilization of calcium from intracellular stores. Prolonged exposure to ET-OH leads to further elevation of $[Ca^{2+}]_i$,[25] supposedly due to increased functioning of Ca^{2+} channels.[26-28]

In the cell membrane, which under normal conditions is practically impermeable to Ca^{2+}, there are at least two kind of channels through which ions penetrate into the cell. Channels of the first type are associated with membrane receptors and are referred to as receptor-operated channels (ROC). Neurotransmitters acting upon these channels include, eg, excitatory amino acids such as glutamic and aspartic acid (see "ET-OH and NMDA-Activated Ion Current" section below). Ca^{2+} channels of the other type open as a result of membrane depolarization and are called voltage-operated channels (VOC).[29-31] A subdivision of VOC into three subtypes has been proposed: T (Transient), N (Neuronal), and L (Low). Only the L type is susceptible to the action of "classic" Ca^{2+} channel inhibitors (see below), although compounds acting on other channel subtypes have been described recently. A better understanding of Ca^{2+} channels has been possible as a result of studies that employed radiolabeled channel inhibitors, especially those belonging to the 1,4-dihydropyridine group (DHP), such as nifedipine.[31]

The chronic administration of ET-OH produces an increase in the number of DHP recognition sites on the VOC-type channels.[32-33] Furthermore, Littleton and Little[33] have shown that the development of ET-OH dependence is associated with an increase in the density of Ca^{2+} channels in the rat brain. This phenomenon is probably responsible for CNS hyperexcitability during ET-OH withdrawal since it has been reported that blockade of Ca^{2+} channels reduces the ET-OH withdrawal symptoms.[33-35] Increased density of VOC is causally associated with biologic effects of chronic ET-OH administration. Since Ca^{2+} ions are intimately involved in transmitter release mechanisms, alterations in Ca^{2+} influx and intracellular concentration may influence the functions of various neurotransmitter systems and brain structures. This mechanism seems to be at least partly responsible for the behavioral and other signs of chronic ET-OH intoxication and for ET-OH withdrawal symptoms such as neuronal excitability (eg, reduced seizure threshold), emotional disturbances (fear and aggressiveness), increased locomotion, etc. There is also evidence that an elevated $[Ca^{2+}]_i$ leads to neuronal damage. It is therefore not surprising that Ca^{2+}-channel inhibitors have beneficial effects in ET-OH intoxicated humans and animals (see below).

Brain Damage Resulting from Chronic ET-OH Abuse: The Role of Calcium and Ca^{2+} Channels

Brain damage is a most dangerous and distructive result of chronic ET-OH intoxication. A persistent elevation of neuronal calcium has been implicated in the neurotoxic effects of ET-OH.[36] Hippocampal neurons appear to be particularly sensitive to the disruptive action of ET-OH. Since chronic ET-OH administration reduces the neurons in this area one may assume that ET-OH-induced impairment of learning and memory is related to this phenomenon.

Ca_i content may be increased by altering the membrane potential and opening Ca^{2+} channels. Therefore inhibition of VOC should be able to limit the increase of $[Ca^{2+}]_i$ and attenuate the neurotoxic effect of ET-OH. Indeed, it has been reported that drugs that reduce calcium entry into the cells through a blockade of the VOC counteract a variety of ET-OH effects, such as the impairment performance and locomotion. We studied the effect of several Ca^{2+}-channel inhibitors on ET-OH-induced suppression of two-way avoidance response in Wistar rats. The administration of ET-OH for three weeks brought about a significant reduction in avoidance acquisition. However, this effect was attenuated in rats treated with Ca^{2+}-channel inhibitors such as diltiazem and verapamil (W. Kostowski, W. Obersztyn, and P. Krzascik, 1993 unpublished data).

Ca^{2+}-Channel Inhibitors and ET-OH Effects

Ca^{2+}-channel inhibitors (CCIs) interact specifically with the L-type channel.[30-31] To date, at least three calcium-binding sites on the channel have been discovered.[29,37] These sites are specific for three different chemical classes of CCIs: 1,4-dihydropyridines (eg, nifedipine, nimodipine, and isradipine), phenylalkylamines (eg, verapamil and gallopamil) and benzothiazepines (eg, diltiazem).

There is evidence that CCIs are capable of influencing many of the behavioral effects of ET-OH. Nifedipine, a DHP derivative, antagonizes the locomotor stimulating effects of low doses of ET-OH in mice,[38] while several CCIs (eg, nifedipine, nimodipine, and verapamil) given systemically potentiate both the hypothermic and the locomotor effects of high doses of ET-OH.[39-40] However, verapamil administered into the cerebral ventricles attenuates the thermolytic action of ET-OH, which suggests that the action on the peripheral blood vessels serves to exacerbate the hypothermic effect of ET-OH. Interestingly, a variety of CCIs have been known to reduce both ET-OH tolerance and ET-OH withdrawal symptoms in laboratory animals. This is in line with findings that show that the number of Ca^{2+} channels is increased in ET-OH-tolerant animals (ie, animals that acquired tolerance to the effects of ET-OH in the course of chronic administration). We have previously shown that nifedipine can significantly attenuate the tolerance to hypothermic effect of ET-OH and reduce the intensity of audiogenic seizures in ET-OH-withdrawn animals.[35] A similar effect of nifedipine on ET-OH withdrawal was observed by Engel and his associates.[38]

More recently we found that isradipine, the new DHP derivative, is capable of reducing the audiogenic seizure response in ET-OH-withdrawn rats, the effect being relatively stronger than that of nifedipine (Figure 20.1).

Considerable evidence suggests that CCIs can attenuate the emotional disturbances that occur during ET-OH withdrawal, in particular fear and anxiety. Indeed, we have shown that diltiazem, well as DHP derivatives (nifedipine and nimodipine), may reduce fear in ET-OH withdrawn rats tested in so-called elevated plus maze apparatus.[41] In another experiment we have demonstrated that diltiazem and other CCIs attenuated aggressiveness in rats undergoing ET-OH withdrawal.[42]

The effect of nifedipine and probably of other CCIs on the action of ET-OH cannot be simply due to pharmacokinetic interactions, eg, altered absorption or elimination.[43] Rather, it appears that a decrease in the

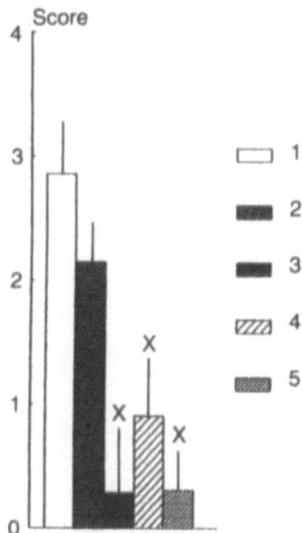

FIGURE 20.1. Effect of CCIs on audiogenic seizures in ET-OH-withdrawn Wistar male rats. The animals received ET-OH intragastrically (5–10 g/kg/24 h in two divided doses) for 5 consecutive days. Sixteen hours after ET-OH withdrawal, audiogenic seizures (evoked by a 100 db electric buzzer) were measured and scored on a scale of 0 to 5. Drugs were injected IP twice daily, 30 min after each ET-OH administration. Legend: 1 = ET-OH + saline, 2 and 3 = ET-OH + isradipine (1.25 or 2.5 mg/kg), 4 and 5 = ET-OH + nifedipine (2.5 or 5.0 mg/kg). Mean values obtained in 10 to 12 rats + SEM. X = $p < 0.005$ vs saline. (W. Dyr, P. Krzascik, and W. Kostowski; Pol J Pharmacol, 1993; in press.)

intracellular calcium may have beneficial therapeutic effect on the development of ET-OH dependence and ET-OH neurotoxicity.

Ca^{2+}-Channel Inhibitors and ET-OH Preference

Changes in neuronal calcium concentration seem to contribute to ET-OH reinforcement, consumption, and preference. The reinforcing effects of addicting drugs, ET-OH included, are mediated, at least in part, through the mesolimbic dopaminergic system. It is possible that one of the mechanism by which CCIs block ET-OH preference involves an interaction with dopaminergic and/or serotonergic (5-HT) neurons in the brain.[22,44–46] In fact, it has been reported that CCIs attenuate the behavioral excitatory effects of dopamine agonists and decrease dopamine turnover in the brain.[47–48] It has also been reported that verapamil modulates 5-HT function in the brain.[49–50]

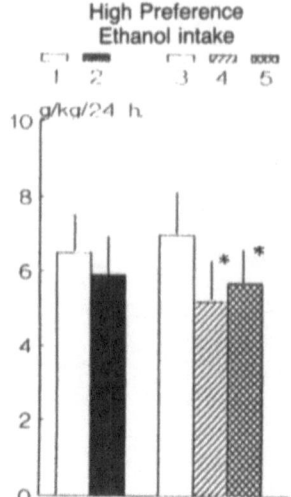
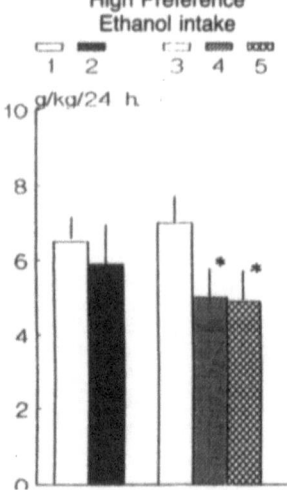

FIGURE 20.2. ET-OH intake (grams per kilo body weight) in rats treated with nifedipine (left) and isradipine (right). Legend: 1 and 3 = baseline consumption preceding saline or drug treatment, 4 and 5 = nifedipine or isradipine (1.25 or 2.5 mg/kg). x = p < 0.05 vs respective baseline value. Mean values obtained in 10 to 12 rats + SEM. (W. Dyr, P. Krzascik, and W. Kostowski; Pol J Pharmacol, 1993; in press.)

Nifedipine and other DHP derivatives, particularly isradipine and other drugs such as verapamil and a novel CCI, GOE 5438, appear to attenuate ET-OH intake and preference in different strains of animals.[25,51-55] We studied the effects of isradipine and nifedipine on ET-OH preference in Wistar rats that developed a high ET-OH consumption after 3-week period of forced ET-OH drinking. Both drugs significantly attenuated ET-OH consumption and preference in these animals (Figure 20.2).

Of special interest are results that show reduced ET-OH consumption in rats selectively bred for voluntary high ET-OH intake (the "P" or *Preferring* rats[52,55]). The observation that nifedipine prevents the ET-OH-induced increase in dopamine turnover in the brain of these rats further supports the notion that the dopaminergic system is intimately involved in the interaction of CCIs with ET-OH-reinforcing mechanism.[38]

In conclusion, considerable evidence suggest that CCIs can attenuate ET-OH preference in ET-OH-drinking laboratory animals. Preliminary clinical results also confirm this effect, but further trials are required. Although the mechanism of action is only partially known, it is possible that intracellular calcium and Ca^{2+} channels may play an important role in ET-OH dependence and craving.

ET-OH and Na^+,K^+-ATPase (NKA)

NKA is a plasma membrane lipoprotein enzyme involved in regulating the transmembrane transport of sodium and potassium, the release and uptake of transmitters, and the resting potential of cell membrane. This enzyme is integrated into the membrane of neuronal cells and, like other membrane enzymes, depends upon the lipid content and physical properties of membranes.[56] ET-OH, which dissolves in lipids, has been shown to inhibit the NKA in vitro.[57] The exact mechanism of this action is poorly understood and several possibilities have been suggested. ET-OH may stabilize the enzyme in one conformation (E_1) or ET-OH may inhibit enzyme dephosphorylation.[58,59] It is also possible that ET-OH may influence the enzyme activity directly by interacting with the enzyme protein or indirectly by interacting with membrane lipids, ie, by altering the microenvironment within which the enzyme resides.[45,60-62]

As mentioned above, ET-OH inhibits NKA in vitro. These studies, however, have yielded conflicting results probably due to methodologic difference such as ET-OH administration, assay conditions or isolation of the membranes.[58,61,63-64] Nevertheless, numerous investigators have shown that tolerance to ET-OH in particular may be more closely related to a reduced enzyme sensitivity to inhibition by ET-OH added in vitro than to the basal activity itself.[61-62] Although some investigators found an increase in basal NKA activity in the brain of rats treated chronically with ET-OH, others failed to demonstrate this phenomenon.[62,65]

The most interesting and important results are those which show resistence to inhibition by high dose of ET-OH when the enzyme is obtained from ET-OH-tolerant animals. This resistance lasted appoximately as long as the behavioral tolerance to ET-OH.[62,66] It has also been reported that noradrenaline sensitizes NKA to the inhibitory action of ET-OH.[58,67] It the presence of noradrenaline, chronic administration of ET-OH leads to a reduction of the inhibitory effect of single ET-OH doses on rat brain ATPase.[59]

It is noteworthy that the reduced inhibitory effect of ET-OH in the presence of noradrenaline was also found in erythrocyte membranes from alcoholic humans. In addition, the basal enzyme activity was significantly reduced in the erythrocytes of ET-OH-dependent patients examined within 24 h of withdrawal.[61]

The results described above suggest that tolerance to ET-OH is due, at least in part, to alterations in the activity of membrane NKA and sodium and potassium transport. However, it should be pointed out that not all studies support this notion. For example, Westcott and Weiner[56] failed to show a difference between the responses of membrane taken from control and from ET-OH-dependent rats. Thus further studies are needed to establish the exact action of ET-OH on NKA.

ET-OH and the Na$^+$ Channel

The inhibitory effect of ET-OH on axonal sodium conduction has been known for many years.[68-71] This phenomenon, however, was first observed using very high ET-OH concentrations (100–500 mM), which would be toxic or lethal in vivo. However, more recent studies have shown that even nontoxic ET-OH concentrations (20 mM) reduce the veratridine-stimulated sodium current of frog nerve.[64] Stokes and Harris[14] reported a substantial inhibitory effect of 100 mM ET-OH on the veratridine-stimulated uptake of calcium by brain synaptosomes, and they provided evidence that the effect was due to an action on the Na^{2+} channel rather than on the Ca^{2+} channels.[14,69,71]

Biochemical measurements of radioactive sodium uptake by isolated brain neurons synaptosomes have demonstrated that ET-OH, like some common anesthetics (halotane and enflurane), inhibits the flux of sodium through the synaptosome membranes.[14,68,72] It has also been reported that ET-OH inhibits the batrachotoxin-stimulated sodium uptake, thus demonstrating suppression of sodium influx through voltage-dependent channels. Interestingly enough, the inhibitory action of tetrodotoxin on sodium uptake was not affected by ET-OH, which indicates that this site of the Na$^+$ channel is not an ET-OH target. It is also noteworthy that the ET-OH effect was selective for certain brain areas, such as the neocortex and the cerebellar cortex[23] (Figure 20.3).

Some effects of ET-OH on veratridine-stimulated sodium uptake can be mimicked by an increase in temperature. A quantitative comparison indicated that the effect of 300 mM ET-OH at 13°C on the K$_d$ was somewhat greater than that of an increase in temperature from 13 to 18°C. This is consistent with results which indicate that an increase in temperature from 13 to 21°C produces similar changes in synaptic membrane as the addition of 300 mM ET-OH at 13°C.[3]

These and other findings[71] provide clear evidence that ET-OH inhibits sodium fluxes in brain synaptosomes, and that the disturbance in membrane lipids caused by ET-OH is probably an important factor in the reduction of Na$^+$ channel function. It is known that the potency of ET-OH and other intoxicant anesthetics is related to their lipid solubility.[68] Evaluation of the physiologic importance of the ET-OH action on sodium fluxes needs further studies. However, it seems reasonable to suggest that the action upon Na$^+$ channels contributes substantially to central depressive and anesthetic effect of alcohol.

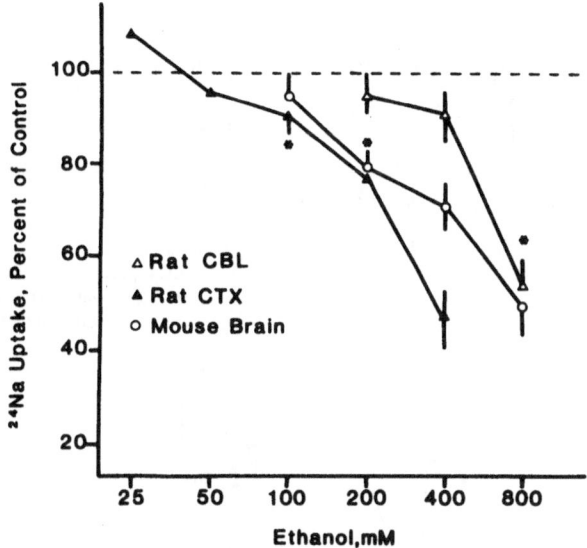

FIGURE 20.3. Effects of ET-OH on sodium uptake by synaptosomes from rat cerebellum (CBL), cortex (CTX), and mouse whole brain. The ordinate represents the veratridine-dependent uptake as a percentage of the uptake in the absence of the drug. Uptake time was 2 s. Values are mean + SEM from 4 to 6 experiments. * = lowest concentration producing a significant (p < 0.05) inhibition of uptake. (From Harris and Bruno.[68])

ET-OH and NMDA-Activated Ion Current

Glutamate, the major excitatory neurotransmitter in the CNS, acts through several receptor subtypes distinguished on the basis of their response to agonists: N-methyl-D-aspartate (NMDA), quisqualate, and kainate.[67] Lovinger, White, and Weight[73] studied the effect of ET-OH on ion current activated by glutamate receptor agonists (NMDA, glutamate, and quisqualate) in voltage-clamped hippocampal neurons of the mouse embryo, and found that the amplitude of the NMDA-activated current was markedly decreased (61% inhibition) in the presence of 50 mM ET-OH (the concentration that produced 50% inhibition was about 30 mM). In contrast ET-OH (50 mM) reduced the amplitude of current activated by kainate and quisqualate by only 18% and 15%, respectively (Figure 20.4). Interestingly, the power to inhibit of the NMDA-activated current of several alcohols (1-butanol, methanol, and isopentanol) was linearly related to their power to intoxicate. Isopentanol was the most potent inhibitor, with a threshold of 0.001 mM.[73]

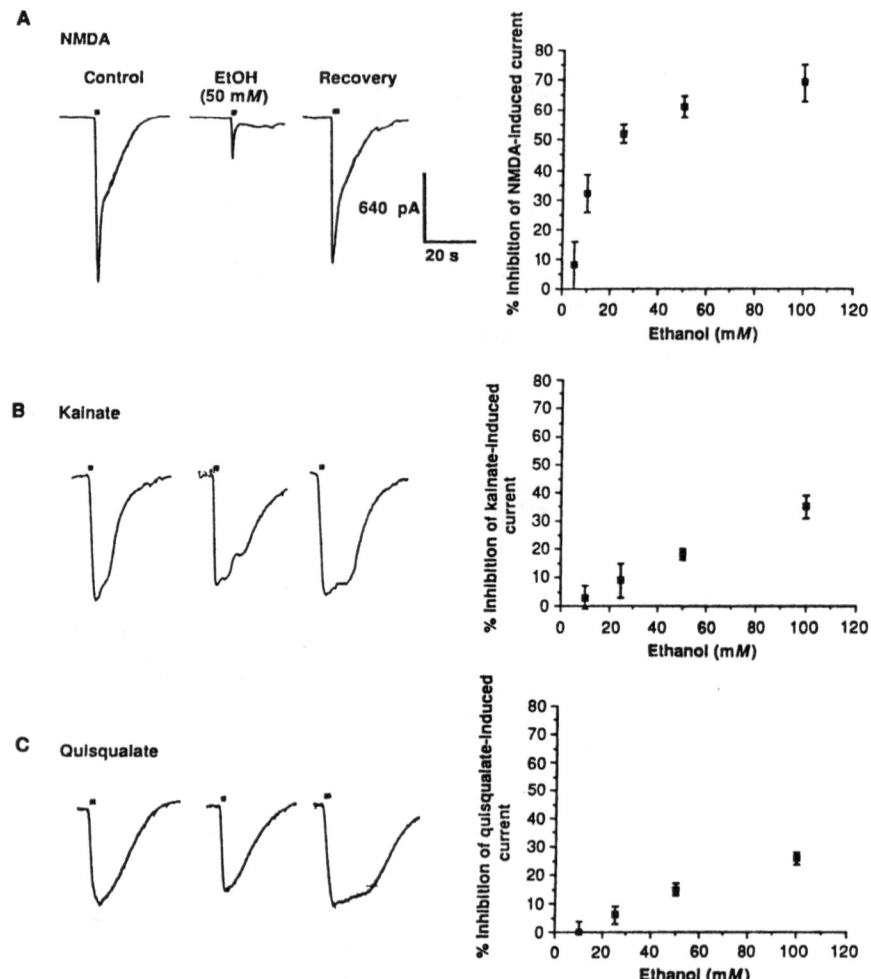

Figure 20.4. ET-OH effects on excitatory amino acid-induced ion currents in hippocampal neurons. (A)Effect of 50 mM ET-OH on ion current induced by NMDA. Control = inward current induced by application to 50mM NMDA. ET-OH = response to NMDA in the presence of ET-OH, Recovery = current induced by NMDA alone (in the absence of ET-OH) 2 min after the end of ET-OH application. Graph on the right shows the average percent inhibition of NMDA-activated current as a function of ET-OH concentration. (B) Effect of 10 μM kainate. (C) Effect of 1 μl quisqualalate. (From Lovinger et al.[73])

The mechanism by which ET-OH reduces the NMDA-activated ion current is poorly understood. Since ET-OH produces changes in membrane fluidity,[3] one may postulate that the ET-OH effect on NMDA receptor is secordary to the changes in the properties of lipids surrounding the receptor protein. It may be also suggested that ET-OH may interact directly with a hydrophobic region of the NMDA receptor ionophore complex. Whatever the mechanism of the ET-OH interaction with NMDA receptor is, the reduction of NMDA-mediated cellular responses contributes subtantially to the biologic action of ET-OH. It is generally accepted that NMDA plays an important role in regulating behavioral processes, such as locomotor activity, learning, and memory. NMDA is involved in synaptic transmission within the hippocampus,[74] the structure responsible for memory processes. In particular, NMDA and intracellular calcium are involved in the so-called long-term potentiation phenomenon (LTP) characteristic of hippocampal cells.[67,75] This phenomenon seems to reflect synaptic mechanism of memory; it is therefore possible that ET-OH-induced memory and learning disturbances may due, at least in part, to an interaction with the NMDA transmission systems.

ET-OH Interaction with GABA-Induced Chloride Current

Two inhibitory neurotransmitters, GABA and glycine, inhibit neuronal activity by opening the Cl^- ion channels, thereby producing an increase in membrane conductance that dampens the depolarizing action of excitatory neurotransmitters.[76] GABA acts on the Cl^- ionophore through the $GABA_A$ receptor, which is activated by muscimol and blocked by the $GABA_A$ antagonist bicuculine. Behavioral studies demonstrate that the effects of ET-OH are potentiated by the $GABA_A$ agonist, muscimol.[77] On the other hand, ET-OH enhances the bioelectric effects of GABA, ie, both the presynaptic and the postsynaptic inhibitory processes.[78] In particular, ET-OH enhances GABA-stimulated ^{36}Cl uptake by rat brain synaptosomes and spinal cord neurons in tissue culture.[79-81] Recently the effect of acute ET-OH on cultures of chick spinal cord neurons were studied using whole-cell, voltage-clamp techniques.[81-82] The results show that ET-OH (20–50 mM) increases the sensitivity of the neurons to GABA and glycine without changing the input resistance or the resting membrane potential. The effect of ET-OH was mimicked by methanol and other alcohols, which suggests that the capacity to alter Cl^- channel function may be a characteristic feature of ET-OH and related organic solvents.[82] Recently some investigators have suggested that alterations in $GABA_A$-Cl^- complex function may be responsible for the greater sensitivity to ET-OH seen in some laboratory animals, such as the ET-OH-sensitive (ANT) rats.[83]

It is now well accepted that the anesthetic and behavioral depressive properties of ET-OH and of other alcohols vary according to their power to disrupt the cell membrane.[40,84] Thus there is a high correlation between the in vitro potencies of alcohols in stimulating ^{36}Cl uptake and their intoxicating properties, while their power to disrupt the membrane is highly correlated with their ability to stimulate GABA-mediated Cl^- uptake.[84]

Contrary to its acute effect, chronic ET-OH administration reduces both the sensitivity and the number (density) of GABA and muscimol binding sites in the brain.[85] The GABA-chloride receptor complex contains several distinct sites, including the so-called benzodiazepine (BZD) receptor.[80] This receptor is responsible for the behavioral actions (eg, anti-anxiety, sedative, and antiepileptic) of BZD such as diazepam, nitrazepam, and others. It has been found that a decrease in GABA-BZD receptor complex is particularly pronounced during ET-OH withdrawal.[18] This phenomenon is probably responsible for the increased brain excitability, reduced seizure threshold, increased anxiety, locomotor disturbances, and other symptoms of withdrawal.

ET-OH and 5-HT-3 Receptor

Serotonergic 5-HT-3 receptors are ligand-gated, cation selective ion channels mediating membrane depolarization and neuronal excitation. Recent evidence from a variety of experimental procedures suggest that the effects of ET-OH are mediated, at least in part, by 5-HT-3 receptors. Low concentrations of ET-OH potentiate 5-HT-induced depolarization at the 5-HT-3 ionophore (E.H. Sellers, G.A. Higgins, and M.B. Sobell; TIPS, 1992; 12:69–75). Since 5-HT-3 receptors have been reported as intimately involved in dopamine release within mesolimbic brain areas and since ET-OH has been shown to release brain dopamine it is conceivable that ET-OH induced reinforcement is mediated through limbic 5-HT-3 receptors. In fact, certain 5-HT-3 receptor antagonists (eg, tropisetron and ondansetron) are capable of influencing ET-OH voluntary drinking and preference (W. Kostowski, W. Dyr, and P. Krzascik; Alcohol, 1993, in press).

Conclusions

Many biologic effects of ET-OH appear to depend upon its action on ion channels and ion currents. The anesthetic and sedative action of ET-OH is related, at least in part, to disturbances in sodium transport, to an action on Ca^{2+} channels, and to a potentiation of the effect of GABA on the Cl^- channel. A reduction in the excitatory action of NMDA and other amino acids also contributes to the depressive effect of ET-OH on the CNS. Tolerance to ET-OH is associated with a variety of changes in ion current function. For example, tolerance to ET-OH is associated with resistance of NKA to inhibition by high single dose of ET-OH. ET-OH-tolerant rats show increased number of voltage-operated Ca^{2+} channels in the brain. This mechanism may be responsible for the increased intracellular Ca^{2+} concentration that occurs during chronic ET-OH administration.

Changes in the function of ion currents and ion channels are intimately associated with alterations in neurotransmission mechanisms.[6] In turn the disturbances in transmitter systems may lead to substantial changes in behavioral processes such as locomotion, emotions, motivation, and others. Disturbances in Ca^{2+} transport are responsible at least in part for ET-OH withdrawal symptoms and ET-OH drinking preference of experimental animals. It is of particular interest that brain damage due to chronic ET-OH intoxication is intimately associated with increased intracellular Ca^{2+} and hence may be attenuated by Ca^{2+}-channel inhibitors. Greater understanding of ET-OH effects on cellular ion processes may lead to more rational approaches to the treatment of ET-OH abuse and intoxication.

References

1. Franks NP, Lieb WR. Do general anaesthetics act by competitive binding to specific receptors? Nature (London) 1984; 310:559–601.
2. Harris RA, Baxter DM, Mitchell MA, Hitzeman RJ. Physical properties and lipid composition of brain membranes from ethanol tolerant-depentent mice. Mol Pharmacol 1981; 30:3209–3215.
3. Harris RA, Schroeder F. Ethanol and the physical properties of brain membranes: Fluorescence studies. Mol Pharmacol 1981; 20:128–137.
4. Erickson CK, Tyler TD, Harris RA. Ethanol: Modification of acute intoxication by divalent cations. Science 1978; 199:1219–1221.
5. Kalant H, Grose W. Effect of ethanol and pentobarbital on release of acetylcholine from cerebral cortical slices. J Pharmacol Exp Ther 1967; 158:386–393.
6. Kelly RB, Deutsch JW, Carson SS, Wagner JA. Biochemistry of neurotransmitter release. Annu Rev Neurosci 1979; 2:399–445.
7. Lynch MA, Littleton JM, Enhanced ^3H noradrenaline release in synaptosomes from ethanol-tolerant animals: The role of nerve terminals calcium concentrations. Alcohol 1985; 20:5–11.
8. Carmichael FJ, Israel Y. Effects of ethanol on neurotransmitter release by rat brain cortical slices. J Pharmacol Exp Ther. 1975; 193:824–834.
9. Wu PH, Naranjo CA, Fan T. Chronic ethanol inhibits rat hippocampal "stimulus secretion" coupling mechanisms for 5-hydroxytryptamine in vitro. Neurochem. Res 1986; 11:801–812.

10. Wu PH, Fan T, Naranjo CA. Increase in the brain regional depolarization-dependent Ca^{2+} uptake in rats preferring ethanol. J Pharmacol Biochem Behav 1987; 27:355–357.
11. Wu PH, Pham T, Naranjo CA. Nifedypine delays the acquisition of tolerance to ethanol. Eur J Pharmacol 1987; 139: 233–236.
12. Garrett KM, Rose DH. Effects of in vivo ethanol administration on Ca^{2+} Mg^{2+} ATPase and ATP-dependent Ca^{2+} uptake activity in synaptosomal brain. Neurochem Res 1983; 8:1013–1028.
13. Lace JW, Schneider CW, Hartline RA. The ethanol sensitivity of calcium taken up by depolarization-dependent process in mouse strains DBA and C57BL. Pharmacol Biochem Behav 1986; 24:1137–1139.
14. Stokes JA, Harris RA. Alcohols and synaptosomal calcium transport. Mol Pharmacol 1982; 22:99–104.
15. Davidson M, Wilce P, Shanley B. Ethanol increases synaptosomal free calcium concentration. Neurosci Lett 1988; 89:165–169.
16. Friedman MB, Erickson CK, Leslie SW. Effects of acute and chronic ethanol administration on whole mouse brain synaptosomal calcium influx. Biochem Pharmacol 1980; 29:1903–1908.
17. Leslie SW, Barr E, Chandler J, Farrar RP. Inhibition of fast and slow-phase depolarization-dependent synaptosomal calcium uptake be ethanol. Pharmacol Exp Ther 1983; 225:571–575.
18. Freund G. Benzodiazepine receptor loss in brains of mice after chronic alcohol consumption. Life Sci 1980; 27:987–992.
19. Lynch MA, Littleton JM. Possible association of alcohol tolerance with increased synaptic calcium sensitivity. Nature (London) 1983; 308:175–176.
20. Michaelis EK, Myers SL. Calcium binding to brain synaptosomes. Biochem Pharmacol 1979; 28:2081–2087.
21. Michaelis ML, Michaelis EK, Tehan T. Alcohol effects on synaptic membrane calcium ion fluxes. Pharmacol Biochem Behav 1983; 18(suppl 1):19–23.
22. Michaelis ML, Michaelis EK, Nunley EW, Galton N. Effects of chronic alcohol administration on synaptic membrane Na^{2+}-Ca^{2+} exchange activity. Brain Res 1987; 414:329–244.
23. Hagiwara S, Byerly L. Calcium channel. Annu Rev Neurosci 1981; 4:69–125.
24. Didly JE, Leslie SW. Are changes in neuronal calcium channels involved in ethanol tolerance? J Pharmacol Exp Ther 1989; 250:985–991.
25. Ross DH. Adaptive changes in Ca^{2+} membrane interactions following chronic exposure to ethanol. In: Gross MM, ed. Alcohol Intoxication and Withdrawal. New York: Plenum Press; 1987:459.
26. Leslie SW, Little HJ. Calcium-channel interaction with ethanol and other sedative-hypnotic drugs. Recent Dev Alcoholism 1987; 5:285–302.
27. Rezvani AH, Crovi SI, Mack CM, Myers RD. Central Ca^{2+} channel blockade reverses ethanol-induced poikilothermia in the rat. Alcohol 1986; 3:273–279.
28. Rezvani AH, Mack CM, De Lacy P, Janovsky D. Verapamil effects on physiological and behavioral responses to ethanol in the rat. Alcohol and Alcoholism 1990; 25:51–58.

29. Ferry DK, Glossman H. Evidence for multiple receptor sites within the putative calcium channels. Naunyn-Schmiedeberg's Arch Pharmacol 1982; 321:80–83.
30. Kostowski W, Pucilowski O. Central action of calcium channel inhibitors: Potential therapeutic uses in psychiatry. New Trends Exp Clin Psychiat 1989; 5:187–196.
31. Miller RJ. Multiple calcium channels and neuronal function. Science 1987; 235:46–52.
32. Dolin SJ, Little HJ, Hudspisth M, Pagonis C, Littleton JM. Increased dihydropyridine calcium channels in rats brain may underlie ethanol physical dependence. Neuropharmacology 1987; 26:270–275.
33. Littleton JM, Little HJ. Dihydropyridine sensitive Ca^{2+} channels in brain are involved in the central nervous system hyperexcitability associated with alcohol withdrawal states. Ann NY Acad Sci 1988; 522:199–202.
34. Little HJ, Dolin SJ, Halsey MJ. Calcium channel antagonists decrease the ethanol withdrawal syndrome. Life Sci 1986; 39:2059–2065.
35. Pucilowski O, Krzascik P, Trzaskowska E, Kostowski W. Different effect of diltiazem and nifedipine on some central actions of ethanol in the rat. Alcohol 1989; 6:165–168.
36. Carlen PL, Wilkinson DA. Alcohol-induced brain damage: confounding variables. Alcohol and Alcoholism 1987; Suppl 1:37–41.
37. Hescheler J, Peltzer D, Traube G, Trautwein W. Does the organic calcium blocker D-600 act from inside or outside on the cardiac membrane? Pfluegers Arch 1982; 393:326–330.
38. Engel JA, Fahlke C, Hulthe P, Hard E, Johanessen K, Snape B, Svensson L. Biochemical and behavioral evidence for an interaction between ethanol and calcium channel antagonists. J Neural Transm 1988; 74:181–193.
39. Isaacson RL, Molina JC, Draski LJ, Johnston JE. Nimodipine's interactions with other drugs: Ethanol. Life Sci 1985; 36:2195–2199.
40. McCreery MJ, Hunt WA. Physico-chemical correlates of alcohol intoxication. Neuropharmacology 1978; 17:451–461.
41. Pucilowski O, Kostowski W. Increased anxiety during ethanol and diazepam withdrawal in rats: Effects of diltiazem and nicardipine. Alcohol 1991; 15:331.
42. Pucilowski O, Kostowski W. Diltiazem suppresses apomorphine-induced fighting and pro-aggressive effect of withdrawal from chronic ethanol or haloperidol in rats. Neurosci Lett 1988; 93:96–100.
43. Gilliani D, Isaacson RL, Burright RG, Johnston J, Fahey J. Nimodipine's effect on alcohol disposition in mice. Alcohol 1988; 5:259–261.
44. Blum K, Noble EP, Sheriden PJ, Montgomery A, Ritchie T, Jageeeswaran P, Nogami H, Briggs A, Cohn JB. Allelic association of human dopamine D-2 receptors gene in alcoholism. JAMA 1990; 263:2055–2060.
45. Swann A. Chronic ethanol and $(Na^+ + K^+)$ adenosine triphosphatase apparent adaptation in cation binding and enzyme conformation. J Pharmacol Exp Ther 1985; 232:275–479.
46. Wong DT, Murphy JM. Serotonergic mechanisms in alcohol intake. In: Sun GH, Rudeen PK, Wei YH, Sum AY, eds. Molecular Mechanisms of Alcohol. New York: Humana Press; 1989:133–146.

47. Fadda F, Gessa GL, Mosca E, Stefanini E. Differential effects of the calcium antagonists nimodipine and flunnarizine on dopamine metabolism in the rat brain. J Neural Transm 1979; 75:195–200.
48. Kostowski W, Krzascik P, Puciłowski O. Effect of calcium channel inhibitors on D-1 receptor mediated responses: SKF 38393-induced grooming and SCH 23390-induced catalepsy in rats. Biogenic Amines 1990; 7:49–44.
49. Brown NL, Sirugua O, Worcell M. The effects of some slow channel blocking drugs on high affinity uptake by rat brain synaptosomes. Eur J Pharmacol 1986; 123:161–165.
50. Rehavi M, Carmi R, Weizman A. Tricyclic antidepressant and calcium channel blockers: Interaction at the (−) desmethoxyverapamil binding site and the serotonin transporter. Eur J Pharmacol 1988; 155:1–9.
51. Fadda F, Mosca E, Colombo G, Gessa GL. Alcohol preferring rats: Genetic sensitivity to alcohol-induced stimulation of dopamine metabolism. Physiol Behav 1990; 47:727–729.
52. Puciłowski O, Rezvani AH, Janowsky DS. Suppression of alcohol and saccharin preference in rats by a novel Ca^{2+} channel inhibitor, Goe 5438. Psychopharmacology 1992; 107:447–452.
53. Rezvani AH, Janovsky D. Decreased ethanol consumption by verapamil in alcohol preferring rats. Prog Neuropsychopharmacol Biol Psychiat 1990; 14:623–631.
54. Rezvani AH, Grady DR, Janovsky D. Effect of calcium channel blockers on alcohol consumption in alcohol-drinking monkeys. Alcohol and Alcoholism 1991; 26:161–167.
55. Rezvani AH, Puciłowski O, Janovsky D. Effects of different Ca^{2+} channel antagonists on alcohol preference in alcohol preferring rats. Alcohol Clin Exp Res 1991; 15:314.
56. Westcott JY, Weiner H. Effect of ethanol on synaptosomal ($Na^+ + K^+$)-ATPase in control and ethanol-dependent rats. Arch Biochem Biophys 1983; 223:51–57.
57. Kalant H, Woo N, Endreny L. Effect of ethanol on the kinetics of rat brain ($Na^+ + K^+$)-ATPase and K^+ dependent phosphatases with different alkali ions. Biochem Pharmacol 1978; 27:1353–1358.
58. Rangaraj N, Kalant H. Interaction of ethanol and catecholamines on rat brain ($Na^+ + K^+$)-ATPase. Can J Physiol Pharmacol 1979; 57:1098–1106.
59. Rangaraj N, Kalant H. Effect of chronic ethanol treatment on temperature dependence and on norepinephrine-sensitization of rat brain ($Na^+ + K^+$) adenosine triphosphatase. J Pharmacol Exp Ther 1982; 223:536–539.
60. Nhamburo PT, Salafsky BI, Hoffman PL, Tabakoff B. Effects of short chain alcohols and norepinephrice on brain ($Na^+ + K^+$) ATPase activity. Biochem Pharmacol 1986; 12:1987–1992.
61. Stibler H, Beaugé F, Borg S. Changes in ($Na^+ + K^+$)-ATPase activity and the composition of surface carbohydrates in erythrocyte membranes in alcoholics. Alcohol Clin Exp Res 1984; 8:522–527.
62. Topel H. Biochemical basis of alcoholism: Statements and hypotheses of present research. Alcohol 1985; 2:711–788.
63. Shanley B, Gurd J, Kalant H. Ethanol tolerance and enhanced calcium-calmodulin-dependent phosphorylation of synaptic membrane proteins. Neurosci Lett 1985; 58:55–59.

64. Tippe A. The effect of n-alkanols on the stationary current voltage behavior and action potential of myelinated nerve. Biochim Biophys Acta 1980; 598:200–205.
65. Levental M, Tabakoff B. Sodium-potassium activated ATPase as a measure of neuronal membrane characteristics in ethanol-tolerant mice. J Pharmacol Exp Ther 1980; 212:316–319.
66. Tabakoff B. Alcohol tolerance in humans and animals. In: Eriksson K, Sinclair JD, Kiianmaa K, eds. Animal Models in Alcohol Research. London: Academic Press; 1980:271–292.
67. Stone TW, Burton NR. NMDA receptors and ligands in the vertebrate CNS. Prog Neurobiol 1988; 30:330–368.
68. Harris RA, Bruno P. Effect of ethanol and other intoxicant anaesthetics on voltage-dependent sodium channels of brain synaptosomes. J Pharmacol Exp Ther 1985; 232:401–406.
69. Harris RA. Differential effects of membrane perturbants on voltage-activated sodium and calcium-dependent potassium channels. Biophys J 1984: 45:132–134.
70. Moore JW, Ulbricht W, Takata W. Effect of ethanol on the sodium and potassium conductance of the squid axon membrane. J Gen Physiol 1964; 48:279–295.
71. Mullin MJ, Hunt WA. Ethanol and pentobarbital inhibit veratrine-stimulated sodium uptake in synaptosomes. Life Sci 1984; 34:287–292.
72. Tamkun MM, Catterall WA. Ion flux studies of voltage sensitive sodium channels in synaptic nerve-ending particles. Mol Pharmacol 1981; 19:78–86.
73. Lovinger DM, White G, Weight FF. Ethanol inhibits NMDA-activated ion current in the hippocampal neurons. Science 1989; 243:1721–1724.
74. Crunelli V, Forda S, Kelly JS. Excitatory aminoacids in the hippocampus: Synaptic physiology and pharmacology. Trends Neurosci 1985; 8:26–30.
75. Collingridge GL. LTP in the hippocampus mechanisms of initiation and modulation by neurotransmitters. Trends Pharmacol Sci 1985; 6:407–411.
76. McBurney RN, Barker JL. GABA-induced conductance fluctuations in cultured spinal neurons. Nature (London) 1978; 274:596–597.
77. Liljequist S, Engel J. Effects of GABA agonist and antagonists on various ethanol-induced behavioral changes. Psychopharmacology 1982; 78:71–75.
78. Davidoff RA. Alcohol and presynaptic inhibition in an isolated spinal cord preparation. Arch Neurol 1973; 28:60–63.
79. Suzdak PD, Schwarz RD, Skolnick P, Paul SM. Ethanol stimulates gamma-aminobutyric acid receptor-mediated chloride transport in rat brain synaptoneurosomes. Proc Natl Acad Sci USA 1986; 83:4071–4075.
80. Ticku MK. Benzodiazepine-GABA receptor ionophore complex: Current concept. Neuropharmacology 1983; 22:1459–1470.
81. Ticku KM, Lowrimore P, Lehoullier P. Ethanol enhances GABA-induced ^{36}Cl-influx in primary spinal cord cultured neurons. Brain Res Bull 1986; 17:123–126.
82. Celentano JJ, Gibbs TT, Farb DH. Ethanol potentiates GABA and glycine-induced chloride currents in chick spinal cord neurons. Brain Res 1988; 445:377–380

83. Korpi ER, Uusi-Oukari M. GABA-A receptor-mediated chloride flux in brain homogenates from rat lines with innate alcohol sensitivities. Neurosci 1989; 32:387–392.
84. Suzdak P, Schwartz RD, Skolnick P, Paul SM. Alcohols stimulate gamma-aminobutyric acid receptor-mediated chloride uptake in brain vesicles: Correlation with intoxication potency. Brain Res 1988; 444:340–345.
85. Volicer L, Biagioni TM. Effect of ethanol administration and withdrawal on GABA-receptor binding in rat cerebral cortex. Subst Alcohol Action Misuse 1982; 3:31–39.

21
Structure and Function of Receptor-Mediated Chloride Channels in the Central Nervous System

KINYA KURIYAMA and MASAAKI HIROUCHI

The central nervous system (CNS) Cl^- channels are typical anion channels, which can be classified as receptor-mediated, voltage-sensitive, and calcium-dependent Cl^- channels.[1] The receptors for GABA and glycine, major inhibitory neurotransmitters in the CNS, contain receptor-mediated Cl^- channels that open upon activation of the receptors, inducing the transduction of an inhibitory signal such as an inhibitory postsynaptic potential (IPSP).

This chapter will discuss recent studies on the molecular nature of these receptor-mediated Cl^- channels and their possible role in the action of hypnotic and anxiolytic drugs.

The GABA$_A$ Receptor-Mediated Cl^- Channel

In general, neurotransmitter receptors have been classified as ionotropic or metabotropic.[1] One group of ionotropic receptors, such as the nicotinic acetylcholine receptor, the N-methyl-D-aspartate (NMDA) and non-NMDA glutamate receptors, and the 5-hydroxytryptamine type$_3$ receptor, form cationic channels (such as the Na^+ and Ca^{2+} channels) and induce excitatory synaptic transmission; another group of receptors, such as the GABA$_A$/benzodiazepine (BZP) and glycine receptors; form anionic channels, eg, the Cl^- channel, and induce inhibitory synaptic transmission by enhancing their permeability.[2]

BZP and barbiturate derivatives are typical drugs that act on these channels.[3,4] Thus, the binding of BZP derivatives activates the GABA receptor and exerts an anxiolytic and/or hypnotic action by potentiating the inhibitory GABAergic pathways. In contrast, β-carboline derivatives, an inverse agonist of the benzodiazepine receptor, inhibits GABA receptor function and induces anxiety (Figure 21.1).[5] GABA receptor function may be also activated by barbiturates, which have an independent binding sites on the Cl^- channel, or may be inactivated by picrotoxin, a CNS stimulant, which in this way induces tonic-clonic

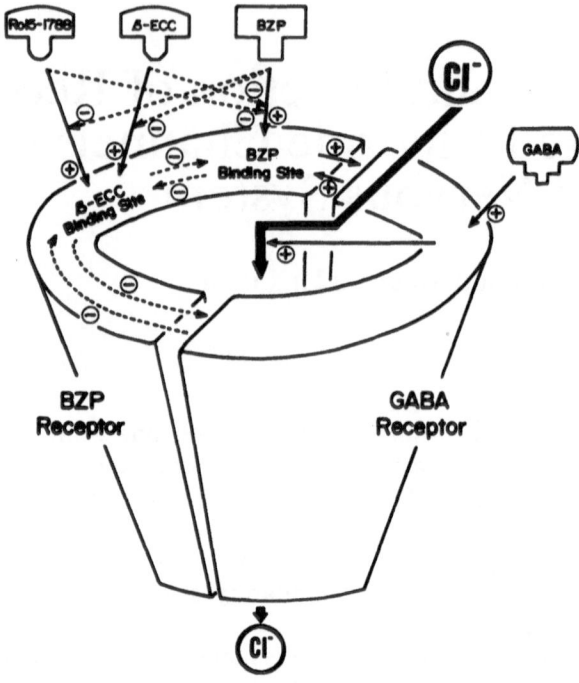

FIGURE 21.1. Hypothetical model for structural components and function of a GABA$_A$ receptor/BZP receptor/Cl$^-$ channel complex.[5]

convulsions. Recently it has been reported that neurosteroid, one of the metabolites of steroid hormones, such as 3α-hydroxy-5-α-dihydroprogesterone, also has binding sites on the Cl$^-$ channel of the GABA$_A$ receptor, which suggests that it may act as an endogenous modulator of stress-induced changes in neuronal function and/or anxiety.[6,7] These are typical examples of how the function of the GABA$_A$ receptor/BZP receptor/Cl$^-$ channel complex may be modulated in the brain.[8]

Recent studies[9] have demonstrated that the GABA receptor consists of five heterogeneous subunits (α, β, γ, δ, ρ; Table 21.1), all of which possess four transmembrane domains in the primary structure (Figure 21.2)[10] similar to those that characterize a superfamily of ionotropic receptors and essential to the formation of the channel. Another characteristic of these GABA$_A$ receptor subunits is the presence of a signal sequence in the N-terminal domain. In addition, there are target sites for phosphorylation by protein kinase A and protein kinase C in the large cytoplasmic loop between TM3 and TM4, and N-glycosylation sites in the extracellular domain.

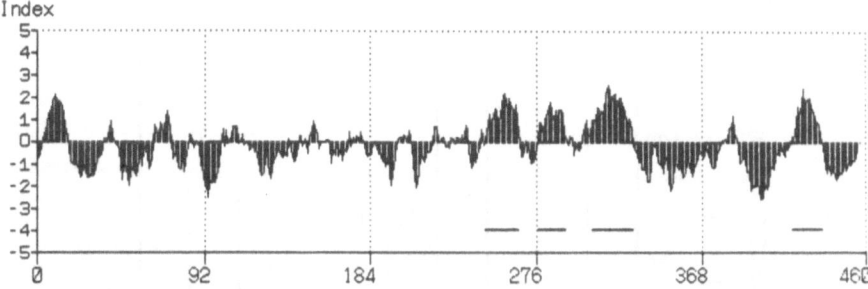

FIGURE 21.2. Hydrophobicity profile of $GABA_A$ receptor α_1 subunit from human brain.[10] Solid bars indicate the positions of the putative membrane-spanning domains.

To clarify the functional roles of $GABA_A$ receptor subunits, the expression of each subunit and of subunit combinations has been examined in *Xenopus* oocytes or in transfected cells. These studies have demonstrated that the formation of a functioning GABA-gated Cl^- channel requires the expression of both the α and the β subunit,[11] and that the expression of the BZP receptor requires the combination of three subunits, α, β, and γ.[12] Although the function of the δ subunits is not clear, it has been suggested that it may be a constituent of the BZP-insensitive GABA receptor.[13] The ρ subunit is abundantly expressed in the retina.[14]

The heterogeneity of α and β subunits seems to be important in establishing the response of the $GABA_A$ receptor/BZP receptor/Cl^- channel complex to drugs. Thus different subunit combinations (eg, α_1–α_6 or β_1–β_4) result in different affinities of the receptor for GABA.[15,16] Heterogeneity is also the basis for the classification of the cerebral BZP receptor into two subtypes with different affinity for ligands such as CL218,872. Thus a combination of α_1, β_1, γ_2 subunits determines the property of the type I central BZP receptor,[17] while a combination of α_2–α_5 subunits determines that of type II.[18] Moreover, it has been found that a combination of α_6, β, and γ subunits produces a BZP receptor having a selective affinity for a behavioral alcohol antagonist, Ro15-

TABLE 21.1. Multiplicity of $GABA_A$ receptor subunits.*

α_1, α_2, α_3, α_4, α_5, α_6
β_1, β_2, β_3, β_4, β_4'
γ_1, γ_2, γ_2', γ_3
δ
ρ

* β_4' and γ_2' are derived from the alternative splicing.

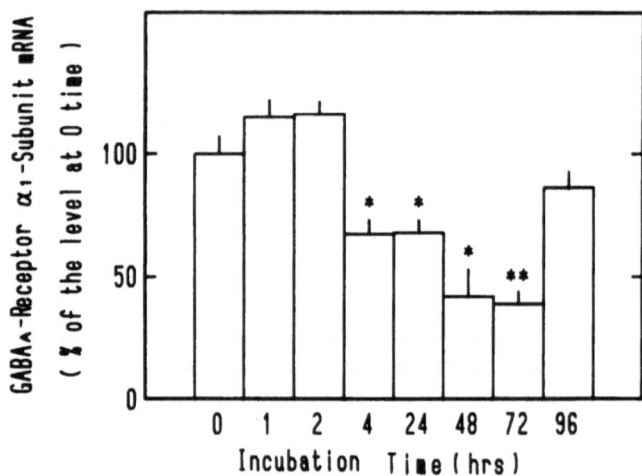

FIGURE 21.3. Time course of changes in the level of GABA$_A$ receptor α_1 subunit mRNA in primary cultured neurons following the treatment with muscimol.[25] *p < 0.05, **p < 0.01, compared with the value obtained at 0 time.

4513.[19] These results indicate that variations in the α subunit are important in determining the affinity of the receptor for GABA as well as the functional properties of Cl$^-$ channels.[20]

It is known that GABA$_A$ receptor function in the brain is regulated by several mechanisms. Thus phosphorylation inhibits the influx of Cl$^-$ ions through the GABA-gated Cl$^-$ channel,[21] a phenomenon probably related to densitization or sequestration of the receptor. Indeed, it has been demonstrated that several subunits of the purified GABA receptor complex can be phosphorylated by protein kinase C and/or by cAMP-dependent protein kinase.[22,23] Furthermore, it has been shown that the prolonged stimulation of the GABA$_A$ receptor results in receptor down-regulation.[24] It is believed that this down-regulation is due to a decreased expression of mRNA. Indeed, in primary cultures of mouse cortical neurons treated with GABA receptor agonists such as muscimol and flunizatrepan, an actual reduction of GABA$_A$ receptor α_1-subunit mRNA can be detected after 4 h of exposure that continues for as long as 72 h (Figure 21.3).[25] The action of muscimol has been found to be

TABLE 21.2. Multiplicity of glycine receptor subunits.*

$\alpha_1, \alpha_1', \alpha_2, \alpha_2'$
β

* α_1' and α_2' are derived from the alternative splicing.

dose-dependent and is antagonized by bicuculline, a $GABA_A$-receptor antagonist. On the other hand, continuous activation of the NMDA receptor, one of the excitatory glutamate receptors, is known to enhance the expression of mRNA for $GABA_A$ receptor subunits.[26]

Glycine Receptor-Mediated Cl^- Channels

Glycine is also known to have an inhibitory action in the CNS similar to that of GABA, although it acts mainly in the spinal cord. This inhibitory action results from the activation of the glycine receptor, which opens the glycine-gated Cl^- channel.[27] Selective ligands for the glycine receptor are few. Among them are β-alanine and taurine, which have low affinities, and the ligands of GABA and BZP receptors. Strychnine and its derivatives are known as selective antagonists of the glycine receptor-mediated Cl^- channel.

The glycine receptor forms the Cl^- channel pore and is known to be constituted by a number of subunits.[28-33] The receptor has four transmembrane domains and structural characteristics are similar to those of the $GABA_A$ receptor. Although the purified glycine receptor is composed of α and β subunits, the injection of mRNA for either subunit into *Xenopus* oocytes results in the expression of receptors with normal channel properties.[34] Therefore it has been suggested that the glycine receptor may be a homo-oligomer, in contrast to the $GABA_A$ receptor, which is know to have a hetero-oligomeric structure.

The expression of the α_1 and α_2 subunits of the glycine receptor in the CNS is known to be uneven.[35] Thus the receptor α_1 subunit is abundant in the spinal cord of the adult, while the α_2 subunit appears in the spinal cord and cerebral tissues, especially at the neonatal stage. In addition, two variants of α_1 and α_2 subunits can be produced by alternative splicing (Table 21.2).[32,33] These results suggest that the two subunits of the glycine receptor may have different physiologic roles and that the full activity of the native glycine receptors in the CNS may be the result of the combined activities of its subunits.

Conclusions

Recent studies on the receptor-mediated Cl^- channels have been reviewed. The function of the CNS is known to be maintained by a balance of excitatory and inhibitory neurons. GABA and glycine are well-known inhibitory neurotransmitters. The activation of their receptors induces the opening of the receptor-mediated Cl^- channels, as well as the appearance of inhibitory postsynaptic potentials. The variety of receptor-mediated Cl^- channels may be essential for the fine regulation of signal transduction within the inhibitory neuronal network. Although studies on the structure and function of receptor-mediated Cl^- channels in the CNS are advancing rapidly, further research on their roles in pathophysiology are urgently needed.

References

1. Eldefrawi AT, Eldefrawi ME. Receptors for γ-aminobutyric acid and voltage-dependent chloride channels as targets for drugs and toxicants. FASEB J 1987; 1:262–271.
2. Strange PG. The structure and mechanism of neurotransmitter receptors. Biochem J 1988; 249:309–318.
3. Braestrup C, Nielsen M, Homore T, Jensen LH, Petersen EN. Benzodiazepine receptor ligands with positive and negative efficacy. Neuropharmacology 1983; 22:1451–1458.
4. Ticku MK. Benzodiazepine-GABA receptor-ionophore complex: Current concepts. Neuropharmacology 1983; 22:1459–1470.
5. Taguchi J, Kuriyama K. Functional modulation of cerebral γ-aminobutyric $acid_A$ receptor/benzodiazepine receptor/chloride ion channel complex with ethyl β-carboline-3-carboxylate: Presence of independent binding site for ethyl β-carboline-3-carboxylate. J Pharmacol Exp Ther 1990; 253:558–566.
6. Puia G, Santi M, Vicini S, Pritchett DB, Purdy RH, Paul SM, Seeburg PH, Costa E. Neurosteroids act on recombinant human $GABA_A$ receptors. Neuron 1990; 4:759–765.
7. Majewska MD, Demirgoren S, Spivak CE, London ED. The neurosteroid dehydroepiandrosterone sulfate is an allosteric antagonist of the $GABA_A$ receptor. Brain Res 1990; 526:143–146.
8. Schumacher M, McEwen BS. Steroid and barbiturate modulation of the $GABA_a$ receptor: Possible mechanisms. Mol Neurobiol 1989; 3:275–304.
9. Burt DR, Kamatchi GL. $GABA_A$ receptor subtypes: From pharmacology to molecular biology. FASEB J 1991; 5:2916–2923.
10. Kuriyama K, Hirouchi M. Structure and function of γ-aminobutyric acid (GABA) receptor: Current state and prospectives. Folia Pharmacol Japon 1989; 94:7–15.
11. Schofield PR, Darlison MG, Fujita N, Burt DR, Stephenson FA, Rodriguez H, Rhee LM, Ramachandran J, Reale V, Glencorse TA, Seeburg PH, Barnard EA. Sequence and functional expression of the $GABA_A$ receptor shows a ligand-gated receptor super-family. Nature 1987; 328:221–227.
12. Pritchett DB, Sontheimer H, Shivers BD, Ymer S, Kettenmann H, Schofield PR, Seeburg PH. Importance of a novel $GABA_A$ receptor subunit for benzodiazepine pharmacology. Nature 1989; 338:582–585.
13. Shivers BD, Killisch I, Sprengel R, Sontheimer H, Kohler M, Schofield PR, Seeburg PH. Two novel $GABA_A$ receptor subunits exist in distinct neuronal subpopulations. Neuron 1989; 3:327–337.
14. Cutting GR, Lu L, O'Hara BF, Kasch LM, Montrose-Rafizadeh C, Donovan DM, Shimada S, Antonarakis SE, Guggino WB, Uhl GR, Kazazian HH. Jr Cloning of the γ-aminobutyric acid (GABA) ρ_1 cDNA: A GABA receptor subunit highly expressed in the retina. Proc Natl Acad Sci USA 1991; 88:2673–2677.

15. Levitan ES, Schofield PR, Burt DR, Rhee LM, Wisden W, Kohler M, Fujita N, Rodriguez HF, Stephenson A, Darlison MG, Barnard EA, Seeburg PH. Structure and functional basis for GABA$_A$ receptor heterogeneity. Nature 1988; 335:76–79.

16. Ymer S, Schofield PR, Draguhn A, Werner P, Kohler M, Seeburg PH. GABA$_A$ receptor β subunit heterogeneity: Functional expression of cloned cDNAs. EMBO J 1989; 8:1665–1670.

17. Pritchett DB, Luddens H, Seeburg PH. Type I and type II GABA$_A$/benzodiazepine receptors produced in transfected cells. Science 1989; 245:1389–1392.

18. Pritchett DB, Seeburg PH. γ-aminobutyric acid$_A$ receptor α$_5$-subunit creates novel type II benzodiazepine receptor pharmacology. J Neurochem 1990; 54:1802–1804.

19. Luddens H, Pritchett DB, Kohler M, Killisch I, Keinanen K, Monyer H, Sprengel R, Seeburg PH. Cerebellar GABA$_A$ receptor selective for a behavioral alcohol antagonist. Nature 1990; 346:648–651.

20. Sigel E, Baur R, Trube G, Mohler H, Malherbe P. The effect of subunit composition of rat GABA$_A$ receptors on channel function. Neuron 1990; 5:703–711.

21. Heuschneider G, Schwartz RD. cAMP and forskolin decrease γ-aminobutyric acid-gated chloride flux in rat brain synaptoneurosomes. Proc Natl Acad Sci USA 1989; 86:2938–2942.

22. Kirkness EF, Bovenkerk CF, Ueda T, Turner AJ. Phosphorylation of γ-aminobutyrate (GABA)/benzodiazepine receptors by cyclic AMP-dependent protein kinase. Biochem J 1989; 259:613–616.

23. Browning MD, Bereau M, Dudek EM, Olsen RW. Protein kinase C and cAMP-dependent protein kinase phosphorylate the β-subunit of the purified γ-aminobutyric acid A receptor. Proc Natl Acad Sci USA 1990; 87:1315–1318.

24. Roca DJ, Rozenberg I, Farrant M, Farb DH. Chronic agonist exposure induces down-regulation and allosteric uncoupling of the γ-aminobutyric acid/benzodiazepine receptor complex. Mol Pharmacol 1990; 37:37–43.

25. Hirouchi M, Ohkuma S, Kuriyama K. Muscimol-induced reduction of GABA$_A$ receptor α$_1$-subunit mRNA in primary cultured neuron. Mol Brain Res 1992; 15:327–331.

26. Memo M, Bovolin P, Costa E, Grayson DR. Regulation of γ-aminobutyric acid$_A$ receptor subunit expression by activation of N-methyl-D-aspartate-sensitive glutamate receptors. Mol Pharmacol 1991; 39:599–603.

27. Betz H, Becker CM. The mammalian glycine receptor: Biology and structure of a neuronal chloride channel protein. Neurochem Int 1988; 13:137–146.

28. Langosch D, Thomas L, Betz H. Concerved quaternary structure of ligand-gated ion channels: The postsynaptic glycine receptor is a pentamer. Proc Natl Acad Sci USA 1988; 85:7394–7398.

29. Grenningloh G, Rienitz A, Schmitt B, Methfessel C, Zensen M, Beyreuther K, Gundelfinger ED, Betz H. The strychnine-binding subunit of the glycine receptor shows homology with nicotinic acetylcholine receptors. Nature 1987; 328:215–220.

30. Grenningloh G, Schmieden V, Schofield PR, Seeburg PH, Siddique T, Mohandas TK, Becker CM, Betz H. Alpha subunit variants of the human

glycine receptor: Primary structures, functional expression and chromosomal localization of the corresponding genes. EMBO J 1990; 9:771–776.

31. Grenningloh G, Pribilla I, Prior P, Multhaup G, Beyreuther K, Taleb O, Betz H. Cloning and expression of the 58 kd β subunit of the inhibitory glycine receptor. Neuron 1990; 4:963–970.

32. Malosio ML, Grenningloh G, Kuhse J, Schmieden V, Schmitt B, Prior P, Betz H. Alternative splicing generates two variants of the α_1 subunit of the inhibitory glycine receptor. J Biol Chem 1991; 266:2048–2053.

33. Kuhse J, Kuryatov A, Maulet Y, Malosio ML, Schmieden V, Betz H. Alternative splicing generates two isoforms of the α subunit of the inhibitory glycine receptor. FEBS Lett 1991; 283:73–77.

34. Schmiden V, Grenningloh G, Schofield PR, Betz H. Functional expression in *Xenopus* oocytes of the strychnine binding 48 kd subunit of the glycine receptor. EMBO J 1989; 8:695–700.

35. Akagi H, Hirai K, Hishinuma F. Cloning of a glycine receptor subtype expressed in rat brain and spinal cord during a specific period of neuronal development. FEBS Lett 1991; 281:160–166.

22
Characterization of Ion Channels in the Central Nervous System: Insights from Radioligand Binding, Autoradiography, and In Situ Hybridization Histochemistry

Andrew L. Gundlach

Ion channels are molecules that form pores in the membrane to allow ion flow and several classes can be distinguished on the basis of their physiological properties.[1] In turn, within each class of ion channel that exists in mammalian systems there is a large diversity of subtypes. I shall present some examples of scientific research, using the techniques of radioligand binding, in vitro receptor autoradiography, and more recently the technique of in situ hybridization histochemistry, which has increased our understanding of ion channel/receptor properties. In the process, I hope to illustrate the uses and advantages of these techniques, the resolution they provide, and their compatability with each other. It should be apparent, therefore, how the application of these techniques has allowed scientists to determine the existence of ion channel/receptor protein complexes and their subtypes, allosteric interactions between different types of ion channel-related receptor molecules, the cellular localization and ontogenic development of ion channel/receptors, and the effect of experimental and pathologic changes in neuronal function on various ion channels. Finally, future directions that research in these areas may take will be discussed along with the challenges that the ever-increasing diversity of ion channels present for neurochemists and neuropharmacologists.

Due to the large number of ion channels and ion channel subtypes that are now known to exist, this review is not intended to be a comprehensive coverage, but it is hoped that the examples cited will adequately convey the possibilities for studies of known or novel classes of channels. Similarly, for reasons of space and brevity, the bibliography is not intended to be comprehensive. The interested reader should consult some of the many more specific and comprehensive reviews.[1-7]

Classes of Ion Channels and Their Roles in the CNS

Ion channels involved in the generation of electrical signals can be grouped into two classes. *Ligand-gated ion channels*, such as the nicotinic acetylcholine receptor, the γ-aminobutyric acid (GABA) receptor, and the glycine receptor, mediate local increases in ion conductance at chemical synapses and thereby depolarize or hyperpolarize the subsynaptic area of the cell.[1,8] In contrast, *voltage-sensitive ion channels* mediate rapid, voltage-gated changes in ion permeability during action potentials in excitable cells.[2]

In most excitable cells, at resting membrane potential excitation is usually generated by Na^+ ion influx, whereas inhibition of neuronal firing results from increased permeability to Cl^- or K^+ ions. In excitable cells, including neurons, the action potential consists of three phases: first a rapid increase in Na^+ permeability mediated by *voltage-sensitive Na^+ channels* causes rapid depolarization during the initial phase. The cell remains depolarized during the plateau phase of the action potential because of the inward movement of Ca^{2+} ions through *voltage-sensitive Ca^{2+} channels*. The action potential is terminated by activation of *voltage-sensitive K^+ channels* that mediate outward movement of K^+ ions, which repolarize the cell. Voltage-sensitive K^+ channels also set the resting membrane potential of the cell and modulate action potential frequency and threshold.[2]

Techniques for the Detection of Ion Channels in the CNS

Our understanding of ion channels has improved dramatically as a result of powerful new techniques such as patch clamping and molecular gene cloning, which have allowed single channel behavior to be monitored, protein amino acid sequences to be derived, and channel function to be related to specific parts of their protein structure. In general, these more recent biochemical and molecular findings confirm earlier indications from pharmacologic studies of a marked diversity of voltage-sensitive and ligand-gated ion channel proteins in the central nervous system and other tissues. Methods similar to those used in the study of neurotransmitter receptors have been used to investigate the biochemical and pharmacologic characteristics of the various types of ion channel in the brain.

The discovery and development of suitable ligands (drugs with high affinity for a given channel or receptor associated with an ion channel) have provided a great deal of information on the characteristics of many types of ion channels. Indeed, a critical first step toward understanding ion channel/receptor function is the identification of tissues, regions, cells, and anatomical systems in which the ion channel/receptor complex(es) exist. This is particularly important in the area of brain research, as cellular interactions are especially complex and their understanding requires the use of techniques with a high degree of anatomical and cellular resolution, such as those based on the use of radiolabeled molecules.

Such experiments using tissue homogenates and different concentrations of radioligand or unlabeled drugs and/or different ions and different incubation times allows the characterization of ion channel-related binding sites, and provides measures of the affinity of the labeled and unlabeled ligands, the density of the binding sites in the tissue, the ionic sensitivity of binding, and the association and dissociation rate constants of binding.

Autoradiography can visualize these binding sites by apposing radiolabeled tissue sections mounted on microscope slides to dry photographic emulsion. The technique is simple and versatile, and under appropriate circumstances it increases the resolution of classical binding techniques to the level of light microscopy, providing valuable anatomic information. Thus autoradiography has been instrumental in identifying subtle differences in the pharmacology of many ion channel/receptor complexes and contributed considerably to the understanding of the multiplicity of ion channel subtypes. Following the development of more selective ligands, autoradiographic techniques allow the identification of brain regions where ion channel subtypes are concentrated. Ion channel receptor autoradiography, however, has limitations. For example, it cannot be used to identify those sites where the ion channel/receptors are synthesized.

The cloning and sequencing of genes or cDNAs coding for ion channel/ receptors means that it is now possible to synthesize nucleic acid probes that allow the detection and visualization of the mRNA coding for these ion channels in tissue sections. This approach, known as in situ hybridization histochemistry, is particularly valuable for studying the distribution of different ion channel subtypes, since very often no specific ligands or antibodies are available for these subtypes. The combined use of ion channel/receptor autoradiography and in situ hybridization histochemistry is particularly useful. Indeed, the visualization of mRNA and ion channel/receptor protein in consecutive brain sections produces results that are both amenable to image analysis and quantitation. Thus such studies have allowed the visualization of the neuronal cell bodies that synthesize ion channel mRNAs, while radioligand autoradiography has revealed the final localization of the ion channel/receptor protein.

Studies of Ion Channel/Receptor Recognition Sites Using Radioligands

Radioligands have been used to successfully identify most of the known classes of ion channels, including both ligand-gated and voltage-operated channels. The following are descriptions of some examples of studies completed for Na^+-, Ca^{2+}-, and K^+-voltage-operated ion channels.

Na$^+$ Channels

Biochemical studies of voltage-sensitive Na^+ channels have utilized specific neurotoxins (that act at several different receptor sites) as molecular probes of channel structure and function. For instance, protein components of Na^+ channels were first identified by photoaffinity labeling with photoreactive derivatives of α-scorpion toxins.[9,10] Na^+ channels have been solubilized from excitable membranes by treatment with non-ionic detergents, detected in solubilized form by high-affinity binding of radiolabeled tetrodotoxin or saxotoxin at their receptor site, and subsequently purified.[11] Functional, purified preparations have also been obtained from human brain. The overall arrangement of α, β_1, and β_2 subunits of the channel protein in the membrane have been inferred from biochemical experiments. The α and β_1 subunits are covalently labeled by neurotoxins that act outside the cell and all three subunits are glycosylated, which indicates that all are exposed to the extracellular surface. Such has been the progress in this area that sufficient information is now available regarding the biochemical properties of the Na^+ channel subunits to formulate models for their disposition in the membrane (eg, Catterall[2]).

Ca^{2+} Channels

There are multiple classes of Ca^{2+} channels in most excitable cells, and there is general agreement that neuronal cells can coexpress different subtypes. The best characterized of these is the high-threshold, slowly-activating, fast-deactivating, voltage-operated Ca^{2+} channels (VOCCs), often subdivided into L, N, and P subtypes. L-type Ca^{2+} channels that mediate long-lasting Ca^{2+} currents are inhibited by three distinct classes of organic Ca^{2+} channel antagonists: dihydropyridines (DHPs), such as nifedipine; phenylalkylamines, such as verapamil; and benzothiazepines, such as diltiazem.[12] These structurally different agents act at three separate allosterically-linked receptor sites, which are reminiscent of the multiple, interacting, neurotoxin receptor sites described for Na^+ channels. In addition, structurally related DHPs that are Ca^{2+}-channel activators (such as Bay K8644) bind at the same DHP receptor site and favor prolonged activation of the Ca^{2+} channel. These Ca^{2+} channel modulators bind with high affinity and have been used as molecular probes to identify and isolate the protein components of DHP-sensitive Ca^{2+} channels (L-type) in the same way that neurotoxins have been used to identify and isolate Na^+ channels (see Catterall[2]).

DHP-Labeled Ca^{2+} Channels

Radiolabeled DHPs, such as [³H]-nitrendipine and [³H]-nimodipine, were the first ligands used to label calcium antagonist receptors, including those present in brain.[13,14] These [³H]-DHPs bind in a specific, saturable, stereoselective, and reversible manner, and receptors for calcium antagonists also show high affinity for various other DHPs. Dissociation constant (K_d) values for the various labeled DHPs are in the range 0.02–0.5 nM in both brain and other tissues (eg, Lee et al.[15]). Linear Scatchard plots derived from binding data imply that a single class of DHP sites exist in most tissues. A variety of evidence suggests that DHP receptors are closely linked to VOCCs. Binding stereoselectivity and order of potency closely parallel calcium flux and transmitter release studies in a variety of cell cultures. DHP ligands and able to discriminate between agonist and antagonist cations.[16] Thus if membranes are treated with a chelator such as EDTA, the number of DHP receptors is greatly reduced. Binding can be restored by the readdition of Ca^{2+}. Binding is facilitated by strontium and barium, which carry current through the channel, but not by cadmium and cobalt, which block Ca^{2+} channels. A variety of calcium antagonists of distinct chemical classes also influence DHP receptors by complex allosteric interactions. Verapamil and methoxyverapamil (D600) only partially inhibit binding of [³H]-nitrendipine. In contrast, diltiazem, fostedil, and other compounds, which share the ability to block Ca^{2+} channels, actually stimulate binding.[17,18] Drug-induced effects on the dissociation of DHP ligands from their binding sites indicate that the interaction between these calcium antagonists and DHP receptors is noncompetitive. Thus verapamil and methoxyverapamil increase the rate of dissociation of [³H]-nitrendipine, which decreases its affinity, while diltiazem slows the dissociation, which increases the affinity of binding.[17,19] Such actions presumably involve a separate but allosterically-linked site.[20]

Modified DHP compounds have been successfully used to isolate the DHP receptor from various tissues. [³H]-azidopine was used effectively as a photoactivated covalent label for the DHP receptor in brain.[21] Other reports indicate that [³H]-nitrendipine, [³H](+)PN200-110, [³H]-diltiazem, and others can also be used to photoaffinity-label calcium antagonist receptors (see Reynolds and Snyder[20]).

Phenylalkylamine-Labeled Ca^{2+} Channels

Phenylalkylamine ligands such as a racemically pure verapamil analogue, $[^3H](-)$desmethoxyverapamil, have also been used in the study of calcium antagonist receptors. This ligand was substantially more potent than verapamil, with a K_d of 0.5 nM in brain membranes. Studies with this ligand confirmed that DHPs interact in a potent and stereoselective fashion with the phenylalkylamine receptor. Like DHP receptors, $[^3H](-)$desmethoxyverapamil binding sites are also regulated by divalent cations. Thus in EDTA-treated brain membranes a number of cations stimulate binding at low concentrations and inhibit binding at higher concentrations. However, agonist cations such as Ca^{2+} and Sr^{2+} and antagonist cations such as Cd^{2+} share this property, which suggests that the phenylalkylamine receptor does not distinguish between agonist and antagonist cations.[22] Diltiazem increases the dissociation of $[^3H](-)$desmethoxyverapamil from the high affinity site in brain, which suggests that diltiazem binds to a distinct receptor that is linked allosterically to the phenylalkylamine receptor.[22] DHPs also regulate the binding of $[^3H](-)$desmethoxyverapamil in a negative, heterotropic, allosteric manner, but not when free divalent cations are chelated by EDTA.

ω-Conotoxin-Sensitive Ca^{2+} Channels

As mentioned above, peptide neurotoxins from different animal species have proven to be powerful tools for the biochemical identification and pharmacologic classification of both voltage-operated and ligand-operated ion channels. In particular, the venom of marine snails of the *Conus* genus contains different peptides (α, μ, and ω), which are highly specific for the nicotinic acetylcholine receptor, Na^+ channels, and Ca^{2+} channels, respectively.[23,24] Thus the introduction of ω-conotoxin fraction GVIA (ω-CTX) into research on VOCCs represented an important step towards better definition of these channel subtypes.

Experiments with iodinated, purified ω-CTX demonstrated specific binding of the toxin to frog and chick brain.[25] At the same time, evidence was reported that L-type VOCCs in brain did not bind ω-CTX and that ω-CTX and nitrendipine bound to different subcellular fractions of rat brain.[26] Distributions of ω-CTX and DHP binding sites in different regions of the CNS have been compared. Striking differences were found in the absolute number of binding sites and in their regional distribution, which suggests that the two ligands were binding to different molecular entities.[27,28] Interestingly, the nervous system appears to be the richest source of ω-CTX binding sites.

Studies that used ω-CTX have been instrumental in achieving greater knowledge of the known VOCC subtypes. In studies of hippocampal and striatal slices, ω-CTX has been shown to block the release of neurotransmitter evoked by electrical stimulation, while DHPs were inactive.[29,30] Only when neurotransmitter release was induced by high KCl concentrations did DHPs have some inhibitory effects. It has also been demonstrated that although ω-CTX blocks most neuronal VOCCs, it does not block neuronal L-type VOCCs (in agreement with radioligand binding data). These studies indicate that a number of different species and/or neuronal cell types possess channels that are insensitive to both ω-CTX and DHPs.[31] Interestingly, more recent studies have indicated that a venom from the funnel-web spider *Halolena curta* is able to potently and persistently inhibit the calcium influx that is maintained in the presence of a DHP and ω-CTX. Thus this venom, or its active constituent, may prove to be a useful tool for the investigation of the role of this novel, high-threshold VOCC in neuronal tissue.[32]

K^+ Channels

At least four or five major classes of K^+ channels exist. Voltage-gated K^+ channels open and close in response to membrane potential change, whereby depolarizing or hyperpolarizing signals can increase channel opening to generate outward or inward rectifications respectively. Ligand-gated K^+ channels are modulated by neurotransmitters, ions such as Ca^{2+}, or nucleotides such as ATP, which act at either intra- or extracellular sites in a positive or negative manner.[33,34] K^+ channels coupled to guanine nucleotide-binding proteins (G proteins) are ligand-sensitive and are coupled to neurotransmitter and hormone receptors by pertussis toxin-sensitive G proteins.[3] Recently it has been noted that the antidiabetic sulfonylureas block specific K^+ channel classes (see Triggle[34]).

Members of the hypoglycemic sulfonylurea class of compounds, including glibenclamide (glyburide) and glypizide, are now known to act via interaction at a specific site associated with an ATP-dependent K^+ channel. Increasing intracellular levels of ATP inhibit the channel, depolarize the cell, activate VOCCs and the entry of Ca^{2+}, and initiate the process of stimulus-secretion coupling. The binding sites for the sulfonylureas have been identified by radioligand binding studies with [^{3}H]-glyburide[35,36] and are particularly rich in brain. Sulfonylurea-sensitive, ATP-gated K^+ channels may also represent the locus of action of another class of compounds known as K^+ activators, which include the vasodilators nicorandil and cromakalim, whose mechanism of action had initially not been clearly understood. Other molecules with activity as potassium activators include the DHP, niguldipine. The ability of a DHP to interact with a K^+ channel is of some interest since this same chemical nucleus interacts with Ca^{2+}, Na^+, and K^+ channels.[34]

Just as studies of Na^+ and Ca^{2+} channels were facilitated by both the use of toxins such as tetrodotoxin, veratridine, and scorpion polypeptide toxins, and the availability of specific synthetic drugs, such as nifedipine, verapamil, and diltiazem, the current availability of both toxins and synthetic agents with specific and potent activities (and the applications of molecular cloning techniques; see below) have made the study of K^+ channels an equally active research area. Venomous toxins, including apamin, charybdotoxin, dendrotoxin, and others, are valuable and potent molecular probes for K^+ channels. Apamin interacts with the widely distributed, small conductance, Ca^{2+}-activated K^+ channel.[37] Charybdotoxin, from the scorpion *Leirus quinquestriatus*, blocks large conductance calcium-activated K^+ channels[38,39] by physical occlusion of the channel mouth. Characterization of high-affinity binding sites for [^{125}I]-charybdotoxin in brain synaptic membranes has been reported.[40] Dendrotoxin, a large 59-residue peptide derived from the venom of the African Mamba snake, interacts with voltage-gated, fast-activating K^+ channels.[41,42] β-bungarotoxin isolated from the venom of the Taiwan Krait *Bungarus multicinctus* possesses calcium-dependent phospholipase A2 activity and K^+ channel-blocking activity,[34] and these latter two toxins have also been used in radioligand binding studies (eg, Awan and Dolly[43] and references therein).

Neuroanatomic Localization of Ion Channel/Receptors by In Vitro Autoradiography

Autoradiographic techniques were first used extensively for the localization of receptors for neurotransmitters, but a great deal of information concerning the distribution of ion channel/receptors in brain is now also available. In the following sections I have given some examples of the use of radioligands to map and characterize various types of ion channel/receptors in normal, experimental, or neuropathologic situations. Notably, autoradiographic studies can also act as a guide to other experiments designed to examine the functional activity of the particular ion channel/receptor, or to isolate the ion channel protein by conveniently identifying areas enriched in binding sites.

Distribution of Ion Channels

The availability of potent and selective radioligands permitted the autoradiographic localization of DHP binding sites in rat and guinea pig brain.[44] DHP receptors are concentrated in areas of the brain rich in synapses. The highest density of DHP receptors occurs in the hippocampus and olfactory bulb, with lower levels in the striatum, cortex, hypothalamus, and midbrain (Figure 22.1). Lesion studies have shown that DHP receptors are localized on granule cells in the dentate gyrus of the hippocampus.[45] Autoradiographic studies reveal that the binding sites

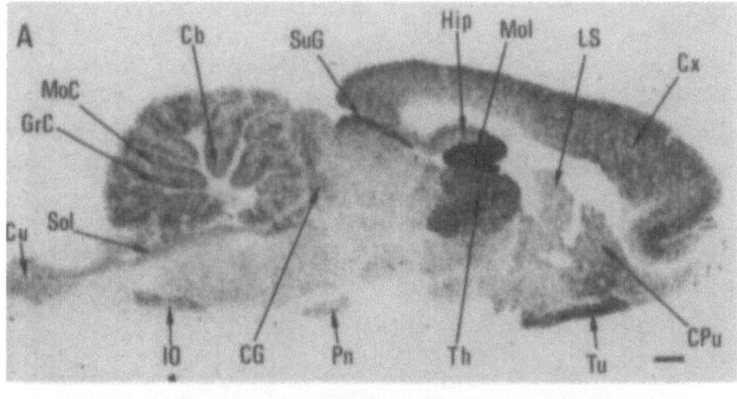

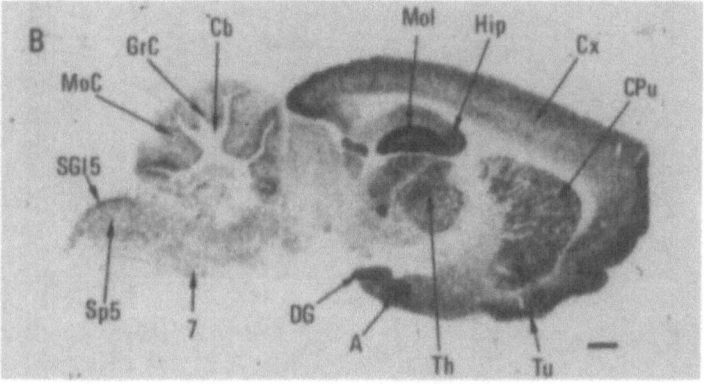

FIGURE 22.1. Autoradiographic distribution of calcium antagonist binding sites in rat brain, labelled with [³H]PN 200-110. Dark areas are regions rich in autoradiographic grains, ie, binding sites. Clear areas are poor in binding sites. The figure shows the distribution of calcium antagonist binding sites in two sagittal sections of rat brain. 7, facial nucleus; A, amygdala; Cb, cerebellum; CG, central (periaqueductal) grey; CPu, caudate putamen; Cu, cuneate nucleus; Cx, cerebral cortex; DG, dentate gyrus; GrC, granular layer of the cerebellum; Hip, hippocampus; IO, inferior olive; LS, lateral septal nucleus; MoC, molecular layer of the cerebellum; Mol, molecular layer of the dentate gyrus; Pn, pontine nuclei; SGI5, substantia gelatinosa of the trigeminum; Sol, nucleus of the solitary tract; Sp5, nucleus of the spinal tract of the trigeminal nerve; SuG, superficial grey layer of the superior colliculus; Th, thalamus; Tu, olfactory tubercle. Bar = 1 mm. (Adapted from an article by Cortés et al.[44] with permission from Springer-Verlag New York Inc.)

for the phenylalkylamine [³H](−)desmethoxyverapamil and of the DHP [³H]-nimodipine have an identical distribution (Figure 22.2).[46,47] The highest specific binding of [³H](−)desmethoxyverapamil also occurs in the molecular layers of the hippocampus and dentate gyrus.[47] These

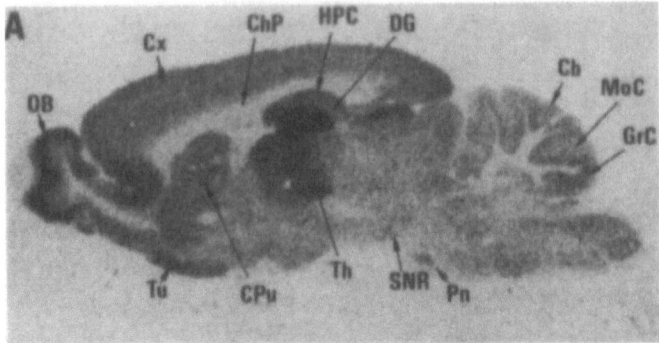

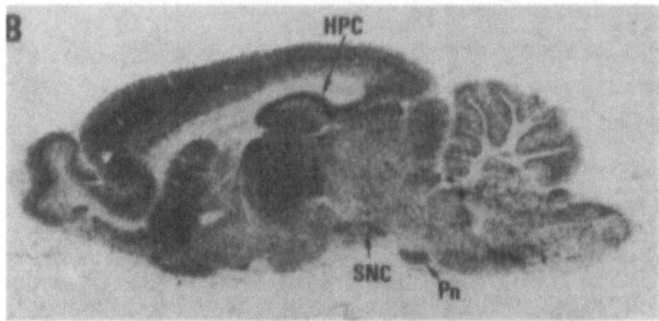

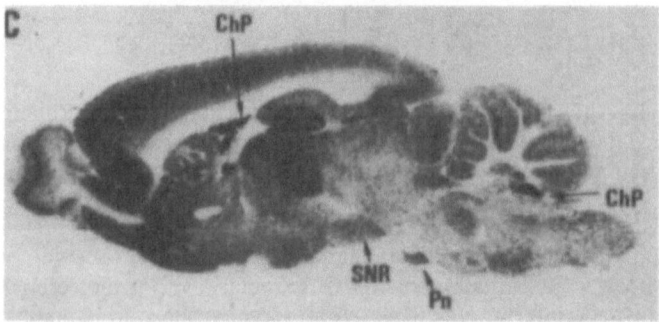

FIGURE 22.2. Autoradiographic distribution of (A) (+)-[³H]PN200-110; (B) (±)-[³H]-methoxyverapamil (D600); and (C) (−)[³H]-desmethoxyverapamil (D888) binding sites in sagittal sections of the rat brain. Abbreviations: Cb, cerebellum; ChP, choroid plexus; CPu, caudate putamen; Cx, cortex; DG, dentate gyrus; GrC, granular layer of the cerebellum; HPC, hippocampus, pyramidal cell layer; MoC, molecular layer of the cerebellum; OB, olfactory bulb; Pn, pontine nuclei; SNC, substantia nigra, pars compacta; SNR, substantia nigra, pars reticulata; Th, thalamus; Tu, olfactory tubercle. (Reprinted with permission from an article by Supavilai et al.[46] in Progress in Brain Research, 1985; 63:89–95, Amsterdam: Elsevier Science Publishers BV.)

findings are in line with the demonstration that these two distinct calcium antagonist binding sites are allosterically coupled. The highly specific anatomic distribution of calcium antagonist binding sites in brain and their almost exclusive localization on neuron terminals[46] led to the suggestion that Ca^{2+}-channel drugs might regulate neuronal functions such as neurotransmitter release (but see below).

Autoradiographic localization studies of $[^{125}I]$-ω-CTX binding sites in rat brain indicated that they too were localized in those areas of the brain rich in synaptic contacts and/or connections.[48,49] The distribution was in part similar to that previously described for $[^3H]$-DHP and $[^3H](-)$desmethoxyverapamil binding sites. However, clear differences were found in some brain areas including the interpeduncular nucleus, pontine nuclei, subiculum, and lateral septal nuclei, these areas being richer in $[^{125}I]$-ω-CTX binding sites that in $[^3H]$-DHP binding sites (see Sher and Clementi[24]).

More recently, both the L-type and the ω-CTX-sensitive VOCCs have been localized more precisely (using antibodies).[50,51] L-type channels have been found at the cell body level and clustered in high density at the base of major dendrites in hippocampal pyramidal cells and in sympathetic neurons. In hippocampal neurons, ω-CTX-sensitive channels were found to be organized in clusters that coincide with the site of synaptic contacts. It has been speculated that the calcium influx through L-type channels located at the cell body may be relevant to the regulation of calcium-dependent intracellular events such as protein phosphorylation or activation of gene expression, while the strategic presynaptic localization of ω-CTX-sensitive channels might be necessary for the rapid generation of large calcium transients able to trigger fast transmitter release in nerve terminals.[24]

Developmental Profile of Ion Channels

Ontogenic studies have been carried out which examine the postnatal development of a number of ion channels. Mourre et al.[52] used labeled tetrodotoxin, apamin, and $(-)$desmethoxyverapamil to study the regional distribution of voltage-dependent Na^+ channels, a class of calcium-dependent K^+ channels, and the phenylalkylamine-sensitive Ca^{2+} channel protein, respectively, in the brain of rats 1 to 15 days old and adult rats.[52] $[^{125}I]$-apamin binding to Ca^{2+}-dependent K^+ channel proteins was detected very early in the germinative zone in association with neuronal cell bodies during their migration and maturation (Figures 22.3 and 22.4). In hippocampus and cerebral cortex, $[^{125}I]$-apamin binding sites were already present at birth and increased to the 20th postnatal day, by which the adult level of expression was established. Development of Ca^{2+} channels ($[^3H](-)$desmethoxyverapamil binding) seemed to follow the development of dendrites. The density of these channel proteins increased steadily until adult age. Na^+ channel proteins ($[^{125}I]$-tetrodotoxin-binding) appear to be absent in the diencephalon at birth, and their appearance and increasing density were strictly correlated to synaptogenesis, particularly in cerebral and cerebellar cortex and hippocampus (Figures 22.3 and 22.4).[52]

Colocalization of Ion Channel/Receptor Subtypes

Autoradiographic techniques can also be readily used to detect species variations in regional and nuclei distribution of ion channel/receptor proteins in brain. Similarly, autoradiography allows the comparison of different populations of ion channel proteins in adjacent brain sections and a contrast of regional distributions to help establish the physical separation or potential coexpression of different populations of ion channels (eg, Olsen et al.[53]). In these studies[53] the regional distribution of binding of various ligands for the different receptor sites of the Cl^- channel/$GABA_A$ receptor complex was investigated, and revealed a significant lack of correspondence in a number of brain regions between the range of ligands compared (Figure 22.5). Thus, in addition to the previously reported discrepancies between high-affinity GABA-agonist binding and benzodiazepine (BZP) receptor distribution, significant differences were observed between particular GABA agonists and antagonists; two different antagonists, bicuculline and SR-95531; and two convulsant ligands, TBPS and TBOB. These findings were most consistent with the existence of at least four $GABA_A$ receptor subtypes that vary in their binding affinities or capabilities, and which correspond to different gene products demonstrated by molecular cloning and protein chemistry.[53,54]

Cellular Localization of Ion Channels

Autoradiographic studies can also readily assess the effect of mutations or lesions that involve the selective loss of a particular cell type and, in turn, the use of physical or genetic lesions allows clarification of the localization of the ion channel ligand binding sites themselves. For example, the autoradiographic distribution of $[^{125}I]$-saxotoxin binding sites associated with voltage-sensitive Na^+ channels has been studied in the cerebellum of different types of neurologic mutant mice—the Weaver, Purkinje Cell Degeneration, Nervous, and Reeler mice.[55] There was a significant decrease in the density of binding sites in the molecular layer and the Purkinje cell layer of Weaver cerebellum, where a large majority of granule cells had disappeared. The hippocampal formation of Reeler mutants presented a homogeneous distribution of the Na^+ channel protein, in contrast to the laminated distribution observed in normal mice. Overall this data suggests that a major proportion of saxotoxin-

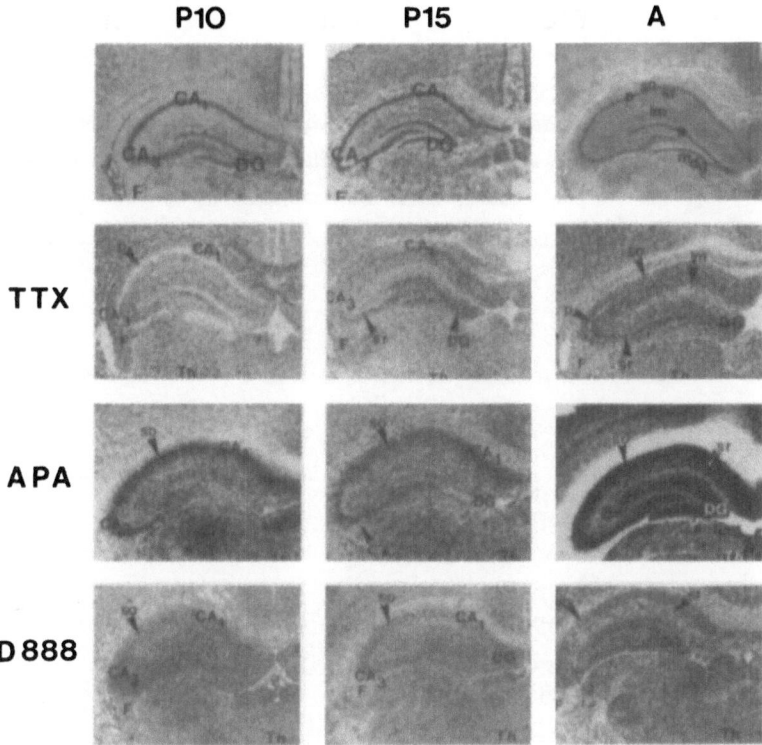

FIGURE 22.3. Localization of ionic channels in the developing hippocampus of the rat brain. In these micrographs, autoradiograms are brightfield images. Dark areas represent zones of high grain density. Grain differences between different images do not necessarily represent quantitative variations. The figure presents different ages, P10, P15, and adult (A), with histological sections in the higher part of the figure and autoradiograms below after tissue section incubation with [³H]-TTX for the Na⁺ channel protein, [¹²⁵I]-apamin (APA) for the calcium-dependent, K⁺-channel proteins and (−)[³H]D888 for the slow Ca²⁺-channel protein. Nonspecific binding for the three labeled ligands in tissue sections incubated in the presence of a large excess of unlabeled ligand was equivalent to film background. The figure clearly shows the different appearances of each ionic channel protein at different times during development. Abbreviations: CA₁ and CA₃, Ammon's horn regions 1 and 3; DG, dentate gyrus; so, stratum oriens; p, stratum pyramidal; sr, stratum radiatum; lm, stratum lacunosum moleculare; m_dg, stratum molecular of the dentate gyrus; g, granular layer of the dentate gyrus; F, fimbria; Ft, fiber tracts; H, hippocampus; Hb, habenula; Th, thalamus. Bar = 1 mm. (Adapted with permission from an article by Mourre et al.[52] in Brain Research, 1987; 417:21–32, Amsterdam: Elsevier Science Publishers BV.)

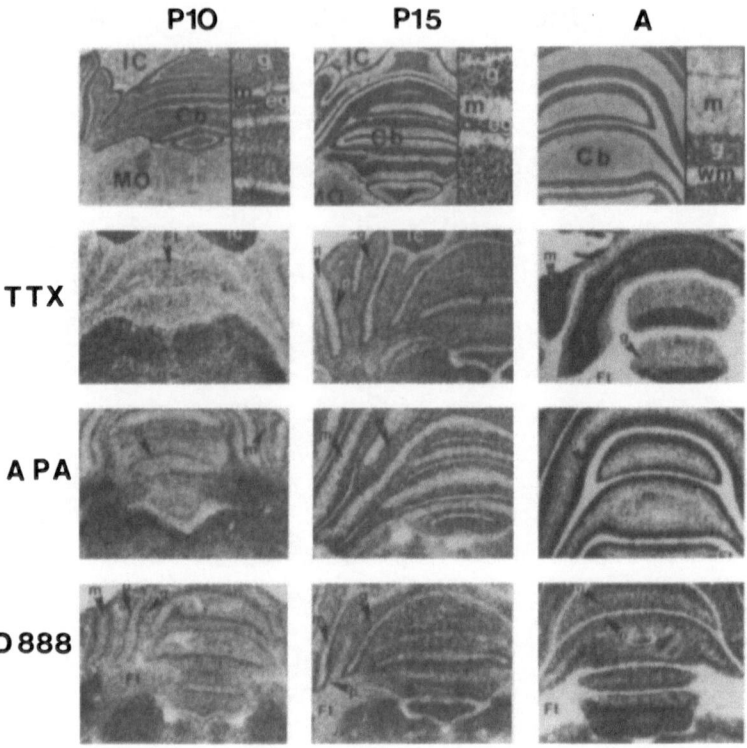

FIGURE 22.4. Localization of ionic channels in the developing cerebellum of the rat brain. Organization is similar to that in Figure 22.3. In the upper row the histological sections correspond to the section level of autoradiograms for a given postnatal age with representations, in insets, of the different layers of cerebellar cortex at a higher power (Bar = 200 μm). Autoradiograms (below) for the different ligands (TTX, APA, (−)D888) used to localize Na$^+$, calcium-dependent K$^+$, and slow Ca^{2+}-channel proteins indicate a different distribution of ionic channel proteins during ontogenesis of the cerebellum of the postnatal rat. Abbreviations: eg, external germinative layer; m, stratum moleculare; g, granular layer; wm, white matter; IC, inferior colliculus; Cb, cerebellum; MO, medulla oblongata; p, Purkinje cell layer; Ft, fiber tracts. Bar = 1mm. (Adapted with permission from an article by Mourre et al.[52] in Brain Research, 1987; 417:21–32, Amsterdam: Elsevier Science Publishers BV.)

sensitive Na$^+$ channels are localized on parallel fibers of granule cells and on axons of basket cells, in a presynaptic position. Furthermore, in contrast to the cell body, the dendritic arborization of Purkinje cells seems to be devoid of [^{125}I]-saxotoxin binding sites.

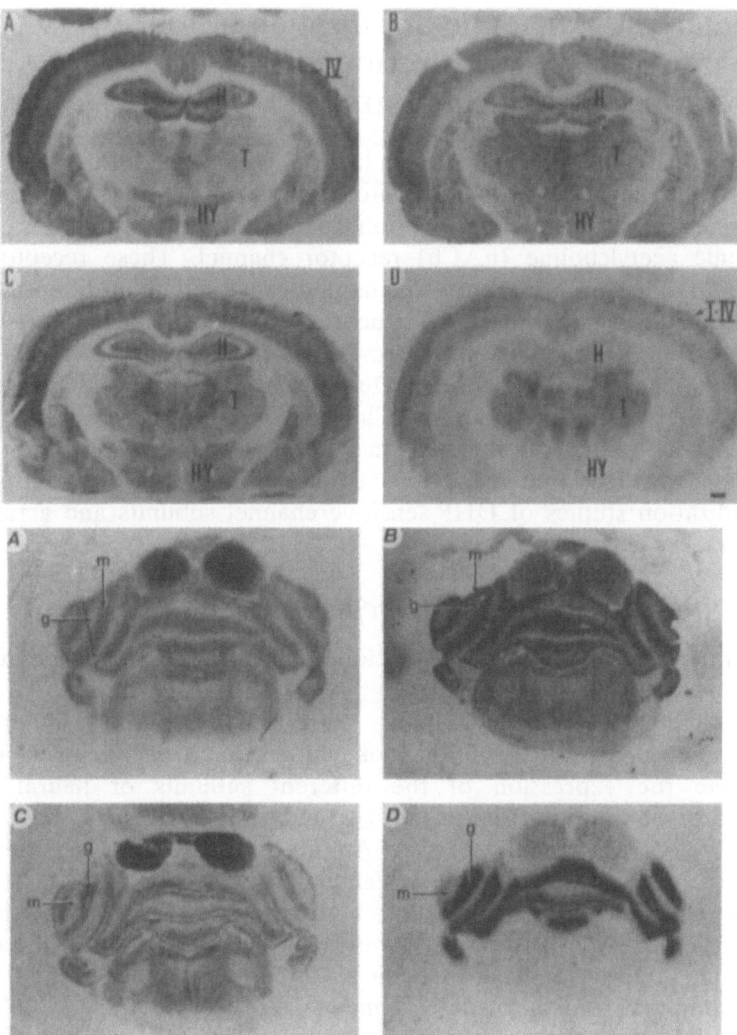

FIGURE 22.5. Photomicrographs illustrating GABA$_A$-benzodiazepine radioligand binding to tissue sections in rat brain. Radioligands used were: (A) 1 nM [^{3}H]-flunitrazepam for BZP receptors; (B) 25 nM [^{3}H]-bicuculline methochloride, a GABA antagonist, for low-affinity agonist sites; (C) 20 nM [^{35}S]TBPS for convulsant sites; and (D) 5 nM [^{3}H]-muscimol for high-affinity GABA agonist sites. Bar = 500 μm. (Top) Brain regions identified in the rostral midbrain include the hippocampus (H), thalamus (T), and hypothalamus (HY). Cerebral cortex layer IV (labeled IV in (A); labeled with arrows in (B), (C) and I–IV (D) are indicated. (Bottom) Cerebellum showing binding to the molecular layer (m) and granule cell layer (g). (Adapted from an article by Olsen et al.[53] in Journal of Chemical Neuroanatomy, 1990; 3:59–76 [Copyright 1990, John Wiley & Sons, Ltd.]. Reprinted by permission of John Wiley & Sons, Ltd.)

Cellular and Anatomic Localization of Ion Channel/Receptor Expression by In Situ Hybridization Histochemistry

Biochemical and molecular studies have yielded much information about the structure of the ligand-activated ion channels. Probably the best characterized example of these ion channel/receptor complexes is the nicotinic acetylcholine (nACh) receptor channel. These receptors are made up of several different subunits that can combine in different ways, which provides great variety of structure. Similar studies of other ligand-activated ion channels have revealed the existence of so-called superfamilies of ion channels. Descriptions are given below of the distribution of the individual nACh, $GABA_A$, and glycine ion channel subunits in brain, and how this can help predict ion channel/receptor location, diversity, and function. Examples are also given of in situ hybridization studies of DHP-sensitive channel subunits and a range of K^+ channel gene transcripts.

Nicotinic Acetylcholine Receptors

A number of cDNA and genomic clones which code for different α and β subunits of the nACh receptor in the CNS have been isolated, which suggests the existence of a family of related receptors in mammalian brain. Several in situ hybridization studies have been carried out to examine the expression of the different subunits of neural nACh receptors in brain (Figure 22.6), and these findings have been summarized in a recent review.[56] Among the α subunits, α_2 shows the most restricted distribution with highest levels of hybridization observed in the interpeduncular nucleus.[57,58] Transcripts for the α_3 and α_4 subunits show a much wider distribution, with high levels detected in the medial habenula and several other thalamic nuclei, as well as the substantia nigra pars compacta and the ventral tegmental area (VTA) (Figure 22.6). α_4 transcripts are also abundant in neocortex and the hippocampal formation. Transcripts for the β_2 subunit present a widespread distribution that overlaps with those of the α_2, α_3, and α_4 subunits, which suggests that β_2 subunits may contribute to the formation of most of the nACh receptors in the brain. In contrast, both β_3 and β_4 display much more restricted distributions with β_3 mRNA is present in the medial and lateral habenula, substantia nigra, and VTA, while β_4 transcripts are detected only in the medial habenula. The overlapping distributions of many of the mRNAs for the different subunits raises the possibility that individual neurons express more than one subtype of α or β subunit.[58] In general, these in situ hybridization studies are in good agreement with those that use nicotinic ligands such as [^{3}H]-acetylcholine and [^{3}H]-nicotine.[59]

GABA$_A$ Receptors

GABA$_A$ receptors that mediate the action of GABA (the major inhibitory neurotransmitter in brain) via a Cl$^-$ channel contain binding sites for a variety of classes of drugs, among them the BZPs (see above). GABA$_A$ receptors are also known to contain at least two classes of subunits, α and β. The α_1 and β_1 subunits show significant homology with each other and with the different subunits of the nACh receptors, which suggests the existence of a superfamily of ligand-gated ion channels.[60]

Expression studies have revealed that a combination of any of the several α subunits ($\alpha_1-\alpha_4$) with the β subunit resulted in a functional GABA$_A$ receptor with many properties of the native receptor.[54] However, these receptors did not display the BZP potentiation observed in native receptors. This suggested the existence of additional components of the receptor necessary for this potentiation, a hypothesis confirmed by the cloning of a novel type of subunit, the γ_2.[61]

Several reports on the localization of the transcripts of the different forms of subunits of GABA$_A$ receptors have been published (eg, Wisden et al.[62,63]). Differential patterns of hybridization observed for the different α subunits also suggest the existence in brain of a multiplicity of GABA$_A$-receptor subtypes (Figure 22.7). The pattern of hybridization for the α_1 subunit shows a significant overlap with the distribution of the pharmacologically defined GABA$_A$-BZP receptors of the so-called type I. Thus it is suggested that this subunit could be present in that type of GABA$_A$ receptor.[63] Expression studies designed to test this hypothesis showed that α_1-containing receptors displayed binding characteristics consistent with those of type I GABA$_A$-BZP receptors, whereas the characteristics of α_2- and α_3-containing receptors were more consistent with type II receptors.[64] γ_2 subunit-containing receptors probably also represent the subset of GABA$_A$ receptors that are coupled to BZP receptor sites (see above), as the distribution of transcripts for the γ_2 subunit overlaps with BZP binding sites (see Vilaró et al.;[56] Shivers et al.;[65] Unnerstall et al.[66]). A fourth subunit class has also been cloned from rat brain (Figure 22.7). Coexpression of this δ subunit with α_1 and β_1 subunits yielded functional receptors that lack BZP binding sites[65] and thus δ subunit-containing receptors could represent the subset of GABA$_A$ receptors present in brain that also lack BZP binding sites and are selectively labeled by [^{3}H]-muscimol.[66,67]

Glycine Receptors

The glycine receptor is a ligand-gated Cl$^-$ channel and has been purified from spinal cord by chromatography, using analogues of its potent natural antagonist, the convulsant strychnine. Purified receptors contain three different polypeptides with molecular weights of 48, 58, and 93 kDa.

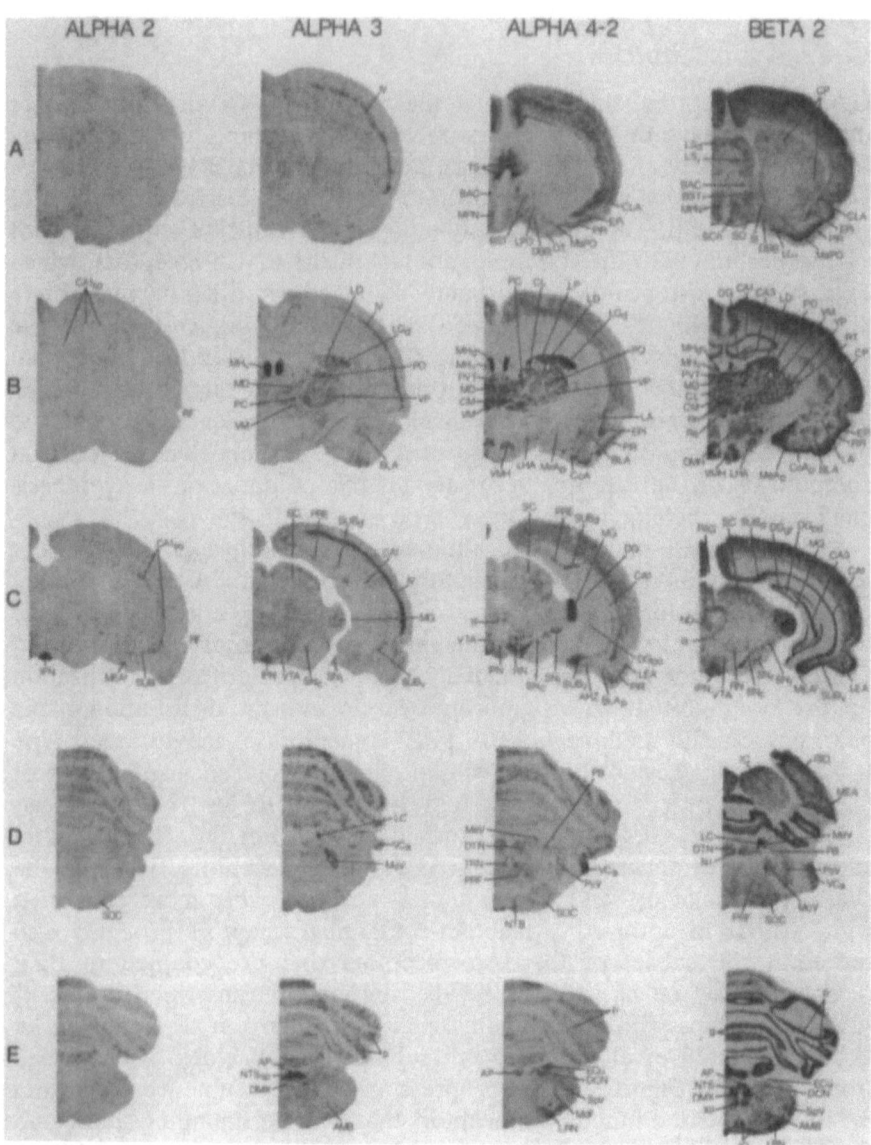

FIGURE 22.6. Comparison of the distribution of nACh receptor alpha2, alpha3, alpha4-2, and beta2 subunit mRNA expression in a rostrocaudal (A–E) series of coronal sections through the rat brain. AHZ, amygdalohippocampal area; AMB, nucleus ambiguus; AP, area postrema; BAC, bed nucleus of the anterior commissure; BLA, basolateral n. (amygdala); BLA$_p$, basolateral n. posterior part (amygdala); BST, bed n. of the stria terminalis; CA1, field CA1 of Ammon's horn; CA1$_{so}$, field CA1 of Ammon's horn, stratum oriens; CA3, field CA3 of Ammon's horn; CL, central lateral n.; CLA, claustrum; CM, central medial n.; CoA, cortical n. (amygdala); CoA$_p$, cortical n. posterior part (amygdala); CP,

Strychnine binding occurs to the 48 kDa subunit (see Betz[68]). A cDNA clone has been isolated for this subunit. The deduced protein is similar both in sequence and in putative transmembrane organisation to the various nACh and $GABA_A$ receptors.

The distribution of cells containing mRNA coding for the strychnine-binding (α) subunit of the glycine receptor in rat brain has been described.[56,69] The hybridization signal clearly shows an increasing rostro-caudal gradient, with the highest levels in spinal cord and several nuclei of the pons and medulla (Figure 22.8). However, significant levels of mRNA are also observed in the red nucleus of the midbrain, the parafascicular nucleus of the thalamus, the zona incerta, and the hypothalamus.[56] In contrast to results for the nACh and $GABA_A$

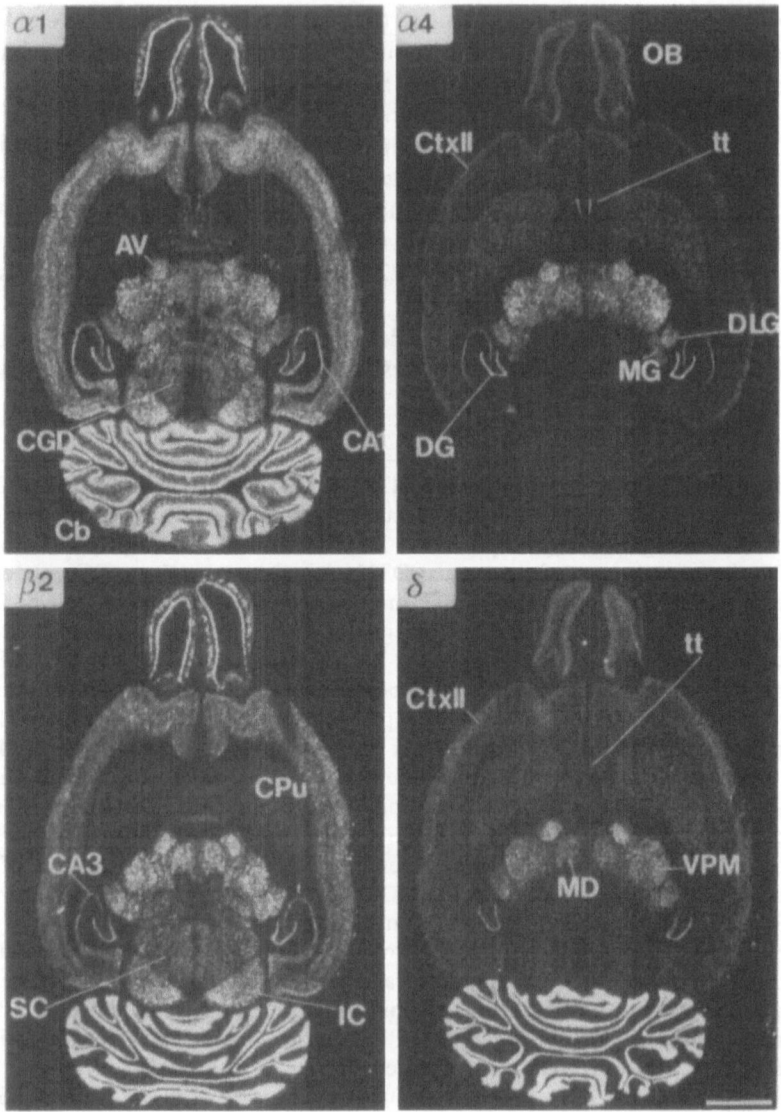

FIGURE 22.7. Comparison of the distribution of mRNAs encoding different subunits of the GABA$_A$ receptor in the rat brain. (Left) Localization of α_1, α_4, β_2, and δ subunits in horizontal sections. (Right) Distribution of α_1–α_5 and δ subunits in coronal sections at the level of the medial preoptic hypothalamic area. Arc, arcuate hypothalamic nucleus; AV, anteroventral thalamic nucleus; CA1–4, fields 1–4 of Ammon's horn; Cb, cerebellum; CGD, central gray, dorsal; CPu caudate putamen; Ctx, neocortex; CtxII, neocortex, layer 2; DG, dentate gyrus; DLG, dorsolateral geniculate thalamic nucleus; Hy, hypothalamus; IC, inferior colliculus; La, lateral amygdaloid nucleus; MD, mediodorsal thalamic nucleus; Me, medial amygdaloid nucleus; MG, medial geniculate nucleus; OB, olfactory

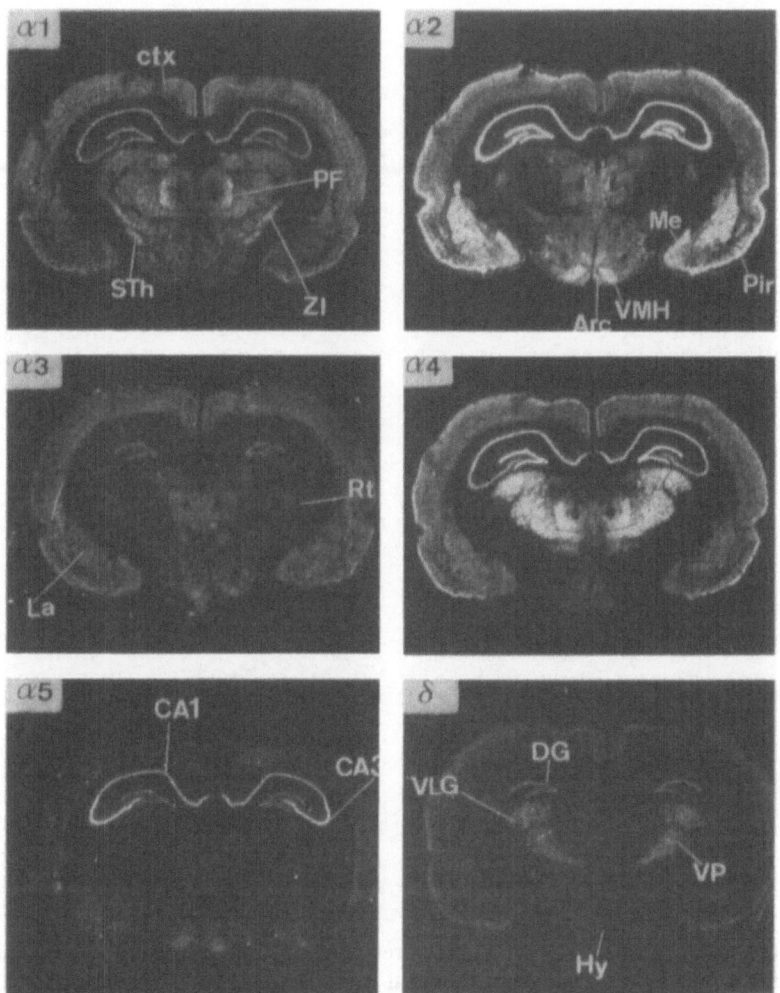

bulb; PF, parafascicular thalamic nucleus; Pir, piriform cortex; Rt, reticular thalamic nucleus; SC, superior colliculus; STh, subthalamic nucleus; tt, tenia tecta; VLG, ventrolateral geniculate nucleus; VMH, ventromedial hypothalamic nucleus; VP, ventroposterior thalamic nucleus; VPM, ventral posteromedial thalamic nucleus; ZI, zona incerta. Bars = 3 mm. (Adapted with permission from an article by Wisden et al.[62] in Journal of Neuroscience, 1992; 12:1040–1062, New York: Oxford University Press.)

receptors, there was an extensive overlap of the distribution of the transcripts for this subunit and that of the [^{3}H]-strychnine binding sites. However, there were indications of diversity in the forms of the strychnine-sensitive glycine receptor that appear to be developmentally-regulated and absent from the adult rat brain.[70]

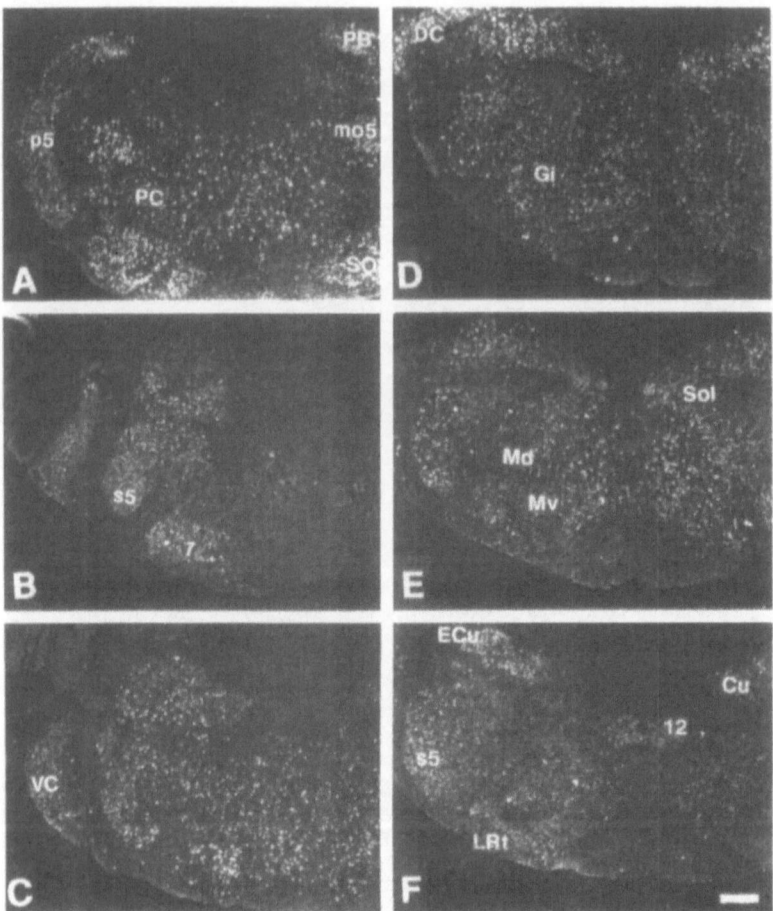

FIGURE 22.8. Localization of glycine receptor α_1 subunit mRNA-containing neurons in the lower brainstem. 7, facial nucleus; 12, hypoglossal nucleus; Cu, cuneate nucleus; DC, dorsal cochlear nucleus; ECu, external cuneate nucleus; Gi, gigantocellular reticular nucleus; LRt, lateral reticular nucleus; Md, medullary reticular field, dorsal; mo5, motor trigeminal nucleus; Mv, medullary reticular field, ventral; p5, principal sensory trigeminal nucleus; PB, parabrachial nucleus; PC, parvocellular reticular nucleus; s5, trigeminal spinal nucleus; Sol, nucleus of the solitary tract; VC, ventral cochlear nucleus. Bar = 0.5 mm. (Reprinted from Neuroscience, 1991; 43:381–395, Sato et al.[69] Localization of glycine receptor α_1 subunit mRNA-containing neurons in the rat brain: An analysis using in situ hybridization histochemistry, Copyright (1991), with permission from Pergamon Press Ltd, Headington Hill Hall, Oxford OX3 OBW, UK.)

Although heterogeneity of receptors for the $GABA_A$ receptor was not unexpected in view of pharmacologic studies with BZPs and GABA agonists, the existence of different glycine receptors and their expression was more surprising. More recent molecular cloning studies have since revealed that several glycine receptor subunits exist. In addition to the originally described glycine receptor α subunit now termed $α_1$, two different $α_2$ cDNAs and an $α_3$ cDNA are expressed in rats and humans (see Betz[68]). All these α-subunit sequences display a high degree of amino acid homology with their $α_1$ counterpart, and correspond to glycine receptor polypeptides, the expression of which is under distinct temporal and regional control.[68] Using sequence-specific oligonucleotides, in situ hybridization experiments reveal that, as described, $α_1$ transcripts are found in the spinal cord, brainstem, and the colliculi, whereas $α_2$ mRNA is also seen in several forebrain regions including the hippocampus, cerebral cortex, and thalamus (Figure 22.9). Low levels of $α_3$ subunit mRNA are detected in the olfactory bulb, the hippocampus, and in particular the cerebellum, whereas β subunit transcripts are abundantly expressed throughout the entire brain and spinal cord.[71] The observation that β subunit mRNA is found in many brain regions where neither [^{3}H]-strychnine binding nor any of the presently known α subunit transcripts are seen at comparable levels suggests the existence of additional glycine receptor α subunits that assemble with the β polypeptide to form novel, strychnine-resistant receptors (Figure 22.9). Alternatively, the β subunit might also be part of other ligand-gated ion channel proteins such as specific $GABA_A$ receptor subtypes or possibly glycine-sensitive glutamate receptors of the N-methyl-D-aspartate (NMDA) type (see Betz;[68] Malosio et al.[71]). During development, $α_2$ mRNA accumulated prenatally and decreased after birth, whereas $α_1$ and $α_3$ subunit transcripts appeared in brain only postnatally. β subunit mRNA has been observed at early embryonic stages and increases continuously to high levels in adult rats. These profiles of expression of different glycine receptor subunits point to novel functions of glycine ion channel/receptor proteins in the mammalian CNS.[71]

These results raise the possibility that both $GABA_A$ and glycine receptors may be present in the same sensory and motor processing systems in higher regions of the brain. Simultaneous use of these receptors may provide better fine-tuning of neuronal activity by exploiting the different kinetic and conductance properties of glycine and $GABA_A$ receptors.[8] Recent studies by Furuyama et al.[72] provide direct evidence of the coexpression of glycine receptor β subunit and $GABA_A$ receptor γ subunit mRNAs in dorsal root ganglion cells of the rat. Interestingly, similar diversity of receptor subunit distribution has been described for the various members of the AMPA/kainate receptor family of ion channel-linked receptors.[73]

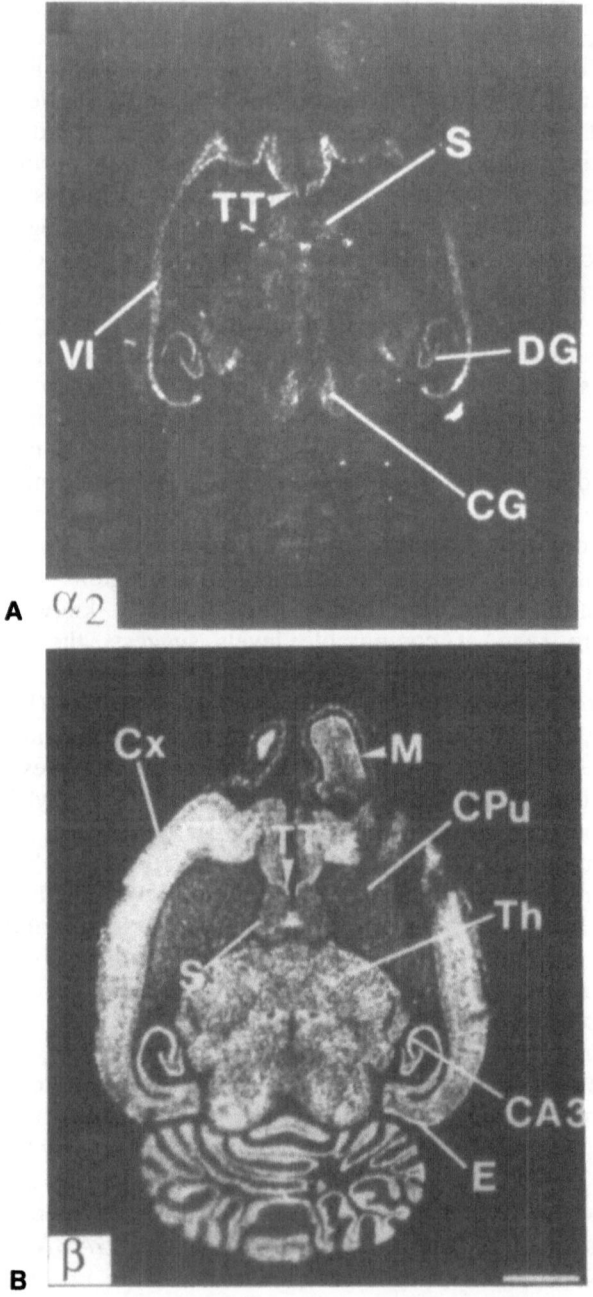

FIGURE 22.9. Localization of GlyR α2- and β-subunit mRNAs in rat brain. Horizontal sections were hybridized to oligonucleotide probes specific for α2 (A) and β (B) transcripts, processed for autoradiography and exposed to autographic

DHP-Sensitive Ca^{2+} Channels

Molecular cloning studies have also identified cDNAs that code for VOCCs. For example, cDNA clones that correspond to multiple forms of the α subunit of the L-type VOCC have been isolated from brain.[74] Chin et al.[75] and others have isolated a brain isoform of the DHP-sensitive, Ca^{2+}-channel α_1 subunit cDNA, whose primary structure is significantly different from its skeletal and cardiac muscle counterparts. These authors have subsequently described the distribution in rat brain of the mRNA that encodes this form of the DHP-sensitive Ca^{2+}-channel α_1 subunit, using in situ hybridization histochemistry (Figure 22.10). Messenger RNA is prominently localized in neuronal cells in the olfactory bulb, dentate gyrus, hippocampus, arcuate nucleus, paraventricular nucleus, ventromedial nucleus, cerebral cortex, superior colliculus, and the cerebellar Purkinje cell layer. Abundance of the DHP channel/receptor mRNA in brain areas responsible for neuroendocrine function, such as the hypothalamus and the pituitary, led these authors to suggest that the DHP-sensitive Ca^{2+} channel (L-type) may play a role in excitation-secretion coupling and in modulating neuroendocrine functions in the CNS.[75]

Of the four main isoforms of brain Ca^{2+} channel α_1 subunit, two are closely related to DHP-sensitive Ca^{2+} channels from other tissues, which suggests that the brain expresses two distinct forms of the L-type Ca^{2+} channel. Recently the primary structure of a third α_1-subunit isoform that is part of an N-type, ω-conotoxin-sensitive Ca^{2+} channel has also been described,[76] although its anatomic distribution within brain has not yet been reported.

Differential Expression of K^+ Channel mRNAs in Rat Brain: Downregulation in the Hippocampus Following Seizures

It is now clear that voltage-sensitive K^+ channels play crucial roles in both mediating and regulating electrical excitability of neurons, including determining resting membrane potential, shaping of the action potential, modulation of neurotransmitter release, and regulation of neuronal firing[6] (see Tsaur et al.[77]). This diversity of function is reflected in the heterogeneity of K^+ channels that have been found in the nervous

---

film. Abbreviations: CA3, CA3 region of the hippocampus; CG, central gray; CPu, caudate putamen; Cx, cortex; DG, dentate gyrus; E, entorhinal cortex; M, mitral cell layer; S, septum; Th, thalamus; TT, tenia tecta; VI, sixth layer of cortex. Bar = 4mm. (Reprinted from an article by Betz,[68] with permission from Elsevier, UK.)

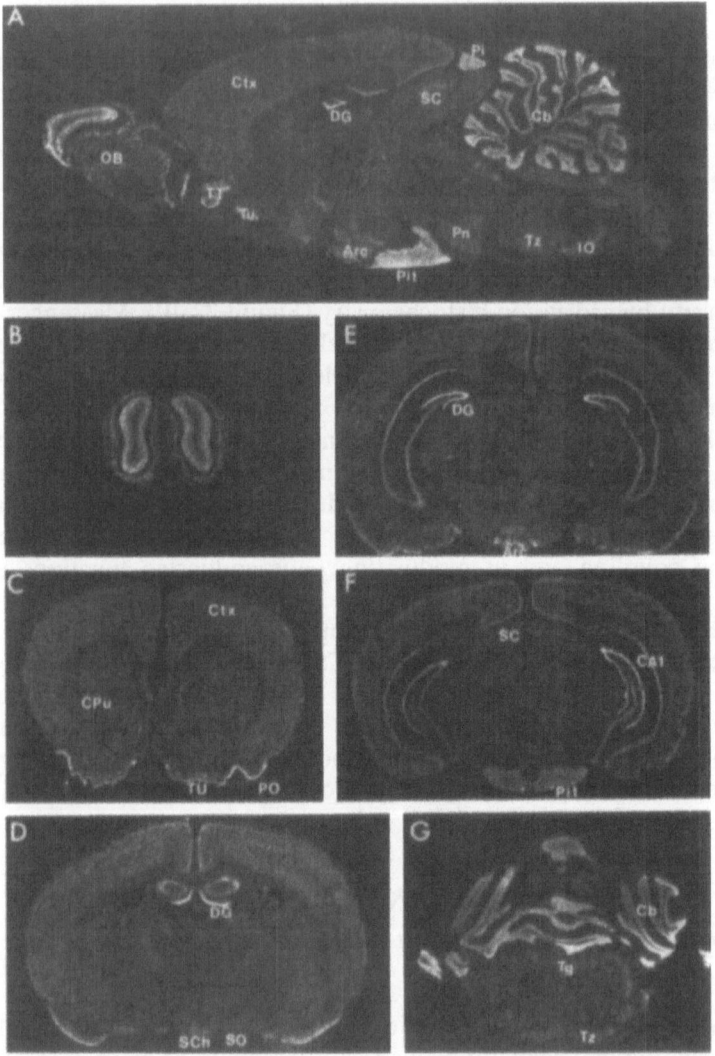

FIGURE 22.10. Localization of the dihydropyridine-sensitive calcium channel α_1 subunit mRNA in adult rat brain by in situ hybridization. Negative film image of sagittal (A) and coronal sections (B–G). Abbreviations: Arc, arcuate nucleus; CA1, CA1 of the hippocampus; Cb, cerebellar cortex; CPu, caudate putamen; Ctx, cerebral cortex; DG, dentate gyrus; IO, inferior olive; OB, olfactory bulb; Pi, pineal gland; Pit, pituitary gland; Pn, pontine nucleus; PO, piriform cortex; SC, superior colliculus; SCh, suprachiasmatic nucleus; SO, supraoptic nucleus; Tg, tegmental nucleus; TT, tenia tecta; Tu, olfactory tubercle; Tz, trapezoid nucleus. (Reprinted with permission from an article by Chin, Smith, Kim, Kim[75] in FEBS Letters 1992; 299:69–74, Amsterdam: Elsevier Science Publishers BV.)

system, both in terms of physiologic properties and in terms of the number of genes that encode them.[1,6] Thus, in a similar way to the ligand-gated channels and the Na^+ channels, molecular cloning studies of the K^+ channels suggest the existence of a large and extended gene family that would allow a "mix and match" association of individual channel subunits expressed from different genes.

K^+ channels are well placed to regulate neuronal and synaptic function and alteration of K^+-channel activity has been implicated in the mechanism of neuronal plasticity (see Tsaur et al.[77]). Despite detailed characterization of various K^+ channels following expression in *Xenopus* oocytes, the physiologic function of specific K^+-channel gene products in the mammalian nervous system has not been extensively examined. The heterogeneity of potassium conductances found in different neurons may be due at least in part to a differential spatial expression of the members of a large K^+-channel gene family.[77]

A recent study examined the pattern of expression of three K^+-channel genes known as Kv1.1, Kv1.2, and Kv4.2. The results of this study showed a striking differential expression of these genes at both the anatomic and cellular level, with each K^+ channel mRNA showing distinctive, though partly overlapping distributions (Figure 22.11).[77] It was also demonstrated that in addition to differential spatial expression, two out of the three K^+ channel mRNAs were also regulated in response to seizure activity.

Kv1.1 and Kv1.2 (encoding delayed-rectifier K^+ channels of the *Shaker*-RCK subfamily) were expressed at high levels in the hippocampus, thalamus, cerebral cortex, and cerebellum. In contrast, Kv4.2 mRNA (an A-type K^+ channel of the *Shal* subfamily) was highly abundant in the cerebellum with intermediate levels in the hippocampus and medial habenula and relatively low levels in cerebral cortex and thalamus. The mRNAs of all three genes were found in the different subregions of the hippocampus, but the relative levels of mRNA were distinct for each gene in the dentate gyrus, CA1, and CA3. Changes in K^+ channel mRNAs were induced by a single dose of the convulsant pentylenetetrazole. Seizure activity produced a reduction of Kv1.2 and Kv4.2 mRNAs in the dentate granule cell layer, while there was no alteration in Kv1.1 mRNA levels in the hippocampus or elsewhere. Administration of the anticonvulsant, diazepam, protected animals from seizure and also prevented the reduction of Kv1.2 and Kv4.2 mRNAs in the dentate gyrus, which indicates that the K^+ channel mRNA reduction was due to the neuronal activity evoked by the seizure. In light of these findings and physiologically important effects of K^+ channels on neuronal excitability and synaptic transmission, it will be important to measure the turnover rate of K^+ channel proteins in vivo and determine the relationship between changes in mRNA level and alterations in K^+ channel density. Tsaur et al.[77] speculated that

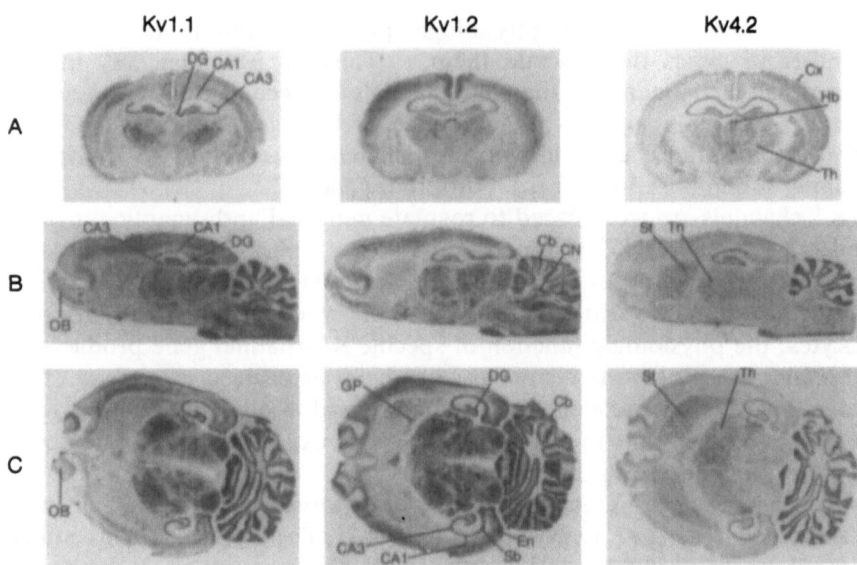

FIGURE 22.11. Differential expression of K$^+$-channel genes Kv1.1, Kv1.2, and Kv4.2 in the rat brain. Coronal (A), sagittal (B), and horizontal (C) sections of adult rat brains were hybridized with ^{35}S-labeled antisense RNA probes specific to the three K$^+$-channel genes, Kv1.1, Kv1.2, and Kv4.2 Abbreviations: CA1, CA3, CA1, and CA3 regions of hippocampus; Cb, cerebellar cortex; CN, deep cerebellar nuclei (where Kv1.1 and Kv1.2 are highly expressed); Cx, cerebral cortex; DG, dentate gyrus; En, entorhinal cortex; GP, globus pallidus (where Kv1.2 is highly expressed); Hb, medial habenular nucleus (where Kv1.2 and Kv4.2 are highly expressed); OB, olfactory bulb; Sub, subiculum; St, striatum; Th, thalamus. (Reprinted with permission from an article by Tsaur, Sheng, Lowenstein, Jan, Jan.[77] Copyright for this material is held by Cell Press, Cambridge, Mass. USA.)

activity-dependent repression of K$^+$-channel gene expression, particularly in excitatory rather than inhibitory neurons, could contribute to the mechanisms of long-term enhancement of synapses and circuits in the nervous system.

In similar studies, Rudy et al.[78] examined the localized expression of mRNA transcripts from a gene encoding components for voltage-gated K$^+$ channels known as KShIIIA. KShIIIA expression was particularly prominent throughout the dorsal thalamus. Transcripts of a closely related gene, NGKII-KV4, were particularly abundant in the cerebellar cortex, where KShIIIA expression was very weak. These results also demonstrate the existence of cell type-specific K$^+$ channel components and again suggest that a reason for the large diversity of K$^+$ channel proteins is the presence of subtypes that participate in specific brain functions. Most nuclei of the dorsal thalamus showed high levels of hybridization. Strong hybridization signals were also seen in the optical layer of the superior colliculus. Cortical areas and parts of the hippocampus show moderate levels of expression. NGKII-KV4 mRNAs were also concentrated in certain areas, but the pattern of expression was quite different from that exhibited by KShIIIA transcripts. NGKII-KV4 mRNA was particularly abundant in the cerebellar cortex, and was also seen in mitral cells of the olfactory bulb. The distribution of NGKII transcripts in the cortex, hippocampus, and thalamus was also quite different from that of KShIIIA mRNAs. Each gene explored in the study was expressed in the hippocampus with distinct patterns. KShIIIA was expressed particularly in the CA3 field, whereas NGKII-KV4 was expressed in the dentate gyrus as well as the CA3. RBKII was expressed throughout the hippocampal formation. The generalized and prominent expression of KShIIIA transcripts throughout, but limited to, the nuclei of the dorsal thalamus, is of particular interest. Expression levels in adjacent thalamic areas are much weaker or nonexistent. The area that gives rise to the dorsal thalamus is embryologically distinct from the areas that give rise to the epithalamus and the ventral thalamic nuclei. The dorsal thalamus is also distinct physiologically, as it corresponds to the parts of the thalamus that project to the cerebral cortex. Thus the results of these in situ hybridization studies provide important clues for the identification of channel function. KShIIIA and NGKII-KV4 proteins could be part of a number of different channel types, which includes high-voltage activating, delayed-rectifier-type channels, or components of inactivating channels.[78]

Overall the results of molecular cloning and subsequent in situ hybridization histochemistry studies clearly indicate the diversity of receptor/ion channel complexes is much larger than suspected and point to the necessity for improved ligands (and antibodies) for detecting the different combinations of receptor channel subunits that exist in receptors, such as the nACh and GABA$_A$ receptors and the various voltage-regulated channels. Studies that use these more specific molecular probes will reveal the correlation between mRNA abundance and densities of channel sites, and allow conclusions about the likely ion channel/receptor turnover rates in different cell populations and/or pathways.

Studies of Ion Channel Receptor Proteins in Human Brain and Clinical Applications of Ion Channel Research

In this monograph series it is relevant to discuss the importance of such a large body of basic research findings to the clinical field, and to detail some of the applications of the important pharmacologic tools and techniques utilized in basic research for the detection of ion channel/

receptor proteins in human brain (or other tissues) and the elucidation of changes in various pathologic and disease states. Since their introduction, radioligand binding and autoradiographic techniques have been used to describe the distribution of a large number of ion channel/receptors in human brain, including Ca^{2+} antagonist-binding sites for the DHPs,[79] voltage-operated Na^+ channels,[80] and the ligand-gated channels, including the nACh receptor, the $GABA_A$ receptor, and the strychnine-sensitive glycine receptor (eg, Palacios et al.[81]). Similar techniques have also been used to describe changes in human disease states such as Alzheimer's disease and Huntington's disease. Such studies have demonstrated a substantial reduction in the density of DHP binding sites in the striatum of patients with Huntington's disease[82] and various specific changes in the density of a number of ion channel-linked binding sites, including apamin binding to the calcium-dependent K^+ channel,[83] and the channel/receptors for the excitatory amino acid glutamate.[84,85] Specific diseases have been associated with direct, specific channel defects, or defects that ultimately affect ion channel expression or function. Among them, inherited myoclonus or spasticity in mice and cattle, which involves an abnormal expression of the Cl^- channel-linked glycine receptor.[86,87].

Obviously these convenient and powerful techniques can be applied to assess the effect of many other parameters on human ion channel number or expression, including such factors as alcohol[88] and epilepsy. Novel clinical applications have been suggested by basic research that examines ion channel receptor characteristics, including the efficacy of Ca^{2+}-channel antagonists in schizophrenia[20] and the ability of blockers of the NMDA receptor-coupled ion channel to prevent excitotoxic damage following ischemic or anoxic episodes. Activators of ATP-sensitive K^+ channels may also fulfill a similiar protective role against the latter brain insults.[89]

The pharmacologic sensitivity of a particular ion channel is relevant to therapeutic drug applications and to the possible treatment of ion channels as pharmacologic receptors. As discussed above, ion channels exist as homologus protein families with specific drug binding sites that exhibit different structure-activity relationships. Antagonists and agonists have been found for most ion channels, some of which are associated with GTP binding proteins, a characteristic that makes them subject to regulation by hormones and drugs.

Future Developments

Radioligand binding studies and autoradiography will continue to have an important role in furthering our understanding of ion channel structure, localization, and function. New ligands are becoming available and their development can, in fact, be aided by the convenient screening of new and existing toxins or experimentally/clinically-used drugs already known to be effective. In addition, newer toxins such as ω-Aga-IIIA, which is isolated from the venom of the funnel-web spider *Agelenopsis aperta* and has been identified in electrophysiologic experiments, can be radiolabeled and used in biochemical studies to identify novel channel binding sites and the existence of novel channels.[90] It is clear that significant progress has been made in the understanding of ion channel proteins and their associated drug receptor binding sites. Advances in this area have been made possible by the use of radioligand binding studies and localization of binding sites in the brain (and other tissues), as well as the differential localization of various ion channel receptor gene products throughout the CNS. Undoubtedly the future will bring about the development of even more selective channel ligands, further genetic analysis of channel structure and function, and, in the long run, development of new therapeutic modalities that possibly involve the identification of ion channel defects in human disease states. Pharmacologic studies will focus attention on the area of drug-induced channel modulation. For example, efforts to elucidate the role of various channel ligands in neuronal cell protection is currently an area of great research interest.

Acknowledgments. I would like to thank Kate Ashley for excellent secretarial assistance, Roy Larkin for reproducing the figures, Richard Loiacono for helpful comments on the manuscript, and Piero Foà for his immense patience.

References

1. Jan LY, Jan YN. Voltage-sensitive ion channels. Cell 1989; 56:13–25.
2. Catterall WA. Structure and function of voltage-sensitive ion channels. Science 1988; 242:50–61.
3. Rudy B. Diversity and ubiquity of K channels. Neuroscience 1988; 25: 729–749.
4. Betz H. Ligand-gated ion channels in the brain: The amino acid receptor superfamily. Neuron 1990; 5:383–392.
5. Gordon D. Ion channels in nerve and muscle cells. Curr Opin Cell Biol 1990; 2:695–707.
6. Hille B. Ionic channels in excitable membranes. 2nd ed. Sunderland, Mass.: Sinauer Associates; 1991.
7. Miller RJ. Voltage-sensitive Ca^{2+} channels. J Biol Chem 1992; 267:1403–1406.
8. Bormann J, Hamill OP, Sakmann B. Mechanism of anion permeation through channels gated by glycine and gamma-aminobutyric acid in mouse cultured spinal neurons. J Physiol (Lond) 1987; 385:243–286.

9. Beneski DA, Catterall WA. Covalent labeling of protein components of the sodium channel with a photoactivable derivative of scorpion toxin. Proc Natl Acad Sci USA 1980; 77:639–643.

10. Sharkey RG, Beneski DA, Catterall WA. Differential labelling of the α and β1 subunits of the sodium channel by photoreactive derivatives of scorpion toxin. Biochemistry 1984; 23:6078–6086.

11. Barchi RL, Cohen SA, Murphy LE. Purification from rat sacrolemma of the saxitoxin-binding component of the excitable membrane sodium channel. Proc Natl Acad Sci USA 1980; 77:1306–1310.

12. Triggle DJ, Janis RA. Calcium channel ligands. Ann Rev Pharmacol Toxicol 1987; 27:347–369.

13. Bellemann P, Ferry D, Lubbecke F, Glossmann H. [³H]Nimodipine and [³H]nitrendipine as tools to directly identify the sites of action of 1,4-dihydropyridine calcium antagonists in guinea pig tissues. Arzneim-Forsch/Drug Res 1982; 32:361–363.

14. Murphy KMM, Snyder SH. Calcium antagonist receptor binding sites labeled with [³H]nitrendipine. Eur J Pharmacol 1982; 77:201–202.

15. Lee HR, Roeske WR, Yamamura HI. High affinity specific [³H](+)PN 200-110 binding to dihydropyridine receptors associated with calcium channels in rat cerebral cortex and heart. Life Sci 1984; 35:721–732.

16. Gould RJ, Murphy KMM, Reynolds IJ, Snyder SH. [³H]nitrendipine-labeled calcium channels discriminate inorganic calcium agonists and antagonists. Proc Natl Acad Sci USA 1982; 79:3656–3660.

17. Yamamura HI, Schoemaker H, Boles RG, Roeske WR. Diltiazem enhancement of [³H]nitrendipine binding to calcium channel-associated drug receptor sites in rat brain synaptosomes. Biochem Biophys Res Commun 1982; 108:640–646.

18. Holck M, Fischli W, Hengartner U. Effects of temperature and allosteric modulators of [³H]nitrendipine binding: Methods for detecting potential Ca²⁺ channel blockers. J Receptor Res 1984; 4:557–695.

19. Boles RG, Yamamura HI, Schoemaker H, Roeske WR. Temperature-dependent modulation of [³H]nitrendipine binding by the calcium channel antagonists verapamil and diltiazem in rat brain synaptosomes. J Pharmacol Exp Ther 1984; 229:333–339.

20. Reynolds IJ, Snyder SH. Calcium antagonist receptors. In: Narahashi T, ed. Ion Channels, Vol 1. New York: Plenum Press; 1988:213–249.

21. Ferry DR, Rombusch M, Goll A, Glossmann H. Photoaffinity labelling of Ca²⁺ channels with [³H]azidopine. FEBS Lett 1984; 169:112–118.

22. Reynolds IJ, Snowman AM, Snyder SH. (−)[³H]Desmethoxyverapamil labels multiple calcium channel modulator receptors in brain and skeletal muscle membranes: Differentiation by temperature and dihydropyridines. J Pharmacol Exp Ther 1986; 237:731–738.

23. Olivera BM, Rivier J, Clark C, Ramilo CA, Corpuz GP, Abogadie FC, Mena EE, Woodward SR, Hillyard DR, Cruz LJ. Diversity of Conus neuropeptides. Science 1990; 249:257–263.

24. Sher E, Clementi F. ω-Conotoxin-sensitive voltage-operated calcium channels in vertebrate cells. Neuroscience 1991; 42:301–307.

25. Cruz LJ, Olivera BM. Calcium channel antagonists. ω-Conotoxin defines a new high affinity site. J Biol Chem 1986; 261:6230–6233.

26. Takemura M, Fukui H, Wada H. Different localization of receptors for ω-conotoxin and nitrendipine in rat brain. Biochem Biophys Res Commun 1987; 149:982–988.
27. Dooley DJ, Lickert M, Lupp A, Osswald H. Distribution of ^{125}I-ω-conotoxin GVIA and ^{3}H-isradipine binding sites in the central nervous system of rats of different ages. Neurosci Lett 1988; 93:318–323.
28. Wagner JA, Snowman AM, Biswas A, Olivera BM, Snyder SH. ω-Conotoxin GVIA binding to a high-affinity receptor in brain: Characterization, calcium sensitivity and solubilization. J Neurosci 1988; 8:3354–3359.
29. Herdon H, Nahorski SR. Investigations of the roles of dihydropyridine and ω-conotoxin-sensitive calcium channels in mediating depolarization-evoked endogenous dopamine release from striatal slices. Naunyn-Schmiedeberg's Arch Pharmacol 1989; 340:36–40.
30. Dooley DJ, Lupp A, Hertting G, Osswald H. ω-Conotoxin GVIA and pharmacological modulation of hippocampal noradrenaline release. Eur J Pharmacol 1988; 148:261–267.
31. Lundy PM, Frew R, Fuller TW, Hamilton MG. Pharmacological evidence of a ω-conotoxin, dihydropyridine-insensitive neuronal Ca^{2+} channel. Eur J Pharmacol Mol Pharmacol Sect 1991; 206:61–68.
32. Lundy PM, Hong A, Frew R. Inhibition of a dihydropyridine, ω-conotoxin insensitive Ca^{2+} channel in rat synaptosomes by venom of the spider *Hololena curta*. Eur J Pharmacol Mol Pharmacol Sect 1992; 225:51–56.
33. Ashcroft FM. Adenosine 5'-triphosphate-sensitive potassium channels. Annu Rev Neurosci 1988; 11:97–118.
34. Triggle DJ. Potassium channels and potassium channel modulators. In: Neumeyer JL, ed. Neurotransmissions, Vol VI. Natick, Mass.: Research Biochemicals; 1990:1–5.
35. Bernardi H, Fosset M, Lazdunski M. Characterization, purification and affinity labeling of brain [^{3}H]glibenclamide-binding protein, a putative neuronal ATP-regulated K^+ channel. Proc Natl Acad Sci USA 1988; 85:9816–9820.
36. Mourre C, Ben AY, Bernardi H, Fosset M, Lazdunski M. Antidiabetic sulfonylureas: Localization of binding sites in the brain and effects on the hyper-polarization induced by anoxia in hippocampal slices. Brain Res 1989; 486:159–164.
37. Hughes M, Romey G, Duval D, Vincent JP, Lazdunski M. Apamin as a selective blocker of the calcium-depend potassium channel on neuroblastoma cells: Voltage-clamp and biochemical characterization of the toxin receptor. Proc Natl Acad Sci USA 1982; 79:1308–1312.
38. Miller C, Moczydlowski E, Latorre R, Phillips M. Charybdotoxin, a high affinity protein inhibitor of single Ca^{2+}-activated K^+ channels of mammalian skeletal muscle. Nature 1985; 313:316–318.
39. Farley J, Rudy B. Multiple types of voltage-dependent Ca^{2+} activated K^+ channels of large conductance in rat brain synaptosomal membrane. Biophys J 1988; 53:919–934.
40. Vázquez J, Feigenbaum P, King VF, Kaczorowski GJ, Garcia ML. Characterization of high affinity binding sites for charybdotoxin in synaptic plasma membranes from rat brain. J Biol Chem 1990; 265:15564–15571.

41. Halliwell JV, Othman IB, Pelchen-Matthews A, Dolly JO. Central action of dendrotoxin: Selective reduction of a transient K^+ conductance in hippocampus and binding to localized receptors. Proc Natl Acad Sci USA 1986; 83:493–497.

42. Penner R, Petersen M, Pierau FK, Dreyer F. Dendrotoxin: A selective blocker of a non-inactivating potassium current in guinea-pig dorsal root ganglion neurones. Pfluegers Arch Gen Physiol 1986; 407:365–369.

43. Awan KA, Dolly JO. K^+ channel sub-types in rat brain: Characteristic locations revealed using β-bungarotoxin, α- and δ-dendrotoxins. Neuroscience 1991; 40:29–39.

44. Cortés R, Supavilai P, Karobath M, Palacios JM. Calcium antagonist binding sites in the rat brain: Quantitative autoradiographic mapping using the 1,4-dihydropyridines [^{3}H]PN 200-110 and [^{3}H]PY 108-068. J Neural Transm 1984; 60:169–197.

45. Cortés R, Supavilai P, Karobath M, Palacios JM. The effects of lesions in the rat hippocampus suggest the association of calcium channel blocker binding sites with specific neuronal population. Neurosci Lett 1983; 42:249–254.

46. Supavilai P, Cortés R, Palacios JM, Karobath M. Calcium entry blockers: autoradiographic mapping of their binding sites in rat brain. Prog Brain Res 1985; 63:89–95.

47. Ferry DR, Goll A, Gadon C, Glossmann H. $(-)$-^{3}H-desmethoxyverapamil labelling of putative calcium channels in brain: autoradiographic distribution and allosteric coupling to 1,4-dihydropyridine and diltiazem binding sites. Naunyn-Schmiedeberg's Arch Pharmacol 1984; 327:183–187.

48. Kerr LM, Filloux F, Olivera BM, Jackson H, Wamsley JK. Autoradiographic localization of calcium channels with [^{125}I]ω-conotoxin in rat brain. Eur J Pharmacol 1988; 146:181–183.

49. Takemura M, Kiyama H, Fukui H, Tohyama M, Wada H. Distribution of the ω-conotoxin receptor in rat brain. An autoradiographic mapping. Neuroscience 1989; 32:405–416.

50. Ahlijanian MK, Westenbroek RE, Catterall WA. Subunit structure and localization of dihydropyridine-sensitive calcium channels in mammalian brain, spinal cord, and retina. Neuron 1990; 4:819–832.

51. Westenbroek RE, Ahlijanian MK, Catterall WA. Clustering of L-type Ca^{2+} channels at the base of major dendrites in hippocampal pyramidal neurons. Nature 1990; 347:281–284.

52. Mourre C, Cervera P, Lazdunski M. Autoradiographic analysis in rat brain of the postnatal ontogeny of voltage-dependent Na^+ channels, Ca^{2+}-dependent K^+ channels and slow Ca^{2+} channels identified as receptors for tetrodotoxin, apamin and $(-)$desmethoxyverapamil. Brain Res 1987; 417:21–32.

53. Olsen RW, McCabe RT, Wamsley JK. $GABA_A$ receptor subtypes: Autoradiographic comparison of GABA, benzodiazepine, and convulsant binding sites in the rat central nervous system. J Chem Neuroanat 1990; 3:59–76.

54. Levitan ES, Schofield PR, Burt DR, Rhee LM, Wisden W, Köhler M, Fujita N, Rodriguez HF, Stephenson A, Darlison MG, Barnard EA, Seeburg PH. Structural and functional basis for $GABA_A$ receptor heterogeneity. Nature 1988; 335:76–79.

55. Mourre C, Widmann C, Lazdunski M. Saxitoxin-sensitive Na$^+$ channels: Presynaptic localization in cerebellum and hippocampus of neurological mutant mice. Brain Res 1990; 533:196–202.
56. Vilaró MT, Martinez-Mir MI, Sarasa M, Pompeiano M, Palacios JM, Mengod G. Molecular neuroanatomy of neurotransmitter receptors: The use of *in situ* hybridization histochemistry for the study of their anatomical and cellular localization. In: Osborne NN, ed. Current Aspects of the Neurosciences, Vol 3. London: Macmillan Press; 1991:1–36.
57. Wada K, Ballivet M, Boulter J, Connolly J, Wada E, Deneris ES, Swanson LW, Heinemann S, Patrick J. Functional expression of a new pharmacological subtype of nicotonic acetylcholine receptor. Science 1988; 240:330–334.
58. Wada E, Wada K, Boulter J, Deneris E, Heinemann S, Patrick J, Swanson L. Distribution of alpha2, alpha3, alpha4 and beta2 neuronal nicotinic receptor subunit mRNAs in the central nervous system: A hybridization histochemical study in the rat. J Comp Neurol 1989; 284:314–335.
59. Clarke PBS, Schwartz RD, Paul SM, Pert CD, Pert A. Nicotinic binding in rat brain: Autoradiographic comparison of [^{3}H]acetylcholine, [^{3}H]nicotine and [^{125}I]-α-bungarotoxin. J Neurosci 1985; 5:1307–1315.
60. Schofield PR, Darlison MG, Fujita N, Burt DR, Stephenson FA, Rodriguez H, Rhee LM, Ramachandran J, Reale V, Glencorse TA, Seeburg PH, Barnard EA. Sequence and functional expression of the GABA$_A$ receptor shows a ligand-gated receptor super-family. Nature 1987; 328:221–227.
61. Pritchett DB, Sontheimer H, Shivers BD, Ymer S, Kettenmann H, Schofield PR, Seeburg PH. Importance of a novel GABA$_A$ receptor subunit for benzodiazepine pharmacology. Nature 1989; 338:582–585.
62. Wisden W, Laurie DJ, Monyer H, Seeburg PH. The distribution of 13 GABA$_A$ receptor subunit messenger RNAs in the rat brain I. Telencephalon, diencephalon, mesencephalon. J Neurosci 1992; 12:1040–1062.
63. Wisden W, Morris BJ, Darlison MG, Hunt SP, Barnard EA. Localization of GABA$_A$ receptor α-subunit mRNAs in relation to receptor subtypes. Mol Brain Res 1989; 5:305–310.
64. Pritchett DB, Lüddens H, Seeburg PH. Type I and Type II GABA$_A$ benzodiazepine receptors produced in transfected cells. Science 1989; 245:1389–1392.
65. Shivers BD, Killisch I, Sprengel R, Sontheimer H, Köhler M, Schofield PR, Seeburg PH. Two novel GABA$_A$ receptor subunits exist in distinct neuronal subpopulations. Neuron 1989; 3:327–337.
66. Unnerstall JR, Kuhar MJ, Niehoff DL, Palacios JM. Benzodiazepine receptors are coupled to a subpopulation of γ-aminobutyric acid (GABA) receptors: Evidence from a quantitative autoradiographic study. J Pharmacol Exp Ther 1981; 218:797–804.
67. Palacios JM, Wamsley JK, Kuhar MJ. High affinity GABA receptors: Autoradiographic localization. Brain Res 1981; 222:285–307.
68. Betz H. Glycine receptors: Heterogeneous and widespread in the mammalian brain. Trends Neurosci 1991; 14:458–461.
69. Sato K, Zhang J-H, Saika T, Sato M, Tada K, Tohyama M. Localization of glycine receptor α$_1$ subunit mRNA-containing neurons in the rat brain: An

analysis using *in situ* hybridization histochemistry. Neuroscience 1991; 43: 381–395.

70. Hoch W, Betz H, Becker C-M. Primary cultures of mouse spinal cord express the neonatal isoforms of the inhibitory glycine receptor. Neuron 1989; 3:339–348.

71. Malosio M-L, Marquèze-Pouey B, Kutse J, Betz H. Widespread expression of glycine receptor subunit mRNAs in the adult and developing rat brain. EMBO J 1991; 10:2401–2409.

72. Furuyama T, Sato M, Sato K, Araki T, Inagaki S, Takagi H, Tohyama M. Co-expression of glycine receptor β subunit and GABA$_A$ receptor γ subunit mRNA in the rat dorsal root ganglion cells. Mol Brain Res 1992; 12:335–338.

73. Keinänen K, Wisden W, Sommer B, Werner P, Herb A, Verdoorn TA, Sakmann B, Seeburg PH. A family of AMPA-selective glutamate receptors. Science 1990; 249:556–560.

74. Snutch TP, Leonard JP, Gilbert MM, Lester HA, Davidson N. Rat brain expresses a heterogeneous family of calcium channels. Proc Natl Acad Sci USA 1990; 87:3391–3395.

75. Chin H, Smith MA, Kim H-L, Kim H. Expression of dihydropyridine-sensitive brain calcium channels in the rat central nervous system. FEBS Lett 1992; 299:69–74.

76. Dubel SJ, Starr TVB, Hell J, Ahlijanian MK, Enyeart JJ, Catterall WA, Snutch TP. Molecular cloning of the α-1 subunit of an ω-conotoxin-sensitive calcium channel. Proc Natl Acad Sci USA 1992; 89:5058–5062.

77. Tsaur M-L, Sheng M, Lowenstein DH, Jan YN, Jan LY. Differential expression of K$^+$ channel mRNAs in the rat brain and down-regulation in the hippocampus following seizures. Neuron 1992; 8:1055–1067.

78. Rudy B, Kentros C, Weiser M, Fruhling D, Serodio P, Vega-Saenz de Miera E, Ellisman MH, Pollock JA, Baker H. Region-specific expression of a K$^+$ channel gene in brain. Proc Natl Acad Sci USA 1992; 89:4603–4607.

79. Peroutka SJ, Allen GS. Calcium channel antagonist binding sites labelled by ^{3}H-nimodipine in human brain. J Neurosurg 1983; 59:933–937.

80. Mourre C, Moll C, Lombet A, Lazdunski M. Distribution of voltage-dependent Na$^+$ channel identified by high affinity receptors for tetrodotoxin and saxitoxin in rat and human brains: Quantitative autoradiographic analysis. Brain Res 1988; 448:128–139.

81. Palacios JM, Probst A, Cortés R. Mapping receptors in the human brain. Trends Neurosci 1986; 9:284–289.

82. Watson DL, Carpenter CL, Marks SS, Greenberg DA. Striatal calcium channel antagonist receptors in Huntington's Disease and Parkinson's Disease. Ann Neurol 1988; 23:303–305.

83. Ikeda M, Dewar D, McCulloch J. Selective reduction of [125I]apamin binding sites in Alzheimer hippocampus: A quantitative autoradiographic study. Brain Res 1991; 567:51–56.

84. Jansen KLR, Faull RLM, Dragunow M, Synek BL. Alzheimer's disease: Changes in hippocampal N-methyl-D-aspartate, quisqualate, neurotensin, adenosine, benzodiazepine, serotonin and opioid receptors—An autoradiographic study. Neuroscience 1990; 39:613–627.

85. Dewar D, Chalmers DT, Graham DI, McCulloch J. Glutamate metabotropic and AMPA binding sites are reduced in Alzheimer's disease: An autoradiographic study of the hippocampus. Brain Res 1991; 553:58–64.

86. Becker C-M. Disorders of the inhibitory glycine receptor: The *spastic* mouse. FASEB J 1990; 4:2767–2771.
87. Gundlach AL. Disorder of the inhibitory glycine receptor: Inherited myoclonus in Poll Hereford calves. FASEB J 1990; 4:2761–2766.
88. Kril JJ, Gundlach AL, Dodd PR, Johnston GAR, Harper CG. Cortical dihydropyridine binding sites are unaltered in human alcoholic brain. Ann Neurol 1989; 26:395–397.
89. Ben-Ari Y, Krnjevic K, Crépel V. Activators of ATP-sensitive K^+ channels reduce anoxic depolarization in CA3 hippocampal neurons. Neuroscience 1990; 37:55–60.
90. Mintz IM, Venema VJ, Adams ME, Bean BP. Inhibition of N- and L-type Ca^{2+} channels by the spider venom toxin ω-Aga-IIIA. Proc Natl Acad Sci USA 1991; 88:6628–6631.

23
Chloride Channels in Cystic Fibrosis

CAROLE M. LIEDTKE

Historical Perspective

Cystic fibrosis (CF), a genetic disease that affects 1 in 2000 Caucasian newborn babies, is often described as a generalized exocrinopathy with altered electrolyte and macromolecular secretion. The syndrome, characterized by chronic obstruction and infection of the respiratory tract, insufficiency of the exocrine pancreas, and elevated levels of sweat electrolytes, was first described in the 1930s,[1] although references to signs and symptoms possibly related to CF have appeared as early as the mid-1600s. Thus some reports refer to cases of steatorrhea and pancreatic insufficiency, while folktales[2] and European folk literature[1,3-6] refer to the ominous significance and early death associated with salty sweat. An 1838 report of meconium ileus and its complications by Rokintansky[3] was followed by more complete descriptions of newborns with symptoms typical of CF.[7] Pancreatic insufficiency received attention before the turn of the century for its characteristic effects on growth[8-10] and was attributed, in some reports, to an inborn error of metabolism.[11]

By the late 1930s, CF was recognized as a disease entity with a defined clinical course. In 1936, Fanconi described two children with clinical features of CF,[1,12] but the first comprehensive description of CF was published by Andersen, who named the disease "cystic fibrosis of the pancreas,"[13] describing its course in infancy, early and late childhood, and relating each stage to the presence of intestinal obstruction, malabsorption, and respiratory complications. In 1946, Andersen and Hodges, recognizing the possible familial basis of CF-related symptoms, proposed that CF was an autosomal recessive inherited disease.[14] The identification of the CF gene and of the protein encoded by this gene soon followed.[15-17]

After the discoveries that sweat duct chloride impermeability played a major role in the high salt content of the sweat[18,19] and that abnormal bioelectric properties of large airway epithelia reflected dysfunctional Na^+ and Cl^- channel activity,[20] the Cl^- channel became a focal point of

CF research efforts. Subsequent studies on sweat secretory coil, nasal epithelium, and small intestine revealed similar abnormalities of chloride permeability, which are still not fully understood. Nevertheless, it has been demonstrated that the product of the CF gene is a cAMP-regulated Cl⁻ channel that is selectively expressed in the apical plasma membrane.[21-23]

Cl⁻ Channels in Epithelia

Cl⁻ channels subserve various functions, including regulation of cell volume,[24-27] stabilization of the membrane potential,[28,29] signal transduction,[30,31] and transepithelial transport.[32,33] Alterations in the Cl⁻ channels themselves or in their regulation result in distinctive phenotypes, the most prevalent of which is CF. In CF transepithelial electrolyte transport is impaired. Transepithelial electrolyte transport is most often thought of as the movement of the electrolytes sodium and chloride, but in actuality also includes the transport of other ions, such H^+, HCO_3^-, and K^+. The focus of this chapter will be Na^+ and Cl^- ions.

In mammalian epithelia, net movement of electrolytes followed by the osmotic flow of water produces either fluid absorption or secretion. Apical or basolateral membrane Cl⁻ channels may play a role in transepithelial transport. In one model of fluid absorption, the current across the epithelial cell layer is carried by sodium as it enters the cell through the apical plasma membrane down the sodium electrochemical gradient (Figure 23.1). Sodium exits in the cell as a result of basolateral

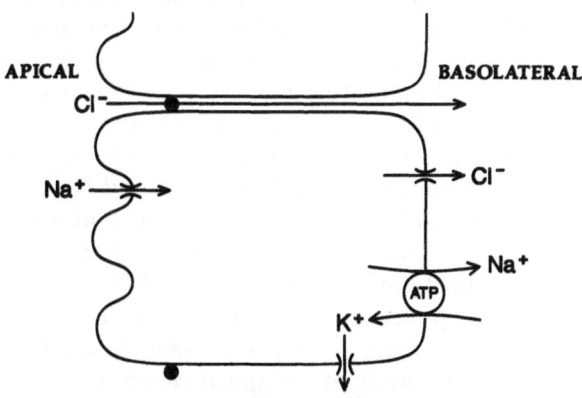

Na Absorption in Airway Epithelium

FIGURE 23.1. Model of chloride channel activity during absorption in airway epithelium. Sodium enters the cell from the lumen through an amiloride-sensitive channel and exits the cell through an $(Na^+ + K^+)$-ATPase in the basolateral membrane. Chloride moves across the epithelium either through the two limiting plasma membranes of the cells or through a paracellular pathway between cells.

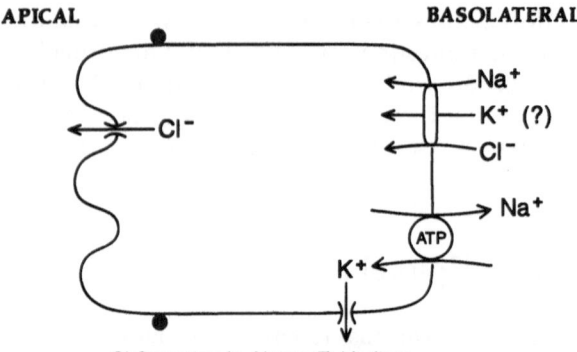

Cl Secretion in Airway Epithelium

FIGURE 23.2. Model for chloride secretion in airway epithelium. Three transporters operate synergistically during chloride secretion. A basolateral NaCl(K) cotransporter provides chloride for secretion by direct coupling of chloride entry to the sodium electrochemical gradient across the basolateral plasma membrane. Basolateral $(Na^+ + K^+)$-ATPase maintains a low sodium electrochemical gradient by pumping sodium from the cell to the serosa. If potassium is transported by the cotransporter, two chloride equivalent enter the cell. Chloride exits the cell at the apical membrane through a channel(s) allowing electrodiffusive movement of chloride. The movement of chloride out of the cell is electrically coupled to the exit of potassium through a channel in the basolateral membrane. Movement of sodium through tight junctions from serosa to lumen maintains electroneutrality.

ATP-dependent $(Na^+ + K^+)$-ATPase activity, which maintains the sodium electrochemical gradient. Charge compensation is achieved by the movement of Cl^- ions through the tight junction from the apical to the basolateral compartment or exiting the cell through a basolateral conductance pathway. In another model, sodium entry at the apical plasma membrane is coupled to a nonelectrolyte amino acid or sugar, and chloride enters through a conductance pathway. During fluid secretion, the current across the epithelial cell layer is carried by chloride (Figure 23.2). Chloride is supplied to the cell by a NaCl(K) cotransport located in the basolateral plasma membrane. The coupled movement of sodium down the electrochemical gradient provides the driving force for chloride and, in some epithelia, potassium entry. In the latter case, electroneutral uptake is maintained by entry of two Cl^- ions. Chloride exits the cell through a conductance pathway at the apical membrane, the charge being compensated by the flow of potassium from the cell to the basolateral compartment through a basolateral potassium conductance pathway. Sodium is pumped from the cell by a $(Na^+ + K^+)$-ATPase. Electroneutrality of the steady state is maintained by the movement of sodium from basolateral to apical compartments through the tight junctions.

In CF, the genetic defect affects apical sodium conductance[32,34] and β-adrenergic activation of apical chloride conductance in airway epithelial cells,[33] small intestine,[35] and sweat gland epithelial cells.[36,37] Other epithelial electrolyte transporters share conductive and/or kinetic properties of normal transporters (Figures 23.1 and 23.2). Thus far there are no indications of abnormalities in the NaCl(K) cotransport,[38] basolateral K$^+$ channel,[39] and basolateral Cl$^-$ channel.[40] A general model for the modulation of airway epithelial transporters that includes a role for adrenergic receptors linked to intracellular second messengers, cAMP, calcium, inositol phosphates (IP$_2$, IP$_3$) and diglycerides, including diacylglycerol (DAG), is shown in Figure 23.3. According to this model, the hydrolysis of PtdIns(4,5)P$_2$ by PtdIns(4,5)P$_2$-specific phospholipase C yields a series of second messengers (IP$_2$, IP$_3$, and DAG).[41] IP$_3$ occupies intracellular receptors, including those linked to the release of calcium to the cytosol and the resulting increase in intracellular calcium levels directly activates the basolateral K$^+$ channels and, in combination with protein kinase C or calcium-calmodulin-dependent protein kinase (CaCdMPK), the apical chloride channels. Calcium and DAG are thought to activate basolateral NaCl(K) cotransport through a calcium-dependent protein kinase C. These responses appear to be normal in CF airway epithelial cells, which suggests that the α-adrenergic receptor-

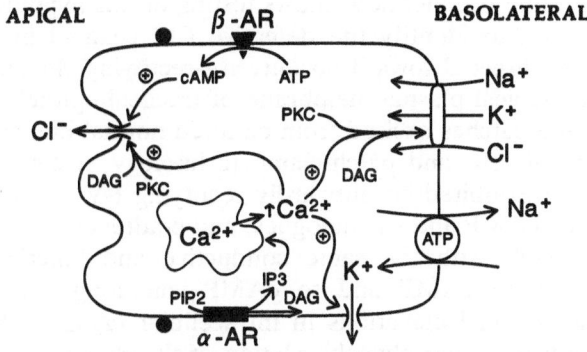

Regulation of Cl$^-$ Secretion in Airway Epithelium

FIGURE 23.3. Modulation of airway epithelial electrolyte transporters through adrenergic receptors. Basolateral α-adrenergic and β-adrenergic receptors mediate activation of PtdIn(4,5-P2)-PLC and adenylate cyclase, respectively, PtdIn(4,5-P2)-PLC produces a rapid breakdown of polyphosphoinositides with subsequent transient accumulation of the inositol phosphates IP2 and IP3 and decrease in inositol lipids. IP3 mobilizes intracellular Ca^{2+}. Elevated Ca^{2+} alone opens the basolateral K$^+$ channel while DAG plus Ca^{2+} activate Ca^{2+}-dependent protein kinase C which stimulates a basolateral NaCl(K) cotransporter and apical chloride channel. Increased cAMP levels directly activate apical chloride channels. AR = adrenergic receptor.

linked intracellular signaling pathway is not affected by the genetic defect.

Respiratory System

The first indication of an abnormal chloride permeability in respiratory epithelium came from the study of CF nasal and large airway epithelia. Conventional electrophysiologic techniques applied to fresh tissue and to cultured epithelial cells provided evidence for an elevated short circuit current (I_{SC}) that was insensitive to β-adrenergic stimulation and to replacement of luminal chloride with an impermeant anion,[42-45] in spite of a normal increase in intracellular cAMP levels in response to the β-adrenergic agonist isoproterenol.[133] This observation led to the conclusion that the defect in electrolyte transport in CF tissue was distal to adenylate cyclase.

Bradykinin induces normal chloride secretion in CF tissue, most likely by elevating intracellular calcium levels via an IP_3-mediated mechanism.[46] Protein kinase C also plays a role in airway epithelial chloride secretion, but is ineffective in CF tissue.[46,47] This suggests that the same Cl^- channel may be regulated by an interaction between cAMP-dependent protein kinase A and protein kinase C pathways. Discovery of the CF gene and its protein product now allows testing of this hypothesis.

The first efforts to identify the defective Cl^- channel in CF airway epithelium were directed toward an outward-rectifying channel of about 25–50 pS in the apical plasma membrane of tracheal epithelial cells.[33,48] Using membrane patches excised from cultured normal human nasal and tracheal epithelial cells and patch-clamp technology, it could be shown that the channel exhibited an outwardly rectifying I-V relationship, and was activated by cAMP and its analogues, and β-adrenergic agents. In CF cells, Cl^- channels with the same conductive and kinetic properties lacked sensitivity to cAMP and to cAMP-generating agents, despite normal agonist-induced elevations in intracellular cAMP.[33] Subsequent experiments indicated that phosphorylation of the channel itself or of an associated regulatory protein had a critical role in the activation of the outward rectifying Cl^- channel. Indeed, the addition of the catalytic subunit of cAMP-dependent protein kinase A and of ATP to excised inside-out membrane patches activated the outward-rectifying Cl^- channel in normal, but not in CF cells.[49,50] Similarly, protein kinase C activators stimulated the outward rectifier in normal cells only.[47] It should be pointed out, however, that this channel occurs infrequently in cell-attached membrane patches.[39,51,52] Thus the question of the role of the outward-rectifying Cl^- channel in transcellular chloride transport in airway epithelium remains unresolved and is further confounded by the discovery that the product of the CF transmembrane conductance

TABLE 23.1. Regulation of chloride channels in human airway epithelium.

Regulated by	Airway segment	Subcellular location	Reference
cAMP	Trachea, nasal polyps	Apical	53–55
		Apical	48
		Apical	133
Ca^{2+}	Trachea, nasal polyps	?	46
		?	57
		Apical	47
		Apical	59, 60
		Apical	58
		Apical	62
Volume	Trachea, nasal polyps	?	57
		Apical	63
		Apical	64
?	Trachea	Basolateral	134

regulator (CFTR) gene is a small-conductance, cAMP-activated Cl⁻ channel.

Small-conductance, chloride-selective channels have been observed in the apical membrane of nasal and tracheal epithelial cells.[48,53–55] At the single-cell level, a nonrectifying 20 pS channel was detected in 24% of successful patches, thus classifying it as the most common Cl⁻ channel in normal cells. However, in CF cells its distribution in the apical membrane and open probability were significantly less than normal.[53] Application of protein kinase A, cAMP, and ATP to excised inside-out membrane patches activated this Cl⁻ channel. Noise analysis of whole cell currents demonstrated that channels of 4.5 pS in human nasal epithelial cells carry most of the chloride current.[55] Single channels of this size were also observed in excised patches. How do the small-conductance Cl⁻ channels relate to CFTR Cl⁻ channels? As of now we do not know. However, the data strongly suggests that most chloride conductance is due to small Cl⁻ channels and hence these may be sites of severe abnormalities.

Other studies using patch-clamp technology confirm the observation that, in addition to the cAMP-regulated Cl⁻ channels, the CF airway epithelial cells contain other Cl⁻ channels characterized by different conductive and kinetic properties and activated by specific stimuli (Table 23.1).[56–58] Thus, as noted above, elevated intracellular calcium activates Cl⁻ channels as well as basolateral K⁺ channels in airway epithelia.[46,59–62] In CF cells, these calcium-activated Cl⁻ channels differ little from normal cells. Whole-cell patch-clamp experiments demonstrate that ionomycin-induced chloride currents in CF cells are normal and are regulated by multifunctional CaCdMPK.[58] In nystatin-permeabilized airway epithelial cells, elevation of intracellular calcium activated apical Cl⁻ channels in normal and CF cells,[56] which indicated again that in CF cells, the function of calcium-activated Cl⁻ channels is normal.

Hypo-osmotic media also induce a large voltage-dependent chloride current in cultured airway epithelial cells.[63,64] The currents generated by volume-regulated Cl^- channels differ significantly from those generated by CFTR Cl^- channels. Thus the volume-regulated Cl^- channel is a 35–50 pS outward-rectifying channel that is inhibited by disulfonic stilbenes (such as 4,4'-diisothiocyanostilbene-2,2'-disulfonic acid [DIDS]), is inactivated by depolarizing voltages, and is expressed normally in CF airway epithelial cells.[64] The observation that K^+ channels are also activated in a hypo-osmotic environment suggests an alternative explanation for the volume-induced activation of Cl^- channels.[65] The opening of basolateral K^+ channels alters the charge distribution within the cell. This promotes the movement of chloride through apical Cl^- channels that are open and provide charge compensation under resting conditions. These studies show that CF Cl^- channels participate in the regulation of cell volume, but the significance of this regulation in the physiology of airway epithelial cells is not clear. The intracellular mechanism of signal transduction of cell swelling and of calcium-activation of apical Cl^- channels have not been clarified. More information is also needed on the contribution of different apical Cl^- channels to chloride secretion.

Cystic Fibrosis Transmembrane Conductance Regulator

The isolation of the gene for CF opened new opportunities to explore the molecular basis of the disease. The CF gene is localized to the long arm of chromosome 7^{66-68} and encodes a protein of 1480 amino acids called the CF transmembrane conductance receptor (CFTR). The predicted protein resembles the traffic ATPase[69] or the ATP-binding cassette (ABC) transporters[70] and has two repeats each consisting of an ATP-binding domain followed by six transmembrane domains (Figure 23.4). A highly charged regulatory domain connects the two halves of CFTR. The regulatory domain has multiple consensus sites for phosphorylation by protein kinase A, protein kinase C, and CaCdMPK.[16] The first mutation observed in CF was a three-base pair deletion in exon 10 of the CFTR gene, which resulted in the loss of a single phenylalanine at codon 508, which was hence designated ΔF508. The ΔF508 deletion is present in 70% of all CF chromosomes. This deletion resides in a region of the CFTR protein that binds adenine nucleotides and, in a synthetic peptide, apparently causes a significant loss of β-sheet structure that is normally conferred by normal CFTR.[71,72] Subsequently, over 100 mutations of CF CFTR have been discovered. However, their presence does not appear to correlate well with the severity of lung disease.[73] Three nonsense mutations (G542X, R553X, and W1282X) are the next most frequent CF mutations,[73,74] but again these mutations do not correlate with the severity of lung disease.[75]

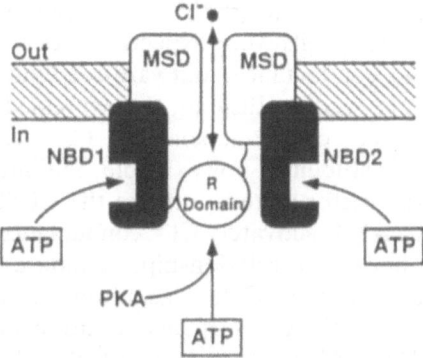

FIGURE 23.4. Model of Cystic Fibrosis Transmembrane Regulator. The model depicts a proposed domain structure and function of CFTR (Anderson, 1992). MSD = membrane-spanning domain; NBD = nucleotide-binding domains; PKA = cAMP-dependent protein kinase; R domain = regulatory domain. The hatched area represents the plasma membrane.

The function of CFTR and its role in abnormal chloride permeability in affected CF epithelia has been explored extensively. Gene transfer of full length, wild-type CFTR into CF cells corrected the defect in chloride permeability.[76,77] Transfection of CF pancreatic cell line (CFPAC-1) by retrovirus-mediated gene transfer[76] or expression of CFTR cDNA in cultured CF airway cells[77] resulted in expression of cAMP-stimulated chloride currents. Forskolin-stimulated current in transfected CFPAC-1 cells displayed a linear I-V relationship that was significantly decreased in chloride-free bathing medium.[76] Transfection with ΔF508-CFTR did not correct the defect in chloride conductance.[77] Thus the deletion in the gene product is related to defective chloride conductance. Direct intratracheal transfer of normal CFTR gene to the lungs of the cotton rat using a replication-deficient recombinant adenovirus vector led to the expression of CFTR mRNA 2 days after infection.[78] Eleven to 14 days after infection, human CFTR protein was detected in infected cells using human CFTR antibody. Introduction of CFTR cDNA into cultured CF airway epithelial cells using a retroviral-mediated gene transfer also corrected the ion-transporting defect.[79] The experiments serve as the basis for considering in vivo recombinant CFTR gene transfer to CF airway epithelial cells. However, despite the success of studies with animals, many questions on the efficacy and targeting of gene transfer remain to be answered.

Introduction of full-length CFTR cDNA into numerous cell lines that do not normally express CFTR induces cAMP-dependent chloride conductance. The cell types used for these studies include mouse fibroblasts[80,81] or L cells,[23] human HeLa cells,[80,81] Chinese hamster ovary

(CHO) cells,[82] *Xenopus* oocytes,[22,83,84] and army worm ovary (Sf9) cells.[85] Studies using these cells were expected to confirm the model of a cAMP-sensitive, large conductance, outward-rectifying Cl⁻ channel, as predicted from previous studies. Instead, the expressed CFTR was a low-(8–10 pS) conductance, cAMP-dependent Cl⁻ channel that was insensitive to disulfonic stilbenes. For example, the infection of a Sf9 cell line with a baculovirus vector system carrying the CFTR gene resulted in the expression of a cAMP-activated Cl⁻ conductance of 8.4 pS.[85] The channel displayed a linear I-V relationship. A mouse fibroblast cell line transfected with plasmid vector pCOF-1 acquired the ability to synthesize CFTR with localization in the cell surface and intracellular membranes.[81] A resulting inducible cAMP-dependent 10 pS chloride conductance closely resembled the linear current-voltage relationship of CFTR expressed in epithelial cells.[86] Although *Xenopus* oocytes may not be the best model for CFTR expression in a nonepithelial cell line, they are widely used to express CFTR. Initial reports indicated expression of cAMP-dependent 7.6 pS chloride conductance only after transfection with wild-type CFTR cDNA.[83] Since then, CFTR expression in oocytes has been found to affect endogeneous potassium conductance.[84] The molecular mechanism that mediates this effect of CFTR is not known, but the results clearly indicate the need for caution in selecting and interpreting data from nonepithelial cell lines.

Further evidence that CFTR is a Cl⁻ channel comes from studies at the molecular level. Site-directed mutations of lysines at positions 95 or 335 to acidic amino acids convert the Cl⁻ channel ion selectivity from Br > Cl > I > F to I > Br > Cl > F.[21] On the other hand, mutation of two arginine residues at positions 347 and 1030 to acidic residues did not affect anion selectivity of cAMP-sensitive Cl⁻ channels. Thus a mutation in the lysine residues is thought to alter anion selectivity. Information derived from hydropathy plots suggests that these sites may be localized in the outer half of the channel.

CFTR expression during cell differentiation has attracted much attention for its implication for the trafficking and/or processing of CFTR by CF epithelial cells. One early study suggested that the structure of CFTR may be a factor in improper intracellular processing and targeting of the ΔF508 CFTR.[87] Wild-type CFTR expressed in heterologous cells produced three polypeptide bands on SDS-PAGE: glycosylated bands A and B with Mr values of 130,000 and 135,000 Da, respectively, and a nonglycosylated band of Mr 150,000. ΔF508 CFTR, as well as several CF-associated mutations, failed to yield a fully glycosylated form of CFTR[87] or to generate cAMP-activated chloride conductance.[88] Although these studies suggest an interference of CFTR mutations in the glycosylation of CFTR in the Golgi apparatus with subsequent effects on trafficking from the endoplasmic reticulum to the plasma membrane, some mutated CFTR associated with severe CF phenotype, eg, G551D, are fully glycosylated.[88] Other evidence also suggests that the role of abnormal protein trafficking as the cause for the CF phenotype should be considered with caution. Thus in *Xenopus* oocytes transfected with mutant CFTR cDNAs, Cl$^-$ channel activation was elicited with high concentrations of phosphodiesterase inhibitor, which suggests that some mutant CFTR still reached the plasma membrane.[22] Using Vero cells transfected with recombinant vaccinia virus, Dalemans et al. demonstrated that less ΔF508 CFTR reached the plasma membrane than normal CFTR.[89] Functional analysis of the defective CFTR showed reduced activity of cAMP-dependent Cl$^-$ channels and longer than normal CFTR closed times. Other kinetic and conductive properties of the ΔF508-CFTR were found to be similar to those of normal CFTR. These studies illustrate the need for more information on the localization of CFTR in normal and CF airway epithelia.

Intestinal Tract

In addition to pancreatic insufficiency and subsequent deficiency of digestive enzymes, the intestinal abnormalities of CF are predominantly those due to obstructions with symptoms of meconium ileus in the newborn and "meconium ileus equivalent" in later life. At first, meconium ileus was attributed to quantitative or qualitative abnormalities in the secretion of macromolecules relative to that of osmotically active solutes.[90] However, the finding that the quantity of osmotically active solutes in the meconium is markedly decreased led to an examination of sodium and chloride secretion and absorption. Since the fluid content of intestinal material reflects the balance between sodium absorption and chloride secretion, problems with either or both electrolyte transport processes may contribute to CF intestinal pathophysiology. Thus excess sodium absorption mediated by amiloride-sensitive Na$^+$ channels in the apical intestinal plasma membrane may lead to dehydration of the intestinal lumen, which results in luminal material that is viscous and sticky. Similarly, altered regulation of intestinal Cl$^-$ channels may decrease the amount of water and ions, which contributes to the dehydration of the intestinal contents. Indeed, studies of ion-transporting intestinal epithelium of the rectum, jejunum, and ileum have provided evidence for abnormal electrolyte transport in CF intestine. The same studies also revealed marked differences in the regulation of electrogenic chloride secretion between intestinal and airway epithelia. In vivo studies of rectal epithelium show that CF is associated with a twofold increase in an amiloride-sensitive component of potential difference (PD)[91] and with loss of responsiveness to cAMP- and calcium-generating secretagogues.[92] This study also demonstrated that the rectal infusion of theophylline produced an increase in I$_{SC}$ in normal rectum, but not in the rectum of patients with CF, despite similar increases in cAMP levels. These results suggest that the β-adrenergic adenylate cyclase system of the rectum epithelium is not altered in CF. Further examination of cAMP-signal transduction showed that in CF, cAMP binds normally to its endogeneous receptors, that is, to the regulatory subunits of cAMP-dependent protein kinase I and II.[92] Thus the failure to respond to cAMP-generating agents is due to a defect distal to cAMP-binding proteins, which is possibly a defect of phosphoprotein that acts as a regulator and/or a transporter.

In vitro bioelectric studies of jejunal biopsy specimens demonstrate that in tissue from CF patients, PD or I_{SC} fail to respond to prostaglandin E_2 (PGE_2) or acetylcholine despite a normal response to mucosal addition of glucose.[93] The latter causes an increase in I_{SC} as a result of electrogenic sodium absorption, comparable to that found in normal tissue. Electrogenic pathways for electrolyte absorption also appear normal in ileal and rectal tissues from CF patients.[94] Thus normal absorption of sodium through amiloride-sensitive Na^+ channels is observed in the rectum, but not in the ileum, which is devoid of amiloride-sensitive Na^+ channels, while the ileum expresses normal glucose-coupled sodium absorption. Basal sodium and chloride fluxes across jejunal tissue appears normal in CF.[95] However, the response of CF intestinal tissue to stimulation differs considerably from that of normal tissue. Thus normal intestinal tissue responds to the phosphodiesterase inhibitor theophylline, PGE_2, cholera toxin, cAMP analogues, and cholinergic agonists with an increase in I_{SC} that corresponds to induction of chloride secretion. In the ileum, histamine, E. coli toxin, and cGMP analogues also stimulate chloride secretion. On the other hand, CF intestinal tissue fails to respond normally to drugs and hormones that induce chloride secretion, even though the tissue is viable, as judged by its response to the mucosal addition of glucose. This lack of response points to a dysfunction in mucosal Cl^- channels at a point distal to the generation of intracellular signals, such as cAMP, calcium, and cGMP. Sinaasappel and co-workers also observed that CF tissue responds to cAMP and calcium, but not to cGMP.[94] As mentioned above, these findings point to a marked difference between intestinal and airway epithelia in CF. CF airway epithelial cells respond to calcium-generating agents, but not to cAMP-generating agents. The inability of CF intestine to respond to agents that generate these second messengers, plus the hypersecretion of potassium suggests that further investigation of chloride conductances, particularly at the primary tissue level, is needed.

Intestinal Cell Lines

Intestinal epithelial cell lines, such as the T84, HT-29, and Caco-2 lines, have been used extensively as models for chloride-secreting epithelia. The adenocarcinoma cell lines Caco-2, when grown to post-confluency,[96] and the HT-29 cells, when deprived of glucose,[97] display morphologic and biochemical differentiation characteristics of fetal colonic cells. Cultured T84 cells, derived from a lung metastasis of a human colonic carcinoma,[98] like Caco-2, form a polarized epithelium with the development of a high transepithelial resistance related to intercellular tight junctions. Each cell line expresses a cAMP-regulated Cl⁻ channel and CFTR protein.

The electrolyte transport properties of intestinal epithelial cell lines have been studied extensively in T84 cells. VIP, a cAMP-generating agent, and the calcium-ionophore A23187 stimulate chloride secretion by activating distinct apical Cl⁻ channels.[99,100] A23187 may mimic the stimulatory response of carbachol,[101] which simultaneously increases intracellular calcium and chloride secretion. Linked to the calcium-stimulated chloride secretion induced by carbachol and by A23187 (as well as histamine and ionomycin) is the role of protein kinase C.[102] Phorbol ester treatment of T84 cells produces a dual effect: initial stimulation of chloride secretion followed by inhibition. Because inhibition of protein kinase C blocks only the inhibitory phase,[102,103] protein kinases C is thought to inhibit chloride secretion by an as yet unknown mechanism. Phorbol ester treatment of T84 cells promotes phosphorylation of a 83 kDa protein within 1 min of cell stimulation.[104] However, the kinetics of phosphorylation (maximal at 30 s) and of induced secretion (maximal at 2 min) indicate a delay that suggests a limiting role of phosphorylation in the cholinergic response. Indeed, a more extensive characterization of carbachol-induced chloride secretion in T84 cells demonstrated a coupling of muscarinic M3 receptors to inositol lipid hydrolysis.[105] The resulting production of diglycerides may activate protein kinase C.

Halm et al.[106] localized an outward-rectifying, 40–45 pS, cAMP-activated Cl⁻ channel to the apical membrane of T84 cells. Measurement of single-channel currents through this channel demonstrated an anion selectivity of SCN > I > Br > Cl > F and an optimal size of permeating anions of 0.4 to 0.5 nm. These findings, together with the concentration dependence of chloride permeation, indicates that the channel behaves as a multi-ion pathway. In this model, the movement of the other permeant anions control the flow of chloride across the apical membrane.[107] The applicability of this model to outward-rectifying Cl⁻ channels in the epithelium of human airways is still not known. Transepithelial responses of T84 cells to hypo-osmotic media or to cAMP-generating agents indicate the presence of multiple Cl⁻ channels.[108,109] The selective sensitivity of cAMP- or hypo-osmotic-induced chloride secretion to Cl⁻-channel blockers and to anion permselectivity provide the most compelling evidence for this conclusion.[109]

Examination of whole-cell currents of T84 cells revealed the existence of chloride conductances other than the 40 pS Cl⁻ channel and distinguished by their I-V relationships, voltage dependence, anion permeability sequence, and stimulatory agents.[27,86,110] These chloride currents are sensitive to calcium, cAMP, or cell swelling. Volume-sensitive chloride conductance is mediated by a 75 pS Cl^- channel that is sensitive to DIDS.[27] Whether this channel is indeed different from the 45 pS outward-rectifying Cl^- channel described above is not clear. Extracellular DIDS and indanyloxyacetic acid (IAA-94) inhibit the outward rectifier, but do not affect a cAMP-induced 8.7 pS Cl^- channel in the apical membrane.[110] This channel is believed to be analogous to the low-conductance airway epithelial Cl^- channel and to be active during cAMP-stimulated secretion.

Caco-2 cells also express a low-conductance Cl^- channel of 8 pS.[111] This nonrectifying anion channel may be activated by increasing cAMP levels with forskolin or by a mixture of phosphodiesterase inhibitor and a cAMP analogue. A second outwardly rectifying anion channel of 62 pS was detected in excised patches from cAMP-stimulated cells,[111] but appears to contribute little to whole-cell currents, and therefore does not appear to have a role in the response of the cells to cAMP. This finding, together with reports of CFTR expression in Caco-2 cells, supports the hypothesis that CFTR functions as a low-conductance, 8 pS, cAMP-regulated Cl^- channel.[111]

Unlike T84 and Caco-2 cell lines, HT-29 cells grow as an unpolarized monolayer in standard medium, but differentiate into a polarized monolayer in glucose-free medium or following treatment with sodium butyrate.[112,113] Different subclones of HT29 cell lines give rise to different cell types, including "mucus" secreting and "epithelial"-like cell types.[113] Characterization of chloride conductances in unpolarized and polarized cells show that calcium and cAMP activate different chloride conductances in HT-29.c119A cells, a stable and irreversibly polarized subclone of HT-29 cells.[114] Calcium-stimulated chloride secretion occurs in unpolarized parent HT-29 cells, but is reduced in the subclone HT-29.c119A.[115] In contrast, forskolin-induced chloride secretion requires cellular polarization. Carbachol mimics the effects of Ca^{2+} ionophore on electrolyte efflux pathways without affecting I_{SC} and hence apparently acts through an intracellular signaling pathway that differs from that stimulated by the ionophore.[116] Chloride conductance is localized in the apical plasma membrane and a separate calcium-stimulated chloride conductance can be found in the basolateral membrane.[115,116] Protein kinase C may also regulate chloride conductance pathway(s) in HT-29.c119A cells as it does in human airway epithelial cells and T84 cells. Acute addition of the phorbol ester PMA stimulates a chloride conductance pathway, whereas long-term exposure inhibits PMA- and cAMP-induced chloride secretion, presumably by downregulating the potassium transporters.[117]

These epithelial cell lines are also good models for the study of the regulation of CFTR gene expression and of its product. In T84 cells,

PMA downregulates CFTR mRNA levels and the responsiveness to cAMP-generating agents in a time-and dose-dependent manner.[118] Whether long-term PMA treatment also affects potassium transporters is not clear. In Caco-2 cells, CFTR protein levels, detected with polyclonal or monoclonal antibodies, remained steady during differentiation despite a ten-fold increase in CFTR mRNA levels, which implied the existence of posttranslational or translational regulation.[119] However, cAMP-activated ^{125}I efflux and whole-cell chloride current were found to be lower in differentitated cells than in undifferentiated cells, which indicates an inverse relationship between function and the expression of CFTR mRNA and protein. In HT-29 cells, prolonged stimulation with cAMP-generating agents leads to a marked increase in CFTR expression with subsequent elevation of Cl⁻ channel activation.[120]

HT-29 cells, grown in a galactose-containing medium, differentiate and their CFTR mRNA levels remain steady, but a subsequent transfer to medium containing glucose results in a ten-fold decrease of CFTR mRNA.[119,121] Differentiated Caco-2 cells have a ten-fold higher level of CFTR mRNA than do undifferentiated cells. However, the higher mRNA levels are accompanied by lower CFTR protein levels, which indicates that the regulation of CFTR may be under translational control. Indeed, differentiation of HT-29 colonic cells causes a 9- to 18-fold increase in CFTR mRNA and 5-fold increase in calcium-induced chloride transport, but does not bring about a detectable sensitivity to forskolin- or dibutyryl cAMP-induced chloride transport.[121] Subclones of differentiated cells with either a mucin-secreting phenotype or an enterocyte-like phenotype also express higher levels of CFTR mRNA when stimulated with cAMP-generating agents. However, the HT-29 subclones have independent patterns of CFTR mRNA regulation, which suggests that final cell phenotype may affect CFTR expression.[121] The finding that increased expression of CFTR mRNA and CFTR protein are apparently not correlated with simultaneous increases of cAMP and calcium regulation of chloride transport clearly points to the complexity of the process of differentiation.

Sweat Gland

During the summer of 1953, di Sant'Agnese and colleagues found abnormally high concentrations of salt in the sweat of CF patients, which they concluded led to excessive loss of salt in the sweat of CF patients suffering from heat prostration.[122] The subsequent observation that all CF patients studied had high sweat sodium and chloride levels lead to the first diagnostic test for CF, the pilocarpine iontophoresis method of sweat testing,[123] and to extensive research into the molecular basis for altered electrolyte transport. Indeed, the sweat gland is an attractive model for studying abnormal electrolyte transport in CF because of its easy availability and because it is not involved in some of the complications of CF, such as chronic bacterial infection. Morphologically, the sweat gland is a continuous tubule separated functionally into a secretory coil (proximal portion) and reabsorptive duct (distal portion), producing a primary fluid that is iso-osmotic with the plasma. The primary fluid moves down the reabsorptive duct, which actively absorbs electrolytes and water. Normally the final sweat is hypotonic.

Secretory Coil

The morphology of the secretory coil is relatively simple, consisting of about equal numbers of interspersed "light" or clear cells and "dark" secretory cells surrounded by a discontinuous layer of myoepithelium. Muscarinic and α- and β-adrenergic agonists are the major regulatory agents of fluid secretion by the secretory coil. Direct cholinergic or α-adrenergic stimulation of isolated, microdissected sweat glands with methacholine or phenylephrine, respectively, increases the maximal sweat rate to the same extent in normal and CF secretory coil cells.[124] However, β-adrenergic stimulation with isoproterenol increases the sweat rate in normal cells, an effect that is potentiated by the phosphodiesterase inhibitor theophylline. This response is diminished, but not absent, in CF cells, even though the intracellular accumulation of cAMP is not significantly different from that occurring in normal cells.[124] This finding rules out a defect in β-adrenergic receptor activation or in adenylate cyclase activity as possible explanations for the abnormal response of the CF cells. Chloride transport in cultured secretory coil cells from normal and CF sweat glands exhibits a fast component sensitive to the Cl⁻ channel blockers DIDS, 9-antracene carboxylate, and diphenylamine 2-carboxylate.[125] β-adrenergic stimulation increases the movement of chloride from normal cells, but not from CF cells, in accordance with the results of electrophysiologic studies that demostrate abnormal chloride permeability and lack of response to β-adrenergic stimulation in the latter.

Identification of an epithelial Cl⁻ channel that corresponds to the chloride permeability of secretory coil cells has proven difficult, most likely due to the variety of anion and cation channels found in these cells. Indeed, Krouse et al.[126] localized four types of anion channels and two types of cation channels in the apical membrane of cultured secretory coil cells. At physiologic voltages, anion channels had chord conductances of 10, 18, 24, and >200 pS. Only the 24-pS channel showed outward rectification, while the others displayed linear current-voltage relationships. Cation channels had chord conductances of 5 and 18 pS and were nonselective for a variety of cations. No differences in channel types and relative abundance in normal and CF cells were found. This implies

TABLE 23.2. Properties of cells of the secretory coil.

Cell type	PD (mV)	Metholyl	Isoproterenol
Myoepithelial	−68.6	39.5	−77.1
Dark cells (β-I)	−57.3	−79.1	−40.2
Light cells (β-S)	−23.6	−39.6	−28.7

that the chloride impermeability of the CF sweat secretory coil, as that of airway epithelia, is most likely due to abnormal regulation of Cl⁻ channel(s) that share the kinetic and conductive properties of the channels found in normal secretory coil cells.

Morphologic and electrophysiologic studies helped pinpoint the cell type with abnormally regulated Cl⁻ channels. Myoepithelial cells are characterized by a high-membrane potential with potassium electrode-like properties.[37] They are localized more superficially than the light and dark cells and contract in response to cholinergic, but not β-adrenergic, stimulation[127] (Table 23.2). These characteristics suggest that myoepithelial cells are nonpolarized muscle cells.[37] Electrophysiologic studies of impaled cells led Jones[128] to a similar conclusion. In this study, cells with spontaneous depolarizing spikes· were identified by injection with Lucifer yellow. Later studies provided evidence for two additional cell types. A β-I (β-adrenergic insensitive) cell was differentiated from a β-S (β-adrenergic sensitive) cell by its V_m and its responsiveness to cholinergic and β-adrenergic stimuli (Table 23.2). The low V_m of the β-I cell and the presence of cholinergic receptors suggest that they be classified as dark cells.[37,128] The β-S cell has a high V_m and responds to both cholinergic and β-adrenergic stimulation, which indicates the presence of both types of receptors.[37] These cells have been classified as light cells and are considered to be the principal fluid-secreting cell of the secretory coil. Because the β-S cells responds to β-adrenergic stimulation and the response of CF secretory coil to β-adrenergic stimuli is normal, this is the likely abnormal cell type in CF. What is apparent from these studies is that the cholinergic and β-adrenergic mechanisms of fluid secretion in the β-S cells do not overlap.[37]

Reabsorptive Duct

Bioelectric properties of acteylcholine-stimulated sweat ducts derived from patients with CF differ markedly from those of normal ducts. An early study showed that the transepithelial potential (V_t) of CF sweat ducts is −66.3 mV, as compared to a normal value of −29.8 mV. [18,19] Replacement of chloride with an impermeant anion did not alter the resting V_t in CF sweat ducts, but did lower normal V_t to approach that of CF ducts perfused with chloride. Although the rate of stimulated sweat secretion was similar in normal and CF sweat ducts, the sweat contained abnormally high levels of sodium (93 mM vs 23 mM) and chloride (97 mM vs 20 mM).[19] Differences in the rates of chloride reabsorption were also noted and were approximately 40% and 20%, respectively, of those noted in normal ducts. The low chloride transport rate in CF ducts, accompanied by a hyperpolarized ductal potential, suggest that chloride permeability is significantly reduced relative to sodium permeability. Electrophysiologic measurements by Bijman and Frömter[129] and Reddy and Quinton[130] showed that the human sweat duct epithelium is capable of generating high electric and osmotic gradients, despite its low resistance due to a high chloride permeability. Indeed, 90% of the total transepithelial conductance is attributed to chloride and 5% to sodium. Electrical potential profiles of normal and CF duct cells also point to significant differences. Thus the apical membrane potential in CF duct cells has a polarity opposite to that of normal duct cells, while the basolateral membrane potential of CF cells is significantly hyperpolarized. Replacement of luminal perfusate chloride with the impermeant anion gluconate in normal duct cells depolarizes the apical membrane potential and hyperpolarizes the basolateral membrane, mimicking the cell electric potential profile of CF ducts in the presence of chloride. These studies demonstate that the electric potential profile of CF duct cells is abnormal, due to the absence of normal chloride conductance.

The chloride impermeability of intact sweat ducts is also observed in cultured sweat duct cells obtained from normal and CF patients.[131,132] Using fluorescent-digital imaging microscopy in combination with the halide-specific fluorescent dye, 6-methoxy-N-(3-sulfopropyl)-quinolinium (SPQ), Ram and colleagues showed that normal duct cells rapidly lose intracellular chloride when chloride in the bathing medium is replaced by gluconate, and that the Cl^--channel blocker diphenylamine-2-carboxylate inhibits this response.[131] The same maneuvers produced a markedly smaller rate of chloride loss in CF duct cells. Replenishment of the bathing medium with chloride induces a more rapid uptake of chloride in normal cells than in CF cells. Thus the reduced chloride permeability of CF sweat ducts may be demonstrated in individual cells of CF origin. Since the cAMP-generating agent PGE_2 increases chloride permeability in normal but not CF cells,[132] it may be concluded that this permeability is due to an abnormality distal to receptor-mediated cAMP generation.

Future Considerations

Despite advances in understanding at the biochemical, pharmacologic, and molecular levels how chloride and sodium traverse an epithelial cell, there remain a number of salient problems to solve. The trafficking of electrolyte transporters through the cell presents a new area of cellular regulation that is as yet not fully realized. Targeting of channels, cotransporters, or exchangers to reach apical or basolateral membranes in proximity to hormone receptors and enzymes that generate mediators requires more intensive investigation, particularly in diseases such as CF, for which there may be an abnormal translational processing of channel or transporter protein(s). The regulation of transporters through receptors other than β-adrenergic receptors need to be more fully investigated. Recent reports from the author's laboratory on α-adrenergic regulation of a NaCl(K) cotransporter and of intracellular mechanisms of signal transduction provides new insights into the complexity of cellular regulation of transepithelial electrolyte transport. However, the contribution of other nonadrenergic receptors, such as imidazol(in)e, muscarinic, and cytokine receptors, are not fully understood. We also need a better understanding of how sodium transport and Na^+ channels are regulated, and how these channels are linked to Cl^- channels, particularly in CF. New information will most likely come from functional and structural investigations combined with new, more advanced physiologic and molecular biology technology.

References

1. Taussig LM. Cystic Fibrosis: An overview. In: Taussig LM, ed. Cystic Fibrosis. New York: Thieme-Straton; 1984:1–9.
2. Home E. On the formation of fat in the intestines of living animals. Roy Soc Land Philosoph Trans 1813; 103:146.
3. Busch R. On the history of mucoviscidosis. Deutche Gesundhs 1978; 33:316.
4. Pfyffer JX. The Dictionary of the Swiss-German Language. 1848.
5. Rochholz EL. Almanac of Children's Songs and Games from Switzerland. Leipzig: J.J. Weber; 1857.
6. Schmidt JD. Book of Folk Philosophy. Chemnitz, Germany: Joh Christoph and Joh David Stoseln; 1729.
7. Landsteiner K. Darmverschuss durch eingedicktes meconium: Pankreatitis. Zentralbe All Pathol Anat 1905; 16:903.
8. Bramwell B. Pancreatic infantilism. Trans Medico-chirur Soc Edinb 1904; 23:162.
9. Clarke C, Hadfield C. Congenital pancreatic disease with infantilism. Quarterly J Med 1923–1924; 17:358.
10. Rentoul JL. Pancreatic infantilism. Br Med J 1904; 2:1694.
11. Garrod AE, Hurtley WH. Congenital family steatorrhea. Quartery J Med 1912; 6:242.
12. Fanconi G, Uehlinger E, Knauer C. Das coeliakiesyndrom bei angeborener zysticher pankreas fibromatose und bronchiektasien. Wein Med Wchnschr 1936; 86:753.
13. Andersen DH. Cystic fibrosis of the pancreas and its relation to celiac disease; a clinical and pathological study. Am J Dis Child 1938; 56:344.
14. Andersen DH, Hodges RG. Celiac syndrome. V. Genetics of cystic fibrosis of the pancreas with a consideration of etiology. Am J Dis Child 1946; 72:62.

518   C.M. Liedtke

15. Kerem BS, Rommens JM, Buchanan JA, Markiewicz D, Cox TK, Chakravart A, Buchwald M, Tsui L-C. Identification of the cystic fibrosis gene: Genetic analysis. Science 1989; 245:1073–1080.
16. Riordan JR, Rommens JM, Kerem B, Alon N, Rozmahel R, Grzelczak Z, Zielenski J, Lok S, Plavsic N, Chou J-L, Drumm ML, Iannuzzi MC, Collins FS, Tsui L-C. Identification of the cystic fibrosis gene: Cloning and characterization of complementary DNA. Science 1989; 245:1066–1072.
17. Rommens JM, Iannuzzi MC, Kerem B-S, Drumm ML, Melmer G, Dean M, Rozmahel P, Cole JL, Kennedy D, Hidaka N, Zsiga M, Buchwald M, Riordan JR, Tsui L-C, Collins FS. Identification of the cystic fibrosis gene: Chromosome walking and jumping. Science 1989; 245:1059–1065.
18. Quinton PM. Chloride impermeability in cystic fibrosis. Nature 1983; 301: 421–422.
19. Quinton PM, Bijman J. Higher bioelectric potentials due to decreased chloride absorption in the sweat glands of patients with cystic fibrosis. N Engl J Med 1983; 308:1185–1189.
20. Knowles M, Gatzy J, Boucher R. Increased bioelectric potential difference across respiratory epithelia in cystic fibrosis. N Engl J Med 1981; 305: 1489–1495.
21. Anderson MP, Gregory RJ, Thompson S, Souza DW, Paul S, Mulligan RC, Smith AE, Welsh MJ. Demonstration that CFTR is a chloride channel by alteration of its anion selectivity. Science 1991; 253:202–205.
22. Drumm ML, Wilkonson DJ, Smit LS, Worrell RT, Strong TV, Frizzell RA, Dawson DC, Collins FS. Chloride conductance expressed by ΔF508 and other mutant CFTRs in *Xenopus* oocytes. Science 1991; 254:1797–1799.
23. Rommens JM, Dho S, Bear CE, Kartner N, Kennedy D, Riordan JR, Tsui L-C, Foskett JK. cAMP-inducible chloride conductance in mouse fibroblasts lines stably expressing the human cystic fibrosis transmembrane conductance regulator. Proc Natl Acad Sci USA 1991; 88:7500–7504.
24. Chamberlin ME, Strange K. Anisosmotic cell volume regulation: A comparative view. Am J Physiol 1989; 257(Cell Physiol 25):C159–C173.
25. Hoffman EK, Simonsen LO. Membrane mechanisms in volume and pH regulation in vertebrate cells. Physiol Rev 1989; 69:315–382.
26. Larson M, Spring KR. Volume regulation in epithelia. Curr Top Membr Transp 1987; 30:105–123.
27. Worrell RT, Butt AG, Cliff WH, Frizzell RA. A volume-sensitive chloride conductance in human colonic cell line T84. Am J Physiol 1989; 256: C1111–C1119.
28. Steinmeyer K, Klocke R, Ortland C, Gronemeier M, Jocskusch H, Gründer S, Jentsch TJ. Inactivation of muscle chloride channel by transposon insertion in myotonic mice. Nature (London) 1991; 354:304–308.
29. Bretag AH. Muscle chloride channels. Physiol Rev 1987; 67:618–724.
30. Grenningloh G, Rienitz A, Schmitt B, Methfessel C, Zensen M, Beyreuther K, Gundelfinger ED, Betz H. The strychnine-binding subunit of the glycine receptor shows homology with nicotinic acetylcholine receptors. Nature (London) 1987; 328:215–220.
31. Schofield PR, Darlison MG, Fujita N, Burt DR, Stephenson FA, Rodriguez H, Rhee LM, Ramachandran J, Reale V, Glencorse TA, Seeburg PH, Barnard EA. Sequence and functional expression of the GABA$_A$ receptor

shows a ligand-gated receptor super-family. Nature (London) 1987; 328: 221–227.

32. Quinton PM. Cystic fibrosis: A disease in electrolyte transport. FASEB J 1990; 4:2709–2717.

33. Welsh MJ. Abnormal regulation of ion channels in cystic fibrosis epithelia. FASEB J 1990; 4:2718–2725.

34. Boucher RC, Cotton CU, Gatzy JT, Knowls MR, Yankaskas JR. Evidence for reduced Cl⁻ and increased Na⁺ permeability in cystic fibrosis human primary cell cultures. J Physiol (Lond) 1988; 405:77–103.

35. DeJonge HR, Bijman Jr, Sinaasappel M. Relation of regulatory enzyme levels to chloride transport in intestinal epithelial cells. Pediatr Pulmonol Suppl 1987; 1:54–57.

36. Sato K, Sato F. Defective beta adrenergic response of cystic fibrosis sweat glands *in vivo* and *in vitro*. J Clin Invest 1984; 73:1763–1771.

37. Reddy MM, Quinton PM. Electrophysiologically distinct cell types in human sweat gland secretory coil. Am J Physiol 1992; 262:C287–C292.

38. Liedtke CM. α_1-Adrenergic regulation of Na-Cl cotransport in human airway epithelia. Am J Physiol 1989; 257:L125–L129.

39. Welsh MJ. An apical membrane chloride channel in human tracheal epithelium. Science 1986; 232:1648–1650.

40. Willumsen NJ, Davis CW, Boucher RC. Intracellular Cl⁻ activity and cellular Cl⁻ pathways in cultured human airway epithelium. Am J Physiol 1989; 256:C1033–C1044.

41. Liedtke CM. α-Adrenergic signaling in human airway epithelial cells involves inositol lipid and phosphate metabolism. Am J Physiol 1992; 262:L183–L191.

42. Cotton CU, Stutts MJ, Knowles MR, Gatzy JT, Boucher RC. Abnormal apical cell membrane in cystic fibrosis respiratory epithelium: An in vitro electrophysiologic analysis. J Clin Invest 1987; 79:80–85.

43. Knowles M, Gatzy J, Boucher R. Relative ion permeability of normal and cystic fibrosis nasal epithelium. J Clin Invest 1983; 71:1410–1417.

44. Knowles MR, Stutts MJ, Spock A, Fischer N, Gatzy JT, Boucher RC. Abnormal ion permeation through cystic fibrosis respiratory epithelium. Science 1983; 221:1067–1070.

45. Widdicombe JH, Welsh MJ, Finkbeiner WE. Cystic fibrosis decreases the apical membrane chloride permeability of monolayers cultured from cells of tracheal epithelium. Proc Natl Acad Sci USA 1985; 82:6167–6171.

46. Boucher RC, Cheng ECH, Paradiso AM, Stutts MJ, Knowles MR, Earp HS. Chloride secretory response of cystic fibrosis human airway epithelia: Preservation of calcium but not protein kinase C- and A-dependent mechanisms. J Clin Invest 1989; 84:1424–1431.

47. Li M, McCann JD, Anderson MP, Clancy JP, Liedtke CM, Nairn AC, Greengard P, Welsh MJ. Regulation of chloride channels by protein kinase C in normal and cystic fibrosis airway epithelia. Science 1989; 244:1353–1356.

48. Frizzell RA, Rechkemmer G, Shoemaker RL. Altered regulation of airway epithelial cell chloride channels in cystic fibrosis. Science 1986; 233:558–560.

49. Li M, McCann JD, Liedtke CM, Nairn AC, Greengard P, Welsh MJ. Cyclic-AMP-dependent protein kinase opens chloride channels in normal but not cystic fibrosis airway epithelia. Nature (London) 1988; 331:358–360.

50. Schoumacher RA, Shoemaker RL, Halm DR, Tallant EA, Wallace RW, Frizzell RA. Phosphorylation fails to activate chloride channels from cystic fibrosis airway cells. Nature (London) 1987; 330:752–754.
51. Jorissen M, Vereecke J, Carmeliet E, Van den Berghe H, Cassiman J-J. Non-selective cation and dysfunctional chloride channels in the apical membrane of nasal epithelial cells cultured from cystic fibrosis patients. Biochim Biophys Acta 1991; 1096:52–59.
52. Wine JJ, Bradyen DJ, Hagiwara G, Krouse ME, Law TC, Muller UJ, Solc CK, Ward CL, Widdicombe JH, Xia Y. Cystic fibrosis, the CFTR, and rectifying Cl⁻ channels. In: Tsui L-C, ed. The Identification of the CF Gene. New York: Plenum; 1991:253–272.
53. Duszyk M, French AS, Man SFP. Cystic fibrosis affects chloride and sodium channels in human airway epithelia. Can J Physiol Pharmacol 1989; 67: 1362–1365.
54. Duszyk M, French AS, Man SFP. The 20-pS chloride channel of the human airway epithelium. Biophys J 1990; 57:223–230.
55. Duszyk M, French AS, Man SFP. Noise analysis and single-channel observations of 4 pS chloride channels in human airway epithelia. Biophys J 1992;61:583–587.
56. Anderson MP, Welsh MJ. Calcium and cAMP activate different chloride channels in the apical membrane of normal and cystic fibrosis epithelia. Proc Natl Acad Sci USA 1991; 88:6003–6007.
57. Chan H-C, Goldstein J, Nelson DJ. Alternate pathways for chloride conductance activation in normal and cystic fibrosis airway epithelial cells. Am J Physiol 1992; 262:C1273–C1283.
58. Wagner JA, Cozens AL, Schulman H, Gruenert DC, Stryer L, Gardner P. Activation of chloride channels in normal and cystic fibrosis airway epithelial cells by multifunctional calcium/calmodulin dependent protein kinase. Nature (London) 1991; 349:793–796.
59. McCann JD, Matsuda J, Garcia M, Kaczorowski G, Welsh MJ. Basolateral K⁺ channels in airway epithelia. I. Regulation by Ca²⁺ and block by charybdotoxin. Am J Physiol 1990; 258:L334–L342.
60. McCann JD, Welsh MJ. Basolateral K⁺ channels in airway epithelia. II. Role in Cl⁻ secretion and evidence for two types of K⁺ channel. Am J Physiol 1990; 258:L343–L348.
61. Welsh MJ, McCann JD. Intracellular calcium regulates basolateral potassium channels in a chloride-secreting epithelium. Proc Natl Acad Sci USA 1985; 82:8823–8826.
62. Willumsen NJ, Boucher RC. Activation of an apical Cl⁻ conductance by Ca²⁺ ionophores in cystic fibrosis airway epithelia. Am J Physiol 1989; 256:C226–C233.
63. McCann JD, Li M, Welsh MJ. Identification and regulation of whole cell chloride currents in airway epithelium. J Gen Physiol 1989; 94:1015–1036.
64. Solc CK, Wine JJ. Swelling-induced and depolarization induced Cl⁻ channels in normal and cystic fibrosis epithelial cells. Am J Physiol 1991; 261:C658–C674.
65. Butt AG, Clapp WL, Frizzell RA. Potassium conductances in tracheal epithelium activated by secretion and cell swelling. Am J Physiol 1990; 258:C630–C638.

66. Knowlton RG, Cohen-Haguenauer O, Van Cong N, Frezal J, Brown VA, Barker D, Braman JC, Schumm JW, Tsui L-C, Buchwald M, Donis-Keller H. A polymorphic DNA marker linked to cystic fibrosis is located on chromosome 7. Nature (London) 1985; 318:380–382.
67. Wainwright BJ, Scambler PJ, Schmidtke J, Watson EA, Law H-Y, Farrall M, Cooke HJ, Eiberg H, Williamson R. Localization of cystic fibrosis locus to human chromosome 7cen-q22. Nature (London) 1985; 318:384–385.
68. White R, Woodward S, Nakamura Y, Leppert M, O'Connell P, Hoff M, Herbst J, Lalouel J-M, Dean M, Vande Woude G. A closely linked genetic marker for cystic fibrosis. Nature (London) 1985; 318:382–384.
69. Ames GF, Mimura CS, Shyamala V. Bacterial periplasmic permeases belong to a family of transport proteins operating from *Escherichia coli* to human: Traffic ATPases. FEMS Microbiol Rev 1990; 6:429–446.
70. Hyde SC, Emsley P, Hartshorn MJ, Mimmack MM, Gileadi U, Pearce SR, Gallagher MP, Gill DR, Hubbard RE, Higgins CF. Structural model of ATP-binding proteins associated with cystic fibrosis, multidrug resistance and bacterial transport. Nature (London) 1990; 346:362–365.
71. Thomas PJ, Shenbagamurthi P, Ysern X, Pedersen PL. Cystic fibrosis transmembrane conductance regulator nucleotide binding to a synthetic peptide. Science 1991; 251:555–557.
72. Thomas PJ, Shenbagamurthi P, Sondek J, Hullihen J, Pedersen PL. The cystic fibrosis transmembrane conductance regulator: Effects of the most common cystic fibrosis-causing mutation on the secondary structure and stability of a synthetic peptide. J Biol Chem 1992; 267:5727–5730.
73. Kerem B-S, Zielinski J, Markiewicz D, Bozon D, Gazit E, Yahav J, Kennedy D, Riordan JR, Collins FS, Rommens JM, Tsui L-C. Identification of mutations in regions corresponding to the two putative nucleotide (ATP) binding folds of the cystic fibrosis gene. Proc Natl Acad Sci USA 1990; 87:8447–8451.
74. Cutting GR, Kasch LM, Rosenstein BJ, Zielenski J, Tsui L-C, Antonarakis SE, Kazazian HH Jr. A cluster of cystic fibrosis mutations in the first nucleotide binding fold of the cystic fibrosis conductance regulator protein. Nature (London) 1990; 346:366–369.
75. Hamosh A, Trapnell BC, Zeitlin PL. Montrose-Rafizadeh C, Rosenstein BJ, Crystal RG, Cutting GR. Severe deficiency of cystic fibrosis transmembrane conductance regulator messenger RNA carrying nonsense mutations R553X andW1316X in respiratory epithelial cells of patients with cystic fibrosis. J Clin Invest 1991; 88:1880–1885.
76. Drumm ML, Pope HA, Cliff WH, Rommens JM, Marvin SA, Tsui L-C, Collins FC, Frizzell RA, Wilson JM. Correction of the cystic fibrosis defect in vitro by retrovirus-mediated gene transfer. Cell 1990; 62:1227–1233.
77. Rich DP, Anderson MP, Gregory RJ, Cheng SH, Paul S, Jefferson DM, McCann JD, Klinger KW, Smith AE, Welsh MJ. Expression of cystic fibrosis transmembrane conductance regulator corrects defective chloride channel regulation in cystic fibrosis airway epithelial cells. Nature (London) 1990; 347:358–363.
78. Rosenfeld MA, Yoshimura K, Trapnell BC, Yoneyama K, Rosenthal ER, Dalemans W, Fukayama M, Bargon J, Stier LE, Stratford-Perricaudet L,

Perricaudet M, Guggino WB, Pavirani A, Lecocq J-P, Crystal RG. In vivo transfer of the human cystic fibrosis transmembrane conductance regulator gene to the airway epithelium. Cell 1992; 68:143–155.

79. Olsen JC, Johnson LG, Stutts MJ, Sarkadi B, Yankaskas JR, Swanstrom R, Boucher RC. Correction of the apical membrane chloride permeability defect in polarized cystic fibrosis airway epithelia following retroviral-mediated gene transfer. Human Gene Therapy 1992; 3:253–266.

80. Anderson MP, Rich DP, Gregory RJ, Smith AE, Welsh MJ. Generation of cAMP-activated chloride currents by expression of CFTR. Science 1991; 251:679–682.

81. Berger HA, Anderson MP, Gregory RJ, Thompson S, Howard PW, Maurer RA, Mulligan R, Smith AE, Welsh MJ. Identification and regulation of the CFTE-generated chloride channel. J Clin Invest 1991; 88:1422–1431.

82. Tabcharani JA, Chang X-B, Riordan JR, Hanrahan JW. Phosphorylation-regulated Cl^- channel in CHO cells stably expressing the cystic fibrosis gene. Nature (London) 1991; 352:628–631.

83. Bear CE, Duguay F, Naismith AL, Kartner N, Hanrahan JW, Riordan JR. Cl^- channel activity in *Xenopus* oocytes expressing the cystic fibrosis gene. J Biol Chem 1991; 266:19142–19145.

84. Cunningham SA, Worrell RT, Benos DJ, Frizzell RA. cAMP stimulated ion currents in *Xenopus* oocytes expressing CFTR cRNA. Am J Physiol 1992; 262:C783–C788.

85. Kartner N, Hanrahan JW, Jensen TJ, Naismith AL, Sun S, Ackerley CA, Reyes EF, Tsui L-C, Rommens JM, Bear CE, Riordan JR. Expression of the cystic fibrosis gene in non-epithelial invertebrate cells produces a regulated anion conductance. Cell 1991; 64:681–691.

86. Cliff WH, Frizzell RA. Separate Cl^- conductances activated by cAMP and Ca^{2+} in Cl^--secreting epithelial cells. Proc Natl Acad Sci USA 1990; 87:4956–4960.

87. Cheng SH, Gregory RJ, Marshall J, Paul S, Souza DW, White GA, O'Riordan CR, Smith AE. Defective intracellular transport and processing of CFTR is the molecular basis of most cystic fibrosis. Cell 1990; 63:827–834.

88. Gregory RJ, Rich DP, Cheng SH, Souza DW, Paul S, Manavalan P, Anderson MP, Welsh MJ, Smith AE. Maturation and function of cystic fibrosis transmembrane conductance regulator variants bearing mutations in putative nucleotide-binding domains 1 and 2. Mol Cell Biol 1991; 11:3886–3893.

89. Dalemans W, Barbry P, Champigny G, Jallat S, Dott K, Dreyer D, Crystal RG, Pavirani A, Lecocq J-P, Lazdunski M. Altered chloride ion channel kinetics associated with the ΔF508 cystic fibrosis mutation. Nature (London) 1991; 354:526–528.

90. Hopfer U. Pathophysiological considerations relevant to intestinal obstruction in cystic fibrosis. In: Quinton PM, Martinez JR, Hopfer U, eds. Fluid and Electrolyte Abnormalities in Exocrine Glands in Cystic Fibrosis. San Francisco: San Francisco Press; 1982:241–251.

91. Orlando RC, Powell DW, Boucher RC, Knowles MR. Colonic and esophageal potential difference measurements in cystic fibrosis. Gastroenterology 1985; 88:1524.

92. Goldstein JL, Nash NT, Al-Bazzaz F, Layden TJ, Rao MC. Rectum has abnormal ion transport but normal cAMP-binding proteins in cystic fibrosis. Am J Physiol 1988; 254:C719–C724.
93. Baxter PS, Wilson AJ, Read NW, Hardcastle J, Hardcastle PT, Taylor CJ. Abnormal jejunal potential difference in cystic fibrosis. Lancet 1989; 2:464–466.
94. Sinaasappel M, Veeze HJ, DeJonge HR. New insights into the pathogenesis of cystic fibrosis. Scand J Gastroenterol 1990; 25(suppl 178):17–25.
95. Berschneider HM, Knowles MR, Azizkhan RG, Boucher RC, Tobey NA, Orlando RC, Powell DW. Altered intestinal chloride transport in cystic fibrosis. FASEB J 1988; 2:2625–2629.
96. Pinto M, Robine-Leon S, Appay MD, Kedinger M, Triadou N, Dussaulx E, Lacroix B, Simon-Assmann P, Haffen K, Fogh J, Zweibaum A. Enterocyte-like differentiation and polarization of the human colon carcinoma cell line Caco-2 in culture. Biol Cell 1983; 47:323–330.
97. Pinto M, Appay M, Simon-Assmann P, Chevalier G, Dracopoli N, Fogh J, Zweibaum A. Entercytic differentiation of cultured human colon cancer cells by replacement of glucose by galactose in the medium. Biol Cell 1982; 44:193–196.
98. Barrett KE, Dharmsathaphorn K. Mechanisms of chloride secretion in an colonic epithelial cell line. In: Lebenthal E, Duffey M, eds. Textbook of Secretory Diarrhea. New York: Raven Press; 1990:59–66.
99. Dharmsathaphorn K, Mandel KG, Masui H, McRoberts JA. Vasoactive intestinal polypeptide-induced chloride secretion by a colonic epithelial cell line: Direct participation of a basolaterally localized Na^+, K^+, Cl^- cotransport system. J Clin Invest 1985; 75:462–471.
100. Mandel KG, Dharmsathaphorn K, McRoberts JA. Characterization of a cyclic AMP-activated Cl^- transport pathway in the apical membrane of a human colonic epithelial cell line. J Biol Chem 1986; 261:704–712.
101. Dharmsathaphorn K, Pandol SJ. Mechanism of chloride secretion induced by carbachol in a colonic epithelial cell line. J Clin Invest 1986; 77:348–354.
102. Kachintorn U, Vongkovit P, Vajanaphanich M, Dinh S, Barrett KE, Dharmsathaphorn K. Dual effects of a phorbol ester on calcium-dependent chloride secretion by T84 epithelial cells. Am J Physiol 1992; 262:C15–C22.
103. Lindeman RP, Chase HS Jr. Protein kinase C does not participate in carbachol's secretory action in T84 cells. Am J Physiol 1992; 263:C140–C146.
104. Cohn JA. Protein kinase C mediates cholinergically regulated protein phosphorylation in a Cl^--secreting epithelium. Am J Physiol 1990; 258:C227–C233.
105. Dickinson KEJ, Frizzell RA, Sekar MC. Activation of T84 cell chloride channels by carbachol involves a phosphoinositide-coupled muscarinic M_3 receptor. Eur J Pharmacol 1992; 225:291–298.
106. Halm DR, Rechkemmer GR, Schoumacher RA, Frizzell RA. Apical membrane chloride channels in a colonic cell line activated by secretory agonists. Am J Physiol 1988; 254:C505–C511.
107. Halm DR, Frizzell RA. Anion permeation in an apical membrane chloride channel of a secretory epithelial cell. J Gen Physiol 1992; 99:339–366.

108. Bell CL, Quinton PM. T84 cells: Anion selectivity demonstrates expression of Cl^- conductance affected in cystic fibrosis. Am J Physiol 1992; 262:C555–C562.

109. McEwan GTA, Brown CDA, Hirst BH, Simmons NL. Hypo-osmolar stimulation of transepithelial Cl^- secretion in cultured human T_{84} intestinal epithelial layers. Biochim Biophys Acta 1992; 1135:180–183.

110. Tabcharani JA, Low W, Elie D, Hanrahan JW. Low-conductance chloride channel activated by cAMP in the epithelial cell line T84. FEBS Lett 1990; 270:157–164.

111. Bear CE, Reyes EF. cAMP-activated chloride conductance in the colonic cell line, Caco-2. Am J Physiol 1992; 262:C251–C256.

112. Augeron C, Maoret JJ, Laboisse CL, Grasset E. Permanently differentiated cell clones isolated from the human colonic carcinoma cell line HT29: Possible models for the study of ion transport and mucus production. In: Alvarado F, Van Os CH, eds. Ion Gradient Coupled Transport. Amsterdam: Elsevier; 1986:363–366.

113. Huet C, Sahuquillo-Merino C, Coudrier E, Louvard D. Absorptive and mucus-secreting subclones isolated from a multipotent intestinal cell line (HT-29) provide new models for cell polarity and terminal differentiation. J Cell Biol 1987; 105:345–357.

114. Augeron C, Laboisse CL. Emergence of permanently differentiated cell clones in a human colonic cancer cell line in culture after treatment with sodium butyrate. Cancer Res 1984; 44:3961–3969.

115. Morris AP, Cunningham SA, Benos DJ, Frizzell RA. Cellular differentiation is required for cAMP but not Ca^{2+}-dependent Cl^- secretion in colonic epithelial cells expressing high levels of cystic fibrosis transmembrane conductance regulator. J Biol Chem 1992; 267:5575–5583.

116. Vaandrager AB, Bajnath R, Groot JA, Bot AGM, deJonge HR. Ca^{2+} and cAMP activate different chloride efflux pathways in HT-29.c119A colonic epithelial cell line. Am J Physiol 1991; 261:G958–G965.

117. Vaandrager AB, van den Berghe N, Bot AGM, deJonge HR. Phorbol esters stimulate and inhibit Cl^- secretion by different mechanisms in a colonic cell line. Am J Physiol 1992; 262(Gastrointest Liver Physiol 25):G249–G256.

118. Trapnell BC, Zeitlin PL, Chu C-S, Yoshimura K, Nakamura H, Guggino WB, Bargon J, Banks TC, Dalemans W, Pavirani A, Lecocq J-P, Crystal RG. Down-regulation of cystic fibrosis gene mRNA transcript levels and induction of the cystic fibrosis chloride secretory phenotype in epithelial cells by phorbol ester. J Biol Chem 1991; 266:10319–10323.

119. Sood R, Bear C, Auerbach W, Reyes E, Jensen T, Kartner N, Riordan JR, Buchwald M. Regulation of CFTR expression and function during differentiation of intestinal epithelial cells. EMBO J 1992; 11:2487–2494.

120. Breuer W, Kartner N, Riordan JR, Cabantchik ZI. Induction of expression of the cystic fibrosis transmembrane conductance regulator. J Biol Chem 1992; 267:10465–10469.

121. Montrose-Rafizadeh C, Guggino WB, Montrose MH. Cellular differentiation regulates expression of Cl^- transport and cystic fibrosis transmembrane conductance regulator mRNA in human intestinal cells. J Biol Chem 1991; 266:4495–4499.

122. di Sant'Agnese PA, Darling RC, Perera GA, Shea E. Abnormal electrolytic composition of sweat in cystic fibrosis of the pancreas. Pediatrics 1953; 12:549–563.
123. Gibson LE, Cooke RE. A test for concentration of electrolytes in sweat in cystic fibrosis of the pancreas utilizing pilocarpine by iontophoresis. Pediatrics 1959; 23:545.
124. Sato K. Differing luminal potential differences of cystic fibrosis and control sweat secretory coils *in vitro*. Am J Physiol 1984; 247:R646–R649.
125. Wood LC, Neufeld EF. A cystic fibrosis phenotype in cells cultured from sweat gland secretory coil: Altered kinetics of ^{36}Cl efflux. J Biol Chem 1990; 265:12796–12800.
126. Krouse ME, Hagiwara G, Chen J, Lewiston NJ, Wine JJ. Ion channels in normal human and cystic fibrosis sweat gland cells. Am J Physiol 1989; 257:C129–C140.
127. Sato K, Nishiyama A, Kobayashi M. Mechanical properties and functions of the myoepithelium in the eccrine sweat gland. Am J Physiol 1979; 237: C177–C184.
128. Jones CJ. Electrophysiological and dye-coupling studies on cell types in eccrine sweat glands isolated from normal subjects and patients with cystic fibrosis: The pusuit of cell signatures. In: Matella G, Quinton PM, eds. Cellular and Molecular Basis of Cystic Fibrosis. San Francisco: San Francisco Press; 1988:115–123.
129. Bijman J, Frömter E. Direct demonstration of high transepithelial chloride-conductance in normal human sweat duct which is absent in cystic fibrosis. Pfluegers Arch 1986; 407:S123–S127.
130. Reddy MM, Quinton PM. Altered electrical potential profile of human reabsorptive sweat duct cells in cystic fibrosis. Am J Physiol 1989; 257: C722–C726.
131. Ram SJ, Kirk KL. Cl⁻ permeability of human sweat duct cells monitored with fluorescence-digital imaging microscopy: Evidence for reduced plasma membrane Cl⁻ permeability in cystic fibrosis. Proc Natl Acad Sci USA 1989; 86:10166–10170.
132. Ram SJ, Weaver ML, Kirk KL. Regulation of Cl⁻ permeability in normal and cystic fibrosis sweat duct cells. Am J Physiol 1990; 259:C842–C846.
133. Welsh MJ, Liedtke CM. Chloride and potassium channels in cystic fibrosis airway epithelium. Nature (London) 1986; 322:467–470.
134. Willumsen NJ, Davis CW, Boucher RC. Cellular Cl⁻ transport in cultured cystic fibrosis airway epithelium. Am J Physiol 1989; 256:C1045–C1053.

24
Cyclic Nucleotide-Activated Channels

Anna Menini and Robert R.H. Anholt

Sensory transduction in both the visual and the olfactory system involves ion channels that are opened directly by cyclic nucleotides. Both the absorption of light and the binding of many, but not all, odorants trigger a chain of enzymatic events that culminate in a change in the concentration of a cyclic nucleotide, a decrease in the concentration of cyclic GMP (cGMP) in the visual system, and an increase in the concentration of cyclic AMP (cAMP) in the olfactory system. Alterations in intracellular levels of cyclic nucleotides in these systems affect the activity of cyclic nucleotide-activated channels located in the plasma membrane. In the visual system closing of these channels leads to hyperpolarization of the cell, whereas in the olfactory system opening of cyclic nucleotide-activated channels results in depolarization. These cyclic nucleotide-induced alterations in the membrane potential represent the first step in neural excitation. Since cyclic nucleotide-activated channels mediate the final steps of the visual and olfactory transduction cascades and initiate the flow of ions that triggers excitation of the cell, they represent the molecular switches that control sensory perception. These observations and recent discoveries of additional members of the cyclic nucleotide-activated channel family in cardiac pacemaker cells[1] and cells of the pineal gland[2] have focused considerable attention on this expanding new family of ion channels that are directly activated by second messengers.

In this chapter we will briefly describe the central role of cyclic nucleotide-activated channels in visual and olfactory transduction. We will then describe in detail the gating properties and mechanisms of ion permeation through cyclic nucleotide-activated channels. Finally, we will describe the structure of these channels and discuss structure-function relationships that are beginning to provide a comprehensive picture of the molecular mechanism by which cyclic nucleotides control the flow of ions through these channels.

Transduction-Excitation Coupling in Sensory Systems

Phototransduction

Phototransduction is the process by which the absorption of a photon of light by a molecule of rhodopsin in vertebrate retinal rods or cones is transformed into an electrical signal. Retinal rods consist of an outer segment and an inner segment connected by a thin, short cilium (Figure 24.1). The rod outer segment is composed of a stack of disk membranes surrounded by the plasma membrane. The visual pigment, rhodopsin, is contained in the disk membrane, whereas cGMP-activated channels are located in the plasma membrane. In darkness there is a circulating current carried by Na^+ and Ca^{2+} ions that enter the outer segment through cGMP-activated channels, and by K^+ ions that flow out of the inner segment. Ionic gradients are maintained by ATP-dependent Na^+-K^+ pumps in the membrane of the inner segment and by Na^+-Ca^{2+}-K^+ exchangers in the plasma membrane of the outer segment. Light causes a reduction in the intracellular concentration of cGMP and results in closure of a fraction of the cGMP-activated channels, thereby causing a hyperpolarization that spreads to the synaptic terminal, where it varies the rate of neurotransmitter release onto postsynaptic cells (Figure 24.1).

The existence of ion channels that are activated directly by cGMP was demonstrated first by patch-clamp experiments on the plasma membrane of rod outer segments.[3] A piece of membrane was excised from the outer segment on the tip of a patch pipette, with the cytoplasmic side facing the bathing solution. The addition of micromolar concentrations of cGMP in the absence of nucleotide triphosphates activated a current, which demonstrated that cGMP directly opened channels without the need for protein phosphorylation. Subsequently, experiments from several laboratories have shown that the cGMP-activated current closely resembles the "light-sensitive" current measured in intact rods, which suggests that this channel indeed mediates phototransduction.

The light-triggered enzymatic reaction cascade that leads to a reduction in the intracellular concentration of cGMP has been reviewed previously[4,5] and we will summarize it here only briefly. Rhodopsin is an integral membrane protein that contains seven transmembrane α-helical domains and is a member of the large superfamily of G protein-coupled receptors.[6] Rhodopsin is activated when light isomerizes a chromophore group, 11-*cis*-retinal, to all-*trans*-retinal.[7] This triggers a series of conformational changes in the protein. One of these conformations, called "metarhodopsin II", can activate about 500 molecules of a heterotrimeric G protein called "transducin,"[8] and catalyzes the release of GDP bound to the α subunit of this G-protein and the binding of GTP.[9,10] After nucleotide exchange, transducin dissociates into two

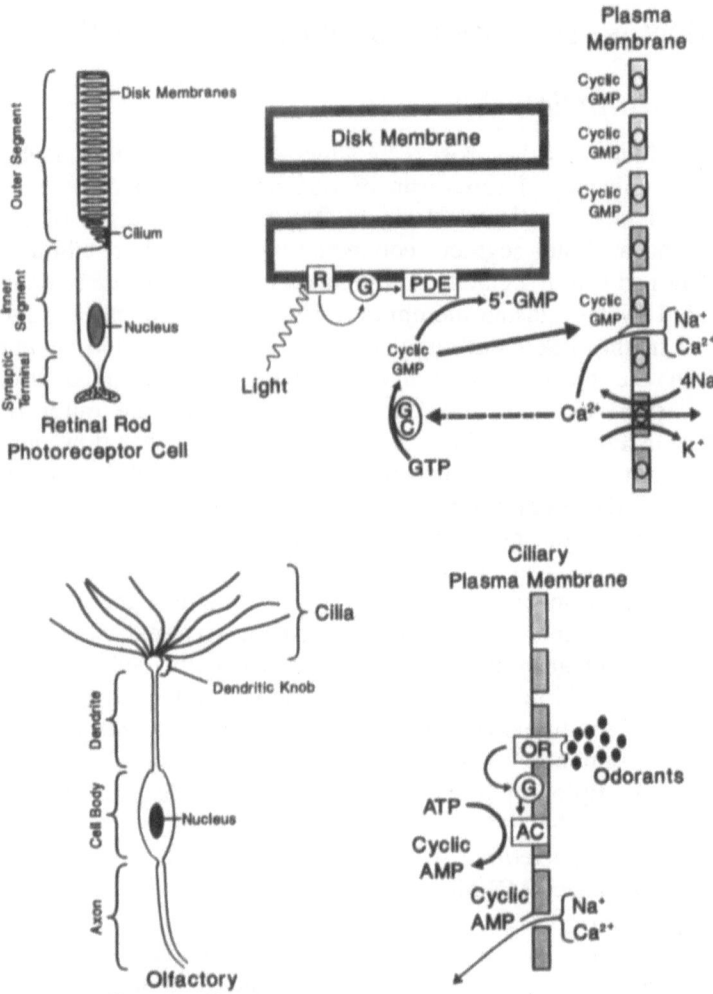

FIGURE 24.1. Schematic representation of the visual (upper diagram) and olfactory (lower diagram) transduction cascades. Visual and olfactory transduction occur at the apical extensions of sensory cells, the rod outer segment and the olfactory cilia, respectively. Both sensory pathways involve activation of primary receptors (R, rhodopsin; OR, odorant receptors) followed by a G protein (G) -mediated enzymatic reaction cascade that leads to a decrease (visual transduction; PDE, phosphodiesterase) or increase (olfactory transduction; AC, adenylate cyclase) in the intracellular concentration of a cyclic nucleotide. Modulation of cyclic nucleotide-activated channels as a result of alterations in the intracellular concentration of cGMP or cAMP generates the electrical signal that activates the cell. The upper scheme also depicts the Na^+-Ca^{2+}-K^+-exchanger that extrudes calcium from the rod outer segment. Calcium is one factor that regulates the synthesis of cGMP by modulating the activity of guanylate cyclase (GC), as described in the text.

subunits, the α subunit that binds GTP, and a βγ complex. In turn, each α subunit can activate a molecule of cGMP phosphodiesterase by removing an inhibitory subunit from this enzyme.[11] Each activated phosphodiesterase rapidly converts about 100 molecules of cGMP into 5′-GMP.[12,13] Thus the absorption of a single photon of light causes the hydrolysis of about 100,000 molecules of cGMP.[12] The reduction in the concentration of cGMP results in closure of some cGMP-activated ion channels in the plasma membrane (Figure 24.1).

The total concentration of cGMP in retinal rod outer segments is $50-60\,\mu M$[14,15] and is determined by the balance between the rate of production of cGMP by guanylate cyclase and its rate of hydrolysis by phosphodiesterase. However, the concentration of free cGMP, estimated from electrophysiologic experiments described below, is only about 5% of the total concentration, ie, only in the low micromolar range, whereas the remaining cGMP is bound to high-affinity binding sites.

Restoration of the dark steady state after light excitation is achieved by turning off each stage in the enzymatic cascade. The isomerized rhodopsin is inactivated through phosphorylation by rhodopsin kinase[16] and subsequent binding of a protein called "arrestin," which competitively inhibits the binding of transducin.[17] The active α subunit of transducin is inactivated due to its own endogenous GTPase activity.[18,19] Following the hydrolysis of GTP to GDP, the α subunit reassociates with the βγ complex. Finally, the phosphodiesterase is likely to be inactivated by recombination with its inhibitory subunit.[20]

The original level of cGMP is restored by guanylate cyclase, which catalyzes the synthesis of cGMP from GTP. Guanylate cyclase is activated by a reduction in the intracellular concentration of calcium.[21-23] Calcium enters the cell through cGMP-activated channels[24-27] and is extruded by the Na^+-Ca^{2+}-K^+ exchanger.[28-31] During illumination some cyclic nucleotide-activated channels close and the influx of calcium decreases. However, the exchanger continues to extrude calcium, which causes a reduction in the internal calcium concentration and consequently stimulation of guanylate cyclase (Figure 24.1).[40] This enzyme is regulated in a cooperative manner by four Ca^{2+} ions, with a concentration of calcium for half maximal activation of about 100 nM,[23] a value close to the estimated free calcium concentration in the dark state of about 300 nM.[32] Moreover, a soluble modulator protein is required for the activity of guanylate cyclase that has been named "recoverin" since it promotes the recovery of the dark state.[33]

Olfactory Transduction

Odorants are recognized and discriminated by olfactory receptor neurons located in the olfactory neuroepithelium. These cells are bipolar neurons. The chemoreceptive region of the cell consists of a dendrite that projects toward the nasal lumen and ends in a dilatation, the "dendritic knob." This knob carries a group of cilia embedded in an extracellular mucous matrix that lines the nasal cavity (Figure 24.1). Electrophysiologic studies have provided conclusive evidence that initial chemosensory events occur at these dendritic cilia.[34-37] The arrival of odorants at the ciliary membrane triggers a cascade of signal transduction events that evoke a generator potential in the cilia that passively spreads to the cell body and axon hillock, where it initiates action potentials through the activation of voltage-sensitive channels. These action potentials travel along the cell's axon to the olfactory bulb of the brain where the first synaptic relay takes place and decoding of the olfactory message is initiated. Olfactory receptor neurons differ in their responsiveness to odorants and different odorants stimulate distinct subpopulations of olfactory neurons.[38,39,109] Thus the molecular structure of an odorant is converted into a spatial and temporal pattern of neuronal activity that is processed by the brain and ultimately perceived as a characteristic odor quality.

Upon arriving at the ciliary membrane odorants are thought to interact with receptors (Figure 24.1). Recently cDNAs that encode a multigene family of putative odorant receptors expressed uniquely in olfactory tissue have been described.[40,109,110] These proteins are members of the superfamily of G protein-coupled receptors and contain the seven-transmembrane-domain structure that is characteristic of this superfamily.[6] They are closely related, forming a distinct subfamily of receptors, but exhibit extensive sequence variations within their third, fourth, and fifth transmembrane domains thought to reflect differences in odorant binding specificities. Although binding of odorants to these proteins or inter-actions between these proteins and second messenger systems have not yet been demonstrated, it is reasonable to expect that this diverse, olfactory tissue-specific array of novel G protein-linked receptors indeed represents the long-sought-after odorant receptors.

Olfactory cilia are rich in adenylate cyclase[41-46] and its associated regulatory G protein (Figure 24.1).[47-49] Both the adenylate cyclase and the G protein have been cloned and are designated adenylate cyclase type III[46] and G_{olf},[49] respectively. When the integrated accumulation of cAMP is measured in the presence of odorants and GTP, stimulation of the enzyme by micromolar-to-millimolar concentrations of some, but not all, odorants is observed.[41-43] Rapid quench-flow measurements reveal transient increases in cAMP in response to submicromolar concentra-tions of odorants, such as isomenthone[44] and citralva.[45] The dose-response curve for the odorant-induced increase in concentration of cAMP was biphasic, which suggests that a rapid, perhaps receptor-mediated, saturable increase occurs at nanomolar-to-micromolar concen-trations of odorant; this is followed by a further increase in cAMP concentration at higher odorant concentrations[45] perhaps due to partitioning of odorants in the membrane.[50] The characteristics of the latter effect account for earlier observations that showed nonsaturating dose-response curves at high concentrations of odorants when the integrated accumulation of cAMP was measured.

On the basis of earlier observations that some odorants fail to activate adenylate cyclase,[42] the notion that additional transduction pathways may participate in olfaction was pursued. Although this line of research needs to be developed, evidence indicates that odorants that fail to stimulate adenylate cyclase, such as pyrazine, may activate phospholipase C and cause the generation of inositol 1,4,5-trisphosphate, which in turn may lead to calcium mobilization.[45,51] In catfish olfactory receptor cells an odorant-activated influx of calcium has been documented using fura-2, a calcium-dependent fluorescent dye.[52] In this system generation of inositol 1,4,5-trisphosphate opens Ca^{2+} channels in the ciliary plasma membrane.[52] The likely involvement of multiple transduction pathways in olfaction suggests that mechanisms through which these pathways can communicate with each other must exist. Indeed, calmodulin mediates crosstalk between the calcium signalling pathway and the cAMP pathway.[53] Calcium bound to calmodulin activates olfactory adenylate cyclase type III via a mechanism that is, at least in part, GTP-dependent and distinct from the mechanism that mediates activation by odorants.[53] Furthermore, isolated olfactory receptor cells from newt fail to undergo inactivation when exposed to high concentrations of odorants in the absence of external calcium or in the presence of internal EGTA suggesting that desensitization to odorants may be calcium-dependent.[54,55] It is possible that following the initial, rapid, transient increase in cAMP concentration, calcium influx that results from the response to odorants may cause a secondary sustained rise in cAMP concentration, and that cAMP-dependent and calcium-dependent phosphorylation may contribute to desensitization.

Soon after it became clear that cAMP plays a central role in olfaction several laboratories investigated its effect on protein phosphorylation[56] and ion channel function. In 1987 Nakamura and Gold discovered a cAMP-gated conductance on the dendritic cilia, olfactory knob, and soma of olfactory receptor cells. Since their studies were made using excised patches, these authors concluded that cAMP opened channels directly.[57] cGMP also opens the olfactory channel and is even more potent than cAMP (Figure 24.2).[57-59] This observation was surprising since cGMP does not appear to be involved in olfactory transduction. Whether activation by cGMP has physiologic significance or is merely coincidental is not clear. Electrophysiologic studies have shown that application of forskolin or analogues of cAMP leads to depolarization of olfactory receptor cells.[36,60,61] These studies consolidated the notion that one

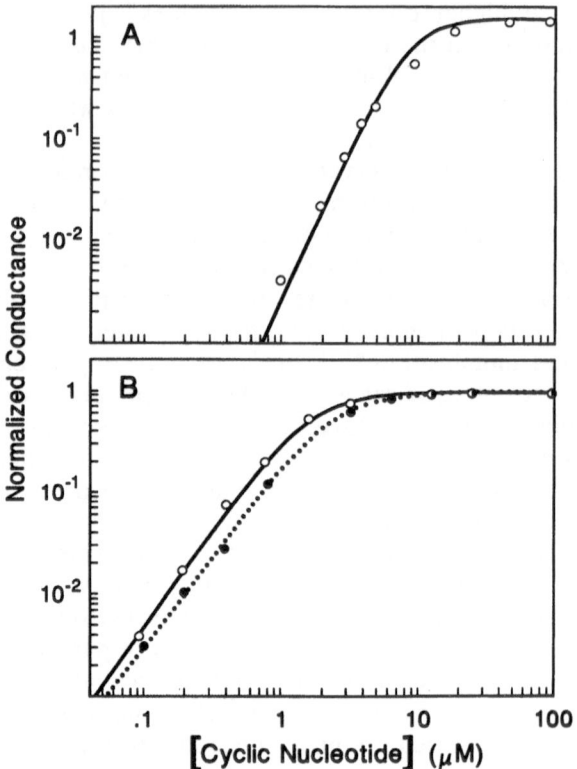

FIGURE 24.2. Activation of currents by cyclic nucleotides in the visual (A) and olfactory (B) systems. Normalized currents as a function of cyclic nucleotide concentration in excised patches from retinal rod outer segment (A) or from olfactory receptor cilia (B) are shown. Note that in the olfactory system cGMP (B; solid line, open circles) is more effective than cAMP (dotted line, filled circles) in activating the cyclic nucleotide-gated conductance, and that cGMP opens channels at lower concentrations in the olfactory system (B) than in the retinal rod outer segment (A). Based on Colamartino, Menini, and Torre[72] (A) and on Nakamura and Gold[57] (B). ((2A) Reproduced with permission, Colamartino et al. (2B) Reproduced with permission, © 1987 Macmillan Magazines Ltd.)

pathway for olfactory transduction consists of odorant-stimulated generation of cAMP via G_{olf} and subsequent opening of cyclic nucleotide-activated channels to evoke the generator potential (Figure 24.1).

Cloning of the cyclic nucleotide-activated channel from rat olfactory tissue,[59] bovine olfactory tissue,[62] and catfish olfactory tissue[63] has shown that it is homologous to its cGMP-activated counterpart of the photoreceptor cells.[64] Thus remarkable similarities exist between visual transduction and olfactory transduction (Figure 24.1). As in the photoreceptor cell, olfactory cyclic nucleotide-activated channels represent the key elements that link stimulus-induced transduction events to excitation of the sensory neuron. Understanding the molecular basis of activation of cyclic nucleotide-activated channels by cAMP and/or cGMP is therefore fundamental to our understanding of vertebrate photoreception and chemoreception.

Ion Permeation Through Cyclic Nucleotide-Activated Channels

Channel Activation by Cyclic Nucleotides

Visual excitation depends on the closing of a fraction of normally open channels,[5,65,66] whereas excitation of olfactory receptor cells is mediated via the opening of channels that are closed at rest. Unlike many other ion channels, cyclic nucleotide-activated channels in retinal rods and in olfactory receptor cells do not show desensitization upon prolonged exposure to cyclic nucleotide.[67,68] The dependence of the current upon cyclic nucleotide concentration shows that half maximal activation of the photoreceptor channel occurs at concentrations between 10 and 50 μM of cGMP (Figure 2A).[3,69-72] The photoreceptor channel can also be activated by cAMP, but here the half maximal concentration for activation by this cyclic nucleotide is larger than 1 mM, which indicates that under physiologic conditions the retinal rod cyclic nucleotide-activated channel is selective for cGMP.[73]

In contrast to the cyclic nucleotide-activated channel in photoreceptor cells, channels of olfactory receptor neurons are activated by micromolar concentrations of both cGMP and cAMP (Figure 2B). Half maximal concentrations for channel activation in amphibian and catfish olfactory receptor cells occurs between 2 and 22 μM for cAMP and between 1 and 4 μM for cGMP.[57,63,68,74,75,106] However, when cDNA that encodes cyclic nucleotide-activated channels from bovine,[76] rat,[59] or catfish[63] olfactory receptor neurons is expressed in transfected cell lines, higher concentrations of cAMP are necessary to activate half of the expressed channels. A subconductance state has been identified in catfish, and it has been suggested that lowered potency of the cyclic nucleotide in eliciting macroscopic currents in transfected cells may reflect a greater probability for opening of the channel to the subconductance state rather than the fully open state.[63]

Activation of the channel by cyclic nucleotides shows cooperativity both in retinal rods and in olfactory receptor cells. Values of Hill coefficients from 1.2 to 3.1 have been reported suggesting that channel opening depends on the cooperative binding of at least three cyclic nucleotide molecules.[3,57,59,63,68-70,72,74-77,106] Analyses of cDNA clones that encode cyclic nucleotide-activated channels from photoreceptor cells and olfactory neurons have shown that each polypeptide contains only a single cyclic nucleotide binding site.[59,62–64] Since cells transfected with cDNA that encodes a single polypeptide express functional channels, a complete channel protein can be formed by several identical monomers that all contribute to the pore-forming domain. Recently, however, a protein sharing a 30% sequence identity with the cyclic nucleotide activated channel protein was obtained from human retina. Coexpression of this protein with cloned human photoreceptor channel protein induced rapid flickering in channel opening characteristic of the native channel, suggesting that the native channel may be a heterooligomere.[105]

Unit Conductance

From the analysis of current noise, the single-channel conductance of photoreceptor channels under physiologic conditions was initially estimated to be 0.1 pS.[78–80] It was suggested that the small size of the unitary conductance could be due to a partial block of the current by divalent cations.[25,79] Indeed, removal of divalent cations produced an increase in the current. The first measurements of single channels in divalent cation-free solutions[70] revealed two conductance states of 24 pS and 8 pS with mean open times of 0.18 and 1 ms, respectively.[69] Recently, when records were obtained at a wider recording bandwidth (10 kHz instead of 1 kHz), it became clear that the channel fluctuates so rapidly between the closed and open state that the mean open time and the single-channel conductance cannot be resolved unequivocally with conventional patch-clamp techniques.[81] This rapid flickering is an intrinsic property of the channel, is largely independent of membrane voltage, and is not affected by intracellular divalent cations such as Ca^{2+} or Mg^{2+}.

The single-channel conductance of the olfactory channel measured at a limited recording bandwidth displays some variability, depending on species. In the absence of divalent cations, values for the unit conductance of 21 pS in bullfrog,[74] 28 pS in newt,[75] 45 pS in salamander,[68,82] and 55 pS in catfish[63] have been reported. In the catfish a subconductance state of 32 pS has also been identified.[63] It is tempting to speculate that records from the bullfrog and newt correspond to the subconductance state observed in catfish olfactory neurons, whereas the conductance measured in the salamander corresponds to the fully open state of the catfish channel.

Density of Channels

cGMP-activated channels are present at high density in the surface membrane of retinal rod outer segments, and only at low density in the inner segment.[83] Channels in the rod outer segment occur at a density of at least 650 per square micron, and their distribution is uniform.[84] This estimate of the channel density was obtained by taking advantage of the observation that excision of a membrane patch from isolated rod outer segments lowered the density of active channels. It has been suggested that this may be due to a physical artifact of patch excision and/or to disruption of enzymatic mechanisms, which normally may regulate the channel's ability to respond to cGMP.

In olfactory receptor neurons the density of channels was estimated from the amplitude of the current activated by saturating concentrations of cAMP divided by the single-channel current and the patch area, which was assumed to be a circular disk covering the pipette tip. These studies showed that cyclic nucleotide-activated channels in olfactory receptor neurons are present at a high density in the ciliary membrane: 920 and 2400 channels per square micron in newt and in toad, respectively.[75] Channels also occur in the dendrite and cell body, but at much lower density: 2 and 6 channels per square micron in newt and toad, respectively.[75] These experiments confirm the notion that olfactory cilia are the sites where chemosensory transduction occurs and in this respect appear analogous to the outer segments of retinal rods that mediate phototransduction.

Ionic Selectivity

The selectivity of cyclic nucleotide-activated channels for monovalent and divalent cations and their interactions were investigated first in intact cells by changing ionic solutions on the extracellular side of the cell.[25-27,36,54,85-87] However, these studies were complicated by the lack of control over intracellular concentrations of ions and cGMP. These problems were subsequently solved with the advent of the excised inside-out patch-clamp technique that allows control over the composition of both the intracellular and extracellular solution.

Both cyclic nucleotide-activated channels in photoreceptor cells and in olfactory receptor neurons are cation-selective, but they do not select strongly between alkali monovalent cations. The ionic selectivity of the photoreceptor channel in excised patches was measured by the permeability ratio, and calculated by changes in the reversal potential and by the sequence of efficacy in carrying macroscopic outward currents.[3,88,89] The permeability ratio sequence for alkali monovalent cations is $Li:Na:K:Rb:Cs = 1.14:1:0.98:0.84:0.58$.[89] A similar selectivity sequence was found for the olfactory cAMP-activated channel. The permeability of monovalent cations decreases with increasing ion radius. This permeability sequence is different from the sequence of mobilities in water, which indicates that cyclic nucleotide-activated channels are not simple water-filled pores, but that cations interact with binding sites within the pore during the permeation process. The selectivity sequence suggests that the energy well in the channel is a "high field strength" site, ie, one with a high negative charge density. The channel is also permeable to a large variety of organic cations and the dimensions of its pore must be at least 0.38×0.5 nm.[90,91]

Both channels are also permeable to divalent cations and under physiologic conditions allow calcium to enter the cell. The permeability to divalent cations has been measured in both intact rods[25-27] and in intact olfactory neurons,[55,92] and indicates that a variety of divalent cations can permeate cyclic nucleotide-activated channels. Recently the permeation properties of divalent cations have also been studied in excised patches from retinal rod outer segments.[72] With only divalent cations at the intracellular side at equiosmolar concentrations (73.3 mM), a small outward current carried by divalent cations was measured. The ratio between the current carried at +60 mV by Na, Sr, Ca, Ba, Mg, and Mn was $83.3:1.4:1:0.58:0.33:0.25$.[72]

Voltage Dependence and Blockage by Divalent Cations

The current-voltage relationships of retinal rod cGMP-activated channels measured in excised patches in the absence of divalent cations rectifies in the outward direction. The properties of the current-voltage relationship of an ion channel can generally be divided in two distinct processes, gating and permeation, where gating is the process that opens or closes the pore and permeation is the ion flux through the open pore. The macroscopic current, due to the opening of many channels, is described by the product of the current through a single open channel (permeation), the number of channels present in the membrane, and the probability of a channel being open (gating). In general, voltage may modulate both the gating and the permeation properties of a channel.

Studies on the kinetic properties of the channel showed that its inter-action with cGMP can be described by a model in which the channel is opened by the binding of three cGMP molecules followed by a rapid open-closed transition of the fully liganded channel.[93] This equilibrium is voltage-dependent and the closed state is favored by hyperpolarization, so that fewer channels will be open at negative potentials. Therefore the gating of the channel is slightly voltage-dependent. Moreover, it has recently been shown that the permeation process is also voltage-dependent[94] (C. Picco and A. Menini, unpublished data). Zimmerman and Baylor[94] suggested that the shape of the current-voltage relation in the steady state is mainly determined by the voltage-dependence of ion permeation, with the voltage-dependence of the channel's open probability making a small additional contribution. Addition of divalent cations blocks the current in a voltage-dependent manner (Figure 24.3).[72,94,95] Moreover, the sequence of their blocking potencies is almost the reverse of their permeability sequence, which suggests that the blocking potency by a divalent cation is inversely related to its ability to carry outward current.[72] This observation, together with the voltage-dependence of blockage, suggests that divalent cations block the cGMP-activated channel by entering the channel and occupying one or more binding sites for

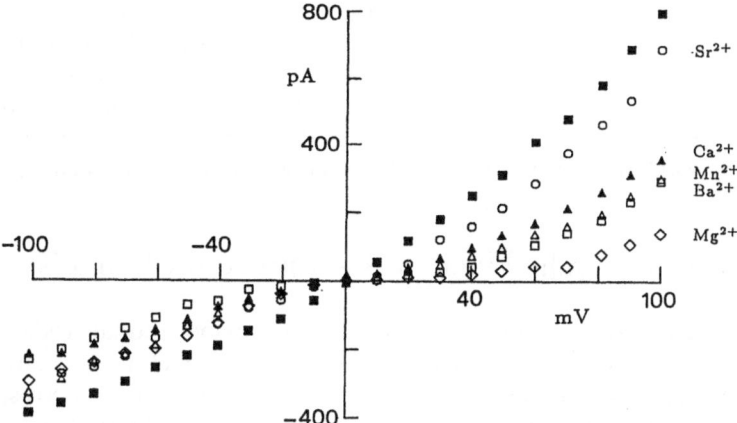

FIGURE 24.3. Voltage-dependent blockage of cyclic nucleotide-activated channels by divalent cations. Current-voltage relations were measured in the presence of 0.1 mM cGMP in excised patches from retinal rod outer segments. The current-voltage relation obtained in symmetrical NaCl solutions in the absence of divalent cations (filled squares) exhibits a small outward rectification. Addition of 1 mM of various salts of divalent cations at the intracellular side caused a reduction of the outward rectification, and in the presence of 1 mM Mg^{2+} the rectification becomes inward. These results indicate that the blocking effect of divalent cations is voltage-dependent.[72]

which cations compete in order to permeate. Several characteristics of ion permeation through the cGMP-activated channel of retinal rods have been accounted for quantitatively by a simple model in which the channel is assumed to contain a single cation binding site located in the channel.[27,89,94,96] This model assumes that only one ion can bind to this site at a time and that divalent cations bind with higher affinity than monovalents. This higher affinity for the binding site in the channel causes divalent cations to move through the channel more slow than monovalents. This model does not exclude other ion binding sites; it suggests that a single site within the channel that sensitive to the transmembrane voltage, dominates the permeation process.

The current-voltage relation of the olfactory channel activated by saturating concentrations of cAMP in the absence of divalent cations is almost linear for channels in the native membrane[57] and for those expressed in transfected cells.[59,63] Addition of divalent cations causes a voltage-dependent blockage of the current. However, at physiologic concentrations of ions channels in the native membrane showed only a small blocking effect,[57] whereas channels expressed in transfected cells displayed a strong outward rectification in the presence of divalent cations[59] similar to that observed in the retinal rod channel.[94,95] Records of single channels also demonstrate that Ca^{2+} and Mg^{2+} ions block the channel.[68,74] Recently, a cyclic nucleotide-activated channel has been identified in cardiac pacemaker cells.[1] Gating of this channel is specific for cAMP and at the same time depends on membrane voltage. The possible regulation of this channel by calcium remains to be investigated.

Structure of Cyclic Nucleotide-Activated Channels

Cyclic nucleotide-activated channels from bovine retinal rods[64] and human[107] and from rat,[59] bovine,[62] and catfish[63] olfactory neurons have been cloned. Their amino acid sequences, deduced from the corresponding nucleotide sequences of their cDNAs, reveal extensive homology (Figure 24.4). Analyses of these sequences indicate that they lack a signal peptide suggesting that both their amino termini and carboxyl termini may be located intracellularly. Recently, it has been demonstrated that the first 92 amino acids of the N terminus of the photoreceptor channel polypeptide are removed after synthesis.[97] This accounts for the observation that cyclic nucleotide-activated channels purified from retinal rods migrate on polyacrylamide gels in SDS with an apparent molecular weight of $63\,kDa$[77] rather than $79\,kDa$, the molecular weight calculated from their amino acid sequences and observed in the case of in vitro synthesized channel polypeptides. Cyclic nucleotide-activated channels from olfactory neurons show similar electrophoretic migration, which suggests that they may undergo similar posttranslational proteolytic modification (R.R.H. Anholt, unpublished data). Cyclic nucleotide-activated channels from bovine rod photoreceptor cells also undergo N-linked glycosylation at a single site.[98] The gene encoding the human retinal rod channel has been localized to chromosome 4 and has a complex structure containing at least 10 exons.[107]

Hydropathicity plots of olfactory and photoreceptor channels show similar patterns. In each case a large, highly charged N-terminal domain is followed by six putative transmembrane domains and an intracellular C-terminal region (Figure 24.4). Sequence divergence between different channels is greatest in the N-terminal and C-terminal domains, whereas homologies are most evident in the transmembrane regions, the cyclic nucleotide binding domain, and a putative pore-forming region, described in detail below. Immunohistochemical studies with site-specific antibodies directed against the N terminus of the mature protein after removal of the first 92 amino acids demonstrated that the N terminus of the mature cGMP-activated channel from retinal rods indeed is located on the intracellular side of the membrane.[97] The intracellular C-terminal domain contains a sequence of amino acids homologous to a similar sequence found in cyclic nucleotide-dependent protein kinases. This sequence is

Met *met* thr glu lys ser asn gly val lys ser ser pro ala asn asn his asn his his   20

pro pro pro ser *ile* lys ala asn gly lys asp asp his arg ala gly ser arg pro gln   40

ser val ala ala asp *asp* asp thr ser pro *glu* leu gln arg leu ala glu met asp thr   60

pro arg arg gly arg *gly* gly phe gln arg ile val arg leu val gly val ile arg asp   80

trp *ala* asn lys asn phe arg glu glu glu pro arg pro asp ser phe leu glu arg phe   100

arg gly pro glu leu gln thr val thr thr his gln gly asp asp *lys* gly gly lys asp   120

gly glu gly *lys* gly thr lys lys *lys* phe *glu* leu phe *val* leu *asp pro* ala *gly* asp   140

trp *tyr tyr* arg *trp leu phe* val *ile* ala met *pro val* leu *tyr* asn *trp* cys leu leu   160

val *ala* arg ala cys phe ser asp *leu* gln arg asn tyr phe val *val trp leu val* leu   180

asp tyr phe *ser asp* thr *val tyr* ile ala *asp* leu ile *ile* arg leu *arg thr* gly phe   200

leu glu gln gly leu leu val lys asp pro lys *lys* leu arg *asp* asn tyr ile *his thr*   220

leu gln phe *lys leu asp val* ala *ser* ile *ile pro thr asp* leu ile *tyr* phe ala val   240

gly ile his ser *pro* glu val arg phe *asn* arg leu leu his phe ala *arg* met *phe* glu   260

phe phe asp arg *thr* glu *thr* arg thr ser tyr pro asn ile phe arg *ile ser asn leu*   280

val leu *tyr* ile leu val *ile ile* his trp asn ala *cys* ile *tyr* tyr val *ile* ser *lys*   300

ser ile gly phe gly val asp thr trp val tyr pro asn ile thr asp pro glu tyr gly   320

tyr leu **ala arg** glu **tyr** ile **tyr** cys **leu tyr trp ser thr leu thr leu thr thr** ile   340

gly glu thr pro pro pro val lys asp glu **glu** *tyr* leu *phe val* ile *phe asp phe leu*   360

ile gly val leu ile phe ala thr ile val gly asn val gly ser met ile ser asn met   380

asn ala thr arg ala glu phe gln *ala* lys ile asp ala val lys his *tyr* met gln phe   400

arg lys *val* ser lys asp met glu ala lys *val* ile lys trp phe asp tyr leu trp thr   420

asn lys lys thr val asp glu arg glu val leu lys asn leu pro ala lys leu arg ala   440

glu ile ala ile asn val his leu ser thr leu lys lys val arg ile phe *gln asp cys*   460

glu ala gly leu leu val glu leu val leu lys leu arg pro gln val **phe ser pro gly**   480

asp tyr ile cys arg lys **gly asp ile gly lys glu met tyr ile ile lys glu gly lys**   500

leu ala val val ala asp asp gly val **thr gln** tyr **ala leu leu ser ala gly ser cys**   520

phe gly glu ile ser ile leu asn ile lys gly ser lys met **gly asn arg *arg thr*** ala   540

asn ile arg ser leu gly tyr ser asp leu phe cys leu ser lys asp asp leu met glu   560

ala val **thr glu tyr pro asp ala** lys **lys val leu glu glu arg gly arg glu *ile* leu**   580

met lys glu **gly leu leu asp glu asn** glu val ala *ala* ser met glu val *asp* val gln   600

glu lys leu glu gln leu *glu* thr asn met *asp* thr *leu* tyr *thr* arg phe ala arg leu   620

leu ala *glu* tyr thr gly ala *gln gln* lys leu *lys* gln arg ile *thr* val leu *glu* thr   640

lys met *lys* gln asn his glu asp asp tyr leu ser asp gly ile asn thr pro *glu* pro   660

thr ala ala glu

FIGURE 24.4. Amino acid sequence of the cyclic nucleotide-activated channel of olfactory receptor neurons. The entire amino acid sequence of the open reading frame deduced from the nucleotide sequence of the cyclic nucleotide-activated channel of rat olfactory receptor neurons, as reported by Dhallan et al.[59] is shown. Amino acids that are identical in the sequences of the channels from bovine retinal rod outer segments and rat olfactory receptor neurons are shown in italics. Six putative transmembrane domains are underlined. A possible channel-forming domain between amino acids 321 and 343 and the cyclic nucleotide binding domain between amino acids 477 and 556 are shown in bold print. Within the latter domain amino acids, which are thought to participate in binding the cyclic nucleotide, are doubly underlined. Note that the presence of an even number of transmembrane domains would place both the N terminus and the C terminus on the cytoplasmic side of the membrane, in agreement with Molday et al.[97]

likely to form the cyclic nucleotide binding site (Figure 24.4).[64,99] It contains a threonine residue, Thr560 in retinal rod cells and Thr537 in the olfactory channel (bovine sequence) that is invariant in the cGMP binding domains of all known cGMP-dependent protein kinases. This threonine is exchanged for an alanine in 23 of 24 cAMP binding sites in cAMP-dependent protein kinases.[64,76,99,100] Indeed, site-directed mutagenesis studies in which this threonine was substituted by an alanine resulted in a change of ligand specificity in cyclic nucleotide-gated channels expressed in *Xenopus* oocytes.[76] Replacement of Thr560 by alanine in photo-receptor channels resulted in an approximately 30-fold increase in the activation constant for cGMP. Similarly, the same substitution in the olfactory channel reduced the sensitivity for cGMP approximately 40-fold, whereas the effect on the sensitivity of the channel to cAMP was minimal. These observations suggest that the hydroxyl group of this threonine is important for recognition of cGMP, but not of cAMP.[76] It has been suggested that this hydroxyl group forms an additional hydrogen bond with the guanine 2-amino group of cGMP (Figure 24.5). Two other highly conserved amino acid residues, an arginine (Arg559) and a glutamate (Glu544), have also been implicated in the formation of the binding site for cyclic nucleotides (Figure 24.5).

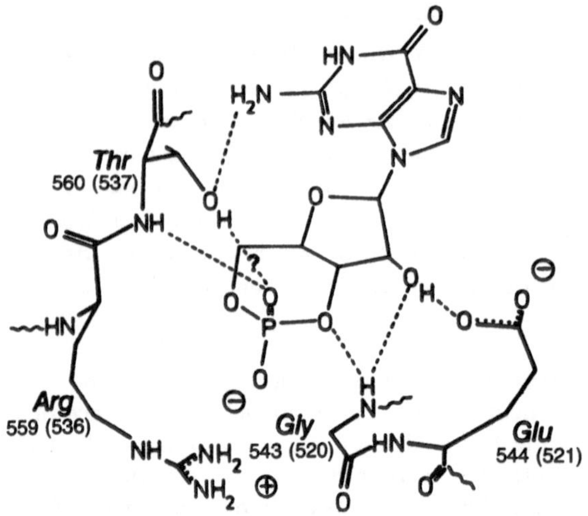

FIGURE 24.5. Formation of the cyclic nucleotide binding site of cyclic nucleotide-gated channels. The dotted lines indicate hypothetical hydrogen bonds, which may interact with cGMP. Amino acid residues that interact with the cyclic nucleotide are numbered according to the sequence of the bovine retinal rods (olfactory) channel polypeptide. The corresponding threonine, arginine, glycine, and glutamate residues of the olfactory cyclic nucleotide-activated channel from rat are indicated by double underlines in Figure 24.4. The question mark indicates a hydrogen bond, which may not be involved in cyclic nucleotide binding. Numbering according to the sequence of bovine retinal rod channel.[76,99]

One region of special interest in the amino acid sequences of cyclic nucleotide-activated channels consists of the fourth and fifth proposed transmembrane domains and the linker region between them (positions 300 to 351 in the olfactory channel and 321 to 372 in the retinal rod channel; Figure 24.4). This region contains an unusually high number of hydroxylated amino acids, mostly tyrosines and threonines, which together comprise 25% of all its amino acids. They are distributed evenly, with an average spacing of two amino acids between consecutive hydroxyls, and are especially concentrated between positions 321 and 343 (numbering is here according to the sequence of the rat olfactory channel), where they occur, on average, 1.4 amino acids apart. This region also contains five invariant prolines, three of which are arranged in a conspicuous tandem array at positions 344 to 346 (Figure 24.4). The region between position 321 and 343 is eminently suitable to form the lining of a cation channel and is highly conserved between retinal rod and olfactory channels with 77% sequence identity over the entire region and 87% sequence identity between positions 321 and 343 (Figure 24.4). It is of particular interest to note that this region shows homologies to the pore-forming hairpin of voltage-gated K^+ channels.[63,101-103] Recently it has been reported that the *Drosophila ether-à-gogo* (*eag*) gene also encodes a channel homologous to cyclic nucleotide-gated ion channels.[102] Similarities between the *eag* channel and cyclic nucleotide-gated channels from retinal rods is greatest in the second, fifth, and sixth putative transmembrane domains, and in segments near the C terminus. However, the linker region between the putative fifth transmembrane α-helical domain and the fourth transmembrane domain of the *eag* channel shows more similarities to the *Shaker* voltage-gated K^+ channel.[102] Alignment of corresponding hairpin sequences of *Shaker* K^+ channels and cyclic nucleotide-activated channels revealed that the former contains a tyrosine and a glycine residue that have no counterparts in the latter. Deletion of these residues abolished the *Shaker* channel selectivity for K^+ and rendered their ion permeation properties similar to those of the cyclic nucleotide-activated channels.[108] These observations indicate that this region of the cyclic nucleotide-gated channel located near the center of the polypeptide may resemble a hairpin structure similar to that proposed to form the transmembrane pore of voltage-dependent K^+ channels.[101,102] Similarities between the family of cyclic nucleotide-activated channels and voltage-gated K^+ channels have also led to the suggestion that the "ball and chain" model proposed for K^+ channel inactivation[103] may be applicable to gating of the cyclic nucleotide-activated channel, in that the cyclic nucleotide binding domain may represent the "ball", which plugs the pore in the absence of cyclic nucleotide. Binding of the cyclic nucleotide would allow this ball to pop out of the channel as a result of a conformational change, and allow ions to flow.[63] Future studies on cyclic nucleotide-activated channels using site-directed mutagenesis are likely to validate or refute these hypotheses.

Thus far a variety of cyclic nucleotide-activated channels has been uncovered: the cGMP-activated channel of retinal rods and of retinal cones, the cAMP-gated channel of olfactory receptor neurons, the voltage-dependent cAMP-gated channel of cardiac pacemaker cells, a cGMP-activated channel in cells from the pineal gland, and the *eag* channel of *Drosophila*. It is likely that future investigations will reveal the existence of additional members of this rapidly expanding family of second messenger-activated channels.

Acknowledgments. The authors are indebted to Drs. A. Zimmerman, D. Baylor, U.B. Kaupp, and S. Firestein for sharing their manuscripts prior to publication. Critical reading of the manuscript and constructive criticism by Dr. Sidney A. Simon, John W. Moore, and V. Torre, and excellent secretarial assistance by Beth Robinson are also gratefully acknowledged. Collaboration between Drs. Menini and Anholt, and preparation of this book chapter, was made possible by a NATO collaborative research grant (CRG-890435). Work in the laboratory of Dr. Robert R.H. Anholt is supported by grants from the National Institutes of Health (DC00394) and the U.S. Army Research Office (DAAL03-86-K-0130).

References

1. DiFrancesco D, Tortora P. Direct activation of cardiac pacemaker channels by intracellular cyclic AMP. Nature 1991; 351:145–147.
2. Dryer SE, Henderson D. A cyclic GMP-activated channel in dissociated cells of the chick pineal gland. Nature 1991; 353:756–758.
3. Fesenko EE, Kolesnikov SS, Lyubarski AL. Induction by cyclic GMP of cationic conductance in plasma membrane of retinal rod outer segment. Nature 1985; 313:310–313.
4. Stryer L. Cyclic GMP cascade of vision. Annu Rev Neurosci 1986; 9:87–119.
5. Pugh EN, Lamb TD. Cyclic GMP and calcium: The internal messengers of excitation and adaptation in vertebrate photoreceptors. Vision Res 1990; 30(12):1923–1948.
6. O'Dowd BF, Lefkowitz RJ, Caron MG. Structure of the adrenergic and related receptors. Ann Rev Neurosci 1989; 12:67–83.
7. Honig B, Ebrey T, Callender RH, Dinur U, Ottolenghi M. Photoisomerization, energy storage, and charge separation: A model for light energy transduction in visual pigments and bacteriorhodopsin. Proc Natl Acad Sci USA 1979; 76:2503–2507.
8. Fung BKK, Stryer L. Photolyzed rhodopsin catalyzes the exchange of GTP for GDP in retinal rod outer segment membranes. Proc Natl Acad Sci USA 1980; 77:2500–2504.
9. Kühn H. Light- and GTP-regulated interaction of GTPase and other proteins with bovine photoreceptor membranes. Nature 1980; 283:587–589.

10. Stryer L, Hurley JB, Fung BK-K. First stage of amplification in the cyclic nucleotide cascade of vision. Curr Top Membr Transp 1981; 15:93–108.

11. Fung BKK, Hurley JB, Stryer L. Flow of information in the light-triggered cyclic nucleotide cascade of vision. Proc Natl Acad Sci USA 1981; 78:152–156.

12. Yee R, Liebman RA. Light-activated phosphodiesterase of the rod outer segment: Kinetics and parameters of activation and deactivation. J Biol Chem 1978; 253:8902–8909.

13. Baehr W, Devlin MJ, Applebury ML. Isolation and characterization of cGMP phosphodiesterase from bovine rod outer segments. J Biol Chem 1979; 254:11669–11677.

14. Cote RH, Biernbaum MS, Nicol GD, Bownds MD. Light-induced decreases in cGMP concentration precede changes in membrane permeability in frog rod photoreceptors. J Biol Chem 1984; 259:9635–9641.

15. Blazynski C, Cohen AI. Rapid decline in cyclic GMP of rod outer segments of intact frog photoreceptors after illumination. J Biol Chem 1986; 261:14142–14147.

16. Kühn H, Dreyer WJ. Light dependent phosphorylation of rhodopsin by ATP. FEBS Lett 1972; 20:1.

17. Kühn H, Hall SW, Wilden U. Light-induced binding of 48-kDa protein to photoreceptor membranes is highly enhanced by phosphorylation of rhodopsin. FEBS Lett 1984; 176:473–478.

18. Wheeler GL, Matuo Y, Bitensky MW. Light-activated GTPase in vertebrate photoreceptors. Nature 1977; 269:822–823.

19. Vuong TM, Chabre M. Subsecond deactivation of transducin by endogenous GTP hydrolysis. Nature 1990; 346:71–74.

20. Deterre P, Bigay J, Forquet F, Robert M, Chabre M. cGMP phosphodiesterase of retinal rods is regulated by two inhibitory subunits. Proc Natl Acad Sci USA 1988; 95:2424–2428.

21. Lolley RN, Racz E. Calcium modulation of cyclic GMP synthesis in rat visual cells. Visual Res 1982; 22:1481–1486.

22. Pepe IM, Panfoli I, Cugnoli C. Guanylate cyclase in rod outer segments of the toad retina. FEBS Lett 1986; 203:73–76.

23. Koch KW, Stryer L. Highly cooperative feedback control of retinal rod guanylate cyclase by calcium ions. Nature (London) 1988; 334:64–66.

24. Capovilla M, Caretta A, Cervetto L, Torre V. Ionic movements through light-sensitive channels of toad rods. J Physiol 1983; 343:295–310.

25. Hodgkin AL, McNaughton PA, Nunn BJ. The ionic selectivity and calcium dependence of the light-sensitive pathway in toad rods. J Physiol 1985; 358:447–468.

26. Nakatani K, Yau KW. Calcium and magnesium fluxes across the plasma membrane of the toad rod outer segment. J Physiol 1988; 395:695–729.

27. Menini A, Rispoli G, Torre V. The ionic selectivity of the light-sensitive current in isolated rods of the tiger salamander. J Physiol 1988; 402:279–300.

28. Yau KW, Nakatani K. Cation selectivity of light-sensitive conductance in retinal rods. Nature 1984a; 309:352–354.

29. Hodgkin AL, McNaughton PA, Nunn BJ. Measurement of sodium-calcium exchange in salamander rods. J Physiol 1987; 391:347–370.

30. Lagnado L, Cervetto L, McNaughton PA. Ion transport by the Na-Ca exchange in isolated rod outer segments. Proc Natl Acad Sci USA 1988; 85:4548–4552.
31. Cervetto L, Lagnado L, Perry RJ, Robinson DW, McNaughton PA. Extrusion of calcium from rod outer segments is driven by both sodium and potassium gradients. Nature 1989; 337:740–743.
32. Ratto GM, Payne R, Owen WG, Tsien RY. The concentration of cytosolic free calcium in vertebrate outer segments measured with fura-2. J Neurosci 1988; 8:3240–3246.
33. Lambrecht HG, Koch KW. A 26 kD calcium binding protein from bovine rod outer segments as modulator of photoreceptor guanylate cyclase. EMBO J 1991; 10:793–798.
34. Frings S, Lindemann B. Single unit recording from olfactory cilia. Biophys J 1990; 57:1091–1094.
35. Firestein S, Shepherd GM, Werbln FS. Time course of the membrane current underlying sensory transduction in salamander olfactory receptor neurons. J Physiol 1990; 430:135–158.
36. Kurahashi T. The response induced by intracellular cyclic AMP in isolated olfactory receptor cells of the newt. J Physiol 1990; 430:355–371.
37. Lowe G, Gold GN. The spatial distribution of odorant sensitivity and odorant-induced currents in salamander olfactory receptor cells. J Physiol 1991; 442:147–168.
38. Duchamp A, Revial MF, Holley A, MacLeod P. Odor discrimination by frog olfactory receptors. Chem Senses Flavour 1974; 1:213–233.
39. Sicard G, Holley A. Receptor cell responses to odorants: Similarities and differences among odorants. Brain Res 1984; 292:283–296.
40. Buck L, Axel R. A novel multigene family may encode odorant receptors: A molecular basis for odor recognition. Cell 1991; 65:175–187.
41. Pace U, Hanski E, Salomon Y, Lancet D. Odorant-sensitive adenylate cyclase may mediate olfactory reception. Nature 1985; 316:255–258.
42. Sklar PB, Anholt RRH, Snyder SH. The odorant-sensitive adenylate cyclase of olfactory receptor cells: Differential stimulation by distinct classes of odorants. J Biol Chem 1986; 261:15538–15543.
43. Shirley SG, Robinson CJ, Dickison K, Aujla R, Dodd GH. Olfactory adenylate cyclase of the rat: Stimulation by odorants and inhibition by Ca^{2+}. Biochem J 1986; 240:605–607.
44. Breer H, Boekhoff I, Tareilus E. Rapid kinetics of second messenger formation in olfactory transduction. Nature 1990; 345:65–68.
45. Boekhoff I, Tareilus E, Strottman J, Breer H. Rapid activation of alternative second messenger pathways in olfactory cilia from rats by different odorants. EMBO J 1990; 9:2453–2458.
46. Bakalyar HA, Reed RR. Identification of a specialized adenylyl cyclase that may mediate odorant detection. Science 1990; 250:1403–1406.
47. Pace U, Lancet D. Olfactory GTP-binding protein: Signal transducing polypeptide of vertebrate chemosensory neurons. Proc Natl Acad Sci USA 1986; 83:4947–4951.
48. Anholt RRH, Mumby SM, Stoffers DA, Girard PR, Kuo JF, Snyder SH. Transduction proteins of olfactory receptor cells: Identification of guanine

nucleotide binding proteins and protein kinase C. Biochemistry 1987; 26:788–795.

49. Jones DT, Reed RR. G$_{olf}$: An olfactory neuron specific G-protein involved in odorant signal transduction. Science 1989; 244:790–795.

50. Anholt RRH. Primary events in olfactory reception. Trends Biochem Sci 1987; 12:58–62.

51. Breer H, Boekhoff I. Odorants of the same odor class activate different second messenger pathways. Chemical Senses 1991; 16:19–29.

52. Restrepo D, Miyamoto T, Bryant BP, Teeter JH. Odor stimuli trigger influx of calcium into olfactory neurons of the channel catfish. Science 1990; 249:1166–1168.

53. Anholt RRH, Rivers AM. Olfactory transduction: Crosstalk between second messenger systems. Biochemistry 1990; 29:4049–4054.

54. Kurahashi T. Activation by odorants of cation-selective conductance in the olfactory receptor cell isolated from the newt. J Physiol 1989; 419:177–192.

55. Kurahashi T, Shibuya T. Ca^{2+}-dependent adaptive properties in the solitary olfactory receptor cell of the newt. Brain Res 1990; 515:261–268.

56. Heldman J, Lancet D. Cyclic AMP-dependent protein phosphorylation in chemosensory neurons: Identification of cyclic nucleotide-regulated phosphoproteins in olfactory cilia. J Neurochem 1986; 47:1527–1533.

57. Nakamura T, Gold GH. A cyclic nucleotide-gated conductance in olfactory receptor cilia. Nature 1987; 325:442–444.

58. Kolesnikov SS, Zhainazarov AB, Kosolapov AV. Cyclic nucleotide-activated channels in the frog olfactory receptor plasma membrane. FEBS Lett 1990; 266:96–98.

59. Dhallan RS, Yau KW, Schrader KA, Reed RR. Primary structure and functional expression of a cyclic nucleotide-activated channel from olfactory neurons. Nature 1990; 347:184–187.

60. Frings S, Lindemann B. Current recording from sensory cilia of olfactory receptor cells in situ. I. The neural response to cyclic nucleotides. J Gen Physiol 1991; 97:1–16.

61. Firestein S, Darrow B, Shepherd GM. Activation of the sensory current in salamander olfactory receptor neurons depends on a G protein-mediated cAMP second messenger system. Neuron 1991; 6:825–835.

62. Ludwig J, Margalit T, Eismann E, Lancet D, Kaupp UB. Primary structure of cAMP-gated channel from bovine olfactory epithelium. FEBS Lett 1990; 270:24–29.

63. Goulding EH, Ngai J, Kramer RH, Colicos S, Axel R, Siegelbaum SA, Chess A. Molecular cloning and single-channel properties of the cyclic nucleotide-gated channel from catfish olfactory neurons. Neuron 1992; 8:45–58.

64. Kaupp UB, Niidome T, Tanabe T, Terada S, Boenigk W, Stuehmer W, Cook NJ, Kangawa K, Matsuo H, Hirose T, Miyata T, Numa S. Primary structure and functional expression from complementary DNA of the rod photoreceptor cyclic GMP-gated channel. Nature 1989; 342:762–766.

65. Yau KW, Baylor DA. Cyclic GMP-activated conductance of retinal photoreceptor cells. Annu Rev Neurosci 1989; 12:289–327.

66. McNaughton PA. Light response of vertebrate photoreceptors. Physiol Rev 1990; 70:847–883.

67. Karpen JW, Zimmerman AL, Stryer L, Baylor DA. Molecular mechanics of the cGMP-activated channel of retinal rods. Cold Spring Harbor Symp Ouant Biol 1988a; 53:325–332.
68. Zufall F, Firestein S, Shepherd GM. Analysis of single cyclic nucleotide-gated channels in olfactory receptor cells. J Neurosci 1991; 11:3573–3580.
69. Haynes LW, Kay AR, Yau KW. Single cyclic GMP-activated channel activity in excised patches of rod outer segment membrane. Nature 1986; 321:66–70.
70. Zimmerman AL, Baylor DA. Cyclic GMP-sensitive conductance of retinal rods consists of aqueous pores. Nature 1986; 321:70–72.
71. Lühring H, Hanke W, Simmoteit R, Kaupp UB. Cation selectivity of the cGMP-gated channel of mammalian rod photoreceptors. In: Borsellino A, Cervetto L, Torre V, eds. Sensory Transduction. New York: Plenum Press; 1990:169–174.
72. Colamartino G, Menini A, Torre V. Blockage and permeation of divalent cations through the cyclic GMP-activated channel from tiger salamander retinal rods. J Physiol 1991; 440:189–206.
73. Tanaka JC, Eccleston JF, Furman RE. Photoreceptor channel activation by nucleotide derivatives. Biochemistry 1989; 28:2776–2784.
74. Suzuki N. Single cyclic nucleotide-activated ion channel activity in olfactory receptor cell soma membrane. Neurosci Res Suppl 1990; 12:S113–S126.
75. Kurahashi T, Kaneko A. High density cAMP-gated channels at the ciliary membrane in the olfactory receptor cell. NeuroReport 1991; 2:5–8.
76. Altenhofen W, Ludwig J, Eismann E, Kraus W, Boenigk W, Kaupp UB. Control of ligand specificity in cyclic nucleotide-gated channels from rod photoreceptors and olfactory epithelium. Proc Natl Acad Sci USA 1991; 88:9868–9872.
77. Cook NJ, Hanke W, Kaupp UB. Identification, purification, and functional reconstitution of the cyclic GMP-dependent channel from rod photoreceptors. Proc Natl Acad Sci USA 1987; 84:585–589.
78. Detwiler PB, Conner JD, Bodoia RD. Gigaseal patch clamp recordings from outer segments of intact retinal rods. Nature 1982; 300:59–61.
79. Bodoia RD, Detwiler PB. Patch-clamp recordings of the light-sensitive dark noise in retinal rods from the lizard and frog. J Physiol (Lond) 1985; 367:183–216.
80. Gray P, Attwell D. Kinetics of light-sensitive channels in vertebrate photoreceptors. Proc R Soc Ser B 1985; 223:379–388.
81. Torre V, Straforini M, Campani M. A quantitative model of phototransduction and adaptation in amphibian rod photoreceptors. Seminars Neurosci 1992; 4:5–13.
82. Zufall F, Shepherd G, Firestein S. Inhibition of the olfactory cyclic nucleotide gated ion channel by intracellular calcium. Proc R Soc Lond 1991; 246:225–230.
83. Watanabe SI, Matthews G. Regional distribution of cGMP-activated ion channels in the plasma membrane of the rod photoreceptor. J Neurosci 1988; 8:2334–2337.

84. Karpen JW, Loney DA, Baylor DA. Cyclic GMP-activated channels of salamander retinal rods: spatial distribution and variation of responsiveness. J Physiol 1992; 448:257–274.
85. Hodgkin AL, McNaughton PA, Nunn BJ, Yau KW. Effect of ions on retinal rods from *Bufo marinus*. J Physiol 1984; 350:649–680.
86. Yau KW, Nakatani K. Electrogenic Na-Ca exchange in retinal rod outer segment. Nature 1984b; 311:661–663.
87. Cervetto L, Menini A, Rispoli G, Torre V. The modulation of the ionic selectivity of the light-sensitive current in isolated rods of the tiger salamander. J Physiol 1988; 406:181–198.
88. Furman RE, Tanaka JC. Monovalent selectivity of the cyclic guanosine monophosphate-activated ion channel. J Gen Physiol 1990; 96:57–82.
89. Menini A. Currents carried by monovalent cations through cyclic GMP-activated channels in excised patches from salamander rods. J Physiol 1990; 424:167–185.
90. Picco C, Menini A. The permeability to organic cations of the cyclic GMP-activated channel from retinal rods. Biophys J 1991; 59:535a.
91. Picco C, Menini A. The permeability of the cGMP-activated channel to organic cations in retinal rods of the tiger salamander. J Physiol 1993; 460:741–758.
92. Suzuki N. Voltage- and cyclic nucleotide-gated currents in isolated olfactory receptor cells. In: Brand JG, Teeter JH, Cagan RH, Kare MR, eds. Chemical Senses: Receptor Events and Transduction in Taste and Olfaction. New York: Marcel Dekker; 1989:469–493.
93. Karpen JW, Zimmerman AL, Stryer L, Baylor DA. Gating kinetics of the cyclic GMP-activated channel of retinal rods: Flash photolysis and voltage jump studies. Proc Natl Acad Sci USA 1986b; 85:1287–1291.
94. Zimmerman AL, Baylor DA. Cation interactions within the cyclic GMP-activated channel of retinal rods from the tiger salamander. J Physiol 1992; 449:759–783.
95. Yau KW, Haynes LW, Nakatani K. Roles of calcium and cyclic GMP in visual transduction. In: Luttgau H, ed. Membrane Control of Cellular Activity. Stuttgart: Gustav Fischer; 1986:343–366.
96. Zimmerman AL, Baylor DA. Ionic permeation in the cGMP-activated channel of retinal rods. Biophys J 1988; 53:472a.
97. Molday RS, Molday LL, Dose A, Clark-Lewis I, Illing M, Cook NJ, Eismann E, Kaupp UB. The cGMP-gated channel of the rod photoreceptor cell: Characterization and orientation of the N-terminus, in press.
98. Wohlfart P, Muller H, Cook NY. Lectin binding and enzymatic deglycosylation of the cGMP-gated channel from bovine rod photo receptors. J Biol Chem 1989; 264:20934–20930.
99. Kaupp UB. The cyclic nucleotide-gated channels of vertebrate photoreceptors and olfactory epithelium. Trends Neurosci 1991; 14:150–157.
100. Weber IT, Shabb JB, Corbin JD. Predicted structures of the cGMP binding domains of the cGMP-dependent protein kinase: A key alanine/threonine difference in evolutionary divergence of cAMP and cGMP binding sites. Biochemistry 1989; 28:6122–6127.

101. Jan LY, Jan YN. A superfamily of ion channels. Nature 1990; 345:672.
102. Miller C. 1990: Annus mirabilis of potassium channels. Science 1991; 252:1092–1096.
103. Hoshi T, Zagotta WN, Aldrich RW. Biophysical and molecular mechanisms of *Shaker* potassium channel inactivation. Science 1990; 250:533–538.
104. Guy HR, Durell SR, Warmke J, Drysdale R, Ganetzky B. Similarities in amino acid sequences of *Drosophila eag* and cyclic nucleotide-gated channels. *Science* 1991; 254:730.
105. Chen TY, Peng YW, Dhallan RS, Ahamed B, Reed RR, Yau KW. A new subunit of the cyclic nucleotide-gated cation channel in retinal rods. Nature 1993; 362:764–767.
106. Frings S, Lynch JW, Lindemann B. Properties of cyclic nucleotide-gated channels mediating olfactory transduction: Activation, selectivity and blockage. J Gen Physiol 1992; 100:1–11.
107. Dhallan RS, Macke JP, Eddy RL, Shows TB, Reed RR, Yau KW, Nathans J. Human rod photoreceptor cGMP-gated channel: Amino acid sequence, gene structure, and functional expression. J Neurosci 1992; 12:3248–3256.
108. Heginbotham L, Abramson T, MacKinnon, R. A functional connection between the pores of distantly related ion channels as revealed by mutant K$^+$ channels. Science 1992; 258:1152–1155.
109. Ngai J, Chess A, Dowling MM, Necles N, Macagno ER, Axel R. Coding of olfactory information: Topography of odorant receptor expression in the catfish olfactory epithelium. Cell 1993; 72:667–680.
110. Ngai J, Dowling MM, Buck L, Axel R, Chess A. The family of odorant receptors in the channel catfish. Cell 1993; 72:657–666.

25
Transport Systems for Arsenic, Antimony, and Cadmium Ions Encoded by Bacterial Plasmids

Anita R. Lynn and Barry P. Rosen

Heavy metal ions are ubiquitous and have beneficial and deleterious actions at the same time.[1,2] For this reason cells frequently have inwardly directed transport systems for their uptake and utilization and outwardly directed systems for extrusion and protection. Salts of arsenic were the first chemotherapeutic agents used in treatment of infectious diseases;[3] indeed, Paul Ehrlich was awarded the 1908 Nobel Prize in Medicine for the development of the antimicrobial arsenical salvarsan, the "magic bullet" that cured syphilis. In his Nobel Award address Ehrlich pointed out that to be effective and arsenical had be taken up into the cells by an arsenical receptor. He also noted that, soon after the start of drug therapy, arsenical resistant organisms appeared, and he postulated that these organisms were resistant because they could no longer take up the toxic arsenical. Our studies in the 1990s expand on Ehrlich's ideas.

Exposure of bacteria to antibiotics, heavy metal ions, and other toxic compounds has led to the evolution of families of resistance determinants. The exposure is sometimes natural or sometimes the result of environmental pollution or intentional use, as in the case of herbicides, pesticides, or chemotherapeutic agents. The mechanisms of resistance are varied.[4] Enzymatic detoxification is widespread, as in the case of the β-lactamases, which produce resistance to penicillin, ampicillin, and other β-lactam antibiotics. Another means for achieving resistance is the modification of the target site, such as methylation of ribosomes to produce erythromycin resistance. Another frequently used stratagem is the expression of new transport systems, which either pump the poisonous agent out of the cell fast enough to prevent toxicity or provide for its uptake to a site of metabolism, as in the case of in mercuric ion detoxification.[5] A wide variety of transport systems are encoded by plasmid genes. These include secondary exchangers such as the tetracycline resistance protein,[6] P-type pumps such as the Cd^{2+} resistance ATPase,[7] and a novel oxyanion-translocating ATPase that provides resistance to arsenicals and antimonials.[8] This chapter focuses on the biochemistry of these two quite different ion-translocating ATPases.

Plasmid-Encoded Transport of Oxyanions of Arsenic and Antimony

There are two plasmid-encoded arsenical resistance operons, found on plasmids of gram-negative and gram-positive organisms. They have some intriguing similarities and homologies, and some remarkable differences. In fact, comparing them suggests how the evolution of an ion-translocating ATPase may have taken place.

The first studies of plasmid-mediated resistance to arsenite and arsenate were made by Novick and Roth[9] using plasmid pI258, isolated from the gram-positive organism *Staphylococcus aureus*. Hedges and Baumberg[10] reported that the transmissible R-factor R773 clinically isolated from the gram-negative *Escherichia coli* also gave rise to arsenate and arsenite resistance. Silver et al.[11] demonstrated that both *S. aureus* and *E. coli* extrude $^{74}AsO_4^{-3}$ when expressing plasmid *ars* genes. The process was shown to be energy-dependent in both gram-negative[12] and gram-positive organisms.[13] Indeed, by depleting the cells of metabolic energy and selectively restoring either the electrochemical proton gradient or the ATP pool we demonstrated that R773-directed efflux of either arsenate[14] or arsenite[15] from *E. coli* was coupled to chemical energy and independent of electrochemical energy, which suggested that the efflux was due to an anion pump rather than to electrophoresis coupled to the outwardly positive membrane potential.

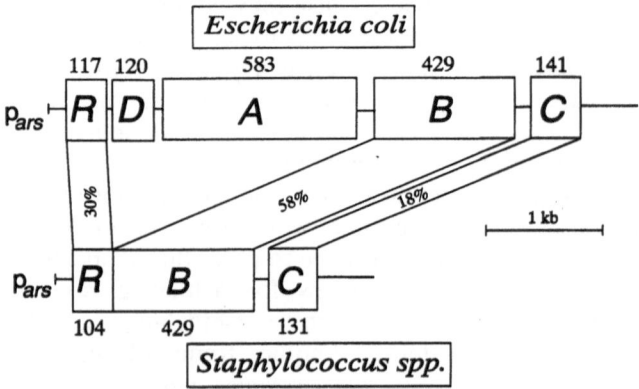

FIGURE 25.1. Physical maps of the *ars* operons: In the top portion the five genes, *arsR*, *arsD*, *arsA*, *arsB*, and *arsC*, of the *ars* operon of the gram-negative plasmid R773 are shown with the direction of transcription from left to right starting at the promoter, *pars*. In the bottom portion the three genes of the *Staphylococcus* plasmids are shown. The homologous regions of the two operons are connected with the percentage similarity indicated. The number of amino acyl residues in each gene product are indicated above or below the corresponding gene. (Modified with permission from Ji and silver.[17])

To determine the mechanism of anion transport we cloned and sequenced the genes for the anion pump from R773.[16] There are five genes in a single operon, two of which are regulatory (*arsR* and *arsD*), two of which form the pump (*arsA* and *arsB*), and the final one, which serves as a modifier of substrate specificity (*arsC*). Two gram-positive determinants have been sequenced, the original plasmid pI258[17] and plasmid pSX267 from *S. xylosus*.[18] Both code for only three genes, *arsR*, *arsB*, and *arsC*. The physical map of the two operons is shown in Figure 25.1. Computer alignment of the protein products indicate a 30% identity between the gram-negative and gram-positive homologues of ArsR, a 58% identity between the ArsB proteins, but between the proteins of ArsC proteins only an 18% identity. The details of the better-known R773 proteins will be discussed first, followed by a discussion of the comparative aspects of the two systems. Only the components of the pump will be reviewed, and not the regulatory aspects of the resistance mechanism.

The Oxyanion Pump Encoded by Plasmid R773

The ArsA Protein

In *E. coli* the *arsA* and *arsB* gene products are sufficient for resistance to arsenite and antimonite, in which the metals are in +III oxidation state.[19] The *arsC* gene is required to produce resistance to arsenate. It is postulated that this gene acts as a modifier subunit that allows the pump to recognize arsenate. As discussed in more detail below, the basic pump is composed of the ArsA protein, the catalytic subunit, and the ArsB protein. The latter is located on the inner membrane in *E. coli*, serves as the membrane anchor for the ArsA protein, and forms the anion-conducting pathway. A model of the oxyanion pump is shown in Figure 25.2.

The physiologic studies described above implicate ATP as the driving force for efflux of both arsenite and arsenate, while analysis of the predicted amino acid sequence of the ArsA protein indicates that the first half of the protein, termed the A1 half, can be aligned with the second or A2 half (Figure 25.3), which suggests that the ArsA protein arose through a duplication and fusion of an ancient gene. Sequence analysis suggests also that the short sequence GKGGVGKT, which is present in both the A1 and A2 halves, is a consensus sequence for the flexible loop portion of a nucleotide binding site.[20] There are no other similarities between the ArsA protein and any other ion pump protein, which suggests that this oxyanion pump is evolutionarily unrelated to any of the reported classes of ion-translocating ATPases and may be the first recognized member of a family of ATP-dependent, anion-translocating systems. Two other superfamilies of ion-translocating ATPases have been identified, and both are cation systems.[21] The F_0F_1 complexes consist of H^+-translocating ATPases, which are found in bacterial, mitochondrial, and chloroplast membranes and form a family with the related vacuolar pumps. The other superfamily includes cation-translocating ATPases of the E_1E_2 or P-type. There are many members of this family, including the cadmium resistance pump discussed below.

As was predicted from both the transport energetics and the primary sequence of the protein, purified ArsA protein is an oxyanion-stimulated ATPase.[22] Although when overexpressed it can be purified as a soluble protein, it is normally bound to the membrane through interactions with the ArsB protein.[23] The ATPase activity of the purified soluble ArsA protein requires the anionic substrate of the pump, arsenite or antimonite. No other anion stimulates it, including arsenate.[24,25] Only ATP will serve as nucleotide substrate. No other nucleotide is hydrolyzed and only the produced ADP inhibits ATP hydrolysis. Inhibitors of other classes of ion-translocating ATPases, including azide, nitrate, vanadate, and N,N'-dicyclohexylcarbodiimide, are without effect on the catalytic activity of the ArsA protein.

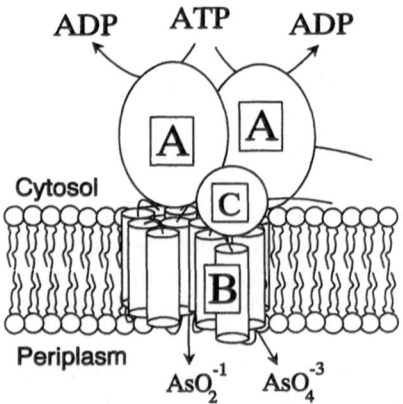

FIGURE 25.2. The plasmid-encoded oxyanion pump: The 63 kDa ArsA and 46 kDa ArsB proteins encoded by the *ars* operon of plasmid R773 interact to form an ATP-coupled extrusion pump for oxyanions of arsenic or antimony of the +III oxidation state. Two ArsA subunits are portrayed in the model, which is consistent with results that demonstrate that ArsA is a functional homodimer in solution. Although the model shows only one copy of the ArsB and ArsC subunits, the stoichiometry of each of these subunits in the pump has not been determined. The ArsA protein is an extrinsic membrane protein with oxyanion-stimulated ATPase activity. The ArsA protein binds to the ArsB protein, an integral membrane protein that functions as an anion channel. Interaction of the ArsC protein with the complex allows recognition and extrusion of oxyanions of arsenic of the +V oxidation state. In the figure the proteins ArsA, ArsB, and ArsC are indicated by A, B, and C.

```
A1:   MQFLQNIPPYLFFTGKGGVGKTSISCATAIRLAEQGKRVLLVSTDPASNVGQVFSQTIGNTIQAIASVPGLSALEIDPQAA
                 .  ::::::::.  :  :.::.  :  : : ..:::  .        . :.  :   :::
A2:   DDIARNEHGLIMLMGKGGVGKTTMAAAIAVRLADMGFDVHLTTSDPAAHL---------SMTLNGSLNNLQVSRIDPHEE

A1:   AQQYRARIVDPIKGVLPDD-VVSSINEQLSGACTTEIAAFDEFTGLLTDASLLTRFDHIIFDTAPTGHTIRLLQLPGAWSS
            ::  ....  ::  :.     ..:.:   ::  :::  :.  :.  .. .:    ::  .. :::::::::. ::.  ::.
A2:   TERYRQHVLE-TKGKELDEAGKRLLEEDLRSPCTEEIAVFQAFSRVIREAGK--RF--VVMDTAPTGHTLLLLLDATGAYHR

A1:   FI-DSNPEGASCLGPMAGLE-KQREQYAYAVEALSDPKRTRLVLVARLQK-STLQEVARTHLELAAIGLKNQYLVINGVLP
          :    :.  ::  :.  .:          .:     .: :  :.::..          :           :    . :
A2:   EIAKKMGEKGHFTTPMMLLQDPPERTKVLLVTLPETTPVLEAANLQADLERAGIHPWGWIINNSLSIADTRSPLLRMRAQQE

A1:   KTEAANDTLAAAIWEREQEALANLPADLAGLPTDTLFLQPVNMVGVSALSRLLSTQPVASPSSDEYLQQRPDIPSLSALV
            :           ::  :  :
A2:   LPQIESVKRQHASRVALVPVLASEPTGIDKLKQLAG
```

FIGURE 25.3. Homology of the A1 and A2 halves of the ArsA protein: The N-terminal (A1) and C-terminal (A2) halves of the ArsA protein are aligned with identical residues (:) and conservative replacements (.) identified. In each pair of lines the top line, designated A1, is from the N-terminal half and the bottom line, designated A2, from the C-terminal half. Underlined are the sequences homologous to the consensus sequence of the flexible glycine-rich loops of nucleotide-binding folds.

The role of the two nucleotide binding sites in ATPase activity and transport is currently under active investigation. Using TNP-ATP, a fluorescent analogue of ATP, we have shown that one mole of ArsA protein binds two moles of nucleotide, which indicates that both the A1 and A2 sites are true ATP binding sites. Binding of ATP to the A1 and A2 sites can be differentiated through binding of α-[^{32}P]-ATP to the ArsA protein by photoreaction with ultraviolet light. Several observations indicate that only the A1 site forms and adduct with α-[^{32}P]-ATP. First, when mutations are inserted into the glycine rich consensus sequence at the A1 site,[26] the mutant proteins bind only one mole of TNP-ATP per mole of protein and no longer form a UV induced adduct with α-[^{32}P]-ATP.[27] Second, similar mutations in the A2 site do not affect adduct formation with α-[^{32}P]-ATP (P. Kaur and B.P. Rosen, unpublished data). Third, when the A1 and A2 sites are expressed as separate polypeptides, the A1 peptide forms a photoadduct; the A2 peptide does not. Mutations in glycine residues in either site leads to loss of catalytic activity. Alteration of K21 or K340 in the consensus sequences in the A1 or A2 sites results in partial loss of resistance (P. Kaur and B.P. Rosen, unpublished data). Purified mutant ArsA protein KE340 retains 20% activity. The K_m for ATP is unaltered, but the V_{max} is reduced fivefold. Thus although the lysyl residues are not absolutely required, both sites appear to be necessary for ATP hydrolysis. Whether both are catalytic remains to be determined.

Genetic analysis of the two nucleotide binding sites has shown that they are independent of each other and capable of fulfilling their function on separate polypeptides. When an *arsA* mutant in the A1 and in the A2 site are expressed in the same cell, the cell becomes arsenite-resistant (P. Kaur and B.P. Rosen, unpublished data). This can only happen if the two ArsA proteins form a complex in which the A1 on one molecule cooperates with the A2 site on another. This genetic complementation suggests that the ArsA protein functions as an oligomer. Biochemical data support this conclusion. In addition to the nucleotide binding sites, the ArsA protein must bind its oxyanionic substrate to be catalytically active.[22] The anion appears to shift the monomer-dimer equilibrium in favor of the dimer, thus stabilizing the latter.[28] Dimerization in the presence of the oxyanion can be demonstrated by light scattering measurements of the size of the ArsA protein in solution. The average mass of the ArsA molecules in solution was 63 kDa in the absence of substrates; that is, all molecules were monomeric. In the presence of antimonite the average value increased to 118 kDa, which is near the mass of the dimer. To demonstrate the existence of the dimer, the ArsA protein was chemically cross-linked using the zero length cross-linker N-ethoxycarbonyl-2-ethoxy-1,2-dihydroquinoline (EEDQ). The 126 kDa cross-linked dimer was detected with SDS polyacrylamide gel electrophoresis only when the ArsA protein was incubated with antimonite or arsenite. Other oxyanions, including arsenate, did not promote dimer formation.

Is dimerization functionally relevant? Since ATP hydrolysis occurs only in the presence of oxyanion, and since dimerization occurs under the same conditions, one can reasonably conclude that binding of oxyanion shifts a monomer-dimer equilibrium in favor of the dimer, and that the dimer is the catalytically competent species. Recently we have found that the affinity of the A1 site for ATP is increased in the presence of antimonite (P. Kaur and B.P. Rosen, unpublished data). In these experiments the amount of photoadduct formation with α-[^{32}P]-ATP at a saturating concentration of ATP is unchanged by antimonite, but the labeling is greatly increased at subsaturating concentrations of ATP. If the role of antimonite is to foster dimer formation, then the dimer would have higher affinity for ATP than the monomer. Thus the function of oxyanion binding is to promote catalysis by interaction of the subunits of the pump (Figure 25.4). The relationship of dimerization of the purified ArsA protein in solution and the conformation of the membrane-bound pump is discussed in more detail below.

The ArsB Protein

The ArsB protein is an inner membrane protein in *E. coli*.[29] Hydropathic profile calculations predict the presence of 10–12 membrane-spanning α helices.[16] A genetic approach was used to determine its topologic arrangement within the membrane.[30] If the gene for a membrane protein is fused with an enzyme that is active only outside of the cell, then fusions

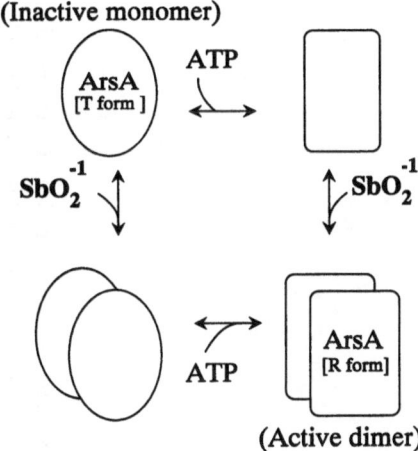

FIGURE 25.4. Conformational states of the ArsA protein: The 63 kDa ArsA has independent binding sites for antimonite and ATP. In the absence of substrates the monomer exists in an inactive R form. Binding of either substrate produces a unique conformational change. The monomer exists in equilibrium with dimer, with binding of oxyanion stabilizing the dimeric form. Binding of both substrates favors the catalytically active T conformation.

with a cytoplasmic enzyme portion will be inactive, while those where the enzyme is exposed to the medium will be active. The *phoA* gene product, alkaline phosphatase, is one such protein. The *blaM* gene product, β-lactamase, is active in the cytosol, but the substrate, ampicillin, is accessible only in the periplasmic space, and so resistance to ampicillin is observed only if the enzyme is translocated outside. The *lacZ* gene product, β-galactosidase, is active only in the cytosol, and thus as a chimeric protein it measures only the portions of membrane proteins exposed to the cytosol. Results of random gene fusions with *lacZ*, *phoA*,[31,32] and *blaM*[33] have led to a topologic model of the ArsB protein with 12 membrane-spanning regions (Figure 25.5).[30]

Although when overexpressed the ArsA can be isolated as a soluble protein, it can also be found as a membrane-bound complex with the ArsB protein.[23] Reconstitution of purified ArsA could be demonstrated using membranes from cells that expressed only the *arsB* gene. Thus the ArsB protein functions as the membrane anchor for the ArsA protein. As discussed above, the active form of the enzyme would be expected to contain two molecules of the ArsA protein. Although the soluble ArsA protein is found as a monomer, it is unlikely that it dissociates from the membrane during the catalytic cycle for two reasons. First, the ArsA protein is so tightly bound to the membrane that only high concentrations of chaotropic agents can remove it[23] and second because, given the

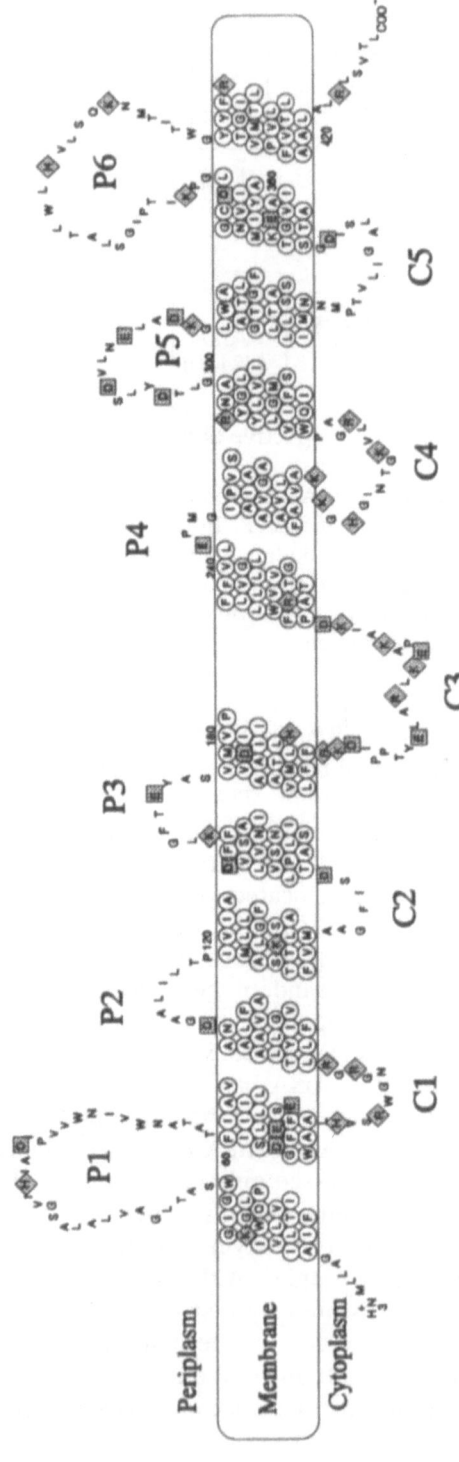

FIGURE 25.5. Topological model of the ArsB protein: The model proposes 12 membrane-spanning α-helices joined by 6 periplasmic loops (P1–P6) and five cytoplasmic loops (C1–C5). The N and C termini are suggested to be cytosolic. The precise placement of each residue cannot be assigned from the data. The suggested placement includes equal numbers of intramembranal positive and negative charges, with glycyl or prolyl residues placed in turn regions where possible. Acid (□) and basic (◊) residues are indicated. (Reprinted with permission from Wu and Rosen.[30])

sigmoidal relationship between the concentration of purified ArsA protein and its binding to the membrane (S. Dey and B.P. Rosen, unpublished data), the monomer-dimer equilibrium would favor the dimer at high protein concentrations. In support of this possibility, the saturation curve for ArsA protein binding to membranes changes from sigmoidal to hyperbolic in the presence of antimonite or arsenite, which indicates that the interaction of ArsA and ArsB proteins is stronger in the presence of oxyanion, that is, under conditions where the dimer is formed. We hypothesize that ArsA subunits move relative to each other to form tense (T) and relaxed (R) forms, according to the nomenclature of Cohen and Monod.[34] The effect of oxyanion is to increase the affinity for ATP and to produce tighter binding of the ArsA protein to membranes, which indicates that the protein, in the absence of anion, is in the T (low affinity) form. Upon oxyanion binding the equilibrium between T and R shifts to the high affinity R form, which favors dimerization in solution or closer association on the membrane (Figure 25.4).

The ArsC Protein

Although the operon provides resistance to both arsenate and arsenite, the two oxyanions are chemically quite different. In the absence of the *arsC* gene, the operon provides resistance to arsenite but not to arsenate.[19] High level arsenate resistance requires all three gene products.[16] Arsenate efflux in cells likewise requires the three gene products. Although the *arsC* is not required for arsenite efflux, it is necessary but not sufficient for the efflux of arsenate.[15,19]

Although the ArsC protein has been purified and crystallized,[35] little is known about the biochemical mechanism of ArsC function. We speculate that the ArsC protein modifies the activity of the oxyanion pump to allow recognition of arsenate.[15,16] Since most phosphate-binding proteins cannot discriminate between arsenate and phosphate, the way in which the ArsC protein specifically recognizes arsenate is of considerable interest, but lack of an in vitro assay hampers studies of the ArsC protein function.

Arsenic Resistance in *Staphylococcus*

A comparison of the nucleotide sequence of the operon from the two staphylococcal plasmids with the E. coli plasmid R773 demonstrates that the two operons encode similar proteins with similar results: resistance to arsenite, antimonite, and arsenate through active anion extrusion. The two ArsC proteins are of similar size, but sequence similarity is only 18%. There are several other genes with products related to the R773 ArsC protein, including *orf3*, an open reading frame in a nitrogenase gene complex,[36] and *amiB*, a gene in a streptococcal operon, which probably codes for a peptide transport system.[37] Another related gene product is an open reading frame, *orf1*, in *Streptomyces viridochromogenes*.[38] The *orf1* product is 28% identical with ArsC, making it the most closely related to all identified ArsC homologues. The function of this open reading frame in *Streptomyces* is unknown. *Streptomyces* is a major antibiotic producer, and many plasmid-encoded antibiotic resistance determinants may have originated in this organism. Since *orf1* is at the start of a cloned chromosomal fragment, it is tempting to speculate that this open reading frame is the last gene in a chromosomal resistance operon, most closely related to the ancestor of the *ars* operons. Since the *ars* operons are plasmid-borne, it is also possible that horizontal transfer of genes between plasmids of gram-negative and gram-positive organisms gave rise to a recent divergence between the two operons. However, the G + C content of the staphylococcal *ars* operon is 31% similar to that of staphylococci, while the G + C content for the R773 *ars* operon is 51% similar to that of gram-negative organisms. Since the G + C content of the two operons are similar to that of their host organisms, a recent divergence is not likely.

In contrast to the limited homology of the ArsC proteins, the 58% sequence identity of the ArsB proteins is striking (Figure 25.6). The structural similarity is even more apparent when their hydropathic profiles are compared. One would expect the biochemical functions of these two ArsB proteins to be the same, which makes it all the more amazing there are no sequences similar to *arsA* gene present on the staphyloccal plasmids. Two possibilities can be considered.

First, both transport systems may be ATP-driven efflux pumps. In the case of plasmid R773, the catalytic ArsA protein subunit is encoded by the *ars* operon. In staphylococci the catalytic subunit could be a chromosomally-encoded, soluble ATPase that normally has another function but can be recruited by the ArsB protein to form a membrane-bound complex. However, both the pI258 and pSX267 arsenical resistance operons have recently been shown to function in *B. subtilis* and *E. coli*.[17,18] Recruitment of a cellular ATPase would require that *S. aureus*, *B. subtilis*, and *E. coli* all produce a protein that can interact with the ArsB protein to yield a functional anion pump. Although there are no data incompatible with such a model, it seems intuitively unlikely that such a generic ATPase would exist in all organisms.

A second hypothesis is that the ArsB protein is sufficient for anion conductance. Although the ArsA-ArsB complex is an anion-translocating ATPase, in the absence of the ArsA protein the ArsB protein may act as an anion channel. Cells produce a membrane potential, positive on the exterior surface, which could serve as the driving force for anion extrusion. This is analogous to the F_0F_1, the H^+-translocating ATPase of bacterial, mitochondria, and chloroplasts.[39] As a whole the complex is an ATP-dependent H^+ pump, but the two parts of the complex can be

```
1    MLLAGAIFILTIVLVIWQPKGLGIGWSATLGAVLALASGVIHIADIPVVWNIVWNATAT
     ***  **;**;;  ********  ** ;*  ;***;*;  ;**;   *;  *  ****** *
1    MTILAIVIFLLTLIFVIWQPKGLDIGITALIGAVVAIITGVVSFSDVLEVTGIVWNATLT

60   FIAVIIISLLLDESGFFEWAALHVSRWGNGRGRLLFTYIVLLGAAVAALFANDGAALILT
     *;***;***;*** ***** *;*   ;   ** *   * ;* **** *** **********
61   FVAVILISLILDEIGFFEWSAIHMVKASNGNGLKMFVFIMLLGAIVAAFFANDGAALILT

120  PIVIAMLLALGFSKSTTLAFVMAAGFISDTASLPLIVSNLVNIVSADFFKLGFTEYASVM
     ***;**;  *** *     *;  *  *** ** ***************;* ;** ** * *
121  PIVLAMVRNLGFNKKVIFPFIIASGFIADTTSLPLIVSNLVNIVSADYFDIGFIEYFSRM

180  VPVDIAAIIATLVMLHLFFRKDIPPTYELARLKEPAKAIKDPATFRTGWVVLLLLLVGFF
     :    :*  :**:;:  * *;*** ** *;;   *  ;*    **** *;   *;** ;****;
181  IIPNIFSLIASILVLWLYFRKSIPKTFNTENLSDPKNVIKDPKLFKLSWIVLAILLVGYL

240  VLEPMGIPVSAIAAVGAAVLFAVAKKGHGINTGKVLRGAPWQIVIFSLGMYLVIYGLRNA
     * *   **** **;; * ;    ;*:*  ;;:  *  *;;****;**;**;*****;;**;*
241  VSEFIQIPVSIIAGIIALIFVILARKSKAVHTKQVIKGAPWNIVVFSIGMYLVVFGLKNV

300  GLTDYLSDVLNELADKGLWAATLGTGFLTALLSSIMNNMPTVLIGALSIDGSTATGVIKE
     *;*  * *;*  ;;    **    * *; ************  * *  *;***;;**
301  GITTILGDILTNISSYGLFSSIMGMGFIAAFLSSIMNNMPTVLIDAIAIGQSSATGILKE

360  AMIYANVIGCDLGPKITPIGSLATLLWLHVLSQKNMTITWGYYFRTGIVMTLPVLFVTLA
     ;*;****** ************    **** ;**  *;** **;***; *;********
361  GMVYANVIGSDLGPKITPIGSLVISPWLHVSTQKGVKISWGTYFKTGIIITIPVLFVTLL

420  ALALRLSVTL
     ;* * * :
421  GLYLTLIIF
```

FIGURE 25.6. Similarity of the ArsB proteins from plasmids R773 and pI257: The ArsB protein from the plasmid R773 is aligned with the ArsB protein from plasmid pI258 with identical residues (*) and conservative replacements (:) indicated. In each pair of lines the top line is the sequence from plasmid R773 and the bottom line is from plasmid pI258.

separated. Purified F_1 is a soluble ATPase. In the absence of F_1 the F_0 is a conducting pathway that transports protons into the cell in response to the membrane potential, $\Delta\psi$. We have recently subcloned the *arsB* gene of the R773 arsenical resistance operon and have shown that it is capable of producing arsenite resistance in *E. coli* in the absence of the *arsA* gene (D. Dou, S. Dey, and B.P. Rosen, unpublished data). Cells expressing the *arsB* gene alone are capable of energy-dependent arsenite extrusion. However, when a nonfunctional mutant *arsA* gene is coexpressed with the *arsB* gene, there is neither resistance nor transport.[26] Similarly the proton permeability of the F_0 is blocked by binding of a nonfunctional F_1. These results suggest that the ArsB protein by itself can catalyze driven arsenite transport. However, binding of an ArsA protein prevents coupling to the $\Delta\psi$ and the ArsA-ArsB complex functions only as an ion-translocating ATPase.

What selective pressure would there be for an already functional resistance system to evolve further? Since the electrochemical proton gradient has an outwardly positive and acid orientation, a uniport should be sufficient for extrusion of the anion by simple electrophoretic movement. If the ArsB protein works solely as an electrophoretic anion uniporter, it could extrude substrate only to its equilibrium potential, the equilibrium between the ratio of ion concentrations on either side of the membrane and the proton motive force,[40] where

$$\log([AsO_2^{-1}]_{out}/([AsO_2^{-1}]_{in}) = \Delta\psi/59.$$

At best, in respiring E. coli $\Delta\psi = -180\,mV$; however, the cell may find itself unable to keep a low cytosolic ion concentration in the face of a high external ion concentration or a low proton-motive force. If the concentration of arsenite outside the cell were $10\,mM$, the intracellular concentration could not be less than $10\,\mu M$ and would probably become considerably higher as the toxic oxyanions interfered with cellular energy metabolism. Thus a uniporter would put the cell at the mercy of its environment. An ATP-linked pump has no such restriction; it can pump out far in excess of the proton equilibrium, maintaining low ion concentration in the cytosol independent of the external concentration.[40]

If this hypothesis is correct, it has important implications about the evolution of ion pumps. Multicomponent ion-translocating ATPases must have evolved in stages. An independent evolution of ion-conducting proteins and soluble ATPases may have a reasonable sequence of events. Their original roles may have been quite different from their present transport function. A complex may then have developed as the subunits evolved affinity for each other, followed by the development of a pump as the interactions became functional. Speculation on evolution is usually unverifiable. Identification of transport systems utilizing highly related proteins with different in vitro coupling mechanisms provides the strongest evidence for an evolutionary hypothesis.

The Oxyanion Pump and Multidrug Resistance

The *mdr* family of pumps, whose human homologue encodes the P glycoprotein, is of considerable medical interest.[41] When amplified it produces the multidrug resistance that leads to the failure of drug treatment in cancer chemotherapy. A homologue of the P glycoprotein appears to confer resistance to arsenite and antimonite in *Leishmania*.[42,43] Other plasmid-encoded relatives of the P glycoprotein provide for resistance to single drugs, such as tylosin or daunorubicin in *Streptomyces*, and probably have a two-component structure, a membrane subunit and extrinsic ATP-utilizing catalytic subunit.[44,45] This structural motif is similar to that of the gram-negative *ars* oxyanion pump.[41]

The P glycoprotein is an ATP-coupled pump that transports a wide variety of unrelated organic compounds out of cells. Both the P glycoprotein and the ArsB protein have 12-membrane-spanning helices arranged in two groups of six. The ArsA protein is an oxyanion-stimulated ATPase with two consensus nucleotide sites. The P glycoprotein also has two domains with the same consensus nucleotide-binding folds. Aside from this limited sequence homology, there is no evolutionary relatedness between the *ars* and *mdr* proteins. Although the P glycoproteins and the *ars* proteins are not homologues, these common structural features may allow them to function similarly to bind and hydrolyze ATP and to couple the ATP-derived energy to the transport of a variety of different chemical species.

As discussed above, the *ars* system extends its range of substrates through the ArsC protein. A gene homologous to the *arsC* gene has been shown to be part of an operon that encodes a P-glycoprotein homologue. The *ami* locus has recently been cloned and sequenced from *Streptococcus pneumoniae*.[37] Sequence analysis suggests this system is related to the oligopeptide permease of *Salmonella typhimurium*, a member of the *mdr* superfamily of solute-translocating ATPases. Mutations in the *ami* locus have pleiotropic effects that range from resistance to aminopterin and methotrexate to sensitivity to branched chain amino acids. Genes in this operon have 27–44% identity with the *opp* operon, depending on the gene. The *ami* locus has an additional gene, *amiB*, which has no counterpart in the *opp* operon. The translation product of this gene has about the same degree of similarity with the R773 ArsC protein as the staphylococcal ArsC proteins. Although the function of this *arsC* homologue is not known, it may operate by expanding the substrate specificity of the *ami* locus. This suggests that an ancestral *arsC*-like gene product may have evolved to serve as a specificity factor that can interact with various protein complexes. Thus the *ars* operon may be a simple model for multidrug specificity and resistance.

Plasmid-Mediated Cadmium Resistance in *Staphylococcus aureus*

The first example of cadmium resistance in bacteria was observed in the staphylococcal penicillinase plasmids described earlier in this chapter. In 1968 Novick and Roth reported that these plasmids contained separate genetic loci for resistance to several toxic ions, including Hg^{2+}, Pb^{2+}, and Bi^{2+} ions, in addition to cadmium and arsenic.[9] Resistance to zinc and to antimony were also noted but appeared to be linked to cadmium and arsenic resistance, respectively. In 1972 Smith and Novick described two loci for cadmium resistance, *cadA*, which conferred a high resistance, and *cadB*, which gave a low resistance.[46]

The *cadB* resistance determinant is found on some, but not all, the staphylococcal penicillinase plasmids that contain *cadA*. It confers resistance to zinc and cadmium, but only at low levels, and it is observable only in the absence of *cadA*.[46] The resistance conferred by *cadB* does not appear to involve efflux, but seems to result from enhanced binding of cadmium to the cell.[47] In addition, a chromosomal-encoded cadmium efflux system has been described in *S. aureus*; however, this resistance is not homologous to the plasmid-mediated system described here.[48]

Resistance to cadmium by the *cadA* determinant is associated with a decreased level of uptake by cells containing the resistance plasmids.[49,50] The decreased uptake is also observed in protoplasts of resistant cells, which suggests that the cell membrane is involved in the exclusion of cadmium.[51] Uptake of cadmium into sensitive cells occurs via an energy-dependent manganese transport system. The suggestion that resistance could be due to an active transport process[52–54] was confirmed by Tynecka, Gos, and Zajac, who demonstrated a specific energy-dependent efflux of cadmium from cells containing the resistance plasmids.[54] The effects of various inhibitors and ionophores on efflux from whole cells led to the conclusion that cadmium transport was mediated by an electroneutral $2H^+/Cd^{2+}$ antiporter that extrudes cadmium in exchange for H^+, where H^+ movement is coupled to the ΔpH.[54]

This model has been revised recently as a result of additional evidence that suggests that cadmium efflux occurs by means of a primary pump. Genetic evidence and more recently biochemical data now indicate that cadmium transport is mediated by an ATPase that is a member of the class of ion-translocating enzymes known as E_1E_2 or P-ATPases.

The *cadA* determinant has been cloned, expressed in *Bacillus subtilis*, and sequenced.[7] Two open reading frames were identified in the 3.5 kb DNA fragment that contains the cadmium resistance determinant. The first, called *cadC*, results in a predicted peptide of 122 amino acids. The second, larger open reading frame, *cadA*, corresponds to a predicted peptide of 727 amino acids.

CadA

Comparison of the predicted amino acid sequence of CadA with known proteins has demonstrated that CadA shares significant sequence homology with some of the P-ATPases. Members of this group of enzymes include the Ca^{2+}-ATPase of sarcoplasmic reticulum, the (Na^+ + K^+)-ATPase, and the K^+-translocating ATPase of *E. coli*. Amino acid sequence homology with the prokaryotic enzymes encoded by *ktr* in *Enterococcus hirae* (formerly *Streptococcus faecalis*) and *kdp* in *E. coli* was 26–30%.[7] Homology was also significant with the eukaryotic enzymes for specific regions of the proteins.

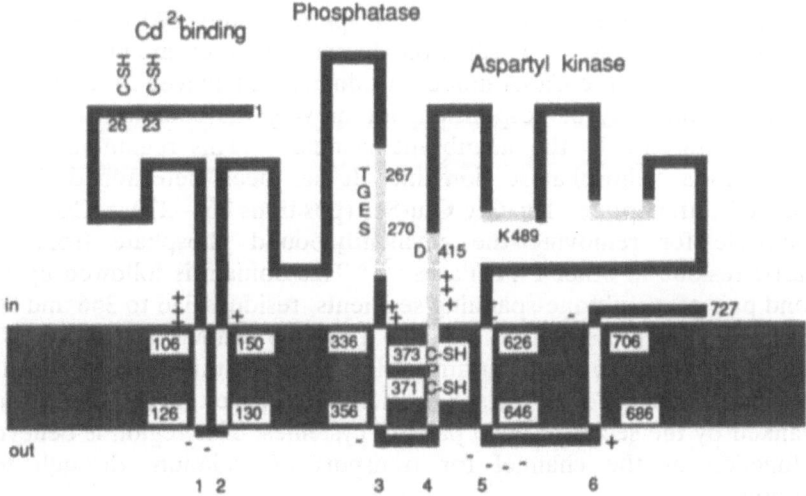

FIGURE 25.7. Model of the CadA cadmium ATPase: The model shows domains that are assigned by homology to other P-ATPases, with attributed functions as described in the text. Amino acids are designated by their standard one-letter abbreviations, except for cysteine (C-SH). Numbers are the positions of the amino acids in the sequence. (Reprinted with permission from Silver and Walderhaug.[56])

The amino terminal of the predicted *cadA* gene product also has homology with the mercuric reductases and with the R100 periplasmic mercury-binding protein.[7,55] The region of homology includes the first two cysteine residues at positions 23 and 26 in CadA. This region in the mercury resistance proteins is believed to be the site for the initial recognition and chelation of Hg^{2+} ions by these proteins. Silver, et al. speculated by analogy that the corresponding region of CadA could be responsible for initial binding of cadmium.[7,55,56]

Using the predicted amino acid sequence of CadA, Silver et al. developed a model of this protein that was derived from models of other members of this group of ATPases, the sarcoplasmic reticulum Ca^{2+}-ATPase and the $(Na^+ + K^+)$-ATPases of animal cell membranes (Figure 25.7).[55,56] Several conserved residues and structural motifs in CadA were identified by analogy with these more-thoroughly-studied enzymes. As described above, the amino-terminal region contains what appears to be the cadmium recognition site, which is analogous to the location of the predicted Ca^{2+} recognition sequence in the Ca^{2+}-ATPase of sarcoplasmic reticulum. This region is followed by the first predicted transmembrane hairpin. The model is drawn with the optimum 20 amino acid candidates for the hydrophobic α-helical membrane-spanning stretches, with only 4 amino acids for the turn between the two membrane segments. The next portion of the sequence is a region of approximately 190 amino acids, which is regarded as a transduction domain in other members of the P-ATPase class. In the CadA model this domain is believed to function as a "funnel" and may be responsible for moving the cadmium from the initial binding site to the membrane channel.[56] This region may also function as a "phosphatase domain." It has been determined that a conserved tetrapeptide, Thr-Gly-Glu-Ser (positions 267–270 in CadA), is responsible for removing the covalently-bound phosphate from the aspartyl residue in other P-ATPases.[57,58] This domain is followed by the second pair of membrane-spanning segments, residues 336 to 356 and 364 to 384. This hairpin structure contains a proline residue at position 372, which corresponds to the invariant proline found in this position in other P-ATPases. In the Cd^{2+}-ATPase-predicted sequence this proline residue is flanked by the second of two pairs of cysteines. This region is believed to function as the channel for transport of cadmium through the membrane.

The next large region of the predicted CadA amino acid sequence in this model is relatively hydrophilic and therefore is believed to be cytosolic.[55,56] This portion of the protein contains several stretches of highly conserved sequence, including the putative phosphorylation site. The sequence Asp-Lys-Thr-Gly-Thr-Leu-Thr, which is completely conserved in all members of the P-ATPases, is found in the Cd^{2+}-ATPase at residues 415 to 421. The Asp of this conserved sequence is the residue that has been shown to be phosphorylated by the γ-phosphate of ATP in several eukaryotic members of this enzyme class. Also in this region is a predicted α-helical loop that is highly conserved among members of the P class of ATPases. This loop, along with a sequence containing an invariant lysine, may be involved in the binding of ATP. Following this ATP-binding region, there is another predicted pair of membrane spanning segments after which the protein ends at Lys_{727}. The eukaryotic members of this class of enzymes generally have one or more additional membrane-spanning hairpins, but the prokaryotic representatives, with the exception of the Mg^{2+}-ATPase MgtB,[59] end with a single pair of membranes spans following the ATP-binding region.[60,61]

CadC

The other predicted protein in the cadmium-resistance determinant, CadC, is a smaller protein of approximately 13.8 kDa.[62] The predicted amino acid sequence of this protein does not yield as much information as does the sequence of CadA. The highly charged CadC protein lacks a stretch of hydrophobic amino acids long enough to span the membrane, and thus it is most likely a soluble protein,[7,15] although neither its cellular location nor any association with CadA have been determined. Significant sequence homologies at the amino acid level with CadC are seen with both *E. coli* and *S. aureus* ArsR and with CadX, a component of the *cadB* cadmium resistance system. As described above, ArsR is a regulatory protein of the arsenic resistance operon, whereas CadX is believed to be a required structural component of the *cadB* system.[56] Initially the sequence homology of ArsR led to the suggestion that CadC is also a regulatory element.[7] Subsequent genetic studies have suggested instead that it is a component of the pump itself and is required for full resistance to cadmium.[62] However, its biochemical function is unknown. Although many P-ATPases are composed of a single polypeptide, a number of members of this class, including two bacterial examples, do contain additional subunits. The potassium transport ATPase of *E. coli* contains two polypeptides, KdpA and KdpC, in addition to the ATP-hydrolyzing protein KdpB.[61] The Mg^{2+}-ATPase of *Salmonella* also contains an additional polypeptide.[59] In both cases the function of these additional proteins is not known.

Biochemistry of Cadmium Transport

Recent biochemical evidence provides support for the model described above in which cadmium transport is mediated by an ATPase. Previous studies of the *cadA* resistance determinant were based on efflux experiments in intact *S. aureus* cells.[52–54,63] More recently, a transport assay that measures the ATP-dependent uptake of $^{109}Cd^{2+}$ into everted membrane vesicles of *B. subtilis* has been developed.[61] No uptake was observed in the presence of NADH or reduced phenazine methosulfate, both of which generate a pH gradient. Thus a proton motive force is insufficient to drive cadmium transport, which indicates that the transporter is unlikely to be a secondary Cd^{2+}/H^+ antiporter, although the possibility that it may exchange the two cations while functioning as an ATP-driven pump cannot be excluded.

If cadmium transport were mediated by an ATPase rather than an antiporter, one would expect that disruption of the proton gradient would have little or no effect on cadmium transport. However, cadmium transport is reduced by about 50% by DCCD, FCCP, or nigericin.[64] Dissipation of the electrical gradient by valinomycin did not inhibit cadmium transport, as anticipated. These data suggest that although ΔpH may have some role in cadmium transport, it is not sufficient to drive transport. The effect of bafilomycin A_1, a potent inhibitor of vacuolar ATPases,[65] is consistent with the hypothesis that CadA is a P-ATPase. This macrolide antibiotic has no effect on the F_1F_0 ATPases and inhibited cadmium transport activity by 50% at a concentration of $10\,\mu M$, a concentration comparable to that required to inhibit other P-type ATPases, but about 50-fold greater than that needed to inhibit vacuolar (V-type) ATPases.

A common characteristic of P-ATPases is the phosphorylation of the enzyme by the γ-phosphate of ATP during the catalytic cycle. The formation of a phosphorylated intermediate of CadA has been demonstrated by the incorporation of label into CadA when *B. subtilis* membranes are incubated with γ-^{32}P-ATP (K-J. Tsai and A.L. Linet, unpublished data). The incorporation of the label is cadmium-dependent and acid-stable and can be removed by treatment with alkali or hydroxylamine. The demonstration of a phosphoenzyme intermediate for CadA provides further support for the classification of the cadmium transporter as a P-ATPase.

The CadA and CadC proteins have been expressed in *E. coli* under control of the T7 phage promoter and identified as [^{35}S]-methionine-labeled bands.[62] In addition, CadA has been identified as a unique protein band on sodium dodecyl sulfate-polyacrylamide gel electrophoresis.[64] The 13,800 Da CadC migrated with an apparent molecular mass of 14 kDa, whereas the CadA protein with a calculated molecular mass of 78,800 migrates with an estimated mass of 68 kDa.

Regulation of the Cadmium Transporter

It has been shown that the *cadA* cadmium resistance operon is an inducible system.[64,66] ATP-dependent cadmium transport in everted membrane vesicles and the appearance of a unique protein band on SDS-polyacrylamide gel electrophoresis increase concomitantly when cells are cultured with as little as $1\,\mu M\,CdCl_2$.[64] Translational and transcriptional fusions of *cadA* and *cadC* with β-lactamase (*bla*) have been used to demonstrate that the operon is inducible by several divalent cations in micromolar concentrations.[66] With these fusions cadmium is the most effective inducer, followed by bismuth and lead. Less effective, but still capable of inducing resistance, were zinc and cobalt. Manganese was totally ineffective. Although zinc is a relatively poor inducer, the CadA system confers resistance of Zn^{2+} when induced with any of the above cations.[66] Early reports of this system noted that resistance to zinc[9] and possibly to bismuth and lead[46,67] were associated with the CadA determinant.

As described above, the similarity of the predicted amino acid sequence of CadC with the *E. coli* and *S. aureus* ArsR regulatory protein first suggested that CadC might be a comparable regulatory element. However, subsequent genetic studies have demonstrated that this is not the case. In the absence of the *cadC* open reading frame, resistance is greatly reduced and is almost fully restored in the presence of the gene on a compatible plasmid *in trans*.[62] In addition to the requirement for CadC for full resistance, fusion studies have shown that the operon is inducible in the absence of a functional *cadC*. A *cadC-bla* transcriptional fusion at the 107th nucleotide of *cadC* exhibited inducible β-lactamase activity.[66]

The start site for mRNA transcription was determined by primer extension to be at residues 576 and 577 of the published sequence. The putative -10 and $+35$ initiation signals, which are comparable to the *E. coli* consensus sequences, precede the start site and are flanked by an inverted repeat sequence. Deletion of the inverted repeat from the *bla* fusion construct leads to a low level constitutive expression of β-lactamase.

Cadmium, Zinc, and Cobalt Resistance in *Alcaligenes eutrophus*

The *czc* resistance determinant, which confers resistance to cadmium, zinc, and cobalt, was found on a large plasmid designated pMOL30 in *Alcaligenes eutrophus* CH34.[68-70] This inducible resistance is due to a decreased accumulation of the cations, mediated by an energy-dependent efflux system. Resting cells of *A. eutrophus* were allowed to accumulate $^{109}Cd^{2+}$, $^{65}Zn^{2+}$, or $^{60}Co^{2+}$. Addition of gluconate as an energy source resulted in efflux of the cation from the resistant cells. Efflux was inhibited by incubation at 4°C or by the addition of 2,4-dinitrophenol, but not by addition of N,N'-dicyclohexylcarbodiimide.[71]

The complete *czc* determinant consists of two operons transcribed in opposite directions (D.H. Nies, personal communication). Sequencing of the operon that contained the structural components revealed four open reading frames identified as *czcC*, *czcB*, *czcA*, and *czcD* with predicted polypeptide sizes of 37.3, 54.5, 116, and 21.2 kDa, respectively.[72] Expression of the operon under control of the T7 phage promoter in *E. coli* resulted in the identification of five polypeptides with approximate molecular masses of 100 (CzcA), 66 (CzcB), 44, 42 (both CzcC) and 21 (CzcD) kDa. Two polypeptides that differed in size by 2 kDa were assigned to *czcC*, but it is not clear whether this is due to proteolytic processing of the peptide or to the presence of two start sites.

Deletion analysis of the operon led to tentative assignments of the peptide functions and to development of a model for the efflux system. Deletion of *czcD* had little or no effect on zinc, cadmium, and cobalt resistance and efflux.[72] Recent studies show that CzcD is required for induction of the operon by extracellular cations, but does not appear to be an element of a two-component regulatory system (D.H. Nies, personal communication).[72] This has led to the speculation that CzcD might function by a specific slow-rate reuptake of the cations to induce the operon. A CzcA deletion mutant that lacked the C-terminal 59 amino acids resulted in a lower Cd^{2+} resistance and reduced efflux of Cd^{2+} and Co^{2+}, while Zn^{2+} resistance and efflux were unaffected. A larger deletion of CzcA led to complete loss of all resistances. Internal deletions of CzcC affected Cd^{2+} and Co^{2+} resistance, but not Zn^{2+}. Mutants in CzcB resulted in loss of efflux and resistance to Zn^{2+} and Cd^{2+}, but retained a residual resistance to Co^{2+}. This analysis allowed the construction of a preliminary model of *czc* function. In this model, CzcA and CzcB together form the core of a resistance and efflux system for Zn^{2+}, with CzcB acting to funnel Zn^{2+} ions to the hydrophobic CzcA transport protein. With the addition of CzcC, the transport activity would attain specificity for Cd^{2+} and Co^{2+}, perhaps through the modification of CzcB by CzcC.[72]

Recent data show that the *czcCBAD* operon is regulated by Zn^{2+}, Cd^{2+}, and Co^{2+} at the level of transcription. The primary transcript of approximately 7000 nucleotides (nt) is unstable and is apparently processed into a more stable 6000-nt intermediate. This processing may occur by deletion of the specific *czcD* transcript from the initial 7000-nt transcript (U. Schwidetzky and D.H. Nies, personal communication).

The second operon contains the regulatory genes *czcS* and *czcR*, both of which code for soluble proteins. Both proteins contain potential metal and DNA binding sites. At least one of the two proteins is required for induction of *czc*, and there is seemingly also a *cis*-acting element required for full induction of the structural operon. The regulatory operon is also regulated at the level of transcription, but apparently not as tightly as the structural region (D.H. Nies, personal communication).

Conclusion

In this chapter we have discussed several unique bacterial, plasmid-mediated ion transport systems. These transport systems have several properties in common and many very different features. All transport metals as a mechanism of resistance to their toxic effects, and in all of these systems transport of these metal ions is energy-dependent. However, these systems belong to several very different classes of ion transporters, particularly with respect to the energetics of transport. Efflux of arsenic or antimony ions is energized either by ATP hydrolysis in the case of gram-negative bacteria, or possibly by the proton motive force in the case of the gram-positives. Cadmium efflux is also energy-dependent, driven by ATP hydrolysis in the case of the *S. aureus* plasmid, but by a completely different type of ATPase than that responsible for arsenic and antimony transport.

The arsenic transport system, with its multiple substrates, provides an excellent model for the study of multidrug resistance systems. In addition, comparison of the differences and similarities of the arsenic transport systems in gram-positive and gram-negative bacteria may provide general insights into the evolutionary development of active transport systems. The cadmium transport ATPase, on the other hand, is an excellent model for the E_1E_2 or P-ATPase class of enzymes. Since it is plasmid rather than chromosomal in origin, the Cd^{2+}-ATPase will be an excellent system in which to study, through the use of mutants, the properties of this important class of enzymes.

Acknowledgments. This work was supported by United States Public Health Service Grants AI19793 and CA54141 to BPR. We thank Drs. D.H. Nies, F. Götz, and S. Silver for supplying the information described in this chapter.

References

1. Foye WO. Antimicrobial activities of mineral elements. In: Weinberg ED, ed. Microorganisms and Minerals. Microbiology Series, Vol 3. New York: Marcel Dekker; 1977:387.
2. Knowles FC, Benson AA. The biochemistry of arsenic. Trends Biochem Sci 1983; 8:178–80.
3. Himmelweit F, ed. On the partial function of the cell. In: Collected Papers of Paul Ehrlich. London: Pergammon Press; 1960:183–194.
4. Foster TJ. Plasmid-determined resistance to antimicrobial drugs and toxic metal ions in bacteria. Microbiol Rev 1983; 47:361–409.
5. Tisa LS, Rosen BP. Plasmid-encoded transport mechanisms. J Bioenerg Biomembr 1990; 22:493–507.
6. McMurry L, Petrucci RE Jr, Levy SB. Active efflux of tetracycline encoded by four genetically different tetracycline resistance determinants in *Escherichia coli*. Proc Natl Acad Sci USA 1980; 77:3974–3977.
7. Nucifora G, Chu L, Misra T, Silver S. Cadmium resistance from *Staphylococcus aureus* plasmid pI258 *cadA* gene results from a cadmium efflux ATPase. Proc Natl Acad Sci USA 1989; 86:3544–3548.
8. Kaur P, Rosen BP. Plasmid-encoded resistance to arsenic and antimony. Plasmid 1992; 27:29–40.
9. Novick RP, Roth C. Plasmid-linked resistance to inorganic salts in *Staphylococcus aureus*. J Bacteriol 1968; 95:1335–1342.
10. Hedges RW, Baumberg S. Resistance to arsenic compounds conferred by a plasmid transmissible between strains of *Escherichia coli*. J Bacteriol 1973; 115:459–460.
11. Silver S, Budd K, Leahy KM, Shaw WV, Hammond D, Novick RP, Willsky GR, Malamy MH, Rosenberg H. Inducible plasmid-determined resistance to arsenate, arsenite and antimony (III) in *Escherichia coli* and *Staphylococcus aureus*. J Bacteriol 1981; 146:983–969.
12. Mobley HLT, Rosen BP. Energetics of plasmid-mediated arsenate resistance in *Escherichia coli*. Proc Natl Acad Sci USA 1982; 79:6119–6122.
13. Silver S, Keach D. Energy-dependent arsenate efflux: The mechanism of plasmid mediated resistance. Proc Natl Acad Sci USA 1982; 79:6114–6118.
14. Mobley HLT, Chen CM, Silver S, Rosen BP. Cloning and expression of R-factor mediated arsenate resistance in *Escherichia coli*. Mol Gen Genet 1983; 191:421–426.

15. Rosen BP, Borbolla MG. A plasmid-encoded arsenite pump produces arsenite resistance in *Escherichia coli*. Biochem Biophys Res Commun 1984; 124:760–765.
16. Chen CM, Misra T, Silver S, Rosen BP. Nucleotide sequence of the structural genes for an anion pump: The plasmid-encoded arsenical resistance operon. J Biol Chem 1986; 261:15030–15038.
17. Ji G, Silver S. Regulation and expression of the arsenic resistance operon from *Staphylococcus aureus* plasmid pI258. J Bacteriol 1992; 174:3684–3694.
18. Rosenstein R, Peschel P, Wieland B, Götz F. Expression and regulation of the *Staphylococcus xylosus* antimonite, arsenite and arsenate resistance operon. J Bacteriol 1992; 174:3676–3683.
19. Chen CM, Mobley HLT, Rosen BP. Separate resistances to arsenate and arsenite (antimonate) encoded by the arsenical resistance operon of R-factor R773. J Bacteriol 1985; 161:758–763.
20. Walker JE, Saraste M, Runswick MJ, Gay NJ. Distantly related sequences in the α- and β-subunits of the ATP synthase, myosin kinases and other ATP-requiring enzymes and a common nucleotide binding fold. EMBO J 1982; 1:945–951.
21. Pedersen PL, Carafoli E. Ion motive ATPases. I. Ubiquity, properties, and significance to cell function. Trends Biochem Sci 1987; 12:146–150.
22. Rosen BP, Weigel U, Karkaria C, Gangola P. Molecular characterization of an anion pump. The ArsA gene product is an arsenite(antimonate) stimulated ATPase. J Biol Chem 1988; 263:3067–3070.
23. Tisa LS, Rosen BP. Molecular characterization of an anion pump: The ArsB protein is the membrane anchor for the ArsA protein. J Biol Chem 1990; 265:190–194.
24. Hsu CM, Rosen BP. Characterization of the catalytic subunit of an anion pump. J Biol Chem 1989; 264:17349–17354.
25. Hsu CM, Rosen BP. Structure of the plasmid-encoded anion translocating ATPase. In: Kotyk A, Skoda J, Paces V, Kostka V, eds. Highlights of Modern Biochemistry. Zeist: VSP International Science Publishers; 1989:743–751.
26. Karkaria CE, Chen CM, Rosen BP. Mutagenesis of a nucleotide binding site of an anion-translocating ATPase. J Biol Chem 1990; 265:7832–7836.
27. Karkaria CE, Rosen BP. Trinitrophenyl-ATP binding to the wild type and mutant ArsA proteins. Arch Biochem Biophys 1991; 288:107–111.
28. Hsu CM, Kaur P, Karkaria CE, Steiner RF, Rosen BP. Substrate-induced dimerization of the ArsA protein, the catalytic component of an anion-translocating ATPase. J Biol Chem 1991; 266:2327–2332.
29. San Francisco MJD, Tisa LS, Rosen BP. Identification of the membrane component of the anion pump encoded by the arsenical resistance operon of R-factor R773. Mol Microbiol 1989; 3:15–21.
30. Wu J, Tisa LS, Rosen BP. Membrane topology of the ArsB protein, the membrane subunit of an anion-translocating ATPase. J Biol Chem 1992; 267:12570–12576.
31. Boyd D, Manoil C, Beckwith J. Determinants of membrane protein topology. Proc Natl Acad Sci USA 1987; 84:8525–8529.
32. Casadaban MJ, Martinez-Arias A, Shapiro SK, Chou J. β-Galactosidase gene fusions for analyzing gene expression in *Escherichia coli* and yeast. Meth Enzymol 1983; 100:293–307.

33. Broome-Smith JK, Spratt BG. A vector for the construction of translational fusions to TEM β-lactamase and the analysis of protein export signals and membrane protein topology. Gene 1986; 49:341–349.
34. Cohen GN, Monod J. Bacterial permeases. Bacteriol Rev 1957; 21:169–194.
35. Rosen BP, Weigel U, Monticello RA, Edwards BPF. Molecular analysis of an anion pump: Purification of the ArsC protein. Arch Biochem Biophys 1991; 284:381–385.
36. Joerger RD, Bishop PE. Nucleotide sequence and genetic analysis of the nifB-nifQ region from Azotobacter vinelandii. J Bacteriol 1988; 170:1475–1487.
37. Alloing G, Trombe M-C, Claverys J-P. The amiB locus of the Gram-positive bacterium Streptococcus pneumoniae is similar to binding protein-dependent transport operons of Gram-negative bacteria. Mol Microbiol 1990; 4:633–644.
38. Behrmann I, Hillemann D, Puehler A, Strauch E, Wohlleben W. Overexpression of a Streptomyces viridochromogenes gene (glnII) encoding a glutamine synthetase similar to those of eucaryotes confers resistance against the antibiotic phosphinothricyl-alanyl-alanine. J Bacteriol 1990; 172:5326–5334.
34. Futai M, Kanazawa H. Structure and function of proton-translocating ATPase (F_0F_1): Biochemical and molecular biological approaches. Microbiol Rev 1983; 47:285–313.
40. Rosen BP, Kashket ER. Energetics of active transport. In: Rosen BP, ed. Bacterial Transport. New York: Marcel Dekker; 1978:559–620.
41. Endicott JA, Ling V. The biochemistry of P-glycoprotein-mediated multidrug resistance. Annu Rev Biochem 1989; 58:136–171.
42. Ouelette M, Fase-Fowler F, Borst P. The amplified H circle of methotrexate-resistant Leishmania tarentolae contains a novel P-glycoprotein gene. EMBO J 1990; 9:1027–1033.
43. Callahan HL, Beverly SM. Heavy metal resistance: A new role for P-glycoproteins in Leishmania. J Biol Chem 1991; 266:18427–18430.
44. Rosteck PR, Reynolds PA, Hershberger CL. Homology between proteins controlling Streptomyces fradiae tylosin resistance and ATP-binding transport. Gene 1991; 102:27–32.
45. Guilfoile PG, Hutchinson CR. A bacterial analog of the mdr gene of mammalian tumor cells is present in Streptomyces peucetius, the producer of daunorubicin and doxorubicin. Proc Natl Acad Sci USA 1991; 88:8553–8557.
46. Smith K, Novick RP. Genetic studies on plasmid-linked cadmium resistance in Staphylococcus aureus. J Bacteriol 1972; 112:761–727.
47. Perry RD, Silver S. Cadmium and manganese transport in Staphylococcus aureus membrane vesicles. J Bacteriol 1982; 150:973–976.
48. Witte W, Green L, Misra TK, Silver S. Resistance to mercury and to cadmium in chromosomally resistant Staphylococcus aureus. Antimicro Agents Chemother 1986; 29:663–669.
49. Chopra I. Decreased uptake of cadmium by a resistant strain of Staphylococcus aureus. J Gen Microbiol 1970; 63:265–267.
50. Tynecka Z, Zajac J, Gos Z. Plasmid dependent impermeability barrier to cadmium ions in Staphylococcus aureus. Acta Microbiol Pol 1975; 7:11–20.
51. Chopra I. Mechanism of plasmid-mediated resistance to cadmium in Staphylococcus aureus. Antimicro Agents Chemother 1975; 7:8–14.

52. Weiss AA, Silver S, Kinscherf TG. Cation transport alteration associated with plasmid-determined resistance to cadmium in *Staphylococcus aureus*. Antimicro Agents Chemother 1978; 14:856–865.
53. Tynecka Z, Gos Z, Zajac J. Reduced cadmium transport determined by a resistance plasmid in *Staphylococcus aureus*. J Bacteriol 1981; 147:305–312.
54. Tynecka Z, Gos Z, Zajac J. Energy-dependent efflux of cadmium coded by a plasmid resistance determinant in *Staphylococcus aureus*. J Bacteriol 1981; 147:313–319.
55. Silver S, Nucifora G, Chu L, Misra T. Bacterial resistance ATPases: Primary pumps for exporting toxic cations and anions. Trends Biochem Sci 1989; 14:76–80.
56. Silver S, Walderhaug M. Gene regulation of plasmid and chromosomal-determined inorganic ion transport in bacteria. Microbiol Rev 1992; 56:195–228.
57. Walderhaug MO, Post RL, Saccomani G, Leonard RT, Briskin DP. Structural relatedness of three ion-transport adenosine triphosphatases around their active sites of phosphorylation. J Biol Chem 1985; 260:3852–3859.
58. Serrano R, Portillo F. Catalytic and regulatory sites of yeast plasma membrane H^+-ATPase studied by directed mutagenesis. Biochim Biophys Acta 1990; 1018:195–199.
59. Snavely MD, Miller CG, Maguire ME. The *mgtB* Mg^{2+} transport locus of *Salmonella typhimurium* encodes a P-type ATPase. J Biol Chem 1991; 266:815–823.
60. Solioz M, Mathews S, Fürst P. Cloning of the K^+-ATPase of *Streptococcus faecalis*. J Biol Chem 1987; 262:7358–7362.
61. Hesse JE, Wieczorek L, Altendorf K, Reicin AS, Dorus E, Epstein W. Sequence homology between two membrane transport ATPases, the Kdp-ATPase of *Escherichia coli* and the Ca^{2+}-ATPase of sarcoplasmic reticulum. Proc Natl Acad Sci USA 1984; 81:4746–4750.
62. Yoon KP, Silver S. A second gene in the *Staphylococcus aureus cadA* cadmium resistance determinant of plasmid pI258. J Bacteriol 1991; 173:7636–7642.
63. Tynecka Z, Skwarek T, Malm A. Anaerobic [109]Cd accumulation by cadmium-resistant and -sensitive *Staphylococcus aureus*. FEMS Microbiol Letts 1990; 69:159–164.
64. Tsai K-J, Yoon KP, Lynn AR. ATP-dependent cadmium transport by the *cadA* cadmium resistance determinant in everted membrane vesicles of *Bacillus subtilis*. J Bacteriol 1992; 174:116–121.
65. Bowman EJ, Siebers A, Altendorf K. Bafilomycins: A class of inhibitors of membrane ATPases from microorganisms, animal cells, and plant cells. Proc Natl Acad Sci USA 1988; 85:7972–7976.
66. Yoon KP, Misra T, Silver S. Regulation of the *cadA* cadmium resistance determinant of *Staphylococcus aureus* plasmid pI258. J Bacteriol 1991; 173:7643–7649.
67. Novick RP, Murphy E, Gryczan TJ, Baron E, Edelman I. Penicillinase plasmids of *Staphylococcus aureus*: Restriction-deletion maps. Plasmid 1979; 2:109–129.

68. Diels L, Faelen M, Mergeay M, Nies D. Mercury transposons from plasmids governing multiple resistance to heavy metals in *Alcaligenes eutrophus* CH34. Arch Int Physiol Biochim 1985; 93:B27–B28.
69. Mergeay M, Nies D, Schlegel HG, Gerits J, Charles P, Van Gijsegem F. *Alcaligenes eutrophus* CH34 is a facultative chemolithotroph with plasmid-bound resistance to heavy metals. J Bacteriol 1985; 162:328–343.
70. Nies D, Mergeay M, Friedrich B, Schlegel HG. Cloning of plasmid genes encoding resistance to cadmium, zinc, and cobalt in *Alcaligenes eutrophus* CH34. J Bacteriol 1987; 169:4865–4868.
71. Nies DH, Silver S. Plasmid-determined inducible efflux is responsible for resistance to cadmium, zinc, and cobalt in *Alcaligenes eutrophus*. J Bacteriol 1989; 171:896–900.
72. Nies DH, Nies A, Chu L, Silver S. Expression and nucleotide sequence of a plasmid-determined divalent cation efflux system from *Alcaligenes eutrophus*. Proc Natl Acad Sci USA 1989; 86:7351–7355.

Index